Dynamic Balancing of Mecha
and Synthesizing of Parallel F

Dan Zhang • Bin Wei
Editors

Dynamic Balancing of Mechanisms and Synthesizing of Parallel Robots

Springer

Editors
Dan Zhang
University of Ontario Institute
of Technology
Oshawa, ON, Canada

Bin Wei
University of Ontario Institute
of Technology
Oshawa, ON, Canada

ISBN 978-3-319-17682-6 ISBN 978-3-319-17683-3 (eBook)
DOI 10.1007/978-3-319-17683-3

Library of Congress Control Number: 2015947930

Springer Cham Heidelberg New York Dordrecht London

Printed on acid-free paper

Springer International Publishing AG Switzerland is part of Springer Science+Business Media (www.springer.com)

Preface

As the emerging technologies of robotics, dynamic balancing for mechanisms and parallel robots has become one of the bottleneck issues for the applications in modern manufacturing industries, space, medical industries, military, and social service. Research and development of various dynamic balancing methods is now being performed more and more actively in every applicable field. This book will introduce state-of-the-art research in these technologies from theory to practice in a systematic and comprehensive way.

Any vibrations in mechanisms will cause inaccuracy while they are in operations. Traditional counterbalance methods make the whole mechanism heavier and have more inertia; the book entitled *Dynamic Balancing of Mechanisms and Synthesizing of Parallel Robots* will be the first book that describes the up-to-date technologies in dynamic balancing of mechanisms and parallel robots. It systematically and thoroughly not only deals with different dynamic balancing principles but also comprises recent advances on dynamic balancing of mechanisms with minimum increase of mass and inertia, synthesizing of parallel robots based on decomposition and integration concept, and finally optimization and control issues for balancing are discussed at length within this book.

We would like to express our deep appreciation to all the authors for their significant contributions to the book. Their commitment, enthusiasm, and technical expertise are what made this book possible. We are also grateful to the publisher for supporting this project and would especially like to thank Ms. Merry Stuber, Editorial Assistant of Springer US, and Ms. Lesley Poliner, Springer US Science and Business Media Project Coordinator, for their constructive assistance and earnest cooperation, both with the publishing venture in general and the editorial details. We hope the readers find this book informative and useful.

This book consists of 20 chapters. Chapter 1 introduces the recent advances on reactionless mechanisms and parallel robots, and the dynamic balancing through reconfiguration concept is proposed. Chapter 2 presents methods and principles used for balancing of planar mechanisms without counterrotations. Chapter 3 considers the shaking moment and shaking force balancing through the use of additional Assur groups mounted on the mechanism to be balanced. Two types of mechanisms are considered, the in-line four-bar linkage and the planar parallel robots with prismatic pairs. Chapter 4 discusses the development of reactionless planar parallel manipulators by using base-mounted counterrotations and inertia flywheel rotating with a prescribed angular velocity. Chapter 5 introduces a new general method to find the dynamic balancing conditions based on the use of natural coordinates for planar mechanisms, and the method has been shown in its application to the design and dynamic balancing of plane mechanisms. Chapter 6 deals with the shaking force and shaking moment balancing of single degree of freedom planar mechanisms by employing the traditional technique of addition of counterweights and counterrotating inertias. Chapter 7 proposes a force balancing method called adjusting kinematic parameters for robotic mechanisms or real-time controllable mechanisms. Chapter 8 proposes a formulation, which can be seen as a tool in selecting appropriate solution(s) according to the expected operation conditions, to address the effects of balancing on mass distributions and dynamic performance. A case of study has been developed by referring to a three degrees-of-freedom spatial parallel manipulator by designing proper counter-rotary counterweights. Chapter 9 focuses on dynamic balancing with respect to a given trajectory for the parallel link robots by modeling control system. Chapter 10 addresses the class of problems that require movement of a dynamic bipedal system according to stringent state-space and temporal requirements despite actuation limits and disturbances. Chapter 11 presents an optimization technique to dynamically balance planar mechanisms by minimizing the shaking forces and shaking moments due to inertia-induced forces. Chapter 12 investigates the dynamic response of mechanism having revolute joints with clearance, and a 4R four-bar mechanism whose two joints have clearances is considered as a model mechanism. Chapter 13 minimizes the shaking force and moment fluctuations at the planar mechanism by employing the genetic algorithm. Chapter 14 presents the optimal balancing for the open-chain robotic system based on the indirect solution of open-loop optimal control problem. Chapter 15 studies the dynamics and control of planar, translation, and spherical parallel manipulators by means of the constraint equations. Chapter 16 deals with the dynamic modeling and control of balanced parallel mechanisms, highlights the importance of the dynamic modeling process, and discusses the impact of the dynamic model, developed in accordance with the methodology, for the control strategy of parallel mechanisms. Chapter 17 describes the control principles necessary for an articulated biped model to accomplish balanced locomotion during walking and climbing. Chapter 18 focuses on the control of a 10-dof biped robot, and a spline-based control system is described in order to generate the servo inputs. Chapter 19 deals with an approach to formulate balancing conditions for the shaking force and shaking moment of planar mechanisms and spatial mechanisms. Chapter 20 addresses the static balancing of six degree-of-freedom articulated wheeled vehicles with multiple leg-wheel subsystem.

Finally, the editors would like to sincerely acknowledge all the friends and colleagues who have contributed to this book.

Oshawa, ON, Canada Dan Zhang
December 2014 Bin Wei

Contents

Contributors

Mario Acevedo Faculty of Engineering, Universidad Panamericana, Mexico City, Mexico

Aliakbar Alamdari The State University of New York at Buffalo, Buffalo, NY, USA

Vigen Arakelian Institut National des Sciences Appliquées (INSA), Rennes, France

Institut de Recherche en Communications et Cybernétique de Nantes (IRCCyN), UMR CNRS 6597, Nantes, France

Department of Mechanical and Control Systems Engineering I.N.S.A. Rennes, 20 avenue des Buttes de Coësmes, CS 14315, F-35043, Rennes, France

IRCCyN, 1 rue de la Noë, BP 92101, F-44321 Nantes Cedex 03, France

Adrian Boeing University of Western Australia, Crawley, WA, Australia

Thomas Bräunl University of Western Australia, Crawley, WA, Australia

Sébastien Briot Institut de Recherche en Communications et Cybernétique de Nantes (IRCCyN), Nantes, France

Daniele Cafolla LARM: laboratory of Robotics and Mechatronics, DICEM, University of Cassino and South Latium Cassino, Cassino, Italy

Giuseppe Carbone LARM: laboratory of Robotics and Mechatronics, DICEM, University of Cassino and South Latium Cassino, Cassino, Italy

Marco Ceccarelli LARM: laboratory of Robotics and Mechatronics, DICEM, University of Cassino and South Latium Cassino, Cassino, Italy

Himanshu Chaudhary Department of Mechanical Engineering, Malaviya National Institute of Technology Jaipur, Jaipur, India

Kailash Chaudhary Department of Mechanical Engineering, Malaviya National Institute of Technology Jaipur, Jaipur, India

Tarcisio Antonio Hess Coelho Department of Mechatronics and Mechanical Systems Engineering, University of Sao Paulo, Sao Paulo, Brazil

André Garnier Coutinho Department of Mechatronics and Mechanical Systems Engineering, University of Sao Paulo, Sao Paulo, Brazil

Nguyen Phong Dien Department of Applied Mechanics, Hanoi University of Science and Technology, Hanoi, Vietnam

Selçuk Erkaya Department of Mechatronics Engineering, Engineering Faculty, Erciyes University, Kayseri, Turkey

Victor Glazunov Institute of Machines sciences of RAS, Moscow, Russia

Andreas Hofmann Massachusetts Institute of Technology (MIT), Cambridge, MA, USA

J Huang Department of Mechanical Engineering, University of Saskatchewan, Saskatoon, SK, Canada

Benjamin Kenwright Edinburgh Napier University, Edinburgh, UK

Nguyen Van Khang Department of Applied Mechanics, Hanoi University of Science and Technology, Hanoi, Vietnam

Sergey Kheylo Moscow State University of Design and Technology, Moscow, Russia

Venkat Krovi The State University of New York at Buffalo, Buffalo, NY, USA

M Moradi Department of Mechanical Engineering, Amirkabir University of Technology, Tehran, Iran

Peddinti Nehemiah Mechanical Engineering Department, Lords Institute of Engineering and Technology, Hyderabad, India

A Nikoobin Faculty of Mechanical Engineering, Semnan University, Semnan, Iran

Renato Maia Matarazzo Orsino Department of Mechanical Engineering, University of Sao Paulo, Sao Paulo, Brazil

P. R. Ouyang Department of Aerospace Engineering, Ryerson University, Toronto, ON, Canada

İbrahim Uzmay Department of Mechatronics Engineering, Engineering Faculty, Erciyes University, Kayseri, Turkey

Vlastimil Votrubec VÚTS, a.s., Liberec, Czech Republic

Bin Wei Robotics and Automation Laboratory, Faculty of Engineering and Applied Science, University of Ontario Institute of Technology, Oshawa, ON, Canada

Taizo Yoshikawa Artificial Intelligence Laboratory Manipulation Group, Stanford University, Stanford, CA, USA

W. J. Zhang Department of Mechanical Engineering, University of Saskatchewan, Saskatoon, SK, Canada

Dan Zhang Robotics and Automation Laboratory, Faculty of Engineering and Applied Science, University of Ontario Institute of Technology, Oshawa, ON, Canada

Chapter 1
Review of Recent Advances on Reactionless Mechanisms and Parallel Robots

Dan Zhang and Bin Wei

Abstract When parallel mechanisms are in motions, because the center of mass (CoM) is not fixed and angular momentum is not constant, vibration is often produced in the system. Shaking force and shaking moment balancing can usually be realized by making the CoM of mechanism fixed and angular momentum constant. There are generally two main ways for shaking force balancing and shaking moment balancing, balancing before kinematic synthesis and balancing at the end of the design process. For the balancing at the end of the design process, addition of counterweights and counter-rotations, addition of active dynamic balancing unit, and addition of auxiliary links are mostly used methods. The advances and problems on dynamic balancing of mechanisms are discussed in detail under the above two main categories here, and balancing through reconfiguration method is proposed, which can reduce the addition of mass and inertia. Fisher's method belongs to the method of balancing before kinematic synthesis.

Keywords Parallel mechanisms • Momentum • Dynamic balancing • Reconfiguration

1.1 Introduction

Parallel mechanisms have been broadly used in the areas of machine tools, telescopes and space, etc., but a problem occurs when they are in operations; it is not dynamic balanced, which affects the accuracy performance when mechanisms are in the process of operations. When mechanisms move, as the center of mass (CoM) of the mechanism is not fixed and angular momentum is not constant, vibration is usually produced in the system. Dynamic balancing can usually be achieved by making the linear and angular momentum of the mechanism constant. The research for dynamic balancing of parallel mechanisms is still in its early stage. Since 2000

D. Zhang • B. Wei (✉)
Robotics and Automation Laboratory, Faculty of Engineering and Applied Science, University of Ontario Institute of Technology, 2000 Simcoe Street North, Oshawa, ON, Canada
e-mail: Dan.Zhang@uoit.ca; Bin.Wei@uoit.ca

D. Zhang, B. Wei (eds.), *Dynamic Balancing of Mechanisms and Synthesizing of Parallel Robots*, DOI 10.1007/978-3-319-17683-3_1

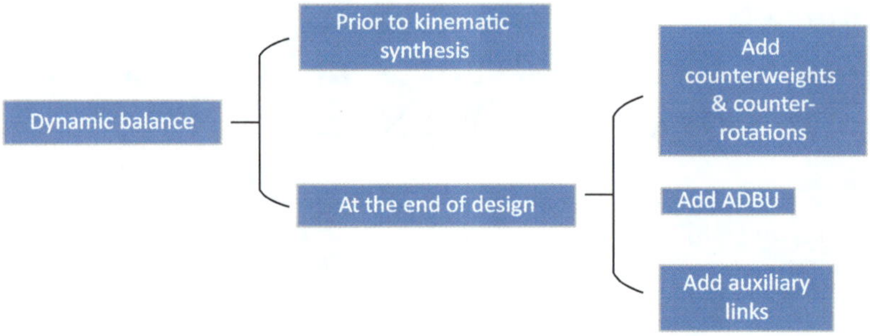

Fig. 1.1 Two main categories for dynamic balancing

when scholars Ricard and Gosselin systematically addressed the dynamic balancing of parallel mechanisms [1], dynamic balancing began to appear more and more in the academic arena. In order to achieve dynamic balance, force balance and moment balance need to be both satisfied at the same time. Force balance is a subset of static balance, which means when the mechanism is force balanced, it is also static balanced, and the mechanisms can remain stable without any actuator forces. Traditionally counterweights are used to achieve force balance, i.e., make the CoM fixed, and counter-rotations are used to make the angular momentum constant. Force balance and moment balance are all about using extra devices (e.g., counterweights, counter-rotations) to counterbalance the shaking force and moment that the original mechanism exerted, but the whole mechanisms will become heavier and have more inertia when using those counterbalancing devices. How to design reactionless mechanisms with minimum increase of mass and inertia has become a common desire. There are generally two main ways for shaking force and shaking moment balancing, i.e., "balancing before kinematic synthesis" and "balancing at the end of the design process" as shown in Fig. 1.1. Here dynamic balancing based on two main categories is discussed in details, and a new balancing principle concept is proposed, the advantage of which is that addition of counterweights can be reduced. For the category of balancing at the end of the design process, addition of counterweights and counter-rotations, addition of active dynamic balancing unit (ADBU), and addition of auxiliary links are mostly used methods; for the category of balancing before kinematic synthesis, Fisher's method is a typical example of this.

For the shaking force balancing, for example, when a link is rotating round a pivot, because CoM of the link is not still, so the link will have a shaking force, when a counterweight is added to the extendable part of that link, then the CoM of the whole link is fixed in the revolute joint, and it is force balanced. If the counterweight is added, the system will become heavy. The second method is to employ ADBU; the ADBU will create a shaking force and shaking moment that the value of which is equal but has opposite direction to the original shaking force and shaking moment so that it can counterbalance those original unbalanced conditions. The third method is to add auxiliary links; the mass of additional link can be used to force balance, for example in [2], all the mass of the moving platform and part of the mass of the links attached to the moving platform for the three-dimensional delta robot. In addition, Fisher's method can also be seen as the method of addition of auxiliary links. Here balance through reconfiguration concept is proposed; for example, we can use screw link so that the link can be moved, the CoM of the link then can be moved to the revolute joint, and then it is balanced; in this method counterweight is not applied but through reconfiguration of the system by moving the screw link the system will not become heavy. For the shaking moment balancing, addition of counter-rotation method, addition of CRCM method, using inherently dynamic balanced 4-bar linkage method, and addition of ADBU method are mostly used principles.

1.2 Prior to Kinematic Synthesis

1.2.1 Dynamic Balanced 4-Bar Linkage

In [3], a 4-bar linkage was proposed to synthesize three degrees of freedom parallel manipulators. By serially connecting two 4-bar linkages, a 2-DOF reactionless serial mechanism was constructed, and the 2-DOF mechanism was used to build the 3-DOF parallel manipulators. The advantage of the above mechanism is that it did not employ counter-rotations, but the drawback is that moving platform is assumed thin, which is not practical. The above 4-bar linkage is actually derived from the principle vector linkage. The three-serial-chain principle vector linkage is evolved to a 4-bar linkage by adding a base link to the ground as shown in Fig. 1.2, and by finding the moment balancing conditions for the 4-bar linkage, the dynamic balanced 4-bar linkage can be derived.

In [4], a 3-DOF serially connected mechanism was derived from two 4-bar mechanisms and one composite mechanism. This 3-DOF mechanism can be used as leg to construct the spatial 6-DOF parallel manipulators. The composite mechanism is derived from a pair of 4-bar mechanisms that orthogonally fixed each other. Because the author wanted to design a spatial 6-DOF parallel manipulator, which requires the 4-bar linkage to move spatially, and due to the fact that the 4-bar linkage is not dynamic balanced when moving spatially, the composite mechanism is developed. Also the synthesized mechanism as shown in Fig. 1.3 is proposed by connecting the 4-bar linkage or composite mechanism to the end bar of the base 4-bar linkage, and the synthesized mechanism was verified that it was dynamic balanced, which is done by the following: if the resulting parameters of the end bar of the base 4-bar linkage and attached mechanism (this attached mechanism can be 4-bar linkage or composite mechanism) meet the balance condition, then the synthesized mechanism will be dynamic balanced.

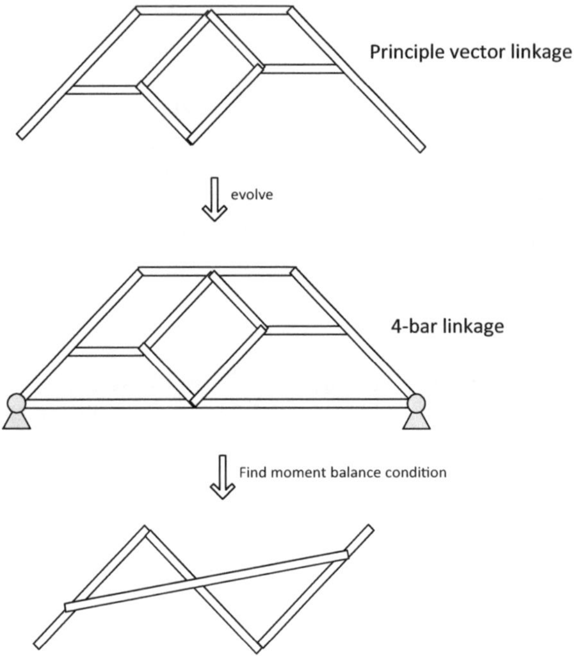

Fig. 1.2 Evolve process for deriving the dynamic balanced 4-bar linkage

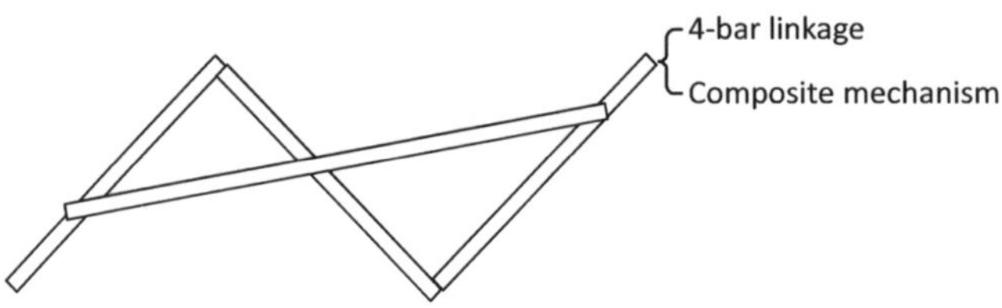

Fig. 1.3 Synthesized mechanism

1.2.2 Fisher's Method

V.D. Wijk has thoroughly investigated this method in his PhD thesis and papers [5–7]; the core content can be concluded as follows: for the shaking force balance, first determine the linear momentum, then determine the force balance condition from the linear momentum, and finally determine the principle dimensions. For the 2-DOF pantograph, first determine linear momentum, then force balance condition, and finally the principle dimensions. Because the 2-DOF pantograph does not have the middle link, it is easier to solve the principle dimensions without using the equivalent linear momentum systems (ELMS). For the 3-DOF and 4-DOF principle vector linkages, they have middle links, so the ELMS is used for the middle links which requires a little more effort to calculate the principle dimensions. For the moment balance, first write the angular momentum, and then substitute the position vectors, position vector derivatives, angle relations, and force balance conditions to the angular momentum equation to obtain the final form of the angular momentum; for linear relations of time-dependent parameters, determine the moment balance condition from the angular momentum and subsequently balance solutions; for nonlinear relations of time-dependent parameters, determine the moment balance condition and subsequently balance solutions. Finally perform synthesis of reactionless mechanisms from the principle vector linkages.

The main content of the Fisher's method that Van der Wijk used in his PhD thesis is to calculate the principle dimensions, and by using the auxiliary links/pantograph links to trace the CoM of the whole mechanism. It is shown that the principle vector linkage architecture is force balanced, and for the moment balance, the relative motions of principle vector linkage architecture have to be constrained by additional elements. The moment balance is achieved mainly through the symmetrical design and constraining the DOF of the mechanism, like adding a slider or something to make the DOF of the mechanism reduced to achieve the moment balance. For the grasping mechanism, it is derived from the 4-DOF principle vector linkage with a slider; the motion of the 4-DOF principle vector linkage (grasping mechanism) is reduced in order to achieve the moment balance. Also the bridge and the roof and wall of house can be derived from the 2-DOF principle vector linkage. The above dynamic balanced mechanisms are all synthesized from the principle vector linkages.

In [8], for the dual-V manipulator, it is derived from two balanced pantographs, and by symmetric designing the structure of legs of 4RRR planar parallel manipulator, shaking moment was balanced out each other when moving along the orthogonal axis, so counter-rotations are no longer needed, only counterweights are used, and the disadvantage is that the manipulator is dynamic balanced only when the manipulator moves in the orthogonal axes with non-rotated moving platform. The idea of the above symmetric designing can also be seen as evolving from pantograph arms with a counter-mass (the arm has a parallelogram shape), and the pantograph arms with a counter-mass was evolved from the normal counter-mass adding in each link, as shown in Fig. 1.4. In the similar paper [9], the author derived the

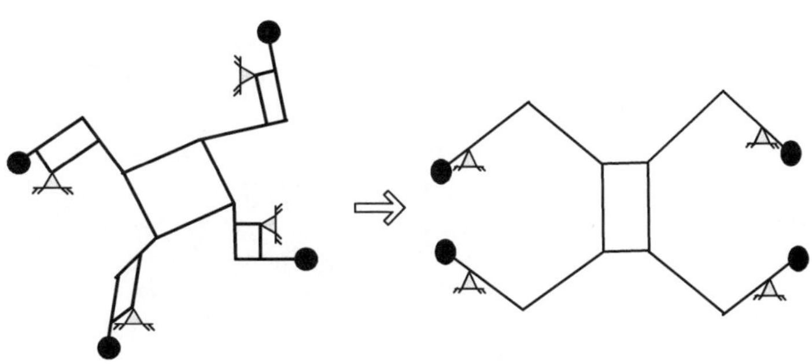

Fig. 1.4 Evolve process for the 4RRR reactionless parallel mechanism

general force balancing conditions of the planar 4RRR parallel manipulator, and the different topologies of 4RRR manipulator from the force balance condition were obtained.

1.3 Balancing at the End of the Design Process

1.3.1 Add Counterweights

1.3.1.1 Normal Counterweights

In [10], a double pendulum was dynamic balanced by using two counterweights and two counter-rotations. The counterweights are placed at the extension of each link like the traditional force balance technique to make the center of mass fixed at revolute joint, and shaking moment balancing is achieved by using planetary gear trains that carry out the counter-rotations. Force balancing condition is derived by using the center of mass formula and making the position of CoM equal to 0; two force balance equations are obtained; from those two static balancing equations it can be seen that the masses and length are both positive; the only way to satisfy the equation is to make the position of the CoM of some links to be negative; to do that, counterweights were added. The shaking moment of upper moving link is balanced by a counter-rotation gear; this counter-rotation gear is mounted on the base, and it is connected to the upper moving link by the following way: there are two gears at base joint (one small gear and one big gear) being fixed together, the counter-rotation gear is connected with the bigger gear, and the small gear is connected to the upper moving link by a belt; through this way, this counter-rotation gear is indirectly connected to the upper moving link and rotates opposite with the upper moving link to achieve the moment balance in order to achieve dynamic balance. For the moment balancing, the author wrote the angular momentum of the whole mechanism, and by making the angular momentum equal to 0, two moment balance equations (moment balance conditions) are derived. The disadvantage of the above force balancing and moment balancing method is that counterweights and planetary gear trains (counter-rotations) are used, which increase the total mass and complexity. In the second part of that paper, the authors also talked about the shaking moment balancing by using flywheel since this solution is constructively more efficient. First the angular momentum of the whole parallel manipulator was derived. In order to achieve shaking moment balance condition for this manipulator, the flywheel was used, and this flywheel needs to have the same and opposite shaking moment so that this flywheel can moment balance the manipulator. This flywheel is driven by another actuator, which belongs to the active dynamic balancing technique. Finally the angular acceleration of this flywheel can be obtained by using the moment formula. But how to link this flywheel to the parallel manipulator was not mentioned.

In [11–13], the idea of dynamic balancing of mechanisms is to use counterweights and counter-rotations (i.e., geared inertia counterweights and planetary-gear-train-inertia counterweight) to force and moment balance linkages, which is quite straightforward. The center of mass formula was used to derive the center of mass of the whole mechanism; then the center of mass was set to be stationary so that the force balance condition can be obtained; subsequently shaking moment of the linkage was described as the time rate of change of the total angular momentum, and the general formula for the total angular momentum of the linkage was used; after that the total angular momentum was set to 0 in order to derive the dynamic balance condition, but later it was found that it was impossible to achieve dynamic balance unless counter-rotations were added. After adding counter-rotations, set the total angular momentum to 0 and the moment balance condition was obtained. The disadvantage of this balance method is that the planetary-gear-train-inertia counterweight was put on the upper moving link rather than the ground.

In [14], the author derived the 3-DOF parallelepiped mechanism (unit) from the basic 1-DOF pivot link as leg to synthesize the spatial parallel manipulator, but this parallelepiped mechanism requires three counter-rotations and six counterweights to achieve dynamic balance condition, which substantially increased the mass, inertia, and complexity of the mechanism. The above parallelepiped mechanism design is not smart because it used the counterweights and counter-rotations. The dynamic balancing condition was directly derived from the center of mass formula and also set angular momentum to 0. Finally, the parallelepiped mechanism was used to construct the spatial parallel manipulators. How to simplify this mechanism has become a future work.

In [15], a parallelogram 5-bar linkage was proposed as a leg for a planar 3-DOF parallel manipulator. Firstly, the moving platform was replaced by two point masses located at the point of attachment of each of the legs to the moving platform; in order to do that, three conditions have to be satisfied: same mass, same inertia, and same center of mass; secondly, for each leg (includes the replaced mass) the static balancing has to be firstly satisfied in order to achieve the dynamic balancing condition, and for the static balancing, the center of mass equation was used and by making the position of CoM equal to 0, two static balance equations are obtained; after obtaining the equations, the next step is to solve it. From those two static balancing equations it can be seen that the masses and length are both positive; the only way to satisfy the equation is to make the position of the CoM of some links to be negative; to do that, counterweights can be added. For the moment balancing, the author wrote the angular momentum of the 5-bar mechanism, and by making the angular momentum equal to 0, three moment balance equations (moment balance conditions) are derived. From the static balancing, two equations were derived, and from the dynamic balancing (angular momentum condition), another three equations were derived; that is, five equations were provided for the dynamic balancing of the leg (5-bar linkage). The novelty of this paper is that the authors proposed the parallelogram 5-bar linkage as a leg of a planar 3-DOF parallel manipulator and analyzed the dynamic balancing of the leg. Future wok is that employ the proposed leg for other kinds of spatial parallel manipulators. The above method is based on the decomposition and integration method; that is, propose a single linkage (leg) first, then dynamic balancing a single linkage, and finally combine those linkages to form the whole parallel manipulator; in other words, decompose first and integrate later. But the disadvantage of the above reactionless mechanism is that the counterweights and counter-rotations were used, which increased the weight, inertia, and complexity. The counterweights are used to keep constant the position of the center of mass while the counter-rotations are used to keep constant the angular momentum.

In [16], the idea of putting the gear, which is used for balancing the shaking moment, on the base can lead to smaller increase of moving masses. This gear is originally mounted on the moving link, so the mass of the counterweight of the base link is needed to force balance this gear as well, but if the gear is put on the mechanism frame, then the counterweight of the base link does not need to force balance this gear, which means the mass of this counterweight of the base link can be decreased. But the disadvantage is that the number of extra devices increased. The balancing method above in which the gear was put on the base of the mechanism is an extension of the method in [11–13].

1.3.1.2 Add CRCM

In [17], it presented the shift modification rules, and the counter-rotary counterweight was evolved from this shift modification rules. In [18], the CRCM was proposed and compared with the separate counter-rotation, and it came to the conclusion that the CRCM principle has reached reduction of added mass and added inertia.

In [19], another three CRCM-based balancing principles were derived, i.e., low inertia configuration balancing principle, one CRCM balancing principle, and only CRCMs near the base balancing principle. According to the paper, the advantage of the first new balancing principle is its low inertia, the advantage of the second new balancing principle is that only one CRCM is necessary for the moment balance of the complete mechanism, and the advantage of the third new balancing principle is its compact construction. Finally several CRCM-based 2-DOF parallel mechanisms were synthesized by using the CRCM-balanced double pendulum. And the 3-DOF planer and spatial parallel manipulators are synthesized by using the balanced double pendulum.

Our perspective is that for the one CRCM configuration, it is not a smart balancing principle because there are two gears on the upper moving link rather than the base frame. For the only CRCMs near the base configuration, the principle is roughly the same with the idler loop or the V. Arakelian and M. Smith mechanism in [10, 16]; that is, the moment of upper moving link is balanced by a CRCM which is connected to the upper moving link through a gear/belt transmission, and the moment of base link is balanced by another CRCM which is connected to a gear that is attached to the base link. But the disadvantage of the only CRCMs near the base configuration is that the CRCM that is used for moment balancing the upper moving link is on the base link, which makes the system heavier. The V. Arakelian and M. Smith mechanism in [10, 16] is that the gear that is used for moment balancing the upper moving link is on the base/ground, which does not affect the system at all.

In [20], the total mass (increase) and reduced inertia of double pendulum were compared within the counter-rotary counter-mass (CRCM), separate counter-rotations (SCR), duplicate mechanisms (DM), and idler loop. Firstly the reduced inertia and total mass of these four balancing principles were derived, and mass-inertia factor was established and this factor was used for judging the additional mass and additional inertia. The comparison results showed that the DM principle had the lowest values for the mass-inertia factor, which means that the DM principle is the most advantageous for low mass and low inertia dynamic balancing, but DM principle requires a larger space. CMCR principle is the second lowest values for the mass-inertia factor, which means CRCM principle is the second most advantageous for low mass and low inertia dynamic balancing, and CRCM principle does not require larger space compared with the DM principle, so the CRCM has more potential to use. The general procedure of the above analysis can be concluded as follows: Step 1: The position vectors of the counter-masses and lump mass were obtained first; then with the derivative of those position vectors, the linear momentum was derived by using linear momentum formula and subsequently making the linear momentum equal to 0, and the force balancing condition was derived. Step 2: The angular momentum about reference point was obtained by using the angular momentum formula, and the relations between the gears were applied to simplify the angular momentum, by making the angular momentum equal to 0; the moment balancing (dynamic balancing) condition was derived. Step 3: When deriving the reduced inertia, either we can determine the kinetic energy first and derive the reduced inertia, or directly obtain the reduced inertia by copying the coefficients of angular velocities in the angular momentum formula but with the transmission ratios squared. Step 4: Determine the total mass. Step 5: The total mass and reduced inertia are compared among those four balancing principles.

Our thought is that it is not necessary to compare the total mass and inertia, because some of the masses and inertia are on the ground, not on the mechanism, so those masses that are on the ground do not really affect the system. In [21] it is the same with the above paper except it compared the total mass and reduced inertia among SCR, CRCM, and DM for a 1-DOF rotatable link rather than a double pendulum.

1.3.1.3 Add Assur Group

In [22], the author used the Assur group and three counterweights to achieve dynamic balance; three counterweights are used to achieve force balance, and Assur group and the counterweights are used to achieve the moment balance. In [23], the paper talked about the shaking force balancing and shaking moment balancing for a planar 3RPR parallel manipulator with prismatic joints; the author proposed two methods for the balancing: the first one is based on the addition of an idler loop between the moving platform and the base; it uses lots of counterweights and counter-rotations, which substantially increase the mass and inertia. The second method is based on the addition of a Scott-Russell mechanism (i.e., special crank-slider mechanism, which belongs to the Assur group) to each leg of the 3RPR parallel manipulator, which can decrease the number of counter-rotations. The second method which is based on the addition of a Scott-Russell mechanism belongs to the passive dynamic balancing; it requires three counter-rotations. It is expected that if we change the passive balancing to active balancing, then the number of counter-rotations can be reduced.

1.3.1.4 Add Active Driven CRCM

In [24], by active driving the CRCM, the double pendulum can be dynamic balanced. The specific angular momentum of ACRCM was derived from the derived angular momentum, then the rotational velocity of the ACRCM was obtained, and the torque of the actuator that actively drove the ACRCM was obtained. Through evaluation, the author found that the ACRCM principle is better than the passive CRCM or with separate counter-rotations mainly in terms of total mass-inertia relation. A 2-DOF ACRCM-balanced parallel manipulator was derived by combining two CRCM to one ACRCM as shown in Fig. 1.5. The 3-DOF planer and spatial parallel manipulators were synthesized by using the ACRCM-balanced double pendulum. For the planer parallel manipulator, it has 1-DOF rotation within a single plane, so only one ACRCM can be used to balance the complete mechanism. For the spatial 3-DOF parallel manipulator, the rotations of the moving platform and links are in two planes; therefore two ACRCM are used to balance the mechanism. It uses the ACRCM; the whole system will still become heavier, because it uses the ACRCM; it belongs to the "consider at the end of the design process" approach.

In the above paper, a 2-DOF ACRCM-balanced parallel manipulator was derived by combining two CRCM to one ACRCM as shown in Fig. 1.5. Inspired by the above design, new 3-DOF planar 3-2RRR and 4-2RRR reactionless parallel manipulators and spatial 3-DOF 3-2RRR and 4-2RRR reactionless parallel manipulators are derived as shown in Fig. 1.6 by employing 2-DOF ACRCM-balanced mechanism.

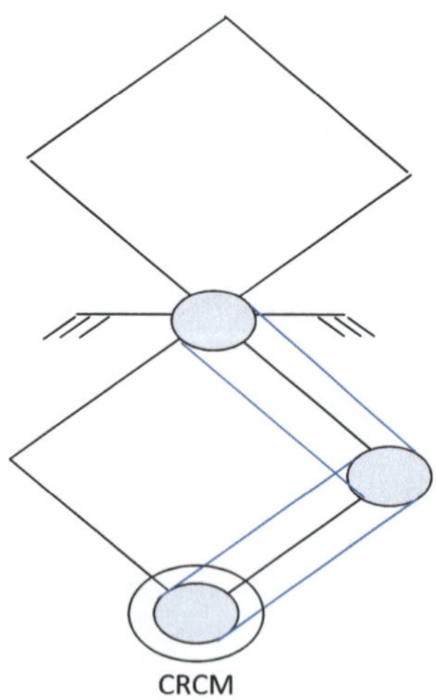

Fig. 1.5 ACRCM-balanced manipulator

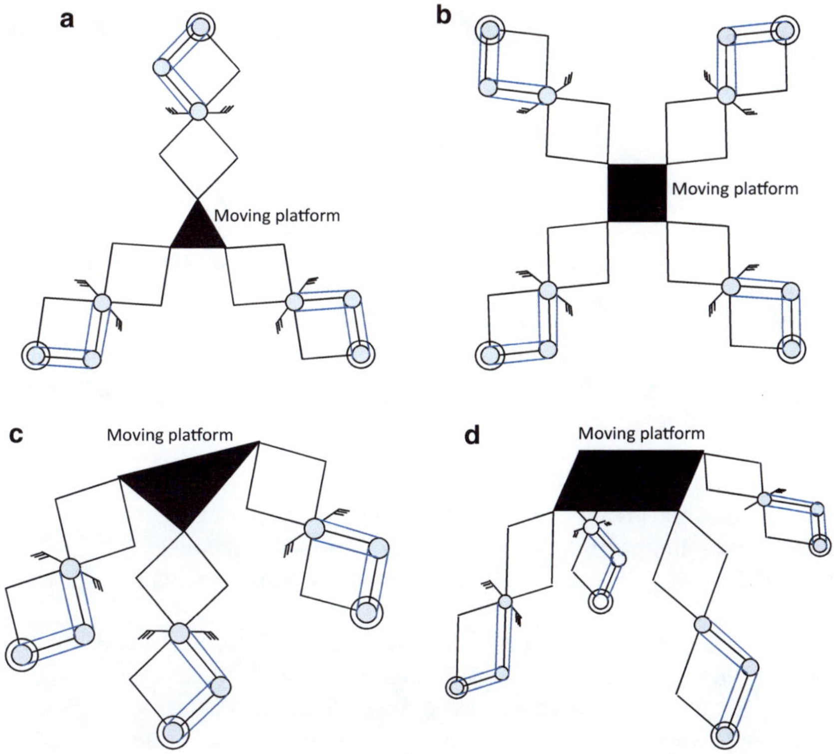

Fig. 1.6 3-DOF planar: (**a**) 3-2RRR and (**b**) 4-2RRR and spatial (**c**) 3-2RRR and (**d**) 4-2RRR reactionless parallel manipulators

1.3.2 Add Auxiliary Links

For dynamic balancing of Clavel's Delta robot, in [2], for the force balance, a solution is proposed that each leg and one-third of the moving platform mass together are balanced with one counter-mass plus an additional link; that is, each leg becomes a 3D pantograph. Furthermore, due to the fact that the moving platform of the Delta robot does not rotate, the above force balance method can be simplified to the following: one leg being a 3D pantograph can balance the complete mass of the moving platform and part of the mass of the links that are attached to the moving platform of the other two legs, and two other counter-masses are attached to the other two legs; that is, the complete Delta robot is force balanced by three counter-masses and additional link. For the moment balance, the author used the active driven method because the angular momentum of the force balanced Delta robot is dependent on the velocity of mechanism; it cannot be made constant by using passive moment balancing methods, for example, geared counter-rotating inertias. It is found that the mass of additional link can be used to force balance all the mass of the moving platform and part of the mass of the links attached to the moving platform of legs 2 and 3. Fisher's method can also be seen as the method of addition of auxiliary links.

1.3.3 Through Reconfiguration

Here force balancing through reconfiguration concept is proposed; for example, we can use screw link as link, the link can be moved so that the CoM of the link can be moved to the revolute joint point, and then it is forced balanced; in this method, counterweight is not used but through reconfiguration of the system by moving the screw link, the system will not become heavy. Figure 1.7 shows such a concept of force balancing through reconfiguration.

The purpose of using counterweight is to move the CoM to the still point, so the question is that can we not use counterweight to achieve the same goal. We can reconfigure the link so that CoM is moved to the still point. We just want to use the function of their links, and in this case it is the rotational function.

For the three link case, if we use counterweights, then it becomes much heavier (Fig. 1.8).

From above, we can see that the function (i.e., rotational function of the links) is not changed at all; the function is still remained. For the force balance by adding counterweight, the whole system becomes much heavier. For the 4R 4-bar linkage, we have the following if the 4R 4-bar linkage is regarded as an open chain of three links in series (Fig. 1.9):

For the crank-slider mechanism, it can be seen as an open chain of three links in series; the third link is a slider that does not rotate and it solely translates. Because link 3 does not rotate, the CoM of the link 3 can be in any point in link 3 (Fig. 1.10).

The above force balanced through reconfiguration crank-slider mechanism maybe can be used as Scott-Russell mechanism, and use the force balanced through reconfiguration crank-slider mechanism to synthesize the planar 3RPR parallel

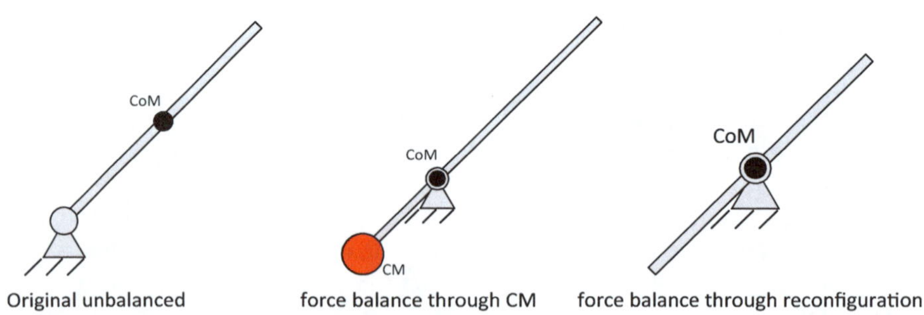

Fig. 1.7 Concept of force balancing through reconfiguration

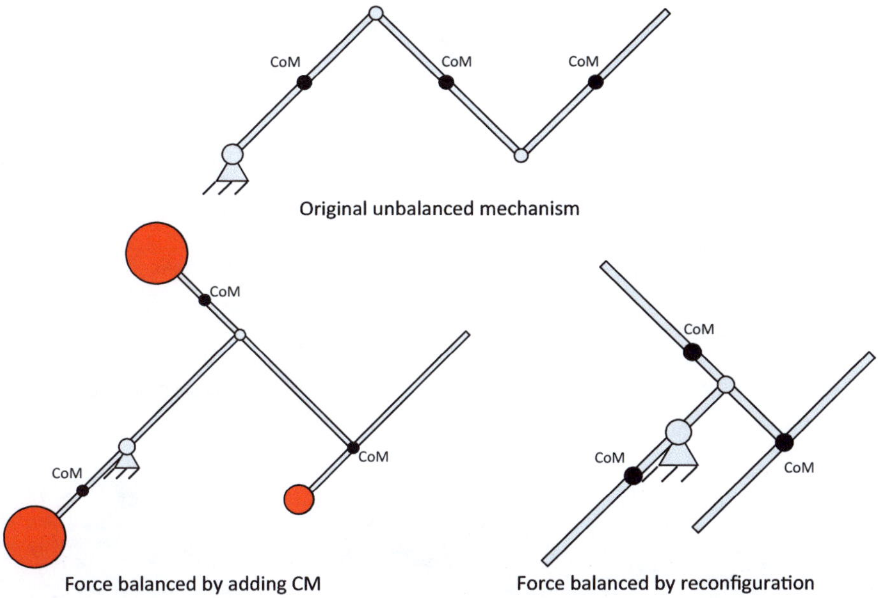

Fig. 1.8 Force balancing of 3-DOF serially connected link through reconfiguration

manipulator. One can see that force balance through reconfiguration does not add any counterweights, and also the function of the crank-slider mechanism remains the same, and does not change at all. If the links of the above crank-slider mechanism have same length, then it is moment balanced as well because it is symmetrical design [8].

In [23], we can use the above force balance through reconfiguration crank-slider mechanism as a Scott-Russell mechanism instead of traditional Scott-Russell mechanism (i.e., an Assur group) and add it in each leg of the 3RPR planar parallel manipulator as shown in Fig. 1.11. And also it is expected that if we change the passive balancing to active balancing, then the number of counter-rotations can be reduced to only one counter-rotation.

Only six counterweights and three counter-rotations are used if it is passive balancing. One can see that by using the force balance through reconfiguration crank-slider mechanism as a Scott-Russell mechanism (i.e., an Assur group), no counterweight is added on the Scott-Russell mechanism; if we stick to the original/traditional Scott-Russell mechanism, two counterweights are added on the Scott-Russell mechanism, which increases the weight. Based on the extension of [22], we can use reconfiguration method to force balance these 4-bar linkage with Assur group instead of adding those three counterweights, and use these through reconfiguration dynamic balanced 4-bar linkage with Assur group to construct the whole parallel robot; that is, decompose first and integrate later. But in [22], what makes the author think to add three CM to those positions to achieve force balance is not explained.

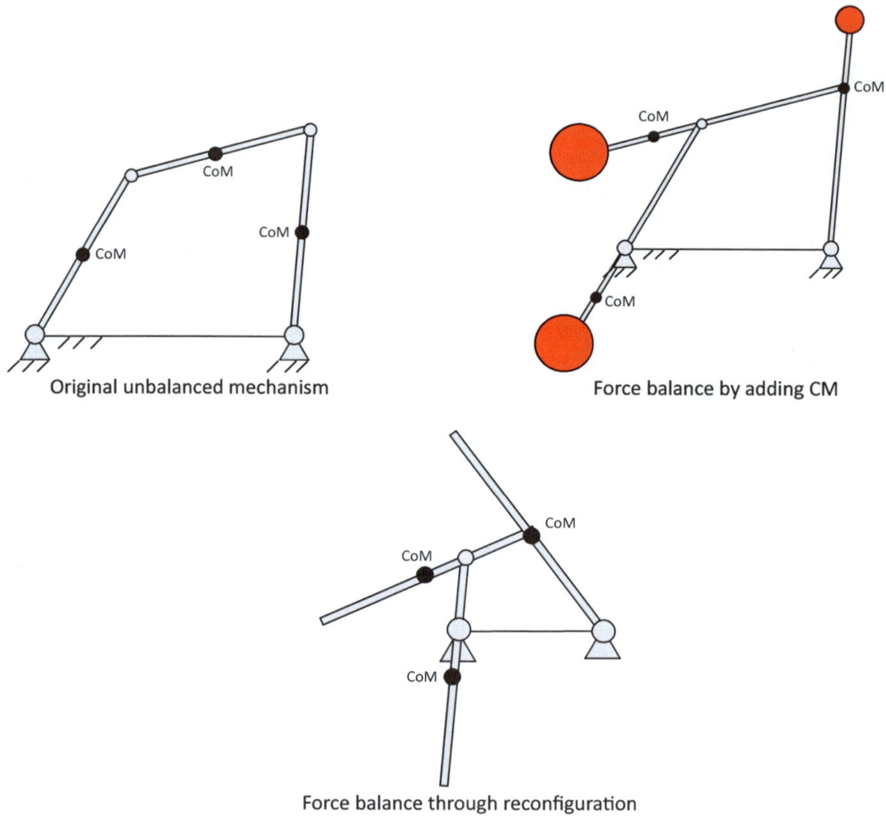

Fig. 1.9 Force balancing of 4R 4-bar linkage through reconfiguration (case I)

The above illustrates the dynamic balancing through reconfiguration method; instead of adding CM, the purpose of which is to move CoM, we can use reconfiguration method to achieve the same goal.

For the SteadiCam, it uses counterweights to achieve force balance, and through adjusting those mass relations dynamic balance is achieved. Here the concept of mass relationship is proposed. There are two links in the bottom acting as the counterweights; it is force balanced. Now if we spin it, it is dynamic balanced. If we move the link 2 up as shown in Fig. 1.12, it is still force balanced, but not dynamic balanced any more. So the question is how we can rearrange the structure, i.e., reconfigure the structure, to regain the dynamic balance.

Imagine that we move an extreme case; that is, let's move the link 2 all the way to the top; it is obvious that if we want to regain the dynamic balance, we need to move the camera counterclockwise direction, so does the mass 1. So we get the same situation; it is just that two masses are in the top and one mass is in the bottom. In other words, if we move the link 2 up a bit, i.e., counterclockwise direction, we need to move the camera counterclockwise as well and so does the mass 1 in order

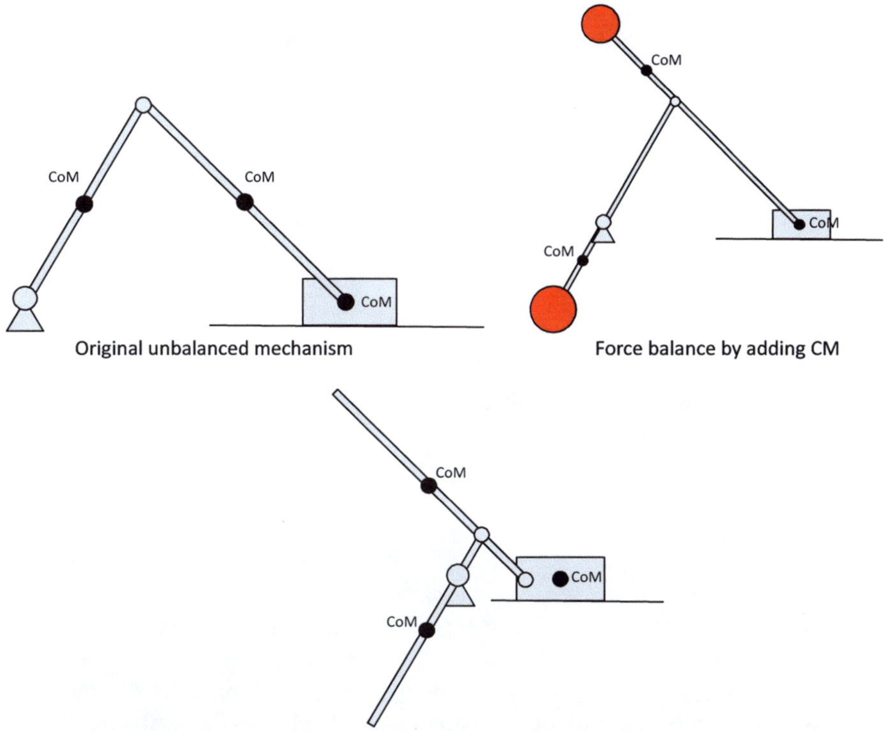

Fig. 1.10 Force balance of crank-slider mechanism through reconfiguration (if the links have same length, it is moment balanced as well)

to regain the dynamic balance. It is all about mass relations; as long as we keep those mass relations, the dynamic balance can be achieved. What is important is the relationship of these three masses.

Figure 1.13 can also be seen as the dynamic balancing through reconfiguration, i.e., through moving the link 2 and mass 2 to achieve dynamic balancing, adapting the position of the link 2 and mass 2.

1.3.4 Active Dynamic Balancing Unit

In [25], the paper deals with the active dynamic balancing. The paper presented an active dynamic balancing unit (ADBU), which is a unit that can be mounted on the base of the unbalanced mechanism and the unit is controlled such that the complete system is dynamically balanced. The goal of the ADBU is to produce balancing forces and balancing moments that are equal and opposite to the total

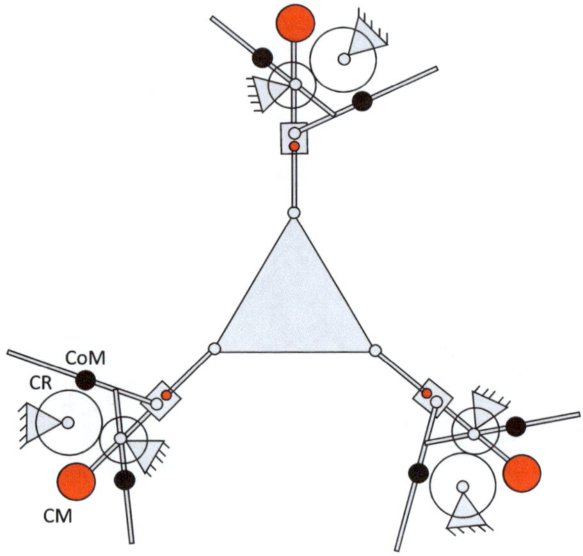

Fig. 1.11 Dynamic balanced 3RPR planar parallel manipulator (passive balancing)

shaking forces and total shaking moments of the machine. The ADBU constitutes of three counter-masses and three counter-rotations; the three counter-masses are used to force balance the shaking force along x, y, and z directions and the three counter-rotations are used to moment balance the shaking moment about x, y, and z directions. Consider the low mass addition aspect; the ADBU is evolved to a new ADBU that the three counter-masses and three counter-rotations are combined. In that paper, the ADBU needs to balance an xy-robot, which means this robot has two shaking forces in the plane, i.e., x and y directions and one shaking moment about z direction, so the ADBU only needs to balance two shaking forces in x and y directions and one shaking moment in z direction. So the ADBU is reduced from the original one to the one that has only two translation motions and one rotation motion. A 2RRR parallel mechanism is used to move the disc in x and y directions; the disc can also rotate; that is, this disc is a CRCM. Future work is to find advanced control strategies for controlling the ADBU.

In [26], a 3-DOF active dynamic balancing mechanism (ADBM) which is attached to the moving platform was proposed, and it is similar to the ADBU. This mechanism not only can balance the moving platform, but also can actuate the moving platform to move in a certain trajectory, but the main function of the ADBM is to balance the shaking force and shaking moment of the moving platform. The counterforces and counter-moments produced by ADBM are equal to the shaking forces and shaking moments plus the actuated force and actuated moment (i.e., one part of the forces and moments produced by ADBM is used to balance the shaking force and shaking moment, the other part of the forces and moments produced by ADBM is used to actuate the moving platform to a certain trajectory).

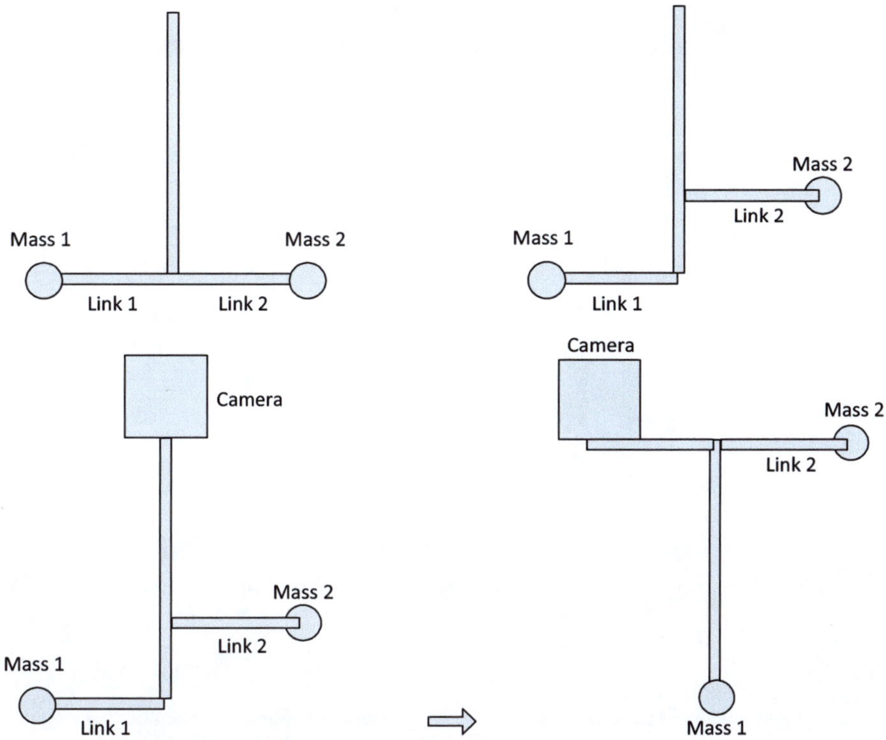

Fig. 1.12 Simplified version of SteadiCam

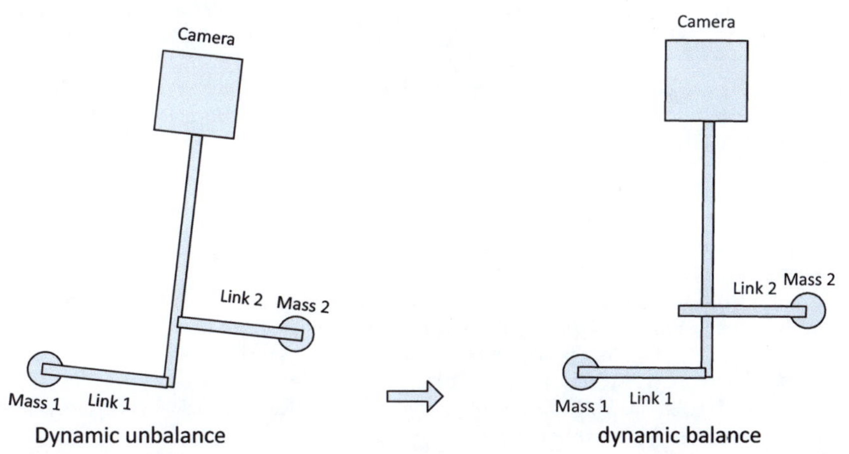

Fig. 1.13 Dynamic unbalance and balance of SteadiCam

1.4 Conclusion

Recent advances and issues on dynamic balancing of mechanisms are reviewed here. How to design reactionless mechanisms with minimum increase of mass and inertia or better yet that does not require counterweights or counter-rotations is the common desire. The advances and problems on dynamic balancing of mechanisms based on two main categories are discussed in details, the concept of dynamic balancing through reconfiguration is proposed, and new reactionless parallel manipulators are derived.

Acknowledgements The authors would like to thank the financial support from the Natural Sciences and Engineering Research Council of Canada (NSERC) and Canada Research Chairs program.

References

1. Ricard, R., Gosselin, C.M.: On the development of reactionless parallel manipulators. Proceedings of ASME Design Engineering Technical Conferences and Computers and Information in Engineering Conference, USA, pp. 1–10 (2000)
2. Wijk, V., Herder, J.: Dynamic balancing of Clavel's Delta robot. In: Computational Kinematics, pp. 315–322. Springer, Berlin (2009)
3. Gosselin, C., Vollmer, F., Cote, G., Wu, Y.N.: Synthesis and design of reactionless three-degree of freedom parallel mechanisms. IEEE Trans. Robot. Autom. **20**(2), 191–199 (2004)
4. Wu, Y.N.: Synthesis and analysis of reactionless spatial parallel mechanisms. Ph.D. dissertation, Laval University (2003)
5. Wijk, V.: Methodology for analysis and synthesis of inherently force and moment-balanced mechanisms—theory and applications. Ph.D. dissertation, University of Twente, Netherlands (2004)
6. Wijk, V., Herder, J.: On the development of low mass shaking force balanced manipulators. In: Advances in Robot Kinematics: Motion in Man and Machine, pp. 411–420. Springer, Dordrecht (2010)
7. Wijk, V., Herder, J.: Synthesis method for linkages with center of mass at invariant link point—pantograph based mechanisms. Mech. Mach. Theory **48**, 15–28 (2012)
8. Wijk, V., Krut, S., Pierrot, F., Herder, J.L.: Design and experimental evaluation of a dynamically balanced redundant planar 4-RRR parallel manipulator. Int. J. Rob. Res. **32**, 744–759 (2013)
9. Wijk, V., Krut, S., Pierrot, F., Herder, J.L.: Generic method for deriving the general shaking force balance conditions of parallel manipulators with application to a redundant planar 4-RRR parallel manipulator. IFToMM 2011 World Congress: The 13th World Congress in Mechanism and Machine Science, Mexico, pp. 1–9 (2011)
10. Arakelian, V.H., Smith, M.R.: Design of planar 3-DOF 3-RRR reactionless parallel manipulators. Mechatronics **18**(10), 601–606 (2008)
11. Gao, F.: Complete shaking force and shaking moment balancing of four types of six-bar linkages. Mech. Mach. Theory **24**(4), 275–287 (1989)
12. Gao, F.: Complete shaking force and shaking moment balancing of 26 types of four-, five- and six-bar linkages with prismatic pairs. Mech. Mach. Theory **25**(2), 183–192 (1990)
13. Gao, F.: Complete shaking force and shaking moment balancing of 17 types of eight-bar linkages only with revolute pairs. Mech. Mach. Theory **26**(2), 197–206 (1991)

14. Wu, Y.N., Gosselin, C.: Design of reactionless 3-DOF and 6-DOF parallel manipulators using parallelepiped mechanisms. IEEE Trans. Robot. **21**(5), 821–833 (2005)
15. Foucault, S., Gosselin, C.: Synthesis, design, and prototyping of a planar three degree-of-freedom reactionless parallel mechanism. ASME J. Mech. Des. **126**(6), 992–999 (2005)
16. Arakelian, V.H., Smith, M.R.: Complete shaking force and shaking moment balancing of linkages. Mech. Mach. Theory **34**(8), 1141–1153 (1999)
17. Herder, J.L., Gosselin, C.: A counter-rotary counterweight (CRCM) for light-weight dynamic balancing. Proceedings of DETC 2004 ASME Design Engineering Technical Conferences and Computers and Information in Engineering Conference, USA, pp. 1–9 (2004)
18. Herder, J.: Reaction-free Systems. Principles, conception and design of dynamically balanced mechanisms. Technical Report, Laval University (2003)
19. Wijk, V., Herder, J.L.: Synthesis of dynamically balanced mechanisms by using counter-rotary counter-mass balanced double pendula. ASME J. Mech. Des. **131**(11), 111003-1–8 (2009)
20. Wijk, V., Demeulenaere, B., Herder, J.L.: Comparison of various dynamic balancing principles regarding additional mass and additional inertia. ASME J. Mech. Rob. **1**(4), 041006-1–9 (2009)
21. Wijk, V., Demeulenaere, B., Gosselin, C., Herder, J.L.: Comparative analysis for low-mass and low-inertia dynamic balancing of mechanisms. ASME J. Mech. Rob. **4**(3), 031008-1–8 (2012)
22. Briot, S., Arakelian, V.: Complete shaking force and shaking moment balancing of in-line four-bar linkages by adding a class-two RRR or RRP Assur group. Mech. Mach. Theory **57**, 13–26 (2012)
23. Briot, S., Bonev, I.A., Gosselin, C.M., Arakelian, V.: Complete shaking force and shaking moment balancing of planar parallel manipulators with prismatic pairs. Proc. Inst. Mech. Eng. K **223**(1), 43–52 (2009)
24. Wijk, V., Herder, J.L.: Dynamic balancing of mechanisms by using an actively driven counter-rotary counter-mass for low mass and low inertia. Proceedings of the Second International Workshop on Fundamental Issues and Future Research Directions for Parallel Mechanisms and Manipulators, France, pp. 241–251 (2008)
25. Wijk, V., Herder, J.: Active dynamic balancing unit for controlled shaking force and shaking moment balancing. ASME 2010 International Design Engineering Technical Conferences and Computers and Information in Engineering Conference, Canada, pp. 1515–1522 (2010)
26. Wang, K., Luo, M., Mei, T.: Dynamics analysis of a three-DOF planar serial-parallel mechanism for active dynamic balancing with respect to a given trajectory. Int. J. Adv. Robot. Syst. **10**, 23 (2013). doi:10.5772/54201

Chapter 2
Design of Reactionless Mechanisms Without Counter-rotations

Vlastimil Votrubec

Abstract This chapter presents methods and principles used for balancing of planar mechanisms without counter-rotations. The fundamentals of balancing are described at first. Balancing only by counterweights provides only the force balance of mechanisms. Several basic methods which balance linkages by internal mass redistribution or adding of counterweights are introduced. These methods are the principal vector method, linearly independent vector method, complex mass method, and linear momentum method. The principles of these methods are explained in the example of the four-bar linkage and some extensions and important outcomes of these methods are added.

Keywords Balancing • Principal vector method • Linearly independent vector method • Complex mass method • Linear momentum method • Four-bar linkage

2.1 Principles of Balancing

During the working process of mechanisms, inertia forces and inertia torques are generated which are exerted to the base as reaction forces and moments. These reactions cause vibrations, inducing noise, wear, fatigue, poor product quality, and other undesired effects. Vibration suppression is usually achieved by applications of damping or other means. However, these solutions do not prevent the origin of vibrations. The balancing compared with, for example, damping eliminates or reduces the inertia forces and moments that cause the vibrations.

The sum of inertia forces that exert on the base of the mechanism is named shaking force and the sum of inertia torques is named shaking moment. The elimination of the shaking force acting on the base for any motion of the mechanism is called force balancing. The elimination of shaking moment is called moment balancing. The combination of force and moment balancing is called dynamic balancing which is synonymous to the terms complete balancing or reactionless.

V. Votrubec (✉)
VÚTS, a.s., Svárovská, 619, Liberec XI - Rüžodol I, 460 01 Liberec, Czech Republic
e-mail: vlastimil.votrubec@vuts.cz

D. Zhang, B. Wei (eds.), *Dynamic Balancing of Mechanisms and Synthesizing of Parallel Robots*, DOI 10.1007/978-3-319-17683-3_2

The research on balancing is very extensive and many principles and methods were described in literature. One of the basic ideas and the most general approach to the dynamic balancing resulted from the conservation of linear momentum law and the conservation of angular momentum law. The first law states that linear momentum p is conserved if the resultant force F is zero, and the second law states that angular momentum h is conserved if the resultant moment M is zero:

$$\frac{dp}{dt} = F = 0 \tag{2.1}$$

$$\frac{dh}{dt} = M = 0 \tag{2.2}$$

The opposites of the terms on the left-hand side of Eqs. (2.1) and (2.2) are the shaking force and shaking moment of the system. The shaking force and the shaking moment of the mechanism will vanish if the linear momentum p and the angular momentum h of a mechanism remain constant for any motion at all times. The linear and angular momentum of a completely balanced mechanism can be written as follows:

$$p = \sum_{i=1}^{n} m_i \dot{r}_i = \text{const} \tag{2.3}$$

$$h = \sum_{i=1}^{n} (I_i \dot{\varphi}_i + r_i \times m_i \dot{r}_i) = \text{const} \tag{2.4}$$

with i being the number of the link of the mechanism, m_i the mass of the link, r_i the position vector of its center of mass, I_i the moment of inertia, and φ_i the angle of rotation. These two constraints are necessary and sufficient conditions for a reactionless mechanism.

From Eq. (2.3), it implies that the center of mass of the force balanced mechanism performs a constant velocity motion or it is stationary. In practical situations, the second possibility is more convenient and the right-hand side of Eq. (2.3) is then set to zero. Similarly, the angular momentum in Eq. (2.4) is usually set to zero.

Constant linear momentum and angular momentum of the system also mean that there are no forces and moments between the system of moving links and the base. Internal forces and moments act within the mechanisms. They include the reaction forces and moments between the links, internal collisions, internal springs, friction, and other. External forces and moments act from outside of the mechanism, for example, gravity force, magnetic force, external springs, and collisions. The sum of all internal forces and moments is zero; therefore, they do not affect the linear and angular momentum and also the dynamic balancing.

The term static balancing is not the same as force balancing. The static balancing implies that the mechanism is in static equilibrium for any motion at all times, which means that the potential energy of a statically balanced mechanism remains constant. A forced balanced mechanism with a stationary center of mass has constant potential energy, so it is also statically balanced. The force balancing, therefore, implies static balancing but the opposite is not true. Constant potential energy can be achieved by using springs; however, there still exists a reaction force and moment on the base.

The positive effects of dynamic balancing are elimination of shaking force and moment and thus reducing vibrations and noise. Balancing has also some undesirable effects that cannot be neglected. The drawbacks of balancing are mainly addition of mass and inertia, influence on input torque, modification of machine design, and costs. Because of these disadvantages, the balancing of mechanisms in practical situations cannot be very often complete and is only partial. The shaking force and shaking moment are then reduced to an acceptable level and the solution is usually obtained by using an optimization procedure.

There are several methods for deriving the conditions for dynamic balancing. The first method is based on calculation of the linear and angular momentum and the conditions for which they are zero (constant). The next method calculates the position of the center of mass of a mechanism and the conditions for which it is stationary. Both methods are convenient for analytic solution of the problem, however, the second only for the force balancing. The conditions for balancing can also be derived using calculation of the shaking force and shaking moment for which they are zero. This method is especially suitable for numeric computations and partial balancing conditions.

There are many principles, methods, and practical solutions for designing reactionless mechanisms. The choice of methods described in this chapter is limited to methods which do not use counter-rotations and methods described in other chapters, for example, duplicate mechanisms, counter-rotations, and input torque balancing mechanisms. The stationary center of mass of a mechanism and thus the force balancing are usually accomplished by the addition of counterweights or redistribution of internal mass. Methods based on these principles are principal vector method [1], method of linearly independent vectors [2], and complex mass method [3]. Balancing of shaking moment without counter-rotations is still challenging. One solution is deriving balancing conditions from the angular momentum equation of the center of mass. Direct balancing of shaking moment is achieved by using noncircular gears or cam mechanisms.

2.2 Principal Vector Method

The principal vector method was published by O. Fisher in 1902 [1] and afterwards it was extended many times. The motion of the center of mass of a mechanism is described analytically and the parameters for which the center of mass is stationary are determined. The position of the center of mass is given by a series of vectors directed along the links of a mechanism. The magnitude of these vectors depends on the mass of each link and its center of mass position. The principal vectors create an augmented mechanism of parallelogram structure which contains the total center of mass.

The position of the total center of mass of the mechanism $\boldsymbol{r_t}$ is given by

$$r_{\mathrm{t}} = \frac{1}{M}\sum_{i=1}^{n} m_i r_{\mathrm{t}i} \tag{2.5}$$

where m_i is the mass of a moving link i, M is the total mass of n moving links, and r_{ti} is the position vector of link center of mass. The position vectors of the first and k center of mass r_{t1} and r_{tk} can be expressed as

$$r_{\mathrm{t1}} = b_1 e_1, \quad r_{\mathrm{tk}} = b_{\mathrm{k}} e_{\mathrm{k}} + \sum_{i=1}^{k-1} a_i e_i \tag{2.6}$$

with e_i being the unit vector directed along the link i, a_i the length of the link, and b_i the distance between the center of mass and the link joint.

This method is, for example, applied on the four-bar linkage as it is shown in Fig. 2.1. The position of the center of mass according to Eqs. (2.5) and (2.6) of this four-bar mechanism is

$$r_t = \frac{1}{M}(m_1 b_1 + m_2 a_1 + m_3 a_1)\, e_1 + \frac{1}{M}(m_2 b_2 + m_3 a_2)\, e_2 + \frac{1}{M} m_3 b_3 e_3 \tag{2.7}$$

This equation can be rewritten to the general formula

$$r_t = \sum_{i=1}^{n} h_i = \sum_{i=1}^{n} h_i e_i \tag{2.8}$$

where the vectors h_i are the *principal vectors*. The absolute values h_i of the last and the kth link are

$$h_{\mathrm{n}} = \frac{1}{M} m_{\mathrm{n}} b_{\mathrm{n}}, \quad h_{\mathrm{k}} = \frac{1}{M}\left(m_{\mathrm{k}} b_{\mathrm{k}} + \sum_{i=\mathrm{k}+1}^{n} m_i a_{\mathrm{k}} \right) \tag{2.9}$$

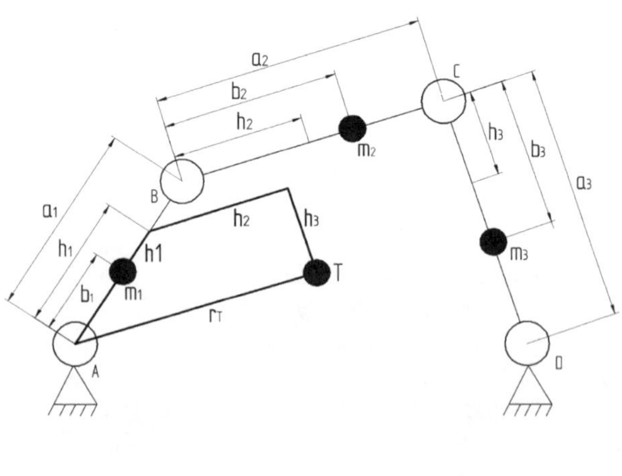

Fig. 2.1 The principle of principal vector method in the example of four-bar linkage

The end point T of the augmented mechanism performs motion of the center of mass of the origin mechanism. This point remains stationary for force balanced mechanisms. The conditions of force balancing are accomplished if the augmented mechanism is geometrically similar to the origin mechanism and if the end point of the augmented mechanism coincides with the fixed point of the origin mechanism. Let λ be the coefficient of geometrical similarity; then the mathematical expression of the first condition is equation

$$\lambda a_i = h_i \tag{2.10}$$

Equations for the four-bar linkage are

$$\lambda a_1 = \frac{1}{M}\left[m_1 b_1 + a_1\left(m_2 + m_3\right)\right] \tag{2.11}$$

$$\lambda a_2 = \frac{1}{M}\left(m_2 b_2 + m_3 a_2\right) \tag{2.12}$$

$$\lambda a_3 = \frac{1}{M} m_3 b_3 \tag{2.13}$$

Excluding λ, the system of equation is reduced to $n-1$ equation with $2n$ variables which create the conditions of a force balanced mechanism. For the given geometry a_i, it is not possible to determine the variables m_i and b_i uniquely. This is an advantage because $n+1$ parameters can be generally chosen, for example, according to the design options. This rule is valid only for an open kinematic chain with revolute joints.

Finally, the conditions of a force balanced four-bar linkage are

$$m_1 \frac{b_1}{a_1} + \frac{m_2\left(a_2 - b_2\right)}{a_2} = 0 \tag{2.14}$$

$$m_2 \frac{b_2}{a_2} + \frac{m_3\left(a_3 - b_3\right)}{a_3} = 0 \tag{2.15}$$

The principal vector method is also usable for mechanisms with more degrees of freedom and loops if the links are connected with revolute joints.

Extensions: The previous example of the force balancing is based on redistribution of the mass of a mechanism. It is also possible to make the total center of mass stationary and to balance the mechanism without changing its properties, using counterweights or an additional mechanism. An augmented pantograph device is a good example of direct balancing [4].

Principal vectors are useful also for the shaking moment balancing [5]. The principal vector linkage is used as a tool for moment balance solutions. Balance conditions are derived from the equation of angular momentum about the center of mass and this equation is written with principal dimensions, total mass, and total inertia radii.

2.3 Linearly Independent Vector Method

An important method of force balance was published in 1969 [2] and it was named the method of linearly independent vectors. It is based on redistribution of link mass, so the total center of mass remains stationary. The total center of mass is stationary when the coefficients of time-dependent terms of the equation describing the position of the center of mass vanish. It is accomplished when the time-dependent unit vectors within the previous equation are linearly independent.

The principle of this method is shown on the four-bar linkage; see Fig. 2.2. The position of the total center of mass corresponds with Eq. (2.5). The position vectors of the individual link centers of mass are expressed in a complex plane using the unit vectors $e^{i\varphi_j}$ with the reference origin at point A as

$$r_{t1} = b_1 e^{i(\varphi_1+\alpha_1)} \tag{2.16}$$

$$r_{t2} = a_1 e^{i\varphi_1} + b_2 e^{i(\varphi_2+\alpha_2)} \tag{2.17}$$

$$r_{t3} = a_4 e^{i\alpha_4} + b_3 e^{i(\varphi_3+\alpha_3)} \tag{2.18}$$

The total center of mass of the four-bar linkage is then

$$\begin{aligned} r_t = \frac{1}{M}\Big[&\left(m_1 b_1 e^{i\alpha_1} + m_2 a_1\right) e^{i\varphi_1} + \left(m_2 b_2 e^{i\alpha_2}\right) e^{i\varphi_2} \\ &+ \left(m_3 b_3 e^{i\alpha_3}\right) e^{i\varphi_3} + m_3 a_4 e^{i\alpha_4}\Big] \end{aligned} \tag{2.19}$$

where M is the sum of link mass m_i. The unit vectors form the loop equation

$$a_1 e^{i\varphi_1} + a_2 e^{i\varphi_2} - a_3 e^{i\varphi_3} - a_4 e^{i\alpha_4} = 0 \tag{2.20}$$

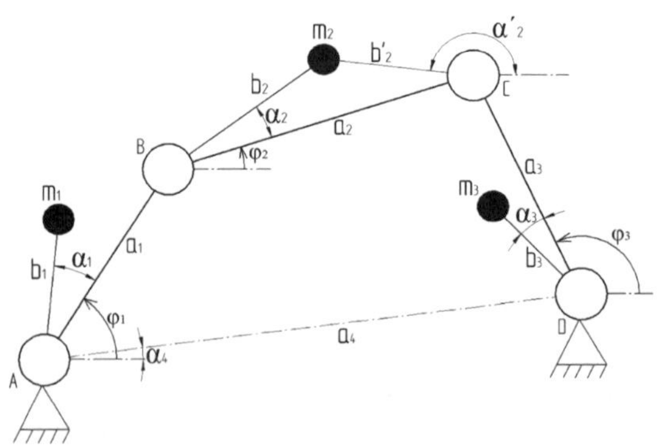

Fig. 2.2 The four-bar linkage

It means that the time-dependent terms in Eq. (2.19) are linearly dependent. If one of the unit vectors from Eq. (2.20) is derived and substituted to Eq. (2.19), then the equation of total center of mass position with linearly independent time-dependent terms is obtained:

$$r_t = \frac{1}{M}\left[\left(m_1 b_1 e^{i\alpha_1} + m_2 a_1 - m_2 \frac{a_1}{a_2} b_2 e^{i\alpha_2}\right) e^{i\varphi_1} \right.$$
$$\left. + \left(m_3 b_3 e^{i\alpha_3} + m_2 \frac{a_3}{a_2} b_2 e^{i\alpha_2}\right) e^{i\varphi_3} + \left(m_3 a_4 + m_2 \frac{a_4}{a_2} b_2 e^{i\alpha_2}\right) e^{i\alpha_4}\right] \tag{2.21}$$

The third term of Eq. (2.21) is constant, so the total center of mass is stationary if the coefficient of the first two time-dependent terms vanishes. The first term can be simplified, using the kinematic identity

$$b_2 e^{i\alpha_2} = a_2 + b_2^{'} e^{i\alpha_2^{'}} \tag{2.22}$$

which finally leads to the conditions of force balancing

$$m_1 b_1 = m_2 b_2^{'} \frac{a_1}{a_2}, \quad \alpha_1 = \alpha_2^{'} \tag{2.23}$$

$$m_3 b_3 = m_2 b_2 \frac{a_3}{a_2}, \quad \alpha_3 = \alpha_2 + \pi \tag{2.24}$$

If another unit vector is substituted to Eq. (2.21), equivalent conditions of balancing would be found. These conditions are also similar to the conditions derived from the principal vector method (Eqs. (2.14) and (2.15)). If the geometry of links is prescribed and cannot be changed, the force balance can be achieved by addition of counterweights which can be mounted on any of the two links. If the condition of static replacement $m_i = m_i^0 + m_i^*$ is satisfied, then the equations for calculation of counterweight parameters are

$$m_i^* b_i^* = \sqrt{(m_i b_i)^2 + \left(m_i^0 b_i^0\right)^2 - 2 m_i m_i^0 b_i b_i^0 \cos\left(\alpha_i - \alpha_i^0\right)} \tag{2.25}$$

$$\tan\alpha_i^* = \frac{m_i b_i \sin\alpha_i - m_i^0 b_i^0 \sin\alpha_i^0}{m_i b_i \cos\alpha_i - m_i^0 b_i^0 \cos\alpha_i^0} \tag{2.26}$$

where m_i^*, b_i^*, α_i^* are the parameters of counterweights, m_i, b_i, α_i are the parameters of balanced linkage resulting from Eqs. (2.23) and (2.24), and m_i^0, b_i^0, α_i^0 are the parameters of the unbalanced linkage.

Extensions: Generalization of the method of linearly independent vectors involves deriving the equation for the position of the total center of mass, eliminating the time-dependent coefficients and equating these coefficients to zero. The solution yields a relation between the link masses and the link geometries which must be fulfilled to obtain force balance. It does not depend on the method of how the balancing conditions are satisfied if the counterweights are added or the links are redesigned. A drawback of counterweight addition is that the other dynamic properties of the mechanism—the input torque, bearing forces, and shaking moment—are greater.

The generalization of this method yields important outcomes known as a *contour theorem* [6]. The first states that a planar mechanism without axisymmetric link groupings can always be fully force balanced by internal mass redistribution or addition of counterweights if from each link there is a contour to the ground by way of revolute joints only. This theorem is equivalent to stating that each equation within a set of independent loop equations cannot contain more than one term with a time-dependent coefficient. The second statement said that n-linked mechanism with one degree of freedom can be fully force balanced with $n/2$ counterweights.

2.4 Complex Mass Method

The complex mass method is derived from the previous method of linearly independent vectors and it was first presented by Walker and Oldham in 1978 [3]. The complex mass method simplifies the theory of linearly independent vectors and it develops a set of general relationships for obtaining the force balancing conditions of multi degrees of freedom, multi bar, and planar linkages. These conditions can be written directly without extracting them from the kinematic equations.

Let a chain of n links connected by revolute joints be pivoted about a frame pivot at one end (Fig. 2.3). The force balance is achieved by adding counterweights to each link. The counterweight on the *k-th* link must satisfy the following condition:

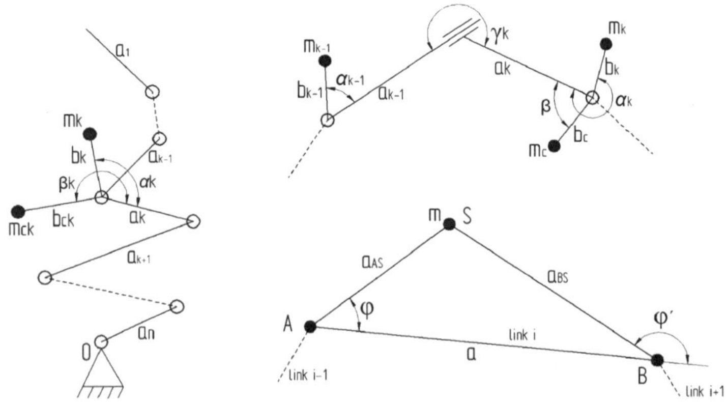

Fig. 2.3 Chain of n links with a counterweight attached to link k (*left*), prismatic joint with a counterweight attached to link k (*above*), and a dependent link i (*below*)

$$m_{\mathrm{ck}}b_{\mathrm{ck}}e^{i\beta_{\mathrm{k}}} + m_{\mathrm{k}}b_{\mathrm{k}}e^{i\alpha_{\mathrm{k}}} + a_{\mathrm{k}}\sum_{i=1}^{k-1}(m_i + m_{\mathrm{c}i}) \tag{2.27}$$

where i is a number of each link starting at the free end, m_i and m_{ci} are the masses of the links and counterweights, a_i the length of a link, and b_i, b_{ci}, α_i, and β_i are the radial and angular coordinates of the mass centers with respect to their lower joint and link.

Similarly, the condition for counterweight balancing of two links connected with prismatic joint is derived. One counterweight for the force balance is sufficient if the next condition is satisfied:

$$m_{\mathrm{c}}b_{\mathrm{c}}e^{i\beta} + m_{\mathrm{k}-1}b_{\mathrm{k}-1}e^{i\alpha_{\mathrm{k}-1}} + m_{\mathrm{k}}b_{\mathrm{k}}e^{i(\alpha_{\mathrm{k}}-\gamma)} = 0 \tag{2.28}$$

Only one prismatic joint in a loop can be contained if a linkage is to be force balanced.

It is not necessary to balance each link to obtain a force balance. If a mechanism contains a loop without prismatic joint, then one arbitrary link can be uncounterweighted. A loop containing one prismatic joint can have one link connected to the prismatic joint uncounterweighted. The loop is then divided into two counterweighted chains. For that reason, the uncounterweighted link, often called a dependent link, has to split its mass into both joints (see Fig. 2.3) according to the rules

$$m_{\mathrm{A}} = m\frac{a_{\mathrm{BS}}}{a}e^{i(\varphi'+\pi)} \tag{2.29}$$

$$m_{\mathrm{B}} = m\frac{a_{\mathrm{AS}}}{a}e^{i\varphi} \tag{2.30}$$

Generally a k-th link of a length a_k which lies in a chain between joints k-1/k and k/k + 1 can contain x masses m_t assigned to a joint as a result of a dependent link or another counterweighted chain being incident at or masses assigned to joint or links higher up the chain. This link can also have u revolute joints with y_{q} masses m_{d} assigned to the *q-th* joint. The *q-th* joint is in offset from the a_{k} link by an angle δ_{q} and in distance l_{k} from the joint k/k + 1. The link has v other links attached by prismatic joints. Masses m_{b} are assigned to the joint k/k + 1 and they are in offset from the a_{k} link by an angle η_{p}. The general force balance condition is

$$m_{\mathrm{k}}b_{\mathrm{k}}e^{i\alpha_{\mathrm{k}}} + m_{\mathrm{ck}}b_{\mathrm{ck}}e^{i\beta_{\mathrm{k}}} + a_{\mathrm{k}}\left[\sum_{t=1}^{x}m_t + \sum_{i=1}^{n}(m_i + m_{\mathrm{c}i})\right]$$
$$+\sum_{q=1}^{u}\left(l_{\mathrm{q}}e^{i\delta_{\mathrm{q}}} + \sum_{d=1}^{y_{\mathrm{q}}}m_{\mathrm{d}}\right) + \sum_{p=1}^{v}\sum_{b=1}^{z}m_b e^{i\eta_{\mathrm{p}}} = 0 \tag{2.31}$$

Eqs. (2.29) and (2.31) form a necessary and sufficient set for establishing the force balance conditions of planar linkages.

The extension of this method defines the minimum number of required counterweights and the most advantageous configuration of counterweights. In Ref. [7], the minimum number of counterweights c required to fully force balance linkages with n moving links and j simple joints is derived, which is given by

$$c = 2\,(n - 1) - j \tag{2.32}$$

This expression applies to any general planar linkage that can be force balanced irrespective of the number of degrees of freedom it has. However, some linkages with special geometries or mass distributions can be balanced with a smaller number of counterweights. If a planar linkage has only one degree of freedom, then Eq. (2.32) is simplified to the expression

$$c = \frac{n}{2} \tag{2.33}$$

which is in agreement with the conclusion in [6].

It was said that the addition of counterweights increases the dynamic parameters of mechanisms. For the given linkage, there can be many combinations of its links which can be counterweighted to give a force balance. The best results with respect to bearing forces, shaking moment, and driving torque are obtained if the chosen counterweighted links are as near as possible to the ground pivots.

The complex mass method was further extended to balance spatial linkages. Ye and Smith [8] developed an equivalence method for complete balancing of planar linkages. By this method, a complex planar linkage can be converted into a number of simple equivalent sub-linkages and cranks.

2.5 Linear Momentum Method

A very general method for deriving balancing conditions is based on equation of linear momentum which is for the force balanced mechanism constant (Eq. (2.3)). This method compared to the previous method requires calculation of derivatives and can be difficult and long, but the equations and conditions can be obtained for any linkage with a proposed balancing device. The principle of this method and derivation of balancing condition are shown again on the four-bar linkage.

The linear momentum of the four-bar linkage with two counterweights (Fig. 2.4) is given by

$$m_1\dot{\boldsymbol{r}}_{t1} + m_2\dot{\boldsymbol{r}}_{t2} + m_3\dot{\boldsymbol{r}}_{t3} + m_{c1}\dot{\boldsymbol{r}}_{tc1} + m_{c2}\dot{\boldsymbol{r}}_{tc2} = 0 \tag{2.34}$$

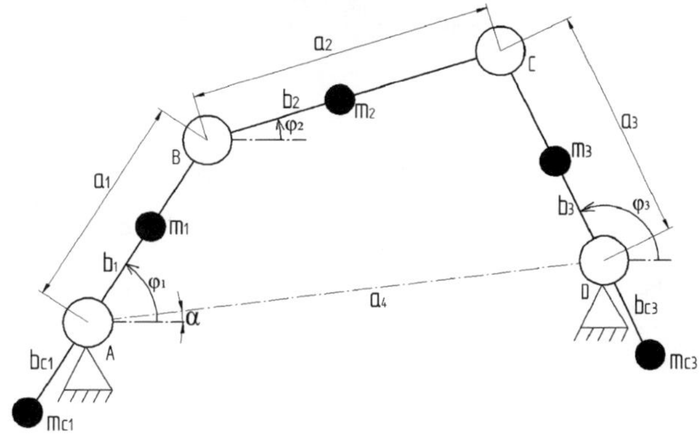

Fig. 2.4 The four-bar linkage with two counterweights attached to the crank and the rocker

where $\boldsymbol{r}_{ti}$ are the velocities of the masses m_i. The position vectors $\boldsymbol{r}_{ti}$ of individual centers of mass with reference origin at point A are

$$\boldsymbol{r}_{t1} = \begin{pmatrix} b_1 \cos\varphi_1 \\ b_1 \sin\varphi_1 \end{pmatrix} \tag{2.35}$$

$$\boldsymbol{r}_{c1} = \begin{pmatrix} -b_{c1} \cos\varphi_1 \\ -b_{c1} \sin\varphi_1 \end{pmatrix} \tag{2.36}$$

$$\boldsymbol{r}_{t2} = \begin{pmatrix} a_1 \cos\varphi_1 + b_2 \cos\varphi_2 \\ a_1 \sin\varphi_1 + b_2 \sin\varphi_2 \end{pmatrix} \tag{2.37}$$

$$\boldsymbol{r}_{t3} = \begin{pmatrix} a_4 \cos\alpha + b_3 \cos\varphi_3 \\ a_4 \sin\alpha + b_3 \sin\varphi_3 \end{pmatrix} \tag{2.38}$$

$$\boldsymbol{r}_{c3} = \begin{pmatrix} a_4 \cos\alpha - b_{c3} \cos\varphi_3 \\ a_4 \sin\alpha - b_{c3} \sin\varphi_3 \end{pmatrix} \tag{2.39}$$

Derivatives of these equations are substituted to Eq. (2.34). After some modifications, the first component of this equation is written as

$$\begin{aligned}&(m_1 b_1 - m_{c1} b_{c1}) \sin\varphi_1 \dot{\varphi}_1 + m_2 \left(a_1 \sin\varphi_1 \dot{\varphi}_1 + b_2 \sin\varphi_2 \dot{\varphi}_2\right) \\ &+ (m_3 b_3 - m_{c3} b_{c3}) \sin\varphi_3 \dot{\varphi}_3 = 0\end{aligned} \tag{2.40}$$

The second component differs from the first only in terms with function *sin* where the second equation has function *cos*. The terms $\sin\varphi_2\dot{\varphi}_2$ are derived from the loop equation

$$\begin{pmatrix} a_1 \cos\varphi_1 + a_2 \cos\varphi_2 \\ a_1 \sin\varphi_1 + a_2 \sin\varphi_2 \end{pmatrix} = \begin{pmatrix} a_4 \cos\alpha + l_3 \cos\varphi_3 \\ a_4 \sin\alpha + l_3 \sin\varphi_3 \end{pmatrix} \tag{2.41}$$

Eq. (2.40) is then expressed as

$$\left(m_1b_1 + m_2a_1 - m_{c1}b_{c1} - \frac{m_2b_2a_1}{a_2}\right)\sin\varphi_1\dot{\varphi}_1 + \left(m_3b_3 - m_{c3}b_{c3} + \frac{m_2b_2a_3}{a_2}\right)\sin\varphi_3\dot{\varphi}_3 = 0 \tag{2.42}$$

and the conditions of balancing are given by

$$\frac{m_1b_1}{a_1} + \frac{m_2(a_2 - b_2)}{a_2} - \frac{m_{c1}b_{c1}}{a_1} = 0 \tag{2.43}$$

$$\frac{m_3b_3}{a_3} + \frac{m_2b_2}{a_2} - \frac{m_{c3}b_{c3}}{a_3} = 0 \tag{2.44}$$

This method works well for other configurations of counterweights and other linkages. The same approach can be applied for moment balancing, only the angular momentum instead of linear momentum is used. Moment balancing usually requires addition of counter-rotations or other balancing methods which are described in the corresponding chapters.

References

1. Fisher, O.: Über die reduzierten Systeme und die Hauptpunkte der Glieder eines Gelenkmechanismus. Z. Math. Phys. **47**, 429–466 (1902)
2. Berkof, R.S., Lowen, G.G.: A new method for completely force balancing simple linkages. J. Eng. Ind. **91B**(1), 21–26 (1969)
3. Walker, M.J., Oldham, K.: A general theory of force balancing using counterweights. Mech. Mach. Theory. **13**(2), 175–185 (1978)
4. Hilpert, H.: Weight balancing of precision mechanical instruments. J. Mech. **3**, 289–302 (1968)
5. van der Wijk, V.: Shaking-moment balancing of mechanisms with principal vectors and momentum. Front. Mech. Eng. **8**(1), 10–16 (2013)
6. Tepper, F.R., Lowen, G.G.: General theorems concerning full force balancing of planar linkages by internal mass redistribution. J. Manuf. Sci. Eng. **94**(3), 789–796 (1972)
7. Walker, M.J., Oldham, K.: Extension to the theory of balancing frame forces in planar linkages. Mech. Mach. Theory **14**(3), 201–207 (1979)
8. Ye, Z., Smith, M.R.: Complete balancing of planar linkages by an equivalence method. Mech. Mach. Theory **29**(5), 701–712 (1994)

Chapter 3
Design of Reactionless Linkages and Robots Equipped with Balancing Assur Groups

Sébastien Briot and Vigen Arakelian

Abstract In the present chapter, we consider the shaking moment and shaking force balancing through the use of additional Assur groups mounted on the mechanism to be balanced. Two types of mechanisms are considered: (1) the in-line four-bar linkage and (2) the planar parallel robots with prismatic pairs. For both types of mechanisms, the proposed solution allows the reduction (or even the cancellation in the case of the four-bar linkage) of the number of counter-rotations used for obtaining the shaking moment balancing, which decreases the design complexity and the inherent problems due to the use of counter-rotations (backlash, noise, vibrations, etc.). All theoretical developments are validated via simulations carried out using ADAMS software. The simulations show that the obtained mechanisms (both in-line four-bar linkages and planar parallel robots) transmit no inertia loads to their surroundings, i.e. the sum of all ground bearing forces and their moments are eliminated.

Keywords Dynamic balancing • Assur groups • Inline four-bar linkage • Planar parallel robots with prismatic pairs

3.1 Introduction

It is known that fast-moving machinery with rotating and reciprocating masses is a significant source of vibration excitation. The high-speed linkages can generate significant fluctuating forces with even small amounts of unbalance. Thus, a primary

S. Briot (✉)
Institut de Recherche en Communications et Cybernétique de Nantes (IRCCyN), UMR CNRS 6597, Nantes, France
e-mail: Sebastien.Briot@irccyn.ec-nantes.fr

V. Arakelian
Institut National des Sciences Appliquées (INSA), Rennes, France

Institut de Recherche en Communications et Cybernétique de Nantes (IRCCyN), UMR CNRS 6597, Nantes, France
e-mail: vigen.arakelian@insa-rennes.fr

D. Zhang, B. Wei (eds.), *Dynamic Balancing of Mechanisms and Synthesizing of Parallel Robots*, DOI 10.1007/978-3-319-17683-3_3

objective of the balancing is to cancel or reduce the variable dynamic loads transmitted to the frame and surrounding structures. The reduction of vibrations leads to the increased accuracy of manipulators [1], which is one of the positive consequences of the balancing. As was mentioned in [2], balancing brings other advantages such as a reduced cycle time [3], reduced noise, wear, and fatigue [4], as well as improved ergonomics [5].

In general, two types of forces must be considered: the externally applied forces and the inertial forces. Inertial forces arise when links of a mechanism are subjected to large accelerations. The inertial force system acting on a given link can be represented as an inertia force acting on a line through the center of mass and an inertia torque about the center of mass. The determination of the inertial forces and torques is well known and it has been disclosed in various handbooks [6]. With regard to the external forces, which are associated with the useful function that the mechanism is to perform, these are often smaller than inertia forces with a much lower variation. On the other hand, when formulating balancing conditions of a mechanism, it is necessary to recognize that, in many cases, external active forces applied to mechanism links constitute internal forces with respect to the mechanism as a whole. Thus, if all external active forces applied to the links of a mechanism are internal forces for the mechanism as whole, then the balance of the mechanism will be ensured under the fulfillment of inertia forces and inertia torque cancellation. Therefore, the balancing of shaking force and shaking moment due to the inertial forces of links acquires a specific importance. The quality of balancing of the moving masses has the influence not only on the level of vibrations but also on the resource, reliability, and accuracy of mechanisms. Besides the mentioned negative effects, vibrations bring to the environments pollution and the loss of energy, and can also provoke various health issues. Consequently, the quality improvement of the mass balancing has not only technical, technological, and economical aspects but also social.

Different approaches and solutions devoted to the shaking force and shaking moment balancing have been developed and documented for one-degree-of-freedom mechanisms [7–9]. Nowadays, a new field for balancing methods applications is the design of mechanical systems for fast manipulation [10], which is a typical problem in advanced robotics [11]. Here also we have similar problems relating to the cancelation or reduction of inertia forces. However, the mechanical systems with multi-degrees of freedom lead to new solutions, such as the shaking force and shaking moment reduction by optimal motions of links, by adding flywheels with prescribed motions, or with the design of new self-balanced manipulators.

For all balancing method, the main challenge is the trade-off between the complexity of the balanced mechanism and the quality of balancing [9]. In the present chapter, we propose a solution which allows the reduction of the balancing complexity by comparison with the usual approaches. The idea is to slightly modify the mechanism design by adding to it Assur groups, i.e. groups which do not add any supplementary degree of freedom into the mechanism [12]. The use of such a solution is detailed in the following of the chapter for the shaking force and shaking moment balancing of:

- the in-line four-bar linkage
- the planar parallel robots with prismatic pairs.

For both types of mechanisms, the proposed solution allows the reduction (or even the cancellation in the case of the four-bar linkage) of the number of counter-rotations used for obtaining the shaking moment balancing, which decreases the design complexity and the inherent problems due to the use of counter-rotations (backlash, noise, vibrations, etc.).

3.2 Complete Shaking Force and Shaking Moment Balancing of In-Line Four-Bar Linkages by Adding a Class-Two RRR or RRP Assur Group

Many high-speed machines contain planar four-bar linkages and the problem of their mass balancing is of continuing interest to machine designers. The previous works on the balancing of planar four-bar linkages may be arranged in the following groups [6]:

1. **Complete shaking force balancing [13–19].** In general, it is carried out by counterweights mounted on the movable links of the linkage. With regard to the several approaches employed for the redistribution of movables masses, the developed methods could be divided into three principal groups:

   (a) The method of "principal vectors" [13]; The aim of this approach was to study the balancing of the mechanism relative to each link and in the determination of those points on the links relative to which a static balance was obtained. These points were called "principal points." Then, from the condition of similarity of the vector loop of the principal points and the structural loop of the mechanism, the necessary conditions of balancing were derived.
   (b) The method of "static substitution of masses"; its aim was to statically substitute the mass of the coupler by concentrated masses, which are balanced thereafter together with the rotating links. Such an approach changes the problem of mechanism balancing into a simpler problem of balancing rotating links. This method was illustrated for four-bar linkage in [14–17].
   (c) The method of "linearly independent vectors" [18], in which the vector equation describing the position of the center of total mass of the mechanism is treated in conjunction with the closed equation of its kinematic chain. The result is an equation of static moments of moving link masses containing single linearly independent vectors. Thereafter following the conditions for balancing the mechanism by reducing the coefficients, which are time-dependent to zero.

   It should be noted that the addition of a counterweighted pantograph device to the planar four-bar linkage has also been used for its complete shaking force balancing [19].

2. **Complete shaking force and partial shaking moment balancing [20–29].** Two principal approaches may be distinguished:
   (a) The shaking moment minimization of fully force balanced linkages [20–26], in which it is shown that the optimum conditions of partial moment balance can be obtained by certain link mass distribution ratios.
   (b) The minimization of the unbalance of shaking moment by transferring the rotation axis of the counterweight mounted on the input crank [27–29]. In the study [27], the first harmonic of the shaking moment is eliminated by attaching the required input link counterweight, not to the input shaft itself, but to a suitable offset one which rotates with the same angular velocity. This approach is original in that, while maintaining the force balance of the mechanism, it is possible to create an additional balancing moment, thereby reducing the shaking moment. This approach has been further developed in works [28, 29].

   It should be noted that optimization algorithms are also widely used in partial balancing of four-bar linkages [30–33].
3. **Complete shaking force and shaking moment balancing [9, 34–43].** The first method of complete shaking force and shaking moment balancing was proposed in study [34], which was extended in [35]. In this approach, the mass of the connecting coupler is substituted dynamically by concentrated masses located at the coupler joints. Thus, the dynamic model of the coupler becomes a weightless link with two concentrated masses. This transforms the problem of four-bar linkage shaking force and shaking moment balancing into a problem of balancing rotating links carrying concentrated masses. The parallelogram structure has also been applied for complete shaking force and shaking moment balancing of four-bar linkages [36]. In the studies [37–41], the authors have proposed methods for complete shaking force and shaking moment balancing by counterweights with planetary gear trains. In [42] a toothed-belt transmission is used to rotate counterweights intended for shaking force balancing, which also allows shaking moment balancing. The disadvantage of these methods is the need for the connection of gears to the oscillating links. The oscillations of the links of the mechanism will create noise unless expensive anti-backlash gears are used.

   Another solution using the copying properties of the pantograph was developed [9, 43], in which the gears driven by the coupler suffer no such sudden reversals so that this problem is almost eliminated. However, it should be noted that the use of the gears for the balancing of four-bar linkages is a drawback for the industrial applications and a fully shaking force and shaking moment balanced four-bar linkage without any gears is more appealing.

   The shaking moment balancing of fully force balanced linkages using a prescribed input speed fluctuation was proposed in [44]. However, such balancing is complicated because it is necessary to use a special type of drive generator. Moreover, it cannot be used for balancing of linkages, which generate the prescribed motions of the output links.

In this section a solution is discussed, which allows the complete shaking force and shaking moment balancing of in-line four-bar linkages with constant input speed by adding a class-two Assur group, i.e. a group which does not add any supplementary degree of freedom into the mechanism [12]. It should be noted that the balancing of the shaking moment without counter-rotations of three particular classes of four-bar linkages was discussed in the studies [45–49]. However, such a method cannot be extended to general four-bar linkages. In this section it is proposed to take advantages of the use of the properties of the four-bar linkage with prescribed geometric parameters [45–49] and to combine it with

1. the principle of the dynamic substitution of link mass by concentrated masses
2. and with the prescription of constant input speed.

It should be mentioned here that the suggested balancing approach can be efficiently applied on the cyclic high-speed machines executing motions in the steady-state regime when the input speed is constant [50, 51].

3.2.1 Complete Shaking Force and Shaking Moment Balancing by Adding a Class-Two RRR Assur Group

3.2.1.1 Theoretical Background Related to the Balanced Four-Bar Linkages with Prescribed Geometrical Parameters

Before considering the suggested balancing concept, let us recall basic notions concerning the balanced four-bar linkages with prescribed geometrical parameters.

In the paper of Berkof and Lowen [21], the angular momentum H and the shaking moment M^{sh}, expressed at point O, of a force balanced in-line four-bar linkage (Fig. 3.1) were expressed as:

$$H = \sum_{i=1}^{3} I_i \dot{\theta}_i, \quad M^{\text{sh}} = \frac{\mathrm{d}H}{\mathrm{d}t} = \sum_{i=1}^{3} I_i \ddot{\theta}_i \tag{3.1}$$

with

$$I_i = m_i \left(k_i^2 + r_i^2 + r_i l_i\right), \quad (i = 1, 3) \tag{3.2}$$

$$I_2 = m_2 \left(k_2^2 + r_2^2 - r_2 l_2\right) \tag{3.3}$$

where r_i is the length of vector $\mathbf{r}_i$ which connects the pivot P_i to the center of mass S_i of link i, l_i is the length of vector $\mathbf{l}_i$ which connects the proximal revolute joint P_i to the distal joint on the same link, and k_i is the radius of gyration with respect to the center of mass of link i, m_i is the mass of link i. Moreover, θ_i is the angular position of link i with respect to the $\mathbf{x}$-axis.

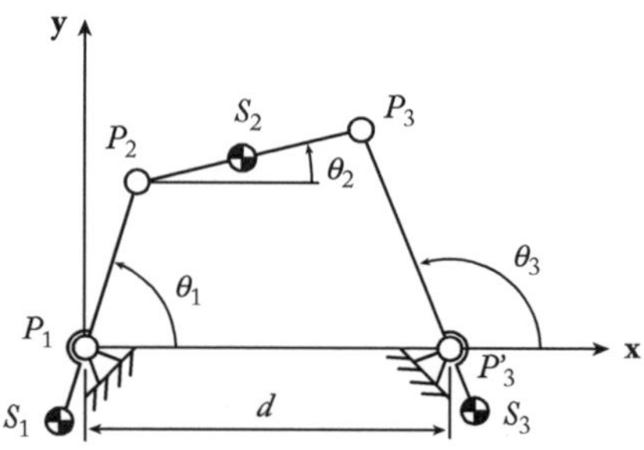

Fig. 3.1 A general in-line four-bar mechanism

With regard to the shaking force balancing, the following expressions were obtained:

$$m_1 r_1 = m_2 l_1 \frac{l_2 - r_2}{l_2} \tag{3.4}$$

$$m_3 r_3 = m_2 r_2 \frac{l_3}{l_2} \tag{3.5}$$

In the works [47, 52], it has been shown that it is possible to cancel the shaking moment of a four-bar mechanism by associating mentioned geometric constraints with an optimal distribution of masses. Three kinds of shaking force and shaking moment balanced four-bar mechanisms were found, which are shown in Fig. 3.2.

In order to illustrate the shaking moment balancing, let us consider the mechanism shown in Fig. 3.2b. The geometrical constraints of this mechanism are the following:

$$l_1 = l_3 \tag{3.6}$$

$$d = l_2 \tag{3.7}$$

where d is the length of the base which is the distance between the two fixed joints on the base.

This leads to the following kinematic relationships:

$$\dot{\theta}_1 - \dot{\theta}_2 + \dot{\theta}_3 = 0 \tag{3.8}$$

Thus, from expressions (3.1) and (3.8), it is easy to see that the shaking moment will be cancelled if $I_1 = I_3 = -I_2$ [see Eqs. (3.2) and (3.3)]. For this purpose the following relationships must be established:

$$k_2^2 = \frac{m_2\left(l_2 r_2 - r_2^2\right) - I_1}{m_2} \tag{3.9}$$

$$k_3^2 = \frac{-m_3\left(l_3 r_3 + r_3^2\right) + I_1}{m_3} \tag{3.10}$$

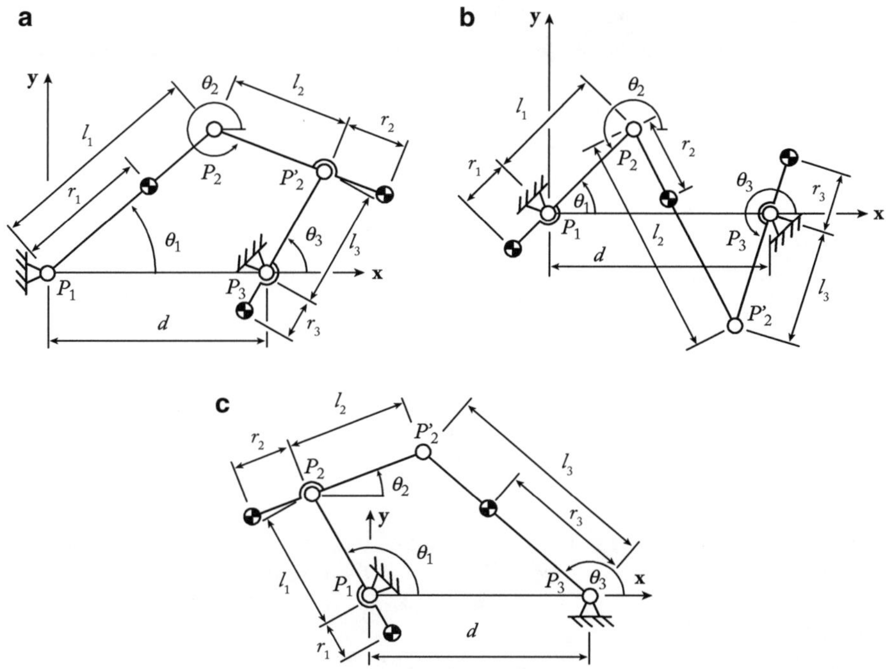

Fig. 3.2 The three kinds of shaking force and shaking moment balanced four-bar mechanisms. (**a**) Case I: $l_1 = d$ and $l_2 = l_3$. (**b**) Case II: $l_2 = d$ and $l_1 = l_3$. (**c**) Case III: $l_3 = d$ and $l_1 = l_2$

where

$$I_1 = m_1 \left(k_1^2 + r_1^2 + r_1 l_1\right) \tag{3.11}$$

It should be noted that similar results have been obtained for the mechanisms shown in Fig. 3.2a, c.

Statement of the Problem The aim of the suggested balancing approach consists of adding a two-link kinematic chain with prescribed geometrical parameters to an in-line four-bar linkage with arbitrary geometrical parameters. It is important to note that the added structure must be an Assur group, i.e. a group which does not add any supplementary degree of freedom into the mechanism [12]. This allows for the modification of the mass redistribution of the obtained six-bar mechanism without perturbation of the kinematic properties of the initial four-bar linkage. We would like to state that this technique allows for the complete shaking force and shaking moment balancing without counter-rotating masses.

Now let us consider the shaking force and shaking moment balancing of an in-line four-bar mechanism using class-two Assur groups with *RRR* kinematic chain.

3.2.1.2 Shaking Force Balancing

Figure 3.3 shows an in-line four-bar linkage with the added class-two *RRR* Assur group. Let us denote the following vectors as: $\mathbf{l}_1 = \mathbf{d}_{OA}$, $\mathbf{l}_2 = \mathbf{d}_{AB}$, $\mathbf{l}_3 = \mathbf{d}_{CB}$, $\mathbf{l}'_3 = \mathbf{d}_{CP_2}$, $\mathbf{l}_4 = \mathbf{d}_{P_2P'_2}$, $\mathbf{l}_5 = \mathbf{d}_{P_3P'_2}$, $\mathbf{r}_1 = \mathbf{d}_{OS_1}$, $\mathbf{r}_2 = \mathbf{d}_{AS_2}$, $\mathbf{r}_3 = \mathbf{d}_{CS_3}$, $\mathbf{r}_4 = \mathbf{d}_{P_2S_4}$, $\mathbf{r}_5 = \mathbf{d}_{P_3S_5}$, $\mathbf{r}_{\mathrm{cw}_1} = \mathbf{d}_{OS_{\mathrm{cw}_1}}$, $\mathbf{r}_{\mathrm{cw}2} = \mathbf{d}_{CS_{\mathrm{cw}_2}}$, $\mathbf{r}_{\mathrm{cw}3} = \mathbf{d}_{P_3S_{\mathrm{cw}_3}}$.

The added class-two *RRR* Assur group has the above-mentioned properties, i.e. it is designed such as:

$$l'_3 = l_5 \tag{3.12}$$

$$e = l_4 \tag{3.13}$$

where l_i $(i = 1 \dots 5)$ is the norm of vector $\mathbf{l}_i$ and e the distance between C and P_3.

As is shown in Fig. 3.3, the Assur group $P_2P'_2P_3$ is attached to the initial linkage *OABC* in such a way that it forms a four-bar linkage with link *BC*, as discussed in Sect. 3.2.1.1.

Let us now derive the expression of the shaking force $\mathbf{f}^{\mathrm{sh}}$ of the obtained mechanism:

$$\mathbf{f}^{\mathrm{sh}} = \sum_{i=1}^{5} m_i \ddot{\mathbf{d}}_{S_i} \tag{3.14}$$

where $\ddot{\mathbf{d}}_{S_i}$ is the translational acceleration of the center of mass S_i and m_i the mass of the link i. Developing and simplifying, one obtains:

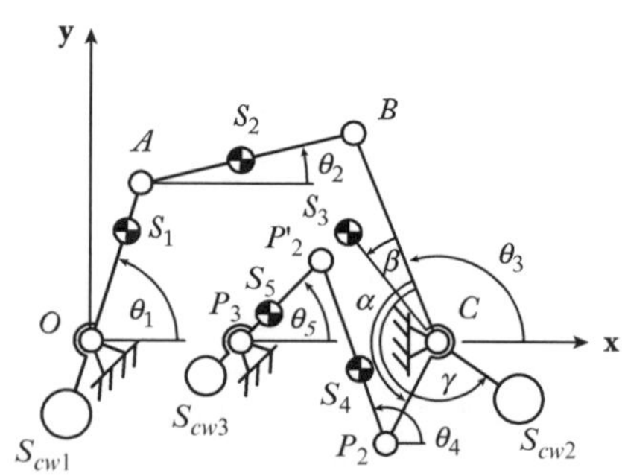

Fig. 3.3 The balanced mechanism with the class-two *RRR* Assur group

$$\mathbf{f}^{sh} = \left(\frac{m_1 r_1}{l_1} + m_2 \frac{l_2 - r_2}{l_2}\right)\ddot{\mathbf{d}}_A + m_2 \frac{r_2}{l_2}\ddot{\mathbf{d}}_B + m_3 \ddot{\mathbf{d}}_{S_3} + m_4 \frac{l_4 - r_4}{l_4}\ddot{\mathbf{d}}_{P_2} + \left(\frac{m_4 r_4}{l_4} + \frac{m_5 r_5}{l_5}\right)\ddot{\mathbf{d}}_{P_2'} \tag{3.15}$$

where r_i is the algebraic values of the norm of vectors $\mathbf{r}_i$, and $\ddot{\mathbf{d}}_A$, $\ddot{\mathbf{d}}_B$, $\ddot{\mathbf{d}}_{S_3}$, $\ddot{\mathbf{d}}_{P_2}$ and $\ddot{\mathbf{d}}_{P_2'}$ represent the acceleration of points A, B, S_3, P_2, and P_2', respectively. Their expressions are:

$$\ddot{\mathbf{d}}_A = l_1 \left(\ddot{\theta}_1 \begin{bmatrix} -\sin\theta_1 \\ \cos\theta_1 \end{bmatrix} - \dot{\theta}_1^2 \begin{bmatrix} \cos\theta_1 \\ \sin\theta_1 \end{bmatrix}\right) \tag{3.16}$$

$$\ddot{\mathbf{d}}_B = l_3 \left(\ddot{\theta}_3 \begin{bmatrix} -\sin\theta_3 \\ \cos\theta_3 \end{bmatrix} - \dot{\theta}_3^2 \begin{bmatrix} \cos\theta_3 \\ \sin\theta_3 \end{bmatrix}\right) \tag{3.17}$$

$$\ddot{\mathbf{d}}_{S_3} = r_3 \left(\ddot{\theta}_3 \begin{bmatrix} -\sin(\theta_3+\beta) \\ \cos(\theta_3+\beta) \end{bmatrix} - \dot{\theta}_3^2 \begin{bmatrix} \cos(\theta_3+\beta) \\ \sin(\theta_3+\beta) \end{bmatrix}\right) \tag{3.18}$$

$$\ddot{\mathbf{d}}_{P_2} = l_3' \left(\ddot{\theta}_3 \begin{bmatrix} -\sin(\theta_3+\alpha) \\ \cos(\theta_3+\alpha) \end{bmatrix} - \dot{\theta}_3^2 \begin{bmatrix} \cos(\theta_3+\alpha) \\ \sin(\theta_3+\alpha) \end{bmatrix}\right) \tag{3.19}$$

$$\ddot{\mathbf{d}}_{P_2'} = l_5 \left(\ddot{\theta}_5 \begin{bmatrix} -\sin\theta_5 \\ \cos\theta_5 \end{bmatrix} - \dot{\theta}_5^2 \begin{bmatrix} \cos\theta_5 \\ \sin\theta_5 \end{bmatrix}\right) \tag{3.20}$$

in which the angles θ_i ($i = 1 \ldots 5$) are defined in Fig. 3.3.

The shaking force $\mathbf{f}^{sh}$ may be cancelled through the addition of three counterweights positioned at points S_{cw_i} (Fig. 3.3), with masses m_{cw_i} ($i = 1, 2, 3$). With such counterweights, the expression of the shaking force becomes:

$$\mathbf{f}^{sh^*} = \mathbf{f}^{sh} + m_{cw_1}\frac{r_{cw_1}}{l_1}\ddot{\mathbf{d}}_A + m_{cw_2}\ddot{\mathbf{d}}_{S_{cw_2}} + m_{cw3}\frac{r_{cw3}}{l_5}\ddot{\mathbf{d}}_{P_2'} \tag{3.21}$$

where r_{cw_i} is the algebraic values of the norm of vectors $\mathbf{r}_{cw_i}$ and

$$\ddot{\mathbf{d}}_{S_{cw_2}} = r_{cw_2} \left(\ddot{\theta}_3 \begin{bmatrix} -\sin(\theta_3+\gamma) \\ \cos(\theta_3+\gamma) \end{bmatrix} - \dot{\theta}_3^2 \begin{bmatrix} \cos(\theta_3+\gamma) \\ \sin(\theta_3+\gamma) \end{bmatrix}\right) \tag{3.22}$$

Thus the shaking force is cancelled if the distribution of the masses is as follows:

$$m_{cw_1} = -\frac{l_1}{r_{cw_1}}\left(m_1\frac{r_1}{l_1} + m_2\frac{l_2 - r_2}{l_2}\right) \tag{3.23}$$

$$\tan\gamma = \frac{m_3 r_3 \sin\beta + m_4\left(1 - r_4/l_4\right) l_3' \sin\alpha}{m_2 l_3 r_2/l_2 + m_3 r_3 \cos\beta + m_4\left(1 - r_4/l_4\right) l_3' \sin\alpha} \tag{3.24}$$

$$m_{\mathrm{cw}_2}^2 = \frac{\left(m_3 r_3 \sin\beta + m_4\left(1 - r_4/l_4\right) l_3' \sin\alpha\right)^2}{r_{\mathrm{cw}_2}^2} + \frac{\left(m_2 l_3 r_2/l_2 + m_3 r_3 \cos\beta + m_4\left(1 - r_4/l_4\right) l_3' \cos\alpha\right)^2}{r_{\mathrm{cw}_2}^2} \tag{3.25}$$

$$m_{\mathrm{cw}_3} = -\frac{l_5}{r_{\mathrm{cw}_3}}\left(m_4\frac{r_4}{l_4} + m_5\frac{r_5}{l_5}\right) \tag{3.26}$$

3.2.1.3 Shaking Moment Balancing

Let us now derive the expression of the shaking moment M^{sh}, expressed at point O, of such a mechanism:

$$M^{\mathrm{sh}} = \sum_{i=1}^{5} m_i\left(x_{S_i}\ddot{y}_{S_i} - y_{S_i}\ddot{x}_{S_i} + k_i^2\ddot{\theta}_i\right) + \sum_{i=1}^{3} m_{\mathrm{cw}_i}\left(x_{S_{\mathrm{cw}_i}}\ddot{y}_{S_{\mathrm{cw}_i}} - y_{S_{\mathrm{cw}_i}}\ddot{x}_{S_{\mathrm{cw}_i}}\right) \tag{3.27}$$

where x_{S_i}, y_{S_i}, $\ddot{x}_{S_i}$, and $\ddot{y}_{S_i}$ are the position and accelerations along **x** and **y** axes of points S_i, respectively, and $x_{S_{\mathrm{cw}_i}}$, $y_{S_{\mathrm{cw}_i}}$, $\ddot{x}_{S_{\mathrm{cw}_i}}$, and $\ddot{y}_{S_{\mathrm{cw}_i}}$ are the position and accelerations along **x** and **y** axes of points S_{cw_i}, respectively, k_i is the radius of gyration of link i.

Now, let us consider that link 2 is a physical pendulum[1] [35], i.e. it can be replaced dynamically by two point masses located at joint centers A and B. This implies that:

$$k_2^2 = r_2\left(l_2 - r_2\right) \tag{3.28}$$

Then, considering that the input speed is constant, i.e. $\ddot{\theta}_1 = 0$ and taking into account (3.28), one can simplify (3.27) as:

$$M^{\mathrm{sh}} = \sum_{i=3}^{5} I_i\ddot{\theta}_i \tag{3.29}$$

where

$$I_3 = m_3\left(k_3^2 + r_3^2\right) + m_2\frac{r_2}{l_2}l_3^2 + m_{\mathrm{cw}_2}r_{\mathrm{cw}_2}^2 + m_4\frac{l_4 - r_4}{l_4}l_3'^2 \tag{3.30}$$

$$I_4 = m_4\left(k_4^2 + r_4^2 - r_4 l_4\right) \tag{3.31}$$

[1] A "physical pendulum" is a link which has such a distribution of masses that it allows the dynamic substitution of link's mass and inertia by two concentrated masses.

$$I_5 = m_5\left(k_5^2 + r_5^2 - r_{\mathrm{cw}_3} r_5\right) + m_4 \frac{r_4}{l_4} l_5 \left(l_5 - r_{\mathrm{cw}_3}\right) \tag{3.32}$$

Thus, this new six-bar mechanism has the same shaking moment as the four-bar mechanism composed of links P_1P_2, P_2P_2', and $P_2'P_3$. Therefore, the initial four-bar linkage balancing problem is transformed in the balancing of the four-bar linkage formed by the added Assur group. Note that the latter has specific geometry and its balancing conditions have been examined in Sect. 3.2.1.1.

Applying these results to the considered mechanism, we obtain:

$$k_4^2 = \frac{-I_3 - r_4^2 + r_4 l_4}{m_4} \tag{3.33}$$

$$k_5^2 = \frac{m_5\left(-r_5^2 + r_{\mathrm{cw}_3} r_5\right) + m_4 r_4/l_4 l_5 \left(l_5 - r_{cw_3}\right) + I_3}{m_5} \tag{3.34}$$

The substitution of (3.33) and (3.34) into (3.29) leads to:

$$M^{\mathrm{sh}} = I_3\left(\ddot{\theta}_3 - \ddot{\theta}_4 + \ddot{\theta}_5\right) \tag{3.35}$$

Taking into account relations (3.6) and (3.6), we have

$$\ddot{\theta}_3 - \ddot{\theta}_4 + \ddot{\theta}_5 = 0 \tag{3.36}$$

and consequently

$$M^{\mathrm{sh}} = 0 \tag{3.37}$$

The proposed balancing technique has been illustrated using the four-bar linkage shown in Fig. 3.2b. However, it can also be achieved via the mechanism of Fig. 3.2a or Fig. 3.2c.

3.2.2 *Complete Shaking Force and Shaking Moment Balancing by Adding a Class-Two RRP Assur Group*

3.2.2.1 Shaking Force Balancing

The second solution, which is proposed for the cancellation of the shaking moment of a four-bar linkage is carried out by adding a class-two *RRP* Assur group (Fig. 3.4). Let us denote the following vectors as: $\mathbf{l}_1 = \mathbf{d}_{OA}$, $\mathbf{l}_2 = \mathbf{d}_{AB}$, $\mathbf{l}_3 = \mathbf{d}_{CB}$, $\mathbf{l}_3' = \mathbf{d}_{CD}$, $\mathbf{l}_4 = \mathbf{d}_{DE}$, $\mathbf{r}_1 = \mathbf{d}_{OS_1}$, $\mathbf{r}_2 = \mathbf{d}_{AS_2}$, $\mathbf{r}_3 = \mathbf{d}_{CS_3}$, $\mathbf{r}_4 = \mathbf{d}_{DS_4}$, $\mathbf{r}_5 = \mathbf{d}_{ES_5}$, $\mathbf{r}_{\mathrm{cw}_1} = \mathbf{d}_{OS_{\mathrm{cw}_1}}$, $\mathbf{r}_{\mathrm{cw}2} = \mathbf{d}_{CS_{\mathrm{cw}_2}}$, $\mathbf{r}_{\mathrm{cw}3} = \mathbf{d}_{DS_{\mathrm{cw}_3}}$.

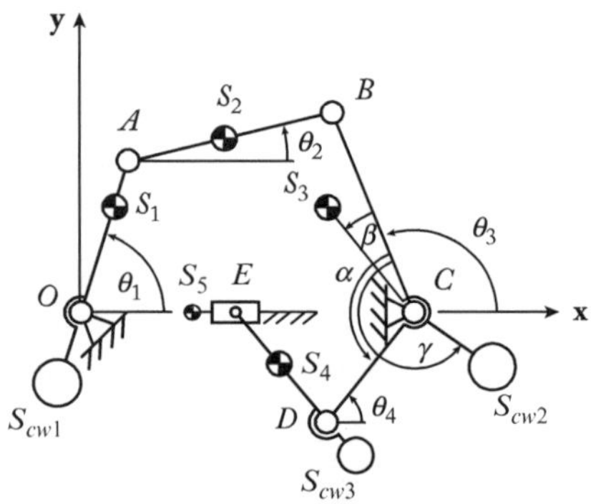

Fig. 3.4 The balanced mechanism with the class-two *RRP* Assur group

In this case, the lengths of added links are the following:

$$l_3' = l_4 \tag{3.38}$$

Thus, the new part created by added links is a Scott–Russell mechanism [53], which is attached to the initial linkage at an angle of α.

The relations between the link accelerations are the following for the Scott–Russell mechanism:

$$\dot{\theta}_3 = -\dot{\theta}_4 \tag{3.39}$$

Let us now derive the expression of the shaking force $\mathbf{f}^{\text{sh}}$ of such a mechanism:

$$\mathbf{f}^{\text{sh}} = \sum_{i=1}^{5} m_i \ddot{\mathbf{d}}_{S_i} \tag{3.40}$$

Developing and simplifying, one obtains:

$$\begin{aligned}\mathbf{f}^{\text{sh}} &= \left(\frac{m_1 r_1}{l_1} + m_2 \frac{l_2 - r_2}{l_2}\right)\ddot{\mathbf{d}}_A + m_2 \frac{r_2}{l_2}\ddot{\mathbf{d}}_B + m_3 \ddot{\mathbf{d}}_{S_3} \\ &+ (m_4 + m_5)\ddot{\mathbf{d}}_D + \left(m_4 \frac{r_4}{l_4} + m_5\right)\ddot{\mathbf{d}}\end{aligned} \tag{3.41}$$

where r_i and l_i are the algebraic values of the norm of vectors $\mathbf{r}_i$, and $\mathbf{l}_i$, respectively,

$$\ddot{\mathbf{d}}_{S_3} = r_3 \left(\ddot{\theta}_3 \begin{bmatrix} -\sin(\theta_3 + \beta) \\ \cos(\theta_3 + \beta) \end{bmatrix} - \dot{\theta}_3^2 \begin{bmatrix} \cos(\theta_3 + \beta) \\ \sin(\theta_3 + \beta) \end{bmatrix}\right) \tag{3.42}$$

$$\ddot{\mathbf{d}} = l_5 \left(\ddot{\theta}_4 \begin{bmatrix} -\sin\theta_4 \\ \cos\theta_4 \end{bmatrix} - \dot{\theta}_4^2 \begin{bmatrix} \cos\theta_4 \\ \sin\theta_4 \end{bmatrix}\right) \tag{3.43}$$

$$\ddot{\mathbf{d}}_D = l'_3 \left(\ddot{\theta}_3 \begin{bmatrix} -\sin(\theta_3 + \alpha) \\ \cos(\theta_3 + \alpha) \end{bmatrix} - \dot{\theta}_3^2 \begin{bmatrix} \cos(\theta_3 + \alpha) \\ \sin(\theta_3 + \alpha) \end{bmatrix} \right) \tag{3.44}$$

in which the angles θ_i $(i = 1 \ldots 4)$ are defined in Fig. 3.4.

The shaking force $\mathbf{f}^{\mathrm{sh}}$ may be cancelled through the addition of three counterweights positioned at points S_{cw_i} (Fig. 3.4), with masses m_{cw_i} $(i = 1, 2, 3)$. With such counterweights, the expression of the shaking force becomes:

$$\mathbf{f}^{\mathrm{sh}^*} = \mathbf{f}^{\mathrm{sh}} + m_{\mathrm{cw}_1} \frac{r_{\mathrm{cw}1}}{l_1} \ddot{\mathbf{d}}_A + m_{\mathrm{cw}_2} \ddot{\mathbf{d}}_{S_{\mathrm{cw}_2}} + m_{\mathrm{cw}_3} \ddot{\mathbf{d}}_D + m_{\mathrm{cw}_3} \frac{r_{\mathrm{cw}3}}{l_4} \ddot{\mathbf{d}} \tag{3.45}$$

where r_{cw_i} is the algebraic values of the norm of vectors $\mathbf{r}_{\mathrm{cw}_i}$ and

$$\ddot{\mathbf{d}}_{S_{\mathrm{cw}_2}} = r_{\mathrm{cw}_2} \left(\ddot{\theta}_3 \begin{bmatrix} -\sin(\theta_3 + \gamma) \\ \cos(\theta_3 + \gamma) \end{bmatrix} - \dot{\theta}_3^2 \begin{bmatrix} \cos(\theta_3 + \gamma) \\ \sin(\theta_3 + \gamma) \end{bmatrix} \right) \tag{3.46}$$

Thus the shaking force is cancelled if the distribution of the masses is as follows:

$$m_{\mathrm{cw}_1} = -\frac{l_1}{r_{\mathrm{cw}_1}} \left(m_1 \frac{r_1}{l_1} + m_2 \frac{l_2 - r_2}{l_2} \right) \tag{3.47}$$

$$\tan\gamma = \frac{m_3 r_3 \sin\beta + (m_4 + m_5)\, l'_3 \sin\alpha}{m_2 l_3 r_2 / l_2 + m_3 r_3 \cos\beta + (m_4 + m_5)\, l'_3 \sin\alpha} \tag{3.48}$$

$$m_{\mathrm{cw}_2}^2 = \frac{\left(m_3 r_3 \sin\beta + (m_4 + m_5)\, l'_3 \sin\alpha\right)^2}{r_{\mathrm{cw}_2}^2} + \frac{\left(m_2 l_3 r_2 / l_2 + m_3 r_3 \cos\beta + (m_4 + m_5)\, l'_3 \cos\alpha\right)^2}{r_{\mathrm{cw}_2}^2} \tag{3.49}$$

$$m_{\mathrm{cw}_3} = -\frac{l_4}{r_{\mathrm{cw}_3}} \left(m_4 \frac{r_4}{l_4} + m_5 \right) \tag{3.50}$$

3.2.2.2 Shaking Moment Balancing

Let us now derive the expression of the shaking moment M^{sh}, expressed at point O, of such a mechanism:

$$M^{\mathrm{sh}} = \sum_{i=1}^{5} m_i \left(x_{S_i} \ddot{y}_{S_i} - y_{S_i} \ddot{x}_{S_i} + k_i^2 \ddot{\theta}_i \right) + \sum_{i=1}^{3} m_{\mathrm{cw}_i} \left(x_{S_{\mathrm{cw}_i}} \ddot{y}_{S_{\mathrm{cw}_i}} - y_{S_{\mathrm{cw}_i}} \ddot{x}_{S_{\mathrm{cw}_i}} \right) \tag{3.51}$$

where x_{S_i}, y_{S_i}, $\ddot{x}_{S_i}$, and $\ddot{y}_{S_i}$ are the position and accelerations along $\mathbf{x}$ and $\mathbf{y}$ axes of points S_i, respectively, and $x_{S_{\mathrm{cw}_i}}$, $y_{S_{\mathrm{cw}_i}}$, $\ddot{x}_{S_{\mathrm{cw}_i}}$, and $\ddot{y}_{S_{\mathrm{cw}_i}}$ are the position and accelerations along $\mathbf{x}$ and $\mathbf{y}$ axes of points S_{cw_i}, respectively, k_i is the radius of gyration of link i.

Now, let us consider as in the previous case, that link 2 is a physical pendulum and that the input speed is constant.

Taking into account (3.28), one can simplify (3.51) as:

$$M^{\mathrm{sh}} = I_3\ddot{\theta}_3 + I_4\ddot{\theta}_4 \tag{3.52}$$

where

$$I_3 = m_3\left(k_3^2 + r_3^2\right) + m_2\frac{r_2}{l_2}l_3^2 + m_{\mathrm{cw}_2}r_{\mathrm{cw}_2}^2 + \left(m_4 + m_5 + m_{\mathrm{cw}_3}\right)l_3'^2 \tag{3.53}$$

$$I_4 = m_4\left(k_4^2 + r_4^2\right) + m_5 l_4^2 + m_{\mathrm{cw}_3}r_{\mathrm{cw}_3}^2 \tag{3.54}$$

Introducing (3.39) into (3.52), we find:

$$M^{\mathrm{sh}} = \left(I_3 - I_4\right)\ddot{\theta}_3 \tag{3.55}$$

Thus, the mechanism will be moment balanced if:

$$I_3 = I_4 \tag{3.56}$$

which can be obtained using a design of link 4 for which the radius of gyration should be equal to

$$k_4 = \sqrt{\frac{I_3 - m_4 r_4^2 - m_5 l_4^2 - m_{\mathrm{cw}_3}r_{\mathrm{cw}_3}^2}{m_4}} \tag{3.57}$$

It should be mentioned that, in order to avoid the singular configurations of the added structure, the value of angle α should be chosen carefully during the design process.

Let us consider two illustrative examples of the proposed balancing technique.

3.2.3 *Illustrative Examples and Numerical Simulations*

3.2.3.1 Balancing by Adding a Class-Two *RRR* Assur Group

Let us carry out the complete shaking force and shaking moment balancing of a four-bar linkage with parameters

- $l_1 = 0.2\,\mathrm{m}$, $l_2 = 0.27\,\mathrm{m}$, $l_3 = 0.25\,\mathrm{m}$, $\beta = 0°$,
- $r_1 = 0.1\,\mathrm{m}$, $r_2 = 0.135\,\mathrm{m}$, $r_3 = 0.125\,\mathrm{m}$,
- $k_1 = 0.056\,\mathrm{m}$, $k_2 = 0.135\,\mathrm{m}$, $k_3 = 0.086\,\mathrm{m}$,
- $m_1 = 1\,\mathrm{kg}$, $m_2 = 1\,\mathrm{kg}$, $m_3 = 1\,\mathrm{kg}$.

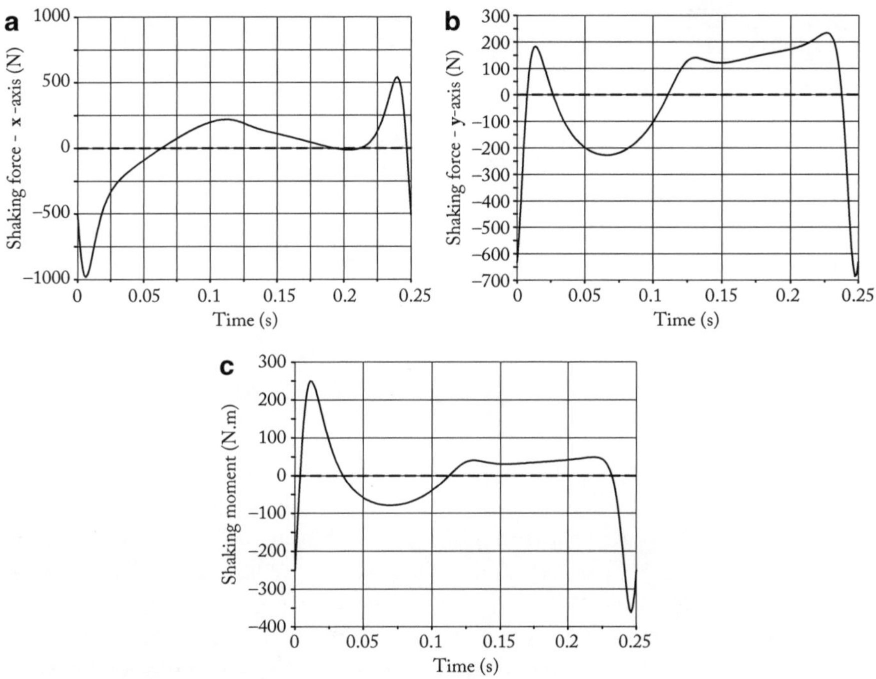

Fig. 3.5 Shaking force and shaking moment of the four-bar mechanism before (*full line*) and after (*dotted line*) balancing. (**a**) Shaking force along **x** axis; (**b**) shaking force along **y** axis; (**c**) shaking moment

The simulations of the proposed mechanism have been carried out using ADAMS software and the obtained results are shown in Fig. 3.5 (full line).

Now we add the *RRR* Assur group with prescribed distribution of the center of mass and inertia. Its geometric and mass properties are:

- $l_4 = 0.8\,\text{m}, l_5 = 0.25\,\text{m}$,
- $r_4 = 0.4\,\text{m}, r_5 = 0.125\,\text{m}$,
- $m_4 = 1.5\,\text{kg}, m_5 = 1\,\text{kg}$.

and the location and mass of the added counterweights are:

- $r_{\text{cw}_1} = -0.1\,\text{m}, r_{\text{cw}_2} = 0.25\,\text{m}, r_{\text{cw}_3} = -0.125\,\text{m}$,
- $m_{\text{cw}_1} = 2\,\text{kg}, m_{\text{cw}_3} = 2.5\,\text{kg}$.

The radii of gyration of elements 4 and 5 and the mass of the counterweight m_{cw_2} are not given as they depend on the value of angle α which is not yet fixed (Fig. 3.3). Their variations as a function of α are shown in Fig. 3.6. In these figures, the values of α are bounded between 60° and 180° in order to avoid the *RRR* Assur group to cross a singularity during the motion. In Fig. 3.5 (dotted line), it is shown that after the addition of the Assur group, the shaking force and shaking moment are cancelled.

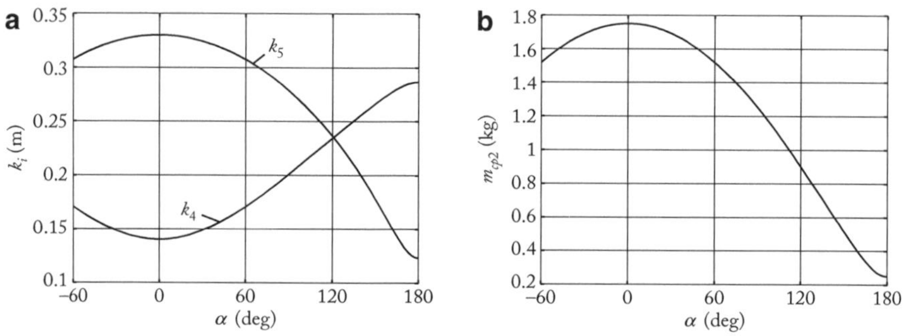

Fig. 3.6 Variation of the values of parameters k_4, k_5 and m_{cw_2} as a function of angle α. (**a**) Variation of k_4 and k_5; (**b**) variation of m_{cw_2}

As angle α is a free parameter that has only influence on the values k_4, k_5, and m_{cw_2}, it can be chosen so that it minimizes one supplementary criterion. In the remainder of the paper, this criterion is chosen to be the linkage input torque [50, 51]. It should however be mentioned that angle α could be used to minimize another criterion such as the power consumption, the energy, etc.

It should also be noted that the input torque τ, i.e. the torque requested by the actuator to move the mechanism, is computed in the two illustrative examples by using the Lagrange equations [51]:

$$\tau = \frac{\mathrm{d}}{\mathrm{d}t}\left(\frac{\partial L}{\partial \dot{\theta}_1}\right) - \frac{\partial L}{\partial \theta_1} \tag{3.58}$$

where $L = T - V$ is the Lagrangian of the system, V is the potential energy (equal to 0 in absence of gravity), and T is the kinetic energy:

$$T = \frac{1}{2}\sum_i m_i\left(\dot{x}_{S_i}^2 + \dot{y}_{S_i}^2\right) + \frac{1}{2}\sum_j I_j\dot{\theta}_j^2 \tag{3.59}$$

$\dot{x}_{S_i}$ and $\dot{y}_{S_i}$ being the velocities along **x** and **y** axes of any center of mass (for links and as well as for counterweights).

In Fig. 3.7, the maximum of the input torque absolute value of as a function of angle α is shown. Thus, it is possible to see that if the value of α is chosen arbitrarily, the input torques can grow up to 2,140 N m (for $\alpha = 0°$). It also appears that the input torque will be minimal if $\alpha = 164°$. In this case, the value of the input torque is 1,010 N m, i.e. about two times less than in the first case.

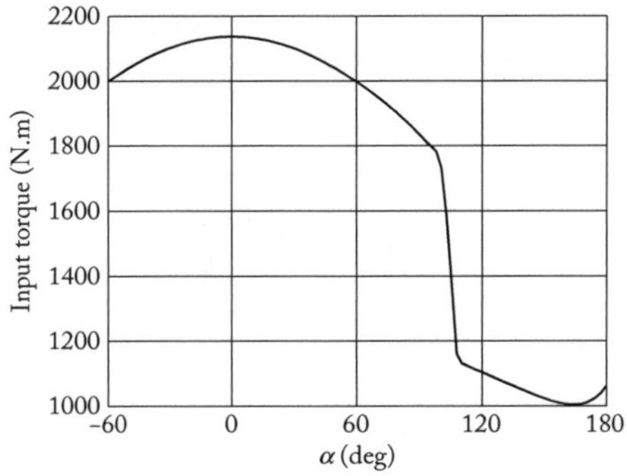

Fig. 3.7 Variation of the maximal input torque absolute value as a function of angle α

3.2.3.2 Balancing by Adding a Class-Two *RRP* Assur Group

We now propose obtaining the complete shaking force and shaking moment balancing of the same mechanism by adding a class-two *RRP* Assur group. Its geometric and mass properties are

- $l_4 = 0.25\,\text{m}$, $l_5 = 0.25\,\text{m}$, $\alpha = -90°$,
- $r_4 = 0.125\,\text{m}$,
- $m_4 = 0.35\,\text{kg}$, $m_5 = 0.1\,\text{kg}$.

and the location and mass of the added counterweights are:

- $r_{\text{cw}_1} = -0.1\,\text{m}$, $r_{\text{cw}_2} = 0.25\,\text{m}$, $r_{\text{cw}_3} = -0.125\,\text{m}$,
- $m_{\text{cw}_1} = 2\,\text{kg}$, $m_{\text{cw}_3} = 0.55\,\text{kg}$.

The radii of gyration of element 4 and the mass of the counterweight m_{cw_2} are not expressed in these tables as they depend on the value of angle α (Fig. 3.4). Their variations as a function of α are shown in Fig. 3.8. In these figures, the values of α are bounded between 25° and 100° or 205° and 280° in order to avoid the *RRP* Assur group to cross a singularity during the motion. The simulations of the proposed mechanism have been carried out using ADAMS software and the results are similar to the previous case shown in Fig. 3.5.

As was mentioned above, the angle α is a not fixed design parameter and it can be found from minimization of the input torque of the mechanism. In Fig. 3.9, maximum of the input torque absolute value as a function of angle α is shown. It is possible to see that if the value of α is chosen arbitrarily, the input torques can grow up to 2,300 N m (for $\alpha = 25°$). It also appears that the input torque will be minimal if $\alpha = 205°$. In this case, the value of the input torque is 1,380 N m, i.e. about 1.7 times less than in the first case.

Let us now consider the balancing of parallel robots by the addition of one or several class-two *RRR* Assur groups.

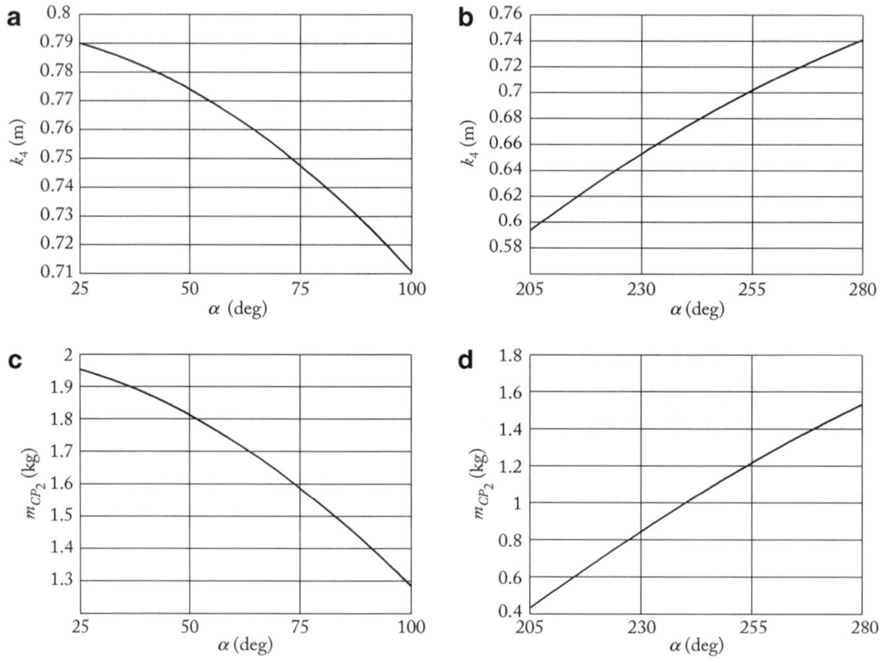

Fig. 3.8 Variation of the values of parameters k_4 and m_{cw_2} as a function of angle α. (**a**) Variation of k_4 for $\alpha \in \{25 \;\; 100\}$ deg. (**b**) Variation of k_4 for $\alpha \in \{205 \;\; 280\}$ deg. (**c**) Variation of m_{cw_2} for $\alpha \in \{25 \;\; 100\}$ deg. (**d**) Variation of m_{cw_2} for $\alpha \in \{205 \;\; 280\}$ deg

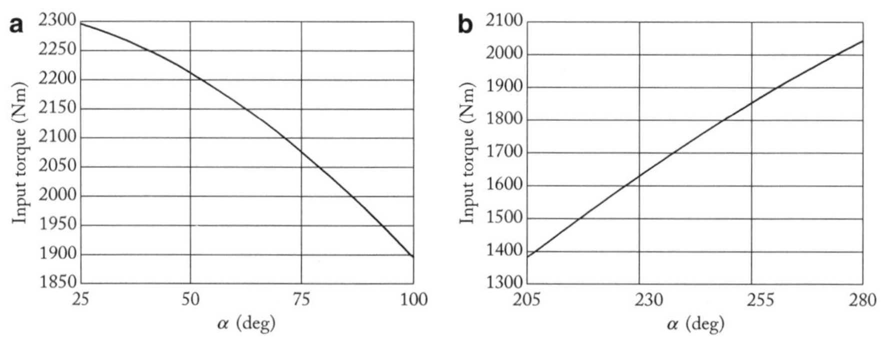

Fig. 3.9 Variation of the maximal input torque absolute value as a function of angle α. (**a**) For $\alpha \in \{25 \;\; 100\}$ deg. (**b**) For $\alpha \in \{205 \;\; 280\}$ deg

3.3 Balancing of Parallel Robots by the Addition of One or Several Class-Two RRR Assur Groups

For robots, shaking force balancing is mostly obtained via an optimal redistribution of movable masses [7–9, 54–60] or adjustment of kinematic parameters [61]. The cancellation of the shaking moment is a more complicated task and can be obtained using three main different methods:

1. Shaking moment balancing using counter-rotations [62, 63],
2. Shaking moment balancing by adding four-bar linkages [45–47, 64, 65] and
3. Shaking moment balancing by optimal trajectory planning [63, 66–69].

Previous works have been devoted to the study of parallel manipulators with revolute joints and few studies have been carried out on complete shaking force and shaking moment balancing of parallel manipulators with prismatic pairs.

In this section, we propose solutions for complete shaking force and shaking moment balancing of planar parallel manipulators with prismatic pairs. We illustrate these solutions via the 3-*RPR* parallel manipulator. All obtained results are validated using ADAMS software simulations

3.3.1 Complete Shaking Moment and Shaking Force Balancing by Adding an Idler Loop Between the Base and the Platform

Inertia force balancing by adding an *idler loop* made of a single *RRR* Assur group is known to be used for 1 degree of freedom (*dof*) mechanisms [19, 36, 70–72]. With regard to planar manipulators, such an approach has only been used in the balancing of gravitational and inertia forces [59, 60, 73, 74].

In this sub-section, the complete shaking force and shaking moment balancing of planar manipulators by adding an idler loop made of a single *RRR* Assur group is discussed. The added balancing loop is mounted between the base and the platform of the mechanism. We illustrate the suggested balancing technique on a 3-*RPR* mechanism (Fig. 3.10). Please note that we do not mention the type of actuation of the mechanism as it has no influence on the balancing.

3.3.1.1 Theoretical Background

Firstly, let us analyze the cancellation of the dynamic reactions of the 3-*RPR* planar parallel mechanism (Fig. 3.10a). Such a mechanism has 3 *dof* (two translations in the (*Oxy*) plane and one rotation of the moving platform around an axis

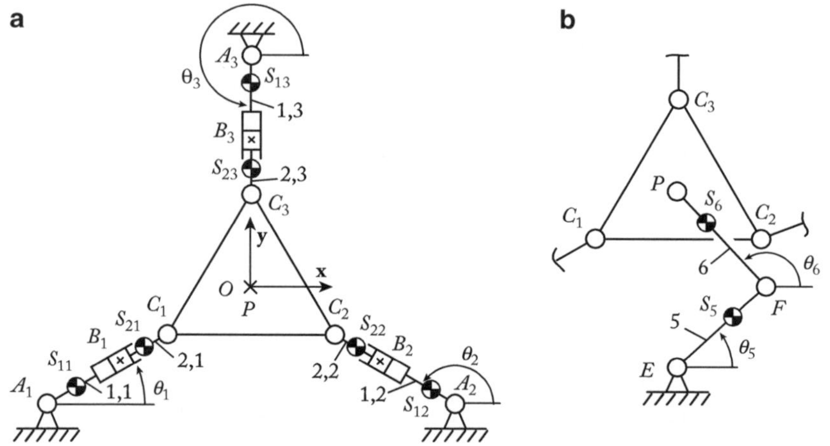

Fig. 3.10 Schematic of the 3-*RPR* robot under study. (**a**) Schematics of the 3-*RPR* mechanism (for this configuration, points O and P coincide). (**b**) Schematics of the added idler loop (*RRR* chain)

perpendicular to (Oxy)) and is composed of three identical legs, each being composed of a revolute joint attached to the base at point A_i (in the following, $i = 1 \ldots 3$), one moving prismatic guide, located at point B_i, and another revolute joint attached to the platform at C_i. The base and platform triangles, denoted $A_1A_2A_3$ and $C_1C_2C_3$, are equilateral. On this manipulator, typically, the actuated joints are the first revolute joint at A_i or the linear guide at B_i.

Considering that the $\mathbf{x}$ axis is directed along the line A_1A_2, the $\mathbf{y}$ axis being perpendicular to the $\mathbf{x}$ axis and the origin of the base frame located at point O, the center of the circumcircle of triangle $A_1A_2A_3$, one can define the coordinate x, y, and ϕ of the platform, as being, respectively, the coordinates of point P along the $\mathbf{x}$ and $\mathbf{y}$ axes and the angle between the line C_1C_2 and A_1A_2.

Let us denote as S_{ij} the center of mass of link ij ($j = 1, 2$), which has a mass m_j and an axial moment of inertia I_{S_j}. The center of mass of the platform is located at point P. The mass of the platform is m_p and its axial moment of inertia I_p.

In order to cancel the shaking forces and shaking moment of the manipulator, an idler loop is added between the base and the platform (Fig. 3.10b). The center of mass of elements 5 and 6 of the idler loop are located at S_5 and S_6, respectively. Their masses are denoted as m_5 and m_6 and their axial moments of inertia as I_{S_5} and I_{S_6}, respectively. The positions of the *coms* are defined such that $\overrightarrow{A_iS_{1i}} = r_1 l_{B_iC_i}\mathbf{u}_i$, $\overrightarrow{C_iS_{2i}} = (r_2 1) l_{B_iC_i}\mathbf{u}_i$, $\overrightarrow{ES_5} = r_5\overrightarrow{EF}$, and $\overrightarrow{FS_{6i}} = r_6\overrightarrow{FP}$, r_1, r_2, r_5, and r_6 being dimensionless coefficients, and $\mathbf{u}_i$ a unit vector directed along $\overrightarrow{B_iC_i}$.

The expression of the shaking force $\mathbf{f}^{sh}$ transmitted by the robot to the ground is:

$$\mathbf{f}^{sh} = \left(\sum_{i=1}^{3} \sum_{j=1}^{2} m_j \ddot{\mathbf{r}}_{S_{ij}} \right) + m_p \ddot{\mathbf{r}}_P + m_5 \ddot{\mathbf{r}}_{S_5} + m_6 \ddot{\mathbf{r}}_{S_6} \tag{3.60}$$

where $\ddot{\mathbf{r}}_{S_{ij}}$, $\ddot{\mathbf{r}}_P$, $\ddot{\mathbf{r}}_{S_5}$, and $\ddot{\mathbf{r}}_{S_6}$ are the accelerations of the *coms* S_{ij}, of P, S_5, and S_6, respectively.

Developing (3.60), it can be demonstrated that the shaking force $\mathbf{f}^{sh}$ can be expressed as:

$$\begin{aligned} \mathbf{f}^{sh} &= (m_1 r_1 - m_2 (1 - r_2)) \sum_{i=1}^{3} \mathbf{a}_i + \left(3m_2 + m_p + m_6 r_6\right) \mathbf{a}_6 \\ &+ \left(3m_2 + m_p + m_5 r_5 + m_6\right) \ddot{\mathbf{r}}_F \end{aligned} \tag{3.61}$$

with

$$\mathbf{a}_i = l_{B_iC_i} \left(\ddot{\theta}_i \begin{bmatrix} -\sin\theta_i \\ \cos\theta_i \end{bmatrix} - \dot{\theta}_i^2 \begin{bmatrix} \cos\theta_i \\ \sin\theta_i \end{bmatrix} \right) \tag{3.62}$$

$$\mathbf{a}_6 = l_{FP} \left(\ddot{\theta}_6 \begin{bmatrix} -\sin\theta_6 \\ \cos\theta_6 \end{bmatrix} - \dot{\theta}_6^2 \begin{bmatrix} \cos\theta_6 \\ \sin\theta_6 \end{bmatrix} \right) \tag{3.63}$$

At this step, only five counterweights are needed in the cancellation of the shaking force, but it could be demonstrated after more derivations that three others are necessary for the cancellation of the shaking moment. Therefore, we propose directly adding three supplementary counterweights (Fig. 3.11). The positions of the eight added counterweights are $\overrightarrow{A_iM_{1i}} = r_{cw_1} l_{B_iC_i} \mathbf{u}_i$, $\overrightarrow{B_iM_{2i}} = (r_{cw_2} - 1) l_{B_iC_i} \mathbf{u}_i$, $\overrightarrow{EM_5} = r_{cw_5} \overrightarrow{EF}$, and $\overrightarrow{FM_6} = r_{cw_6} \overrightarrow{FP}$, r_{cw_1}, r_{cw_2}, r_{cw_5}, and r_{cw_6} being dimensionless coefficients. Their masses are, respectively, denoted m_{cw_1}, m_{cw_2}, m_{cw_5}, and m_{cw_6}. With the addition of the counterweights, the shaking force becomes:

$$\begin{aligned} \mathbf{f}^{sh^*} &= \mathbf{f}^{sh} + (m_{cw_1} r_{cw_1} - m_{cw_2} (1 - r_{cw_2})) \sum_{i=1}^{3} \mathbf{a}_i + m_{cw_6} r_{cw_6} \mathbf{a}_6 \\ &+ (m_{cw_5} r_{cw_5} + m_{cw_6}) \ddot{\mathbf{r}}_F \end{aligned} \tag{3.64}$$

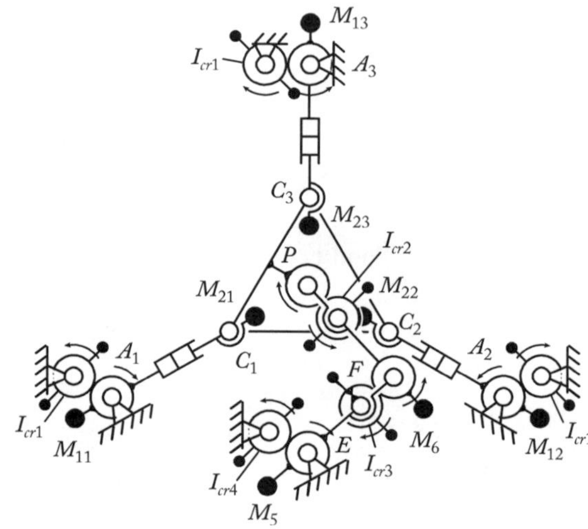

Fig. 3.11 Schematics of the 3-RPR mechanism with the added RRR chain used for the cancellation of the shaking force and shaking moment

Thus, the shaking force is cancelled if:

$$\begin{aligned} m_{\mathrm{cw}_1} &= -\frac{m_1 r_1}{r_{\mathrm{cw}_1}} \\ m_{\mathrm{cw}_2} &= -\frac{m_2(1-r_2)}{1-r_{\mathrm{cw}_2}} \\ m_{\mathrm{cw}_6} &= -\frac{3\left(m_2+m_{\mathrm{cw}_2}\right)+m_{\mathrm{p}}+m_6 r_6}{r_{\mathrm{cw}_6}} \\ m_{\mathrm{cw}_5} &= -\frac{3\left(m_2+m_{\mathrm{cw}_2}\right)+m_{\mathrm{p}}+m_5 r_5+m_6+m_{\mathrm{cw}_6}}{r_{\mathrm{cw}_5}} \end{aligned} \tag{3.65}$$

The expression of the shaking moment M_O^{sh} of the modified structure (expressed at point O) can be written as:

$$M_O^{\mathrm{sh}} = \frac{\mathrm{d}}{\mathrm{d}t} H_O \tag{3.66}$$

where H_O is the angular momentum of the leg (expressed at point O). Thus, in order to cancel the shaking moment, the angular momentum is held constant over time.

The expression of the angular momentum H_O is equal to:

$$\begin{aligned} H_O = & \sum_{i=1}^{3}\sum_{j=1}^{2}\left(m_j\left(x_{S_{ij}}\dot{y}_{S_{ij}} - y_{S_{ij}}\dot{x}_{S_{ij}}\right) + m_{\mathrm{cw}_j}\left(x_{M_{ij}}\dot{y}_{M_{ij}} - y_{M_{ij}}\dot{x}_{M_{ij}}\right) + I_{S_j}\dot{\theta}_i\right) \\ & + I_{\mathrm{p}}\dot{\phi} + \sum_{j=5}^{6}\left(m_j\left(x_{S_j}\dot{y}_{S_j} - y_{S_j}\dot{x}_{S_j}\right) + m_{\mathrm{cw}_j}\left(x_{M_j}\dot{y}_{M_j} - y_{M_j}\dot{x}_{M_j}\right) + I_{S_j}\dot{\theta}_j\right) \end{aligned} \tag{3.67}$$

where x_Q, y_Q, $\dot{x}_Q$, and $\dot{y}_Q$ are the position and velocities of any point Q along **x** and **y** axes, respectively (Q being either point S_{ij}, M_{ij}, ($j = 1, 2$), S_j or M_j ($j = 5, 6$)).

Developing and introducing (3.65) into (3.67) yields

$$\begin{aligned} H_O = &\sum_{i=1}^{3} \left(I_{S_1} + I_{S_2} + \left(m_1 r_1^2 + m_{\text{cw}_1} r_{\text{cw}_1}^2 + m_2 (1 - r_2)^2 + m_{\text{cw}_2} (1 - r_{\text{cw}_2})^2 \right) l_{B_i C_i}^2 \right) \dot{\theta}_i \\ &+ \left(I_{S_6} + \left(m_6 r_6^2 + m_{\text{cw}_6} r_{\text{cw}_6}^2 + m_{\text{p}} + 3 (m_2 + m_{\text{cw}_2}) \right) l_{FP}^2 \right) \dot{\theta}_6 \\ &+ \left(I_{S_5} + \left(m_5 r_5^2 + m_{\text{cw}_5} r_{\text{cw}_5}^2 + m_6 + m_{\text{cw}_6} + m_{\text{p}} + 3 (m_2 + m_{\text{cw}_2}) \right) l_{EF}^2 \right) \dot{\theta}_5 \\ &+ \left(I_{\text{p}} + 3 (m_2 + m_{\text{cw}_2}) l_{C_i P}^2 \right) \dot{\phi} \end{aligned} \tag{3.68}$$

After such modifications of the *RRR* chain, the angular momentum of the legs of the mechanism and of the *RRR* chain can be balanced using six counter-rotations (Fig. 3.11), which have an axial moment of inertia equal to:

$$\begin{aligned} I_{\text{cr}_1} &= I_{S_1} + I_{S_2} + \left(m_1 r_1^2 + m_{\text{cw}_1} r_{\text{cw}_1}^2 + m_2 (1 - r_2)^2 + m_{\text{cw}_2} (1 - r_{\text{cw}_2})^2 \right) l_{B_i C_i}^2 \\ I_{\text{cr}_2} &= I_{\text{p}} + 3 (m2 + m_{\text{cw}_2}) l_{C_i P}^2 \\ I_{\text{cr}_3} &= I_{S_6} + (m_6 r_6^2 + m_{\text{cw}_6} r_{\text{cw}_6}^2 + m_{\text{p}} + 3 (m_2 + m_{\text{cw}_2})) l_{FP}^2 + 2 I_{\text{cr}_2} \\ I_{\text{cr}_4} &= I_{S_5} + (m_5 r_5^2 + m_{\text{cw}_5} r_{\text{cw}_5}^2 + m_6 + m_{\text{cw}_6} + m_{\text{p}} + 3 (m_2 + m_{\text{cw}_2})) l_{EF}^2 + 2 I_{\text{cr}_3} \end{aligned} \tag{3.69}$$

3.3.1.2 Illustrative Examples and Numerical Simulations

Let us illustrate the suggested balancing approach using numerical simulations carried out with ADAMS software. For this purpose, non-balanced and balanced 3-*RPR* parallel manipulators will be compared.

The chosen trajectory for simulations is a straight line of the controlled point of the platform, achieved in $t_f = 0.25$ s, between $P_0 = [x_0 \;\; y_0]^T = [0.05\,\text{m} \;\; 0\,\text{m}]^T$ and $P_f = \left[x_f \;\; y_f\right]^T = [0.2\,\text{m} \;\; 0\,\text{m}]^T$ with a rotation of the platform from $\phi_0 = 0°$ to $\phi_f = 30°$. For the displacement of the mechanism, fifth order polynomial laws are used and therefore the trajectory is defined by the following expressions:

$$\begin{cases} x(t) = x_0 + s(t)(x_f - x_0) \\ y(t) = 0 \\ \phi(t) = \phi_0 + s(t)(\phi_f - \phi_0) \end{cases} \tag{3.70}$$

with

$$s(t) = \frac{10}{t_f^3} t^3 - \frac{15}{t_f^4} t^4 + \frac{6}{t_f^5} t^5 \tag{3.71}$$

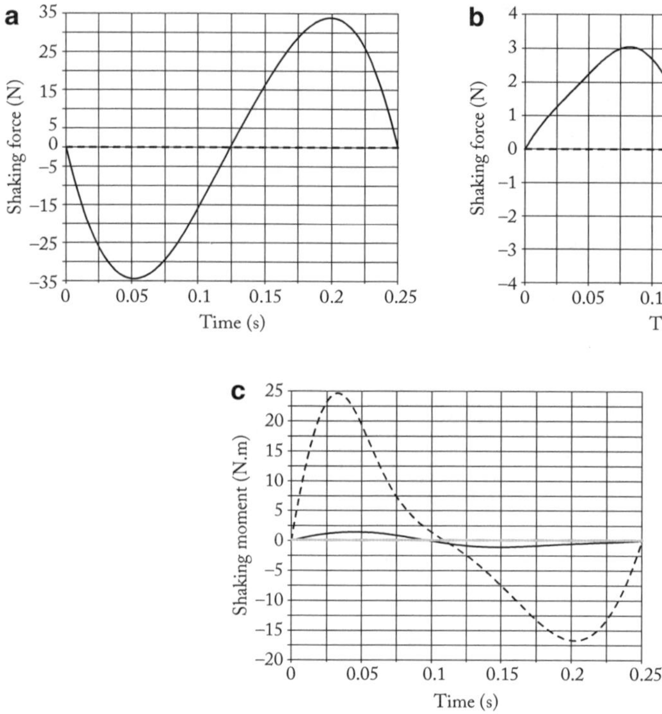

Fig. 3.12 Shaking force and shaking moment before (*solid line*) and after (*dashed line*) the addition of the counterweights, and after the addition of the counter-rotations (*gray line*). (**a**) Shaking force along **x**-axis; (**b**) shaking force along **y**-axis; (**c**) shaking moment around **z**-axis

The parameters used for the simulations are the followings

- $l_{OE} = 0\,\text{m}$, $l_{OA_i} = 0.35\,\text{m}$, $l_{B_iC_i} = 0.05\,\text{m}$, $l_{C_iP} = 0.1\,\text{m}$, $l_{EF} = 0.15\,\text{m}$ and $l_{FP} = 0.1581\,\text{m}$,
- $r_1 = 2$, $r_2 = r_5 = r_6 = 0.5$,
- $m_1 = 0.75\,\text{kg}$, $m_2 = 0.37\,\text{kg}$, $m_5 = 0.42\,\text{kg}$, $m_6 = 0.47\,\text{kg}$; $m_\text{p} = 1\,\text{kg}$,
- $I_{S_1} = 0.00344\,\text{kg m}^2$, $I_{S_2} = 0.00025\,\text{kg m}^2$, $I_{S_5} = 0.00122\,\text{kg m}^2$, $I_{S_6} = 0.00146\,\text{kg m}^2$, $I_\text{p} = 0.00436\,\text{kg m}^2$.

For such parameters and such a trajectory, the shaking force and shaking moment are computed using ADAMS software and are presented in Fig. 3.12 (solid line). Then, we add the counterweights and the idler loop *EFP* to the mechanism. The position coefficients of the counterweights are all equal to $r_{\text{cw}_j} = -0.5$ ($j = 1, 2, 5, 6$). Therefore, the added masses are equal to $m_{\text{cw}_1} = 0.75\,\text{kg}$, $m_{\text{cw}_2} = 0.37\,\text{kg}$, $m_{\text{cw}_5} = 6.92\,\text{kg}$, $m_{\text{cw}_6} = 21.66\,\text{kg}$. The new values of the shaking force and moment are presented in Fig. 3.12 (dashed line). It is possible to see that with the added counterweights the shaking efforts are cancelled, while the maximal value of shaking moment is increased by a factor 17. Finally, we add the counter-rotations. Their values are equal to $I_{\mathrm{cr}_1} = 0.01917\,\mathrm{kg\,m^2}$, $I_{\mathrm{cr}_2} = 0.02665\,\mathrm{kg\,m^2}$, $I_{\mathrm{cr}_3} = 0.18169\,\mathrm{kg\,m^2}$, $I_{\mathrm{cr}_4} = 0.72781\,\mathrm{kg\,m^2}$. With such counter-rotations, the shaking moment is balanced (in gray line in Fig. 3.12c).

3.3.2 Complete Shaking Force and Shaking Moment Balancing Using Scott–Russell Mechanism

In this sub-section, another approach for complete shaking force and shaking moment balancing is developed, which consists of adding Scott–Russell mechanisms (mechanisms made of *RRR* Assur groups) to each leg of the initial architecture of a manipulator. This approach enables a reduction in the number of counter-rotations.

3.3.2.1 Properties of the Scott–Russell Mechanism

Let us observe a simple slider-crank mechanism (Fig. 3.13). The center of mass of link i ($i = 1, 2, 3$) is denoted as S_i. Link i has a mass m_i and an axial moment of inertia I_{S_i}. The positions of the centers of mass are $\overrightarrow{AS_1} = r1\overrightarrow{AB}$, $\overrightarrow{BS_2} = r_2\overrightarrow{BC}$, $\overrightarrow{CS_3} = l_3 r_3 \mathbf{x}$, r_1, r_2 and r_3 being dimensionless coefficients, and l_3 being a constant length.

It is known that the complete shaking force and shaking moment balancing of a general slider-crank mechanism can be obtained by adding two counterweights mounted on the links and two pairs of counter-rotations. However, it is possible to balance this mechanism without counter-rotation if it has specific geometrical parameters, as in Scott–Russell mechanisms ($a = 0\,\mathrm{m}$, $l_{AB} = l_{BC}$—Fig. 3.13).

Let us consider the balancing of this mechanism. The expression of the shaking force $\mathbf{f}^{sh}$ of a slider-crank mechanism can be written as:

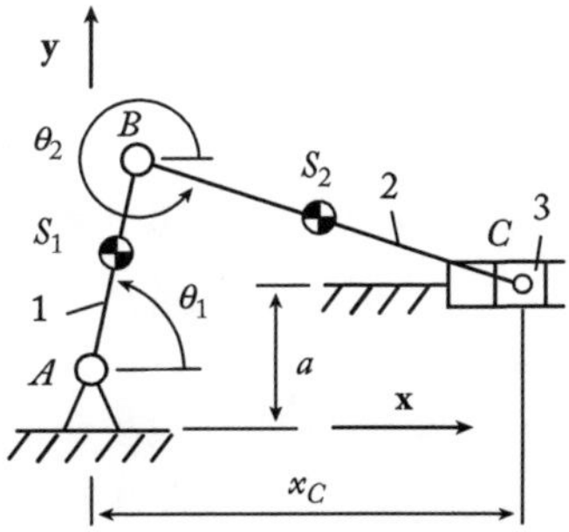

Fig. 3.13 A general slider-crank mechanism

$$\mathbf{f}^{sh} = \sum_{i=1}^{3} m_i \ddot{\mathbf{r}}_{S_i} = (m_1 r_1 + m_2)\,\ddot{\mathbf{r}}_B + (m_2 r_2 + m_3)\,\mathbf{a} \tag{3.72}$$

with

$$\mathbf{a} = l_{BC}\left(\ddot{\theta}_2 \begin{bmatrix} -\sin\theta_2 \\ \cos\theta_2 \end{bmatrix} - \dot{\theta}_2^2 \begin{bmatrix} \cos\theta_2 \\ \sin\theta_2 \end{bmatrix}\right) \tag{3.73}$$

$\ddot{\mathbf{r}}_B$ being the acceleration of point B.

The constant terms of (3.72) can be cancelled by the addition of two counterweights located at M_j, ($j = 1, 2$) (Fig. 3.13), whose masses are denoted as m_{cw_j}. Their positions are equal to: $\overrightarrow{AM_1} = r_{cw_1}\overrightarrow{AB}$, $\overrightarrow{BM_2} = r_{cw_1}\overrightarrow{BC}$, r_{cw_1} and r_{cw_2} being dimensionless coefficients. With the addition of the counterweights, the shaking force becomes:

$$\mathbf{f}^{sh^*} = \mathbf{f}^{sh} + (m_{cw_1} r_{cw_1} + m_{cw_2})\,\ddot{\mathbf{r}}_B + m_{cw_2} r_{cw_2}\mathbf{a} \tag{3.74}$$

Thus, the shaking force is balanced if:

$$m_{cw_2} = -\frac{m_2 r_2 + m_3}{r_{cw_2}} \text{ and } m_{cw_1} = -\frac{m_1 r_1 + m_2 + m_{cw_2} + m_3}{r_{cw_1}} \tag{3.75}$$

The expression of the angular momentum H_A (expressed at point A) is:

$$H_A = \sum_{j=1}^{3} \left(m_j \left(x_{S_j}\dot{y}_{S_j} - y_{S_j}\dot{x}_{S_j}\right)\right) + \sum_{j=1}^{2} \left(m_{cw_j}\left(x_{M_j}\dot{y}_{M_j} - y_{M_j}\dot{x}_{M_j}\right) + I_{S_j}\dot{\theta}_j\right) \tag{3.76}$$

where x_Q, y_Q, $\dot{x}_Q$, and $\dot{y}_Q$ are the position and velocities of any point Q along $\mathbf{x}$ and $\mathbf{y}$ axes, respectively [Q being either point S_j or M_j, ($j = 1, 2, 3$)].

Developing and introducing (3.75) into (3.76),

$$\begin{aligned} H_A = &\left(I_{S_1} + \left(m_1 r_1^2 + m_{cw_1} r_{cw_1}^2 + m_2 + m_{cw_2} + m_3\right) l_{AB}^2\right)\dot{\theta}_1 \\ &+ \left(I_{S_2} + \left(m_2 r_2^2 + m_{cw_1} r_{cw_1}^2 + m_3\right) l_{BC}^2\right)\dot{\theta}_2 \end{aligned} \tag{3.77}$$

with

$$\dot{\theta}_2 = -\frac{\dot{y}_B\,(x_C - x_B) + (a - y_B)\,(\dot{x}_C - \dot{x}_B)}{l_{BC}^2} \tag{3.78}$$

where x_B, y_B, x_C are the coordinates along $\mathbf{x}$ and $\mathbf{y}$ axes of points B and C, respectively, and $\dot{x}_B$, $\dot{y}_B$, $\dot{x}_C$ their velocities.

In order to cancel the shaking moment M_A^{sh}, the angular momentum has to be constant or null. Developing (3.77), M_A^{sh} can be cancelled if $a = 0$ and $l_{AB} = l_{BC}$ (in such a case, $\dot{\theta}_1 = -\dot{\theta}_2$) and if:

$$\begin{aligned} &I_{S_1} + \left(m_1 r_1^2 + m_{cw_1} r_{cw_1}^2 + m_2 + m_{cw_2} + m_3\right) l_{AB}^2 \\ &\quad - I_{S_2} - \left(m_2 r_2^2 + m_{cw_1} r_{cw_1}^2 + m_3\right) l_{BC}^2 = 0 \end{aligned} \tag{3.79}$$

3.3.2.2 Balancing of a Manipulator's Leg Using a Scott–Russell Mechanism

Now let us consider a manipulator's leg with an added Scott–Russell mechanism (i.e., an additional *RRR* Assur group—Fig. 3.14). Let us denote as S_4 the center of mass of link 4, which has a mass m_4 and an axial moment of inertia I_{S_4}. The position of S_4 is such that: $\overrightarrow{AS_4} = l_3 r_4 \mathbf{u}$, r_4 being a dimensionless coefficient and $\mathbf{u}$ a unit vector along $\overrightarrow{CS_3}$.

Now the shaking force becomes:

$$\begin{aligned} \mathbf{f}^{sh} = &\left(m_1 r_1 + m_{cw_1} r_{cw_1} + m_2 + m_{cw_2} + m_3\right) \ddot{\mathbf{r}}_B + \left(m_3 r_3 + m_4 r_4\right) \mathbf{a}_1 \\ &+ \left(m_2 r_2 + m_{cw_2} r_{cw_2} + m_3\right) \mathbf{a}_2 \end{aligned} \tag{3.80}$$

with

$$\begin{aligned} \mathbf{a}_1 &= l_3 \left(\ddot{\theta}_{03} \begin{bmatrix} -\sin\theta_{03} \\ \cos\theta_{03} \end{bmatrix} - \dot{\theta}_{03}^2 \begin{bmatrix} \cos\theta_{03} \\ \sin\theta_{03} \end{bmatrix}\right) \\ \mathbf{a}_2 &= l_{AB} \left(\left(\ddot{\theta}_{03} - \ddot{\theta}_{41}\right) \begin{bmatrix} -\sin(\theta_{03} - \theta_{41}) \\ \cos(\theta_{03} - \theta_{41}) \end{bmatrix} - \left(\dot{\theta}_{03} - \dot{\theta}_{41}\right)^2 \begin{bmatrix} \cos(\theta_{03} - \theta_{41}) \\ \sin(\theta_{03} - \theta_{41}) \end{bmatrix}\right) \end{aligned} \tag{3.81}$$

At this step, only one supplementary counterweight is necessary for the cancellation of the shaking force, but it could be demonstrated after more derivations that another is necessary for the cancellation of the shaking moment. Therefore, we

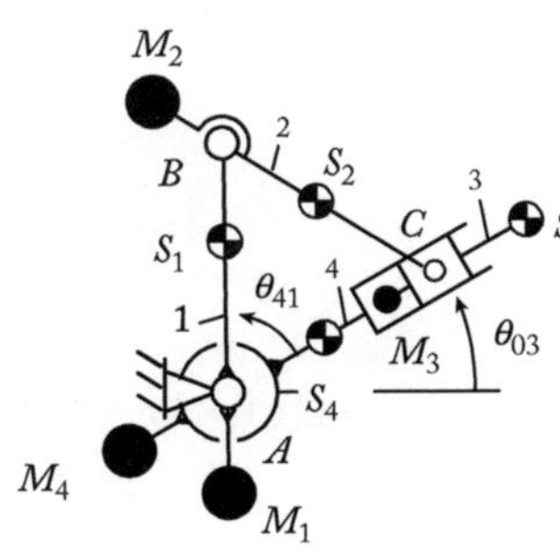

Fig. 3.14 A manipulator leg with added Scott–Russell mechanism

propose adding this additional counterweight directly. The two counterweights are located at points M_3 and M_4, defined such that: $\overrightarrow{CM_3} = r_{cw_3} l_3 \mathbf{u}$, $\overrightarrow{AM_4} = r_{cw_4} l_3 \mathbf{u}$, r_{cw_3} and r_{cw_4} being dimensionless coefficients. Their masses are respectively denoted m_{cw_3} and m_{cw_4}. With the addition of the counterweights, the shaking force becomes:

$$\mathbf{f}^{sh^*} = \mathbf{f}^{sh} + m_{cw_3} \ddot{\mathbf{r}}_B + m_{cw_3} \mathbf{a}_1 + \left(m_{cw_3} r_{cw_3} + m_{cw_4} r_{cw_4} \right) \mathbf{a}_2 \tag{3.82}$$

Thus, the shaking force is cancelled if:

$$\begin{aligned} m_{cw_4} &= -\frac{m_4 r_4}{r_{cw_4}} \\ m_{cw_3} &= -\frac{m_3 r_3}{r_{cw_3}} \\ m_{cw_2} &= -\frac{m_2 r_2 + m_3 + m_{cw_3}}{r_{cw_2}} \\ m_{cw_1} &= -\frac{m_1 r_1 + m_2 + m_{cw_2} + m_3 + m_{cw_3}}{r_{cw_1}} \end{aligned} \tag{3.83}$$

Developing and simplifying, the expression of the angular momentum is equal to:

$$H_A = I_{eq_1} \dot{\theta}_{03} + I_{eq_2} \dot{\theta}_{41} \tag{3.84}$$

with

$$\begin{aligned} I_{eq_1} &= \sum_{i=1}^{4} I_{S_i} + \left(m_1 r_1^2 + m_{cw_1} r_{cw_1}^2 + m_2 (1 - r_2)^2 + m_{cw_2} (1 - r_{cw_2})^2 \right) l_{AB}^2 \\ &\quad + \left(m_3 r_3^2 + m_{cw_3} r_{cw_3}^2 + m_4 r_4^2 + m_{cw_4} r_{cw_4}^2 \right) l_3^2 \end{aligned} \tag{3.85}$$

$$\begin{aligned} I_{eq_2} &= I_{S_1} + \left(m_1 r_1^2 + m_{cw_1} r_{cw_1}^2 + m_2 + m_{cw_2} + m_3 \right) l_{AB}^2 \\ &\quad - I_{S_2} - \left(m_2 r_2^2 + m_{cw_2} r_{cw_2}^2 + m_3 \right) l_{BC}^2 \end{aligned} \tag{3.86}$$

From (3.79), $I_{eq_2} = 0$. Therefore, the shaking moment of the leg can be cancelled using a simple counter-rotation I_{cr} with an axial moment of inertia equal to I_{eq_1}.

3.3.2.3 Shaking Moment and Shaking Force Balancing of the 3-*RPR* Manipulator

Now, let us apply such an approach to the 3-*RPR* mechanism. First of all, let us substitute the platform mass by three points masses located at C_1, C_2, and C_3, with the values of mass equal to m_{p_1}, m_{p_2}, and m_{p_3}, respectively [38, 73, 75]. Such a condition can be obtained if:

$$m_{p_i} = m_p/3 \text{ and } I_p = 3m_{p_i} l_{C_iP}^2 \tag{3.87}$$

Such a decomposition of the platform enables us to consider the shaking force and shaking moment balancing of each leg of the mechanism. Then, modifying each leg in order to obtain a mechanism similar to a Scott–Russell linkage, the shaking force and shaking moment are cancelled if:

$$\begin{aligned}
m_{cw_4} &= -\frac{m_4 r_4}{r_{cw_4}} \\
m_{cw_3} &= -\frac{m_3 r_3 + m_{p_i}}{r_{cw_3}} \\
m_{cw_2} &= -\frac{m_2 r_2 + m_3 + m_{cw_3} + m_{p_i}}{r_{cw_2}} \\
m_{cw_1} &= -\frac{m_1 r_1 + m_2 + m_{cw_2} + m_3 + m_{cw_3} + m_{p_i}}{r_{cw_1}} \\
0 &= I_{S_1} + \left(m_1 r_1^2 + m_{cw_1} r_{cw_1}^2 + m_2 + m_{cw_2} + m_3\right) l_{A_iB_i}^2 \\
&\quad - I_{S_2} - \left(m_2 r_2^2 + m_{cw_2} r_{cw_2}^2 + m_3\right) l_{B_iC_i}^2, \text{ and} \\
I_{cr} &= \sum_{i=1}^{4} I_{S_i} + \left(m_1 r_1^2 + m_{cw_1} r_{cw_1}^2 + m_2 (1 - r_2)^2 + m_{cw_2} (1 - r_{cw_2})^2\right) l_{A_iB_i}^2 \\
&\quad + \left(m_3 r_3^2 + m_{cw_3} r_{cw_3}^2 + m_4 r_4^2 + m_{cw_4} r_{cw_4}^2 + m_{p_i}\right) l_3^2
\end{aligned} \tag{3.88}$$

taking into account that I_{cr} is the axial moment of inertia of the counter-rotations (Fig. 3.15).

Thus, with this approach it is possible to create a fully balanced shaking force and shaking moment 3-*RPR* mechanism with only three counter-rotations (Fig. 3.15), i.e., this method enables a reduction in the number of counter-rotations by a factor of two.

3.3.2.4 Illustrative Examples and Numerical Simulations

The parameters used for the simulations are the followings

- $l_{OA_i} = 0.35$ m, $l_{A_iB_i} = l_{B_iC_i} = 0.25$ m, $l_{C_iP} = 0.1$ m and $l_3 = 0.025$ m,
- $r_1 = r_2 = 0.5$, $r_3 = 0$ and $r_4 = 4$,
- $m_1 = 1.09$ kg, $m_2 = 1.1$ kg, $m_3 = 0.37$ kg, $m_4 = 0.75$ kg; $m_p = 1$ kg,
- $I_{S_1} = 0.00738$ kg m^2, $I_{S_2} = 0.58389$ kg m^2, $I_3 = 0.00344$ kg m^2, $I_{S_4} = 0.00025$ kg m^2, and $I_p = 0.01$ kg m^2.

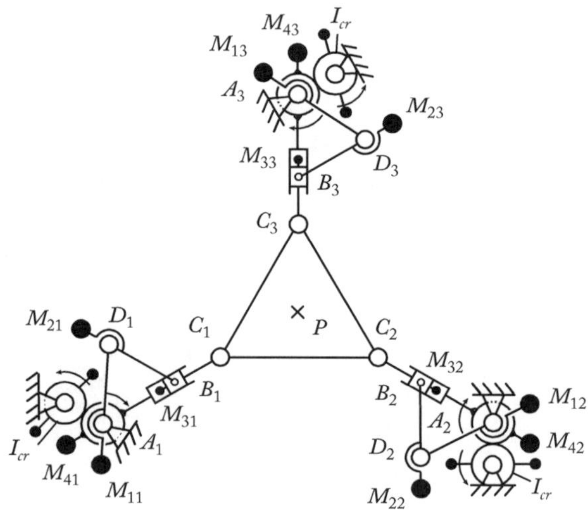

Fig. 3.15 Schematics of a shaking force and shaking moment balanced 3-RPR mechanism

For these new parameters and for the trajectory used in Sect. 3.3.1, taking into account that the position coefficients of the counterweights are equal to $r_{\mathrm{cw}_j} = -0.5$ $(j = 1, 3, 4)$, $r_{\mathrm{cw}_2} = -1$, the new values of the counterweights are: $m_{\mathrm{cw}_1} = 3.17\,\mathrm{kg}$, $m_{\mathrm{cw}_2} = 11.71\,\mathrm{kg}$, $m_{\mathrm{cw}_3} = 0.33\,\mathrm{kg}$, $m_{\mathrm{cw}_4} = 0.75\,\mathrm{kg}$. The shaking force and shaking moment are then computed (dashed line in Fig. 3.16). Finally, we add the counter-rotations. Their values are equal to $I_{\mathrm{cr}} = 1.56907\,\mathrm{kg\,m^2}$. With such counter-rotations, the shaking moment is balanced (gray line in Fig. 3.16c).

Finally, it should be noted that the combination of the proposed two techniques of balancing enables the creation of fully balanced parallel manipulators with modified legs. As examples, different structures of balanced manipulators are presented in Fig. 3.17 (3-*RPR*, 3-*PRR* and 3-*PRP*) in which one leg with a prismatic pair is replaced by a leg with only revolute joints. Such a modification allows the displacement of the center of mass of the manipulator to C_3 and then to balance the manipulator via the modified leg $C_3B_3A_3$.

In the same way, it is possible to balance a parallel manipulator with prismatic pairs by adding fewer Scott–Russell mechanisms. The balancing schemes for several parallel manipulators are presented in Fig. 3.18.

Thus, it has been presented new balancing schemes for the shaking force and shaking moment of planar parallel manipulators whose legs are made of prismatic pairs.

Usually, the balancing of parallel manipulators with prismatic pairs is only attained via a considerably complicated design. This section showed that it is possible to balance planar parallel mechanisms using Scott–Russell mechanisms. Such an approach enables a division of the number of counter-rotations by two. Numerical simulations carried out using ADAMS software validated the obtained results and illustrated that the suggested balancing enables the creation of a parallel manipulator transmitting no inertia load to its base.

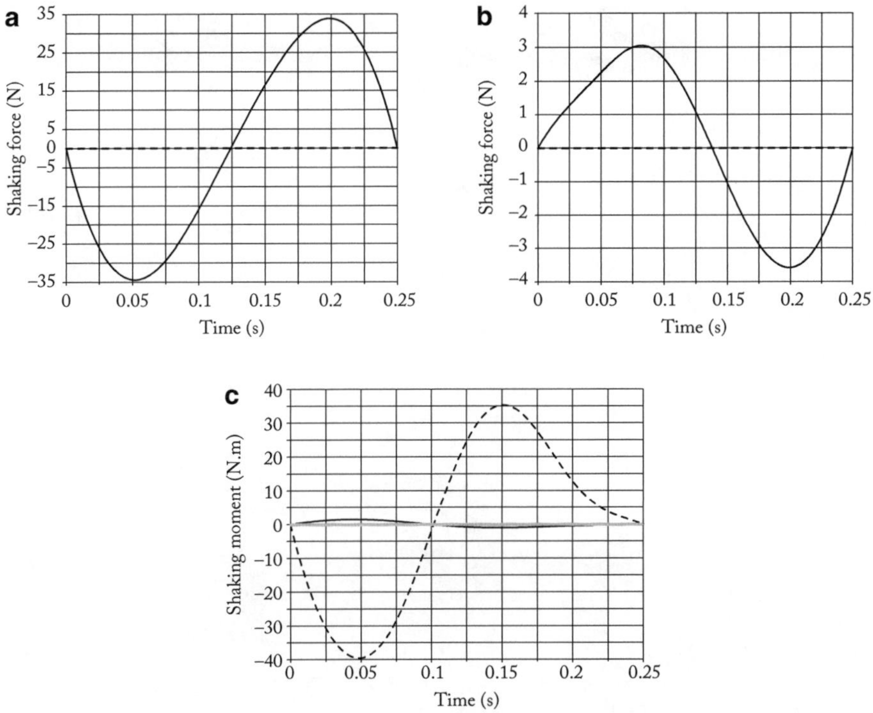

Fig. 3.16 Shaking force and shaking moment before (*solid line*) and after (*dashed line*) the addition of the counterweights, and after the addition of the counter-rotations (*gray line*). (**a**) Shaking force along **x**-axis; (**b**) shaking force along **y**-axis; (**c**) shaking moment around **z**-axis

3.4 Conclusions

For all balancing method, the main challenge is the trade-off between the complexity of the balanced mechanism and the quality of balancing. In the present chapter, we have proposed a solution which allows the reduction of the balancing complexity by comparison with the usual approaches. The idea was to slightly modify the mechanism design by adding to it Assur groups, i.e. groups which do not add any supplementary degree of freedom into the mechanism. The use of such a solution was detailed for the shaking force and shaking moment balancing of:

- the in-line four-bar linkage
- the planar parallel robots with prismatic pairs.

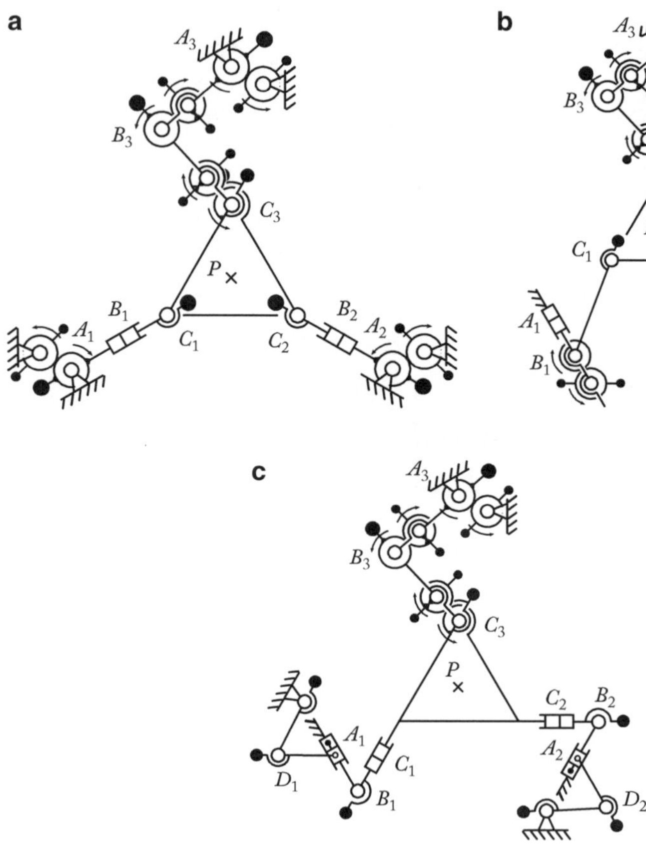

Fig. 3.17 Complete shaking force and shaking moment balancing of planar manipulators with prismatic pairs via structural modification of one leg. (**a**) Balanced 3-*RPR* parallel manipulator; (**b**) balanced 3-*PRR* parallel manipulator; (**c**) balanced 3-*PRP* parallel manipulator

For both types of mechanisms, the proposed solution allowed the reduction (or even the cancellation in the case of the four-bar linkage) of the number of counter-rotations used for obtaining the shaking moment balancing, which decreases the design complexity and the inherent problems due to the use of counter-rotations (backlash, noise, vibrations, etc.).

All theoretical developments were validated via simulations carried out using ADAMS software. The simulations showed that the obtained mechanisms transmitted no inertia loads to their surroundings, i.e. the sum of all ground bearing forces and their moments were eliminated.

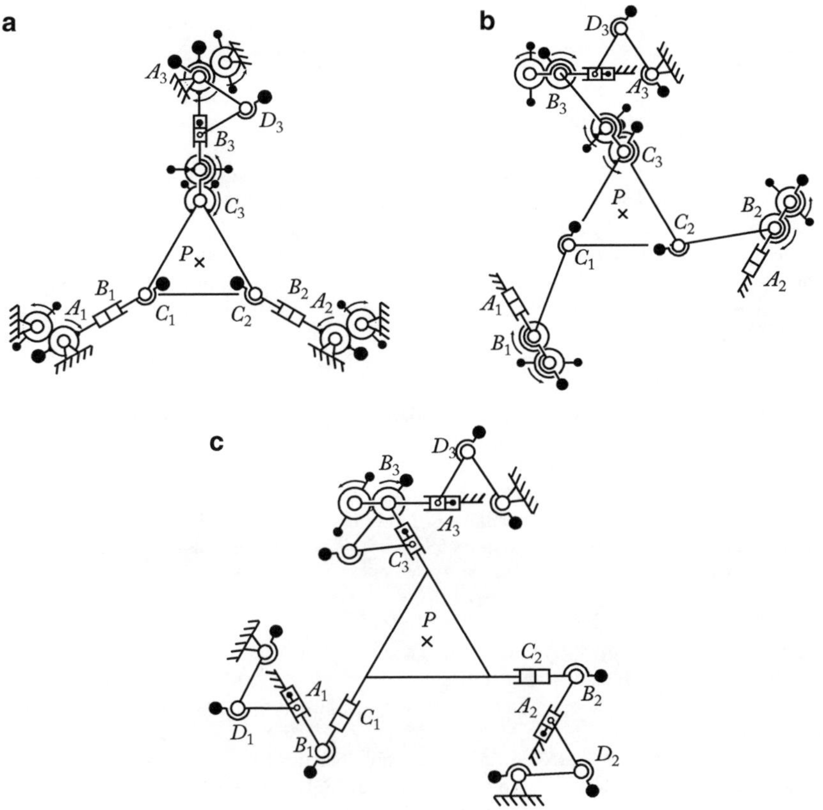

Fig. 3.18 Complete shaking force and shaking moment balancing of planar manipulators with prismatic with reduced number of Scott–Russell mechanisms. (**a**) Balancing of 3-*RPR* parallel manipulator; (**b**) balancing of 3-*PRR* parallel manipulator; (**c**) balancing of 3-*PRP* parallel manipulator

References

1. Foucault, S., Gosselin, C.M.: On the development of a planar 3-dof reactionless parallel mechanism. In: Proceedings of the ASME Design Engineering Technical Conferences and Computers and Information in Engineering Conference (IDECT/CIE 2002), Montreal, September 2002
2. van der Wijk, V., Demeulenaere, B., Gosselin, C.M., Herder, J.L.: Comparative analysis for low-mass and low-inertia dynamic balancing of mechanisms. ASME J. Mech. Robot. **4** (2012)
3. Raaijmakers, R.: Besi zoekt snelheidslimiet pakken en plaatsen op (besi attacks the speedlimit for pick and place motion). In: Mechatronica nieuws (Dutch Magazine), pp. 26–31 (2007)
4. Lowen, G.G., Berkof, R.S.: Survey of investigations into the balancing of linkages. ASME J. Mech. **3**, 221–231 (1968)
5. Ishida, K., Matsuda, T.: Performance characteristics and working comfortableness of forest workers of a new non-vibrating chain saw utilizing perfectly balanced rotation-reciprocation device. In: Proceedings of the Fifth World Congress of Theory of Machines and Mechanisms, Montreal, pp. 951–954 (1979)
6. Arakelian, V., Briot, S.: Balancing of Linkages and Robot Manipulators – Advanced Methods with Illustrative Examples. Springer, Cham (2014)

7. Lowen, G.G., Tepper, F.R., Berkof, R.S.: Balancing of linkages – an update. Mech. Mach. Theory **18**(3), 213–230 (1983)
8. Arakelian, V., Dahan, M., Smith, M.R.: A historical review of the evolution of the theory on balancing of mechanisms. In: Ceccarelli, M. (ed.) Proceedings of the International Symposium on History of Machines and Mechanisms (HMM 2000), pp. 291–300. Kluwer Academic, Dordrecht (2000)
9. Arakelian, V., Smith, M.R.: Shaking force and shaking moment balancing of mechanisms: A historical review with new examples. ASME J. Mech. Des. **127**, 334–339 (2005)
10. Nabat, V., Pierrot, F., Mijangos, M.R., Arteche, J.M.A., Zabalo, R.B., Company, O., Perez De Armentia, K.F.: High-speed parallel robot with four degrees of freedom, patent (2006)
11. van der Wijk, V., Krut, S., Pierrot, F., Herder, J.L.: Design and experimental evaluation of a dynamically balanced redundant planar 4-RRR parallel manipulator. Int. J. Robot. Res. **32**, 744–759 (2013)
12. Leinonen, T.: Terminology for the theory of machines and mechanisms. Mech. Mach. Theory **26**, 435–539 (1991)
13. Fischer, O.: über die reduzierten Systeme und die Hauptpunkte der Glieder eines Gelenkmechanismus. Zeit. für Math. Phys. **47**, 429–466 (1902)
14. Artobolevsky, I.I., Edelshtein, B.V.: Methods of Inertia Calculation for Mechanisms of Agricultural Machines. Selkhozizdate, Moscow (1935) [in Russian]
15. Crossley, F.R.E.: Dynamics in Machines. Roland Press, New York (1954)
16. Talbourdet, G.L., Shepler, P.R.: Mathematical solution of 4-bar linkages - IV. Balancing of linkages. Mach. Des. **13**, 73–77 (1941)
17. Smith, M.R., Maunder, L.: Inertia forces in a four-bar linkage. Mech. Eng. Sci. **9**(3), 218–225 (1967)
18. Berkof, R.S., Lowen, G.G.: A new method for completely force balancing simple linkages. ASME J. Eng. Ind. **91**(1), 21–26 (1969)
19. Hilpert, H.: Weight balancing of precision mechanical instruments. Mechanisms **3**(4), 289–302 (1968)
20. Lowen, G.G., Berkof, R.S.: Determination of force-balanced four-bar linkages with optimum shaking moment characteristics. ASME J. Eng. Ind. **93**(1), 39–46 (1971)
21. Lowen, G.G., Berkof, R.S.: Theory of shaking moment optimization of force-balanced four-bar linkages. ASME J. Mech. Des. **12**, 53–60 (1970)
22. Berkof, R.S., Lowen, G.G.: Theory of shaking moment optimization of force-balanced four-bar linkages. ASME J. Eng. Ind. **93**(1), 53–60 (1971)
23. Wiederrich, J.L., Roth, B.: Momentum balancing of four-bar linkages. ASME J. Mech. Des. **98B**(4), 1289–1295 (1976)
24. Elliot, J.L., Tesar, D.: The theory of torque, shaking force and shaking moment balancing of four link mechanisms. ASME J. Eng. Ind. **99**(3), 715–722 (1977)
25. Carson, W.L., Stephens, J.M.: Feasible parameter design spaces for force and root-mean-square moment balancing on in-line 4R 4-bar linkage synthesized for kinematic criteria. Mech. Mach. Theory **13**(6), 649–658 (1978)
26. Haines, R.S.: Minimum RMS shaking moment or driving torque of a force-balanced 4-bar linkage using feasible counterweights. Mech. Mach. Theory **16**, 185–190 (1981)
27. Shchepetilnikov, V.A.: The determination of the mass centers of mechanisms in connection with the problem of mechanism balancing. Mechanisms **3**, 367–389 (1968)
28. Arakelian, V., Dahan, M.: Partial shaking moment balancing of fully force balanced linkages. Mech. Mach. Theory **36**(11–12), 1241–1252 (2001)
29. Arakelian, V., Dahan, M.: Complete shaking force and partial shaking moment balancing of planar four-bar linkages. Proc. Inst. Mech. Eng. K J. Multibody Dyn. **15**, 31–34 (2001)
30. Zhang, S.: A constitutive method of objective function for the dynamic optimum balance of shaking force in linkage. Mech. Mach. Theory **29**(6), 829–835 (1994)
31. Zhang, S., Chen, J.: The optimum balance of shaking force and shaking moment of linkages. Mech. Mach. Theory **30**(4), 589–597 (1995)

32. Qi, N.M., Pennestri, E.: Optimum balancing of four-bar linkages. Mech. Mach. Theory **26**(3), 337–348 (1991)
33. Chaudhary, H., Saha, S.K.: Balancing of four-bar linkages using maximum recursive dynamic algorithm. Mech. Mach. Theory **42**(2), 216–232 (2007)
34. Kamenski, V.A.: On the question of the balancing of planar linkages. Mechanisms **3**(4), 303–322 (1968)
35. Berkof, R.S.: Complete force and moment balancing of inline four-bar linkages. Mech. Mach. Theory **8**(3), 397–410 (1973)
36. Bagci, C.: Complete shaking force and shaking moment balancing of link mechanisms using balancing idler loops. ASME J. Mech. Des. **104**, 482–493 (1982)
37. Ye, Z., Smith, M.R.: Complete balancing of planar linkages by an equivalence method. Mech. Mach. Theory **29**(5), 701–712 (1991)
38. Arakelian, V., Smith, M.R.: Complete shaking force and shaking moment balancing of linkages. Mech. Mach. Theory **34**(8), 1141–1153 (1999)
39. Gao, F.: Complete shaking force and shaking moment balancing of 17 types of eight-bar linkages only with revolute pairs. Mech. Mach. Theory **26**(2), 197–206 (1991)
40. Berestov, L.B.: Comparative analysis of the reactions in the kinematic pairs of the four-bar linkages for the different methods of balancing. J. Mech. Mach. 61–70 (1977) [in Russian]
41. Dresig, H., Naake, S., Rockausen, L.: Vollständiger und harmonischer Ausgleich ebener Mechanismen. VDI, Düsseldorf (1994)
42. Esat, I., Bahai, H.: A theory of complete force and moment balancing of planar linkage mechanisms. Mech. Mach. Theory **34**(6), 903–922 (1999)
43. Arakelian, V., Dahan, M.: Balanced four-bar articulated mechanism has output member extended beyond axis of second pivot and meshing gears. FR, 2817008 (2002)
44. Kochev, I.S.: Full shaking moment balancing of planar linkages by a prescribed input speed fluctuation. Mech. Mach. Theory **25**(4), 459–466 (1990)
45. Ricard, R., Gosselin, C.M.: On the development of reactionless parallel manipulators. In: Proceedings of the ASME Design Engineering Technical Conferences and Computers and Information in Engineering Conference (IDETC/CIE 2000), Baltimore (2000)
46. Wu, Y., Gosselin, C.M.: Synthesis of reactionless spatial 3-dof and 6-dof mechanisms without separate counter-rotations. Int. J. Robot. Res. **23**(6), 625–642 (2004)
47. Gosselin, C.M., Vollmer, F., Côté, G., Wu, Y.: Synthesis and design of reactionless three-degree-of-freedom parallel mechanisms. IEEE Trans. Robot. Autom. **20**(2), 191–199 (2004)
48. Gosselin, C.M., Moore, B., Schicho, J.: Dynamic balancing of planar mechanisms using toric geometry. J. Symb. Comput. **44**(9), 1346–1358 (2009)
49. Jiang, Q., Gosselin, C.M.: Dynamic optimization of reactionless 4-bar linkages. ASME J. Dyn. Syst. Meas. Control **132**, 041006 (2010)
50. Demeulenaere, B., Berkof, R.: Improving machine drive dynamics: a structured design approach towards balancing. ASME J. Mech. Des. **130**(8), (2008)
51. Berkof, R.S.: The input torque in linkages. Mech. Mach. Theory **14**(1), 61–73 (1979)
52. Wu, Y., Gosselin, C.M.: On the dynamic balancing of multi-dof parallel mechanisms with multiple legs. ASME J. Mech. Des. **129**, 234–238 (2007)
53. Martin, G.H.: Kinematics and Dynamics of Machines, 3rd edn. McGraw-Hill, Columbus (2002)
54. Agrawal, S.K., Fattah, A.: Reactionless space and ground robots: Novel design and concept studies. Mech. Mach. Theory **39**, 25–40 (2004)
55. Wang, J., Gosselin, C.M.: Static balancing of spatial three-degree-of-freedom parallel mechanisms. Mech. Mach. Theory **34**, 437–452 (1999)
56. Newman, W.S., Hogan, N.: The optimal control of balanced manipulators. In: Proceedings of the ASME Winter Annual Meeting, California (1986)
57. Laliberté, T., Gosselin, C.M., Jean, M.: Static balancing of 3-DOF planar parallel mechanisms. IEEE/ASME Trans. Mechatron. **4**(4), 363–377 (1999)
58. Fujikoshi, K.: Balancing apparatus for jointed robot. JP, pp. 51–12 (1976)

59. Wang, J., Gosselin, C.M.: Static balancing of spatial four-degree-of-freedom parallel mechanisms. Mech. Mach. Theory **35**(4), 563–592 (2000)
60. Russo, A., Sinatra, R., Xi, F.: Static balancing of parallel robots. Mech. Mach. Theory **40**(2), 191–202 (2005)
61. Ouyang, P.R., Zhang, W.J.: Force balancing of robotic mechanisms based on adjustment of kinematic parameters. ASME J. Mech. Des. **127**, 433–440 (2005)
62. Herder, J.L., Gosselin, C.M.: A counter-rotary counterweight for light-weight dynamic balancing. In: Proceedings of the ASME Design Engineering Technical Conferences and Computers and Information in Engineering Conference (IDETC/CIE 2004), Salt Lake City, pp. 659–667 (2004)
63. Fattah, A., Agrawal, S.K.: On the design of reactionless 3-dof planar parallel mechanisms. Mech. Mach. Theory **41**(1), 70–82 (2006)
64. Foucault, S., Gosselin, C.M.: Synthesis, design, and prototyping of a planar three degree-of-freedom reactionless parallel mechanism. ASME J. Mech. Des. **126**, 992–999 (2004)
65. Wu, Y., Gosselin, C.M.: Design of reactionless 3-dof and 6-dof parallel manipulators using parallelepiped mechanisms. IEEE Trans. Robot. **21**(5), 821–833 (2005)
66. Papadopoulos, E., Abu-Abed, A.: Design and motion planning for a zero-reaction manipulator. In: Proceedings of the IEEE International Conference on Robotics and Automation (ICRA 1994), San Diego, pp. 1554–1559 (1994)
67. Arakelian, V., Briot, S.: Dynamic balancing of the scara robot. In: Proceedings of 17th CISM-IFToMM Symposium on Robot Design, Dynamics, and Control (RoManSy 2008), Tokyo (2008)
68. Briot, S., Arakelian, V.: Complete shaking force and shaking moment balancing of the position-orientation decoupled PAMINSA manipulator. In: Proceedings of the IEEE/ASME International Conference on Advanced Intelligent Mechatronics (AIM2009), Singapore (2009)
69. Briot, S., Arakelian, V., Le Baron, J.P.: Shaking force minimization of high-speed robots via centre of mass acceleration control. Mech. Mach. Theory **57**, 1–12 (2012)
70. Frolov, K.V.: Theory of Mechanisms and Machines. Vishaya Shkola, Moscow (1987)
71. Doronin, V., Pospelov, A.: Balanced slider-crank mechanism. SU, 1627769 (1991)
72. Arakelian, V.: équilibrage dynamique complet des mécanismes. Mech. Mach. Theory **33**(4), 425–436 (1998)
73. Baradat, C., Arakelian, V., Briot, S., Guegan, S.: Design and prototyping of a new balancing mechanism for spatial parallel manipulators. ASME J. Mech. Des. **130**(7) (2008)
74. Leblond, M., Gosselin, C.M.: Static balancing of spatial and planar parallel manipulators with prismatic actuators. In: Proceedings of the ASME 1998 DETC Conference (1998)
75. Seyferth, W.: Massenersatz duch punktmassen in räumlichen getrieben. Mech. Mach. Theory **9**, 49–59 (1974)

Chapter 4
Design of Reactionless Planar Parallel Manipulators with Inertia Flywheel or with Base-Mounted Counter-rotations

Vigen Arakelian

Abstract This chapter discusses the development of reactionless planar parallel manipulators, which apply no reaction forces or moments to the mounting base during motion. Design equations and techniques are proposed which allow for the dynamic substitution of the mass of the moving platform of a parallel manipulator by three concentrated masses. The dynamic model of the moving platform consequently represents a weightless link with three concentrated masses. This allows for the transformation of the problem of the design of a reactionless manipulator into a problem of balancing pivoted legs carrying concentrated masses. The total angular momentum of the manipulator can be reduced to zero using two approaches: (1) on the basis of counter-rotations, and (2) using an inertia flywheel rotating with a prescribed angular velocity. The suggested solutions are illustrated through 3-DOF 3-RRR planar parallel manipulators. Computer simulations and the results verified by showing that the manipulator are indeed reactionless, there being no forces or moments transmitted to the base during motion of the moving platform.

Keywords Shaking force • Shaking moment • Balancing • Reactionless manipulators • parallel manipulators

4.1 Introduction

In high-speed mechanical systems, mass balancing of the moving links brings about a reduction of the variable dynamic loads on the frame and, as a result, a reduction of vibrations. Different approaches and solutions have been developed

V. Arakelian (✉)
Department of Mechanical and Control Systems Engineering I.N.S.A. Rennes, 20 avenue des Buttes de Coësmes, CS 14315, F-35043 Rennes, France

IRCCyN, 1 rue de la Noë, BP 92101, F-44321 Nantes Cedex 03, France
e-mail: vigen.arakelyan@insa-rennes.fr; vigen.arakelyan@irccyn.ec-nantes.fr

D. Zhang, B. Wei (eds.), *Dynamic Balancing of Mechanisms and Synthesizing of Parallel Robots*, DOI 10.1007/978-3-319-17683-3_4

and documented [1–3] but, despite its long history, mechanism balancing theory continues to develop and new approaches and solutions are constantly being reported. A new field for their application is the design of fast parallel manipulators, which are very efficient for advanced robotic applications. Previous work on the problem of balancing of parallel manipulators may be arranged in the following groups.

(a) Shaking force balancing by counterweights mounted on the movable links of the parallel manipulator [4–9]: The aim of these balancing methods is the redistribution of movable masses by adding counterweights to the links, which allows the fixation of the common centre of mass of the moving links of the manipulator. After such a redistribution of the masses, the gravitational and inertia forces are cancelled.
(b) Gravitational force balancing by springs mounted on the movable links of the parallel manipulator [8–13]: Such a balancing can be defined as when the weights of the links do not produce any force on the actuators for any configuration of the manipulator; that is, potential energy of the parallel manipulator is constant for all possible configurations. It should be noted that many results in the field of balancing of robotic arms and linkages [14–19] can be successfully applied to the balancing problems of parallel manipulators.
(c) Gravitational force balancing by secondary mechanisms coupled with the parallel manipulator [20–26]: In this case the balancing element, which can be a spring [20], a counterweight [21] or an actuating power cylinder [22–24], is mounted on the links of the secondary mechanism. In these studies the added system is a pantograph linkage which allows the gravitational forces to be balanced.

These approaches have been developed for inertia or gravitational force balancing of parallel manipulators. In the case of the shaking force balancing the mentioned methods allow the cancellation of the resultant of all reaction forces at the frame. However, the unbalanced angular moments create a moment on the frame, which can also be significant.

Among several works on this subject, studies devoted to the design of reactionless parallel manipulators [25, 26] should be highlighted. These manipulators are of interest because the inertia forces are cancelled together with the total angular momentum of the manipulator. Such a design enables the cancellation of the reaction forces and torques at the frame of the parallel manipulator.

In this study thought the design of reactionless 3-DOF 3-RRR planar parallel manipulators and the proposed balancing technique are disclosed.

4.2 3-DOF 3-RRR Planar Parallel Manipulator and Dynamic Model with Concentrated Masses (A Special Shape of the Moving Platform)

The moving platform of a planar 3-DOF 3-RRR parallel manipulator is connected to its legs by three revolute joints P_i (i = 1, 2, 3) (Fig. 4.1).

Each leg comprises two links connected by revolute joints $A_i (i = 1, 2, 3)$ and they are mounted on the frame by revolute joints $O_i (i = 1, 2, 3)$. The input parameters of such a manipulator are defined by the joint angles $\theta_i (i = 1, 2, 3)$ of each leg and the output parameters by the pose of the moving platform, i.e. its orientation φ and position of one point of the moving platform, by example, the centre of mass of the moving platform (x_O, y_O).

Note that all axes of revolute joints are parallel; that is, this is a mechanism in which all points of the links describe paths located in parallel planes.

Thus the conditions for dynamic substitution of the mass of the platform (Fig. 4.2) by three concentrated masses situated on the axis of joints $P_i (i = 1, 2, 3)$ are the following:

$$\sum_{i=1}^{3} m_i = m_{\mathrm{pl}} \tag{4.1}$$

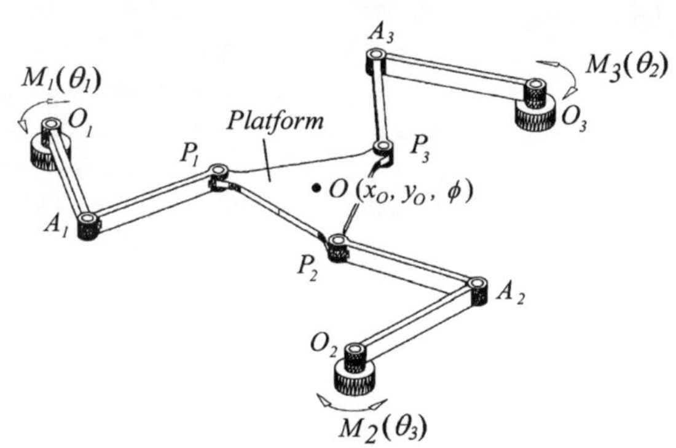

Fig. 4.1 Planar 3-DOF 3-RRR parallel manipulator

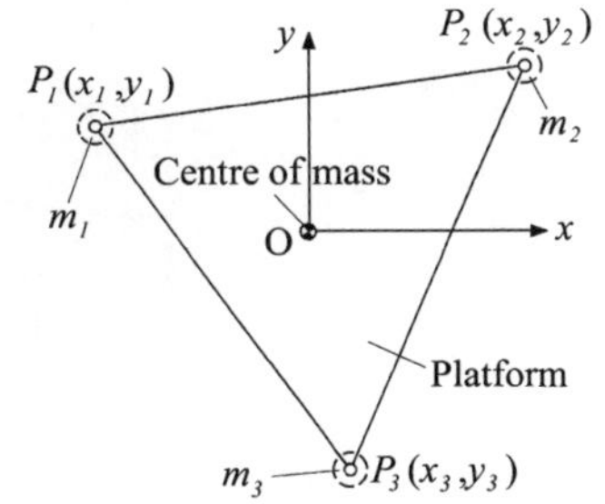

Fig. 4.2 Moving platform of the parallel manipulator and the point masses $m_i (i = 1,\ 2,\ 3)$

$$\sum_{i=1}^{3} m_i x_i = 0 \quad (4.2)$$

$$\sum_{i=1}^{3} m_i y_i = 0 \quad (4.3)$$

$$\sum_{i=1}^{3} m_i \left(x_i^2 + y_i^2\right) = I_{\mathrm{pl(zz)}} \quad (4.4)$$

where m_i are point masses located on the joint axis; x_i and y_i are coordinates of point masses with respect to the platform frame xOy; m_{pl} is the mass of the moving platform and $I_{\mathrm{pl(zz)}}$ is the axial moment of inertia of the moving platform with respect to the centre of mass.

Thus, if we have a platform with such a redistribution of masses, when the conditions (4.1)–(4.4) are fulfilled, the mass of the platform can be dynamically substituted by three concentrated masses; that is, the platform can be considered as a weightless link with three point masses attached.

A design example of such a platform is now considered. Figure 4.3 shows a platform of a parallel manipulator, which represents a cylinder of radius R. The axial moment of inertia of this platform with respect to the centre of the mass is equal to $J_{\mathrm{pl(zz)}} = m_{\mathrm{pl}} R^2/2$. If it is desired to substitute dynamically the mass of the platform by three point masses disposed on the vertices of an isosceles triangle $P_1P_2P_3$, it is necessary to situate the revolute joints P_i at distances $r = R/\sqrt{2}$ from the centre O. In this case the axial moment of inertia of the three point masses and the moving platform are identical. This is an example of conceivable shape, but it is obviously possible to find several examples which allow the dynamic substitution of the mass of the platform by three concentrated masses.

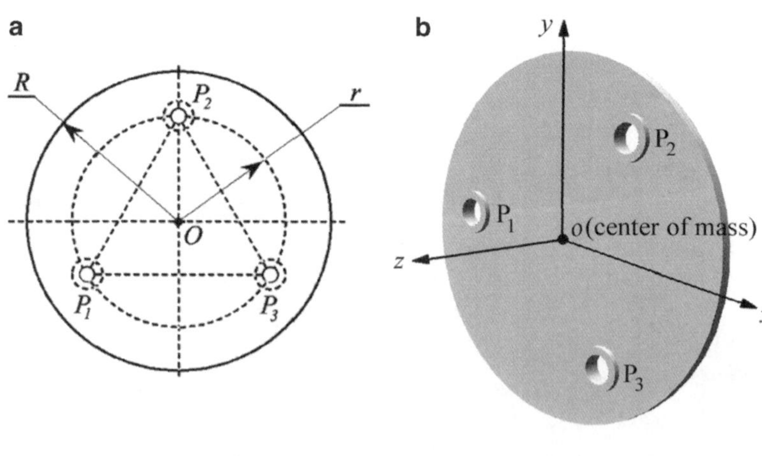

Fig. 4.3 A special shape of the moving platform: (**a**) drawing in 2D; (**b**) CAD model in 3D

Thus, we can replace the effect of mass and inertia of the moving platform by three point masses, which are at the end of each leg. This model allows the transformation of the manipulator balancing problem into one of balancing the legs. The latter is much simpler than the former.

4.3 Balancing of Legs

4.3.1 Balancing by Counter-rotations

The suggested balancing method is based on balancing of the inertia forces by means of counterweights mounted on the links and balancing of the total angular momentum by means of counterweights with planetary gear trains to generate the counter-rotations.

4.3.1.1 Shaking Force Balancing

In order to achieve the dynamic balancing of the manipulator, we first have to ensure that it is force-balanced, i.e. statically balanced. As mentioned above, the mass of the moving platform is substituted by three equivalent point masses located at the legs; that is, each leg of the manipulator can be balanced independently.

The centre of mass of each leg relative to its base O_i (Fig. 4.4) can be found by the expressions

$$x_{Si} = (m_{2i}x_{2i} + m_{3i}x_{3i} + m_i x_{Pi}) \,/\, (m_{2i} + m_{3i} + m_i) \tag{4.5}$$

$$y_{Si} = (m_{2i}y_{2i} + m_{3i}y_{3i} + m_i y_{Pi}) \,/\, (m_{2i} + m_{3i} + m_i) \tag{4.6}$$

where

$$x_{2i} = r_{S2i} \cos\theta_i \tag{4.7}$$

$$y_{2i} = r_{S2i} \sin\theta_i \tag{4.8}$$

$$x_{3i} = l_{2i} \cos\theta_i + r_{S3i} \cos\theta_{i+3} \tag{4.9}$$

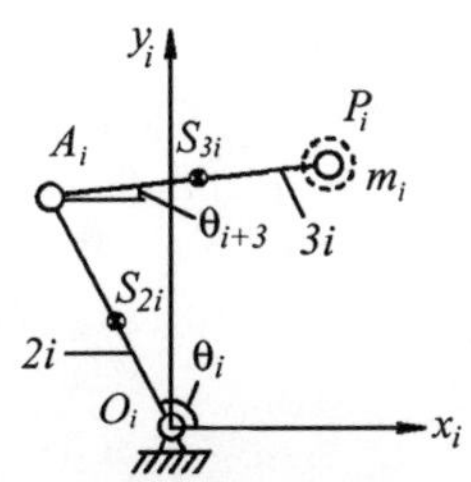

Fig. 4.4 Modeling of leg i of the planar parallel manipulator ($i = 1, 2, 3$)

$$y_{3i} = l_{2i} \sin \theta_i + r_{S3i} \sin \theta_{i+3} \quad (4.10)$$

$$x_{Pi} = l_{2i} \cos \theta_i + l_{3i} \cos \theta_{i+3} \quad (4.11)$$

$$y_{Pi} = l_{2i} \sin \theta_i + l_{3i} \sin \theta_{i+3} \quad (4.12)$$

m_{2i} and m_{3i} are the masses of links *2i* and *3i*; m_i is the point mass obtained from dynamic substitution of masses of the platform; $r_{S2i} = l_{O_iS_{2i}}$ is the distance of the centre of mass S_{3i} of the link 3*i* from the joint centre A_i; $r_{S3i} = l_{A_iS_{3i}}$ is the distance of the centre of mass S_{2i} of the link 2*i* from the joint centre O_i; $l_{2i} = l_{O_iA_i}$ and $l_{3i} = l_{A_iP_i}$ are the lengths of the links *2i* and *3i*.

It is clear that the motion of the centre of mass of each leg is generated by the two angles θ_i and θ_{i+3}. Thus, for the position of the centre of mass to remain constant and located at the axis O_i, it is sufficient that the coefficients of the variables θ_i and θ_{i+3} be equal to zero, i.e.

$$m_{3i}r_{S3i} + m_i l_{3i} = 0 \quad (4.13)$$

$$m_{2i}r_{S2i} + (m_i + m_{3i})\, l_{2i} = 0 \quad (4.14)$$

The conditions (4.13) and (4.14) can be satisfied by adding two counterweights mounted on links *2i* and *3i*, which produce negative values of radii r_{S2i} and r_{S3i}.

After such a redistribution of masses, all moving masses of the manipulator can be replaced by three fixed masses $m_{Oi} = m_{2i} + m_{3i} + m_i$ $(i = 1, 2, 3)$ located at the axis O_i and the centre of mass of the manipulator is located at the centre of these three fixed masses. Thus, the centre of mass of the manipulator remains motionless for any motion of links and hence, the manipulator transmits no inertia loads to its base.

4.3.1.2 Shaking Force and Shaking Moment Balancing

Now that the inertia force balancing is achieved, we have to consider the cancellation of the shaking moment. As in the first case, we consider the balancing of the manipulator legs. There are several approaches for complete shaking moment balancing of articulated dyads with two revolute joints. The balancing method applied in this case is based on the shaking moment balancing by means of counterweights with planetary gear trains carrying out the counter-rotations. The dynamic balancing scheme of each leg is designed in the following manner. The gear 3_{GRi} (Fig. 4.5) is mounted on the rotation axis $O_i$of input link *2i* and is linked kinematically with *3i* through belt transmission *6i*. It meshes also with gear *4i*

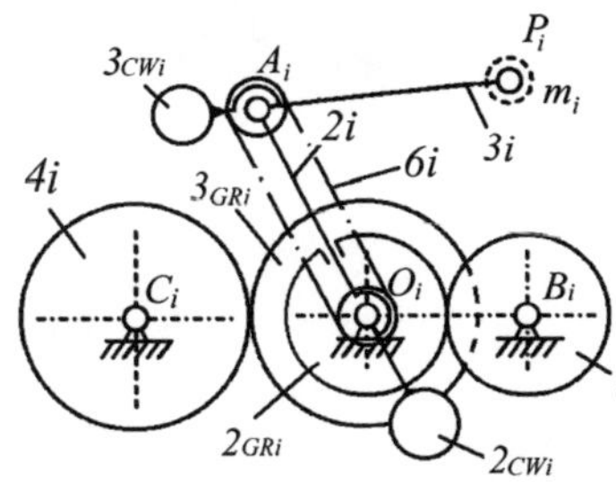

Fig. 4.5 Dynamic balancing scheme of legs ($i = 1, 2, 3$)

mounted on the base. The gear 2_{GRi} is mounted on input link *2i* and meshes with gear *5i* mounted on the rotation axis B_i. It should be noted that the joint between the gear 3_{GRi} and the frame is different from the joint O_i and it will be designated O_{3RGi}.

Thus the shaking moment may be balanced by the moment of inertia of gears *4i* and *5i* taking into account that the angular velocities of links are the following: $\dot{\theta}_{4i} = -\dot{\theta}_{i+3}$ and $\dot{\theta}_{5i} = -\dot{\theta}_i$.

After shaking force balancing, the shaking moment applied on the base is constant relative to any point; that is, for a given position of the manipulator it has the same value for any point of the base and can be expressed as

$$M^{sh} = \sum_{i=1}^{3} M_i^{sh} = \sum_{i=1}^{3} \frac{dH_{Oi}}{dt} \quad (i = 1, 2, 3) \tag{4.15}$$

where H_{Oi} is the angular momentum of the moving links of each leg with respect to O_i.

In order to have a shaking moment equal to zero for all trajectories, the sum of the angular momentum of the legs must be constant over time.

The angular momentum for each leg with added planetary gear trains can be written as

$$\begin{aligned} H_{Oi} &= m_{2i}\left(x_{2i}\dot{y}_{2i} - y_{2i}\dot{x}_{2i}\right) + I_{2i}\dot{\theta}_i + I_{2GRi}\dot{\theta}_i + I_{5i}\dot{\theta}_{5i} \\ &\quad + m_{3i}\left(x_{3i}\dot{y}_{3i} - y_{3i}\dot{x}_{3i}\right) + I_{3i}\dot{\theta}_{i+3} + I_{3RGRi}\dot{\theta}_{i+3} + I_{4i}\dot{\theta}_{4i} \\ &\quad + m_i\left(x_{Pi}\dot{y}_{Pi} - y_{Pi}\dot{x}_{Pi}\right), \quad (i = 1, 2, 3) \end{aligned} \tag{4.16}$$

where I_{2i} and I_{3i} are the moments of inertia of links *2i* and *3i* about the centres of mass of the links (axial moment of inertia), I_{2GRi} and I_{3GRi} are the axial moments of inertia of gears *2GRi* and *3GRi* and I_{4i} and I_{5i} are the axial moments of inertia of the added gears.

We substitute Eqs. (4.7)–(4.12) and their derivatives into Eq. (4.16) and, taking into account condition (4.13), we obtain the following expression of the angular momentum for each leg:

$$H_{Oi} = \left(I_{3GRi} + I_{3i} + m_{3i}r_{3Si}^2 + m_i l_{3i}^2\right)\dot{\theta}_{i+3} + I_{4i}\dot{\theta}_{4i}$$
$$+ \left(I_{2GRi} + J_{2i} + m_{2i}r_{2Si}^2 + (m_{3i} + m_i)\, l_{2i}^2\right)\dot{\theta}_i + I_{5i}\dot{\theta}_{5i}, \quad (i = 1,\ 2,\ 3) \tag{4.17}$$

from which we obtain the conditions of shaking moment balancing:

$$I_{4i} = I_{3GRi} + I_{3i} + m_{3i}r_{3Si}^2 + m_i l_{3i}^2 \tag{4.18}$$

$$I_{5i} = I_{2GRi} + I_{2i} + m_{2i}r_{2Si}^2 + (m_{3i} + m_i)\, l_{2i}^2 \tag{4.19}$$

Hence, any 3-DOF 3-RRR parallel manipulator satisfying Eqs. (4.1)–(4.4), (4.13), (4.14), (4.18) and (4.19) will be dynamically balanced, i.e. reactionless.

The disadvantage of the suggested balancing scheme is the need for the connection of gears to the oscillating links. The oscillations of the links of the manipulator will create noise unless expensive anti-backlash gears are used.

Anti-backlash gears are devices that bias the gear always to favour one side of the tooth through spring action. Regardless of the direction of movement, they should always "push" up against the same side of the tooth. They are basically comprised of two gears that are spring-loaded in opposing directions. One gear is attached to the mechanism being moved, and the other simply "floats" to provide the bias.

4.3.2 Numerical Example and Simulation Results

We shall now examine the ground bearing forces and the ground bearing moments of a 3-DOF 3-RRR parallel manipulator which is fully force and moment balanced. The geometry and mass distribution parameters of the links are listed in Table 4.1.

The platform of the examined parallel manipulator represents a cylinder with radius $R = 0.082$ m and $m = 3$ kg. As a result, $I_{\mathrm{pl(zz)}} = 0.01$ kgm^2 (so that $m_i = 1$ kg and $r_i = 0.058$ m $(i = 1, 2, 3)$). The drivers are given by the expressions $\theta_i = a_i\pi + b_i\,(2\pi t/T - \sin(2\pi t/T))$, where $a_1 = 1/3$, $a_2 = 4/3$, $a_3 = 10/3$, $b_1 = 1/6$, $b_2 = -1/6$, $b_3 = 1/12$ and $T = 0.3$sec. The angles of the input links θ_i are measured with respect to the global X-axis. The driver functions give zero velocity and acceleration at the start and end of the motion.

Figures 4.6 and 4.7 show the resultant bearing forces of the balanced planar 3-DOF 3-RRR manipulator along the X- and Y-axes.

In Fig. 4.8 are presented the variations of the moment of the ground bearing forces and the reactions of the input torques.

Table 4.1 Parameters of the balanced manipulator

Parameter	Legs 1	2	3
X_{O_i}(m)	0	0.46	0.22
Y_{O_i}(m)	0	0	0.4
l_{2i} (m)	0.18	0.18	0.18
l_{3i} (m)	0.18	0.18	0.18
x_{P_i} $(t=0)$, (m)	0.18	0.28	0.22
y_{P_i} $(t=0)$, (m)	0	0	0.087
m_i (kg)	1	1	1
r_i (m)	0.058	0.058	0.058
m_{2i} (kg)	7.2	7.2	7.2
r_{2Si} (m)	0.09	0.09	0.09
m_{3i} (kg)	2.6	2.6	2.6
r_{3Si} (m)	0.07	0.07	0.07
J_{2i} (kg m^2)	0.02	0.02	0.02
J_{3i} (kg m^2)	0.017	0.017	0.017
J_{4i} (kg m^2)	0.068	0.068	0.068
J_{5i} (kg m^2)	0.2	0.2	0.2

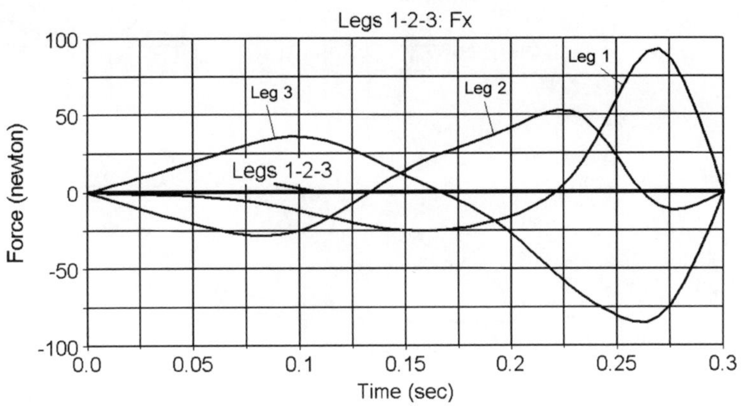

Fig. 4.6 Variations of the ground bearing forces of the balanced manipulator along the X-axis

4.4 Balancing by Inertia Flywheel

In this section we consider the shaking moment cancellation of the fully force balanced 3-DOF 3-RRR parallel manipulator by an inertia flywheel with prescribed rotation. It is evident that this solution is constructively more efficient.

Figure 4.9 shows the fully force-balanced 3-DOF 3-RRR parallel manipulator and balancing inertia flywheel, which is mounted on the base of the manipulator. The conditions for balancing the shaking moment of the manipulator are determined from the following consideration.

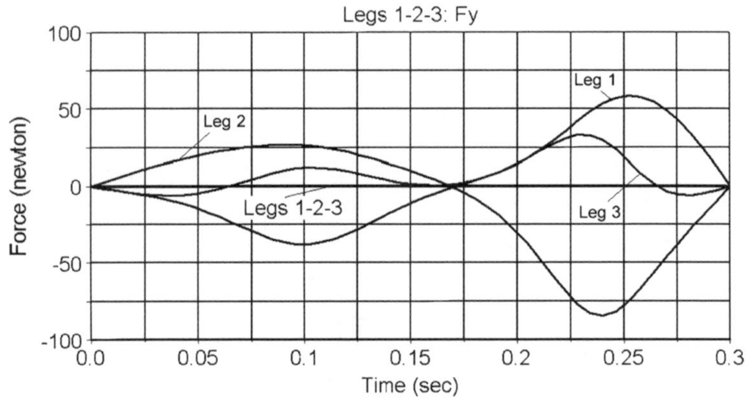

Fig. 4.7 Variations of the ground bearing forces of the balanced manipulator along the Y-axis

Fig. 4.8 Variations of the moment of the ground bearing forces and the reactions of the input torques of the balanced manipulator

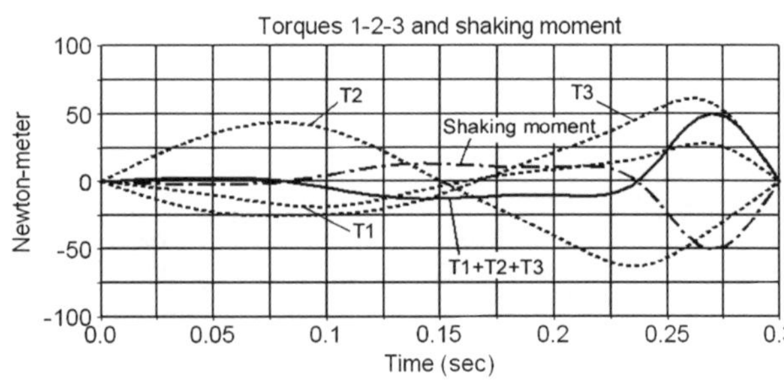

Fig. 4.9 Shaking moment balancing of fully force-balanced 3-DOF 3-RRR parallel manipulator by an inertia flywheel

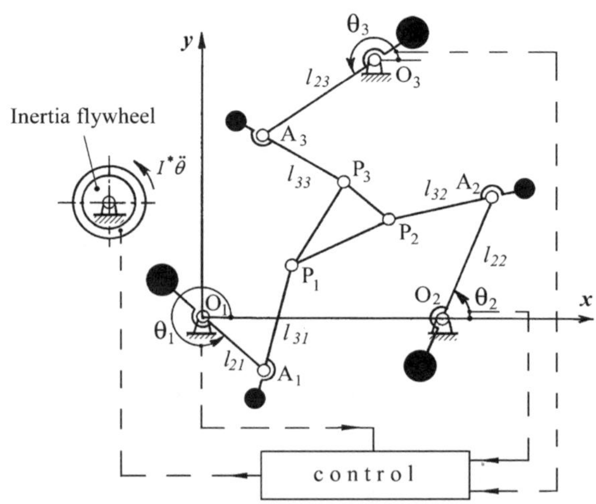

Please note that in this case the platform has axial inertia moment, which cannot be dynamically substituted by three concentrated masses, i.e. $\sum_{i=1}^{3} m_i \left(x_i^2 + y_i^2\right) \neq I_{\mathrm{pl(zz)}}$.

Thus, the angular momentum for the fully force-balanced manipulator can be written as

$$H = \sum_{i=1}^{3} H_{Oi} + \Delta I_{pl}\dot{\phi} = \sum_{i=1}^{3} \Big[\left(J_{2i} + m_{2i} r_{2Si}^2 + (m_{3i} + m_i)\, l_{2i}^2\right) \dot{\theta}_i + \left(J_{3i} + m_{3i} r_{3Si}^2 + m_i l_{3i}^2\right) \dot{\theta}_{i+3} \Big] + \Delta I_{pl}\dot{\phi} \tag{4.20}$$

where $\Delta I_{\mathrm{pl(zz)}} = I_{\mathrm{pl(zz)}} - \sum_{i=1}^{3} m_i \left(x_i^2 + y_i^2\right)$.

Hence, the shaking moment is the following:

$$M^{sh} = \frac{d}{dt}\left(\sum_{i=1}^{3} H_{Oi} + \Delta I_{\mathrm{pl(zz)}}\dot{\phi}\right) = \sum_{i=1}^{3}\left(K_i \ddot{\theta}_i + K_{i+3}\ddot{\theta}_{i+3}\right) + \Delta I_{\mathrm{pl(zz)}}\ddot{\phi} \tag{4.21}$$

where $K_i = I_{2i} + m_{2i} r_{2Si}^2 + (m_{3i} + m_i)\, l_{2i}^2$ and $K_{i+3} = I_{3i} + m_{3i} r_{3Si}^2 + m_i l_{3i}^2$.

To balance the shaking moment, an inertia flywheel with axial inertia moment I^* can be used. The angular acceleration of this inertia flywheel driven by a complementary actuator 4 is the following:

$$\ddot{\theta} = M^{\mathrm{sh}}/I^* = \sum_{i=1}^{3}\left(K_i \ddot{\theta}_i + K_{i+3}\ddot{\theta}_{i+3}\right) + \Delta I_{\mathrm{pl(zz)}}/I^* \tag{4.22}$$

It should be noted that the axial inertia moment of the flywheel must be selected in such a manner that its rotation with prescribed acceleration will be feasible. Therefore, the reaction of the balancing inertia flywheel on the frame cancels the shaking moment due to the parallel manipulator. In other words, the actuator, which rotates the balancing inertia flywheel with a prescribed angular acceleration $\ddot{\theta}$, has a reaction on the frame which is similar but opposite to the shaking moment of the parallel manipulator. Thus, full shaking moment is annulled.

The angular velocity $\dot{\theta}(t)$ and angular displacements $\theta(t)$ can be determined by simple integration of the obtained values of $\ddot{\theta}(t)$.

4.4.1 Numerical Example and Simulation Results

Let us consider a numerical example for computer simulation. As a model we could use the previous example with the link parameters given in Table 4.1. However, for the best illustration of the suggested balancing approach, we change the value of the axial moment of inertia of the platform: $J_{\mathrm{pl(zz)}} = 0.01\ 5\mathrm{kgm}^2$; that is, the mass of the platform cannot be dynamically substituted by three concentrated masses. It should

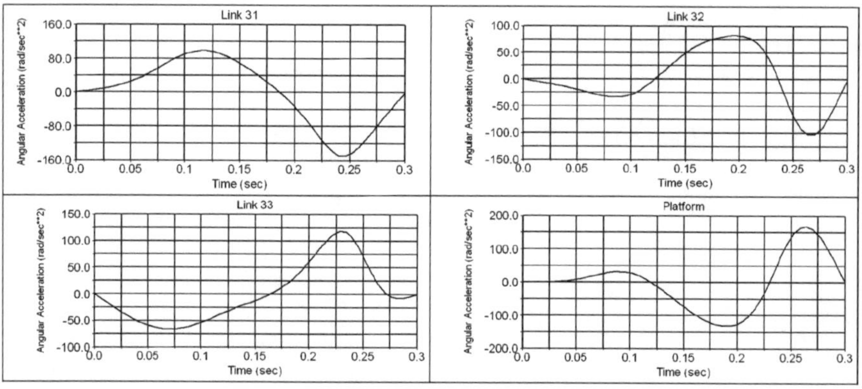

Fig. 4.10 Angular accelerations of links

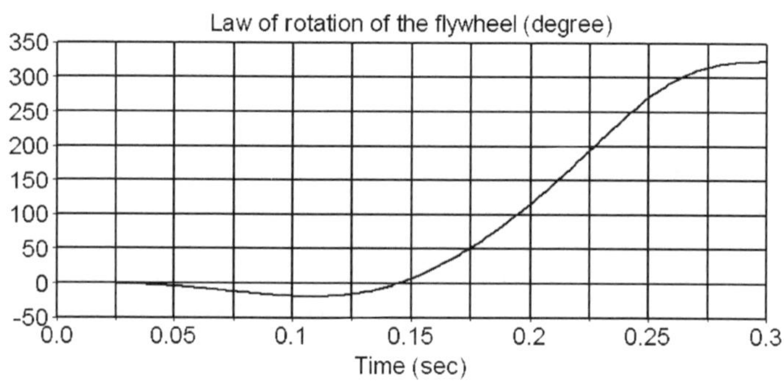

Fig. 4.11 Law of rotation of the balancing flywheel

be noted that in this case we do not need this condition. Thus, by substituting statically the mass of the platform by three concentrated masses, the shaking force balancing is carried out in the same way as in the previous case.

Then the angular accelerations of the movable links are determined (Fig. 4.10) taking into account that the drivers are given by the expressions $\theta_i = a_i\pi + b_i(2\pi t/T - \sin(2\pi t/T))(i = 1, 2, 3)$, where $a_1 = 1/3$, $a_2 = 4/3$, $a_3 = 10/3$, $b_1 = 1/6$, $b_2 = -1/6$, $b_3 = 1/12$ and $T = 0.3$sec.

Now, by determining the shaking moment from Eq. (4.21) and taking as the axial moment inertia of the flywheel $I^* = 0.01$ kgm^2, we determine the angular acceleration of the balancing flywheel, which gives complete shaking moment balancing of the manipulator. Figure 4.11 shows the obtained law of rotation of the flywheel, which produces complete shaking moment balancing.

4.5 Conclusions

A new field for shaking force and shaking moment balancing is the design of fast parallel manipulators, which are very efficient for advanced robotic applications. In this study, the shaking force and shaking moment balancing approach is developed for planar parallel manipulators which is illustrated via 3-DOF 3-RRR parallel manipulators. It is based on the dynamic substitution of the mass of the platform by three concentrated masses situated at the axes of the revolute joints of the legs. By application of this approach the dynamic model of the platform represents a weightless link with three concentrated masses attached. This allows for the transformation of the problem of shaking force and shaking moment balancing of the manipulator into a problem of balancing legs carrying concentrated masses. A design example of a platform with point masses is examined. However this approach requires the use of counter-rotations which increase the mass and inertia of the system. For this reason, a second approach is proposed, which allows the cancellation of the shaking moment of the system by means of a flywheel. It should be noted that such a solution can also be applied on planar serial manipulators. Numerical examples confirm that after such a balancing, the manipulator transmits no inertia loads to its surroundings; that is, the sum of all ground bearing forces and their moments are eliminated.

References

1. Lowen, G.G., Tepper, F.R., Berkof, R.S.: Balancing of linkages – an update. Mech. Mach. Theory **18**(3), 213–230 (1983)
2. Arakelian, V.H., Dahan, M., Smith, M.R.: A historical review of the evolution of the theory on balancing of mechanisms. Proceedings of the international symposium on history of machines and mechanisms, pp. 291–300. Kluwer Academic, Dordrecht (2000)
3. Arakelian, V.H., Smith, M.: Shaking force and shaking moment balancing of mechanisms: a historical review with new examples. J. Mech. Des. **127**, 334–339 (2005)
4. Bolotin, L.M., Glazunov VA: Spatial mechanism, Patent SU 1757873, August 30, 1992
5. Jean, M., Gosselin, C.M.: Static balancing of planar manipulators. Proceedings of the 1996 IEEE international conference on robotics and automation, Minneapolis, Minnesota, April 1996. pp. 3732–3737
6. Wang, J., Gosselin, C.M.: Static balancing of spatial six-degree-of-freedom parallel mechanisms with revolute actuators. Proceedings of DETC'98, Atlanta, Georgia, USA, September 1998 (Paper DECH/MECH-5961)
7. Arakelian, V., Le Baron, J.-P., Mottu, P.: Torque minimisation of the 2-DOF serial manipulators based on minimum energy consideration and optimum mass redistribution. Mechatronics **21**(1), 310–314 (2011)
8. Laliberté, T., Gosselin, C.M.: Static balancing of 3-DOF planar parallel mechanisms. IEEE. Trans. Mechatron. **4**(4), 363–377 (1999)
9. Ebert-Uphoff, I., Gosselin, C.M., Laliberté, T.: Static balancing of spatial parallel platform mechanisms – revisited. J. Mech. Des. **122**, 43–51 (2000)

10. Leblond, M., Gosselin C.M.: Static balancing of spatial and planar parallel manipulators with prismatic actuators. Proceedings of DETC'98, Atlanta, Georgia, USA, September 1998 (Paper DECH/MECH-5963)
11. Arakelian, V., Ghazaryan, S.: Improvement of balancing accuracy of robotic systems: application to leg orthosis for rehabilitation devices. Mech. Mach. Theory **43**(5), 565–575 (2008)
12. Johnson, K., Ebert-Uphoff, I.: Development of a spatial statically-balanced parallel platform mechanism. Proceedings of the year 2000 parallel kinematic machines international conference and second European-American PKM forum, Ann Arbor, MI, September 2000. pp. 143–159
13. Nathan, R.H.: A constant force generation mechanism. J. Mech. Des. **107**, 508–512 (1985)
14. Minotti, P., Pracht, P.: Ressort et mécanismes : une solution aux problèmes d'équilibrage. Mech. Mach. Theory **23**(2), 157–168 (1988)
15. Streit, D.A., Gilmore, B.J.: Perfect spring equilibrators for rotatable bodies. J. Mech. Transmiss. Automat. Des. **111**, 451–458 (1989)
16. Streit, D.A., Shin, E.: Equilibrator for planar linkages. J. Mech. Des. **115**, 604–611 (1993)
17. Walsh, G.J., Streit, D.A., Gilmore, B.J.: Spatial spring equilibrator theory. Mech. Mach. Theory **26**(2), 115–170 (1991)
18. Simionescu, I., Ciupitu, L.: The static balancing of the industrial robot arms; part I: discrete balancing. Mech. Mach. Theory **35**(9), 1287–1298 (2000)
19. Simionescu, I., Ciupitu, L.: The static balancing of the industrial robot arms; part II: continuous balancing. Mech. Mach. Theory **35**(9), 1299–1311 (2000)
20. Herder, J.L.: Some consideration regarding statically balanced parallel mechanisms. Proceedings of the workshop on fundamental issues and future research directions for parallel mechanisms and manipulators, Quebec City, Quebec, Canada, 3–4 October 2002. pp. 40–45
21. Russo, A., Sinatra, R., Xi, F.: Static balancing of parallel robots. Mech. Mach. Theory **40**(2), 191–202 (2005)
22. Baradat, C., Arakelian, V., Maurine, P.: Parallel robot including the load compensation system. Patent FR2880575, 14 juillet 2006
23. Baradat C., Arakelian V., Briot S., Guegan S.: Design and prototyping of a new balancing mechanism for spatial parallel manipulators. Transactions of the ASME. J. Mech. Des. **130**(7), 072305 (13 pages) (2008)
24. Briot, S., Baradat, C., Guegan, S., Arakelian, V.: Contribution to the mechanical behavior improvement of the robotic navigation device Surgiscope®. Proceedings of the 2007 ASME International Design Engineering Technical Conferences (DETC 2007), Las Vegas, Nevada, USA, September 4–7 2007
25. Foucault, S., Gosselin, C.M.: Synthesis, design and prototyping of a planar three degree-of-freedom reactionless parallel mechanism. J. Mech. Des. **126**, 992–999 (2004)
26. Fattah, A., Agrawal, S.K.: On the design of reactionless 3-DOF planar parallel mechanisms. Mech. Mach. Theory **41**, 70–82 (2006)

Chapter 5
Design of Reactionless Mechanisms with Counter-Rotary Counter-Masses

Mario Acevedo

Abstract In this chapter a new method to find the force and moment balancing conditions based on Natural Coordinates is introduced. The method is simple and can be highly automated, it is very prone to be used in combination with a system for the manipulation of symbolic expressions. These conditions can be interpreted and used for the creation of dynamic balanced linkages by design. The application of the method is demonstrated through the dynamic balancing of a simple pendulum (open-loop linkage) and a general four-bar mechanism (closed-loop linkage), particularly by the design of counter-rotary counter-masses applying optimization. The resulting designs are presented and their virtual prototypes simulated using a general multibody dynamics simulation software (ADAMS), specifying the resulting geometry (dimensions), shaking force, shaking moment, and driving torque.

Keywords Dynamic balancing • Counter rotary counter-masses • Optimization • Planar mechanisms • Natural coordinates

5.1 Introduction

Force and moment balancing (dynamic balancing) of rigid body linkages with constant mass links is a traditional but still very active area of research in mechanical engineering. Its benefits are well known as machine vibrations often occur due to dynamic unbalance inducing noise, wear, fatigue problems [1], limiting the full potential of many machines. Mechanisms that are dynamically balanced do not transmit vibrations to the base, a useful property in hand tools, in objects and vehicles moving in free space, and in robotics.

But dynamic balancing of linkages has some difficulties and drawbacks. First finding the balancing conditions may be complicated [2] and second a substantial

M. Acevedo (✉)
Faculty of Engineering, Universidad Panamericana, Augusto Rodin No. 498, Insurgentes Mixcoac, 03920 Mexico City, Mexico
e-mail: mario.acevedo@up.edu.mx

D. Zhang, B. Wei (eds.), *Dynamic Balancing of Mechanisms and Synthesizing of Parallel Robots*, DOI 10.1007/978-3-319-17683-3_5

amount of mass and inertia must generally be added [3, 4]. A complete overview of dynamic balancing techniques and methods can be found in [5–7].

An increment in mass or inertia implies more power to drive the mechanism, so research has therefore been focused on reducing these disadvantages. A way to reduce the necessary mass and inertia to balance the mechanisms can be to use the counter-masses, necessary for the force balance of the linkage, also for balancing the moment. This principle, compared with other balancing principles, has shown effective in the reduction of the additional mass and inertia, see [4, 8], and used effectively to synthesize different dynamic balanced mechanisms using the double pendulum as building element [9].

In this work a completely new general method to find the dynamic balancing conditions, based on the use of Natural Coordinates [10], is introduced. The method is direct and very easy to automate, and can be used to obtain the shaking force and the shaking moment balancing conditions for the linkages in the plane and in space, although at this time is presented only for planar mechanisms. Once the balancing conditions are found, these are used to the effective design of reactionless mechanisms with counter-rotary counter-masses.

The chapter is organized as follows: in Sect. 5.2 the new method is introduced, showing how Natural Coordinates are very useful to directly obtain the balancing conditions. In Sects. 5.3 and 5.4, the application of the method to the dynamic balancing of a single pendulum and of a four-bar mechanism are presented. In Sect. 5.5 some numerical examples are solved to obtain specific mechanisms designs, and the results obtained from dynamic simulations made with ADAMS are presented. Finally some concluding remarks are made in Sect. 5.6.

5.2 Balancing Conditions Using Natural Coordinates

For the effective design of reactionless mechanisms with counter-rotary counter-masses it is necessary first to obtain the dynamic balancing conditions. In this section a new method based on Natural Coordinates is presented. The method is straightforward and can be easily automated, mainly it has the advantage of being suitable for the application of a computer algebra system.

A dynamic balanced mechanism must be completely force and moment balanced. In fact the mechanism that is not balanced by force first, cannot be balanced by moments.

A mechanism is force balanced if its linear momentum, $\mathbf{l}_m$, is conserved. This condition in general can be expressed as:

$$\mathbf{l}_m = \sum_{i=1}^{n} \mathbf{l}_i = cnt. \tag{5.1}$$

where n is the number of total moving elements in the linkage.

When working in reference-point coordinates (Cartesian Coordinates) the linear momentum of body i can be calculated as:

$$\mathbf{l}_i = \sum_{i=1}^{n} m_i \mathbf{v}_i \tag{5.2}$$

where $\mathbf{v}_i$ and m_i are the velocity of the center of mass and the mass of the ith moving body, respectively.

On the other hand, a mechanism is moment balanced if its angular momentum, $\mathbf{h}_m$, is conserved. This condition is expressed by:

$$\mathbf{h}_m = \sum_{i=1}^{n} \mathbf{r}_i \times (\mathbf{l}_i) + \mathbf{h}_i = cnt. \tag{5.3}$$

where $\mathbf{r}_i$ is the position vector of the center of mass. Again when working in reference-point coordinates the angular momentum of body i can be calculated as:

$$\mathbf{h}_i = \mathbf{I}_i \boldsymbol{\omega}_i \tag{5.4}$$

where $\mathbf{I}_i$ and $\boldsymbol{\omega}_i$ are the inertia tensor with respect to the center of mass and the angular velocity of the ith body, respectively.

So it is necessary to find equivalent expressions to Eqs. (5.2) and (5.4) in natural coordinates to calculate the linear and angular momentum of the mechanism. In the next subsections these expressions are developed.

5.2.1 Linear Momentum of a Body Using Natural Coordinates

Equations (5.1) and (5.3) are in general well recognized when using reference-point coordinates (Cartesian Coordinates), but in this work we are interested on its form in Natural Coordinates. So the first step to find this form is to develop each part of the equations according to our goal.

When dealing with mechanical systems in the plain, Natural coordinates introduce a set of points to define a body, the basic points, see [10] for a detailed explanation. So a body can be modeled in Natural Coordinates with a pair of points, i and j, as seen in Fig. 5.1. In this figure it can be noted an inertial fixed reference frame **XY**, a local reference frame **xy** attached to the moving body at the basic point i, $(0, 0)$. It is also noted that the second basic point j has its position in local coordinates at $(l, 0)$, and that the center of mass of the body, point g, at (x, y).

Using this pair of points and considering that the body has a total mass concentrated at g equal to m, and a moment of inertia I_i with respect to the origin of the local reference frame (point i), the constant mass matrix of a body can be expressed as (see [10]):

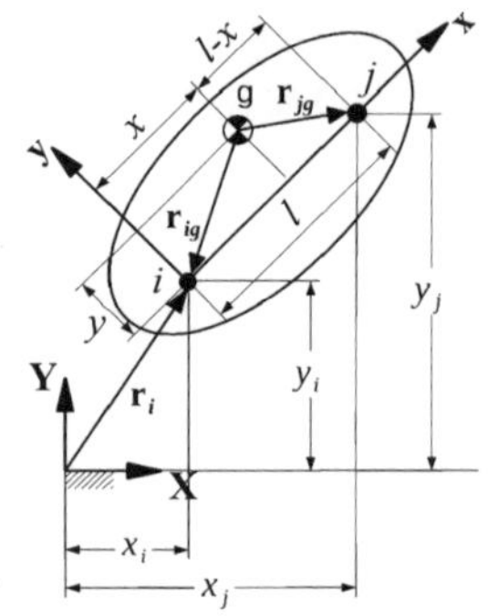

Fig. 5.1 A general model of a body using Natural Coordinates. Two basic points, *i* and *j*, with a local moving reference frame attached to the body at point *i*, the origin

$$\mathbf{M} = \begin{bmatrix} m - \frac{2mx}{l} + \frac{I_i}{l^2} & 0 & \frac{mx}{l} - \frac{I_i}{l^2} & -\frac{my}{l} \\ 0 & m - \frac{2mx}{l} + \frac{I_i}{l^2} & \frac{my}{l} & \frac{mx}{l} - \frac{I_i}{l^2} \\ \frac{mx}{l} - \frac{I_i}{l^2} & \frac{my}{l} & \frac{I_i}{l^2} & 0 \\ -\frac{my}{l} & \frac{mx}{l} - \frac{I_i}{l^2} & 0 & \frac{I_i}{l^2} \end{bmatrix} \tag{5.5}$$

The pair of basic points introduce a vector of four coordinates represented by $\mathbf{q}_k$, the positions, and its time derivative $\dot{\mathbf{q}}$, the velocities:

$$\mathbf{q} = \begin{bmatrix} x_i \; y_i \; x_j \; y_j \end{bmatrix}^T \tag{5.6}$$

$$\dot{\mathbf{q}} = \begin{bmatrix} \dot{x}_i \; \dot{y}_i \; \dot{x}_j \; \dot{y}_j \end{bmatrix}^T \tag{5.7}$$

So it is possible to calculate a set of the linear momentum vectors associated with the basic points in the body as:

$$\mathbf{M}\dot{\mathbf{q}} = \begin{bmatrix} m - \frac{2mx}{l} + \frac{I_i}{l^2} & 0 & \frac{mx}{l} - \frac{I_i}{l^2} & -\frac{my}{l} \\ 0 & m - \frac{2mx}{l} + \frac{I_i}{l^2} & \frac{my}{l} & \frac{mx}{l} - \frac{I_i}{l^2} \\ \frac{mx}{l} - \frac{I_i}{l^2} & \frac{my}{l} & \frac{I_i}{l^2} & 0 \\ -\frac{my}{l} & \frac{mx}{l} - \frac{I_i}{l^2} & 0 & \frac{I_i}{l^2} \end{bmatrix} \begin{bmatrix} \dot{x}_i \\ \dot{y}_i \\ \dot{x}_j \\ \dot{y}_j \end{bmatrix} = \begin{bmatrix} \mathbf{l}_i \\ \mathbf{l}_j \end{bmatrix} \tag{5.8}$$

where $\mathbf{l}_i$ and $\mathbf{l}_j$ are the linear momentum associated with the points i and j, respectively. They can be expressed as:

$$\mathbf{l}_i = \begin{bmatrix} \left(\frac{I_i}{l^2} - \frac{2mx}{l} + m\right)\dot{x}_i + \left(\frac{mx}{l} - \frac{I_i}{l^2}\right)\dot{x}_j - \frac{my}{l}\dot{y}_j \\ \left(\frac{I_i}{l^2} - \frac{2mx}{l} + m\right)\dot{y}_i + \frac{my}{l}\dot{x}_j + \left(\frac{mx}{l} - \frac{I_i}{l^2}\right)\dot{y}_j \end{bmatrix} \tag{5.9}$$

$$\mathbf{l}_j = \begin{bmatrix} \left(\frac{mx}{l} - \frac{I_i}{l^2}\right)\dot{x}_i + \frac{my}{l}\dot{y}_i + \left(\frac{I_i}{l^2}\right)\dot{x}_j \\ -\frac{my}{l}\dot{x}_i + \left(\frac{mx}{l} - \frac{I_i}{l^2}\right)\dot{y}_i + \left(\frac{I_i}{l^2}\right)\dot{y}_j \end{bmatrix} \tag{5.10}$$

Then the total linear momentum of the body is:

$$\mathbf{l} = \mathbf{l}_i + \mathbf{l}_j = \begin{bmatrix} \left(m - \frac{mx}{l}\right)\dot{x}_i + \left(\frac{my}{l}\right)\dot{y}_i + \left(\frac{mx}{l}\right)\dot{x}_j - \left(\frac{my}{l}\right)\dot{y}_j \\ \left(\frac{my}{l}\right)\dot{x}_i + \left(m - \frac{mx}{l}\right)\dot{y}_i + \left(\frac{my}{l}\right)\dot{x}_j - \left(\frac{mx}{l}\right)\dot{y}_j \end{bmatrix} \tag{5.11}$$

5.2.2 Angular Momentum of a Body Using Natural Coordinates

The angular momentum of the body, represented by a pair of masses on points i and j, with respect to its center of mass g can be calculated as:

$$\mathbf{h}_g = \mathbf{r}_{ig} \times \mathbf{l}_i + \mathbf{r}_{jg} \times \mathbf{l}_j \tag{5.12}$$

where $\mathbf{r}_{ig} = [-x \;\; -y]^T$ and $\mathbf{r}_{jg} = [(x-l) \;\; -y]^T$ are the position vector of points i and j with respect to the center of mass of the body expressed in the global fixed reference frame, and can be calculated by:

$$\mathbf{r}_{ig} = \mathbf{A}\bar{\mathbf{r}}_{ig}; \; \mathbf{r}_{jg} = \mathbf{A}\bar{\mathbf{r}}_{jg} \tag{5.13}$$

where $\bar{\mathbf{r}}_{ig}$ and $\bar{\mathbf{r}}_{jg}$ are the position vector of i and j expressed in local coordinates, while $\mathbf{A}$ is the rotation matrix:

$$\mathbf{A} = \frac{1}{l}\begin{bmatrix} (x_j - x_i) & (y_i - y_j) \\ (y_j - y_i) & (x_j - x_i) \end{bmatrix} \tag{5.14}$$

So the general form of the angular momentum of the body with respect to the global fixed reference frame can be calculated as:

$$\mathbf{h} = \mathbf{r}_g \times \mathbf{l} + \mathbf{h}_g \tag{5.15}$$

where $\mathbf{r}_g = \mathbf{r}_i - \mathbf{r}_{ig}$.

5.3 Dynamic Balancing of a Single Pendulum

5.3.1 Linear and Angular Momentum

As an example of the application of the previous developed equations consider a general single pendulum, Fig. 5.2, rotating at constant angular velocity ω.

In this case points A and B can be identified as basic points, so the linear momentum of the system can be obtained substituting the corresponding values in Eq. (5.11):

$$\mathbf{l} = \mathbf{l}_A + \mathbf{l}_B = \begin{bmatrix} \left(m - \frac{mx}{l}\right)\dot{x}_A + \left(\frac{my}{l}\right)\dot{y}_A + \left(\frac{mx}{l}\right)\dot{x}_B - \left(\frac{my}{l}\right)\dot{y}_B \\ \left(\frac{my}{l}\right)\dot{x}_A + \left(m - \frac{mx}{l}\right)\dot{y}_A + \left(\frac{my}{l}\right)\dot{x}_B - \left(\frac{mx}{l}\right)\dot{y}_B \end{bmatrix} \tag{5.16}$$

But the point A is fixed, so its velocity is zero. Substituting this result in the previous equation the linear momentum of the single pendulum expressed in natural coordinates can be obtained:

$$\mathbf{l} = \begin{bmatrix} \left(\frac{mx}{l}\right)\dot{x}_B - \left(\frac{my}{l}\right)\dot{y}_B \\ \left(\frac{my}{l}\right)\dot{x}_B - \left(\frac{mx}{l}\right)\dot{y}_B \end{bmatrix} \tag{5.17}$$

This result clearly indicates that to have a constant (invariant) linear momentum for the pendulum, the location of the center of mass must be the origin of the local reference system, point A, that coincides with the origin of the global fixed coordinate system. This means $(x, y) = (0, 0)$. In practice it is necessary to add a counterweight.

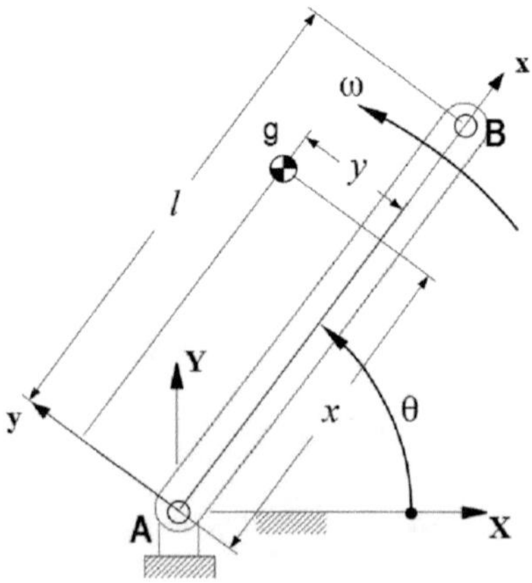

Fig. 5.2 A general model of a single pendulum rotating at constant angular velocity

On the other hand, the angular momentum of the pendulum can be obtained by substituting the corresponding values in Eq. (5.15), considering that the point A is fixed, so finally obtaining:

$$\mathbf{h} = \frac{I_A}{l^2}\left(\dot{y}_B\, x_B - \dot{x}_B\, y_B\right) \tag{5.18}$$

By substituting the corresponding values of the coordinates of point B and its derivatives:

$$\begin{aligned} x_B &= l\,\cos(\theta) \\ y_B &= l\,\sin(\theta) \\ \dot{x}_B &= -\omega\, l\,\sin(\theta) \\ \dot{y}_B &= \omega\, l\,\cos(\theta) \end{aligned}$$

in Eq. (5.18) it is possible to obtain the angular momentum of the single pendulum expressed in reference-point coordinates. For example, in the case of a uniform bar with its center of mass at the middle of its length:

$$\mathbf{h} = \omega\, I_A \tag{5.19}$$

but $I_A = I_g + m\left(\frac{l}{2}\right)^2$ so:

$$\mathbf{h} = \omega\left[I_g + m\left(\frac{l}{2}\right)^2\right] \tag{5.20}$$

This last results can indicate that to obtain a moment balanced pendulum, it is necessary to add a counter-inertia moving with an opposite angular velocity.

It can also be seen that the equations of linear momentum and angular momentum in Natural Coordinates have very simple forms and, as should be expected, they are correct only if the positions and velocities used are consistent with the kinematic constraints of the system.

5.3.2 Dynamic Balancing

For the dynamic balancing of a single pendulum it is necessary the addition of a counterweight, Eq. (5.17), and the addition of a counter-inertia, Eq. (5.18). Both the counterweight and the counter-inertia can be added as single counter-rotary counter-mass that rotates with an opposite angular velocity with respect to the angular velocity of the pendulum, as can be seen in Fig. 5.3. In Fig. 5.3, body one is the extended bar that works as the pendulum, defined by two basic points, A and B. The counter-rotary counter-mass used to dynamic balance the pendulum is the second

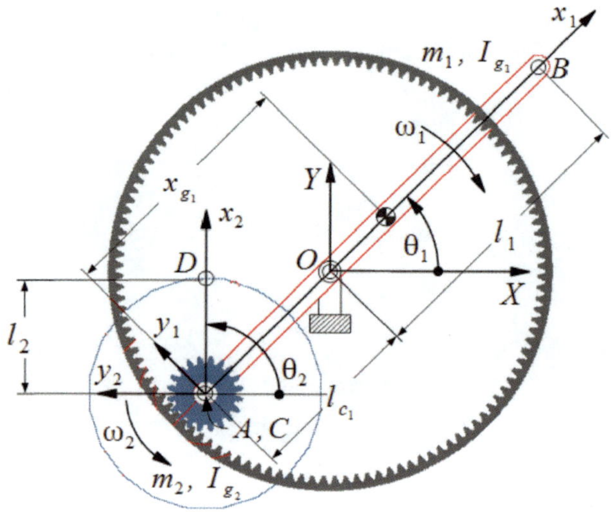

Fig. 5.3 A single pendulum, dynamic balanced by a counter-rotary counter-mass

body, defined by two basic points, C and D. It is important to note that points A and C have the same position and velocity, so strictly they are the same point.

Using Eq. (5.1) and applying Eq. (5.11) for each body, the linear momentum of the system in Fig. 5.3 can be calculated. The following result can be obtained, taking into account that in a general solution, the center of mass of the system must be in the line defined by points A and B ($y_{g_1} = 0$):

$$\mathbf{l}_m = \begin{bmatrix} \left(-\frac{m_1\, x_{g1}}{lc_1 + l_1} + m_2 + m_1\right) \dot{x}_A + \frac{m_1\, x_{g1}}{lc_1 + l_1}\, \dot{x}_B \\ \\ \left(-\frac{m_1\, x_{g1}}{lc_1 + l_1} + m_2 + m_1\right) \dot{y}_A + \frac{m_1\, x_{g1}}{lc_1 + l_1}\, \dot{y}_B \end{bmatrix} \tag{5.21}$$

where x_{g_1} is the location of the center of mass of the bar with respect to the local reference frame of body one, with the origin at A.

As mentioned before, Eq. (5.21) will be correct as long as the correct values of the velocities are substituted. This velocities should come from the solution of velocities problem, this means to solve the time derivative of the constraint equations of the system (generally the closed-loop equations). Then in this case the corresponding values of the velocities are:

$$\begin{aligned} \dot{x}_A &= lc_1\, \omega_1\, \sin(\theta_1) \\ \dot{y}_A &= -lc_1\, \omega_1\, \cos(\theta_1) \\ \dot{x}_B &= -l_1\, \omega_1\, \sin(\theta_1) \\ \dot{y}_B &= l_1\, \omega_1\, \cos(\theta_1) \end{aligned} \tag{5.22}$$

so the linear momentum of this system is:

$$\mathbf{l}_m = \begin{bmatrix} \{\omega_1 \sin(\theta_1)\} \left\{lc_1 \left(-\frac{m_1 x_{g1}}{lc_1+l_1} + m_2 + m_1\right) - \frac{l_1 m_1 x_{g1}}{lc_1+l_1}\right\} \\ -\{\omega_1 \cos(\theta_1)\} \left\{lc_1 \left(-\frac{m_1 x_{g1}}{lc_1+l_1} + m_2 + m_1\right) - \frac{l_1 m_1 x_{g1}}{lc_1+l_1}\right\} \end{bmatrix} \tag{5.23}$$

which is invariant if:

$$lc_1 \left(-\frac{m_1 x_{g1}}{lc_1 + l_1} + m_2 + m_1\right) - \frac{l_1 m_1 x_{g1}}{lc_1 + l_1} = 0$$

meaning that the force balancing condition of the system is:

$$x_{g_1} = lc_1 \left(\frac{m_2}{m_1} + 1\right) \tag{5.24}$$

On the other hand, using Eq. (5.3) and applying Eq. (5.18) for each body, the angular momentum of the system can be formulated. And substituting the force balancing condition, Eq. (5.24), the velocities at Eq. (5.22), and the corresponding positions:

$$\begin{aligned} x_A &= -lc_1 \cos(\theta_1) \\ y_A &= -lc_1 \sin(\theta_1) \\ x_B &= l_1 \cos(\theta_1) \\ y_B &= l_1 \sin(\theta_1) \end{aligned} \tag{5.25}$$

the final form of the angular momentum can be obtained as:

$$\mathbf{h}_m = \omega_1 \left[I_1 - lc_1^2 (m_1 + m_2)\right] + \omega_2 I_2 \tag{5.26}$$

where I_1 is the inertia moment with respect to the local coordinate system of body one at point A, and I_2 is the inertia moment with respect to the local coordinate system of body two at point C. In this case:

$$\begin{aligned} I_1 &= I_{g_1} + lc_1^2 \left(\frac{m_2^2}{m_1} + 2\,m_2 + m_1\right) \\ I_2 &= I_{g_2} \end{aligned}$$

where I_{g_1} and I_{g_2} are the moments of inertia with respect to the corresponding center on mass of each body. Additionally it is known that:

$$\omega_2 = -\left(\frac{lc_1}{R_2}\right)\omega_1 \tag{5.27}$$

so the angular momentum of the system is:

$$\mathbf{h}_m = \omega_1\left[I_{g_1} + lc_1^2\left(\frac{m_2^2}{m_1} + m_2\right) - \left(\frac{lc_1}{R_2}\right)I_{g_2}\right] \tag{5.28}$$

Then to obtain an invariant angular momentum the following moment balancing condition must be maintained:

$$\left[I_{g_1} + lc_1^2\left(\frac{m_2^2}{m_1} + m_2\right) - \left(\frac{lc_1}{R_2}\right)I_{g_2}\right] = 0 \tag{5.29}$$

In this way Eqs. (5.24) and (5.29) are the design conditions to obtain a dynamic balanced single pendulum.

5.4 Dynamic Balancing of a Four-Bar Mechanisms

Consider now a general four-bar mechanism as the one shown in Fig. 5.4, that is modeled in Natural Coordinates. In this case body one is defined with points *A* and *B*, body two is defined with points *B* and *C*, and body three is defined with points *C* and *D*. The origin of the local reference frames is also indicated, being at *A*, *B*, and *C*, respectively.

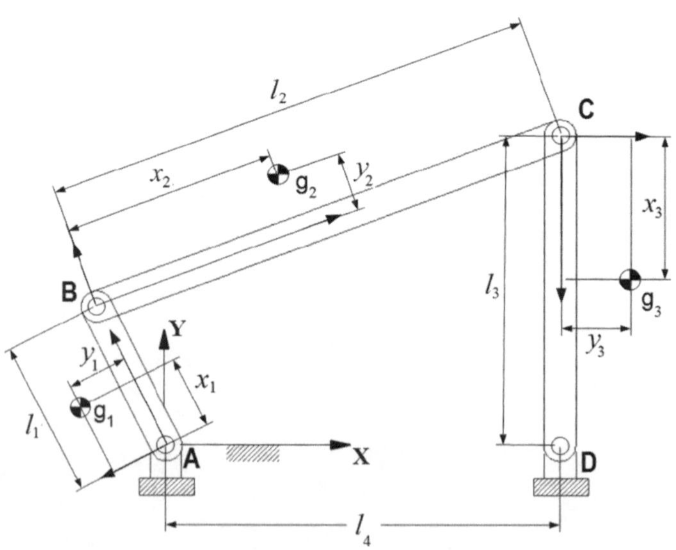

Fig. 5.4 A four-bar mechanism modeled in Natural Coordinates

The total linear momentum of this mechanism can be obtained by calculating the linear momentum of each body and then applying the Eq. (5.1) to get:

$$\mathbf{l}_m = \begin{bmatrix} a\,\dot{x}_B + b\,\dot{y}_B + c\,\dot{x}_C + d\,\dot{y}_C \\ \\ a\,\dot{y}_B + b\,\dot{x}_B + c\,\dot{y}_C + d\,\dot{x}_C \end{bmatrix} \tag{5.30}$$

where:

$$\begin{aligned} a &= \left(-\frac{m_2\,x_2}{l_2} + \frac{m_1\,x_1}{l_1} + m_2\right) \\ b &= \left(\frac{m_2\,y_2}{l_2} - \frac{m_1\,y_1}{l_1}\right) \\ c &= \left(-\frac{m_3\,x_3}{l_3} + \frac{m_2\,x_2}{l_2} + m_3\right) \\ d &= \left(\frac{m_3\,y_3}{l_3} - \frac{m_2\,y_2}{l_2}\right) \end{aligned} \tag{5.31}$$

The Eq. (5.31), equated to zero, are the force balancing conditions of the four-bar mechanism:

$$\begin{aligned} \left(-\frac{m_2\,x_2}{l_2} + \frac{m_1\,x_1}{l_1} + m_2\right) &= 0 \\ \left(\frac{m_2\,y_2}{l_2} - \frac{m_1\,y_1}{l_1}\right) &= 0 \\ \left(-\frac{m_3\,x_3}{l_3} + \frac{m_2\,x_2}{l_2} + m_3\right) &= 0 \\ \left(\frac{m_3\,y_3}{l_3} - \frac{m_2\,y_2}{l_2}\right) &= 0 \end{aligned} \tag{5.32}$$

This result is exactly the same as the one obtained in [11].

On the other hand, the angular momentum of the mechanisms can be obtained using Eq. (5.3), calculating previously the angular momentum of each body applying Eq. (5.15). In this case the angular momentum of the system is:

$$\begin{aligned} \mathbf{h}_m = {} & \dot{x}_B\,(-e\,y_B - h\,x_C + f\,y_C) \\ & + \dot{y}_B\,(e\,x_B + f\,x_C - h\,y_C) \\ & + \dot{x}_C\left(h\,x_B + f\,y_B - g\,y_C - \frac{l_4}{l_3} m_3\,y_3\right) \\ & + \dot{y}_C\left(e\,x_B + h\,y_B + g\,x_C - \frac{l_4}{l_3^2} I_3 + \frac{l_4}{l_3} m_3\,x_3\right) \end{aligned} \tag{5.33}$$

where

$$e = \left(\frac{I_2}{l_2^2} + \frac{I_1}{l_1^2} - \frac{2\,m_2\,x_2}{l_2} + m_2 \right) \tag{5.34}$$

$$f = \left(\frac{I_2}{l_2^2} - \frac{m_2\,x_2}{l_2} \right) \tag{5.35}$$

$$g = \left(\frac{I_3}{l_3^2} + \frac{I_2}{l_2^2} - \frac{2\,m_3\,x_3}{l_3} + m_3 \right) \tag{5.36}$$

$$h = \left(\frac{m_2\,y_2}{l_2} \right) \tag{5.37}$$

5.4.1 Dynamic Balancing of the Parallel Mechanism

A special case of a four-bar mechanisms is when the crank and the rocker have the same length and move parallel to each other, a parallel four-bar mechanisms, as the one in Fig. 5.5.

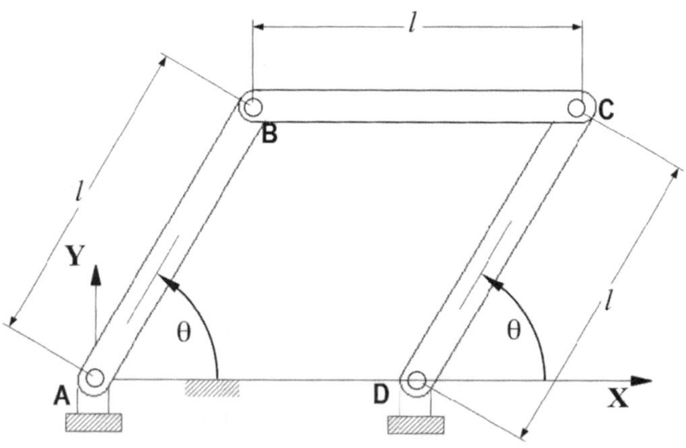

Fig. 5.5 A parallel four-bar mechanism

This is an inline[1] four-bar mechanism, and some of its geometric characteristics can be used to simplify force and moment balancing conditions to get an efficient design for the dynamic balanced system.

Considering that in a parallel inline four-bar mechanism $y_1 = y_2 = y_3 = 0$ and $l_1 = l_2 = l_3 = l$, the linear momentum equation, Eq. (5.30), can be expressed as:

$$\mathbf{l}_m = \begin{bmatrix} \dot{x}_B\left(-\frac{m_2 x_2}{l} + \frac{m_1 x_1}{l} + m_2\right) + \dot{x}_C\left(-\frac{m_3 x_3}{l} + \frac{m_2 x_2}{l} + m_3\right) \\ \dot{y}_B\left(-\frac{m_2 x_2}{l} + \frac{m_1 x_1}{l} + m_2\right) + \dot{y}_C\left(-\frac{m_3 x_3}{l} + \frac{m_2 x_2}{l} + m_3\right) \end{bmatrix} \tag{5.38}$$

but $\dot{x}_B = \dot{x}_C$ and $\dot{y}_B = \dot{y}_C$, so Eq. (5.38) can be expressed in terms of the velocity of point C as:

$$\mathbf{l}_m = \begin{bmatrix} \dot{x}_C\left(\frac{m_1 x_1}{l} - \frac{m_3 x_3}{l} + m_2 + m_3\right) \\ \dot{y}_C\left(\frac{m_1 x_1}{l} - \frac{m_3 x_3}{l} + m_2 + m_3\right) \end{bmatrix} \tag{5.39}$$

This equation indicates that this system can be force balanced by a single counterweight at body three (body one could be chosen in the same way), and the balancing condition is:

$$\frac{m_1 x_1}{l} - \frac{m_3 x_3}{l} + m_2 + m_3 = 0 \tag{5.40}$$

On the other hand, taking into account the angular momentum of a four-bar mechanism, Eq. (5.33), and considering that $y_1 = y_2 = y_3 = 0$ and $l_1 = l_2 = l_3 = l$, as in the case of the linear momentum, and that $x_C = x_B + l$, $y_C = y_B$, $\dot{x}_C = \dot{x}_B$, and $\dot{y}_C = \dot{y}_B$, the angular momentum of a parallel inline four-bar mechanism can be expressed as:

$$\begin{aligned} \mathbf{h}_m = &- \dot{x}_B\, y_B \left(\frac{I_3}{l^2} + \frac{I_1}{l^2} - \frac{2\, m_3\, x_3}{l} + m_3 + m_2\right) \\ &+ \dot{y}_B\, x_B \left(\frac{I_3}{l^2} + \frac{I_1}{l^2} - \frac{2\, m_3\, x_3}{l} + m_3 + m_2\right) \\ &- m_3\, x_3 + m_2\, x_2 + l\, m_3 \end{aligned} \tag{5.41}$$

Equations (5.40) and (5.41) can help us to the design of different dynamic balanced parallel four-bar mechanisms. This can be done by assigning different values to x_1, x_2, and x_3, the location of the center of mass of each moving link.

[1]The term "inline" means that the centers of mass of the links must lie on the line connecting the pivots (which can be extended beyond the pivots). The links need not be symmetrical in any way [12].

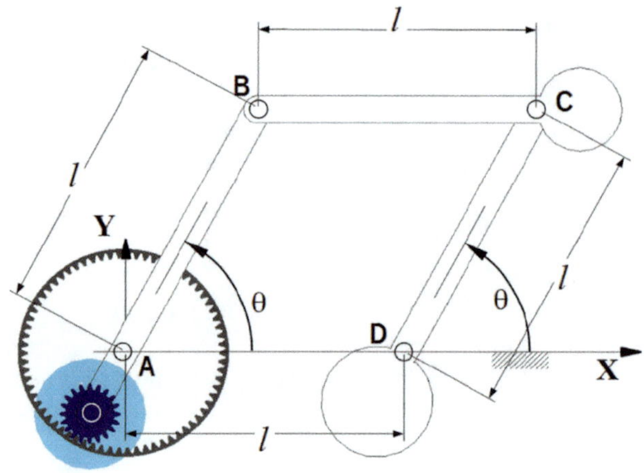

Fig. 5.6 A proposed design of an inline parallel four-bar mechanism with two counter-masses and one counter-rotary counter-mass

Consider first a mechanism with a counter-rotary counter-mass at the crank, a counterweight at the coupler, and a counterweight at the rocker. Figure 5.6 is a representation of this design, previously reported also in [13].

In this case $x_1 = 0$ and $x_2 = l$ were chosen, meaning that the center of mass of link one is at joint A and the center of mass of link two is at joint C, see Fig. 5.5. Substituting these values in Eq. (5.40), the corresponding value of x_3 can be found:

$$x_3 = l\left(\frac{m_2}{m_3} + 1\right)$$

Substituting the previous values for x_1, x_2, and x_3, and considering that:

$$\begin{aligned} x_B &= l\cos(\theta) \\ y_B &= l\sin(\theta) \\ \dot{x}_B &= -l\sin(\theta)\omega \\ \dot{y}_B &= l\cos(\theta)\omega \end{aligned} \tag{5.42}$$

the angular momentum, Eq. (5.41), finally results in:

$$\mathbf{h}_m = \left(\frac{I_3}{l^2} + \frac{I_1}{l^2} - m_3\,l^2 - m_2\,l^2\right)\omega$$

where ω is the angular velocity of the crank and the rocker. Then for the shaking moment balancing of the system it is necessary to add a counter-rotary counter-mass with the same magnitude but in opposite direction to the value of $\mathbf{h}_m$.

An alternative more efficient design could be a mechanism with a counter-rotary counter-mass at the crank and a counterweight at the rocker. Figure 5.7 is a representation of this design, also reported in [14].

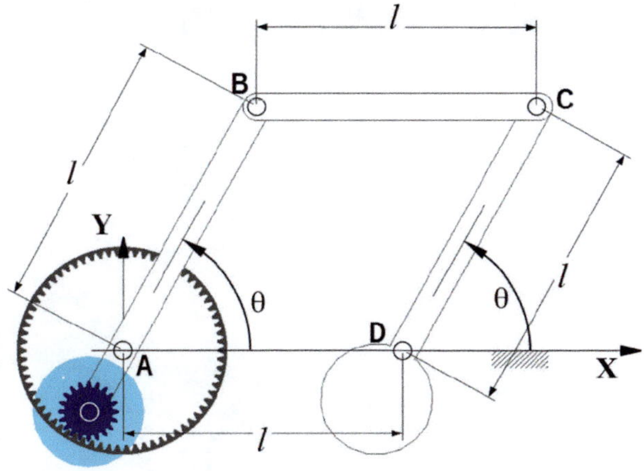

Fig. 5.7 A proposed design of an inline parallel four-bar mechanism with one counter-mass and one counter-rotary counter-mass

In this case $x_1 = -\frac{l\,m_2}{2\,m_1}$ and $x_2 = l/2$ were chosen. Substituting these values in Eq. (5.40), the corresponding value of x_3 can be found as:

$$x_3 = l + \frac{l\,m_2}{2\,m_3}$$

Substituting the previous values for x_1, x_2, and x_3, and considering the positions and velocities of point B, Eq. (5.42), the angular momentum, Eq. (5.41), finally results in:

$$\mathbf{h}_m = \left(\frac{I_3}{l^2} + \frac{I_1}{l^2} - m_3\,l^2\right)\omega$$

where ω is the angular velocity of the crank and the rocker. Again for the shaking moment balancing of the system it is necessary to add a counter-rotary counter-mass with the same magnitude but in opposite direction to the found value of $\mathbf{h}_m$.

5.4.2 Dynamic Balancing of the Inline Four-Bar Mechanism

Another special case is the inline four-bar mechanisms balanced by two counter-rotary counter-masses. This case has been previously studied in detail in [12] and the proposed design is similar to the one presented in Fig. 5.8.

In an inline four-bar mechanism $y_1 = y_2 = y_3 = 0$, so the linear momentum equation, Eq. (5.30), can be expressed as:

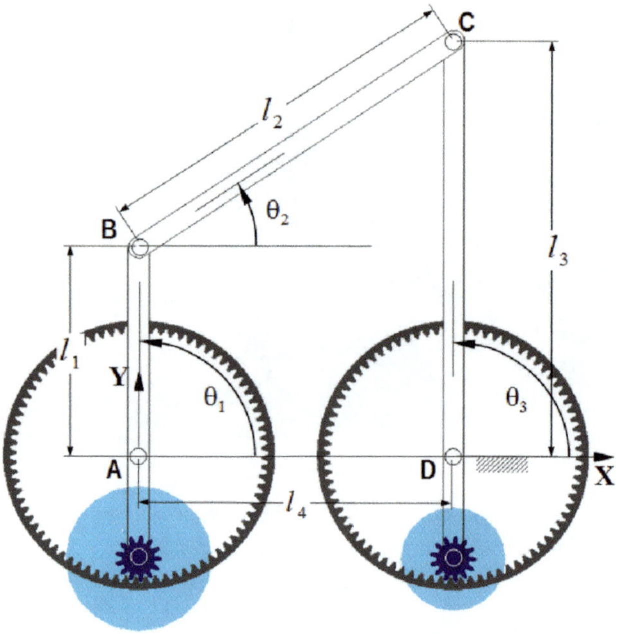

Fig. 5.8 Design proposal for the dynamic balancing of an inline four-bar mechanism

$$\mathbf{l}_m = \begin{bmatrix} \dot{x}_B\left(-\frac{m_2 x_2}{l_2} + \frac{m_1 x_1}{l_1} + m_2\right) + \dot{x}_C\left(-\frac{m_3 x_3}{l_3} + \frac{m_2 x_2}{l_2} + m_3\right) \\ \dot{y}_B\left(-\frac{m_2 x_2}{l_2} + \frac{m_1 x_1}{l_1} + m_2\right) + \dot{y}_C\left(-\frac{m_3 x_3}{l_3} + \frac{m_2 x_2}{l_2} + m_3\right) \end{bmatrix} \tag{5.43}$$

If the coupler is left without change, this equation indicates that this system can be force balanced by a two counterweights, one at body one and one at body three. And their location can be determined by the following balancing conditions:

$$\left(-\frac{m_2 x_2}{l_2} + \frac{m_1 x_1}{l_1} + m_2\right) = 0$$
$$\left(-\frac{m_3 x_3}{l_3} + \frac{m_2 x_2}{l_2} + m_3\right) = 0 \tag{5.44}$$

On the other hand taking into account the angular momentum of a four-bar mechanism, Eq. (5.33), and considering that $y_1 = y_2 = y_3 = 0$, the angular momentum of an inline four-bar mechanism can be expressed as:

$$
\begin{aligned}
\mathbf{h}_m = & -\dot{x}_B\left\{y_B\left(\frac{I_2}{l_2^2}+\frac{I_1}{l_1^2}-\frac{2\,m_2\,x_2}{l_2}+m_2\right)+y_C\left(\frac{I_2}{l_2^2}-\frac{2\,m_2\,x_2}{l_2}\right)\right\}\\
& +\dot{y}_B\left\{x_B\left(\frac{I_2}{l_2^2}+\frac{I_1}{l_1^2}-\frac{2\,m_2\,x_2}{l_2}+m_2\right)+x_C\left(\frac{I_2}{l_2^2}-\frac{2\,m_2\,x_2}{l_2}\right)\right\}\\
& -\dot{x}_C\left\{y_C\left(\frac{I_3}{l_3^2}+\frac{I_2}{l_2^2}-\frac{2\,m_3\,x_3}{l_3}+m_3\right)+y_B\left(\frac{I_2}{l_2^2}-\frac{2\,m_2\,x_2}{l_2}\right)\right\}\\
& +\dot{y}_C\left\{x_B\left(\frac{I_3}{l_3^2}+\frac{I_2}{l_2^2}-\frac{2\,m_3\,x_3}{l_3}+m_3\right)+x_C\left(\frac{I_2}{l_2^2}-\frac{2\,m_2\,x_2}{l_2}\right)\right\}\\
& +\dot{y}_C\left(-\frac{l_4\,I_3}{l_3^2}+\frac{l_4\,m_3\,x_3}{l_3}\right)
\end{aligned}
\tag{5.45}
$$

but if the coupler is considered a physical pendulum (as done in [12]) then $x_2 = l_2/2$ and $I_2 = l_2\,m_2\,x_2$, and substituting the force balancing conditions, Eq. (5.44), the angular momentum of this system is:

$$
\begin{aligned}
\mathbf{h}_m = & -\dot{x}_B y_B\left(\frac{I_1}{l_1^2}-\frac{m_2}{2}\right)+\dot{y}_B x_B\left(\frac{I_1}{l_1^2}-\frac{m_2}{2}\right)\\
& -\dot{x}_C y_C\left(\frac{I_3}{l_3^2}+\frac{m_2}{2}+m_3\right)+\dot{y}_C x_C\left(\frac{I_3}{l_3^2}+\frac{m_2}{2}+m_3\right)\\
& +\dot{y}_C\left(-\frac{l_4\,I_3}{l_3^2}+\frac{l_4\,m_2}{2}+l_4\,m_3\right)
\end{aligned}
\tag{5.46}
$$

Finally considering the positions and velocities of the points B and C:

$$
\begin{aligned}
x_B &= l_1\cos(\theta_1); & y_B &= l_1\sin(\theta_1)\\
\dot{x}_B &= -\,l_1\sin(\theta_1)\omega_1; & \dot{y}_B &= l_1\sin(\theta_1)\omega_1\\
x_C &= l_3\cos(\theta_3)+l_4; & y_C &= l_3\sin(\theta_3)\\
\dot{x}_C &= -\,l_3\sin(\theta_1)\omega_3; & \dot{y}_C &= l_3\sin(\theta_3)\omega_3
\end{aligned}
$$

and substituting in Eq. (5.46), after some reductions the resulting expression for the angular momentum is:

$$
\mathbf{h}_m = \left(I_1+\frac{l_1^2\,m2}{2}\right)\omega_1+\left(I_3-\frac{l_3^2\,m2}{2}-l_3^2\,m_3\right)\omega_3 \tag{5.47}
$$

This equations clearly show that an inline four-bar mechanism, with a physical pendulum as coupler, can be dynamic balanced just with two counter-rotary counter-masses, as reported in [12].

5.5 Design Examples and Simulation Results

5.5.1 *Dynamic Balancing of a Single Pendulum*

Let us suppose that it is desired the dynamic balancing of a single pendulum as the one shown in Fig. 5.9, using the results given in Eqs. (5.24) and (5.29).

The pendulum is made of aluminum with density, mass, and moment of inertia as indicated in Fig. 5.9. The force balancing of this system implies to comply with the balancing condition, the Eq. (5.24).

Applying a solution similar to the one proposed in Fig. 5.3, the bar *OB* is modified to get the bar *AB*, made also of aluminum. Setting $l_{c_1} = 15\,\text{cm}$, $x_{g_1} = 32.5\,\text{cm}$, both with respect to the local reference frame of the pendulum at point *O*. Then this new bar has a total mass $m_1 = 0.36467\,\text{kg}$ and a moment of inertia $I_{g_1} = 0.013471\,\text{kg}\,\text{m}^2$. Substituting this values at Eq. (5.24) the corresponding value of m_2 can be found:

$$m_2 = m_1\left(\frac{x_{g_1}}{l_{c_1}} - 1\right)$$

$$m_2 = 0.36467\left(\frac{0.325}{0.15} - 1\right)$$

$$m_2 = 0.425445\,\text{kg} \tag{5.48}$$

Note that this change implies an increment of near 1.8 times the original mass.

The resulting mass, m_2, has to be distributed between the gear and the disk that form the counter-rotary counter-mass, maintaining the moment balancing condition in Eq. (5.29).

The gear material (the density), diameter, and thickness depend on the mechanical design rules, while the physical characteristics of the disk depend on the convenience of the designer. On the other hand, a high value for the angular velocity

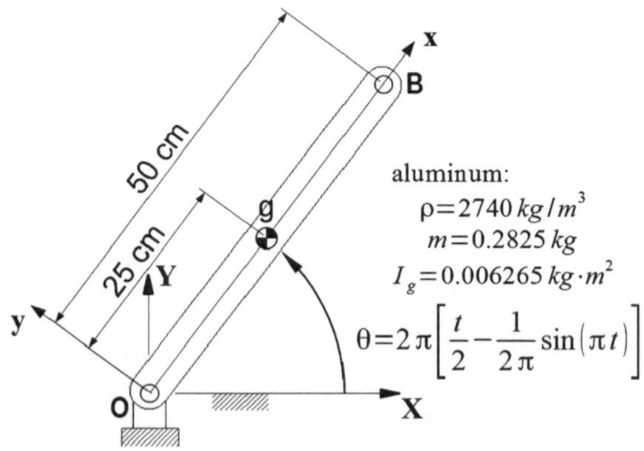

Fig. 5.9 A general single pendulum. In the equation of θ, t stands for time

ω_2, see Eq. (5.27), helps to reduce the inertia required in the counter-rotary counter-mass. Taking these factors into account and using a modulus $m = 2$, a gear with 20 teeth and $R_2 = 2\,\text{cm}$ is chosen. A 170 teeth crown gear results.

All previous values are substituted in Eq. (5.29) to obtain an appropriate value for I_{g_2} :

$$\left[I_{g_1} + lc_1^2\left(\frac{m_2^2}{m_1} + m_2\right) - \left(\frac{lc_1}{R_2}\right) I_{g2}\right] = 0$$

$$\left[0.013471 + 0.15\left(\frac{(0.425445)^2}{0.36467} + 0.425445\right) - \left(\frac{0.15}{0.02}\right) I_{g_2}\right] = 0$$

then,

$$I_{g_2} = 0.004562\,\text{kg}\,\text{m}^2$$

This inertia moment corresponds to both the gear and the disk, so:

$$I_{g_2} = I_g + I_d \tag{5.49}$$

where I_g is the moment of inertia of the gear and I_d is the moment of inertia of the disk. In the same way the mass of the counter-rotary counter-mass should be:

$$m_2 = m_g + m_d \tag{5.50}$$

where m_g and m_d are the mass of the gear and the disk, respectively.

The gear is chosen made of steel, and its moment of inertia determined by its design, in this case $m_g = 0.09803\,\text{kg}$ and $I_g = 0.00001961\,\text{kg}\,\text{m}^2$. On the other hand, the mass and the moment of inertia of the disk can be calculated by:

$$m_d = \pi\, R_d^2\, t_d\, \rho_d$$

$$I_d = \frac{1}{2} m_d R_d^2$$

where R_d, t_d, and ρ_d are the radius, the thickness, and the density of the disk, respectively.

Substituting the previous values in Eqs. (5.50) and (5.50), considering that the disk is made of brass (8,545 kg m^2), the following two equations are obtained:

$$m_2 = 0.09803 + 8545\,\pi\, t_d\, R_d^2$$

$$0.425445 = 0.09803 + 8545\,\pi\, t_d\, R_d^2 \tag{5.51}$$

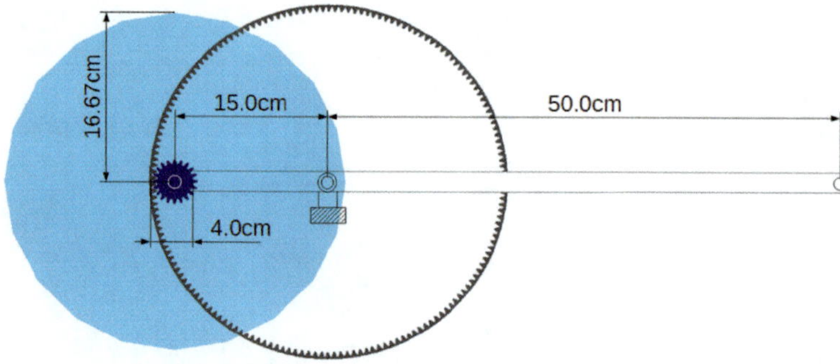

Fig. 5.10 Resulting design in the dynamic balancing of a pendulum

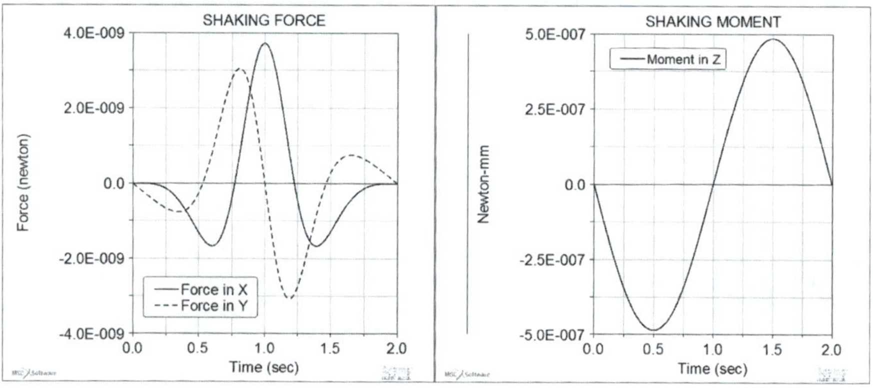

Fig. 5.11 Resulting shaking force and shaking moment in the dynamic balancing of a pendulum

and

$$I_{g_2} = \frac{1}{2} 8545\, \pi\, t_d\, R_d^4$$
$$0.004562 = \frac{1}{2} 8545\, \pi\, t_d\, R_d^4 \tag{5.52}$$

that can be solved simultaneously to get $R_d = 166.566$ mm and $t_d = 0.43961$ mm.

The final design is shown in Fig. 5.10. The resulting shaking force and shaking moment are shown in Fig. 5.11. In Fig. 5.12 the shaking force of the not balanced pendulum is shown, additionally the driving torque required to move the original pendulum and the driving torque necessary to move the balanced pendulum are compared.

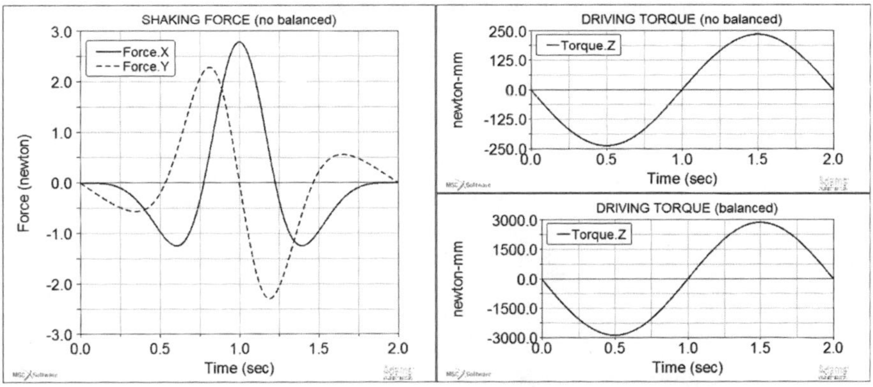

Fig. 5.12 Resulting reaction force and driving torque in the non-balanced pendulum. The driving torque of the balanced system is also included for comparison

5.5.2 Dynamic Balancing of a Four-Bar Mechanism

Frequently, when designing a dynamic balanced mechanism, it is important to maintain the counterweights near to the fixed joints (fixed pivots) attached to ground. This practice reduces the total additional inertia introduced in the balancing process and helps to reduce the increment in the driving torque.

In this example the dynamic balancing of a four-bar mechanisms is solved, by applying the design proposed in Fig. 5.8, an inline four-bar mechanism. This solution complies with the conditions exposed in the previous paragraph and works fine with the application of two counter-rotary counter-masses near the base, one at the crank and one at the rocker.

Consider the mechanism in Fig. 5.13. All elements are made of aluminum and their cross section is equal for all (2×1 cm), the values for the corresponding mass and moment of inertia are indicated in the figure. Its motion is defined by the function specified for angle θ_1.

To get the full dynamic balancing of this mechanism it is necessary first to balance it by forces, imposing the force balancing conditions expressed by Eq. (5.44). These conditions assure that the center of mass of the system will be stationary at the origin of the fixed reference frame, point A. Note that in this case the coupler of the mechanism (element 2) will be changed to be a physical pendulum, so x_2 and m_2 are completely determined.

On the other hand, at the same time it is necessary to impose the moment balancing condition. This can be obtained by taking into account the total angular momentum of the system, Eq. (5.47). Meaning that in order to moment balancing this system it is necessary to make the total angular momentum equal to zero. This clearly can be done by adding two counter-rotary counter-masses, one at the crank and one at the rocker.

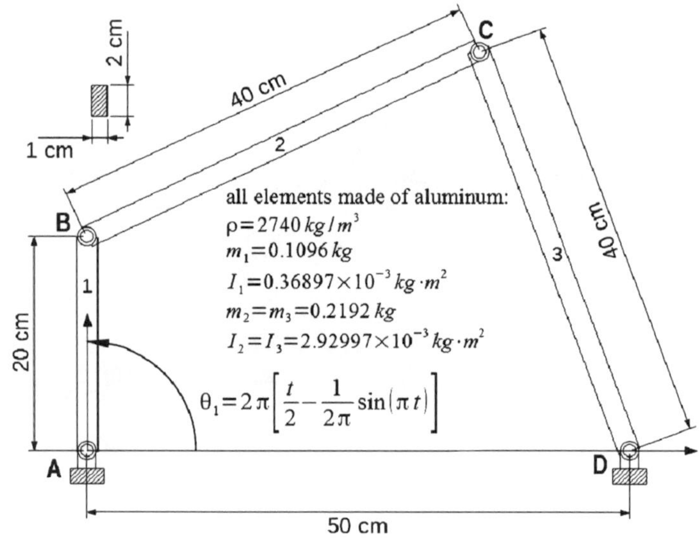

Fig. 5.13 Original four-bar mechanism to be dynamic balanced

In this way the four dynamic balancing conditions are:

$$\left(-\frac{m_2 x_2}{l_2} + \frac{m_1 x_1}{l_1} + m_2\right) = 0 \tag{5.53}$$

$$\left(-\frac{m_3 x_3}{l_3} + \frac{m_2 x_2}{l_2} + m_3\right) = 0 \tag{5.54}$$

$$\left(I_1 + \frac{l_1^2 m2}{2}\right)\omega_1 - I_{PC_1}(k_1\omega_1) = 0 \tag{5.55}$$

$$\left(I_3 - \frac{l_3^2 m2}{2} - l_3^2 m_3\right)\omega_3 - I_{PC_3}(k_3\omega_3) = 0 \tag{5.56}$$

where I_{c_1} and I_{c_3} are the moment of inertia of the counter-rotary counter-masses attached to the crank and the rocker, respectively. As can be noted these counter-rotary counter-masses must rotate with an angular velocity in the opposite direction with respect to their associated elements. The counter-rotation can be achieved by introducing gears, belts, etc., in this case a set of gears are chosen, giving a design similar to the one presented in Fig. 5.13, so

$$k_1 = \frac{d_1}{R_{P_1}}; \quad k_3 = \frac{d_3}{R_{P_3}}$$

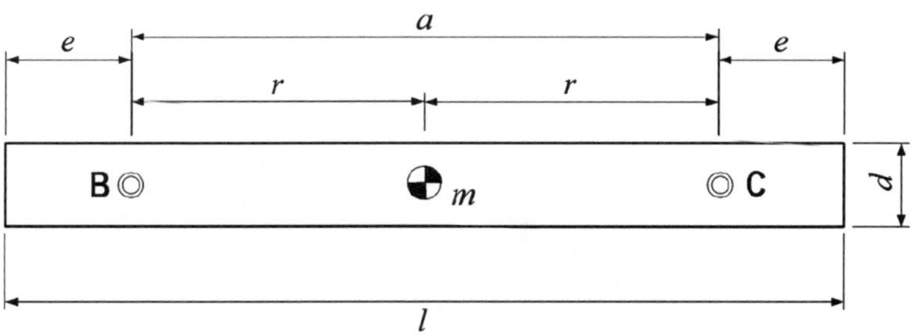

Fig. 5.14 Rectangular bar predesigned to be a physical pendulum

It can also be seen from the figure that the coupler has been changed to be a physical pendulum. So the original coupler has been modified to have its moment of inertia with respect to its center of mass equal to the moment of inertia generated by two equal punctual masses located at points B and C, respectively:

$$I_{CM_2} = m_2 \frac{l_2^2}{2}$$

In this case the physical pendulum is made by extending the original rectangular bar satisfying the following equation, [12], (see Fig. 5.14):

$$\frac{e}{h} = \frac{1}{2}\sqrt{3\left(\frac{a}{h}\right)^2 - 1} - \frac{a}{2h}$$

In this case the original coupler has $a = 400\,\text{mm}$ and $h = 20\,\text{mm}$, then $e = 146.266\,\text{mm}$, and the new coupler will have a total length $l = 692.532\,\text{mm}$, and a mass $m_2 = 0.3795\,\text{kg}$.

The new coupler is dynamically equivalent to a pair of masses located at B and at C, then the crank and the rocker can be balanced independently.

5.5.2.1 Balancing of the Crank

For the dynamic balancing of the crank, Eqs. (5.54) and (5.56) must be solved simultaneously to find the appropriate counter-rotary counter-mass that balance for forces and moments.

The counter-rotary counter-mass, I_{C_1}, is made by a gear (the pinion) with radius R_{P_1} and thickness t_{P_1}, and a disk with radius R_{C_1} and thickness t_{C_1}. On the other hand, the length of the crank should be increased by a distance d_1 to connect to these new elements, as seen in Fig. 5.15. All these variables form a set of five unknowns, so three of them must be set by election.

From the mechanical design point of view the pinion diameter and its thickness are determined by the general design rules for the gears. So choosing a modulus for

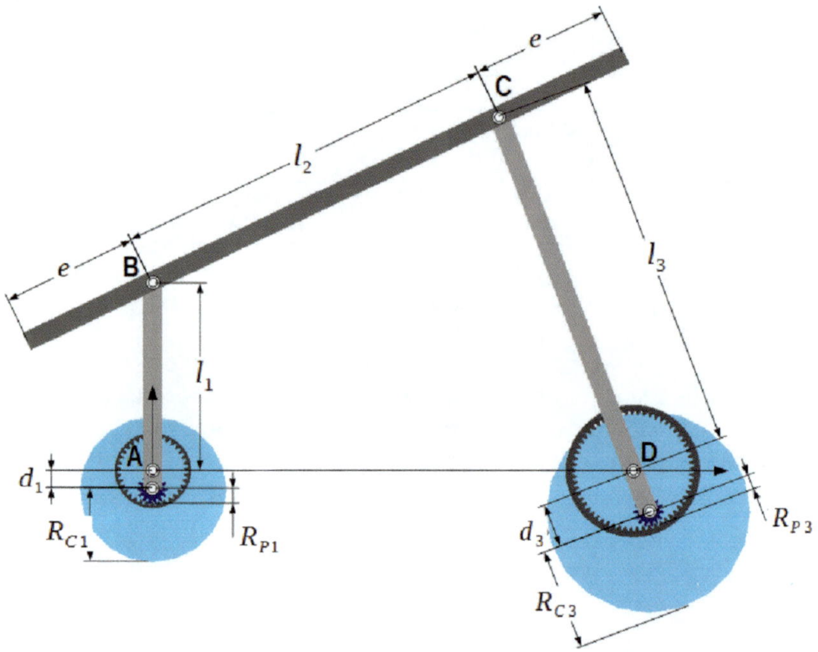

Fig. 5.15 Proposed design for an inline dynamic balanced four-bar mechanism

the crown-pinion set automatically sets the radius, thickness, and number of teeth. In this case a modulus $m = 2$ has been chosen, $R_{P_1} = 14\,\text{mm}$, and $t_{P_1} = 10\,\text{mm}$.

To determine the remaining variables: d_1, R_{C_1}, and t_{C_1}, an optimization problem is solved. It has been chosen to minimize Eq. (5.56) subject to the following constraints: Eq. (5.54), $R_{C_1} > 0$, $t_{C_1} > 0$ and $d_1 > 0$. But to do this the corresponding values for lengths, masses, and moments of inertia were substituted:

$$m_2 = 0.3795\,\text{kg};\; l_2 = 400\,\text{mm}$$

$$m_1 = m_c + m_{d_1} + m_{P_1} + m_{C_1}$$

$$I_1 = I_c + \frac{1}{4}m_c l_1^2 + I_{d_1} + \frac{1}{4}m_{d_1} d_1^2 + I_{P_1} + m_{P_1} d_1^2 + I_{C_1} + m_{C_1} d_1^2$$

$$I_{PC_1} = I_{P_1} + I_{C_1}$$

where m_c, m_{d_1}, m_{P_1}, and m_{C_1} are the corresponding mass of the crank, the added piece of bar, the pinion, and the disk at the crank, respectively. In similar way I_c, I_{d_1}, I_{P_1}, and I_{C_1} are the moments of inertia of the crank, the added piece of bar, the pinion, and the disk, respectively. I_{PC_1} is precisely the counter-inertia at the crank.

All these masses and moments of inertia are determined by the geometry and density of the bodies. In this case the bar is considered made of aluminum ($\rho_a = 2{,}740\,{}^{\text{kg}}/_{\text{m}^3}$), the pinion of steel ($\rho_s = 7{,}801\,{}^{\text{kg}}/_{\text{m}^3}$), and the disk of brass ($\rho_b = 8{,}545\,{}^{\text{kg}}/_{\text{m}^3}$), so:

$$m_c = htl_1\rho_a; \quad I_c = \frac{1}{12}ht\rho_a l_1^3$$

$$m_{d_1} = htd_1\rho_a; \quad I_{d_1} = \frac{1}{12}ht\rho_a d_1^3$$

$$m_{P_1} = \pi R_{P_1}^2 t_{P_1}\rho_s; \quad I_{P_1} = \frac{1}{2}\pi R_{P_1}^4 t_{P_1}\rho_s$$

$$m_{C_1} = \pi R_{C_1}^2 t_{C_1}\rho_b; \quad I_{C_1} = \frac{1}{2}\pi R_{C_1}^4 t_{C_1}\rho_b$$

where $h = 20$ mm and $t = 10$ mm as specified in Fig. 5.13.

Substituting all values in Eqs. (5.54) and (5.56), the force and moment balancing conditions finally are:

$$-8545\,\pi\, d_1\, R_{C_1}{}^2\, t_{C_1} - 0.274\, d_1^2 - 0.01529\,\pi\, d_1 + 0.048911 = 0 \tag{5.57}$$

$$\begin{aligned} &-305178.57\,\pi\, d\, Rc^4\, tc + 8545\,\pi\, d^2\, Rc^2\, tc + 0.1827\, d^3 \\ &+ 0.01529\,\pi\, d^2 - 1.0702972 \times 10^{-4}\,\pi\, d + 0.0090515 = 0 \end{aligned} \tag{5.58}$$

Finally solving the optimization problem, using an open source code implementation of the method introduced in [15], the following results are obtained:

$$R_{C_1} = 76.59\,\text{mm}, \quad t_{C_1} = 15.191\,\text{mm}, \quad d_1 = 20.0\,\text{mm}$$

5.5.2.2 Balancing of the Rocker

For the dynamic balancing of the rocker, Eqs. (5.55) and (5.56) must be solved simultaneously to find the appropriate counter-rotary counter-mass that balance for forces and moments.

The counter-rotary counter-mass, I_{C_3}, is made by a gear (the pinion) with radius R_{P_3} and thickness t_{P_3}, and a disk with radius R_{C_3} and thickness t_{C_3}. On the other hand, the length of the crank should be increased by a distance d_3 to connect to these new elements, as seen in Fig. 5.15. All these variables form a set of five unknowns, so three of them must be set by election.

From the mechanical design point of view the pinion diameter and its thickness are determined by the general design rules for the gears. So choosing a modulus for the crown-pinion set automatically sets the radius, thickness, and number of teeth. In this case a modulus $m = 2$ has been chosen, $R_{P_3} = 14$ mm, and $t_{P_3} = 10$ mm.

To determine the remaining variables: d_3, R_{C_3}, and t_{C_3}, an optimization problem is solved. It has been chosen to minimize Eq. (5.56) subject to the following constraints: Eq. (5.55), $R_{C_3} > 0$, $t_{C_3} > 0$ and $d_3 > 0$. But to do this the corresponding values for lengths, masses, and moments of inertia were substituted:

$$m_2 = 0.3795\,\text{kg};\; l_2 = 400\,\text{mm}$$

$$m_3 = m_r + m_{d_3} + m_{P_3} + m_{C_3}$$

$$I_1 = I_r + \frac{1}{4} m_r\, l_1^2 + I_{d_3} + \frac{1}{4} m_{d_3}\, d_3^2 + I_{P_3} + m_{P_3}\, d_3^2 + I_{C_3} + m_{C_3}\, d_3^2$$

$$I_{PC_3} = I_{P_3} + I_{C_3}$$

where m_r, m_{d_3}, m_{P_3}, and m_{C_3} are the corresponding mass of the rocker, the added piece of bar, the pinion, and the disk at the rocker, respectively. In similar way I_r, I_{d_3}, I_{P_3}, and I_{C_3} are the moments of inertia of the rocker, the added piece of bar, the pinion, and the disk, respectively. I_{PC_3} is precisely the counter-inertia at the rocker.

All these masses and moments of inertia are determined by the geometry and density of the bodies. The bar is considered made of aluminum($\rho_a = 2,740\,^{\text{kg}}/_{\text{m}^3}$), the pinion of steel($\rho_s = 7,801\,^{\text{kg}}/_{\text{m}^3}$), and the disk of brass ($\rho_b = 8,545\,^{\text{kg}}/_{\text{m}^3}$), so:

$$m_r = htl_3\rho_a; \quad I_r = \frac{1}{12} ht\rho_a l_3^3$$

$$m_{d_3} = htd_3\rho_a; \quad I_{d_3} = \frac{1}{12} ht\rho_a d_3^3$$

$$m_{P_3} = \pi R_{P_3}^2 t_{P_3}\rho_s; \quad I_{P_3} = \frac{1}{2}\pi R_{P_3}^4 t_{P_3}\rho_s$$

$$m_{C_3} = \pi R_{C_3}^2 t_{C_3}\rho_b; \quad I_{C_3} = \frac{1}{2}\pi R_{C_3}^4 t_{C_3}\rho_b$$

where $h = 20\,\text{mm}$ and $t = 10\,\text{mm}$ as specified in Fig. 5.13.

Substituting all values in Eqs. (5.55) and (5.56), noting that the origin of the local reference frame of the rocker is at point C, so all distances have to be taken with respect to this point, the force and moment balancing conditions finally are:

$$-21362.5\,\pi\, d_3\, R_{C_3}^2\, t_{C_3} - 0.685\, d_3^2 - 0.038225\,\pi\, d_3 + 0.2994 = 0 \tag{5.59}$$

$$\begin{aligned} &-305178.57\,\pi\, d_3\, Rc^4\, tc + 8545.0\,\pi\, d_3^2\, Rc^2\, tc \\ &+ 6836.0\,\pi\, d\, Rc^2\, tc + 0.1827\, d_3^3 + 0.01529\,\pi\, d_3^2 \\ &+ 0.2192\, d_3^2 + 0.012125\,\pi\, d_3 - 0.05374192 = 0 \end{aligned} \tag{5.60}$$

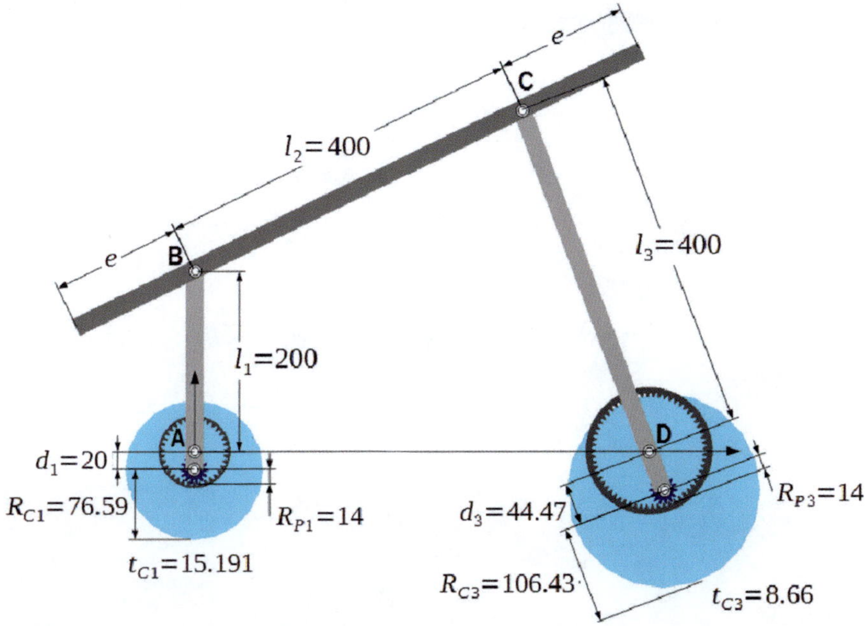

Fig. 5.16 Resulting design for an inline dynamic balanced four-bar mechanism. All measures in mm

Finally solving the optimization problem, using an open source code implementation of the method introduced in [15], the following results are obtained:

$$R_{C_3} = 106.43\,\text{mm}, \quad t_{C_3} = 8.66\,\text{mm}, \quad d_3 = 44.47\,\text{mm}$$

5.5.2.3 Resulting Inline Four-Bar Mechanism

The final design of the proposed inline four-bar mechanisms can be seen at Fig. 5.16. This results in an increment of 9.7 times the original mass of the system.

The comparison of the shaking force, the shaking moment and the driving torque, can be seen in Figs. 5.17, 5.18, and 5.19, respectively.

5.6 Concluding Remarks

In this chapter a completely new general method to find the dynamic balancing conditions based on the use of Natural Coordinates has been introduced. The method can be used for linkages in the plane and in space, although it is presented for planar mechanisms.

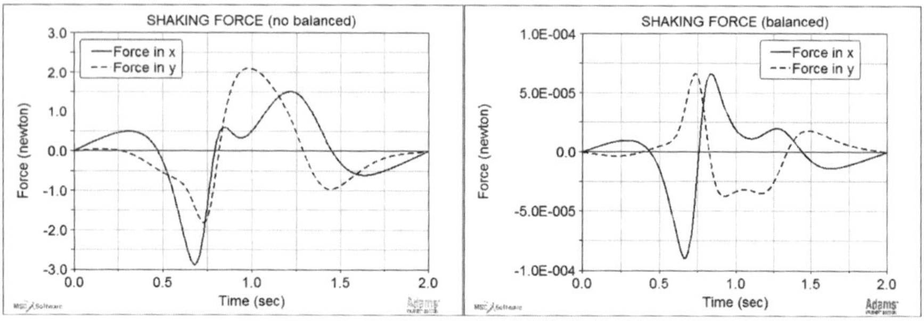

Fig. 5.17 Resulting shaking force in the inline dynamic balanced four-bar mechanism

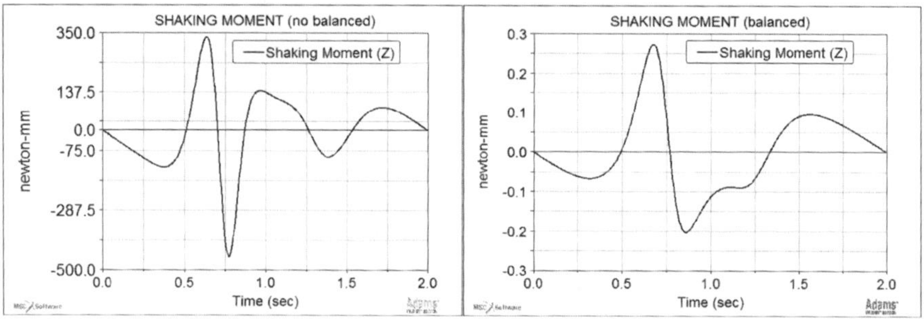

Fig. 5.18 Resulting shaking moment in the inline dynamic balanced four-bar mechanism

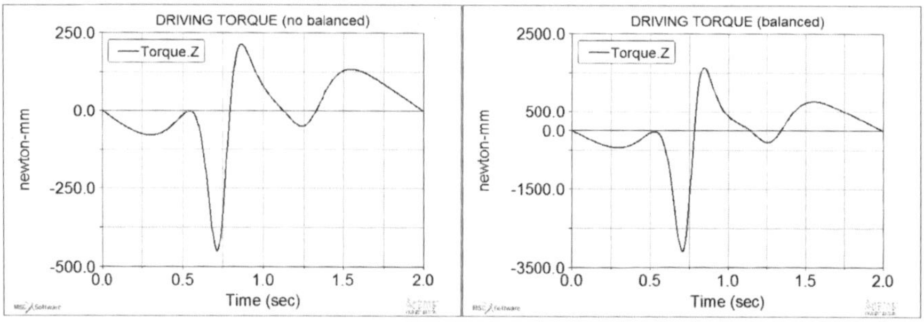

Fig. 5.19 Resulting driving torque in the inline dynamic balanced four-bar mechanism

The method is simple and can be highly automated, and it has been shown in its application to the design of dynamic balanced planar mechanisms using counter-rotary counter-masses. In particular the resulting equations of a general simple pendulum and of a general four-bar mechanism are presented. It is shown that these equations must be solved simultaneously in order to obtain a feasible design, and that in a more general case the use of optimization could be better from the mechanical design point of view. Detailed results obtained from the dynamic simulations made using virtual prototypes defined in ADAMS are included, showing the validity and applicability of the method, and helping the reader to deeply understand the concepts, maybe repeating the examples and proposing variations of the presented designs.

References

1. Lowen, G.G., Berkoft, R.S.: Survey of investigations into the balancing of linkages. J. Mech. **3**, 221–231 (1968)
2. Dresig, H., Rockhausen, L., Naake, S.: Balancing conditions for planar mechanisms, flexible mechanisms, dynamics and analysis. In: 22nd ASME Mechanism Conference, DE, vol. 127, pp. 67–73. ASME, New York (1992)
3. Kochev, I.S.: General theory of complete shaking moment balancing of planar linkages: A critical review. Mech. Mach. Theory **35**, 1501–1514 (2000)
4. van der Wijk, V., Herder, J.L., Demeulenaere, B.: Comparison of various dynamic balancing principles regarding additional mass and additional inertia. ASME J. Mech. Robot. **1**, 21–30 (2009)
5. Arakelian, V., Dahan, M., Smith, M.: A historical review of the evolution of the theory on balancing of mechanisms. In: Ceccarelli, M. (ed.) International Symposium on History of Machines and Mechanisms - Proceedings (HMM2000), pp. 291–300. Springer, Berlin (2000)
6. Arakelian, V.G., Smith, M.R.: Shaking force and shaking moment balancing of mechanisms: A historical review with new examples. ASME J. Mech. Des. **127**, 334–339 (2005)
7. Arakelian, V.G., Smith, M.R.: Erratum: Shaking force and shaking moment balancing of mechanisms: A historical review with new examples. ASME J. Mech. Des. **127**, 1035 (2005)
8. Herder, J.L., Gosselin, C.M.: A counter-rotary counterweight (CRCW) for light-weight dynamic balancing. In: Proceedings of the DETC 2004, Paper No. DETC2004-57246. ASME, New York (2004)
9. van der Wijk, V., Herder, J.L.: Synthesis of dynamically balanced mechanisms by using counter-rotary countermass balanced double pendula. ASME J. Mech. Des. **131**, 111003 (2009)
10. García de Jalón, J., Bayo, E.: Kinematic and Dynamic Simulation of Multibody Systems: The Real-Time Challenge. Springer, Berlin (1994)
11. Berkof, R.S., Lowen, G.G.: A new method for complete force balancing simple linkages. ASME J. Manuf. Sci. Eng. **91**, 21–26 (1969)
12. Berkof, R.S.: Complete force and moment balancing of inline four-bar linkages. Mech. Mach. Theory **8**, 397–410 (1973)
13. van der Wijk, V., Herder, J.L.: Double pendulum balanced by counter-rotary counter-masses as useful element for synthesis of dynamically balanced mechanisms. In: Proceedings of the DETC 2008, Paper No. DETC2008-49402. ASME, New York (2008)
14. Acevedo, M., Carbone, G., Ceccarelli, M.: Application of counter-rotary counterweights to the dynamic balancing of a spatial parallel manipulator. Appl. Mech. Mater. **162**, 224–233 (2012)
15. Powell, M.J.D.: A direct search optimization method that models the objective and constraint functions by linear interpolation. In: Gomez, S., Hennart J.P. (eds.) Advances in Optimization and Numerical Analysis, pp. 51–67. Kluwer Academic, Dordrecht (1994)

Chapter 6
Shaking Force and Shaking Moment Balancing of Six- and Eight-Bar Planar Mechanisms

Peddinti Nehemiah

Abstract This chapter presents the dynamic balancing technique for shaking force and shaking moment balancing of six- and eight-bar planar mechanisms. Shaking force is balance by the method of redistribution of mass and shaking moment by geared inertia elements. The planetary gears used to balance shaking moment of links not directly connected to the frame in earlier methods are mounted on the base of the mechanism which is constructively more efficient. The proposed method is illustrated by numerical examples and it is observed that better results are obtained than those of the previous method.

Keywords Shaking force • Shaking moment • Dynamic balancing • Watt mechanisms • Self-balanced Slider-crank mechanism

6.1 Introduction

Mechanisms, particularly those which run at high speeds, generate variable forces on their foundations. These forces may cause noise, vibration, and unnecessary wear and fatigue. If these devices were balanced they would run more smoothly due to a reduction in these undesirable qualities. The balancing of a linkage would eliminate the vibration and noise and maintains a peaceful and productive environment, and it also minimizes the alternating components of the dynamic forces acting on the frame of the mechanism and machine. Therefore, the problems of shaking force and shaking moment balancing have attracted the attention of the machine and mechanism designers for a long time. Balancing of shaking force and shaking moment in high-speed mechanisms/machines reduces the forces transmitted to the frame. In effect, this reduces the noise and wear, improves the dynamic performance, and extends the fatigue life of the mechanisms. A considerable amount of research on balancing of shaking force and shaking moment in planar mechanisms has been carried out in the past.

P. Nehemiah (✉)
Mechanical Engineering Department, Lords Institute of Engineering and Technology, Hyderabad, India
e-mail: prof.nehemiah@gmail.com

D. Zhang, B. Wei (eds.), *Dynamic Balancing of Mechanisms and Synthesizing of Parallel Robots*, DOI 10.1007/978-3-319-17683-3_6

6.1.1 Shaking Force in a Mechanism

Of special interest to the designer are the forces transmitted to the frame or foundation of the machine owing to the inertia of moving links and other machine members. When these forces vary in magnitude or direction, they tend to shake or vibrate the machine, and consequently they are called shaking forces. Thus the shaking forces are the forces, which act upon the frame of a machine owing only to the inertia forces of the moving parts. Shaking forces and shaking moments are the unbalanced forces and moments generated when the planar and spatial mechanisms are in motion. The study of these forces and moments is important when they run at high speeds. These undesirable qualities of the mechanism reduce the performance of the mechanism. The shaking force generated by the mechanism can be determined as follows:

If a four-bar linkage is considered, as an example, with links 2, 3, and 4 as the moving members and link 1 as the frame, then the inertia forces associated with the moving members are $-m_2A_{G_2}, -m_3A_{G_3}, -m_4A_{G_4}$. Therefore, taking the moving members as a free body, it can be immediately written as

$$\sum F = F_{12} + F_{14} + (-m_2A_{G_2}) + (-m_3A_{G_3}) + (-m_4A_{G_4}) = 0$$

Using "F_S" as a symbol for the resulting shaking force, it is defined as equal to the resultant of all the reaction forces on the ground link 1,

$$F_S = F_{21} + F_{41}$$

Therefore, from the previous equation, it can be written as

$$F_S = -\left(m_2A_{G_2} + m_3A_{G_3} + m_4A_{G_4}\right)$$

Thus a general equation for the shaking forces in any machine is

$$F_S = -\sum_{2}^{n} m_nA_{G_n}$$

where it is understood that link 1 is always the frame and where "n" is the number of members making up the machine.

6.1.2 Shaking Moment of the Mechanism

The shaking moment of a linkage can be described as the time rate of change of the total angular momentum with respect to the reference origin "O." It is

$M = I\alpha$, where M is the shaking moment w.r.t. point "O"; α is the angular acceleration; and I is the mass moment of inertia.

6.1.3 Methods of Balancing

Balancing of linkages is an important step in the design of machinery. When shaking forces and shaking moments of the whole mechanisms are to be balanced then balancing of sub-linkages is considered. The linkages consist of different sub-linkages; this study considers two sub-linkages as most of the mechanisms are formed by them. Many methods [1–75] have been developed for the balancing of shaking force and shaking moment of planar linkages:

1. Method of redistribution of mass [1–6]
2. Method of double crank with symmetrical properties [7]
3. Method of active balancing [8–16]
4. Methods of balancing by planetary systems attached to the coupler [17–30]
5. Method of balancing by minimizing vibration [31–41]
6. Computational methods of optimization for balancing [42–68]
7. Methods for the minimization of shaking moments [69–74]
8. Balancing by opposite movements [75]

This chapter deals with the shaking force and shaking moment balancing of single degree of freedom planar mechanisms. Specifically, the author employs the traditional technique of addition of counterweights and counter-rotating inertias in order to balance six- and eight-bar linkages through the development of analytical expressions. This chapter is the extension of the work carried out by the authors [18–23].

6.2 Articulation Dyad

6.2.1 Complete Shaking Force and Shaking Moment Balancing of an Articulation Dyad

An open kinematic chain of two binary links and one joint is called a dyad. When two links are articulated by a joint so that movement is possible that arrangement of links is known as articulation dyad. The well-known scheme of complete shaking force and shaking moment balancing of an articulation dyad [18–23] is shown in Fig. 6.1.

For shaking force balancing link 2 is dynamically replaced by two point masses. A counterweight $m_{\mathrm{CW}_2} = (m_2 l_{\mathrm{AS}_2}) / r_{\mathrm{CW}_2}$ is added to link 2 which permits the displacement of the center of mass of link 2 to joint A. Then, by means of a counterweight with mass $m_{\mathrm{CW}_1} = [(m_2 + m_{\mathrm{CW}_2}) \, l_{\mathrm{OA}} + m_1 l_{\mathrm{OS}_1}] / r_{\mathrm{CW}_1}$ a complete balancing of shaking force is achieved. A complete shaking moment balance is realized through four gear inertia counterweights 3–6, one of them being of the planetary type and mounted on link 2.

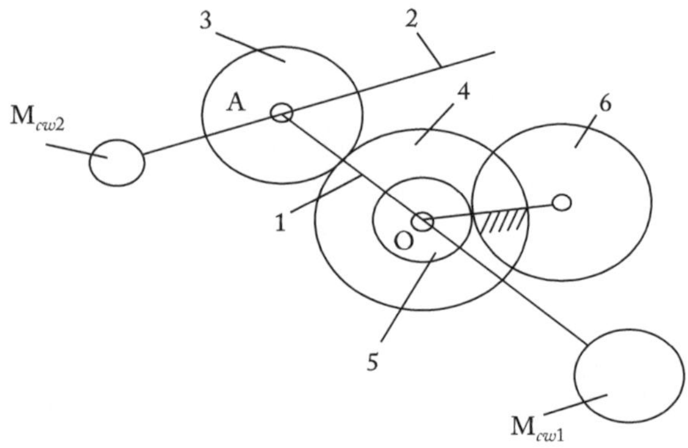

Fig. 6.1 Complete shaking force and shaking moment balancing of an articulation dyad

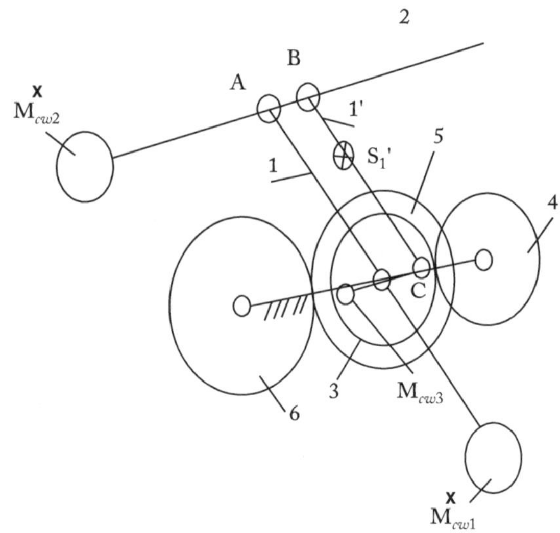

Fig. 6.2 Complete shaking force and shaking moment balancing of an articulation dyad by gear inertia counterweights mounted on the base

6.2.2 Complete Shaking Force and Shaking Moment Balancing of an Articulation Dyad by Gear Inertia Counterweights Mounted on the Base

The scheme used in this work (Fig. 6.2) is distinguished from the earlier scheme by the fact that gear 3 is mounted on the base and is linked kinematically with link 2 through link $1'$.

To prove the merits of such a balancing, the application of the new system with the mass of link $1'$ not taken into account is considered. In this case (compared to the usual method in Fig. 6.1), the mass of the counterweight of link 1 will be reduced by an amount

$$\delta m_{\mathrm{cw}_1} = m_3 l_{\mathrm{OA}} \Big/ r_{\mathrm{cw}_1} \qquad (6.1)$$

where m_3 is the mass of gear 3.

l_{OA} is the distance between the centers of hinges O and A.

r_{cw_1} is the rotation radius of the center of mass of the counterweight.

It is obvious that the moment of inertia of the links is correspondingly reduced. If the gear inertias are made in the form of heavy rims in order to obtain a large moment of inertia, the moments of inertia of the gear inertia counterweights may be presented as

$$I = \frac{m_i D_i^2}{4} \, (i = 3 \ldots 6).$$

Consequently, the mass of gear 6 will be reduced by an amount

$$\delta m_6 = 4 \left(m_3 l_{\mathrm{OA}}^2 + \delta m_{\mathrm{cw}_1} r_{\mathrm{cw}_1}^2 \right) \frac{T_6}{D_6^2 T_5} \qquad (6.2)$$

where

T_5 and T_6 are the numbers of teeth of the corresponding gears. Thus, the total mass of the system will be reduced by an amount

$$\delta m = \delta m_{\mathrm{cw}_1} + \delta m_6 \qquad (6.3)$$

Here the complete shaking force and shaking moment balancing of the articulation dyad with the mass and inertia of link $1'$ taken into account are considered. For this purpose initially, statically replace mass m_1' of link $1'$ by two point masses m_{B} and m_{c} at the centers of the hinges B and C:

$$\begin{aligned} m_{\mathrm{B}} &= m_{1'} l_{\mathrm{CS}_{1'}} / l_{\mathrm{BC}} \\ m_C &= m_{1'} l_{\mathrm{BS}_{1'}} / l_{\mathrm{BC}} \end{aligned} \qquad (6.4)$$

where

l_{BC} is the length of link 1.

$l_{\mathrm{CS}_{1'}}$ and $l_{\mathrm{BS}_1'}$ are the distances between the centers of joints C and B and the center of mass S_1' of link $1'$, respectively.

After such an arrangement of masses the moment of inertia of link $1'$ will be equal to

$$I_{S_1'}^* = I_{S_1'} - m_{1'} l_{\mathrm{BS}_1'} l_{\mathrm{CS}_1'} \qquad (6.5)$$

where

$I_{S_1'}$ is the moment of inertia of link $1'$ about the center of mass S_1' of the link.

Thus a new dynamic model of the system is obtained, where the link $1'$ is represented by two point masses $m_{\mathrm{B}}, m_{\mathrm{C}}$ and has a moment of inertia $I_{S_1'}^*$.

This fact allows for an easy determination of the parameters of the balancing elements as follows:

$$m^*_{\mathrm{CW}_2} = (m_2 l_{\mathrm{AS}_2} + m_{\mathrm{B}} l_{\mathrm{AB}}) / r_{\mathrm{CW}_2} \tag{6.6}$$

where

m_2 is the mass of link 2.

l_{AB} is the distance between the centers of the hinges A and B

l_{AS_2} is the distance of the center of hinge A from the center mass of S_2 of link 2.

r_{CW_2} is the rotation radius of the center of mass of the counterweight with respect to A and

$$m^*_{\mathrm{CW}_1} = [(m_2 + m_{\mathrm{CW}_2} + m_{\mathrm{B}})\, l_{\mathrm{OA}} + m_1 l_{\mathrm{OS}_1}] / r_{\mathrm{CW}_1} \tag{6.7}$$

where m_1 is the mass of link 1.

l_{OS_1} is the distance of the joint center O from the center of mass S_1 of link 1.

Also,

$$m_{\mathrm{CW}_3} = m_{\mathrm{C}} l_{\mathrm{OC}} / r_{\mathrm{CW}_3} \tag{6.8}$$

where

$$l_{\mathrm{OC}} = l_{\mathrm{AB}}$$

r_{CW_3} is the rotation radius of the center of mass of the counterweight.

Taking into account the mass of link 1′ brings about the correction in Eq. (6.3) in this case,

$$\delta m = \delta m_{\mathrm{CW}_1} + \delta m_6 - \delta m_1' \tag{6.9}$$

where $\delta m_1'$ is the value characterizing the change in the distribution of the masses of the system links resulting from the addition of link 1′.

6.3 Asymmetric Link with Three Rotational Pairs

A link with three nodes is called ternary link, where nodes are points for attachment to other links. In previous work by Gao Feng [18] relating to balancing of linkages with a dynamic substitution of the masses of the link by three rotational pairs shown in Fig. 6.3 two replacement points A and B are considered. This results in the need to increase the mass of the counterweight. However, such a solution may be avoided by considering the problem of dynamic substitution of link masses by three point masses. Usually the center of mass of such an asymmetric link is located inside a triangle formed by these points.

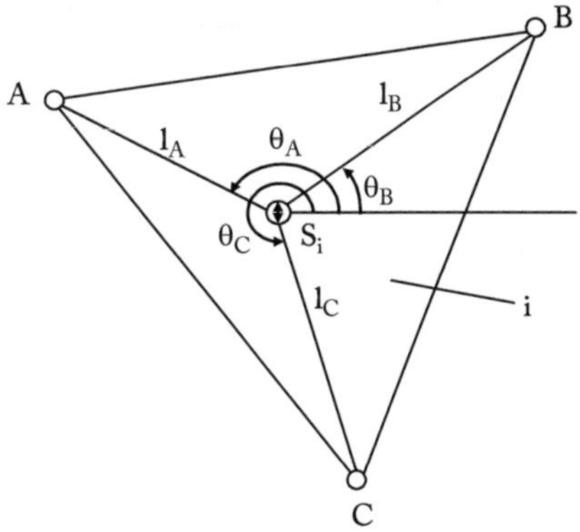

Fig. 6.3 Dynamic substitution of the masses of the link by three rotational pairs

The conditions for dynamic substitution of masses are the following:

$$\begin{bmatrix} 1 & 1 & 1 \\ l_A e^{i\theta_A} & l_B e^{i\theta_B} & l_C e^{i\theta_C} \\ l_A^2 & l_B^2 & l_C^2 \end{bmatrix} \begin{bmatrix} m_A \\ m_B \\ m_C \end{bmatrix} = \begin{bmatrix} m_i \\ 0 \\ I_{S_i} \end{bmatrix}$$

where m_A, m_B, and m_C are point masses.

l_A, l_B and l_C are the moduli of radius vectors of corresponding points.

θ_A, θ_B and θ_C are angular positions of radius vectors; m_i is the mass of link.

I_{S_i} is the moment of inertia of the link about an axis through S_i (axial moment of inertia of link).

From this system of equations the masses are obtained:

$$m_A = D_A/D_i; m_B = D_B/D_i; m_C = D_C/D_i \tag{6.10}$$

where D_A, D_B, D_C and D_i are determinants of the third order obtained from the above system of equations.

6.4 Summary

The complete shaking force is balanced by the method of redistribution of mass and making the total mass center of the mechanism stationary. The complete shaking moment is balanced by geared inertia counterweights. The planetary gears which are mounted on the links not directly connected to the frame in earlier method are mounted on the frame of the mechanism by connecting the planetary gear and the corresponding link by a link of known mass, center of mass, and mass moment of inertia. This arrangement makes the balanced mechanism constructively more efficient and compact and yields better results over the Gao Feng method.

6.5 Watt Mechanism with Three Fixed Points Linkage

Watt mechanism consists of six links; out of them two are ternary and the remaining four are binary links. In Watt mechanism two ternary links are directly connected to one another. This mechanism is obtained when one of the ternary links in the basic Watt chain is fixed. This is a simple mechanism as the radii of path curvature of all motion transfer points are known. This mechanism is used in steam engines and is also used to oscillate the agitator in some washing machines. In the Watt mechanism with three fixed points shown in Fig. 6.4, link 1 and 3 are ternary links and all other links are binary links. The balanced Watt mechanism with three fixed points is shown in Fig. 6.5.

6.5.1 Shaking Force Balancing of the Mechanism

For shaking force balancing link 3 is dynamically replaced by three point masses m_{B3}, m_{C3} and m_{D3} and then the problems of sub-linkages OAB and DEF are considered.

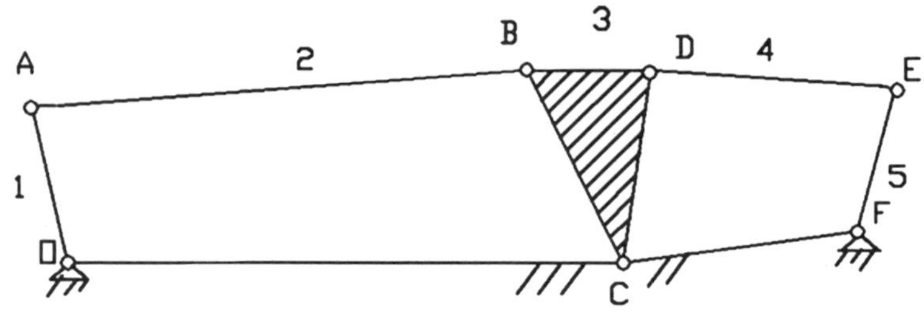

Fig. 6.4 Watt mechanism with three fixed points

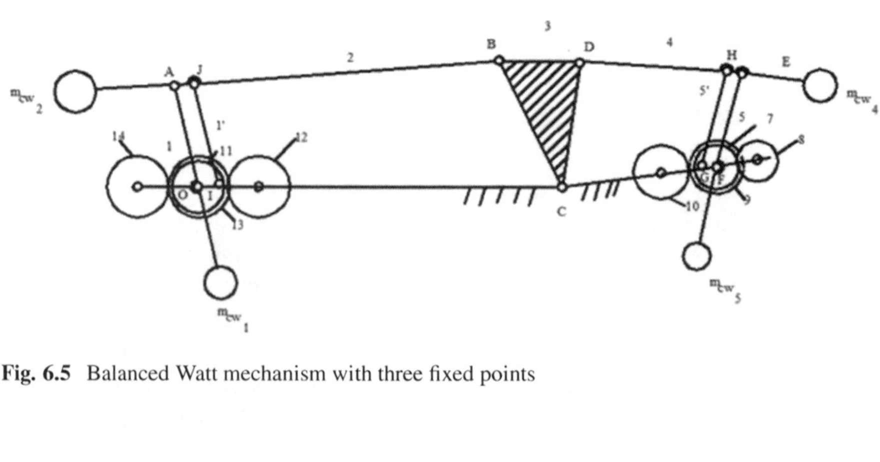

Fig. 6.5 Balanced Watt mechanism with three fixed points

The dynamic conditions for link 3 to be replaced by three point masses are

$$\begin{bmatrix} 1 & 1 & 1 \\ l_B e^{i\theta_B} & l_C e^{i\theta_C} & l_D e^{i\theta_D} \\ l_B^2 & l_C^2 & l_D^2 \end{bmatrix} \begin{bmatrix} m_{B3} \\ m_{C3} \\ m_{D3} \end{bmatrix} = \begin{bmatrix} m_3 \\ 0 \\ I_{S_3} \end{bmatrix}$$

$$m_{B3} = D_B \Big/ D_3 \; ; m_{C3} = D_C \Big/ D_3 \; ; m_{D3} = D_D \Big/ D_3 \tag{6.11}$$

where

l_B, l_C, l_D are the moduli of radius vectors of corresponding points.

$\theta_B, \theta_C, \theta_D$ are the angular positions of radius vectors.

m_3 is the mass of link 3.

I_{S_3} is the mass moment of inertia link 3 about its center of mass.

D_B, D_C, D_D and D_3 are the third-order determinants obtained from the system of equations.

For sub-linkage DEF link 4 is dynamically replaced by two point masses m_{D4} and m_{P4} and then kinematically linked link 4 and its corresponding gear inertia counterweight 7 by link 5′ and link 5′ is statically replaced by two point masses m_G and m_H and attached a counterweight m_{CW_4} against link 4. Then link 5 has been dynamically replaced by two point masses m_{E5}, m_{P5} and attached a counterweight m_{CW_5} against it.

For link 4 to be dynamically replaced by two point masses the condition to be satisfied is $k_4^2 = l_{DS_4} l_{P_4S_4}$

where

k_4 is the radius of gyration of link 4 about its center of mass.

l_{DS_4} is arbitrarily fixed.

$l_{P_4S_4}$ is obtained from the above condition:

$$m_{D4} = m_4 l_{P_4S_4} \Big/ (l_{DS_4} + l_{P_4S_4})$$

$$m_{P4} = m_4 l_{DS_4} \Big/ (l_{DS_4} + l_{P_4S_4})$$

For link 5 to be dynamically replaced by two point masses the condition to be satisfied is

$$k_5^2 = l_{ES_5} l_{P_5S_5}$$

where

k_5 is the radius of gyration of link 5 about its center of mass.

l_{ES_5} is arbitrarily fixed.

$l_{P_5S_5}$ is obtained from the above condition:

$$m_{E5} = m_5 l_{P_5S_5} \Big/ (l_{ES_5} + l_{P_5S_5})$$

$$m_{P5} = m_5 l_{ES_5} \Big/ (l_{ES_5} + l_{P_5S_5})$$

and counterweight mass against "G" is equal to

$$m_{\mathrm{CW}_7} = {}^{m_{\mathrm{G}} l_{\mathrm{FG}}} \Big/ {}_{r_{\mathrm{CW}_7}} \tag{6.12}$$

$$\begin{aligned} m_{\mathrm{G}} &= {}^{m'_5 l_{\mathrm{HS}_5}} \Big/ {}_{l_{\mathrm{GH}}} \\ m_{\mathrm{H}} &= {}^{m'_5 l_{\mathrm{GS}_5}} \Big/ {}_{l_{\mathrm{GH}}} \\ I'^{*}_{\mathrm{S}_5} &= I'_{\mathrm{S}_5} - m'_5 l_{\mathrm{GS}_5} l_{\mathrm{HS}_5} \\ m_{\mathrm{CW}_4} &= {}^{(m_4 l_{\mathrm{ES}_4} + m_{\mathrm{D3}} l_{\mathrm{DE}} + m_{\mathrm{H}} l_{\mathrm{EH}})} \Big/ {}_{r_{\mathrm{CW}_4}} \\ m_{\mathrm{CW}_5} &= {}^{((m_4 + m_{\mathrm{D3}} + m_{\mathrm{H}} + m_{\mathrm{CW}_4}) \, l_{\mathrm{EF}} + m_5 l_{\mathrm{FS}_5})} \Big/ {}_{r_{\mathrm{CW}_5}} \end{aligned} \tag{6.13}$$

where $r_{\mathrm{CW}_4} = (l_{\mathrm{P}_4\mathrm{S}_4} - l_{\mathrm{ES}_4})$ is the radius of rotation of counterweight m_{CW_4} and $r_{\mathrm{CW}_5} = (l_{\mathrm{P}_5\mathrm{S}_5} - l_{\mathrm{FS}_5})$ is the radius of rotation of counterweight m_{CW_5}.

For sub-linkage OAB link 2 is dynamically replaced by two point masses $m_{\mathrm{B2}}, m_{\mathrm{P2}}$ and then kinematically linked link 2 and its corresponding gear inertia counterweight 11 by link $1'$ and link $1'$ is statically replaced by two point masses $m_{\mathrm{I}}, m_{\mathrm{J}}$ and attached a counterweight m_{CW_2} against link 2. Then link 1 is dynamically replaced by two point masses $m_{\mathrm{A1}}, m_{\mathrm{P1}}$ and attached a counterweight m_{CW_1} against it.

For link 2 to be dynamically replaced by two point masses the condition to be satisfied is $k_2^2 = l_{\mathrm{BS}_2} l_{\mathrm{P}_2\mathrm{S}_2}$

where k_2 is the radius of gyration of link 2 about its center of mass.

l_{BS_2} is arbitrarily fixed and $l_{\mathrm{P}_2\mathrm{S}_2}$ is obtained from the above condition:

$$\begin{aligned} m_{\mathrm{B2}} &= {}^{m_2 l_{\mathrm{P}_2\mathrm{S}_2}} \Big/ {}_{(l_{\mathrm{BS}_2} + l_{\mathrm{P}_2\mathrm{S}_2})} \\ m_{\mathrm{P2}} &= {}^{m_2 l_{\mathrm{BS}_2}} \Big/ {}_{(l_{\mathrm{BS}_2} + l_{\mathrm{P}_2\mathrm{S}_2})} \end{aligned}$$

For link 1 to be dynamically replaced by two point masses the condition to be satisfied is

$$k_1^2 = l_{\mathrm{AS}_1} l_{\mathrm{P}_1\mathrm{S}_1}$$

where

k_1 is the radius of gyration of link 1 about its center of mass.

l_{AS_1} is arbitrarily fixed.

$l_{\mathrm{P}_1\mathrm{S}_1}$ is obtained from the above condition:

$$m_{A1} = {m_1 r_{P_1S_1}} \Big/ (l_{AS_1} + l_{P_1S_1})$$

$$m_{P1} = {m_1 l_{AS_1}} \Big/ (l_{AS_1} + l_{P_1S_1})$$

$$m_I = {m'_1 l_{IS_5}} \Big/ l_{IJ}$$

$$m_J = {m'_1 l_{JS_5}} \Big/ l_{IJ}$$

$$m_{CW_{11}} = {m_I l_{OI}} \Big/ r_{CW_{11}}$$

$$I'^{*}_{S_1} = I'_{S_1} - m'_1 l'_{JS_1} l_{IS_1}$$

$$m_{CW_2} = (m_2 l_{AS_2} + m_{B3} l_{AB} + m_J l_{AJ}) \Big/ r_{CW_2}$$

$$m_{CW_1} = ((m_2 + m_{B3} + m_J + m_{CW_2})\, l_{OA} + m_1 l_{OS_1}) \Big/ r_{CW_1}$$

where $m_{CW_{11}}$ is the counterweight attached against point mass m_I.

$r_{CW_2} = (l_{P_2S_2} - l_{AS_2})$ is the radius of rotation of counterweight m_{CW_2}, and $r_{CW_1} = (l_{P_1S_1} - l_{OS_1})$ is the radius of rotation of counterweight m_{CW_1}.

6.5.2 Shaking Moment Balancing of the Mechanism

The shaking moments generated by links 1, 2, 4, and 5 are given in Eq. (6.14). The links 2 and 4 are not directly connected to the frame, and the geared inertia counterweights required to balance the shaking moments of these two links are mounted on the base of the mechanism, by kinematically linking them to the corresponding links by links of known mass and center of mass.

The shaking moment generated by the linkage is determined by the sum

$$
\begin{aligned}
M^{int} &= M_1^{int} + M_5^{int} + M_2^{int} + M_4^{int} \\
M_1^{int} &= \left(I_{S_1} + m_1 l_{OS_1}^2 + (m_{CW_2} + m_J + m_2 + m_{B3})\, l_{OA}^2 + m_{CW_1} r_{CW_1}^2 + I'^{*}_{S_1} + m'_1 l_{OS_1}^2\right) \alpha_1 \\
M_5^{int} &= \left(I_{S_5} + m_5 l_{FS_5}^2 + (m_{CW_4} + m_H + m_4 + m_{D3})\, l_{EF}^2 + m_{CW_5} r_{CW_5}^2 + I'^{*}_{S_5} + m'_5 l'^2_{FS_5}\right) \alpha_5 \\
M_2^{int} &= \left(2 m_I l_{OI}^2\right) \alpha_2 \\
M_4^{int} &= \left(2 m_G l_{FG}^2\right) \alpha_4
\end{aligned}
\tag{6.14}
$$

where

M_1^{int}, M_5^{int} are the shaking moments of rotating links 1 and 5, respectively.

I_{S_1}, I_{S_5} are mass moments of inertia of links 1 and 5 about their centers of masses, respectively.

$I'^*_{S_1}, I'^*_{S_5}$ are the changed moments of inertia of links 1′, 5′, respectively.

$\alpha_1, \alpha_2, \alpha_4, \alpha_5$ are the angular accelerations of links 1, 2, 4, and 5, respectively.

For shaking moment balancing eight gear inertia counterweights are used, four at F and four at O.

6.6 Watt Mechanism with Two Fixed Points

The Watt mechanism with two fixed points is obtained when one of the binary links in the basic Watt chain is fixed. This mechanism is generally used in steam engines. In the Watt mechanism with two fixed points shown in Fig. 6.6, links 2 and 3 are ternary links and all other links are binary links. The balanced Watt mechanism with two fixed points is shown in Fig. 6.7.

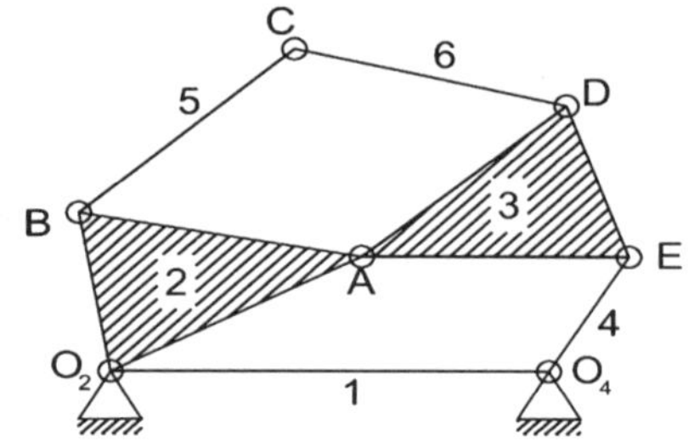

Fig. 6.6 Watt mechanism with two fixed points

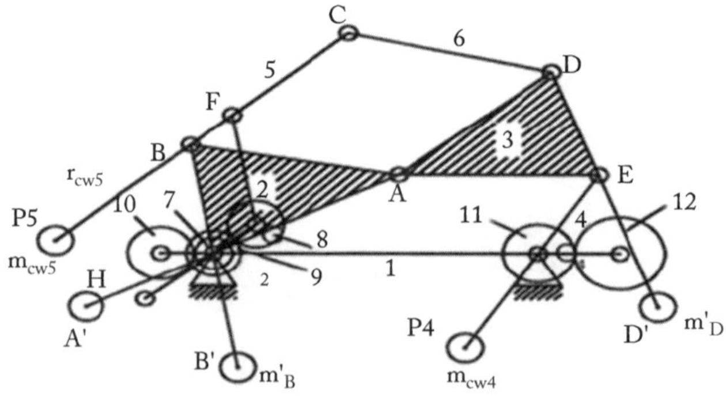

Fig. 6.7 Balanced Watt mechanism with two fixed points

6.6.1 Shaking Force Balancing of the Mechanism

Here the link 2 is dynamically replaced by three point masses $m_{A_2}, m_{B_2}, m_{O_2}$ by using the following conditions:

$$\begin{bmatrix} 1 & 1 & 1 \\ l_{O_2}e^{i\theta_{O_2}} & l_A e^{i\theta_A} & l_B e^{i\theta_B} \\ {l_{O_2}}^2 & {l_A}^2 & {l_B}^2 \end{bmatrix} \begin{bmatrix} m_{O_2} \\ m_{A_2} \\ m_{B_2} \end{bmatrix} = \begin{bmatrix} m_2 \\ 0 \\ I_{S_2} \end{bmatrix}$$

$$m_{O_2} = \frac{D_{O_2}}{D_2}, \; m_{A_2} = \frac{D_{A_2}}{D_2}, \; m_{B_2} = \frac{D_{B_2}}{D_2}$$

where $l_{O_2},\ l_A,\ l_B$ are the moduli of radius vectors of corresponding points.

$\theta_{O_2},\ \theta_A,\ \theta_B$ are the angular positions of radius vectors.

m_2 is the mass of link 2.

I_{S_2} is the mass moment of inertia of link 2 about its center of mass.

$D_{O_2},\ D_{A_2},\ D_{B_2}$ and D_2 are the third-order determinants obtained from the system of equations.

For link 6 to be statically replaced by the point masses m_{C_6} and m_{D_6}

$$m_{C_6} = \frac{m_6 l_{DS_6}}{l_{CD}}$$

$$m_{D_6} = \frac{m_6 l_{CS_6}}{l_{CD}}$$

Changed mass moment of inertia $I_{S_6}^{\prime *} = I_{S_6}^{\prime} - m_6 l_{DS_6} l_{CS_6}$

For link 5 to be dynamically replaced by two point masses m_{C_5} and m_{P_5} the condition to be satisfied is

$${k_5}^2 = l_{CS_5} l_{P_5S_5}$$

where l_{CS_5} is arbitrarily taken and $l_{P_5S_5}$ is obtained from the above condition:

$$m_{C_5} = \frac{m_5 l_{P_5S_5}}{(l_{P_5S_5} + l_{CS_5})}$$

$$m_{P_5} = \frac{m_5 l_{CS_5}}{(l_{P_5S_5} + l_{DS_5})}$$

After link 5 is dynamically replaced by two point masses it is kinematically connected to its corresponding gear inertia counterweight 8 by link 2′; moreover link 2′ is statically replaced by two point masses m_G and m_F:

$$m_G = \frac{m_2^{\prime} l^{\prime}_{FS_2}}{l_{FG}}$$

$$m_{\mathrm{F}} = \frac{m'_2 l'_{\mathrm{GS}_2}}{l_{\mathrm{FG}}}$$

Counterweight m_{CW_5} can be obtained as

$$m_{\mathrm{CW}_5} = \frac{(m_{\mathrm{C6}} l_{\mathrm{BC}} + m_{\mathrm{F}} l_{\mathrm{BF}} + m_5\, l_{\mathrm{BS}_5})}{r_{\mathrm{CW}_5}} \tag{6.15}$$

where $r_{\mathrm{CW}_5} = l_{\mathrm{P_5S_5}} - l_{\mathrm{CS}_5}$ is radius of rotation of counterweight m_{CW_5}.

Link 3 is dynamically replaced by three point masses $m_{\mathrm{A3}}, m_{\mathrm{D3}}, m_{\mathrm{E3}}$ by using the following conditions:

$$\begin{bmatrix} 1 & 1 & 1 \\ l_{\mathrm{A}} e^{i\theta_{\mathrm{A}_3}} & l_{\mathrm{A}} e^{i\theta_{\mathrm{D}}} & l_{\mathrm{B}} e^{i\theta_{\mathrm{E}}} \\ l_{\mathrm{A}}{}^2 & l_{\mathrm{D}}{}^2 & l_{\mathrm{E}}{}^2 \end{bmatrix} \begin{bmatrix} m_{\mathrm{A}_3} \\ m_{\mathrm{D}_3} \\ m_{\mathrm{E}_3} \end{bmatrix} = \begin{bmatrix} m_3 \\ 0 \\ I_{\mathrm{S}_3} \end{bmatrix}$$

$$m_{\mathrm{A3}} = \frac{D_{\mathrm{A3}}}{D_3},\ m_{\mathrm{D3}} = \frac{D_{\mathrm{D3}}}{D_3},\ m_{\mathrm{E3}} = \frac{D_{\mathrm{E3}}}{D_3} \tag{6.16}$$

where $l_{\mathrm{A}}, l_{\mathrm{D}}, l_{\mathrm{E}}$ are the moduli of radius vectors of corresponding points.

$\theta_{\mathrm{A3}}, \theta_{\mathrm{D}}, \theta_{\mathrm{E}}$ are the angular positions of radius vectors.

m_3 is the mass of link 3.

I_{S3} is the mass moment of inertia of link 2 about its center of mass.

$D_{\mathrm{A3}}, D_{\mathrm{D3}}, D_{\mathrm{E3}}$ and D_2 are the third-order determinants obtained from the system of equations.

Counterweight against point B of link 2 can be obtained as

$$m'_{\mathrm{B}} = {(m_{\mathrm{CW}_5} + m_{\mathrm{F}} + m_5 + m_{\mathrm{C6}})\, l_{\mathrm{O_2B}}} \Big/ {l'_{\mathrm{O_2B}}}$$

where $l'_{\mathrm{O_2B}}$ is arbitrarily fixed.

Counterweight against point A of link 3 can be obtained as

$$m'_{\mathrm{A}} = {(m_{\mathrm{A2}} + m_{\mathrm{A3}})\, l_{\mathrm{O_2A}}} \Big/ {l'_{\mathrm{O_2A}}}$$

where $l'_{\mathrm{O_2A}}$ is arbitrarily chosen.

Counterweight against point D of link 3 can be obtained as t

$$m'_{\mathrm{D}} = {(m_{\mathrm{D3}} + m_{\mathrm{D6}})\, l_{\mathrm{DE}}} \Big/ {l'_{\mathrm{DE}}}$$

where l'_{DE} is arbitrarily chosen.

For link 4 to be dynamically replaced by two point masses m_{E4}, m_{P4} the condition to be satisfied is $k_4^2 = l_{ES_4} l_{P_4S_4}$, where l_{ES_4} is arbitrarily is chosen and $l_{P_4S_4}$ is obtained from the above condition:

$$m_{E4} = m_4 l_{P_4S_4} \Big/ (l_{P_4S_4} + l_{ES_4}); \quad m_{P4} = m_4 l_{ES_4} \Big/ (l_{P_4S_4} + l_{ES_4})$$

Counterweight against link 4 can be obtained as

$$m_{CW_4} = \frac{\left(m_{E3} + m_{D6} + m'_D\right) l_{O_4E}}{r_{CW_4}}$$

where $r_{CW_4} = l_{P_4S_4} - l_{O_4S_4}$ is the radius of rotation of counterweight m_{CW_4}.

6.6.2 Shaking Moment Balancing of the Mechanism

The shaking moments generated by links 2, 4, and 5 are given in Eq. (6.17).

The shaking moment generated by the mechanism can be determined by the sum

$$M^{int} = M_2^{int} + M_5^{int} + M_4^{int} \tag{6.17}$$

where

$$M_2^{int} = \Big(I_{S_2} + I'^{*}_{S_2} + m'_2 l'_{GS_2} l'_{FS_2} + (m_{A2} + m_{A3})\, l^2_{O_2A} + (m_{CW_5} + m_F + m_5 + m_{C6})\, l^2_{O_2B} + m'_B l'^2_{O_2B}\Big)\, \alpha_2$$

$$M_4^{int} = \left(I_{S_4} + m_4 l^2_{O_4S_4} + m_{CW_4} r^2_{CW_4} \left(m'_D + m_{D3} + m_{D6} + m_{E3}\right) l^2_{O_4E}\right) \alpha_4$$

$$M_5^{int} = \left(2 m_G l^2_{O_2G}\right) \alpha_5$$

$M_2^{int}, M_4^{int}, M_5^{int}$ are the shaking moments of rotating links 2, 4, and 5, respectively.

I_{S_2}, I_{S_4} are the mass moment of inertia of links 2 and 4, respectively.

$\alpha_{2,}, \alpha_4, \alpha_5$ are the angular accelerations of links 2, 4, and 5, respectively.

For shaking moment balancing six gear inertia counterweights are used, four at O_2 and two at O_4.

Shaking force of the mechanism by the proposed method:

$$F_{Proposed} = -\left(m_2 A_{G2} + m_3 A_{G3} + m_4 A_{G4} + m_5 A_{G5} + m_6 A_{G6} + m'_2 A'_{G2}\right)$$

Shaking moment of the mechanism by the proposed method:

$$M^{int}_{proposed} = M_2^{int} + M_4^{int} + M_5^{int}$$

Shaking force of the mechanism by Gao Feng's method:

$$F_{\text{Gaofeng}} = -(m_2A_{G2} + m_3A_{G3} + m_4A_{G4} + m_5A_{G5} + m_6A_{G6} + m_{G8}A_{G8})$$

Shaking moment of the mechanism by Gao Feng's method:

$$M_{\text{Gaofeng}}^{\text{int}} = M_2^{\text{int}} + M_4^{\text{int}} + M_5^{\text{int}} + \left(I_{S8} + 2m_{G8}l_{FG}^2\right)\alpha_2$$

Numerical example: The Watt mechanism with two fixed points shown in Fig. 6.6 has the following parameters:

$m_2 = 2$ kg, $k_2 = 0.1198$ m, $m_3 = 1.8$ kg, $k_3 = 0.1178$ m, $m_4 = 7$ kg, $k_4 = 0.237$m, $m_5 = 2.8$ kg, $k_5 = 0.934$ m, $m_6 = 3$ kg, $k_6 = 0.369$ m, $l_{A3} = 3.7$ m, $l_E = 5.8$ m, $l_D = 5.6$ m, $\theta_A = 0°$, $\theta_B = 117°$, $\theta_{O_2} = 262°$, $l_{O_2} = 5$ m, $l_A = 2.6$ m, $l_B = 3.7$, $l_{O_2B} = 5$, $l_{O_2A} = 2.3$ m, $l_{AB} = 9$ m, $l_{BC} = 8$ m, $l_{CD} = 6$ m, $l_{BF} = 2.1$ m, $l_{O_4E} = 9$ m, $l_{DE} = 7$ m, $l_{AE} = 5$ m, $l_{AD} = 2.5$ m, $\theta_E = 0°$, $\theta_{A\#} = 208°$, $\theta_D = 147°$, $m_2' = 0.5$ kg, $\omega_2 = 10$ rad/s, $\alpha_2 = 10$ rad/s^2

6.6.3 Comparison Between the Results of Proposed and Gao Feng Methods

The results of shaking force and shaking moment by proposed method and Gao Feng method for Watt mechanism with two fixed points are shown in Tables 6.1 and 6.2.

The shaking forces in Watt mechanism with two fixed points are determined at intervals of 90°. At all positions better results are produced by proposed method. Shaking force of the mechanism is maximum 2,726.43 N, at 0°, and minimum 791.96 N, at 180° in the proposed method. The shaking force gradually decreases from maximum at 0° to minimum at 180° and again gradually increases to

Table 6.1 Shaking force comparison of Watt mechanism with two fixed points

Crank angle(deg)	Shaking force generated in proposed method (N)	Shaking force generated in Gao Feng's method (N)
0	2,726.43	15,285.24
90	1,840.32	14,399.16
180	791.96	13,350.82
270	923.45	13,482.31
360	2,726.43	15,285.22

Table 6.2 Shaking moment comparison of Watt mechanism with two fixed points

Crank angle(deg)	Shaking moment generated in proposed method $\times 10^5$ N m	Shaking moment generated in Gao Feng's method $\times 10^5$ N m
0	−468.22	−468.19
90	−5.23	−5.18
180	33.17	33.23
270	39.41	39.46
360	−468.25	−468.19

maximum at 360°. The shaking moment of Watt mechanism with two fixed points is maximum 468.2×10^5 N m, at 0°, and minimum -5.23×10^5 N m, at 90°. The shaking moment gradually decreases from 0° to 90° and again increases to maximum at 360°. It can be observed that shaking forces by proposed method are very much less at all intervals of crank angle than those by Gao Feng's method. As there is only one planetary gear 8 to be mounted on the base of the mechanism, there is a little improvement in the shaking moment balancing, but the shaking forces have been substantially reduced. Though the results of a numerical example are not available in the literature to make a comparison in Tables 6.1 and 6.2, the balanced mechanisms of both the proposed and Gao Feng methods can be compared construction-wise. It can be observed that the balanced mechanism of proposed method is constructively more efficient and compact and occupies less space.

6.7 Self-Balanced Slider-Crank Mechanism

In the two identical slider-crank mechanism shaking forces are automatically balanced as the movements of the two slider-crank mechanisms are opposite to each other, so it is called as self-balanced slider-crank system. These mechanical systems find successful applications in engines, agricultural machines, mills, and various automatic machines (Fig. 6.8).

6.7.1 Self-Balanced Slider-Crank System with an Imagined Articulation Dyad

Figure 6.9 shows a self-balanced slider-crank system with an imagined articulation dyad B′D′E, which forms a pantograph with the initial system. The similarity factor of the formed pantograph is $k = {}^{l_{AD}}\!/_{l_{AB}} = 1$ and $l_{BB'} = l_{DD'}, l_{B'D'} = l_{AD} + l_{AB}$.

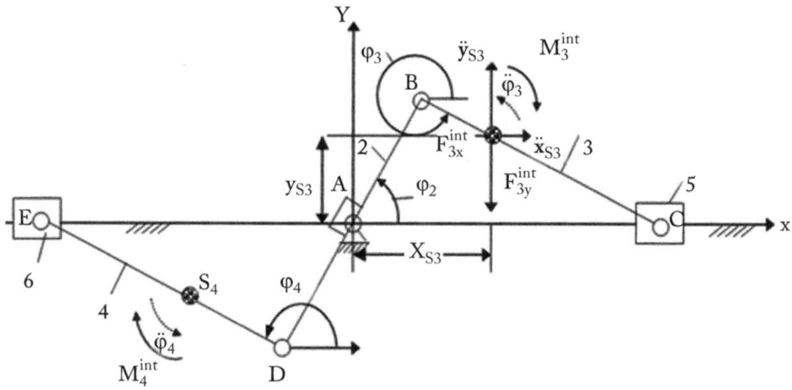

Fig. 6.8 Self-balanced slider-crank system

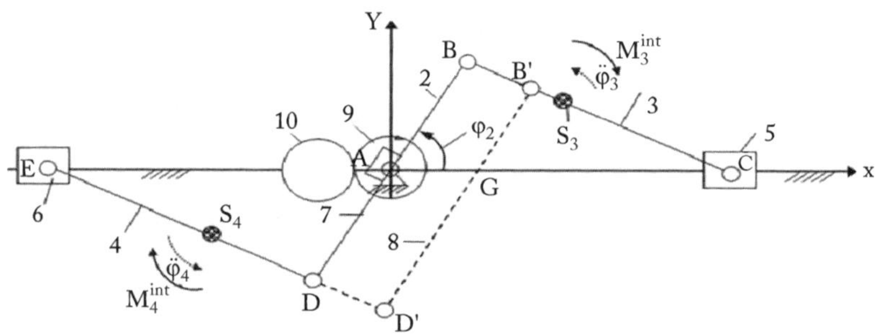

Fig. 6.9 Self-balanced slider-crank system with an imagined articulation dyad B′D′E

By substituting dynamically the mass m_3 of the connecting coupler 3 by point masses at the centers B, B′ and C and using the following condition

$$\begin{bmatrix} 1 & 1 & 1 \\ l_{BS_3} & -l_{CS_3} & l_{B'S_3} \\ l_{BS_3}{}^2 & l_{CS_3}{}^2 & l_{B''S_3}{}^2 \end{bmatrix} \begin{bmatrix} m_B \\ m_C \\ m_{B'} \end{bmatrix} = \begin{bmatrix} m_3 \\ 0 \\ I_{S_3} \end{bmatrix}$$

where $l_{BS_3}, l_{CS_3}, l_{B'S_3}$ are the distances of joint centers B, C, and B' from the centers of masses S_3 of the link 3.

I_{S_3} is the axial moment of inertia of link 3; we determine the value of the point masses

$$m_B = D_B/D_3; m_C = {}^{D_C}\!\big/_{D_3}\,; m_{B'} = D_{B'}/D_3 \tag{6.18}$$

where $D_B, D_C, D_{B'}, D_3$ are determinants of the third order obtained from the system of equations.

We now require imagined link B′D′ to be balanced about point G of the pantograph, i.e.,

$$m_{D'} = m_{B'} l_{B'G} / l_{D'G}$$

The concentrated point masses m_G, m_C, m_E to be balanced about center A, i.e.,

$$m_E = \left(m_G l_{BB'} + m_C l_{BC} \right) / l_{DE}$$

where $l_{BB'}, l_{BC}$ are the distances of joint centers B′, C from the joint center B, and l_{DE} is the distance of joint center D from the joint center E:

$$m_G = m_{B'} + m_{D'} \tag{6.19}$$

Finally the concentrated point masses m_B, m_D are also to be balanced about center A, i.e.,

$m_D = m_B l_{AB} / l_{AD}$. Thus we obtain the values of three concentrated point masses $m_{D'}, m_D, m_E$ which allow the determination of the mass and inertia parameters of the connecting coupler 4

where $l^*_{DS_4} = l_{DE} - l_{ES_4}; l_{D'S_4} = l_{D'E} - l_{ES_4}$.

6.7.2 *Shaking Moment Balancing of the Mechanism*

The shaking moment transmitted to the frame by links 2 and 7 is calculated using the angular acceleration of link 2. The shaking moment transmitted to the frame by connecting rods 3 and 4 is calculated using angular acceleration of link 8, as their point masses are brought to link 8:

$$M_2^{int} + M_7^{int} = \left(I_{S_2} + m_2 l_{AS_2}^2 + m_B l_{AB}^2 + m_D l_{AD}^2 + m_7 l_{AS_7}^2 + I_{S_7} \right) \alpha_2$$

$$M_8^{int} = \left(I_{S_8} + m_8 l_{GS_8}^2 + m_{D'} l_{D'G}^2 + m_{B'} l_{B'G}^2 \right) \alpha_8 \tag{6.20}$$

Total shaking moment generated by the mechanism:

$$M^{int} = M_2^{int} + M_7^{int} + M_8^{int} \tag{6.21}$$

The shaking moment generated by the mechanism is balanced by addition of gear inertia counterweights 9 and 10.

For any mechanism with the given numerical values of link mass, length, mass moment of inertia, and radius of gyration, the shaking force and shaking moment can be calculated using the above equations. To balance the shaking moment generated by the mechanism geared inertia counterweights with the equal amount of inertia moment can be mounted on the frame of the mechanism.

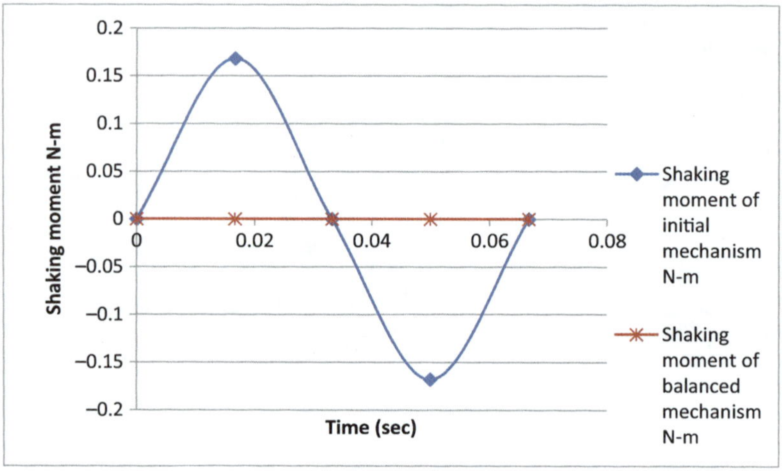

Fig. 6.10 Time vs. shaking moment

Numerical example:
The parameters of the self-balanced slider-crank system are the following:

$$l_{AB} = l_{AD} = 0.05m; l_{BC} = l_{DE} = 0.2m; l_{CS_3} = l_{ES_4} = 0.1m; m_3 = m_4 = 0.35 \text{ kg};$$

$$m_5 = m_6 = 2 \text{ kg}; I_{S_3} = I_{S_4} = 0.005 \text{ kg} - m^2; \omega_{AB} = 30\pi/s; \alpha_{AB} = 15 \text{ rad}/s^2;$$

$$m_2 = m_7 = 0.3 \text{ kg}; I_{S_2} = I_{S_7} = 0.003 \text{ kg}; l_{AS_2} = l_{AS_7} = 0.025m;$$

$$I_{S_8} = 0.006 \text{ kg} - m^2; m_8 = 0.6 \text{ kg}$$

Figure 6.10 shows the variations of the shaking moment of the initial mechanical system. For cancellation of the shaking moment it is necessary to redistribute the masses of the second connecting coupler. By dynamically substituting the mass m_3 of the connecting coupler 3 by point masses at centers B, B′, C and taking into account conditions $m'_D; m_E; m_D$, we calculate the mass and inertia parameters of the connecting coupler 4. Figure 6.10 illustrates the obtained results. So by mounting geared inertia counterweights the shaking moment is cancelled. The shaking moment of initial mechanism was +0.168 N m at 90° and −0.168 N m at 270°. It has been observed that the shaking moment of self-balanced slider-crank mechanism has been zero at all angular positions of the crank.

If the driving torque and time are plotted for both the unbalanced and balanced linkages then it can be observed that the driving torque is slightly higher for the balanced mechanism, as the inertia elements are mounted on the base of the mechanism for shaking moment balancing. On the other hand the performance of the balanced mechanism improves considerably and it also increases the fatigue life of the mechanism.

6.8 Eight-Bar Mechanism with Three Fixed Points and Three Ternary Links (Mechanism with Low Degree of Complexity)

The eight-bar mechanism with three fixed points and three ternary links shown in Fig. 6.11 has eight links, four ternary links and four binary links; one of the ternary links is fixed. It has ten binary joints. The degree of freedom of this mechanism is one. It is a mechanism with low degree of complexity as the path curvature of motion transfer point "E" is not known. Links 2, 4, 6, and 7 are binary links and 1, 3, 5, and 8 are four ternary links; among them link 8 is fixed link. The links 2, 4, 6, and 7 are not directly connected to the frame. The geared inertia counterweights required to balance the shaking moments generated by links 2, 4, 6, and 7 are mounted on the frame of the mechanism by kinematically linking geared inertia counterweights and the corresponding links by links of known mass and center of mass. The balanced eight-bar mechanism with three fixed points and three ternary links is shown in Fig. 6.12.

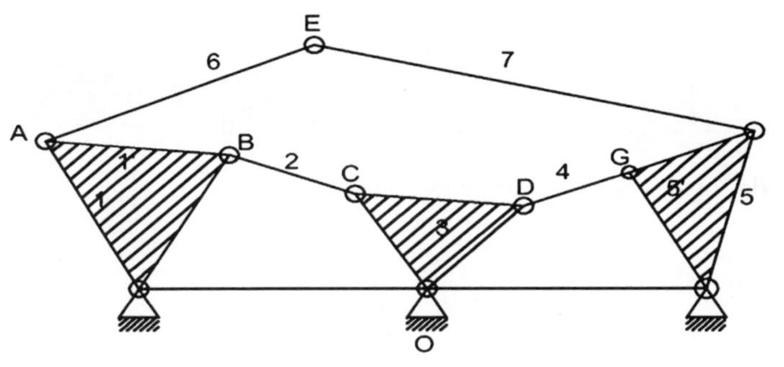

Fig. 6.11 Eight-bar mechanism with three fixed points and three ternary links

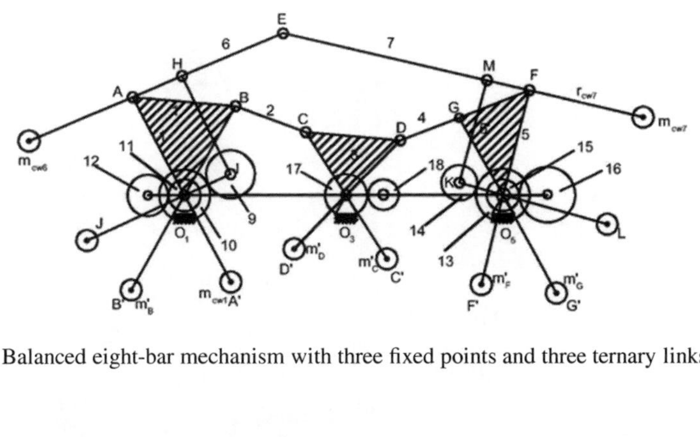

Fig. 6.12 Balanced eight-bar mechanism with three fixed points and three ternary links

6.8.1 Shaking Force Balancing of the Mechanism

Here link 1 is dynamically replaced by three point masses m_{O_1}, m_{A_1}, m_{B_1} by using the following conditions:

$$\begin{bmatrix} 1 & 1 & 1 \\ l_{O_1}e^{i\theta_{O_1}} & l_A e^{i\theta_A} & l_B e^{i\theta_B} \\ {l_{O_1}}^2 & {l_A}^2 & {l_B}^2 \end{bmatrix} \begin{bmatrix} m_{O_1} \\ m_{A_1} \\ m_{B_1} \end{bmatrix} = \begin{bmatrix} m_1 \\ 0 \\ I_{S_1} \end{bmatrix}$$

$$m_{O_1} = \frac{D_{O_1}}{D_1},\ m_{A_1} = \frac{D_{A_1}}{D_1},\ m_{B_1} = \frac{D_{B_1}}{D_1} \tag{6.22}$$

where l_{O_1}, l_A, l_B are the moduli of radius vectors of corresponding points.

θ_{O_1}, θ_A, θ_B, are the angular positions of radius vectors.

m_1 is the mass of link 1.

I_{S_1} is the mass moment of inertia of link 1 about its center of mass.

D_{O_1}, D_{A_1}, D_{B_1} and D_1 are the third-order determinants obtained from the system of equations.

Link 2 is statically replaced by two point masses m_{B_2} and m_{C_2}:

$$m_{B_2} = \frac{m_2\, l_{CS_2}}{l_{BC}};\ m_{C_2} = \frac{m_2\, l_{BS_2}}{l_{BC}}$$

Changed mass moment of inertia $I'^{*}_{S_2} = I_{S_2} - m_2 l_{CS_2} l_{BS_2}$.

For link 6 to be dynamically replaced by two point masses m_{E_6} and m_{P_2} the condition to be satisfied is

$${k_6}^2 = l_{ES_6} l_{P_6S_6}$$

where k_6 is the radius of gyration of link 6 about its center of mass.

l_{ES_6} is arbitrarily fixed and $l_{P_6S_6}$ is obtained from the above condition:

$$m_{E_6} = \frac{m_6\, l_{P_6S_6}}{(l_{ES_6} + l_{P_6S_6})};\ m_{P_6} = \frac{m_6\, l_{ES_6}}{(l_{ES_6} + l_{P_6S_6})}$$

After dynamically replacing link 6 by two point masses, it is kinematically connected to its corresponding gear 9 by link 1′ and it is statically replaced by two point masses m_H and m_I:

$$m_H = \frac{m'_1 l_{IS'_1}}{l_{HI}};\ m_I = \frac{m'_1 l_{HS'_1}}{l_{HI}}$$

Changed mass moment of inertia $I'^{*}_{S_1} = I'_{S_1} - m'_1 l'_{IS_1} l'_{HS_1}$

where I'_{S_1} is the original mass moment of inertia link 1′.

Counterweight m_{CW_6} against link 6 is calculated by using the formula

$$m_{CW_6} = \frac{(m_6 l_{AS_6} + m_H l_{AH})}{r_{CW_6}} \tag{6.23}$$

where $r_{CW_6} = l_{P_6S_6} - l_{AS_6}$ is the radius of rotation of counterweight m_{CW_6}.

Counterweight against point mass B can be obtained as

$$m_{B'} = \frac{(m_{B1} + m_{B2})\, l_{BO_1}}{l_{B'O_1}}$$

Counterweight against link 1 is calculated by using the formula

$$m_{CW_1} = \frac{(m_{CW_6} + m_H + m_6)}{r_{CW_1}} \tag{6.24}$$

where r_{CW_1} is radius of rotation of counterweight against link 1 which is arbitrarily taken.

For link 3 to be dynamically replaced by three point masses m_{C_3}, m_{D_3}, m_{O_3} the conditions to be satisfied are

$$\begin{bmatrix} 1 & 1 & 1 \\ l_{O_3}e^{i\theta_{O_3}} & l_C e^{i\theta_C} & l_D e^{i\theta_D} \\ {l_{O_3}}^2 & {l_C}^2 & {l_D}^2 \end{bmatrix} \begin{bmatrix} m_{O_3} \\ m_{C_3} \\ m_{D_3} \end{bmatrix} = \begin{bmatrix} m_3 \\ 0 \\ I_{S_3} \end{bmatrix}$$

$$m_{O_3} = \frac{D_{O_3}}{D_3},\ m_{C_3} = \frac{D_{C_3}}{D_3},\ m_{D_3} = \frac{D_{D_3}}{D_3} \tag{6.25}$$

where l_{O_3}, l_C, l_D are the moduli of radius vectors of corresponding points.

θ_{O_3}, θ_C, θ_D are the angular positions of radius vectors.

m_3 is the mass of link 3.

I_{S_3} is the mass moment of inertia of link 3 about its center of mass.

D_{O_3}, D_{C_3}, D_{D_3}, D_3 are the third-order determinants obtained from the system of equations.

Counterweight against point C can be obtained as

$$m_{C'} = \frac{(m_{C_2} + m_{C_3})\, l_{O_3c}}{l_{C'O_3}}$$

For link 4 to be statically replaced by two point masses m_{D_4} and m_{G_4} the conditions to be satisfied are

$$m_{D_4} = \frac{m_4 l_{GS_4}}{l_{DG}}; \quad m_{G_4} = \frac{m_4 l_{DS_4}}{l_{DG}}$$

Changed mass moment of inertia of link 4 can be obtained as $I'^{*}_{S_4} = I_{S_4} - m_4 l_{DS_4} l_{GS_4}$

Counterweight against point D can be obtained as

$$m_{D'} = \frac{(m_{D_3} + m_{D_4})\, l_{DO_3}}{l_{D'O_3}}$$

For link 5 to be dynamically replaced by three point masses m_{O_5}, m_{G_5}, m_{F_5} the conditions to be satisfied are

$$\begin{bmatrix} 1 & 1 & 1 \\ l_{O_5} e^{i\theta_{O_5}} & l_G e^{i\theta_G} & l_F e^{i\theta_F} \\ {l_{O_5}}^2 & {l_G}^2 & {l_F}^2 \end{bmatrix} \begin{bmatrix} m_{O_5} \\ m_{G_5} \\ m_{F_5} \end{bmatrix} = \begin{bmatrix} m_5 \\ 0 \\ I_{S_5} \end{bmatrix}$$

$$m_{O_5} = \frac{D_{O_5}}{D_5}, \; m_{G_5} = \frac{D_{G_5}}{D_5}, \; m_{F_5} = \frac{D_{F_5}}{D_5} \tag{6.26}$$

where l_{O_5}, l_G, l_F are the moduli of radius vectors of corresponding points.

θ_{O_5}, θ_G, θ_F are the angular positions of radius vectors.

m_5 is the mass of link 5.

I_{S_5} is the mass moment of inertia of link 5 about its center of mass.

D_{O_5}, D_{G_5}, D_{F_5}, D_5 are the third-order determinants obtained from the system of equations.

For link 7 to be dynamically replaced by two point masses m_{E_7} and m_{P_7} the condition to be satisfied is

$${k_7}^2 = l_{FS_7} l_{P_7S_7}$$

where k_7 is the radius of gyration of link 7 about its center of mass.

l_{FS_7} is the arbitrarily fixed and $l_{P_7S_7}$ is obtained from the above condition:

$$m_{E_7} = \frac{m_7 l_{P_7S_7}}{l_{FS_7} + l_{P_7S_7}}; \quad m_{P_7} = \frac{m_7 l_{FS_7}}{l_{FS_7} + l_{P_7S_7}}$$

After link 7 is dynamically replaced by two point masses, it is kinematically linked to its corresponding gear 13 by link 5′ and moreover link 5′ is statically replaced by two point masses m_M and m_K:

$$m_M = \frac{m'_5 l_{KS'_5}}{l_{KM}}; \quad m_K = \frac{m'_5 l_{MS'_5}}{l_{KM}}$$

Counterweight m_{CW_7} against link 7 is calculated using the formula

$$m_{\mathrm{CW}_7} = \frac{(m_{\mathrm{M}} l_{\mathrm{FM}} + m_7 l_{\mathrm{FS}_7})}{r_{\mathrm{CW}_7}}$$

where $r_{\mathrm{CW}_7} = l_{\mathrm{P}_7\mathrm{S}_7} - l_{\mathrm{FS}_7}$ is the radius of rotation counterweight.

Counterweight against point G can be obtained as

$$m'_{\mathrm{G}} = \frac{(m_{\mathrm{G}_4} + m_{\mathrm{G}_5})\, l_{\mathrm{O}_5\mathrm{G}}}{l_{\mathrm{O}_5\mathrm{G}'}}$$

where $l_{\mathrm{O}_5\mathrm{G}'}$ is arbitrarily taken.

Counterweight against point F can be obtained as

$$m'_{\mathrm{F}} = \frac{(m_7 + m_{\mathrm{CW}_7} +)\, l_{\mathrm{O}_5\mathrm{F}}}{l_{\mathrm{O}_5\mathrm{F}'}}$$

where $l_{\mathrm{O}_5\mathrm{F}'}$ is arbitrarily taken

6.8.2 Shaking Moment Balancing of the Mechanism

The shaking moments of links 1, 3, 5, 6, and 7 are given as follows:

The shaking moment generated by the linkage is determined by the sum

$$M^{\mathrm{int}} = M_1^{\mathrm{int}} + M_6^{\mathrm{int}} + M_3^{\mathrm{int}} + M_5^{\mathrm{int}} + M_7^{\mathrm{int}} \tag{6.27}$$

$$M_1^{\mathrm{int}} = \left(I_{\mathrm{S}_1} + m_1 l^2_{\mathrm{O}_1\mathrm{S}_1} + I'^{*}_{\mathrm{S}_1} + m'_1 l^2_{\mathrm{IS}'_1} + m_{\mathrm{B}2} l^2_{\mathrm{O}_1\mathrm{B}} + m'_{\mathrm{B}} l^2_{\mathrm{O}_1\mathrm{B}'} + m_{\mathrm{CW}_1} r^2_{\mathrm{CW}_1}\right)\alpha_1$$

$$M_6^{\mathrm{int}} = \left(I_{\mathrm{S}_6} + m_{\mathrm{CW}_6} r^2_{\mathrm{CW}_6} + m_6 l^2_{\mathrm{AS}_6} + m_{\mathrm{H}} l^2_{\mathrm{AH}} + 2 m_{\mathrm{J}} l^2_{\mathrm{O}_1\mathrm{J}}\right)\alpha_6$$

$$M_3^{\mathrm{int}} = \left(I_{\mathrm{S}_3} + m_3 l^2_{\mathrm{O}_3\mathrm{S}_3} + m_{\mathrm{D}_4} l^2_{\mathrm{O}_3\mathrm{D}} + m'_{\mathrm{D}} l^2_{\mathrm{O}_3\mathrm{D}'} + m'_{\mathrm{C}} l^2_{\mathrm{O}_3\mathrm{C}'}\right)\alpha_3$$

$$M_5^{\mathrm{int}} = \left(I_{\mathrm{S}_5} + m_5 l^2_{\mathrm{O}_5\mathrm{S}_5} + m_{\mathrm{G}4} l^2_{\mathrm{o}5\mathrm{G}} + m'_{\mathrm{G}} l^2_{\mathrm{O}_5\mathrm{G}'} + m'_{\mathrm{F}} l^2_{\mathrm{O}_5\mathrm{F}'} + + m'_5 l^2_{\mathrm{KS}'_5}\right)\alpha_5$$

$$M_7^{\mathrm{int}} = \left(I_{\mathrm{S}_7} + m_7 l^2_{\mathrm{FS}_7} + m_{\mathrm{CW}_7} r^2_{\mathrm{CW}_7} + m_{\mathrm{M}} l^2_{\mathrm{FM}} + 2 m_{\mathrm{K}} l^2_{\mathrm{O}_5\mathrm{K}}\right)\alpha_7$$

where M_1^{int}, M_3^{int}, M_5^{int}, M_6^{int}, M_7^{int} are the shaking moments generated by links 1, 3, 5, 6, and 7, respectively.

I_{S_1}, I_{S_3}, I_{S_5}, I_{S_6}, and I_{S_7} are the mass moment of inertias of links 1, 3, 5, 6, and 7, respectively.

α_1, α_3, α_5,α_6 and α_7 are the angular accelerations of links 1, 3, 5, 6, and 7, respectively.

For shaking moment balancing ten geared inertia counterweights are used, four at O_5, two at O_3, and four at O_1.

6.9 Summary

Self-balanced slider-crank mechanism has been studied with numerical example and it is observed that shaking moment is completely balanced. Shaking force and shaking moment balancing expressions are developed for eight-bar mechanism with three fixed points and three ternary links.

References

1. Berkof, R.S.: Complete force and moment balancing of inline four-bar linkage. Mech. Mach. Theory **8**, 397–410 (1973)
2. Esat, I., Bahai, H.: A theory of complete force and moment balancing of planar linkage mechanisms. Mech. Mach. Theory **34**, 903–922 (1999)
3. Berkof, R.S., Lowen, G.G.: A new method for completely force balancing simple linkages, Trans. ASME J. Eng. Ind. **91**(1), 21–26 (1969)
4. Sherwood, A.A., Hokey, B.A.: The Optimization of mass distribution in mechanisms using dynamically similar systems. J. Mech. **4**, 243–260 (1969)
5. Sherwood, A.A.: The Optimum distribution of mass in the coupler of a plane four-bar linkage. J. Mech. **1**, 229–234 (1966)
6. Sherwood, A.A.: The dynamics of the harmonic space slider-crank mechanisms. J. Mech. **1**, 203–208 (1966)
7. Kochev, I.S.: Theory of symmetrical mechanisms. Int. J. Mech. Mach. Theory **25**(4), 467–478 (1990)
8. Kochev, I.S.: Full shaking movement balancing of planar linkages by a prescribed input speed fluctuation. Int. J. Mech. Mach. Theory **25**(4), 459–466 (1990)
9. Kochev, I.S.: Active balancing of the frame shaking moment in high speed planar machines. Int. J. Mech. Mach. Theory **27**(1), 53–58 (1992)
10. Hong-sen, Y., Guo-jhih, Y.: Integrated control and mechanism design for the variable input speed servo four-bar linkages. Int. J. Mechatron. **19**, 274–285 (2009)
11. Yan, H.S., Hsu, M.H., Fong, M.K., Hsieh, W.H.: A kinematic approach for eliminating the discontinuity of motion characteristics of cam-follower systems. J. Appl. Mech. Robotics **1**(2), 1–6 (1994)
12. Yan, H.S., Tsai, M.C., Hsu, M.H.: A variable speed method for improving motion characteristics of cam-follower systems. ASME Trans. J. Mech. Design **118**(1), 250–258 (1996)
13. Yan, H.S., Tsai, M.C., Hsu, M.H.: An experimental study of the effects of cam speed on cam-follower systems. Int. J. Mech. Mach. Theory **31**(4), 397–412 (1996)
14. Yan, H.S., Fong, M.K.: An Approach for reducing the peak acceleration of cam-follower systems using a B-spline representation. J. China Soc. Mech. Eng. (Taiwan) **15**, 48–55 (1994)
15. Yan, H.S., Chen, W.R.: On the output motion characteristics of variable input speed servo-controlled slider-crank mechanisms. Int. J. Mech. Mach. Theory **35**(4), 541–561 (2000)

16. Kochev, I.S.: General method for active balancing of combined shaking moment torque fluctuations in planar linkages. Int. J. Mech. Mach. Theory **25**(6), 679–687 (1990)
17. Ye, Z., Smith, M.R.: Complete balancing of planar linkages by an equivalence method. Mech. Mach. Theory **29**(5), 701–712 (1994)
18. Feng, G.: Complete shaking force and shaking moment balancing of 26 types of four-, five-, six-bar linkages with prismatic pairs. Int. J. Mech. Mach. Theory **25**(2), 183–192 (1990)
19. Feng, G.: Complete shaking force and shaking moment balancing of four types of six-bar linkages. Int. J. Mech. Mach. Theory **24**(4), 275–287 (1989)
20. Feng, G.: Complete shaking force and shaking moment balancing of 17 types of eight-bar linkages only with revolute pairs. Int. J. Mech. Mach. Theory **26**(2), 197–206 (1991)
21. Arakelian, V.: Complete shaking force and shaking moment balancing of linkages. Int. J. Mech. Mach. Theory **33**(4), 425–436 (1998)
22. Arakelian, V.H., Smith, M.R.: Complete shaking force and shaking Moment balancing of planar linkages. Int. J. Mech. Mach. Theory **34**, 1141–1153 (1999)
23. Nehemiah, P., Sundarasiva Rao, B.S.K., Ramji, K.: Shaking force and Shaking moment balancing of planar mechanisms with high degree of complexity. Jordan J. Mech Ind. Eng. (JJMIE) **6**(1), 17–24 (2012)
24. Yan, H.S., Soong, R.C.: kinematic and dynamic design of four-bar linkage by counterweighing with variable input speed. Int. J. Mech. Mach. Theory **36**, 1050–1071 (2001)
25. Chiou, S.T., Bai, G.J., Chang, W.K.: Optimum balancing design of the drag-link drive of mechanical presses for precision cutting. Int. J. Mach. Tools Manuf. **38**(3), 131–141 (1998)
26. Barker, C.R., Tso, P.L.: Characteristics surface for three position function generation with planar four-bar mechanism. ASME Trans. J.Mech. Transm. Automat. Des. **111**, 104–109 (1989)
27. Bagci, C.: Complete shaking force and shaking moment balancing of link mechanisms using balancing idler loops. ASME Trans. J. Mech. Des. **104**, 482–493 (1982)
28. Wu, Y., Gosselin, C.M.: Synthesis of reaction less spatial 3-dof and 6-dof mechanisms without separate counter rotations. Int. J. Robotics Res. **23**(6), 625–642 (2004)
29. Bagci, C.: Shaking force balancing of planar linkages with force transmission irregularities using balancing idler loops. Int. J. Mech. Mach. Theory **14**(4), 267–284 (1979)
30. Kochev, I.S.: A new general method for full force balancing of planar linkages. Int. J. Mech. Mach. Theory **23**(6), 475–480 (1988)
31. Zhang, S., Chen, J.: The optimum balance of shaking force and shaking moment of linkages. Int. J. Mech. Mach. Theory **38**(4), 589–597 (1995)
32. Li, C.-H., Tso, P.-L.: The study of dynamic balancing for high speed presses. JSME Int J Series C **49**(3), 657–662 (2006)
33. Stevensen Jr., E.N.: Balancing of machines. Trans ASME J Eng. Ind. **95**, 650–656 (1973)
34. Hocky, B.A.: An improved technique for reducing the fluctuation of kinetic energy in plane mechanisms. J. Mech. **6**, 405–418 (1971)
35. Spreitzer, G.: The path of the inertia pole of a rigid body in constrained motion. Int. J. Mech. Mach. Theory **7**, 231–245 (1972)
36. Zhe, L., Shixian, B.: Optimum balancing of linkages with clearances. Int. J. Mech. Mach. Theory **27**(5), 535–541 (1992)
37. Wiederrich, J.L., Roth, B.: Momentum balancing of four-bar linkages. Trans. ASME J. Eng. Ind. Series B **98**, 1289–1295 (1976)
38. Porter, B., Sanger, D.J.: Synthesis of dynamically optimal four-bar linkages. Institution of mechanical engineers conference on mechanisms, Londan, paper C 69, (1972)
39. Kochev, I.S.: Qualitative theory of the reactions and stress in high speed planar linkages. Int. J. Mech. Mach. Theory **27**(1), 59–68 (1992)
40. Ogawa, K., Funabashi, H.: On the balancing of the fluctuations of input torques caused by inertia forces in the crank and rocker mechanisms. Trans. ASME J. Eng. Ind. **91**(1), 97–102 (1969)
41. Hocky, B.A.: The minimization of the fluctuation of input shaft torque in plane mechanisms. Int. J. Mech. Mach. Theory **7**, 335–346 (1972)

42. Smith, M.R., Maunder, L.: Inertia forces in a four-bar linkage. J. Mech. Eng. Sci. **9**(3), 218–225 (1967)
43. Smith, M.R.: Dynamic analysis and balancing of linkages with interactive computer graphics. Comput. Aided Des. **7**(1), 15–19 (1975)
44. Ettefgh, M.M., Abbasidoust, F., Milanchian, H., Yazdanian Asr, M.: Complex crank-slider mechanism dynamic balancing by binary genetic algorithm (BGA). Proceedings of the 2011 international symposium on innovations in intelligent systems and applications. Istanbul, Turkey, IEEE Trans. 277–281 (2011)
45. Carson, W.L., Stephens, J.M.: Feasible parameter design space for force and root-mean-square momentum balancing on in-line 4R four bar linkage synthesized for kinematic criteria. Int. J. Mech. Mach. Theory **13**(6), 649–658 (1978)
46. Elliott, J.L., Tesar, D.: The theory of torque, shaking force, shaking moment. Trans. ASME J. Eng. Ind. **99**(3), 715–722 (1977)
47. Elliott, J.L., Tesar, D.: A general mass balancing method for complex planar mechanisms. Int. J. Mech. Mach. Theory **17**(2), 153–172 (1982)
48. Haines, R.S.: Minimum r.m.s shaking moment or driving torque of a force balanced four bar linkage using feasible counterweights. Int. J. Mech. Mach. Theory **16**(3), 185–190 (1981)
49. Himanshu, C., Saha, S.K.: Balancing of shaking force and shaking moments for planar mechanisms using the equimomental systems. Int. J. Mech. Mach. Theory **43**, 310–334 (2008)
50. Himanshu, C., Saha, S.K.: Balancing of four bar linkages using maximum recursive dynamic algorithm. Int. J. Mech. Mach. Theory **42**, 216–232 (2007)
51. Ilia, D., Sinatra, R.: A novel formulation of the dynamic balancing of five-bar linkages with applications to link optimization. Int. J. Multi-Body Syst. Dyn. **21**, 193–211 (2009)
52. Sadler, J.P., Mayne, R.W.: Balancing of mechanisms by non-linear programming. proceedings of 3rd Applied mechanics conference, Oklahoma state university, paper 29, (1973)
53. Walker, M.J., Smith, M.R.: Force and moment reduction in linkage mechanisms by computerized optimization. Proceedings of 2nd IFToMM International symposium linkages and computer-aided design methods, Bucharest, June 1977
54. Nikravsh, P.E., Gim, G.: Systematic construction of the equation of motion for multi-body systems containing closed kinematic loops. ASME J. Mech. Des. **115**, 143–149 (1993)
55. Saha, S.K.: Dynamics of serial multi-body systems using the decoupled natural orthogonal complement matrices. ASME J. Appl. Mech. **66**, 986–996 (1999)
56. Angeles, J., Lee, S.: The formulation of dynamical equations of holonomic mechanical systems using a natural orthogonal complement. ASME J. Appl. Mech. **55**, 243–244 (1988)
57. Lee, T.W., Cheng, C.: Optimum balancing of combined shaking force, shaking moment and torque fluctuations in high speed linkages. ASME J. Mech. Transm. Automat. Des. **106**(2), 242–251 (1984)
58. Feng, B., Morita, N., Torii, T.: A new optimization method for dynamic design of planar linkage with clearance at joints- optimization the mass distribution of links to reduce the change of joint forces. ASME J. Mech. Des. **124**, 68–73 (2000)
59. Denavit, J., Hartenberg, R.S.: A kinematic notation for lower pair mechanisms based matrices. ASME J. Appl. Mech. **77**, 215–221 (1955)
60. Freudenstein, F.: Approximate synthesis of four bar linkages. Trans. ASME J. Eng. Ind. **77**, 853–861 (1955)
61. Han, C.Y.: Balancing of high speed machinery. Trans. ASME J. Eng. Ind. Series B **89**(1), 111–118 (1967)
62. Mewes, E.: Unbalanced inertia forces in slider crank mechanisms of large eccentricity. Trans. ASME J. Appl. Mech. **25**, 225–232 (1958)
63. Qi, N.M., Pennestri, E.: Optimum balancing of four bar linkages-A refined algorithm. Int. J. Mech. Mach. Theory **26**(3), 337–348 (1991)
64. Lowen, G.G., Berkof, R.S.: Survey of investigations into the balancing of linkages. J. Mech. **3**, 221–231 (1968)
65. Tepper, F.R., Lowen, G.G.: On the distribution of the R.M.S shaking moment of unbalanced planar mechanisms: Theory of Isomomental Ellipses. ASME paper no.72-Mech-4,Mechanisms conference, October 8–12, 1972

66. Tepper, F.R., Lowen, G.G.: Shaking force optimization of four-bar linkage with adjustable constraints on ground bearing forces. Trans. ASME J. Eng. Ind. **97 (series B)**, 643–651 (1975)
67. Himanshu, C., Saha, S.K.: constraint wrench formulation for closed loop systems using two-level recursions. ASME J. Mech. Des. **129**, 1234–1242 (2007)
68. Zhang, S.: A constituting method of objective function for the dynamic optimum balance of shaking forces in linkage. Int. J. Mech. Mach. Theory **29**(6), 829–835 (1994)
69. Lowen, G.G., Berkof, R.S.: Determination of force balanced four-bar linkages with optimum shaking moment characteristics. Trans. ASME J. Eng. Ind. **93**(B (1)), 39–46 (1971)
70. Berkof, R.S., Lowen, G.G.: Theory of shaking moment optimization of force balanced four bar linkages. Trans. ASME J. Eng. Ind. **93**(B (1)), 53–60 (1971)
71. Arakelian, V.H., Dahan, M., Smith, M.R.: Complete shaking force and partial shaking moment balancing of planar four bar linkages. J. Multi-Body Dynam. Institut. Mech. Eng. **215, part K**, 31–34 (2001)
72. Lowen, G.G., Tepper, F.R., Berkof, R.S.: The quantitative influence of complete four balancing on the forces and moments of certain families of four bar linkages. Int. J. Mech. Mach. Theory **9**(3/4), 299–323 (1974)
73. Haines, R.S.: PhD thesis, university of Newcastle upon Tyne, England, 156–157, 1982
74. Kochev, I.S.: General theory of complete shaking moment balancing of planar linkages: A critical review. Int. J. Mech. Mach. Theory **35**, 1501–1514 (2000)
75. Turbin, B.I., Koropetz, A.A., Koropetz, Z.A.: The possibility of the shaking force balancing in the system with oscillating links. Int. J. Mech. Mach. Theory **7**, 87–90 (1978)

Chapter 7
Synthesizing of Parallel Robots Using Adjusting Kinematic Parameters Method

P.R. Ouyang, W.J. Zhang, and J. Huang

Abstract Force balancing is a very important issue in mechanism design and has only recently been introduced to the design of robotic mechanisms. In this chapter, a force balancing method called adjusting kinematic parameters (AKP) for robotic mechanisms or real-time controllable (RTC) mechanisms is proposed, as opposed to existing force balancing methods, e.g., the counterweights (CW) method. Both the working principle of the AKP method and the design equation are described in detail. A particular implementation of the AKP method for the RTC mechanisms where two pivots on a link are adjustable is presented. After that, a hybrid approach to force balancing of robotic mechanisms is proposed, and this hybrid approach is to combine AKP and counterweights (CW) approaches, called AKP+CW in short. The main motivation of the AKP+CW approach is that CW and AKP each has its own advantage and disadvantage, and thus a combined one may strengthen both. This chapter presents the force balancing principles and equations for the AKP+CW approach. Software called ADAMS is employed as a tool for the simulated experiment to verify the effectiveness of the proposed approach. The joint forces and torques are calculated for the trajectory tracking of the RTC mechanisms. The implication of the work to the balancing of mechanisms in general is that many different force balancing methods may be combined based on the hybridization principle proposed in this chapter to become a novel one. Simulation results show that the AKP method and AKP+CW method are consistently better than the CW method in terms of the reduction of the joint forces and the torques in the servomotors, and the smoothing of the fluctuation of the joint force.

P.R. Ouyang (✉)
Department of Aerospace Engineering, Ryerson University, Toronto, ON, Canada
e-mail: pouyang@ryerson.ca

W.J. Zhang • J. Huang
Department of Mechanical Engineering, University of Saskatchewan, Saskatoon, SK, Canada

D. Zhang, B. Wei (eds.), *Dynamic Balancing of Mechanisms and Synthesizing of Parallel Robots*, DOI 10.1007/978-3-319-17683-3_7

7.1 Introduction

Mechanisms driven by real-time controllable (RTC) motors, or servomotors, are called RTC mechanisms. In general, RTC mechanisms are multi-degrees of freedom systems. RTC mechanisms are also called *mechatronic mechanisms*. A generic task of an RTC mechanism is to generate trajectory tracking motion. RTC mechanisms are fundamental building blocks in many machine tools and advanced robots due to their flexibility in terms of adapting to different applications without the need of redesign of their physical structures. Since there is no well-developed guidance available to design an RTC mechanism system with consideration of both control structures and mechanical structure designs, it is therefore significant to develop methodologies for this purpose.

For an RTC mechanism, the mechanical properties, such as shaking force balancing, shaking moment balancing, machine cooling, and vibration, are highly coupled with the characteristics of the controller. One challenging issue in the RTC mechanism design is that in order to achieve the optimal performance from the overall system viewpoint, both the controller design and the mechanical structure design need to be considered simultaneously. This chapter only takes the property of force balancing into consideration in addressing this issue. Force balancing can be categorized into static balancing problem and dynamic balancing problem. Static balancing is defined as a set of conditions under which the weight of the links of a mechanism does not produce any torque or force at the actuators under static conditions for any configuration of the manipulator or mechanism [1]. This is a mechanical solution which does not include solution by control method. This chapter focuses on static balancing. In this chapter, a parallel robot with two degrees of freedom is used as an example for the force balancing.

Many studies on force balancing of spatial mechanisms, i.e., robots, are performed at Laval University. These studies are limited within the scope of mechanical structure design only. When attached with a controller and programmed to follow different trajectories, however, mechanisms designed with complete force balancing property in this manner may not generate satisfactory dynamic performance. This consequence was revealed by the recent studies carried out at the Advanced Engineering Design Laboratory (AEDL) of the University of Saskatchewan [2] and other research [3].

At AEDL, a novel method called the adjusting kinematic parameters (AKP) method for force balancing of real-time controllable RTC mechanisms was developed [2]. This method suggests that adjusting the kinematic parameters can achieve force balancing of RTC mechanisms. Although showing advantages, there are some problems with this method. *First*, the initial development assumed that the center of mass of each link in a mechanism was in line with its axis; see Fig. 7.1a. However, in general, the center of mass of a link is likely off-line with its axis; see Fig. 7.1b. In mechanisms, a link may also take the ternary form instead of the binary form; see Fig. 7.1c. *Second*, the physical implementation was not considered in Wang's study. It should be noted that when adjusting the kinematic parameters of a link, extra

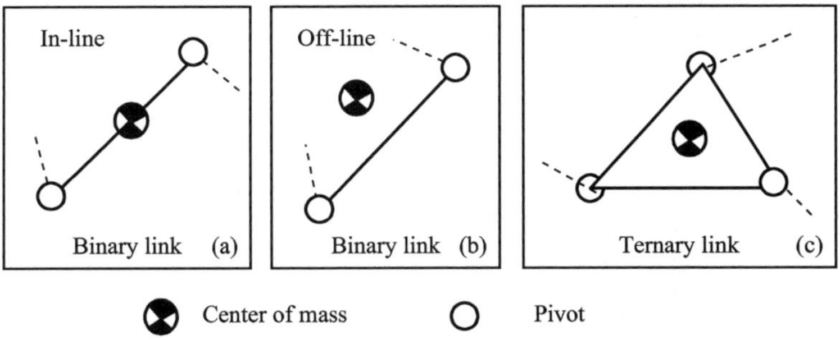

Fig. 7.1 Different situations of mass centers and different forms of links

masses are included in the system, and adjustment of kinematic parameters related to these masses may involve a change of the mass distribution of the system. *Third*, the exploration of how control methods affect trajectory tracking performance with respect to different force balancing methods was not well addressed. In particular, the control method used was a simple PD law with gains selected in a trial-and-error manner. *Finally*, trajectory planning was not well studied in spite of its importance in the AKP method. In fact, whenever the kinematic parameters are adjusted, the geometry of the mechanism may be varied; hence the trajectory must be re-planned to achieve the desired motion task.

7.2 Principles for Complete Force Balancing

There are two principles for complete force balancing: (1) making the total mass center of a mechanism stationary [4, 5], and (2) making the total potential energy of a mechanism stationary [1, 6]. *Principle* (1) is explained as follows.

Consider a mechanism that transmits forces to its base at point O. Let f_0 stand for the sum of the reacting forces the mechanism imposes on the base. The application of the Newton's second law leads to

$$\mathbf{f}_0 = -\frac{d}{dt}\{M\dot{\mathbf{r}}_c\} + M\mathbf{g} \tag{7.1}$$

where M is the total mass of a mechanism, $\mathbf{g}$ the gravitational acceleration, and $\dot{r}_c$ the velocity of the *mass center* (MC) of the mechanism. It can be seen from Eq. (7.1) that the undesired shaking force results from changes in the system's linear momentum. This dynamic component becomes zero if the system MC does not change in any configuration, i.e., $\mathbf{r}_c = \mathbf{constant}$, during a period of motion. Therefore, *Principle (1)* can be stated as follows: to transmit zero shaking force, the mechanism's MC has to be stationary or configuration invariant. The property of configuration invariance of MC can be obtained in several ways, depending on the

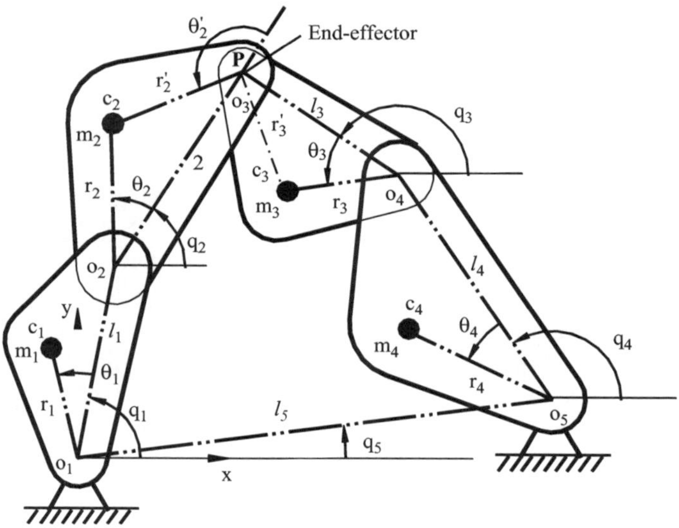

Fig. 7.2 A 2 DOF parallel robotic mechanism with arbitrary mass distribution

types of the mechanisms. For instance, for any mechanism containing only revolute kinematic pairs, the mass distribution of the mechanism to achieve the force balance can be obtained by using the *linearly independent vector* (LIV) approach [4].

The LIV approach is explained by taking a 2 DOF parallel robotic mechanism as an example; see Fig. 7.2. This mechanism consists of two kinematic chains connecting the fixed base and the end effector of the mechanism. The two kinematic chains are O_1, O_2, O_3, P and O_5, O_4, O_3, P, respectively. Two revolute actuators are mounted at joints O_1 and O_5 and described by joint variables q_1 and q_4, respectively. Joints O_2, O_3, and O_4 are passive revolute joints. Point P (overlapping with O_3) is the position of the end effector of the mechanism.

The stationary total mass center condition can be expressed by the following equation:

$$\mathbf{r}_c = \frac{1}{M}\sum_{i=1}^{4} m_i \mathbf{r}_i = \mathbf{constant} \tag{7.2}$$

where m_i and $\mathbf{r}_i$ are the mass and the position vector of mass center of link i, respectively, and $M = \sum_{i=1}^{4} m_i$. According to the LIV approach, the position vector $\mathbf{r}_i$ in Eq. (7.2) can be expressed as

$$\begin{cases} \mathbf{r}_1 = r_1 \mathrm{e}^{\mathbf{i}(q_1+\theta_1)} \\ \mathbf{r}_2 = l_1 \mathrm{e}^{\mathbf{i}q_1} + r_2 \mathrm{e}^{\mathbf{i}(q_2+\theta_2)} \\ \mathbf{r}_3 = l_5 \mathrm{e}^{\mathbf{i}q_5} + l_4 \mathrm{e}^{\mathbf{i}q_4} + r_3 \mathrm{e}^{\mathbf{i}(q_3+\theta_3)} \\ \mathbf{r}_4 = l_5 \mathrm{e}^{\mathbf{i}q_5} + r_4 \mathrm{e}^{\mathbf{i}(q_4+\theta_4)} \end{cases} \tag{7.3}$$

Substituting Eq. (7.3) into Eq. (7.2) leads to

$$\begin{aligned} M\mathbf{r}_c = {} & \left(m_3 l_5 \mathrm{e}^{\mathbf{i}q_5} + m_4 l_5 \mathrm{e}^{\mathbf{i}q_5}\right) + \left(m_1 r_1 \mathrm{e}^{\mathbf{i}\theta_1} + m_2 l_1\right) \mathrm{e}^{\mathbf{i}q_1} + \left(m_2 r_2 \mathrm{e}^{\mathbf{i}\theta_2}\right) \mathrm{e}^{\mathbf{i}q_2} \\ & + \left(m_3 r_3 \mathrm{e}^{\mathbf{i}\theta_3}\right) \mathrm{e}^{\mathbf{i}q_3} + \left(m_4 r_4 \mathrm{e}^{\mathbf{i}\theta_4} + m_3 l_4\right) \mathrm{e}^{\mathbf{i}q_4} \end{aligned} \tag{7.4}$$

The unit vectors $\mathrm{e}^{\mathbf{i}q_1}$, $\mathrm{e}^{\mathbf{i}q_2}$, $\mathrm{e}^{\mathbf{i}q_3}$, and $\mathrm{e}^{\mathbf{i}q_4}$ are constrained by the kinematic closed-loop equation, i.e., $l_1 \mathrm{e}^{\mathbf{i}q_1} + l_2 \mathrm{e}^{\mathbf{i}q_2} - l_3 \mathrm{e}^{\mathbf{i}q_3} - l_4 \mathrm{e}^{\mathbf{i}q_4} - l_5 \mathrm{e}^{\mathbf{i}q_5} = 0$. Substituting this constraint equation into Eq. (7.4) leads to

$$\begin{aligned} M\mathbf{r}_c = {} & \left(m_3 l_5 \mathrm{e}^{\mathbf{i}q_5} + m_4 l_5 \mathrm{e}^{\mathbf{i}q_5} + \lambda_{25} m_2 r_2 \mathrm{e}^{\mathbf{i}\theta_2} \mathrm{e}^{\mathbf{i}q_5}\right) \\ & + \left(m_1 r_1 \mathrm{e}^{\mathbf{i}\theta_1} + m_2 l_1 - \lambda_{21} m_2 r_2 \mathrm{e}^{\mathbf{i}\theta_2}\right) \mathrm{e}^{\mathbf{i}q_1} + \left(m_3 r_3 \mathrm{e}^{\mathbf{i}\theta_3} + \lambda_{23} m_2 r_2 \mathrm{e}^{\mathbf{i}\theta_2}\right) \mathrm{e}^{\mathbf{i}q_3} \\ & + \left(m_4 r_4 \mathrm{e}^{\mathbf{i}\theta_4} + m_3 l_4 + \lambda_{24} m_2 r_2 \mathrm{e}^{\mathbf{i}\theta_2}\right) \mathrm{e}^{\mathbf{i}q_4} \end{aligned} \tag{7.5}$$

where $\lambda_{ij} = {}^{l_j}\!\big/_{l_i}$, $i, j = 1, \ldots, 5$. In order to make the total mass center stationary, all the terms with the time-varying quantities (q_1, q_3 and q_4) in Eq. (7.5) should vanish. This will result in the following equations:

$$m_1 r_1 \mathrm{e}^{\mathbf{i}\theta_1} + m_2 l_1 - \lambda_{21} m_2 r_2 \mathrm{e}^{\mathbf{i}\theta_2} = 0 \tag{7.6}$$

$$m_3 r_3 \mathrm{e}^{\mathbf{i}\theta_3} + \lambda_{23} m_2 r_2 \mathrm{e}^{\mathbf{i}\theta_2} = 0 \tag{7.7}$$

$$m_4 r_4 \mathrm{e}^{\mathbf{i}\theta_4} + m_3 l_4 + \lambda_{24} m_2 r_2 \mathrm{e}^{\mathbf{i}\theta_2} = 0 \tag{7.8}$$

As long as Eqs. (7.6) to (7.8) are satisfied in the design process, the shaking force of the parallel robotic mechanism is cancelled. Such a mechanism is called a force balanced mechanism.

Principle (2) is that if the total potential energy of a mechanism in any configuration is kept constant (i.e., the weight of the mechanism has no effect on the actuators), then the mechanism is force balanced. The expression of the total potential energy of the mechanism can be written as

$$V = V_w + V_s \tag{7.9}$$

where V_w and V_s are, respectively, the gravitational potential energy and the elastic potential energy stored in the springs. The way of implementing this principle is to eliminate the effect of the potential energy through properly adding springs into the original mechanism [1].

As shown in Fig. 7.1, two relationships $r_2 e^{i\theta_2} = l_2 + r_2' e^{i\theta_2'}$ and $r_3 e^{i\theta_3} = l_3 + r_3' e^{i\theta_3'}$ can be readily obtained. By using these relationships, Eqs. (7.6)–(7.8) can be rewritten as

$$m_1 r_1 l_2 = l_1 m_2 r_2' \quad \text{and} \quad \theta_1 = \theta_2' \tag{7.10}$$

$$m_3 r_3 l_2 = l_3 m_2 r_2 \quad \text{and} \quad \theta_3 = \pi + \theta_2 \tag{7.11}$$

$$m_4 r_4 l_3 = l_4 m_3 r_3' \quad \text{and} \quad \theta_4 = \theta_3' \tag{7.12}$$

From the above three equations, it can be seen that whenever the mass distribution of one of the links is given, the mass distributions of the remaining three links can be determined. The equations to calculate the additional masses can be derived as (assume that link 2 is unchanged)

$$m_i^* r_i^* e^{i\theta_i^*} = m_i r_i e^{i\theta_1} - m_i^0 r_i^0 e^{i\theta_i^0} \quad (i = 1, 3, 4) \tag{7.13}$$

where m_i^0, r_i^0, and θ_i^0 are the parameters of the original link; m_i^*, r_i^*, and θ_i^* are the parameters of the counterweights (CW); and m_i, r_i, and θ_i are the parameters after adding or deducting the counterweights to the original mechanism.

Apparently, in general, either m_i^* or r_i^* is arbitrarily selected, then the other one can be obtained from Eq. (7.13).

There are several problems associated with the CW method. The first problem is that both joint forces and actuator output torques might increase. The second problem is that the vibration behavior of the mechanism may be degraded [3]. The third problem is that the balanced mechanism may have a poor trajectory tracking performance, and consume more energy when running at high speeds [7, 8]. Although a careful design of the mass redistribution may help in solving these problems to some degree, new methods are needed for further improvement. It should be further noted that the CW method is only applicable to pivot joint mechanisms.

7.3 New Force Balancing Condition Equations

When implementing the extended AKP method, the masses of sliding blocks, which are used to adjust the pivots, must be taken into account in the force balancing condition equations. The implementation of the pivot adjustment is illustrated in Fig. 7.3 [7, 8]. When sliding block 3 is adjusted along link 1 or sliding block 4 adjusted along link 2, the mass distribution of this group of links will vary.

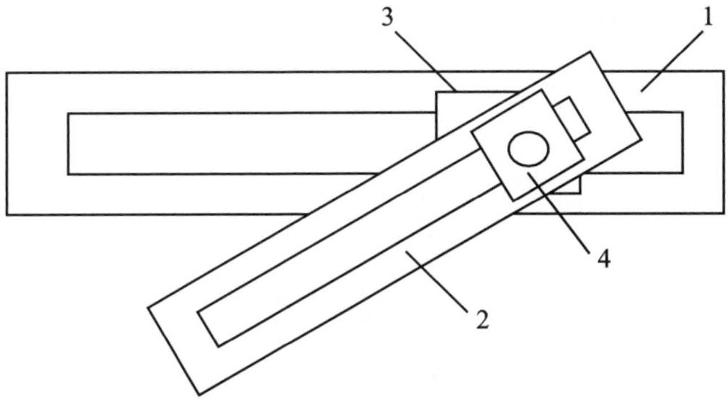

Fig. 7.3 Illustration of pivot adjustment

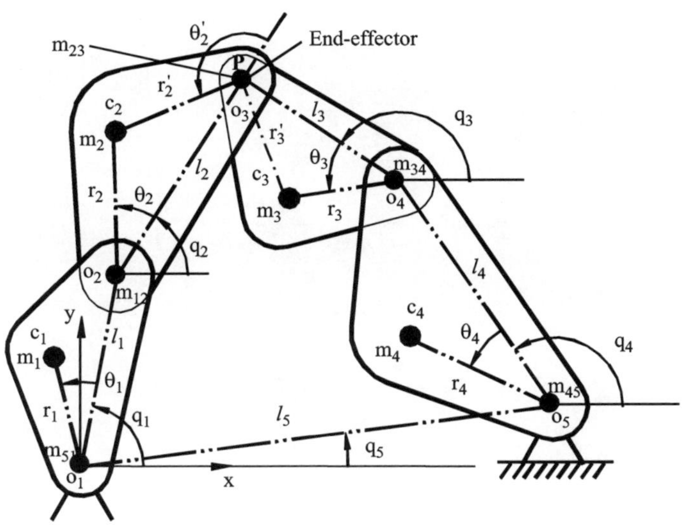

Fig. 7.4 Scheme of parallel robotic mechanism with the point mass at the pivot

Assume that the sliding block is a point mass acting on the pivot, denoting m_{ij} as the mass of the sliding block between link i and link j. The schematic diagram of the parallel robotic mechanism with consideration of these point masses, m_{ij}, is shown in Fig. 7.4.

The new force balancing condition equations can be derived by following the LIV approach:

$$m_1 r_1 e^{i\theta_1} + (m_2 + m_{12})\, l_1 - l_1 m_2 r_2 e^{i\theta_2}/l_2 = 0 \tag{7.14}$$

$$m_3 r_3 e^{i\theta_3} + m_{23} l_3 + l_3 m_2 r_2 e^{i\theta_2}/l_2 = 0 \tag{7.15}$$

$$m_4 r_4 e^{i\theta_4} + (m_3 + m_{23} + m_{34})\, l_4 + l_4 m_2 r_2 e^{i\theta_2}/l_2 = 0 \tag{7.16}$$

Using the relationship $r_2 e^{\mathbf{i}\theta_2} = l_2 + r_2' e^{\mathbf{i}\theta_2'}$, Eq. (7.14) can be rewritten as

$$l_2 m_1 r_1 e^{i\theta_1} = l_1 m_2 \tilde{r}_2' e^{i\tilde{\theta}_2'} \tag{7.17}$$

where $\tilde{r}_2' = \sqrt{\left(r_2' \cos\theta_2' - m_{12} l_2/m_2\right)^2 + \left(r_2' \sin\theta_2'\right)^2}$

$$\tilde{\theta}_2' = \tan^{-1}\left(\frac{r_2' \sin\theta_2'}{r_2' \cos\theta_2' - m_{12} l_2/m_2}\right)$$

Similarly, Eq. (7.15) can be rewritten as

$$l_2 m_3 r_3 e^{i\theta_3} + l_3 m_2 \tilde{r}_2 e^{i\tilde{\theta}_2} = 0 \tag{7.18}$$

where $\tilde{r}_2 = \sqrt{\left(r_2 \cos\theta_2 + m_{23} l_2/m_2\right)^2 + \left(r_2 \sin\theta_2\right)^2}$

$$\tilde{\theta}_2 = \tan^{-1}\left(\frac{r_2 \sin\theta_2}{r_2 \cos\theta_2 + m_{23} l_2/m_2}\right)$$

Likewise, Eq. (7.16) can be rewritten as

$$l_2 m_4 r_4 e^{i\theta_4} + l_4 m_2 \widehat{r}_2 e^{i\widehat{\theta}_2} = 0 \tag{7.19}$$

where $\widehat{r}_2 = \sqrt{\left(r_2 \cos\theta_2 + (m_3 + m_{23} + m_{34})\, l_2/m_2\right)^2 + \left(r_2 \sin\theta_2\right)^2}$

$$\widehat{\theta}_2 = \tan^{-1}\left(\frac{r_2 \sin\theta_2}{r_2 \cos\theta_2 + (m_3 + m_{23} + m_{34})\, l_2/m_2}\right)$$

Eqs. (7.17) to (7.19) are the new force balancing equations and can be rearranged as

$$l_2 m_1 r_1 = l_1 m_2 \tilde{r}_2' \quad \text{and} \quad \theta_1 = \tilde{\theta}_2' \tag{7.20}$$

$$l_2 m_3 r_3 = l_3 m_2 \tilde{r}_2 \quad \text{and} \quad \theta_3 = \tilde{\theta}_2 + \pi \tag{7.21}$$

$$l_2 m_4 r_4 = l_4 m_2 \widehat{r}_2 \quad \text{and} \quad \theta_4 = \widehat{\theta}_2 + \pi \tag{7.22}$$

Compared with the original force balancing Eqs. (7.10)–(7.12), it can be seen that the effect due to the sliding blocks is reflected by the augmented parameters, i.e., $(\tilde{r}_2, \tilde{r}_2', \widehat{r}_2)$ and $(\tilde{\theta}_2, \tilde{\theta}_2', \widehat{\theta}_2)$, which are only related to link 2, the masses of the sliding blocks, and link 3. The force balancing condition equations, i.e., Eqs. (7.20)–(7.22), imply that the mass distribution of a force balanced mechanism should satisfy Eqs. (7.20)–(7.22). It is noted that the force balancing condition equations above are derived by following principle (1), and therefore, they are applicable to the AKP method.

7.4 The Extended AKP Method

Examining the original and the new force balancing equations shown in Eqs. (7.10) to (7.12) and Eqs. (7.20) to (7.22), respectively, it is clear that these equations can also be satisfied by changing the kinematic parameters, l_i ($i = 1, 2, 3, 4$), whilst maintaining the total mass of a mechanism unchanged. It should be noted that when l_i is varied, the parameters r_i and θ_i will be changed accordingly. Both the original and the extended AKP method are developed based on this observation. In particular, the derivation of the extended AKP method based on the new force balancing equations is given as follows [9]. A general design case will be presented first, followed by a special design case.

In the general design situation, assume that the mass centers of the links are arbitrarily distributed, with $\theta_i \neq 0$ for $i = 1$ to 4, as shown in Fig. 7.5. Let l_i^0 and (r_i^0, θ_i^0) represent the length and the mass center of link i, respectively, where superscript "0" indicates the parameters prior to the adjustment of pivot o_i^0. It is observed that in order to satisfy the force balancing condition equations specified in Eqs. (7.20) to (7.22), adjusting of only one pivot on a link is not sufficient; instead, two pivots must be adjusted. The extended AKP method is thus accomplished in two steps. The first step is to adjust a pivot from o_i^0 to o_i so that the angle relationship between link i and link $i + 1$ can be satisfied. The second step is to adjust o_{i+1}^0 to o_{i+1} so that all Eqs. (7.20) to (7.22) can be satisfied. Now, let l_i and (r_i, θ_i) represent the new length and the new mass center of link i, respectively, and let v_i and w_i represent the adjusted amounts of the two pivots o_i and o_{i+1}, respectively. From Fig. 7.5, the following equations can be obtained:

$$l_i = l_i^0 - v_i - w_i \tag{7.23}$$

$$r_i^0 \cos\theta_i^0 = v_i + r_i \cos\theta_i \tag{7.24}$$

$$r_i^0 \sin\theta_i^0 = r_i \sin\theta_i \tag{7.25}$$

for $i = 1, 2, 3, 4$.

Equation (7.25) can be rewritten as

$$r_i = \frac{r_i^0 \sin\theta_i^0}{\sin\theta_i} \tag{7.26}$$

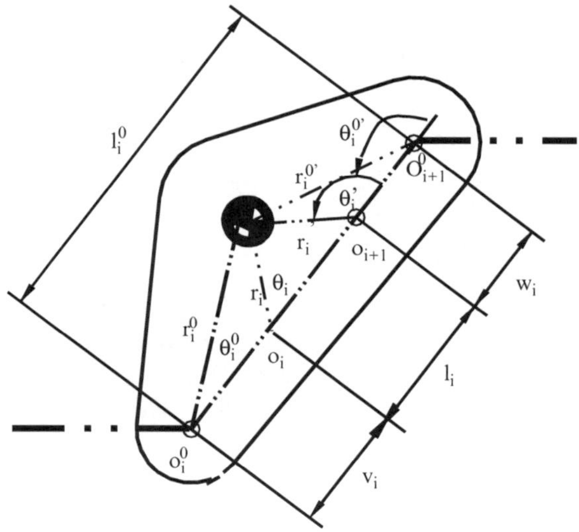

Fig. 7.5 Two-step kinematic parameter adjustment in the extended AKP method

Substituting Eq. (7.26) into Eq. (7.24) yields

$$v_i = r_i^0 \sin\left(\theta_i - \theta_i^0\right) / \sin\theta_i \tag{7.27}$$

Based on the force balancing condition equations given in Eqs. (7.20) to (7.22) and the results given in Eqs. (7.23) to (7.27), the implementation of the extended AKP method can be illustrated using the following example. Referring to Fig. 7.4, assume that the pivots of link 2 are unchanged, and the force balancing conditions can be achieved by adjusting the pivots of the other three links. The detailed procedure is illustrated below.

For link 1:

Pivot O_1 on link 1 is adjusted using the following equation derived from Eqs. (7.27) and (7.20):

$$v_1 = r_1^0 \sin\left(\tilde{\theta}_2' - \theta_1^0\right) / \sin\tilde{\theta}_2' \tag{7.28}$$

Pivot O_2 on link 1 is adjusted using Eq. (7.20). To determine the amount of adjustment for pivot O_2 on link 1, i.e., w_1, substituting Eq. (7.28) and Eqs. (7.23) to (7.25) into Eq. (7.20) yields

$$w_1 = l_1^0 - \frac{r_1^0\left(m_2\tilde{r}_2' \sin\left(\tilde{\theta}_2' - \theta_1^0\right) + m_1 l_2 \sin\theta_1^0\right)}{m_2\tilde{r}_2' \sin\tilde{\theta}_2'} \tag{7.29}$$

For link 3:
Substituting Eq. (7.21) into Eq. (7.27) yields

$$v_3 = r_3^0 \sin\left(\tilde{\theta}_2 - \theta_3^0\right) / \sin\tilde{\theta}_2 \tag{7.30}$$

Substituting Eq. (7.30) and Eqs. (7.23) to (7.25) into Eq. (7.21) yields

$$w_3 = l_3^0 - \frac{r_3^0\left(m_2\tilde{r}_2 \sin\left(\tilde{\theta}_2 - \theta_3^0\right) - m_3 l_2 \sin\theta_3^0\right)}{m_2\tilde{r}_2 \sin\tilde{\theta}_2} \tag{7.31}$$

For link 4:
Substituting Eq. (7.22) into Eq. (7.27) yields

$$v_4 = r_4^0 \sin\left(\widehat{\theta}_2 - \theta_4^0\right) / \sin\widehat{\theta}_2 \tag{7.32}$$

Substituting Eq. (7.32) and Eqs. (7.23) to (7.25) into Eq. (7.22) yields

$$w_4 = l_4^0 - \frac{r_4^0\left(m_2\widehat{r}_2 \sin\left(\widehat{\theta}_2 - \theta_4^0\right) - m_4 l_2 \sin\theta_4^0\right)}{m_2\widehat{r}_2 \sin\widehat{\theta}_2} \tag{7.33}$$

For the special design case where mass distributions of all the links are in line with their kinematic axes, adjusting of only one pivot for each link is sufficient. In this case, since conditions $\theta_i^0 = 0$ and $\theta_i = 0$ hold, Eq. (7.25) thus becomes meaningless. Furthermore, if the sliding blocks are not considered in the implementation, the force balancing equations given in Eqs. (7.23) to (7.25) are simplified as the original ones given in Eqs. (7.10) to (7.12). Substituting Eqs. (7.23) and (7.24) into Eqs. (7.10) to (7.12), the adjusting amounts for links 1, 3, and 4, respectively, can be obtained as follows:

$$v_1 = \left(m_2 r_2' l_1^0 \cos\theta_1 - m_1 r_1^0 l_2 \cos\theta_1^0\right) / \left(m_2 r_2' \cos\theta_1 - m_1 l_2\right) \tag{7.34}$$

$$v_3 = \left(m_2 r_2 l_3^0 \cos\theta_3 - m_3 r_3^0 l_2 \cos\theta_3^0\right) / \left(m_2 r_2 \cos\theta_3 - m_3 l_2\right) \tag{7.35}$$

$$v_4 = \left(k l_4^0 \cos\theta_4 - m_4 r_4^0 \cos\theta_4^0\right) / \left(k \cos\theta_4 - m_4\right) \tag{7.36}$$

with $k = \sqrt{(m_2 r_2/l_2)^2 + m_3^2 + (2 m_2 r_2 m_3 \cos\theta_2/l_2)}$.

Moreover, following the above discussion, if link 3 is selected to be unchanged, the adjusting amounts for other three links can be determined as follows:

$$v_1 = \left(h l_1^0 \cos\theta_1 - m_1 r_1^0 \cos\theta_1^0\right) / \left(h \cos\theta_1 - m_1\right) \tag{7.37}$$

$$v_2 = \left(m_3 r_3 l_2^0 \cos\theta_2 - m_2 r_2^0 l_3 \cos\theta_2^0\right) / \left(m_3 r_3 \cos\theta_2 - m_2 l_3\right) \tag{7.38}$$

$$v_4 = \left(m_3 r_3' l_4^0 \cos\theta_4 - m_4 r_4^0 l_3 \cos\theta_4^0\right) / \left(m_3 r_3' \cos\theta_4 - m_4 l_3\right) \tag{7.39}$$

with $h = \sqrt{(m_3 r_3/l_3)^2 + m_2^2 + (2 m_3 r_3 m_2 \cos\theta_3/l_3)}$.

When designing an RTC mechanism, for a given task, i.e., a set of trajectory points that the end effector should follow, through the inverse kinematics, the corresponding joint angles are readily determined. However, if the extended AKP method is adopted for force balancing design, the kinematic parameters of the mechanism will vary. This will in turn change the inverse kinematics of the mechanism. In order to enable the end effector of the mechanism to follow the same trajectory, the motion profiles of the joints must be adjusted according to the new inverse kinematics. This adjustment is only possible through the implementation of programmable actuators, i.e., RTC actuators. Therefore, the application of the extended AKP method is only limited to RTC mechanisms.

To verify the extended AKP method, a parallel robotic mechanism prototype was built using LEGO blocks, as shown in Fig. 7.6. Figure 7.6a shows the unbalanced system. After applying the extended AKP method, the kinematic parameters were changed and the shaking force of the mechanism was cancelled. Figure 7.6b, c illustrates two configurations of the force balanced mechanism. The mechanism was stabilized at these two positions, and in fact, it was stable at any other positions as well. That is, the mechanism is force balanced.

7.5 Comparison of the Extended AKP Method with the CW Method: Joint Reaction Force

Two examples are used here to verify the effectiveness of the extended AKP method. Comparison is made between the extended AKP method and the CW method in terms of reduction of the joint forces.

Example 1 The kinematic parameters of the original mechanism without considering force balancing are listed in Table 7.1. Using the CW method and the extended AKP method, the kinematic parameters of the force balanced mechanism are computed and listed in the same table. In the CW case, assume that link 3 is unchanged and the other three links are subject to additional masses. In the extended AKP case, assume that all the movable links are subject to pivot adjustments.

The mechanism is supposed to fulfill the following task: the end effector is requested to move from point A (0.3, 0.2) to point B (0.2, 0.3) within 1 s, and subsequently to point C (0.1, 0.2) within 2 s. The unit of the coordinates is meter for all examples. Furthermore, for each segment of the trajectories, (1) the velocities of the end effector at these three points are zero, and (2) the accelerations of the end

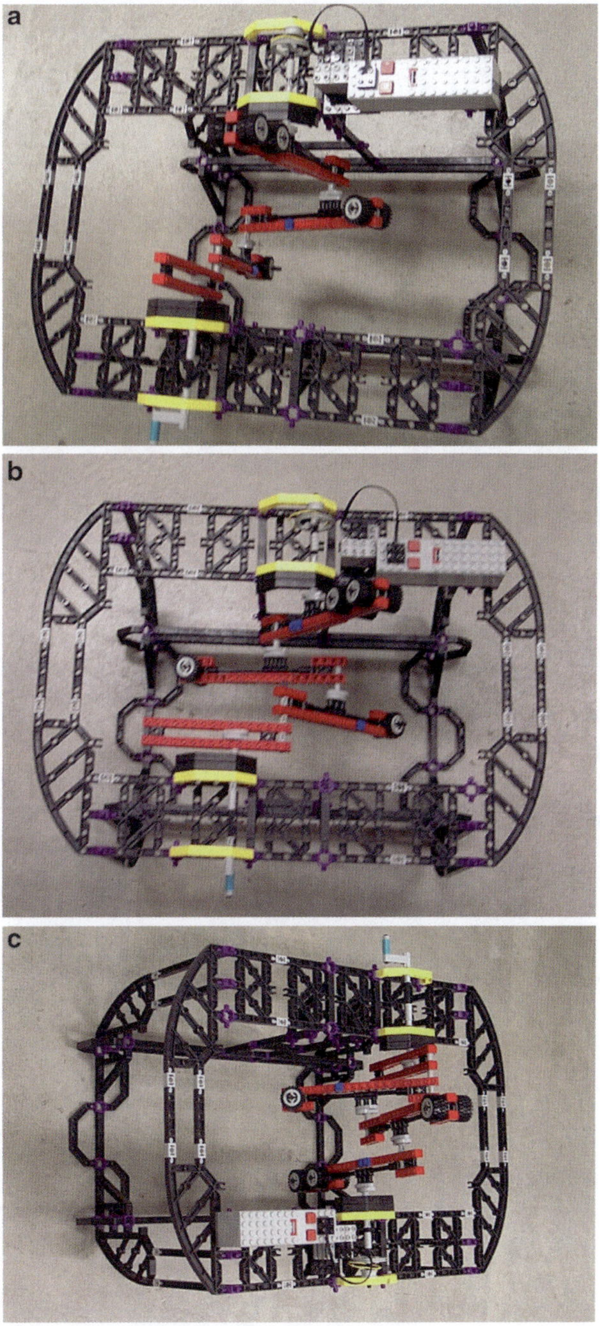

Fig. 7.6 (**a**) Unbalanced mechanism. (**b**) Balanced mechanism using the extended AKP method (1). (**c**) Balanced mechanism using the extended AKP method (2)

Table 7.1 Parameters for different mechanisms (in-line case 1)

Parameters	Unbalanced linkage	CW linkage	AKP linkage
l_1 (m)	0.2	0.2	0.0977
l_2 (m)	0.3	0.3	0.28
l_3 (m)	0.4	0.4	0.2445
l_4 (m)	0.3	0.3	0.1009
l_5 (m)	0.3	0.3	0.3
r_1 (m)	0.05	0.1758	0.0523
r_2 (m)	0.15	0.0506	0.13
r_3 (m)	0.08	0.08	0.0756
r_4 (m)	0.1	0.103	0.0991
m_1 (kg)	0.25	0.5912	0.25
m_2 (kg)	0.25	0.4445	0.25
m_3 (kg)	0.375	0.375	0.375
m_4 (kg)	0.5	0.8742	0.5
$I_1\left(\mathrm{kg\cdot m^2}\right)$	0.004	0.0133	0.004
$I_2\left(\mathrm{kg\cdot m^2}\right)$	0.01	0.0229	0.01
$I_3\left(\mathrm{kg\cdot m^2}\right)$	0.02	0.02	0.02
$I_4\left(\mathrm{kg\cdot m^2}\right)$	0.02	0.0475	0.02
θ_1(rad)	0	π	π
θ_2(rad)	0	π	0
θ_3(rad)	0	0	π
θ_4(rad)	0	π	π

effector at the initial and final points are zero. The trajectories at the two actuators can be determined based on the inverse kinematic and the motion planning method that will be described in Chap. 4. In particular, the trajectories at the two actuators for the unbalanced mechanism and the force balanced mechanism using the CW method are, respectively, expressed as follows:

For actuator 1:

$$q_1(t) = 90.0 + 187.2701\times t^3 - 261.4406\times t^4 + 96.7905\times t^5 \quad \text{if } t \le 1$$

$$\begin{aligned} q_1(t) = {} & 112.6199 - 38.9289\times(t-1)^2 + 109.4122\times(t-1)^3 \\ & - 67.4608\times(t-1)^4 + 12.5189\times(t-1)^5 \quad \text{if} \quad 1 < t \le 3 \end{aligned}$$

For actuator 2:

$$q_4(t) = -13.688 + 657.9647\times t^3 - 894.1248\times t^4 + 320.5211\times t^5 \quad \text{if } t \le 1$$

$$\begin{aligned} q_4(t) = {} & 27.5324 - 87.5649\times(t-1)^2 + 157.4044\times(t-1)^3 \\ & - 85.2165\times(t-1)^4 + 14.8542\times(t-1)^5 \quad \text{if} \quad 1 < t \le 3 \end{aligned}$$

As mentioned in Sect. 4, the trajectory of the force balanced mechanism using the extended AKP method must be re-planned due to the change of the kinematic parameters. After re-planning, the expressions of the new trajectories at the two actuators are as follows:

For actuator 1:

$$q_1(t) = 63.4541 + 201.7492 \times t^3 - 290.3993 \times t^4 + 111.27 \times t^5 \quad \text{if } t \leq 1$$

$$q_1(t) = 86.0739 - 24.4488 \times (t-1)^2 + 152.8515 \times (t-1)^3 \\ - 105.4703 \times (t-1)^4 + 20.4828 \times (t-1)^5 \quad \text{if} \quad 1 < t \leq 3$$

For actuator 2:

$$q_4(t) = -13.688 + 657.9647 \times t^3 - 894.1248 \times t^4 + 320.5211 \times t^5 \qquad \text{if } t \leq 1$$

$$q_4(t) = 70.6729 - 185.6443 \times (t-1)^2 + 286.676 \times (t-1)^3 \\ - 145.3904 \times (t-1)^4 + 24.437 \times (t-1)^5 \quad \text{if} \quad 1 < t \leq 3$$

Using the software called "Working Model 2D" (Knowledge Revolution, 1999), the joint forces in the five pivots can be calculated. Figure 7.7 shows the joint forces in the two actuators.

It can be seen from Fig. 7.7 that, when the mechanism runs at low speeds (about 5 rpm), the forces in the two actuators for the force balanced mechanism do not change very much, and the extended AKP method insures smaller forces at both the x-direction and the y-direction than the CW method. Furthermore, the joint forces at the y-direction using the CW method are significantly larger than those of the extended AKP method and the unbalanced mechanism. This phenomenon agrees with one of the weaknesses associated with the CW method; that is, joint forces will be increased. It is interesting to observe that the variations of joint forces using both the CW and the extended AKP methods are small, while the joint forces of the unbalanced mechanism vary considerably, especially in the x-direction. The reason for this phenomenon can be explained as follows. After the shaking force is balanced, the mass center of the system is stationary during operation. Furthermore, the inertia term is small at low speeds. Therefore, the variation of the forces in the two actuators is small. On the other hand, for the unbalanced mechanism, the mass center of the mechanism changes with the system configurations. Therefore, variation of the forces in the two actuators is large.

It should be noted that, for the force balanced mechanism, the total forces at the two actuators should be zero in the x-direction and should be equal to the total weight of the mechanism in the y-direction. This observation is confirmed with the results shown in Fig. 7.7.

Figure 7.8 shows the results when the mechanism runs at high speeds (about 50 rpm). Performance totally different from the low-speed motion is observed.

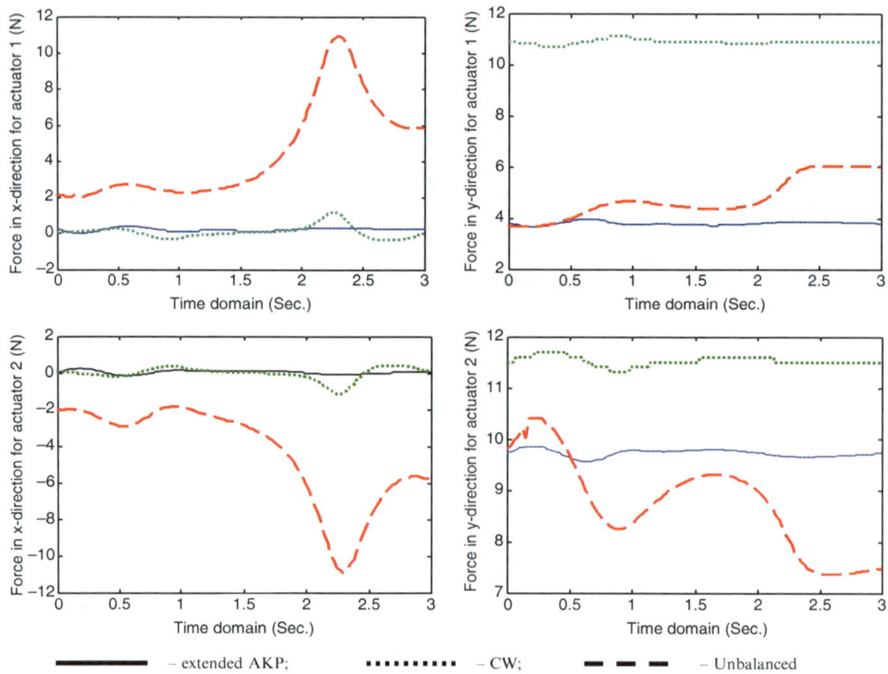

Fig. 7.7 Joint forces in the two actuators at low speeds

First, variations of the forces are very large for all three cases. Second, the CW method generates the worst performance in the x-direction; although the total forces in the two actuators in the x-direction are still maintained at zero, the sharp variation of forces exhibits. Nevertheless, the extended AKP method remains to produce the best performance.

The results above are expected. When the mechanism runs at high speeds, the inertia forces becomes the dominant term in the dynamics. Since the CW method adopts the adding mass approach for force balancing purpose, its inertia force takes more weight. While with the extended AKP method, the total mass of the system is unchanged; therefore the inertia force does not differ significantly from the unbalanced mechanism.

To further illustrate the effectiveness of the extended AKP method, Table 7.2 lists the minimum and maximum joint forces in the x-direction and the y-direction for the cases of applying the AKP method and the CW method, respectively. It is observed that the joint forces generated by using the extended AKP method have a smaller variation range than those using the CW method. The reduction of the forces in the x-direction is more remarkable by using the AKP method than by using the CW method.

Example 2 In this example, a mechanism with the mass center off-line of the kinematic axis is studied. The kinematic parameters of the original mechanism

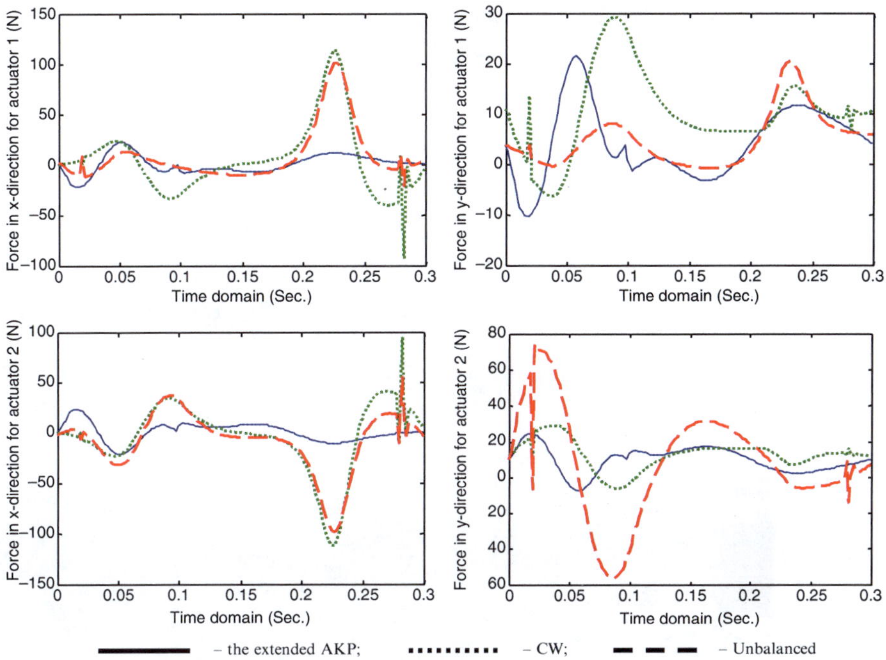

Fig. 7.8 Joint forces in the two actuators at high speeds

Table 7.2 Joint forces (min and max) using different force balancing methods

		Joint forces in x-direction				Joint forces in y-direction			
		AKP method		CW method		AKP method		CW method	
Speeds	Pivot no.	Min	Max	Min	Max	Min	Max	Min	Max
Low speeds	1	−0.205	0.204	−0.405	1.130	3.640	3.930	10.70	11.10
	2	−0.208	0.228	−1.210	0.338	−1.470	−1.200	−5.350	−4.870
	3	−0.233	0.164	−1.130	0.387	0.930	1.320	−0.869	−0.595
	4	−0.514	0.444	−1.140	0.487	4.320	5.300	2.430	3.420
	5	−0.204	0.205	−1.130	0.405	9.560	9.840	11.30	11.70
High speeds	1	−22.80	21.60	−93.50	113.0	−10.50	21.40	−6.40	29.20
	2	−21.30	25.00	−121.0	81.00	−18.10	11.80	−31.10	17.30
	3	−24.20	18.30	−113.0	89.40	−21.90	22.30	−14.10	13.30
	4	−56.90	42.10	−114.0	110.0	−50.10	60.70	−48.30	51.90
	5	−21.60	22.80	−113.0	93.50	−7.870	23.90	−6.770	28.80

without force balancing and the modified mechanisms using the CW method and the extended AKP method are listed in Table 7.3, respectively. In the redesign of the mechanism, link 2 is assumed to be unchanged.

In this example, the mechanism is supposed to fulfill the following task: the end effector is requested to move from point A (0.3, 0.25) to point C (0.1, 0.2)

Table 7.3 Parameters for different mechanisms (off-line)

Parameters	Unbalanced linkage	CW linkage	AKP linkage
l_1 (m)	0.15	0.15	0.0866
l_2 (m)	0.26	0.26	0.26
l_3 (m)	0.26	0.26	0.2078
l_4 (m)	0.14	0.14	0.08838
l_5 (m)	0.30	0.30	0.30
r_1 (m)	0.075	0.0866	0.10
r_2 (m)	0.15	0.15	0.15
r_3 (m)	0.08485	0.10	0.12
r_4 (m)	0.115	0.1415	0.135
m_1 (kg)	1	2	1
m_2 (kg)	2	2	2
m_3 (kg)	2	3	2
m_4 (kg)	2	4	2
$I_1\ (\text{kg}\cdot\text{m}^2)$	0.01	0.03	0.01
$I_2\ (\text{kg}\cdot\text{m}^2)$	0.05	0.05	0.05
$I_3\ (\text{kg}\cdot\text{m}^2)$	0.04	0.06	0.04
$I_4\ (\text{kg}\cdot\text{m}^2)$	0.02	0.04	0.02
θ_1(deg)	90	150	150
θ_2(deg)	30	30	30
θ_3(deg)	225	210	210
θ_4(deg)	192.83	169.11	188.21

and intermediate point B (0.2, 0.2). The time duration between two neighboring points is 2 s at low speeds (about 5 rpm) and 0.1 s at high speeds (about 100 rpm). Furthermore, for each segment of the trajectories, (1) the velocity of the end effector at the intermediate point B is determined by the method that will be discussed in Chap. 4, and (2) the accelerations of the end effector at the initial and final tracking points are zero.

Figures 7.9 and 7.10 show the total forces in the two actuators for the unbalanced mechanism and the balanced mechanism using the extended AKP method and the CW method at low speeds and high speeds, respectively. From these two figures, it can be seen that the extended AKP method is the best in terms of the reduction of forces at both low speeds and high speeds, and the CW method is the worst. So it is demonstrated that the extended AKP method is better than the CW method in the reduction of the joint forces for an off-line mechanism.

From the force profiles shown in these figures, it can be seen that, in order to accomplish the same motion task, the extended AKP method needs the least amount of forces at both low speeds and high speeds among all the three design cases. The CW method, however, demands the highest forces. The extended AKP method is thus demonstrated to be better than the CW method in terms of the joint force reductions.

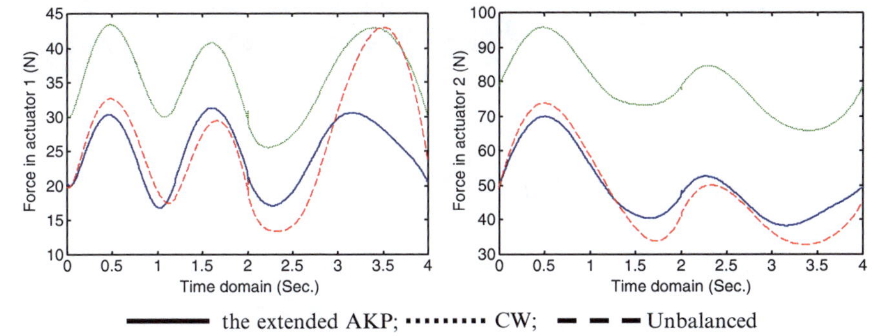

Fig. 7.9 Total forces in the two actuators at low speeds

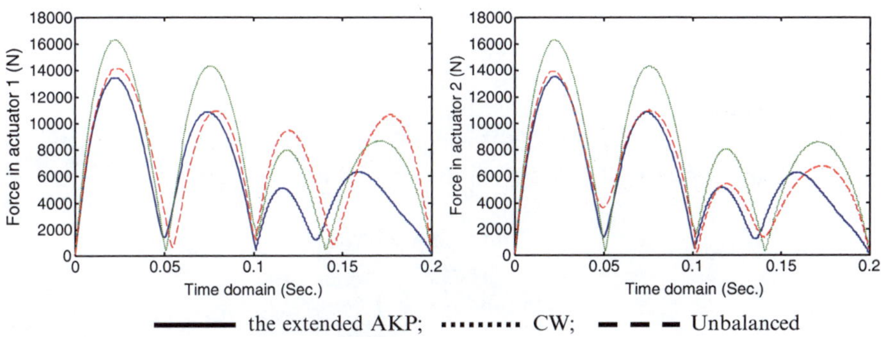

Fig. 7.10 Total forces in the two actuators at high speeds

However, the adjustment of the kinematic parameters may not always work in some applications. One case of the failure with the AKP approach is such that the trajectory may be out of the new workspace which results from the change of the kinematic parameters of the mechanism [9]. The other case is that a required change in the kinematics parameter may be out of the physical region. Another shortcoming with AKP is an increase in complexity of the mechanism structure. In the next section, a new solution will be introduced, that is, the hybridization of AKP and CW method.

7.6 Hybrid AKP+CW Method for Force Balancing

To overcome the limitations of both the CW method and the AKP method, we propose a force balancing strategy which combines these two methods in the pursuit of taking advantages of both the methods while alleviating their shortcomings. The generalized one of this strategy is called "hybridization engineering" [10]. Through such a hybrid approach, it is possible that a good performance and a limited workspace modification of the designed RTC mechanism can be achieved. The same configuration shown in Fig. 7.4 will be considered as an example for the implementation of the hybrid AKP+CW method [11].

7.6.1 *Hybrid AKP+CW Method*

In the following discussion, we also consider that link 2 in Fig. 7.4 is unchanged without the loss of generality. We use the following notation for the particular schemes of the CW+AKP method: "1CW+3/4AKP" means CW is applied to link 1, and AKP is applied to link 3 and link 4, respectively. We consider two particular hybrid schemes of the AKP+CW method and present their design equations accordingly.

7.6.1.1 Hybrid 1: 1AKP+3/4CW Method

For this scheme, the AKP method will be applied to link 1 and the CW method applied to links 3 and 4, respectively.

For link 1, following the AKP approach, we have

$$v_1 = r_1^0 \sin\left(\theta_2' - \theta_1^0\right) / \sin\theta_2' \tag{7.40}$$

$$w_1 = l_1^0 - \frac{r_1^0\left(m_2 r_2' \sin\left(\theta_2' - \theta_1^0\right) + m_1 l_2 \sin\theta_1^0\right)}{m_2 r_2' \sin\theta_2'} \tag{7.41}$$

For link 3, by using the CW method from Eq. (7.13), we get

$$\theta_3 = \theta_2 + \pi \tag{7.42}$$

$$m_3 r_3 = l_3 m_2 r_2 / l_2 \tag{7.43}$$

$$m_3{}^* r_3{}^* e^{i\theta_3^*} = m_3 r_3 e^{i\theta_3} - m_3{}^0 r_3{}^0 e^{i\theta_3^0} \tag{7.44}$$

For link 4, following the same procedure as for link 3, we can obtain

$$r_3{}' e^{i\theta_3'} = r_3 e^{i\theta_3} - l_3 \tag{7.45}$$

$$m_4 r_4 = l_4 m_3 r_3{}' / l_3 \tag{7.46}$$

$$\theta_4 = \theta_3' \tag{7.47}$$

$$m_4{}^*r_4{}^*\mathrm{e}^{i\theta_4^*} = m_4r_4\mathrm{e}^{i\theta_4} - m_4{}^0r_4{}^0\mathrm{e}^{i\theta_4^0} \tag{7.48}$$

7.6.1.2 Hybrid 2: 1CW+3/4AKP Method

Following the same procedure as mentioned above, we can obtain the design equations.

For link 1,

$$\theta_1 = \theta_2' \tag{7.49}$$

$$m_1r_1 = l_1m_2r_2'/l_2 \tag{7.50}$$

$$m_1{}^*r_1{}^*\mathrm{e}^{i\theta_1^*} = m_1r_1\mathrm{e}^{i\theta_1} - m_1{}^0r_1{}^0\mathrm{e}^{i\theta_1^0} \tag{7.51}$$

For link 3,

$$v_3 = r_3^0\sin\left(\theta_2 - \theta_3^0\right)/\sin\theta_2 \tag{7.52}$$

$$w_3 = l_3^0 - \frac{r_3^0\left(m_2r_2\sin\left(\theta_2 - \theta_3^0\right) - m_3l_2\sin\theta_3^0\right)}{m_2r_2\sin\theta_2} \tag{7.53}$$

For link 4,

$$r_3{}'\mathrm{e}^{i\theta_3'} = r_3\mathrm{e}^{i\theta_3} - l_3 \tag{7.54}$$

$$v_4 = r_4^0\sin\left(\theta_3' - \theta_4^0\right)/\sin\theta_3' \tag{7.55}$$

$$w_4 = l_4^0 - \frac{r_4^0\left(m_3r_3'\sin\left(\theta_3' - \theta_4^0\right) - m_4l_3\sin\theta_4^0\right)}{m_3r_3'\sin\theta_3'} \tag{7.56}$$

7.6.2 *Design Examples and Illustrations*

We used ADAMS for calculating all joint forces and torques in the motor. We considered three situations for the motors: *situation 1*: both motors are regular constant velocity (CV motor for short) motors; *situation 2*: both motors are servomotors which have a prescribed trajectory; *situation 3*: one motor is a CV motor and the other is a servomotor.

Table 7.4 gives the detailed designed parameters of the parallel robotic mechanism of Fig. 7.4. The second column is associated with the original design, and the

Table 7.4 Parameters of the designed mechanisms

Parameters	Unbalanced	CW	AKP	Hybrid 1	Hybrid 2
l_1 (m)	0.15	0.15	0.13	0.15	0.13
l_2 (m)	0.26	0.26	0.26	0.26	0.26
l_3 (m)	0.26	0.26	0.208	0.208	0.26
l_4 (m)	0.14	0.14	0.0884	0.0884	0.14
l_5 (m)	0.3	0.3	0.3	0.3	0.3
r_1 (m)	0.075	0.0866	0.15	0.0866	0.15
r_2 (m)	0.15	0.15	0.15	0.15	0.15
r_3 (m)	0.08485	0.1	0.12	0.12	0.1
r_4 (m)	0.115	0.141	0.135	0.135	0.141
m_1 (kg)	1	2	1	2	1
m_2 (kg)	2	2	2	2	2
m_3 (kg)	2	3	2	2	3
m_4 (kg)	2	4	2	2	4
I_1 (kg·m^2)	0.01	0.0328	0.01	0.0328	0.01
I_2 (kg·m^2)	0.05	0.05	0.05	0.05	0.05
I_3 (kg·m^2)	0.04	0.0605	0.04	0.04	0.0605
I_4 (kg·m^2)	0.02	0.077	0.02	0.02	0.077
θ_1 (deg)	90	150	150	150	150
θ_2 (deg)	30	30	30	30	30
θ_3 (deg)	225	210	210	210	210
θ_4 (deg)	192.83	188.19	190.89	190.89	188.19

third is the modified design based on CW, and so forth for the remaining columns. For situation 1, we further considered two cases: low speed ($\pi/6$ rad/s) and high speed ($10\pi/3$ rad/s). The results are reported in the following.

7.6.2.1 Results for Situation 1

In all the figures, F_{X1} and F_{X2} represent the X-axis forces (horizontal here) on two motors, respectively; F_{Y1} and F_{Y2} represent the Y-axis forces (vertical) on two motors, respectively, and T_1 and T_2 represent torques on two motors, respectively.

Figures 7.11 and 7.12 show the fluctuation of the driving forces in two directions and the control torques of the parallel robotic mechanism in two motors at a low constant speed of $\pi/6$ rad/s (i.e., the period is 12 s). In Fig. 7.3, the big driving force fluctuations of the unbalanced mechanism are clearly shown. As listed in Table 7.5, the driving force for the unbalanced mechanism rises from the minimum 1.6187 N at time 0.78 s to the maximum 4.957 N at time 8.06 s, followed by a sharp decline to 1.86 N at the end of the period. At this point, a cycle has been completed and a new one begins. Despite far less significant changes, similar trends are seen in all

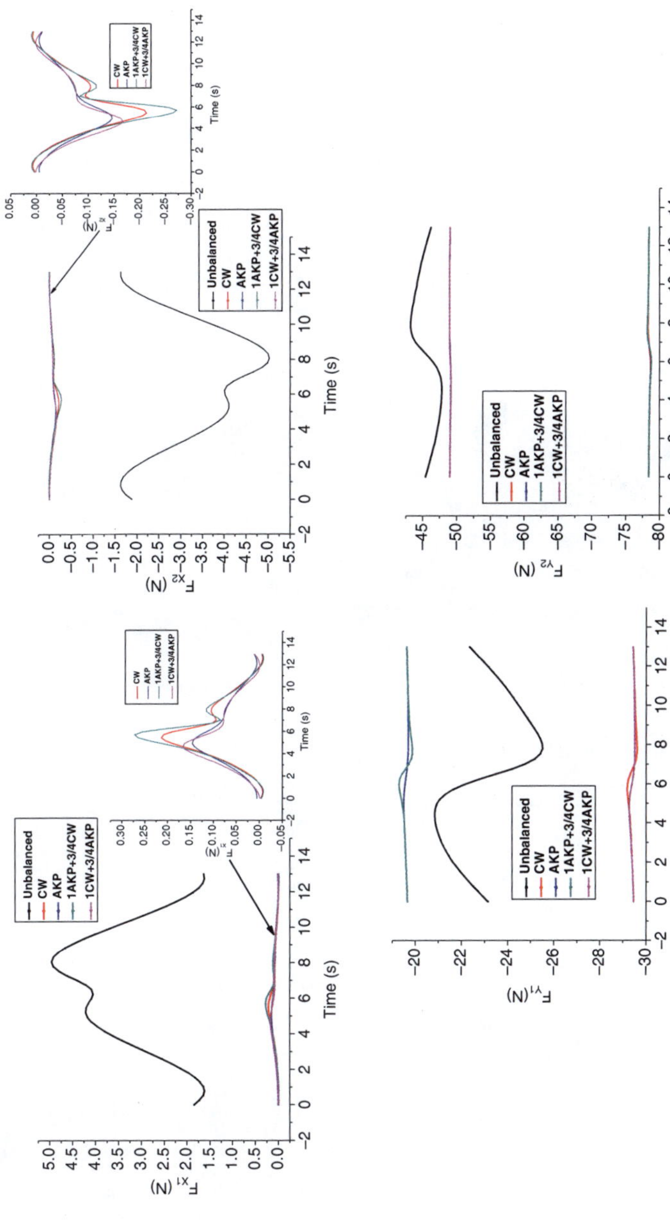

Fig. 7.11 Driving forces for two motors at low speed

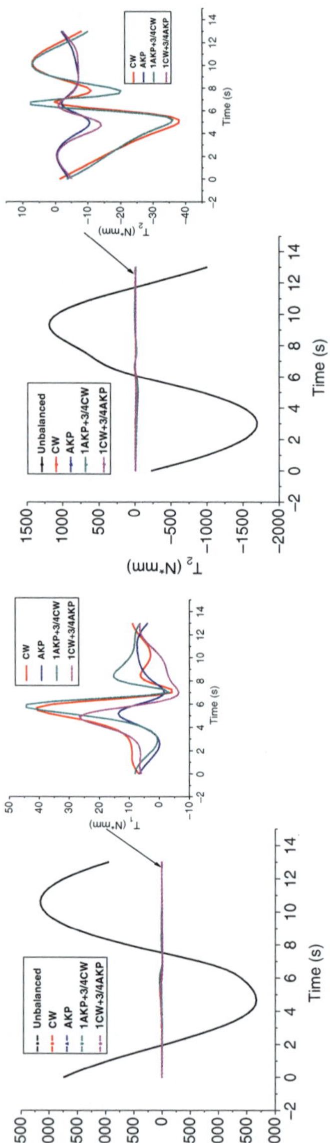

Fig. 7.12 Control torques for two motors at low speed

Table 7.5 Minimum and maximum driving forces of motors at low speed

Force	Range	Unbalanced	CW	AKP	Hybrid 1	Hybrid 2
F_{X1}	Min (N)	1.619	−0.007	0.005	−0.009	−0.001
	Max (N)	4.957	0.214	0.146	0.272	0.167
F_{X2}	Min (N)	−5.030	−0.214	−0.146	−0.272	−0.167
	Max (N)	−1.625	0.007	−0.005	0.009	0.001
F_{Y1}	Min (N)	−25.515	−29.611	−19.699	−19.877	−29.512
	Max (N)	−20.853	−29.186	−19.501	−19.293	−29.275
F_{Y2}	Min (N)	−47.817	−78.687	−49.145	−78.774	−49.178
	Max (N)	−43.219	−78.262	−48.947	−78.190	−48.941

Table 7.6 Minimum and maximum control torques of motors (N mm) at low speed

Force	Range	Unbalanced	CW	AKP	Hybrid 1	Hybrid 2
T_1	Min	−1666.969	−4.135	−1.627	−3.203	−6.373
	Max	2156.143	40.821	13.838	44.473	26.539
T_2	Min	−1693.543	−38.028	−10.568	−36.108	−14.020
	Max	1190.208	7.094	−0.423	7.612	−0.644

the other balanced mechanisms, rising from a minimum to a maximum, followed by a sharp decline at the end of the period. Table 7.5 summarizes the minimum and maximum values of the forces for the unbalanced mechanism and the four balanced mechanisms with different approaches at a low speed.

In general, at low-speed simulation, the unbalanced mechanism fluctuates further more than the balanced mechanisms in terms of force and torque, as shown in Table 7.6. Among the balanced mechanisms, the best performances go to AKP and hybrid 2 (1CW+3/4AKP).

Figures 7.13 and 7.14 describe the fluctuation of the driving forces in the X direction and control torques in two motors at a constant high speed of $10\pi/3$ rad/s. Tables 7.7 and 7.8 list the minimum and maximum driving forces and control torques for two motors, respectively. Comparing Fig. 7.13 with Fig. 7.11, one can see that the distributions of driving forces in the X direction are totally changed when the system operated in a high speed. The differences of the fluctuations of driving forces among all five designs become nonsignificant. The same change trends also appeared for the control torques, as shown in Fig. 7.14 and Table 7.8.

Overall, at high speed, forces and torques considerably differ from those at low speed. The forces nearly increase by 20 times and torques by 100 times. From the simulation results, we can see that the most remarkable phenomenon is that CW and 1AKP+3/4CW are poorer even than the unbalanced mechanism, but still AKP and hybrid 2 (1CW+3/4AKP) are the best.

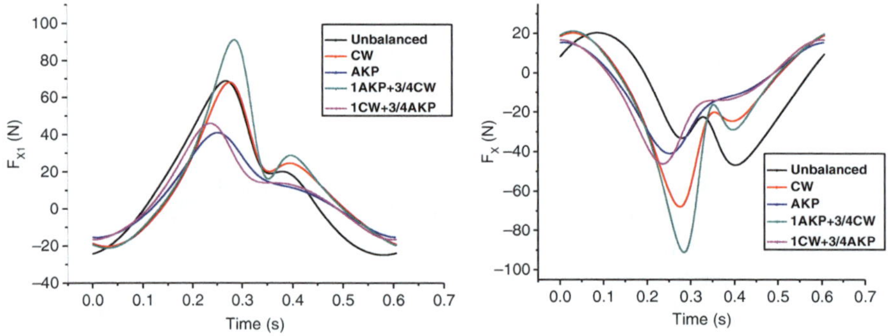

Fig. 7.13 Driving forces for two motors in the *X* direction at high speed

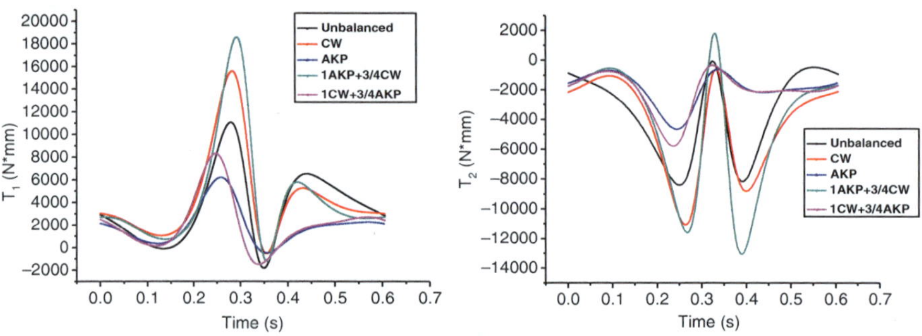

Fig. 7.14 Control torques for two motors at high speed

Table 7.7 Minimum and maximum driving forces of the motors at high speed

Force	Range	Unbalanced	CW	AKP	Hybrid 1	Hybrid 2
F_{X1}	Min (N)	68.803	68.083	40.940	91.156	46.049
	Max (N)	−24.898	−20.328	−15.404	−21.143	−16.676
F_{X2}	Min (N)	−46.863	−67.926	−40.955	−91.127	−46.209
	Max (N)	20.268	20.215	15.400	21.168	16.791
F_{Y1}	Min (N)	−116.555	−102.772	−50.914	−122.878	−62.135
	Max (N)	47.815	66.602	27.916	110.357	28.929
F_{Y2}	Min (N)	−141.116	−174.532	−96.543	−208.603	−107.366
	Max (N)	12.878	−5.260	−17.725	24.701	−16.160

7.6.2.2 Results for Situation 2

For situation 2, the operation duration is 4 s for a full rotation of both motors with varying speeds, and the prescribed trajectories of two motors are defined as follows:

Motor 1:

$$q_{1d} = q_{1d0} + \left(6 \times t^5/4^5 - 15 \times t^4/4^4 + 10 \times t^3/4^3\right) \times 2\pi \quad t \in [0, 4]$$

Table 7.8 Minimum and maximum torques of motors (N mm) at high speed

Force	Range	Unbalanced	CW	AKP	Hybrid 1	Hybrid 2
T_1	Min	−1831.242	−546.258	−472.298	−1215.42	−1485.57
	Max	11066.674	15604.21	6208.644	18605.1	8313.735
T_2	Min	−8406.293	−11073.7	−4660.89	−13051.9	−5791.45
	Max	−71.239	−607.81	−612.595	1807.899	−341.812

Motor 2:

$$q_{2d} = q_{2d0} + \left(6 \times t^5/4^5 - 15 \times t^4/4^4 + 10 \times t^3/4^3\right) \times 2\pi \quad t \in [0, 4]$$

Figure 7.15 shows the fluctuation of the driving forces and control torques in two motors. From this figure, it demonstrated that the AKP and hybrid 2 (1CW+3/4AKP) are the best in terms of low driving forces and less fluctuation of the control torques.

7.6.2.3 Results for Situation 3

For situation 3, the operation duration is 4 s and the trajectories of two motors are different and defined as follows:

Motor 1:

$$q_{1d} = 0.5\pi \times t \quad t \in [0, 4]$$

Motor 2:

$$q_{2d} = \left(6 \times t^5/4^5 - 15 \times t^4/4^4 + 10 \times t^3/4^3\right) \times 2\pi, \quad t \in [0, 4]$$

Figure 7.16 shows the fluctuation of the driving forces and control torques in two motors under different trajectories, respectively.

Based on these three situations where trajectories are given, AKP and hybrid 2 (1CW+3/4AKP) are the best in terms of less fluctuation of driving forces and less control torques, and the unbalanced mechanism is the worst. However, CW and hybrid 1 (1AKP+3/4CW) appear to be no better than the unbalanced mechanism at high-speed situation.

7.7 Conclusions

The extended AKP method is developed in this chapter. One of the important contributions of this extension lies in the new idea on the adjustment of two pivots on each link. With this design method, the original AKP method was extended to

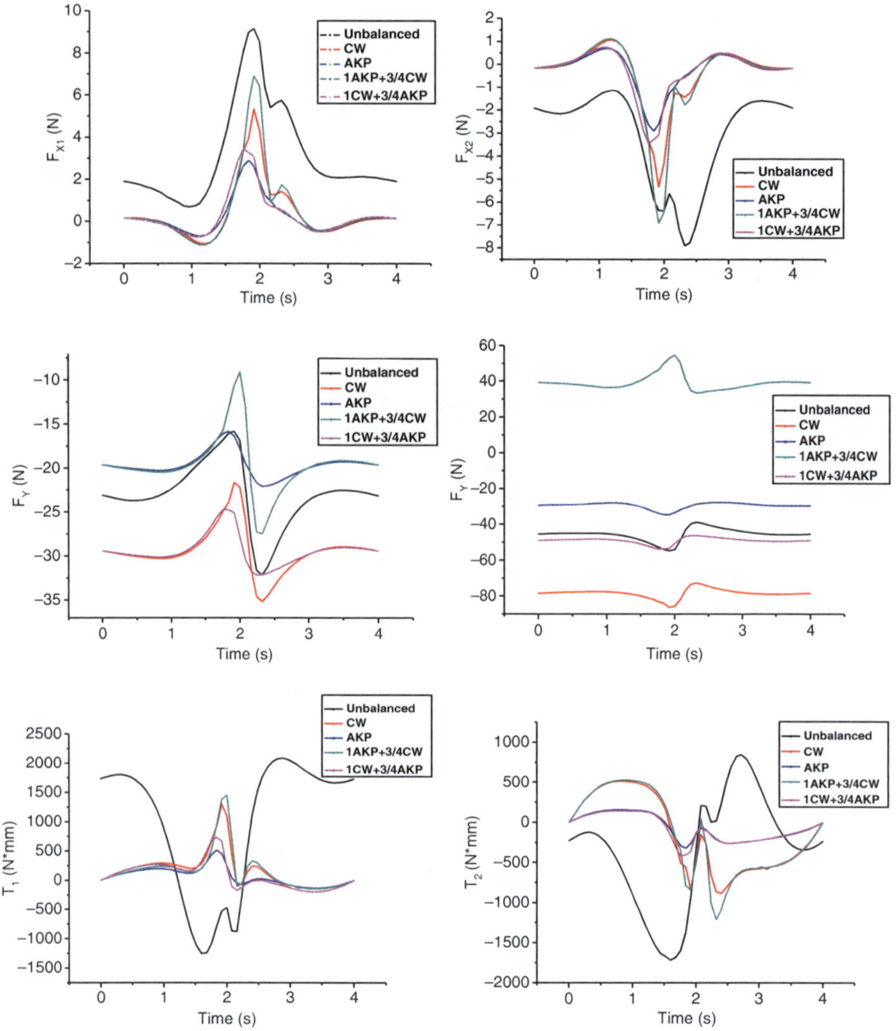

Fig. 7.15 Driving forces and control torques for two motors for the same trajectories

any planar mechanism with "off-line" mass centers. The derived design equations of the extended AKP method are in a general form, from which the special design case with "in-line" mass centers can be readily derived. Two different configurations for three cases: the force unbalanced mechanism, and the force balanced mechanism using the extended AKP method and the CW method, respectively, are studied to demonstrate the effectiveness of the extended AKP method. The joint forces of the individual pivots are calculated at both low speeds and high speeds for these

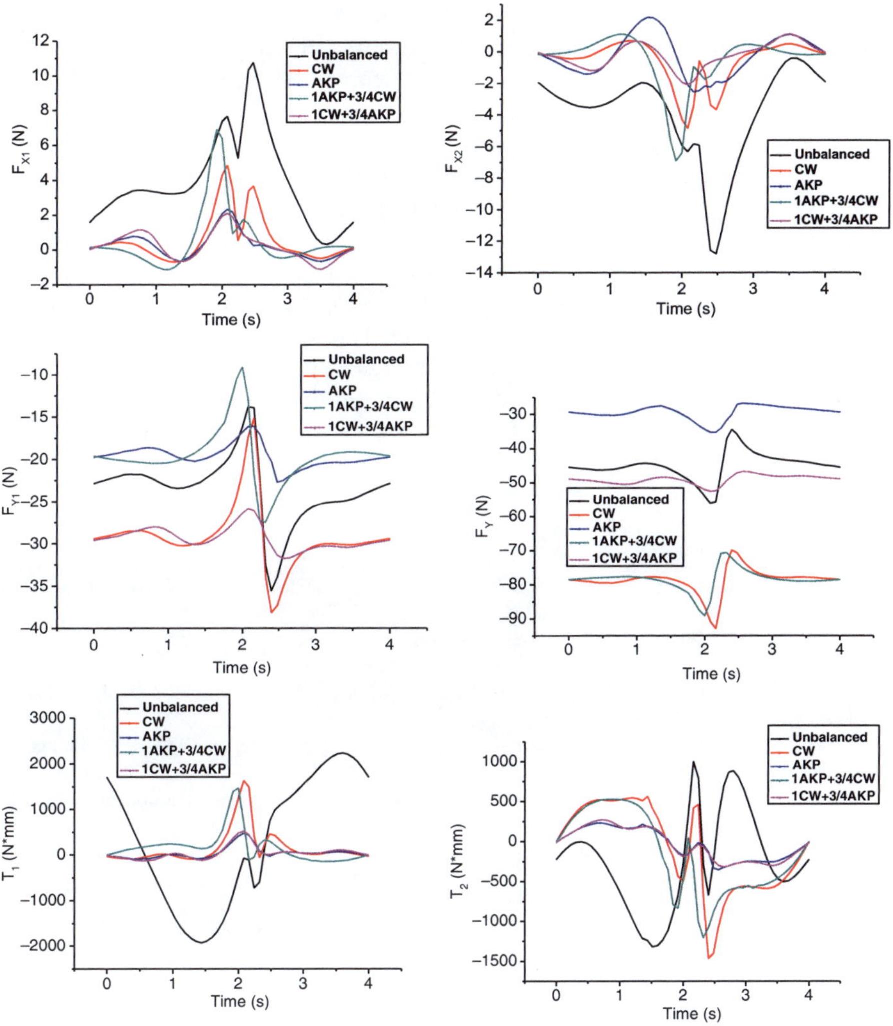

Fig. 7.16 Driving forces and control torques for two motors for different trajectories

configurations. All the results have shown that the extended AKP method is better than the CW method in terms of the joint force reductions and the variation decrease at high speeds.

It should be noticed that the static balancing solution has some demerit when the mechanism runs at high speeds. In this case, the inertia forces becomes the dominant term in the dynamics. Since the CW method adopts the adding mass approach for force balancing purpose, its inertia force takes more weight. While with the extended AKP method, the total mass of the system is unchanged; therefore the inertia force does not differ significantly from the unbalanced mechanism.

In addition, following the hybridization principle, the hybrid AKP+CW method was proposed for the force balancing. Two different hybrid schemes are designed for the parallel robotic mechanism and their performances are compared with AKP, CW, and unbalanced mechanism. Simulation results show that AKP and hybrid 2 (1CW+3/4AKP) can achieve good performance in all situations. As the operating speed increases, CW and hybrid 1 (1AKP+3/4CW) get worse. When the operating speed goes up to $10\pi/3\ \mathrm{s}^{-1}$, CW and hybrid 1 (1AKP+3/4CW) are even worse than the unbalanced mechanism.

It should be noted that, in the force balancing with the CW+AKP approach, certain parameters need to be selected. In this study, the optimal selection of these parameters has not been considered. The optimal selection of design parameters in the CW+AKP approach will be addressed in future. Another work is planned in the future on moment balancing and torque balancing with this hybrid approach.

References

1. Wang, J.G., Gosselin, C.M.: Static balancing of spatial four-degree-of-freedom parallel mechanisms. Mech. Mach. Theory **35**(4), 563–592 (2000)
2. Wang, Z.H.: Mechatronic design to real-time controllable mechanical systems: force balancing and trajectory tracking. M.Sc. thesis, University of Saskatchewan (2000)
3. Xi, F.F., Sinatra, R.: Effect of dynamic balancing on four-bar linkage vibrations". Mech. Mach. Theory **32**(6), 715–728 (1997)
4. Berkof, R.S., Lowen, G.G.: A new method for completely force balancing simple linkages". Trans. ASME J. Eng. Ind. **91**(B), 21–26 (1969)
5. Bagci, C.: Shaking force balancing of planar linkages with force transmission irregularities using balancing idler loops. Mech. Mach. Theory **14**(4), 267–284 (1979)
6. Gosselin, C.M.: Static balancing of spherical 3-DoF parallel mechanisms and manipulators". Int. J. Rob. Res. **18**(8), 819–829 (1999)
7. Ouyang, P.R., Zhang, W.J., Wu, F.X.: Nonlinear PD control for trajectory tracking with consideration of the design for control methodology. IEEE IRAC **4**, 4126–4131 (2002)
8. Ouyang, P.R.: Force balancing design and trajectory tracking control of real-time controllable mechanisms. M.Sc. thesis, University of Saskatchewan (2002)
9. Ouyang, P.R., Zhang, W.J.: Force balancing of robotic mechanisms based on adjustment of kinematic parameters. ASME J. Mech. Des. **127**(3), 433–440 (2005)
10. Zhang W.J., Ouyang, P.R., Gupta, M.M., Sun, Z.H.: A novel hybridization design principle for intelligent mechatronics systems. ICAM 2010, Japan (2010)
11. Huang, J., Ouyang, P.R., Cheng, L., Zhang, W.J.: A Hybrid Approach To Force Balancing Of Robotic Mechanisms. Proceedings of the ASME 2010 International Mechanical Engineering Congress & Exposition (IMECE 2010), pp. 735–744 (2010)

Chapter 8
Balancing of a 3-DOFs Parallel Manipulator

D. Cafolla, G. Carbone, and M. Ceccarelli

Abstract This chapter gives an overview on static and dynamic balancing. Basic approaches are discussed for achieving the design of mechanisms having a fully balanced behavior under different operation conditions. A formulation is proposed to address the effects of balancing on mass distributions and dynamic performance. The proposed formulation is applied for the dynamic balancing of a three DOFs (degrees-of-freedom) spatial parallel manipulator, namely CaPaMan 2bis (Cassino Parallel Manipulator 2bis). This parallel manipulator has three identical legs, where each leg is composed by a four-bar mechanism, an orthogonal revolute joint, and a spherical joint that is attached on the mobile platform. The proposed solution for achieving the balancing of CaPaMan 2bis is based on the use of counter-rotary counterweights. The obtained results are validated by simulations by using a general-purpose software for multi-body dynamics analysis.

Keywords Parametric • Design • Optimization • Parallel manipulator

8.1 Introduction

Inertia forces exist wherever parts having mass are accelerated [1]. A careful attention to inertia forces must be given, since the first design steps. In fact, each moving part must be designed to perform satisfactorily under all combinations of inertia, payloads, and externally acting forces. Inertia forces are also important, since any resulting external or shaking force becomes a disturbing force on the supporting frame and associated parts. In both cases varying forces acting on elastic bodies can produce serious, even destructive, vibrations of the parts or complete machine and adjacent structures and equipments, as outlined for example in [1, 2]. The presence of vibration and the accompanying noise can provide serious problems to other machines and to human operators as discussed for example in [3].

D. Cafolla (✉) • G. Carbone • M. Ceccarelli
LARM: laboratory of Robotics and Mechatronics, DICEM, University of Cassino and South Latium Cassino, Cassino, Italy
e-mail: cafolla@unicas.it; carbone@unicas.it; ceccarelli@unicas.it

D. Zhang, B. Wei (eds.), *Dynamic Balancing of Mechanisms and Synthesizing of Parallel Robots*, DOI 10.1007/978-3-319-17683-3_8

Several applications in many different fields can get significant advantages from balancing. Very well known is, for example, the balancing of rotors or tyres that has been even coded in international standards such as [4, 5]. In the robotic field the balancing of manipulators with large payloads is often considered for the static and/or dynamic balancing of industrial robots with high payloads, since it may significantly reduce the power consumption [6, 7]. Balancing is also important to reduce wear for machine tools or to improve the operation comfort (by reducing vibrations) such as reported for example in [8]. The operation comfort is a key issue in applications such as flight simulators [9] or in any application requiring human-robot interaction, such as the rehabilitation of patients as proposed in [10]. Careful attention to dynamic balancing is usually addressed, for example, when designing and operating single-wheel robots such as in [11] or biped and humanoid robots such as reported in [12].

A general approach for minimization of the magnitude of the inertial shaking forces can be formulated by introducing another shaking force that is equal in magnitude and opposite in direction to the original shaking force. This process is called balancing [1–6]. Based on this concept, several balancing techniques have been developed and they can be found in a very rich literature such as reported in [13–23]. Further investigations are still undergoing for identifying balancing techniques that can better fit to specific applications such as reported in [24–33] or for implementing specific optimal design procedures such as proposed by [34, 35].

In this chapter we propose a formulation to address the effects of balancing on mass distributions and dynamic performance. The proposed formulation can be seen as a tool for designers in selecting the most appropriate solution(s) according to the expected operation conditions. The proposed formulation can be conveniently applied for the balancing of parallel manipulators. A specific case of study has been developed by referring to a three-degrees-of-freedom (3-DOFs) spatial parallel manipulator by designing proper counter-rotary counterweights (CRCW).

8.2 Problems and Requirement for Balancing

8.2.1 Static Balancing

A mechanism is statically balanced if its potential energy is constant for all possible configurations. Static balance of a body occurs when the center of gravity of the object is laying on its axis of rotation and the object can remain stationary without the application of any force [1].

A simple example of static balancing can be made by referring to a disk as shown in the schemes of Fig. 8.1. In particular, Fig. 8.1a shows a disk having an irregular shape. Due to this shape the center of mass of the disk is not laying on the axis of rotation. The disk is not statically balanced and it has a tendency to rotate due to the

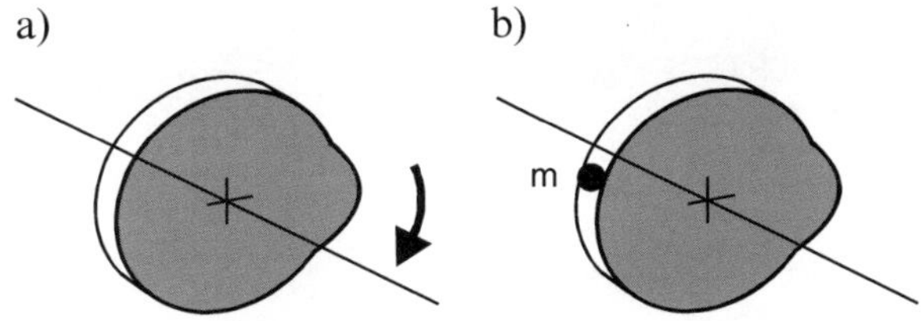

Fig. 8.1 Schemes of a disk: (**a**) The lack of static balancing lets the disk turn to the force of gravity; (**b**) a counterweight mass m is added to balance the disk

force of gravity. Static balancing is usually achieved by using additional mechanical elements like elastic components or counterweight masses, either directly mounted on the links of the mechanism or by using auxiliary components [1, 6, 7].

The static balancing of the disk in Fig. 8.1a can be easily achieved by using a counterweight mass m as shown in Fig. 8.1b. This counterweight mass will move the center of mass to let it coincide with the axis of rotation. Accordingly, the disk will no more have a tendency to rotate about its axis of rotation.

The calculation of the counterweight mass and location is made by considering the centrifugal force F_c of the disk in the form

$$F_c = M\,\omega^2\,r \tag{8.1}$$

where M is the mass of the disk, ω is the angular speed of rotation of the disk, and r is the radial distance of the center of mass from the rotation axis.

One can add a mass m at a radius r_m to the disk so that the resulting centrifugal force will become zero such as shown in the scheme of Fig. 8.1b. Accordingly one can write Eq. (8.1) in the forms

$$F_c = 0 = M\,\omega^2\,r - m\omega^2\,r_m \tag{8.2}$$

$$M\,\omega^2\,r = m\,\omega^2\,r_m \tag{8.3}$$

If the angular speed ω is nonzero, one can divide both sides of Eq. (8.3) by ω so that one can obtain the following expression of the required mass for letting the centrifugal force become zero as

$$m \;= M\;r/r_m \tag{8.4}$$

8.2.2 Dynamic Balancing

A rotating system of mass is in dynamic balance when the rotation does not produce any resultant centrifugal force or couple. Thus, the system rotates without requiring the application of any external force or couple, other than that required to support its weight [1]. Statically balanced disks such as in Fig. 8.1b may still be dynamically unbalanced due to the presence of centrifugal effects.

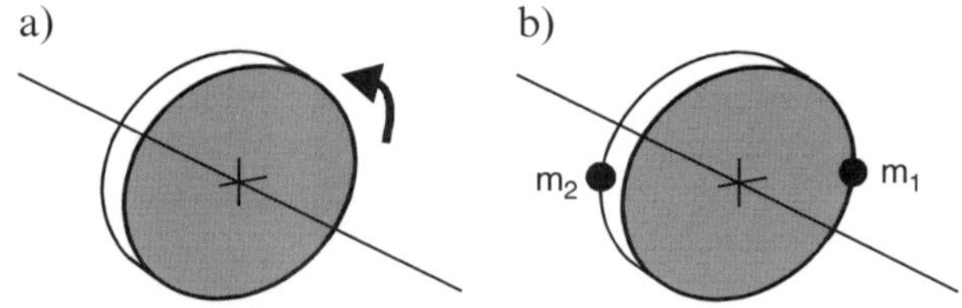

Fig. 8.2 Schemes of a disk: (**a**) The lack of dynamic balancing produces a couple acting on the rotating axis; (**b**) counterweight masses m_1 and m_2 are added to dynamically balance the disk

Dynamically unbalanced rotating shafts are usually balanced by adding two identical weights m_1 and m_2, as shown in the schemes of Fig. 8.2.

The two masses m_1 and m_2 will not modify the static balancing while they will produce a counterclockwise centrifugal effect that will achieve the dynamic balancing.

Several approaches can be used for dynamic balancing of mechanisms. The classical method to obtain statically and dynamically balanced mechanisms consists on adding mass and inertia elements to the system so that the center of mass remains unchanged (statically balanced) while the angular momentum becomes zero for any motion. The dynamic balancing can be achieved using several methods or "principles." For example, [19, 20] propose the following methods:

- Using counter-rotary counter-masses
- Using separate counter-rotations
- Using idler loops
- Using a duplicated mechanism

Similarly, [14, 15] propose the following methods:

- Balancing by counterweights mounted on the movable links
- Harmonic balancing by two counter-rotating masses
- Balancing by opposite movements

The balancing by counterweights is based on adding counterweight masses that can have the same weight and opposite dynamic effects of the links being part of the mechanism to be balanced. In the case of complete shaking force balancing this approach is generally limited to simple mechanisms having only revolute joints.

The harmonic balancing by two counter-rotating masses is based on harmonic analysis. The reduction of inertia effects is accomplished by the balancing only of certain harmonics of the shaking forces and shaking moments. Unbalanced forces and moments are approximated by Fourier series (or Gaussian least-square formulation) and then each frequency component is studied. This approach has been used successfully for engine balancing. For example, the balancing shafts are used for balancing of a second harmonic of the shaking force.

The balancing by opposite movements requires the addition of an axially symmetric duplicate mechanism that will produce opposite motions and dynamic effects as compared with the mechanism to be balanced. In this case, shaking force and shaking moment can become both zero.

The balancing by added dyads is achieved when adding links (dyads) to a mechanism to make a parallelogram chain (consisting of the initial links of the mechanism and the added dyad). In this way, the dyad transfers the motion of the coupler link to a shaft on the frame, where it is connected to a counterweight of considerably reduced mass as reported for example in [28]. In this way, it is possible to create an additional balancing moment for reducing the shaking moment while maintaining the static balance of the mechanism.

The major drawback of most of the abovementioned approaches is that a considerable amount of mass and inertia is added to the system, and: "The price paid for shaking force and shaking moment balancing is discouraging" [16]. Another possible approach is to generate trajectories which minimize or eliminate reaction forces and torques [32]. This approach is however quite restrictive and applicable only for special cases.

8.2.3 A Procedure for Balancing

In Fig. 8.3 a general procedure is outlined for solving a problem of balancing.

The first step of the procedure is the definition of proper kinematic and dynamic. It is to note that identifying general formulations for the solution of kinematic

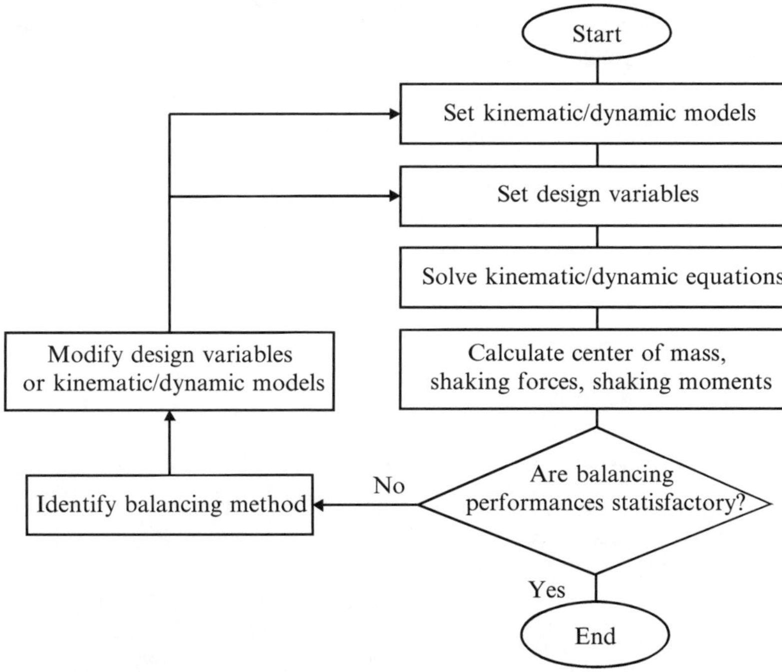

Fig. 8.3 The proposed procedure for balancing

and dynamic models of parallel manipulators is a quite hard task. This topic is widely addressed in the literature as mentioned for example in [36–40]. Close-form solutions of kinematics/dynamics have been identified only for a subset of the feasible kinematic architectures of parallel manipulators. In other cases, simplified and/or approximate and/or iterative approaches should be considered. Then, it is necessary to identify the main characteristics of the system under investigation including masses and inertia properties. The key design variables and sizes should be identified with proper numerical values. Among the design variables special care should be given to the expected input/output motions as function of time.

The solution of kinematics and dynamics equations allows the calculation of the coordinates of center of mass as well as the values of shaking forces and shaking moments. The obtained values depend on the robot architecture but also on the expected input/output motions. Accordingly, coordinates of center of mass as well as the values of shaking forces and shaking moments will be numerically calculated as function of time. The results of these calculations will provide direct information on the fulfilling of static and dynamic balancing conditions.

In general, a mechanism will be statically and dynamically balanced only if—in any operating condition—the calculated coordinates of center of mass coincide with the axis of rotation and the value of shaking forces and shaking moments are zero. However, a designer may desire that a mechanism is just statically balanced or that a mechanism is dynamically balanced only for a specific set of input/output motions. Specific conditions may lead a designer to tolerate a certain amount of unbalancing or even to consider not necessary a static/dynamic balancing. The desired balancing performances should be carefully defined in a case-by-case manner by considering specific design constraints such as construction cost limitations, complexity limitations, size limitations, lightweight requirements, and power consumption improvements.

If the balancing performances are not satisfactory, the procedure will search for a suitable method for fulfilling the desired static/dynamic balancing among those identified in the previous section. Design variables and kinematic/dynamic models should be updated accordingly. Often, it can be necessary to repeat the above procedure in an iterative manner until the desired performances are achieved. A suitable iterative search of proper design variables can be achieved by means of optimal search algorithms while the choice of different balancing methods usually requires a reengineering process.

8.3 A Case of Study for Balancing

A case of study is herewith proposed in order to show the key steps of a balancing design process as referring to a parallel manipulator having three active DOFs.

At LARM in Cassino significant research activity has been devoted to the design of parallel manipulators such as the CaPaMan (Cassino Parallel Manipulator) series. A prototype of CaPaMan 2bis built at LARM is shown in Fig. 8.4. It has been

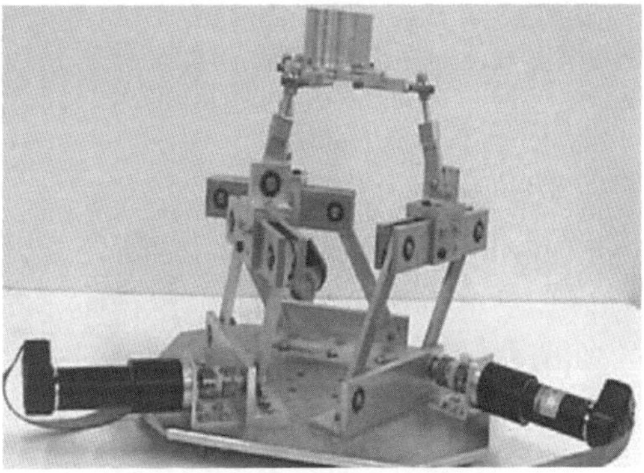

Fig. 8.4 A photo of CaPaMan 2bis

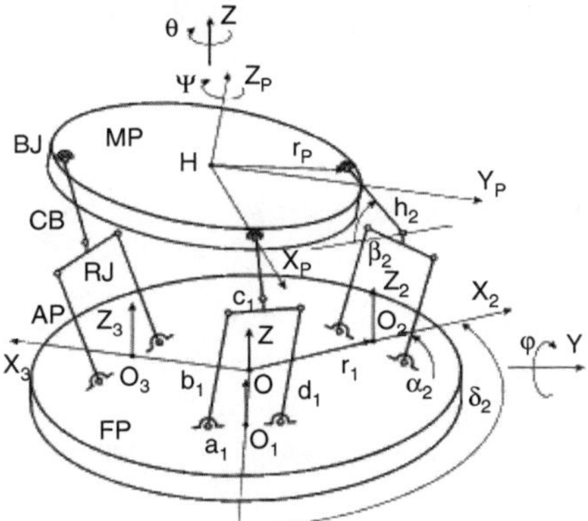

Fig. 8.5 A kinematic scheme of CaPaMan 2bis

implemented as a part of a hybrid robotic architecture for surgical tasks in the work, as well as a trunk module in a humanoid robot design named as CALUMA (CAssino Low-cost hUMAnoid robot) [41–44].

8.3.1 Definition of a Kinematic Model

A kinematic scheme of CaPaMan 2bis is shown in Fig. 8.5. It is composed of a movable platform (MP) connected to a fixed base (FP) through three leg mechanisms. Each leg mechanism is composed of an articulated parallelogram (AP) whose coupler carries a revolute joint (RJ), a connecting bar (CB) that transmits the motion from AP to MP through RJ, and a spherical joint (BJ), which is installed on MP at point J. Revolute joint RJ installed on the coupler of AP has the rotation axis coinciding with the parallelogram plane. Each leg mechanism is rotated $2\pi/3$ with respect to the neighboring one so that the leg planes lie along two vertices of an equilateral triangle, giving symmetry properties to the mechanism. They can be identified for the k-th leg mechanism ($k = 1, 2, 3$) as ak is the length of the frame link; bk is the length of the input crank; ck is the length of the coupler link; dk is the length of the follower crank; and hk is the length of the connecting bar. The sizes of movable platform MP and fixed base FP are given by distances rp and rf, as shown in Fig. 8.1b. Points H and O are the center points of MP and FP, respectively. Points Ok are the middle point of frame link ak, Jk are the connecting points between. An inertial frame O-XYZ has been assumed to be fixed to base FP. A moving frame HXPYPZP has been attached to platform MP. O-XYZ has been fixed with Z-axis orthogonal to the FP plane and X-axis as coincident with the line joining O to O1. Moving frame HXPYPZP has been fixed to platform MP with ZP orthogonal to the MP plane and XP-axis as coincident to the line joining H to J and YP to give a Cartesian frame. Angle δk is the structure rotation angle between OX1 and OXk as well as between HJ1 and HJk. They are equal to $\delta 1 = 0$, $\delta 2 = 2\pi/3$, and $\delta 3 = 4\pi/3$, the k-th leg mechanism, and platform MP. The design parameters of the CaPaMan 2bis are listed in Table 8.1.

8.3.2 Definition of a Dynamic Model

The CaPaMan 2bis parallel manipulator has been modeled using mixed coordinates. So in general terms for a given kinematic chain, the position of each body is described by a set of coordinates that combines reference-point coordinates defined at the center of mass of each body and relative coordinates defined at each joint. In this way a set of dependent coordinates p was defined to describe the manipulator as

$$p = \begin{pmatrix} g_1 \\ e_1 \\ \vdots \\ g_n \\ e_n \end{pmatrix} \tag{8.5}$$

Table 8.1 Mechanical design parameters and inertial properties of CaPaMan 2bis

$a_k = c_k$ (mm)	$b_k = d_k$ (mm)	r_p (mm)	r_f (mm)	h_k (mm)	$I_{cxy} = I_{cxy} = I_{cy}$ (kg mm^2)
100	100	65	65	50	0
α_k (deg)	β_k (deg)	m (kg)	τ (N/m)	I_{czz} (Kg mm^2)	$I_{cxx} = I_{cxy}$ (kg mm^2)
45–135	30–120	2.3	2	24,600	12,400

where g_i are the positions of the center of mass of the body i defined as

$$g_i^{\mathrm{T}} = \left(x_i \; y_i \; z_i\right) \tag{8.6}$$

and e_i are the Euler angles of body i defined in the form

$$\mathrm{e}_i^{\mathrm{T}} = \left(\alpha_i \; \beta_i \; \gamma_i\right) \tag{8.7}$$

In particular for CaPaMan 2bis, 13 bodies were modeled and described by the corresponding set of coordinates, giving a total of 78 coordinates in vector p.

The dynamic equations for parallel architectures are generally difficult to formulate in closed form because of the high nonlinearity existing in the kinematics. A simplification in the dynamic analysis consists of neglecting the inertia of leg mechanisms in comparison with the inertia of the movable plate. This neglecting can be justified when you consider that the leg motion is smoother than that one correspondingly obtained for the movable plate. In fact, the motion and mass of the movable plate are more significant with respect to the corresponding leg properties in most cases. Further details on this matter can be found in [43, 44].

The mobile platform has been transformed in an equivalent model. The equivalence is obtained with three identical point masses arranged symmetrically, shown in Fig. 15, with

$$m_1 = m_2 = m_3 = \tfrac{1}{3}m \tag{8.8}$$

where m is the original total mass of the mobile platform. The only solution to this equivalence is to have

$$y_2 = y_3, \quad x_2 = -x_3, \quad y_1 = -2y_2 \tag{8.9}$$

$$I_{xx} = \frac{1}{2}mr_{\mathrm{pm}}, \quad I_{yy} = I_{xx}, \quad I_{zz} = \frac{1}{2}mr_{\mathrm{pm}}, \quad I_{xy} = 0 \tag{8.10}$$

The distance between the center of the mobile platform and the point of attachment to each leg (at the spherical joint) is 65 mm. This means that $r_{\mathrm{pm}} = 65$ mm and using Eq. (8.10), the circular platform results with a diameter of 183.848 mm.

Focusing on the dynamic balancing of CaPaMan 2bis these three point masses are located at the corresponding spherical joints, the points attaching to the legs, so these legs can be balanced independently as shown in Fig. 8.6.

8.3.3 Selection of Design Variables

A 3D CAD model has been built as shown in Fig. 8.7. The model has been simplified so that the mass coming from the equivalence of the moving plate is concentrated at

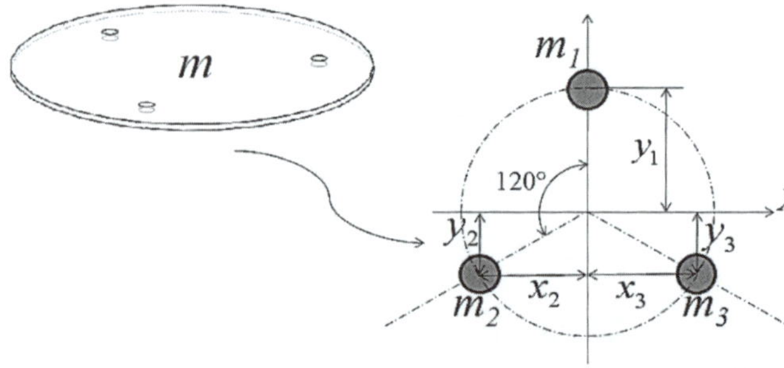

Fig. 8.6 Dynamic equivalence of the mobile platform with three point masses

Fig. 8.7 Basic architecture of CaPaMan 2bis indicating the two subsystems that have been used

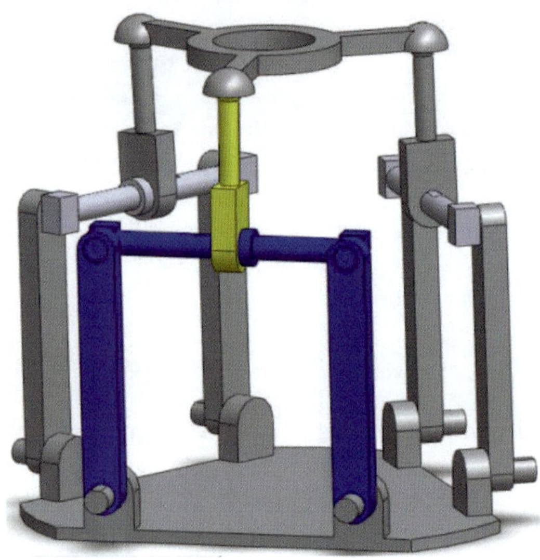

Fig. 8.8 A scheme of a connecting bar CB with main design variables

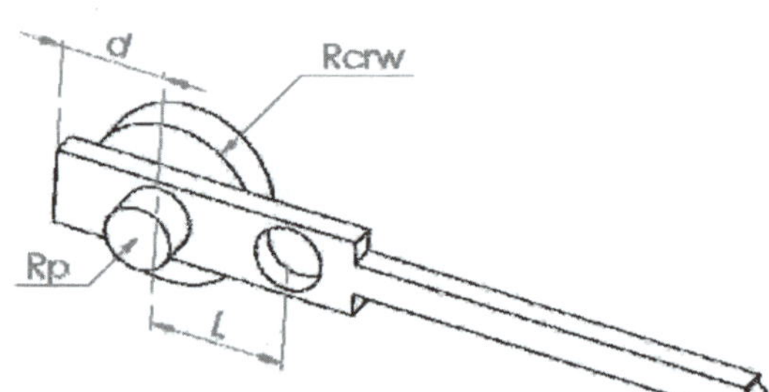

the spherical joints. Main attention for the balancing will be given to the connecting bars CB (one of them is shown in yellow in Fig. 8.7) and to the coupler-link in the four-bar mechanism (one of them is shown in blue in Fig. 8.7).

The desired balancing condition can be defined so that one can achieve a stationary center of mass. The parameters taken in consideration for the balancing study are shown in Fig. 8.8. The design variables are listed with their properties in Table 8.2.

Table 8.2 The considered feasible ranges of the main design variables

Name	Values	Units
L	Min:10 Max:30	mm
Dp	Min:2 Max:30	mm
d	Min:2 Max:30	mm

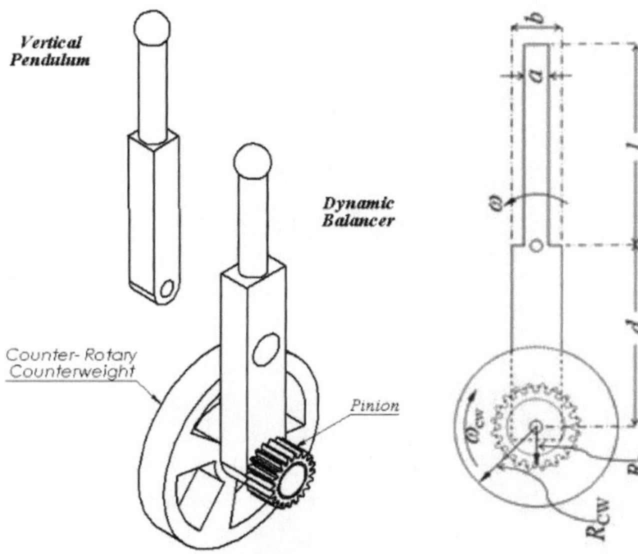

Fig. 8.9 A design for the dynamic balancer: (**a**) Mechanical design layout. (**b**) Design parameters

8.3.4 Choice of Balancing Method

Considering the characteristics of CaPaMan 2bis it can be convenient to select the balancing method that is based on CRCW and counterweights. Given the architecture of CaPaMan 2bis one can use a total of six CRCW and three counterweights.

Once the mobile platform has been split into three point masses, it is possible to balance each leg independently. The original vertical pendulum is modified to become a dynamic balancer. All the geometry of the elements has been simplified, as shown in Fig. 8.9a, in order to facilitate the generation of the balancing equations. The pendulum is a bar with rectangular cross section and thickness $t = 4$ mm. The pinion and CRCW have been considered as solid disks of steel $\left(\rho = 7,800\ {}^{\mathrm{kg}}\!/_{\mathrm{m}^3}\right)$. The CRCW and the pinion has a thickness of 8.355 mm and are considered of aluminum $\left(\rho = 2,740\ {}^{\mathrm{kg}}\!/_{\mathrm{m}^3}\right)$. The dynamic balancer is designed in such a way that linear and angular momentum are conserved, getting zero shaking force and zero shaking moment and obtaining a reactionless system.

A constant linear momentum can be obtained if the total center of mass is stationary. Considering **r** as the position vector of the global center of mass one can write

$$\frac{d\mathbf{r}}{dt} = 0 \tag{8.11}$$

that can be expressed as

$$\mathbf{r} = \frac{1}{M}\sum_{i=1}^{n} m_i \mathbf{r}_i; \quad M = \sum_{i=1}^{n} m_i \tag{8.11}$$

where m_i is the mass of the *i*th body, $\mathbf{r}_i$ is the position vector of the center of mass of the *i*th body, and M is the total mass of the system.

For the dynamic balancer shown in Fig. 8.9b the global position of the center of mass can be calculated as

$$\mathbf{r} = \frac{1}{M}\left[\frac{1}{3}m_{\mathrm{MP}}l + \frac{1}{2}m_{\mathrm{VP}}l - \frac{1}{2}m_{\mathrm{SB}}d - m_{\mathrm{P}}d - m_{\mathrm{CW}}d\right] \tag{8.12}$$

where $m_{\mathrm{VP}} = \rho_a atl$ is the mass of the vertical pendulum, $m_{\mathrm{SB}} = \rho_a btl$ is the mass of the supporting bar, $m_{\mathrm{P}} = \rho_s \pi t R_{\mathrm{P}}^2$ is the mass of the pinion, and $m_{\mathrm{CW}} = \rho_s \pi t R_{\mathrm{CW}}^2$ is the mass of the CRCW. A point mass of $m_{\mathrm{PM}} = 1.2123$ g, corresponding to one-third of the total mass of the mobile platform, is taken into account.

To impose a stationary center of mass, Eq. (8.12) can be written as

$$\mathbf{r} = \frac{1}{M}\left[1.2123(l) + \frac{1}{2}\mathrm{at}\,(l-d)\,\rho - \frac{1}{2}\pi t d\rho R_P^2 - \frac{3}{2}\pi t d\rho R_{\mathrm{CW}}^2\right] = 0 \tag{8.13}$$

Considering ω as the angular velocity of the bar and ω_{CW} the angular velocity of the pinion-CRCW set, and noting that $\omega_{\mathrm{CW}} = -(d/R_P)\,\omega$, the total angular momentum $\mathbf{H}_{\mathrm{tot}}$ of the system can be calculated as

$$\mathbf{H}_{\mathrm{tot}} = \begin{bmatrix} 1.2123\left(l^2\right) + \frac{1}{3}a\rho t\left(l^3 + d^3\right) + \pi\rho t d^2\left(\frac{1}{2}R_P^2 + \frac{3}{2}R_{\mathrm{CW}}^2\right) \\ +\frac{1}{4}\pi\rho t d\left(R_P + 3\frac{R_{\mathrm{CW}}^2}{R_P}\right) \end{bmatrix}\omega = 0 \tag{8.14}$$

where the general equation to calculate the moment of inertia of a bar ($I = \frac{1}{12}ml^2$) and the general equation to calculate the moment of inertia of a disk ($I = \frac{1}{2}mR^2$) have been used.

Table 8.3 Results from the solution to get a valid dynamic balancer

l (mm)	a (mm)	b (mm)	t (mm)	T_c (mm)	d (mm)	R_P (mm)	R_{CW} (mm)
64	3	8	4	8.355	40	0.332	7.066

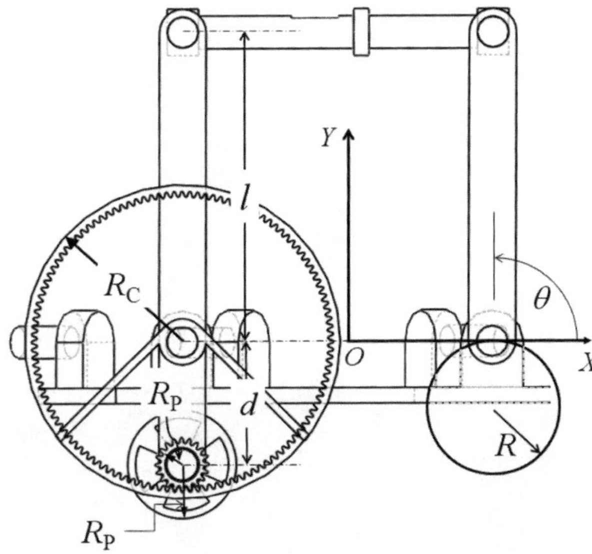

Fig. 8.10 Articulated parallelogram balanced by a single dynamic balancer

Equations (8.13 and 8.14) can be solved simultaneously to obtain a valid dynamic balancer, using an ordinary Newton–Raphson method. The parameters l and a are already determined as they define the pendulum to be balanced, and b has been chosen equal to a (in fact the geometry and dimensions of this part of the dynamic balancer could be a topic to investigate). Different values of d may be chosen, leaving $\mathbf{R_P}$ and as unknowns. Table 8.3 shows representative results choosing $d = 40$ mm, as this value warranties enough distance between the CRCW and the coupler axes of rotation.

The obtained dynamic balancer has been mounted on the coupler of the four-bar linkage (the articulated parallelogram) of the limb, next balanced following a similar procedure as the one used for the vertical pendulum.

The articulated parallelogram can be balanced by a single dynamic balancer and a single counterweight, as in Fig. 8.10, to preserve the linear and angular momentums, obtaining a reactionless system.

The conservation of the linear momentum **L** can be obtained by keeping it constant (zero). Linear momentum can be calculated as in Eq. (8.15); two densities are used: ρ_1 is associated to all the elements of the four-bar mechanism, and ρ_2 is associated to the CRCW and to the counterweight. The thickness (t) has been considered the same for all the elements. Additionally take note that a point mass of 0.00121 g is considered, and added to the coupler of the four-bar mechanism.

This point mass comes by considering the basic balancer obtained from the balancing of the vertical pendulum in the form

$$\mathbf{L} = \left\{ \begin{array}{l} \omega \sin(\theta) \begin{pmatrix} \pi \rho_2 t R^3 + \pi d \rho_2 t R_P^2 + \pi d \rho_2 t R_{\mathrm{CW}}^2 \\ -2al^2 t \rho_1 - (0.07844) l + \frac{1}{2} a d^2 \rho t \end{pmatrix} \\ \omega \cos(\theta) \begin{pmatrix} -\pi \rho_2 t R^3 - \pi d \rho_2 t R_P^2 - \pi d \rho_2 t R_{\mathrm{CW}}^2 \\ +2al^2 t \rho_1 + (0.07844) l - \frac{1}{2} a d^2 \rho t \end{pmatrix} \end{array} \right\} = \text{const.} \tag{8.15}$$

Constant linear momentum can be achieved if the sum of terms in the parenthesis of Eq. 8.15 is zero; this is the necessary condition to get null shaking force. Therefore, one can write

$$\begin{aligned} &\pi \rho_2 t R^3 + \pi d \rho_2 t R_P^2 + \pi d \rho_2 t R_{\mathrm{CW}}^2 \\ &- 2al^2 t \rho_1 - (0.07844) l + \frac{1}{2} \mathrm{ad}^2 \rho t = 0 \end{aligned} \tag{8.16}$$

On the other hand to get a null shaking moment it is necessary to make the total angular momentum of the system constant or zero. For the articulated parallelogram shown in Fig. 20 the total angular momentum can be calculated as

$$\begin{aligned} H_Z = \omega &\begin{pmatrix} \frac{3}{2} \pi \rho_2 t R^4 - \frac{1}{2} \pi d \rho_2 t R_P^3 + \pi d^2 \rho_2 t R_P^2 - \frac{1}{2} \frac{\pi d \rho_2 t R_{\mathrm{CW}}^4}{R_P} \\ + \pi d^2 \rho_2 t R_{CW}^2 + \frac{5}{3} a l^3 t \rho_1 + \frac{1}{3} \mathrm{ad}^3 t \rho_1 + (0.07844) l^2 \end{pmatrix} \\ + \omega \cos(\theta) &\begin{pmatrix} \frac{1}{2} \pi l \rho_2 t R^3 - \frac{1}{2} \pi d l \rho_2 t R_P^2 \\ - \frac{1}{2} \pi d l \rho_2 t R_{CW}^2 - \frac{1}{4} d^2 l t \rho_1 \end{pmatrix} = 0 \end{aligned} \tag{8.17}$$

ω is the angular velocity of both cranks (the coupler never rotates); it has also been considered that $\omega_{\mathrm{CW}} = -(d/R_P)\,\omega$ is the angular velocity of the CRCW-pinion set, just as in the vertical pendulum case. To get a constant angular momentum both terms in parenthesis in Eq. (8.17) must be zero, and then the conditions to accomplish can be summarized as

$$\begin{aligned} &\frac{3}{2} \pi \rho_2 t R^4 - \frac{1}{2} \pi d \rho_2 t R_P^3 + \pi d^2 \rho_2 t R_P^2 - \frac{1}{2} \frac{\pi d \rho_2 t R_{\mathrm{CW}}^4}{R_P} \\ &+ \pi d^2 \rho_2 t R_{\mathrm{CW}}^2 + \frac{5}{3} a l^3 t \rho_1 + \frac{1}{3} a d^3 t \rho_1 + (0.07844) l^2 = 0 \end{aligned} \tag{8.18}$$

$$\frac{1}{2} \pi l \rho_2 t R^3 - \frac{1}{2} \pi d l \rho_2 t R_P^2 - \frac{1}{2} \pi d l \rho_2 t R_{\mathrm{CW}}^2 - \frac{1}{4} d^2 l t \rho_1 = 0 \tag{8.19}$$

Table 8.4 Balancing results

l (mm)	a (mm)	t (mm)	R_P (mm)	d (mm)	R (mm)	R_{CW} (mm)
70	5	10	5	2.5	17.2	44.2

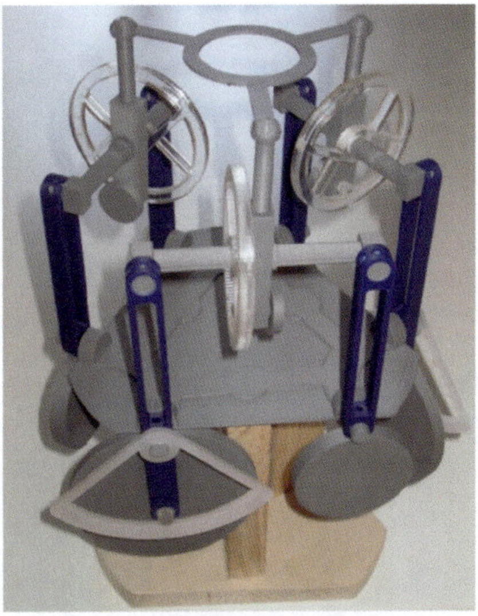

Fig. 8.11 Modified CaPaMan 2bis concept with counter-rotary counterweights and counterweights [28]

The latter can be solved simultaneously using the Newton–Raphson algorithm for a set of nonlinear equations. In this case the length (l) of both cranks and the coupler (considered made of aluminum with $\rho_1 = 2740\ ^{kg}/_{m^3}$) of the articulated parallelogram is known, $l = 70$ mm, and the width and the thickness are set to $a = 10$ mm and $t = 10$ mm, respectively, for all elements. The CRCW and the counterweight are considered made of steel, with $\rho_2 = 7,800\ ^{kg}/_{m^3}$. Finally the radius of the pinion is set to $R_P = 5$ mm, even though it is possible to choose any other value, taking into account that the lesser value the bigger the radius of the CRCW. Solutions are shown in Table 8.4. The final design with the modified parts is shown in Fig. 8.11.

8.3.5 Numerical Validation of Balancing

The balancing computation verified has been with a dynamic simulation using MSC ADAMS imposing some general motions to the DOFs associated to the cranks in the articulated parallelograms. Three generic cubic functions have been used to guide the 3-DOFs in the form

$$\alpha_i = \alpha_{i0} + \frac{3\Delta\alpha_i}{t_{\text{if}}^2}t^2 - \frac{2\Delta\alpha_i}{t_{\text{if}}^3}t^3; \quad \Delta\alpha_i = \alpha_{\text{if}} - \alpha_{i0} \tag{8.20}$$

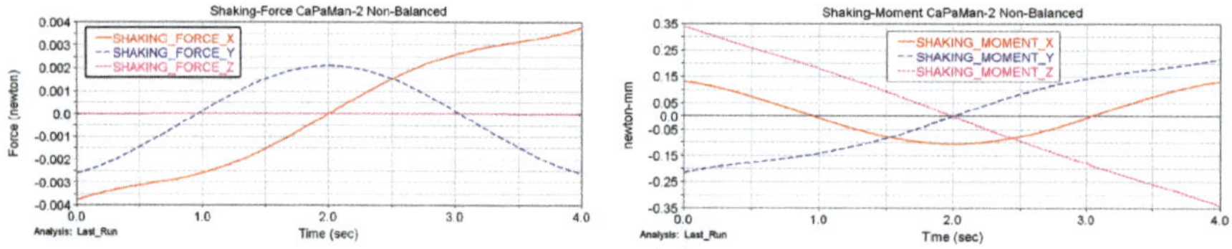

Fig. 8.12 Computed results from the simulation of the ADAMS non-balanced model: (**a**) Shaking forces, (**b**) shaking moments

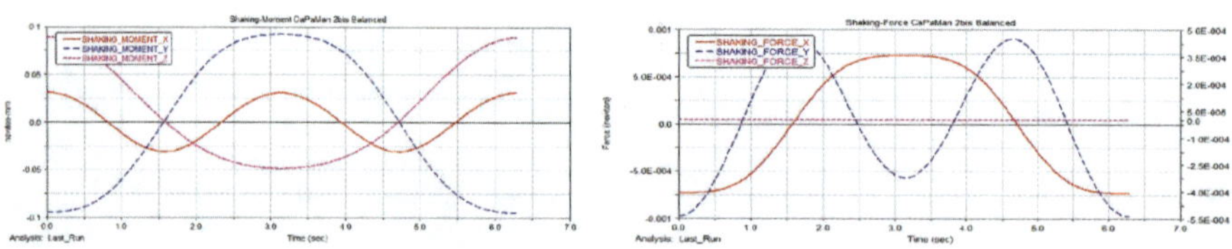

Fig. 8.13 Computed results from the simulation of the ADAMS balanced model: (**a**) Shaking forces, (**b**) shaking moments

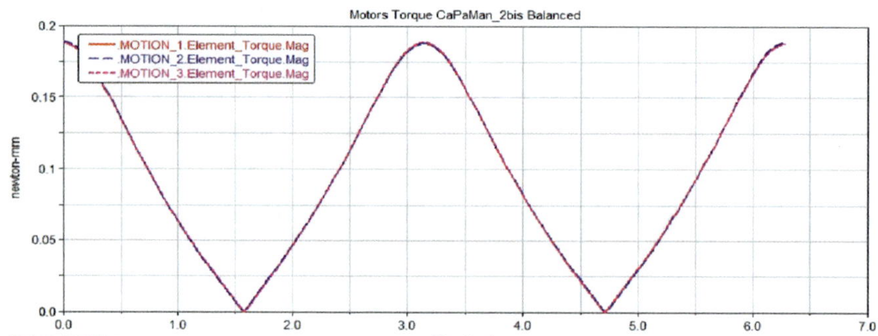

Fig. 8.14 Needed motor torque

Applying the function in Eq. 39 to one leg and keeping the other ones fixed the results obtained in the non-balanced mechanism are presented in Fig. 8.12, while the results obtained in the balanced mechanism are presented in Fig. 8.13. With the balancing procedure shaking forces in X and Y drop by an order of magnitude and shaking forces in Z remain constantly zero. A significant improvement is noticed in the shaking moments especially in Y and Z, the most important axis concerning vibrations due to the nature of the model motion. Analyzing the motor torques shown in Fig. 8.14 it can be noticed that also if more mass has been added the needed torque to move the manipulator is very low.

The application of CRCW to the dynamic balancing of a spatial 3-DOFs parallel manipulator, given the special architecture of the mechanism, shows that this procedure leads to very good results in shaking force and to an interesting reduction in shaking moment.

8.4 Conclusions

This chapter has addressed the key issues for achieving static and dynamic balancing of a robotic system. A general procedure has been outlined in order to clarify the main steps that should be considered in the design process for achieving suitable balancing performances. The proposed procedure has been considered specifically for manipulators having parallel architecture. A case of study has been reported in full details in order to show the feasibility and effectiveness of the proposed procedure. In particular, the proposed case of study refers to a 3-DOFs parallel manipulator, whose name is CaPaMan 2bis. The main result of the proposed balancing procedure has been the improvement of CaPaMan 2bis design for dynamically balanced operation. A numerical simulation has been carried out in MSC.ADAMS environment for validating the expected dynamic performances of the improved CaPaMan 2bis design.

Acknowledgements Authors wish to acknowledge the significant contribution and inspiration to this chapter given by Prof. Mario Acevedo, Universidad Panamericana, Mexico City.

References

1. Uicker, J.J., Pennock, G.R., Shigley, J.E.: Theory of machines and mechanisms. Oxford University Press, New York (2003)
2. Angeles, J.: Dynamic response of linear mechanical systems, modeling, analysis and simulation. Springer, (2012)
3. Norton, M.P., Karczub D.G.: Fundamentals of noise and vibration analysis for engineers. Cambridge University Press, (2003)
4. International Organization for Standardization (ISO): Mechanical vibration—Balancing—Guidance on the use and application of balancing standards. ISO 19499, (2007)
5. International Organization for Standardization (ISO): Test methods for measuring tyre uniformity. ISO 13326, (1998)
6. Ceccarelli, M.: Fundamentals of mechanics of robotic manipulation. Kluwer Academic Publishers, (2004)
7. Arakelian, V.H., Briot, S.: Dynamic balancing of the Scara robot. Proceedings of the 17-th CISM-IFToMM Symposium on Robot Design, Dynamics and Control (RoManSy 2008), Tokyo, pp. 167–174 (2008)
8. Rivin, E.I.: Mechanical design of robots. McGraw Hill, (1987)
9. Kemmerling, P.T Jr.: Dynamic characteristics of flight simulator motion systems. Agard Conference Proceedings No. 249, London, p. 20, 1978, available on-line at http://www.dtic.mil/dtic/tr/fulltext/u2/a063850.pdf
10. Agrawal, A., Agrawal, S.K.: Design of gravity balancing leg orthosis using non-zero free length springs. Mech. Mach. Theory **40**, 693–709 (2005)
11. Xu, Y., Au, K.W., Nandy, G.C., Brown H.B.: Analysis of actuation and dynamic balancing for a single-wheel robot. IEEE/RSJ International Conference on Intelligent Robots and Systems. Vol. 3 pp. 1789–1794, Victoria, (1998)
12. Park, J., Haan, J., Park, F.C.: Convex optimization algorithms for active balancing of humanoid robots. IEEE Trans. Robot. **23**(4), 817–822 (2007)

13. Berkof, R.S., Lowen, G.G.: A new method for completely force balancing simple linkages". J. Eng. Ind. **91**(B), 21–26 (1969)
14. Arakelian, V.H., Smith, M.R.: Complete shaking force and shaking moment balancing of linkages. Mech. Mach. Theory **34**(8), 1141–1153 (1999)
15. Arakelian, V.H., Smith, M.R.: Shaking force and shaking moment balancing of mechanisms: a historical review with new examples. ASME Trans. J. Mech. Des. **127**, 334–339 (2005)
16. Kochev, I.S.: General theory of complete shaking moment balancing of planar linkages: a critical review". Mech. Mach. Theory **35**, 1501–1514 (2000)
17. Herder J.L., Gosselin C.M.: A Counter-Rotary Counterweight (CRCW) for Light-Weight Dynamic Balancing. Proceedings of the ASME Design Engineering Technical Conferences and Computers and Information in Engineering Conference, Salt Lake City, Utah, USA, September 28-October 2, (2004)
18. Russo, A., Sinatra, R., Xi, F.: Static balancing of parallel robots". Mech. Mach. Theory **40**, 191–202 (2005)
19. Van der Wijk V., Herder, J.: Synthesis of dynamically balanced mechanisms by using counter-rotary countermass balanced double pendula. ASME J. Mech. Des., **131**(11), (2009)
20. Van der Wijk, V.L., Herder, J., Demeulenaere, B.: Comparison of various dynamic balancing principles regarding additional mass and additional inertia. J. Mech. Robot., JMR-08-1193, (2009)
21. Wu, Y., Gosselin, C.M.: On the Dynamic Balancing of Multi-DOF Parallel Mechanisms With Multiple Legs". ASME J. Mech. Des. **129**(2), 234–238 (2007)
22. Lim, T.G., Cho, H.S., Chung, W.K.: A parameter identification method for robot dynamic models using a balancing mechanism. Robotica Int. J. **7**(4), 327–337 (1989)
23. Rahman, T., Ramanathan, R.: A simple technique to passively gravity-balance articulated mechanisms. ASME J. Mech. Des. **117**(4), 655–658 (1995)
24. Deepak S.R., Ananthasuresh G.K.: Static balancing of a four-bar linkage and its cognates. Mech. Mach. Theory. (2011)
25. Ebert-Uphoff, I., Gosselin, C.M., Laliberté, T.: Static balancing of spatial parallel platform mechanisms-revisited". J. Mech. Des. **122**, 43–51 (2000)
26. Elliot J.L., Tesar D.: The theory of torque, shaking force and shaking moment balancing of four link mechanisms. ASME J. Eng. Ind. **99**, (1977)
27. Gosselin, C.M., Vollmer, F., Côté, G., Wu, Y.: Synthesis and design of reactionless three-degree-freedom parallel mechanisms. IEEE Trans Robot. Automat. **20**(2), (2004)
28. Acevedo, M., Ceccarelli, M., Carbone, G., Cafolla, D.: Complete dynamic balancing of a 3-DOFs spatial parallel mechanisms by the application of counter-rotary counterweights. EUROMECH Colloquium 524, University of Twente, Enschede, Netherlands, February 27 March 1, (2012)
29. Gosselin, C.M., Vollmer, F., Côté, G., Wu, Y.: Synthesis and Design of Reactionless Three-Degree-of-Freedom Parallel Mechanisms". IEEE Trans. Robot. Automat. **20**(2), 191–199 (2004)
30. Gosselin, C.M., Moore, B., Schicho, J.: Dynamic balancing of planar mechanisms using toric geometry. J. Symbolic Comput. In Press, Accepted Manuscript (2009)
31. Laliberté, T., Gosselin, C.M., Jean, M.: Static balancing of 3-DOFs planar parallel mechanisms. IEEE/ASME Trans. Mechatron. **4**, 363–377 (1999)
32. Moore, B., Schicho, J., Gosselin, C.M.: Dynamic balancing of spherical 4r linkages. J. Mech. Robot. Submitted, paper number JMR-08-1176, (2009)
33. Xi, F.: Dynamic balancing of hexapods for high-speed applications. Robotica Int. J. **17**(3), 335–342 (1999)
34. Alici, G., Shirinzadeh, B.: Optimum synthesis of planar parallel manipulators based on kinematic isotropy and force balancing. Robotica Int. J. **22**(1), 97–108 (2004)
35. Alici, G., Shirinzadeh, B.: Optimum dynamic balancing of planar parallel manipulators based on sensitivity analysis. Mech. Mach. Theory **41**(12), 1520–1532 (2006)
36. Angeles, J.: Fundamentals of robotic mechanical systems: theory, methods, and algorithms. Springer, Dordrecht (2006)

37. Ceccarelli, M.: Fundamentals of mechanics of robotic manipulation. Kluwer Academic Publisher, Dordrecht (2004)
38. Merlet, J.P.: Parallel robots. Springer, Dordrecht (2006)
39. García de Jalón, J., Bayo, E.: Kinematic and dynamic simulation of multibody systems: the real-time challenge. Springer, New York (1994)
40. Rolland, L.: Certified solving of the forward kinematics problem with an exact method for the general parallel manipulators. Int. J. Robot. Soc. Japan **19**(9), 995–1025 (2005)
41. Aguirre, G., Acevedo, M., Carbone, G., Ceccarelli, M.: Kinematic and dynamic analyses of a 3-DOFs parallel manipulator by symbolic formulations. Thematic Conference on Advances in Computational Multibody Dynamics, ECCOMAS Lisbon, Paper MB2003-010, (2003)
42. Hernandez-Martinez, E.E., Conghui, L., Carbone, G., Ceccarelli, M., Lopez-Cajun, C.S.: Experimental and Numerical Characterization of CaPaMan 2bis Operation". J. Appl. Res. Technol. **8**(1), 101–119 (2010)
43. Carbone, G., Ceccarelli, M., Oliveira, P.J., Saramago, S.F.P., Carvalho, J.C.M.: Optimum Path Planning of CaPaMan (Cassino Parallel Manipulator) by Using Inverse Dynamics". Robotica: Int. J. **26**(2), 229–239 (2008)
44. Carvalho, J.C.M., Ceccarelli, M.: A closed form formulation for the inverse dynamics of cassino parallel manipulator. J. Multibody Syst. Dynamics **5**, 185–210 (2001). Vol. 6, pp.303

Chapter 9
Dynamic Balancing with Respect to a Given Trajectory

Taizo Yoshikawa

Abstract To control the parallel link robots with better performance in terms of high rigidity, high degree of accuracy, high speed or acceleration, high load-carrying capacity, static balancing, and dynamic balancing are important factors. Generally, static balancing can be obtained by using counterweights or springs, and no computer control is involved. On the other hand, dynamic balancing utilizes control system to coordinate the motions of balancing elements. In this chapter, we persist on dynamic balancing with respect to a given trajectory for the parallel link robots by modeling control system. This chapter is organized in the following manner. In section "Modeling of Kinematics," geometric feature of Stewart Platform is introduced and modeling process of kinematics and Jacobian matrices is introduced. In section "Jacobian Analysis," modeling process of Jacobian matrix is presented. In section "Dynamics," modeling process of dynamics equations of a six DOF Stewart Platform is presented. In section "The Operational Space Formulation," the Operational Space Formulation is presented to control task dynamics at the end-effector. In section "Trajectory Generation," modeling method of smooth trajectory is presented. In section "Trajectory Tracking Control," control method to realize stable trajectory tracking motion is presented. Finally, control method to realize dynamic balancing with respect to a given trajectory is presented in section "Dynamic Balancing with Respect to a Given Trajectory."

9.1 Introduction

Originally, parallel link robot was designed as a flight simulator. Since then, many variants of the parallel link robots have experienced and a wide variety of applications have benefited from this mechanism. Currently, parallel link robots have been widely used several areas of industry, surgical operations, flight simulators, helicopter runway, throwing platform of missiles, surgical operations, 3D printer,

T. Yoshikawa (✉)
Artificial Intelligence Laboratory Manipulation Group, Stanford University,
Stanford, CA 94305, USA
e-mail: taizouy.94040@gmail.com

D. Zhang, B. Wei (eds.), *Dynamic Balancing of Mechanisms and Synthesizing of Parallel Robots*, DOI 10.1007/978-3-319-17683-3_9

etc. Parallel link robots demonstrate several advantages over serial link robots. Unlike serial robots, parallel link robots are composed of closed kinematic chain. This architecture provides high rigidity and high payload-to-weight-ratio, which allows the parallel link robots to handle larger and heavier loads than serial link robots. Under heavy loads, serial robots cannot perform precision positioning and oscillate at high-speeds. On the other hand, positioning accuracy of parallel link robots is high because the load is shared by several parallel kinematic chains and the positioning error of the platform cannot exceed the average error of the legs positions.

However, the parallel link robots have some drawbacks. When the base plate and mobile plate are parallel to the ground and a center of the mobile plate and a center of the base plate are in same position, the mass of the plate is divided into each leg equally. In this case, same torque needs to be applied to each motor. On the other hand, when position and orientation of the end-effector in Cartesian space changes, the motion needs to be transferred into leg dynamics and joint torque command needs to be generated precisely accounting for position error and quick response. To realize quick motion with less position error, high power motor and gear system are desired but it also causes high power consumption, high cost, less safety, large design, and relatively small workspace in comparison with serial link robots.

To control the parallel link robots with better performance in terms of high rigidity, high degree of accuracy, high speed or acceleration, high load-carrying capacity, static balancing, and dynamic balancing are important factors. Generally, static balancing can be obtained by using counterweights or springs and no computer control is involved. On the other hand, dynamic balancing utilizes control system to coordinate the motions of balancing elements. In this chapter, we persist on dynamic balancing with respect to a given trajectory for the parallel link robots by modeling control system. The aim of this chapter is to model the dynamic formulation of a six DOF Stewart Platform and discuss about dynamic balancing with respect to a given trajectory.

9.2 Modeling of Kinematics

9.2.1 Stewart Platform

Parallel link robots consist of (1) fixed base plate and (2) mobile plate which a gripper or a tool is mounted, and (3) several legs or rods to connect the fixed plate and the mobile plate. The individual leg is designed to rotate along one joint which is fixed to the base plate, which generates motion of the leg. Both side of the rod is connected by ball joint or any joint which enables free rotation in work space. Each legs and rods are connected and constrained at the mobile plate, which enables translation and orientation of the mobile plate at the end-effector. This

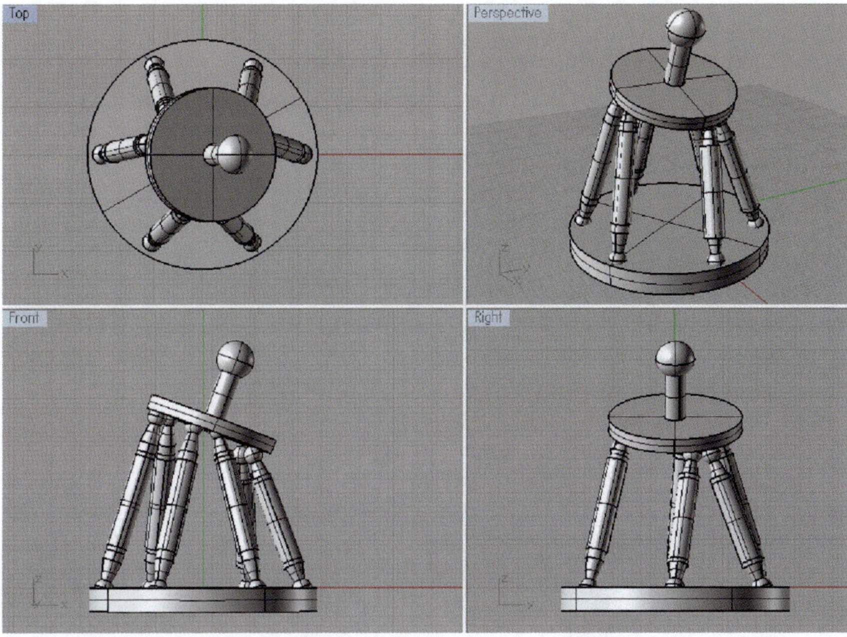

Fig. 9.1 Image of the Stewart Platform

parallel configuration consists of a closed kinematics chain which allows a higher performance. Therefore, this configuration enables precise position and orientation control of the end-effector in Cartesian space by controlling each joint position of the leg. General actuation system can be applied to control joint position of the leg such as motor and gear or linear actuator. However, the parallel link robots have some drawback of relatively small workspace in comparison with serial link robots.

The most widely used structure of parallel link robots is the Stewart Platform. The early version of the Stewart Platform was designed by V. Eric Gough in 1954. The Stewart Platform was invented as a flight simulator by Stewart in 1965. A Stewart Platform is a parallel manipulator device that is used in many applications for positioning objects [1, 2]. The Stewart Platform is a parallel mechanism that consists of a rigid body mobile plate and a fixed base plate joined by six adjustable legs that allow it to be precisely adjusted and controlled. The six legs attach the base plate to the adjustable mobile plate. Both ends of the legs are connected using universal joints that allow for a wide range of motion. The legs are designed to be varied in length and depending on the lengths, the position and orientation of the mobile platform is controlled (Fig. 9.1).

9.2.2 Kinematics

The kinematic and dynamic modeling of Stewart Platform is extremely complicated in comparison with serial robots [3–7]. For parallel link robots, inverse kinematics is straight forward and there is no complexity deriving Equations. On the other hand, forward kinematics of Stewart Platform is generally said to be very complicated and difficult to solve. It is because it contains the solution of many nonlinear equations and generally has more than one solution. In order to overcome this problem, accurate kinematic and dynamic identification is needed.

The Stewart Platform consists of two rigid frames, the base platform and mobile platform, connected by six variable length legs. These legs are identical kinematics chains, couple the moveable upper and the fixed lower platform by universal joints. We now need to express the three tool plate vertices with respect to the Base frame. To do this, we need to define a homogenous transformation matrix which represents the position and orientation of the Tool frame embedded within the Base frame. The used notations to describe the parallel link robots are defined in the following.

9.2.3 Required Notation

- The base platform has orthogonal axes $X_{\mathrm{B}}, Y_{\mathrm{B}}, Z_{\mathrm{B}}$ attached to the center of mass of the base plate.
- The mobile platform has orthogonal axes $X_{\mathrm{M}}, Y_{\mathrm{M}}, Z_{\mathrm{M}}$ located at the center of mass of the mobile plate. This transform will be a function of the $X_{\mathrm{B}}, Y_{\mathrm{B}}, Z_{\mathrm{B}}$ position values of the tool and three orientation parameters represented in the Base frame.
- R_{B} is the absolute frame, tied to the base platform.
- R_{M} is the mobile frame, tied to the mobile platform.
- Let O_{B} be the origin of the absolute coordinate system of the base platform, R_{B}.
- Let O_{M} be the origin of the absolute coordinate of the mobile platform, R_{M}, whose coordinates are in the absolute frame R_{B}:
  $\overrightarrow{O_{\mathrm{B}}O_{\mathrm{M}}} = [X_{\mathrm{M}}, Y_{\mathrm{M}}, Z_{\mathrm{M}}]^T$
- ,P_i^{B} is the center of the joint position between the segment i and the fixed base plate on the base:
  $O_{\mathrm{B}}P_i^{\mathrm{B}} = \left[a_{x(i)}, a_{y(i)}, a_{z(i)}\right]^T \ (i = 1, \ldots, 6)$
- P_i^{M} is the center of the joint between the segment i and the mobile part:
  $O_{\mathrm{M}}P_i^{\mathrm{M}} = \left[b_{x(i)}, b_{y(i)}, b_{z(i)}\right]^T \ (i = 1, \ldots, 6)$
- $R_{\mathrm{B}}^{\mathrm{M}}(\alpha, \beta, \gamma)$ is the rotation matrix expressing the transformation of orientation from orthogonal axes $X_{\mathrm{B}}, Y_{\mathrm{B}}, Z_{\mathrm{B}}$ at the center of mass of the base plate to orthogonal axes $X_{\mathrm{M}}, Y_{\mathrm{M}}, Z_{\mathrm{M}}$ located at the center of mass of the mobile plate.

- X is the task coordinate vector: $X = [X_T, Y_T, Z_T, \alpha, \beta, \gamma]^T$
- R_M and R_B are the radius of the moving platform and fixed base respectively.
- The length of each leg is L_i.
- X_i^B is the center of the joint position between the segment i and the fixed base plate in global framework that is equal to the base absolute frame R_B.
- X_i^M is the center of the joint position between the segment i and the mobile plate in global framework.

9.2.4 Coordinate Transformation

Euler angles are a means of representing the spatial orientation of any reference frame as a composition of three elemental rotations. There are several conventions for Euler angles, depending on the axes about which the rotations are carried out. The three rotational angles are defined as roll-pitch-yaw angles α, β and γ. The angle values represent the rotation about the x, y, and z axis, as R_x, R_y, R_z, respectively:

$$R_x(\alpha) = \begin{bmatrix} 1 & 0 & 0 \\ 0 & \cos\alpha & -\sin\alpha \\ 0 & \sin\alpha & \cos\alpha \end{bmatrix}, \tag{9.1}$$

$$R_y(\beta) = \begin{bmatrix} \cos\beta & 0 & \sin\beta \\ 0 & 1 & 0 \\ -\sin\beta & 0 & \cos\beta \end{bmatrix}, \tag{9.2}$$

$$R_z(\gamma) = \begin{bmatrix} \cos\gamma & -\sin\gamma & 0 \\ \sin\gamma & \cos\gamma & 0 \\ 0 & 0 & 1 \end{bmatrix}. \tag{9.3}$$

Using a Z-Y-X Euler angle to represent the rotation matrix associated with this parameterization of orientation, we have coordinate transformation matrix for orientation in 3×3 form and in homogeneous coordinates as:

$$R_B^M(\alpha, \beta, \gamma) = R_z R_y R_x$$
$$= \begin{bmatrix} \cos\beta\cdot\cos\gamma & \cos\alpha\cdot\sin\gamma + \cos\alpha\cdot\sin\beta\cdot\cos\gamma & \sin\alpha\cdot\sin\gamma + \cos\alpha\cdot\sin\beta\cdot\cos\gamma \\ \cos\beta\cdot\sin\gamma & \cos\alpha\cdot\cos\gamma + \sin\alpha\cdot\sin\beta\cdot\sin\gamma & \sin\alpha\cdot\cos\gamma + \cos\alpha\cdot\sin\beta\cdot\sin\gamma \\ \sin\beta & \sin\alpha\cdot\cos\beta & \cos\alpha\cdot\cos\beta \end{bmatrix} \tag{9.4}$$

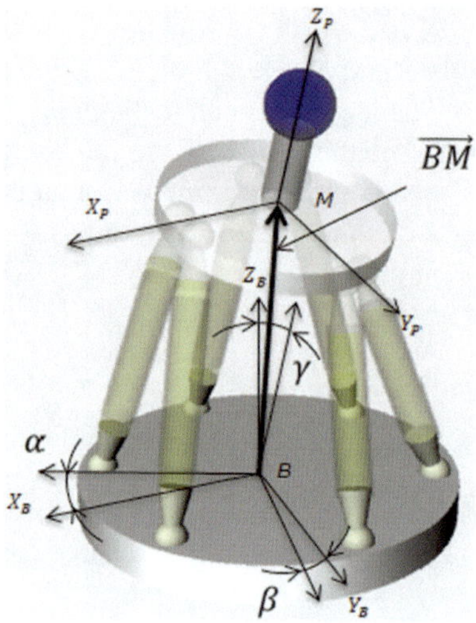

Fig. 9.2 Coordinate transformation of the Stewart Platform

$$T_{\mathrm{B}}^{\mathrm{M}}\,(X_{\mathrm{M}}, Y_{\mathrm{M}}, Z_{\mathrm{M}}, \alpha, \beta, \gamma) = \begin{bmatrix} \cos\beta\cdot\cos\gamma & \cos\alpha\cdot\sin\gamma+\cos\alpha\cdot\sin\beta\cdot\cos\gamma & \sin\alpha\cdot\sin\gamma+\cos\alpha\cdot\sin\beta\cdot\cos\gamma & x \\ \cos\beta\cdot\sin\gamma & \cos\alpha\cdot\cos\gamma+\sin\alpha\cdot\sin\beta\cdot\sin\gamma & \sin\alpha\cdot\cos\gamma+\cos\alpha\cdot\sin\beta\cdot\sin\gamma & y \\ \sin\beta & \sin\alpha\cdot\cos\beta & \cos\alpha\cdot\cos\beta & z \\ 0 & 0 & 0 & 1 \end{bmatrix} \tag{9.5}$$

The coordinate transformation matrix from the base platform to the mobile platform (Fig. 9.2) is

$$\begin{bmatrix} X_{\mathrm{M}} \\ Y_{\mathrm{M}} \\ Z_{\mathrm{M}} \end{bmatrix} = R_{\mathrm{B}}^{\mathrm{M}} \begin{bmatrix} X_{\mathrm{B}} \\ Y_{\mathrm{B}} \\ Z_{\mathrm{B}} \end{bmatrix} + \begin{bmatrix} x \\ y \\ z \end{bmatrix} \quad \text{or} \quad \begin{bmatrix} X_{\mathrm{M}} \\ Y_{\mathrm{M}} \\ Z_{\mathrm{M}} \\ 1 \end{bmatrix} = T_{\mathrm{B}}^{\mathrm{M}} \begin{bmatrix} X_{\mathrm{B}} \\ Y_{\mathrm{B}} \\ Z_{\mathrm{B}} \\ 1 \end{bmatrix}. \tag{9.6}$$

9.2.5 Kinematic Constraints

In Fig. 9.3, point P_i^{B} is the connecting points placed on base platform and point P_i^{M} is the connecting points placed on mobile platform. The separation angles between points, P_1^{B} and P_2^{B}, P_3^{B} and P_4^{B} and P_5^{B} and P_6^{B}, are denoted by 2φ. The separation angles between center points are $\theta = 60$ [deg]. Then we have points P_i^{B} locally on the base platform as

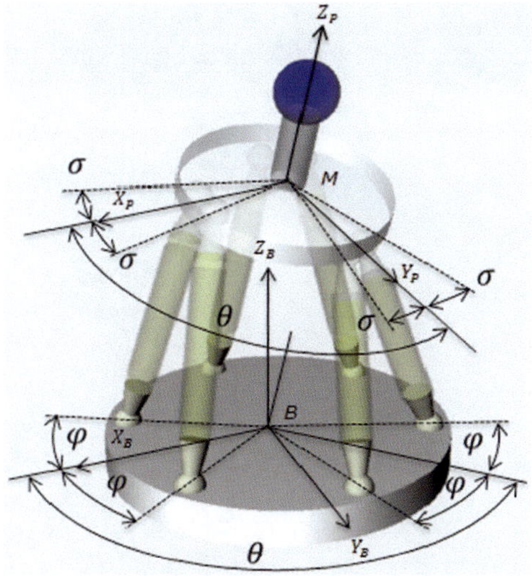

Fig. 9.3 The separation angles between anchor points

$$P_1^{\mathrm{B}} = \begin{bmatrix} \cos(-\varphi) & -\sin(-\varphi) & 0 \\ \sin(-\varphi) & \cos(-\varphi) & 0 \\ 0 & 0 & 1 \end{bmatrix} \begin{bmatrix} R_{\mathrm{B}} \\ 0 \\ 0 \end{bmatrix}, \tag{9.7}$$

$$P_2^{\mathrm{B}} = \begin{bmatrix} \cos\varphi & -\sin\varphi & 0 \\ \sin\varphi & \cos\varphi & 0 \\ 0 & 0 & 1 \end{bmatrix} \begin{bmatrix} R_{\mathrm{B}} \\ 0 \\ 0 \end{bmatrix}, \tag{9.8}$$

$$P_3^{\mathrm{B}} = \begin{bmatrix} \cos(\theta-\varphi) & -\sin(\theta-\varphi) & 0 \\ \sin(\theta-\varphi) & \cos(\theta-\varphi) & 0 \\ 0 & 0 & 1 \end{bmatrix} \begin{bmatrix} R_{\mathrm{B}} \\ 0 \\ 0 \end{bmatrix}, \tag{9.9}$$

$$P_4^{\mathrm{B}} = \begin{bmatrix} \cos(\theta+\varphi) & -\sin(\theta+\varphi) & 0 \\ \sin(\theta+\varphi) & \cos(\theta+\varphi) & 0 \\ 0 & 0 & 1 \end{bmatrix} \begin{bmatrix} R_{\mathrm{B}} \\ 0 \\ 0 \end{bmatrix}, \tag{9.10}$$

$$P_5^{\mathrm{B}} = \begin{bmatrix} \cos(2\theta-\varphi) & -\sin(2\theta-\varphi) & 0 \\ \sin(2\theta-\varphi) & \cos(2\theta-\varphi) & 0 \\ 0 & 0 & 1 \end{bmatrix} \begin{bmatrix} R_{\mathrm{B}} \\ 0 \\ 0 \end{bmatrix}, \tag{9.11}$$

$$P_6^{\mathrm{B}} = \begin{bmatrix} \cos(2\theta+\varphi) & -\sin(2\theta+\varphi) & 0 \\ \sin(2\theta+\varphi) & \cos(2\theta+\varphi) & 0 \\ 0 & 0 & 1 \end{bmatrix} \begin{bmatrix} R_{\mathrm{B}} \\ 0 \\ 0 \end{bmatrix}. \tag{9.12}$$

The separation angles between points, P_1^{M} and P_2^{M}, P_3^{M} and P_4^{M} and P_5^{M} and P_6^{M}, are denoted by 2σ. The separation angles between a center point of P_1^{M} and P_2^{M} and a center point of P_3^{M} and P_4^{M} is $\theta = 60$ [deg]. The separation angles between a middle point of P_3^{M} and P_4^{M} and a middle point of P_5^{M} and P_6^{M} is $\theta = 60\,[\mathrm{deg}]$. Then we have points P_i^{M} locally on the base platform as

$$P_1^{\mathrm{M}} = \begin{bmatrix} \cos(-\sigma) & -\sin(-\sigma) & 0 \\ \sin(-\sigma) & \cos(-\sigma) & 0 \\ 0 & 0 & 1 \end{bmatrix} \begin{bmatrix} R_{\mathrm{M}} \\ 0 \\ 0 \end{bmatrix}, \tag{9.13}$$

$$P_2^{\mathrm{M}} = \begin{bmatrix} \cos\sigma & -\sin\sigma & 0 \\ \sin\sigma & \cos\sigma & 0 \\ 0 & 0 & 1 \end{bmatrix} \begin{bmatrix} R_{\mathrm{M}} \\ 0 \\ 0 \end{bmatrix}, \tag{9.14}$$

$$P_3^{\mathrm{M}} = \begin{bmatrix} \cos(\theta-\sigma) & -\sin(\theta-\sigma) & 0 \\ \sin(\theta-\sigma) & \cos(\theta-\sigma) & 0 \\ 0 & 0 & 1 \end{bmatrix} \begin{bmatrix} R_{\mathrm{M}} \\ 0 \\ 0 \end{bmatrix}, \tag{9.15}$$

$$P_4^{\mathrm{M}} = \begin{bmatrix} \cos(\theta+\sigma) & -\sin(\theta+\sigma) & 0 \\ \sin(\theta+\sigma) & \cos(\theta+\sigma) & 0 \\ 0 & 0 & 1 \end{bmatrix} \begin{bmatrix} R_{\mathrm{M}} \\ 0 \\ 0 \end{bmatrix}, \tag{9.16}$$

$$P_5^{\mathrm{M}} = \begin{bmatrix} \cos(2\theta-\sigma) & -\sin(2\theta-\sigma) & 0 \\ \sin(2\theta-\sigma) & \cos(2\theta-\sigma) & 0 \\ 0 & 0 & 1 \end{bmatrix} \begin{bmatrix} R_{\mathrm{M}} \\ 0 \\ 0 \end{bmatrix}, \tag{9.17}$$

$$P_6^{\mathrm{M}} = \begin{bmatrix} \cos(2\theta+\sigma) & -\sin(2\theta+\sigma) & 0 \\ \sin(2\theta+\sigma) & \cos(2\theta+\sigma) & 0 \\ 0 & 0 & 1 \end{bmatrix} \begin{bmatrix} R_{\mathrm{M}} \\ 0 \\ 0 \end{bmatrix}. \tag{9.18}$$

The connecting points P_i^{M} in global framework are defined by X_i^{M}

$$X_i^{\mathrm{M}} = X_{\mathrm{M}} + R_{\mathrm{B}}^{\mathrm{M}} P_i^{\mathrm{M}}. \tag{9.19}$$

Defining tool position $(\Delta x_{\mathrm{T}}, \Delta y_{\mathrm{T}}, \Delta z_{\mathrm{T}})$ on the local coordinate system of the mobile platform O_{M}, the tool position X_{T} in global coordinate system is defined as

$$X_{\mathrm{T}} = X_{\mathrm{M}} + R_{\mathrm{B}}^{\mathrm{M}} (\Delta x_{\mathrm{T}}, \Delta y_{\mathrm{T}}, \Delta z_{\mathrm{T}})^T. \tag{9.20}$$

Deleting X_{M} from Eqs. (9.17) and (9.18), we have the connecting points X_i^{M} in global framework according to Tool position and orientation

$$X_i^{\mathrm{M}} = X_{\mathrm{T}} + R_{\mathrm{B}}^{\mathrm{M}} P_i^{\mathrm{M}} - R_{\mathrm{B}}^{\mathrm{M}} (\Delta x_{\mathrm{T}}, \Delta y_{\mathrm{T}}, \Delta z_{\mathrm{T}})^T. \tag{9.21}$$

To represent three translations and three orientation of the mobile platform, anchor points P_i^{M} of the mobile platform, which are the end-effector of each three variable legs, are controlled. If the position and orientation of the legs are achieved by

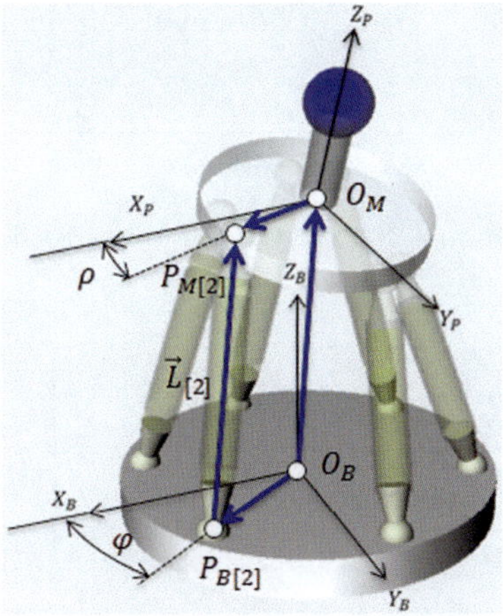

Fig. 9.4 The separation angles between anchor points

linear motion of the leg, geometrical relationship is defined for each leg (Fig. 9.4). Applying the transformation matrix, we have Equations of motion for each leg motion

$$L_i = X_i^{\mathrm{M}} - X_i^{\mathrm{B}} = \overrightarrow{O_{\mathrm{B}}O_{\mathrm{M}}} + R_{\mathrm{B}}^{\mathrm{M}} P_i^{\mathrm{B}} - X_i^{\mathrm{B}} \ (i = 1, \ldots, 6). \tag{9.22}$$

When the position and orientation of the end-effector is given, taking the magnitudes of the vector differences between corresponding base plate and mobile plate vertices, the length of each leg is computed as the following

$$L_i = \left| \overrightarrow{O_{\mathrm{B}}O_{\mathrm{M}}} + R_{\mathrm{B}}^{\mathrm{M}} P_i^{\mathrm{B}} - X_i^{\mathrm{B}} \right| \ (i = 1, \ldots, 6). \tag{9.23}$$

9.3 Jacobian Analysis

For the Stewart platform, we have seen that a closed-form solution of the inverse kinematics can be obtained.

$$L_i = F_i\,(x, y, z, \alpha, \beta, \gamma) \quad (i = 1, \ldots, 6) \tag{9.24}$$

According to the chain rule, functions of the differentials as

$$\delta L_i = \frac{\partial F_i}{\partial x}\delta x + \frac{\partial F_i}{\partial y}\delta y + \frac{\partial F_i}{\partial z}\delta z + \frac{\partial F_i}{\partial \alpha}\delta \alpha + \frac{\partial F_i}{\partial \beta}\delta \beta + \frac{\partial F_i}{\partial \gamma}\delta \gamma \ (i = 1, \ldots, 6). \tag{9.25}$$

Dividing both sides by the differential time element δt, we have

$$\frac{\delta L_i}{\delta t} = \frac{\partial F_i}{\partial x}\frac{\delta x}{\delta t} + \frac{\partial F_i}{\partial y}\frac{\delta y}{\delta t} + \frac{\partial F_i}{\partial z}\frac{\delta z}{\delta t} + \frac{\partial F_i}{\partial \alpha}\frac{\delta \alpha}{\delta t} + \frac{\partial F_i}{\partial \beta}\frac{\delta \beta}{\delta t} + \frac{\partial F_i}{\partial \gamma}\frac{\delta \gamma}{\delta t}. \tag{9.26}$$

In matrix form, Eq. (9.24) becomes,

$$\begin{bmatrix} \dot{L}_{[1]} \\ \dot{L}_{[2]} \\ \dot{L}_{[3]} \\ \dot{L}_{[4]} \\ \dot{L}_{[5]} \\ \dot{L}_{[6]} \end{bmatrix} = \begin{bmatrix} \frac{\delta F_1}{\partial x} & \frac{\delta F_1}{\partial y} & \frac{\delta F_1}{\partial z} & \frac{\delta F_1}{\partial \alpha} & \frac{\delta F_1}{\partial \beta} & \frac{\delta F_1}{\partial \gamma} \\ \frac{\delta F_2}{\partial x} & \frac{\delta F_2}{\partial y} & \frac{\delta F_2}{\partial z} & \frac{\delta F_2}{\partial \alpha} & \frac{\delta F_2}{\partial \beta} & \frac{\delta F_2}{\partial \gamma} \\ \vdots & \vdots & \vdots & \vdots & \vdots & \vdots \\ \frac{\delta F_6}{\partial x} & \frac{\delta F_6}{\partial y} & \frac{\delta F_6}{\partial z} & \frac{\delta F_6}{\partial \alpha} & \frac{\delta F_6}{\partial \beta} & \frac{\delta F_6}{\partial \gamma} \end{bmatrix} \begin{bmatrix} \frac{\delta x}{\delta t} \\ \frac{\delta y}{\delta t} \\ \frac{\delta z}{\delta t} \\ \frac{\delta \alpha}{\delta t} \\ \frac{\delta \beta}{\delta t} \\ \frac{\delta \gamma}{\delta t} \end{bmatrix} \tag{9.27}$$

$$\begin{bmatrix} \dot{L}_{[1]} \\ \dot{L}_{[2]} \\ \dot{L}_{[3]} \\ \dot{L}_{[4]} \\ \dot{L}_{[5]} \\ \dot{L}_{[6]} \end{bmatrix} = \begin{bmatrix} \frac{\delta F_1}{\partial x} & \frac{\delta F_1}{\partial y} & \frac{\delta F_1}{\partial z} & \frac{\delta F_1}{\partial \alpha} & \frac{\delta F_1}{\partial \beta} & \frac{\delta F_1}{\partial \gamma} \\ \frac{\delta F_2}{\partial x} & \frac{\delta F_2}{\partial y} & \frac{\delta F_2}{\partial z} & \frac{\delta F_2}{\partial \alpha} & \frac{\delta F_2}{\partial \beta} & \frac{\delta F_2}{\partial \gamma} \\ \vdots & \vdots & \vdots & \vdots & \vdots & \vdots \\ \frac{\delta F_6}{\partial x} & \frac{\delta F_6}{\partial y} & \frac{\delta F_6}{\partial z} & \frac{\delta F_6}{\partial \alpha} & \frac{\delta F_6}{\partial \beta} & \frac{\delta F_6}{\partial \gamma} \end{bmatrix} \begin{bmatrix} \dot{x} \\ \dot{y} \\ \dot{z} \\ \dot{\alpha} \\ \dot{\beta} \\ \dot{\gamma} \end{bmatrix}. \tag{9.28}$$

Equation (9.26) shows commonly used expression of the inverse of the conventional Jacobian matrix, which relates the velocities of the active joints to the generalized velocity of the mobile platform. It can be described as

$$\dot{L}_i = J^{-1}\dot{X} \qquad (i = 1, \ldots, 6) \tag{9.29}$$

where $\dot{L}_i$ and X are the velocities of the leg and the mobile platform respectively. The Jacobian matrix can be derived by formulating a velocity loop-closure equation for the ith leg can be written as

$$\overline{O_B O_M} + \overline{O_M P_i^M} = \overline{O_B P_i^B} + \overline{P_i^B P_i^M} \ (i = 1, \ldots, 6)\,, \tag{9.30}$$

The center of the joint between the leg i and the mobile part P_i^M on the moving platform with reference to the base coordinate system is obtained as

$$X_i^M = X_M + R_B^M P_i^M = P_i^B + \left(P_i^M - P_i^B\right). \tag{9.31}$$

Differentiating the above equation with respect to time yields

$$\dot{X}_i^M = \dot{X}_M + \omega_B \times m_i = L_i \omega_i \times d_i + \dot{L}_i d_i. \tag{9.32}$$

Where ω_B is angular velocity of the mobile platform with reference to the base platform and is represented as $\omega_B = \left(\dot{\omega}_x, \dot{\omega}_y, \dot{\omega}_z\right)^T = \left(\dot{\alpha}, \dot{\beta}, \dot{\gamma}\right)^T$, m_i is the vector $P_M P_i^M$ and d_i is the vector $P_i^B P_i^M$. Equation (9.28) can be written in matrix form, which yields a scalar equation as

$$J_x \dot{X}_M = J_l \dot{L}_i. \tag{9.33}$$

According to Eq. (9.29), we have Jacobian matrix and the inverse of Jacobian matrix as

$$\dot{X}_M = J_x^{-1} J_l \dot{L}_i = J \dot{L}_i. \tag{9.34}$$

$$\dot{L}_i = J_l^{-1} J_x \dot{X}_M = J^{-1} \dot{X}_M. \tag{9.35}$$

$$J^{-1} = J_l^{-1} J_x = J_l \dot{L}_i = \begin{bmatrix} d_1{}^T \; (m_1 \times d_1)^T \\ d_2{}^T \; (m_2 \times d_2)^T \\ d_3{}^T \; (m_3 \times d_3)^T \\ d_4{}^T \; (m_4 \times d_4)^T \\ d_5{}^T \; (m_5 \times d_5)^T \\ d_6{}^T \; (m_6 \times d_6)^T \end{bmatrix}. \tag{9.36}$$

We have the Jacobian matrix and the inverse of Jacobian matrix shown in Eq. (9.34).

Stiffness is an important factor in the design of the parallel link robot, since parallel link robots are required to perform with high-accuracy positioning, high-speed machining, lower mass/inertia properties, and higher structural rigidity. Stiffness can be defined as the capacity of a mechanical system to sustain loads without excessive changes on its geometry. The stiffness varies with the posture of the mechanism, geometry, the applied forces, deformations, or compliant displacements. Generally, stiffness is defined through the "stiffness matrix" K, which relates the translational/rotational displacement of the end-effector and the static forces/torques. The "compliance matrix" k is defined by the inverse of *K*. If the legs are of the same type and the spring constants associated with all the legs are of the same value, the stiffness matrix for the Stewart platform (Fig. 9.5) is given by

$$K = kJ^T J \tag{9.37}$$

9.4 Dynamics

This section presents the dynamics of the parallel link robots. The dynamic modeling of parallel link robots is quite complicated because of their closed-loop structure, coupled relationship between system parameters, high nonlinearity in

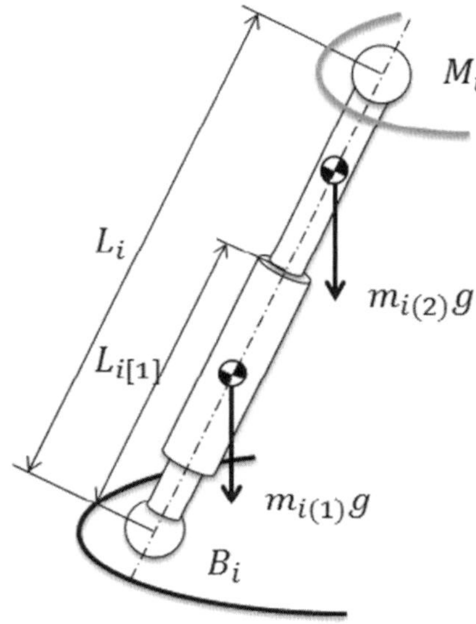

Fig. 9.5 Schematic view of the ith leg of the parallel link robot

system dynamics and kinematic constraints. Nevertheless, the dynamic modeling of the parallel link robots is quite important because precise positioning and good dynamic performance under high load are required. To obtain the dynamic model of parallel link robots, there are many valuable studies published by many researches. Generally, two methods, the Newton Euler formulation, and the Lagrangian formulation, can be used in order to obtain Equations of motion. The Newton Euler formulation is derived by the direct interpretation of Newton's Second Law of Motion, which describes that the net force on an object is equal to the rate of change of its linear momentum in an inertial reference frame. Using the Newton-Euler approach, calculations of constraint forces are required but are eliminated to obtain the final equations of motion. On the other hand, Lagrangian formulation relies on the energy properties of mechanical systems to compute Equations of motion. In this method, all the workless forces and constraint forces are automatically eliminated. The resulting equations can be computed in closed form expression in terms of joint torques and joint displacements. In this work, inverse dynamics of the Stewart platform has been formulated by Lagrangian formulation.

9.4.1 Lagrange's Equations

To form Equations of motion, we define the Lagrangian, L, as the difference between the kinetic energy K and potential energy U of the system. The kinetic energy and potential energy for both of these parts are computed and then the dynamic equations are derived using these energies. Lagrangian equations of motion is

$$\frac{d}{\mathrm{d}t}\left(\frac{\partial L}{\partial \dot{q}}\right) - \frac{\partial L}{\partial q} = \tau \tag{9.38}$$

where q is the generalized coordinate vector, $\dot{q}$ is the generalized velocity vector and τ is the generalized force vector. Appling $L = K - U$ to Eq. (9.33), the Lagrangian equations of motion is defined as

$$\frac{d}{\mathrm{d}t}\left(\frac{\partial K}{\partial \dot{q}}\right) - \frac{\partial K}{\partial q} + \frac{\partial U}{\partial q} = \tau \tag{9.39}$$

here, the first two terms and third term in the left side of Eq. (9.34) represents Inertial forces and Gravity vector respectively.

9.4.2 Kinetic Energy and Potential Energy

Using the generalized velocity vector $\dot{q}$, the kinetic energy is defined as

$$K = \frac{1}{2}\dot{q}^T M(q)\dot{q}. \tag{9.40}$$

Applying Eq. (9.35) to each component of Eq. (9.34), we have

$$\frac{d}{\mathrm{d}t}\left(\frac{\partial K}{\partial \dot{q}}\right) = \frac{d}{\mathrm{d}t}\left(\frac{\partial}{\partial \dot{q}}\left(\frac{1}{2}\dot{q}^T M(q)\dot{q}\right)\right) = \frac{d}{\mathrm{d}t}\left(M(q)\dot{q}\right) = M(q)\ddot{q}, \tag{9.41}$$

$$\frac{\partial K}{\partial q} = \frac{\partial}{\partial q}\left(\frac{1}{2}\dot{q}^T M(q)\dot{q}\right) = \frac{1}{2}\dot{q}^T \begin{bmatrix} \frac{\partial M(q)}{\partial q_1} \\ \vdots \\ \frac{\partial M(q)}{\partial q_6} \end{bmatrix} \dot{q}^T - \dot{M}(q)\dot{q}. \tag{9.42}$$

Therefore, the inertial forces are

$$\frac{d}{\mathrm{d}t}\left(\frac{\partial K}{\partial \dot{q}}\right) - \frac{\partial K}{\partial q} = M(q)\ddot{q} + \frac{1}{2}\dot{q}^T \begin{bmatrix} \frac{\partial M(q)}{\partial q_1} \\ \vdots \\ \frac{\partial M(q)}{\partial q_6} \end{bmatrix} \dot{q}^T - \dot{M}(q)\dot{q}. \tag{9.43}$$

The parallel link robot is considered as a multi-robot system with i serial robots moving according to a common end-effector. Figure 9.6 shows the links, the frames, position vector, angular velocity, force, and inertia for the leg i ($i = 1, \ldots, 6$). In Fig. 9.6, v_{c_i} is the translational velocity vector for the ith body and ω_i is the angular velocity vector, m_i is mass and I_{C_i} is inertia for the ith link. The kinetic energy for the link i can be defined by the translational motion and rotational motion

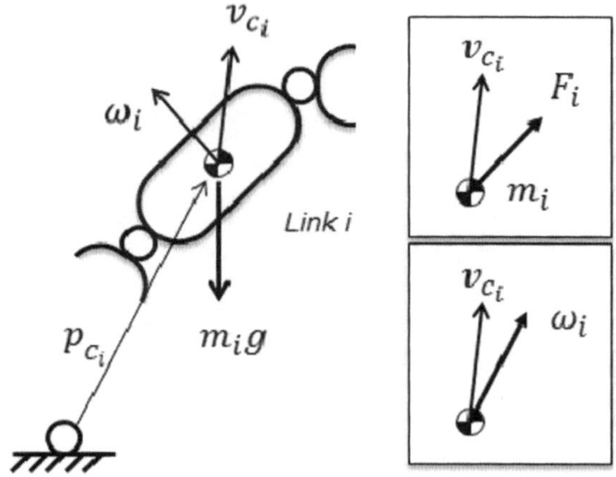

Fig. 9.6 Modeling of kinetic energy for link i

$$K_i = \frac{1}{2}\left(m_i {v_{c_i}}^T v_{c_i} + {\omega_i}^T I_{C_i}\omega_i\right). \tag{9.44}$$

Total Kinetic energy can be derived by adding each Kinetic energy of the link

$$K = \sum_{i=1}^{N} K_i = \frac{1}{2}\sum_{i=1}^{N}\left(m_i {v_{c_i}}^T v_{c_i} + {\omega_i}^T I_{C_i}\omega_i\right). \tag{9.45}$$

Appling transitional Jacobian matrix J_{v_i} which relates the generalized coordinate vector and rotational Jacobian matrix J_{ω_i} which relates the generalized coordinate vector and angular velocity vector, we have relationship as follows:

$$v_{c_i} = J_{v_i}\dot{q}, \tag{9.46}$$

$$\omega_{c_i} = J_{\omega_i}\dot{q}. \tag{9.47}$$

Here,

$$J_{v_i} = \left[\begin{array}{ccc} \frac{\partial p_{c_i}}{\partial q_1} \cdots \frac{\partial p_{c_i}}{\partial q_i} & 0 \cdots 0 \end{array}\right], \tag{9.48}$$

$$J_{v_i} = \left[\bar{\varepsilon}_1 z_1 \ldots \bar{\varepsilon}_i z_i \quad 0 \cdots 0\right]. \tag{9.49}$$

Equation (9.40) can be modified by Eqs. (9.41) and (9.42) as follows:

$$\begin{aligned} \frac{1}{2}\dot{q}^T M(q)\dot{q} &= \frac{1}{2}\sum_{i=1}^{N}\left(m_i\dot{q}^T J_{v_i}^T J_{v_i}\dot{q} + \dot{q}^T J_{\omega_i}^T I_{C_i} J_{\omega_i}\dot{q}\right) \\ &= \frac{1}{2}\dot{q}^T\left[\sum_{i=1}^{N}\left(m_i J_{v_i}^T J_{v_i} + J_{\omega_i}^T I_{C_i} J_{\omega_i}\right)\right]\dot{q}. \end{aligned} \tag{9.50}$$

Therefore, we have

$$M(q)\dot{q} = \sum_{i=1}^{N} \left(m_i J_{v_i}^T J_{v_i} + J_{\omega_i}^T I_{C_i} J_{\omega_i} \right). \tag{9.51}$$

For the link i, the Potential energy is defined by mass of the link i, that is $G = \partial U / \partial q$

$$G = J_{v_i} G = \begin{bmatrix} J_{v_1}^T & J_{v_2}^T & \cdots & J_N^T \end{bmatrix} \begin{bmatrix} m_1 g \\ m_2 g \\ \vdots \\ m_N g \end{bmatrix}. \tag{9.52}$$

9.4.3 Dynamics Equation

Finally, the Lagrange's equations of motion shown in Eq. (9.33) is clearly defined by the inertial forces shown in Eq. (9.38) and Gravity vector shown in Eq. (9.47) respectively

$$\begin{aligned} \frac{d}{dt}\left(\frac{\partial K}{\partial \dot{q}}\right) - \frac{\partial K}{\partial q} + \frac{\partial U}{\partial q} &= M(q)\ddot{q} + \dot{q}^T \begin{bmatrix} \frac{\partial M(q)}{\partial q_1} \\ \vdots \\ \frac{\partial M(q)}{\partial q_6} \end{bmatrix} \dot{q}^T - \dot{M}(q)\dot{q} + G \\ &= M(q)\ddot{q} + V(q, \dot{q}) + G. \end{aligned} \tag{9.53}$$

Generally, the joint space dynamics of a robot are described by Eq. (9.48) where q is the $n \times 1$ generalized vector in joint space, M(q) is the $n \times n$ mass/inertia matrix, $V(q, \dot{q})$ is the Coriolis and centrifugal torque and $g(q)$ is gravity torque. In order to derive the dynamic equations of the Stewart Platform, the whole system is separated into two parts: the mobile platform and the legs.

9.4.4 Dynamics Equation of Mobile Platform

The kinetic energy for the mobile platform can be defined by the translational motion and rotational motion

$$K_{\mathrm{MP}} = \frac{1}{2}\dot{q}^T M(q)\dot{q} = \frac{1}{2}\left(m_{\mathrm{MP}} \left(X_{\mathrm{M}}^2 + Y_{\mathrm{M}}^2 + Z_{\mathrm{M}}^2 \right) + \omega_{\mathrm{MP}}^T I_c \omega_{\mathrm{MP}} \right). \tag{9.54}$$

where, m_{MP} is mass of the mobile platform, $(X_{\mathrm{M}}, Y_{\mathrm{M}}, Z_{\mathrm{M}})$ is the absolute coordinate of the mobile platform, ω_{MP} is the angular velocity of the moving platform and I_c is the rotational inertia mass. Here, the generalized coordinate vector can be defined as $\dot{q} = \left(\dot{X}_{\mathrm{M}}, \dot{Y}_{\mathrm{M}}, \dot{Z}_{\mathrm{M}}, \dot{\alpha}, \dot{\beta}, \dot{\gamma}\right)$. Applying this definition into Eq. (9.49) yields

$$K_{\mathrm{MP}} = \frac{1}{2}\left(\dot{X}_{\mathrm{M}}, \dot{Y}_{\mathrm{M}}, \dot{Z}_{\mathrm{M}}, \dot{\alpha}, \dot{\beta}, \dot{\gamma}\right) M_{\mathrm{MP}}(q)\left(\dot{X}_{\mathrm{M}}, \dot{Y}_{\mathrm{M}}, \dot{Z}_{\mathrm{M}}, \dot{\alpha}, \dot{\beta}, \dot{\gamma}\right)^T \tag{9.55}$$

where, $M_{\mathrm{MP}}(q)$ is the 6×6 mass diagonal matrix of the moving platform with carrying load

$$M_{\mathrm{MP}}(q) = \begin{bmatrix} m_{\mathrm{MP}} & 0 & 0 & 0 & 0 & 0 \\ 0 & m_{\mathrm{MP}} & 0 & 0 & 0 & 0 \\ 0 & 0 & m_{\mathrm{MP}} & 0 & 0 & 0 \\ 0 & 0 & 0 & I_{c[x]} & 0 & 0 \\ 0 & 0 & 0 & 0 & I_{c[y]} & 0 \\ 0 & 0 & 0 & 0 & 0 & I_{c[z]} \end{bmatrix}. \tag{9.56}$$

And the potential energy of the mobile platform is

$$U_{\mathrm{MP}} = m_{\mathrm{MP}} g Z_{\mathrm{M}} q = \left[0\ 0\ m_{\mathrm{MP}} g\ 0\ 0\ 0\right]\left[\dot{X}_{\mathrm{M}}, \dot{Y}_{\mathrm{M}}, \dot{Z}_{\mathrm{M}}, \dot{\alpha}, \dot{\beta}, \dot{\gamma}\right]^T, \tag{9.57}$$

$$\frac{\partial U}{\partial q} = \frac{\partial}{\partial q}\left(m_{\mathrm{MP}} g Z_{\mathrm{M}} q\right) = G_{\mathrm{MP}}, \tag{9.58}$$

$$V_{\mathrm{MP}}(q, \dot{q}) = \frac{1}{2}\dot{q}^T \begin{bmatrix} \frac{\partial M_{\mathrm{MP}}(q)}{\partial q_1} \\ \vdots \\ \frac{\partial M_{\mathrm{MP}}(q)}{\partial q_6} \end{bmatrix} \dot{q}^T - \dot{M}_{\mathrm{MP}}(q)\dot{q}. \tag{9.59}$$

Then we have Equation of motion according to Eq. (9.48) as

$$M_{\mathrm{MP}}(q)\ddot{q} + V_{\mathrm{MP}}(q, \dot{q}) + G_{\mathrm{MP}} = \tau_{\mathrm{MP}}. \tag{9.60}$$

9.4.5 Dynamics Equation of Legs

Each leg consists of two parts: the moving part and the fixed part. Therefore, N in Eqs. (9.39)–(9.47) equals to 2 and $i = 1, 2, \ldots, 6$. The kinetic energy for the mobile platform can be defined by the translational motion and rotational motion

$$K_{\mathrm{LEG}(i)} = \sum_{j=1}^{2} K_j = \frac{1}{2}\sum_{j=1}^{2}\left(m_i {v_{c_j}}^T v_{c_j} + {\omega_j}^T I_{C_j} \omega_j\right). \tag{9.61}$$

And the potential energy of the mobile platform is

$$G = J_{v_j} G = \left[J_{v_1}^T \; J_{v_2}^T \right] \begin{bmatrix} m_1 g \\ m_2 g \end{bmatrix}. \tag{9.62}$$

The mass and inertia matrix is

$$M_i(q) = \sum_{j=1}^{2} \left(m_j J_{v_j}^T J_{v_j} + J_{\omega_j}^T I_{C_j} J_{\omega_j} \right). \tag{9.63}$$

The Coriolis and centrifugal torque is

$$V_{\mathrm{LEG}(i)}\left(q, \dot{q}\right) = \frac{1}{2} \dot{q}^T \begin{bmatrix} \frac{\partial M_{\mathrm{LEG}}(q)}{\partial q_1} \\ \vdots \\ \frac{\partial M_{\mathrm{LEG}}(q)}{\partial q_6} \end{bmatrix} \dot{q}^T - \dot{M}_{\mathrm{LEG}(i)}(q)\dot{q}. \tag{9.64}$$

Then we have Equation of motion according to Eq. (9.48) as

$$M_{\mathrm{LEG}(i)}(q)\ddot{q} + V_{\mathrm{LEG}(i)}\left(q, \dot{q}\right) + G_{\mathrm{LEG}(i)} = \tau_{\mathrm{LEG}(i)}. \tag{9.65}$$

Since the platform is divided into two parts, inertia, Coriolis-Centrifugal and gravity matrix in Eqs. (9.54) and (9.59)

$$M(q) = M_{\mathrm{MP}}(q) + \sum_{j=1}^{6} \left(M_{\mathrm{LEG}(i)}(q) \right), \tag{9.66}$$

$$V\left(q, \dot{q}\right) = V_{\mathrm{MP}}\left(q, \dot{q}\right) + \sum_{j=1}^{6} \left(V_{\mathrm{LEG}(i)}\left(q, \dot{q}\right) \right), \tag{9.67}$$

$$G(q) = G_{\mathrm{MP}}(q) + \sum_{j=1}^{6} \left(G_{\mathrm{LEG}(i)}(q) \right). \tag{9.68}$$

9.5 The Operational Space Formulation

The dynamics of a robot must be considered in order to execute fast, accurate, and compliant motion. Khatib [8] presented the *Operational Space Formulation* (1987), which was applied to end-effector dynamics for rigid-body robot manipulators. Generally, task of the end-effector is specified by means of a position in Cartesian coordinates and an orientation in terms of Euler angles. On the other hand, the robot configuration is measured through sensors located in the joints compensating dynamic effects of the link. The operational space formulation handles with these two dynamics decoupling tasks from redundant null space dynamics, which results in any additional forces applied within the null space remain dynamically consistent with the tasks. The Operational Space Formulation provides different solution such as resolved-motion rate control, resolved-acceleration control, and force-based control. If the robot model is accurately known, dynamically consistent framework can be extended more into multiple-task control with prioritization, constrained dynamics, etc. In this point of view, the Operational space control is one of the most efficient approaches to control robots from end-effector control of manipulators up to humanoid robots [9] due to its potential for dynamically consistent control, compliant control, force control, and hierarchical control. In this section, the Operational Space Formulation by task and posture is introduced to control the end-effector of the Stewart Platform dynamically efficient method [10].

Corresponding to the instantaneous linear/angular velocity, ϑ, in task space, the following relationship is defined by the Jacobian, $J(q)$

$$\vartheta = J(q)\dot{q}, \tag{9.69}$$

Task dynamic behavior is obtained by projecting the joint space dynamics into the space associated with the task:

$$\Lambda(q)\dot{\vartheta} + \mu\,(q,\dot{q}) + p(q) = F \tag{9.70}$$

here, $\Lambda(q)$, $\mu\,(q,\dot{q})$, and $p(q)$ are the inertia matrix, the vector of Coriolis/centrifugal forces and the vector of gravity forces mapped into the operational space and are defined as follows;

$$\Lambda(q) = \left(JA^{-1}J^{T}\right)^{-1}, \tag{9.71}$$

$$\mu\,(q,\dot{q}) = \Lambda\left(JA^{-1}B - \dot{J}\dot{q}\right)^{-1}, \tag{9.72}$$

$$p(q) = \Lambda JA^{-1}g. \tag{9.73}$$

The control force, F, in Eq. (9.68) provides a decoupled control structure by

$$F = \hat{\Lambda}(q)f^{*} + \hat{\mu}\,(q,\dot{q})\,\mu + \hat{p}(q) \tag{9.74}$$

where $\hat{.}$ represents estimates of the model parameters. f^{*} is the command to the unit mass system. When the estimates are perfect, the following decoupled equations of motion for the end-effector are obtained

$$\dot{\vartheta} = f^{*}. \tag{9.75}$$

The Operational Space Formulation provides decomposition of joint forces into two control vectors; (1) the joint torque corresponding to forces acting at the task and (2) joint torque that only affects the posture behavior in the null space

$$\Gamma = \Gamma_{\text{Task}} + \Gamma_{\text{Posture}} = J^T F + N^T(q)\Gamma. \tag{9.76}$$

Here, $N^T(q)$ is the dynamically consistent null space projection matrix.

$$N^T(q) = I - J^T \bar{J}^T, \tag{9.77}$$

$$\bar{J}^T = \Lambda J A^{-1}. \tag{9.78}$$

And Γ_{Posture} is joint space dynamics defined by Eq. (9.51)

$$\Gamma_{\text{Posture}} = M(q)\ddot{q} + V(q, \dot{q}) + G. \tag{9.79}$$

The term, $N^T(q)$, guarantees that the null space control torque will not generate any force on the task control.

9.5.1 Trajectory Tracking Task

The Operational Space Command in Cartesian Space is applied to the position and orientation control of the end-effector as Eq. (9.66). $\dot{\vartheta}$ is defined as a simple PD control

$$\dot{\vartheta} = f^* = K_{px}\left(X_{\text{Trajectory}[i]} - X_{\text{T}}\right) + K_{vx}\left(\dot{X}_{\text{Trajectory}[i]} - \dot{X}_{\text{T}}\right), \tag{9.80}$$

K_{px} and K_{vx} are the PD gains in operational space, which are selected for the unit-mass system $\ddot{X} + K_{vx}\dot{X} + K_{px}X = 0$. The term $X_{\text{Trajectory}[i]}$ is a position command of the leg i following designed trajectory and X_{T} is position of the tool. For a simple positioning tracking task, the applied force to the trajectory point is

$$F = \Lambda f^* = \Lambda\left(K_{px}\left(X_{\text{Trajectory}[i]} - X_{\text{T}}\right) + K_{vx}\left(\dot{X}_{\text{Trajectory}[i]} - \dot{X}_{\text{T}}\right)\right). \tag{9.81}$$

The joint torque corresponding to forces acting at the task is calculated by Eq. (9.72). For given trajectory, accuracy of trajectory tracking is decided by accuracy of dynamics model and design of PD control. Generally, if we increase the position tracking gain K_{px}, it will reduce the position tracking error, however if K_{px} is too high, it will cause instability of the system. At the same time, if we increase the velocity gain K_{vx}, the response to track the trajectory will increase, however if it is too high, it will cause overshoot while tracking and will cause instability of the system. We need to design the best position gain and velocity gain according to the application.

9.6 Trajectory Generation

9.6.1 Trajectory Generation in Cartesian Space

In this section, we focus on methods of computing a trajectory that describes the desired motion of a manipulator in Cartesian space [11]. Here, trajectory refers to a time history of position, velocity, and acceleration for six degrees of freedom. When we specify paths as motions of the tool or the end-effector, relative to the base coordinate system, the basic problem is to move the manipulator from an initial position to final position in time. To design the motion in more detail representation than simply defining the initial and final positions, a sequence of desired intermediate points must be specified. Generally, *Path Points* simply includes the initial and final points, however in more detail representation, the path points include all the desired intermediate points from the initial position to final position in time. These paths can be planned from the user's design of path points in Cartesian space. Many works have been conducted on path planning, however, among such functions, the polynomial splines are often used for planning trajectories that involve multiple segments. In this section, we consider methods of trajectory generation such as cubic spline and parabolic blending, in which the path shapes are interpolated between data points.

9.6.2 Cubic Polynomials

Among polynomial splines, the cubic spline [11, 15] is the lowest-degree polynomial that can provide a smooth trajectory with continuous velocity, acceleration, and jerk profile. Cubic polynomials interpolation takes several data points as the input and interpolates the values for the segments between them, in which each segment is represented by a cubic polynomial function of time. A cubic polynomial is a polynomial of degree 3. The cubic polynomial has the form

$$X(t) = a_0 + a_1 t + a_2 t^2 + a_3 t^3. \tag{9.82}$$

Here, we suppose that the initial position and velocity are, $X(0) = x_0, \dot{X}(0) = v_0$ and the final position and velocity are, $X(t_f) = x_f, \dot{X}(t_f) = v_f$. According to Eq. (9.80), the velocity and acceleration will have the following forms

$$\dot{X}(t) = a_1 + 2a_2 t + 3a_3 t^2, \tag{9.83}$$

$$\ddot{X}(t) = 2a_2 + 6a_3 t, \tag{9.84}$$

Using initial constraints, we can solve the coefficients in Eq. (9.80)

$$a_0 = x_0, \tag{9.85}$$

$$a_1 = 0, \tag{9.86}$$

$$a_2 = \frac{3}{{t_\mathrm{f}}^2}\left(x_\mathrm{f} - x_0\right), \tag{9.87}$$

$$a_3 = \frac{2}{{t_\mathrm{f}}^3}\left(x_\mathrm{f} - x_0\right). \tag{9.88}$$

Figure 9.7 shows a cubic polynomial trajectory, velocity, and acceleration of $a_0 = 0.25, a_2 = 8, a_3 = 1.5$.

It is more practical if the trajectory derived passes through intermediate via points with continuous velocity and continuous acceleration without stops. In this case, velocities at the intermediate points need not to be zero. That is $X(t_i) = x_i$, $\dot{X}(t_i) = v_i$ and $X(t_{i+1}) = x_{i+1}$, $\dot{X}(t_{i+1}) = v_{i+1}$, where $X(t_i)$ is an intermediate point at $T = t_i$ and $X(t_{i+1})$ is an intermediate point at $T = t_{i+1}$ between the initial position and final position (Fig. 9.8). Then we have

$$a_0 = x_i, \tag{9.89}$$

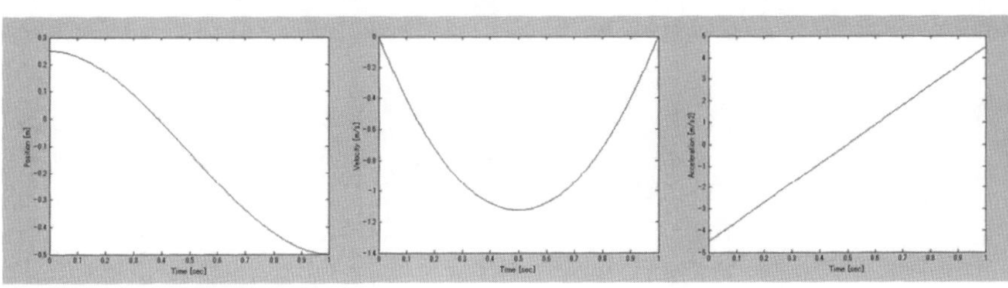

Fig. 9.7 Cubic polynomials

Fig. 9.8 Cubic polynomials with via points

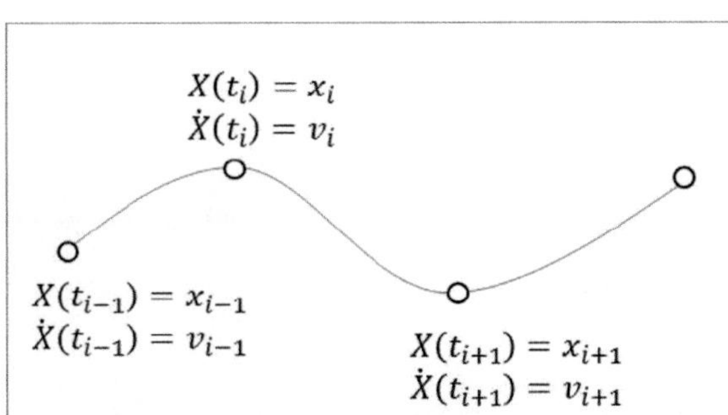

$$a_1 = \dot{x}_i, \tag{9.90}$$

$$a_2 = \frac{3}{{t_{i+1}}^2}\left(x_{i+1} - x_i\right) - \frac{2}{t_{i+1}}\dot{x}_i - \frac{2}{t_{i+1}}\dot{x}_{i+1}, \tag{9.91}$$

$$a_3 = \frac{2}{{t_\mathrm{f}}^3}\left(x_\mathrm{f} - x_0\right). \tag{9.92}$$

$$a_3 = \frac{2}{{t_{i+1}}^3}\left(x_{i+1} - x_i\right) + \frac{1}{{t_{i+1}}^2}\left(\dot{x}_{i+1} + \dot{x}_i\right). \tag{9.93}$$

9.6.3 *Linear Interpolation with Parabolic Blend*

If we were to simply just connect the desired intermediate points with a linear function in Cartesian space, it would cause the velocity to be discontinuous at the beginning and end of the motion. One way to create a smooth and continuous trajectory and velocity taking linear trajectory and interpolate it linearly is to add parabolic blend region in the trajectory. In this section, we specify linear interpolation with parabolic blend [11, 12]. We assume that the acceleration and deceleration will not change during the blend region, having same duration and can be instantaneously generated. In Fig. 9.9, the end of the blend region at $t = t_\mathrm{b}$ has to be the same as the linear region from $t = t_\mathrm{b}$ to $t = t_\mathrm{f} - t_\mathrm{b}$ which yields

$$\ddot{X}t_\mathrm{b} = \frac{x_\mathrm{h} - x_\mathrm{b}}{t_\mathrm{h} - t_\mathrm{b}}. \tag{9.94}$$

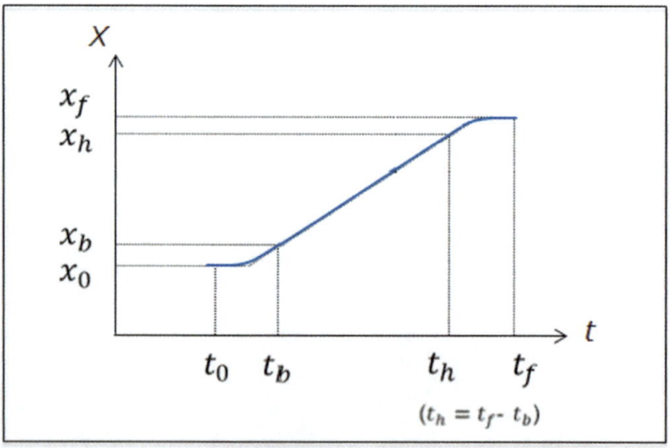

Fig. 9.9 Linear interpolation with parabolic blend

Since x_b is position at $t = t_b$, it is given by

$$X_b = x_0 + \frac{1}{2}\ddot{X}{t_b}^2 \tag{9.95}$$

then we have

$$\ddot{X}{t_b}^2 - \ddot{X}tt_b + (x_f - x_0) = 0. \tag{9.96}$$

where $t = 2t_b$. Equation (9.94) is the second-order equation of t_b and can be solved as

$$t_b = \frac{t}{2} - \frac{\sqrt{\ddot{X}t^2 - 4\ddot{X}\left(x_f - x_0\right)}}{2\ddot{X}} = 0. \tag{9.97}$$

Figure 9.9 shows image of the linear interpolation with parabolic blend. Depending on the acceleration, blend region will change. For example, if the acceleration is high the blend region will be shorter. In Fig. 9.10, MatLab simulation result of Cubic polynomials and Linear interpolation with parabolic blend is shown. The red lines show position, green lines show velocity, and blue lines show acceleration. We can see different characteristics in these two trajectory generation.

It is more practical if the trajectory derived passes through intermediate via points with continuous velocity and continuous acceleration without stops. Figure 9.11 shows the linear interpolation with parabolic blend with several segments. The followings are given constraints to specify the trajectory.

- Positions x_i, x_j, x_k, x_l, x_m
- Time durations t_{dij}, t_{djk}, t_{dkl}, t_{dlm}
- Segment times t_{ij}, t_{jk}, t_{kl}, t_{lm}
- Segment velocities $\dot{x}_{ij}$, $\dot{x}_{jk}$, $\dot{x}_{kl}$, $\dot{x}_{lm}$
- Blend times t_i, t_j, t_k, u_l, t_m

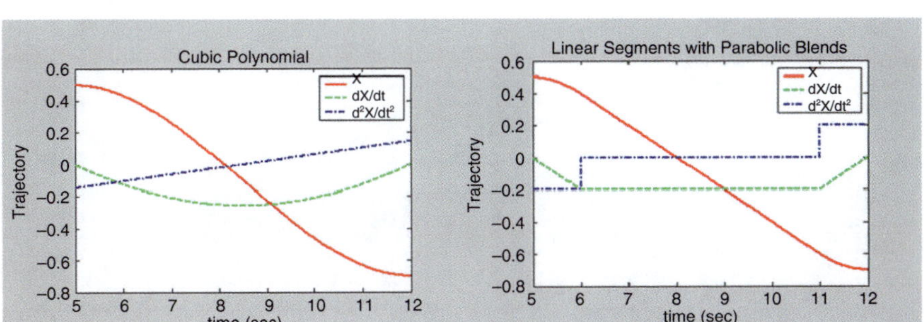

Fig. 9.10 MatLab simulation result of cubic polynomials and linear interpolation with parabolic blend

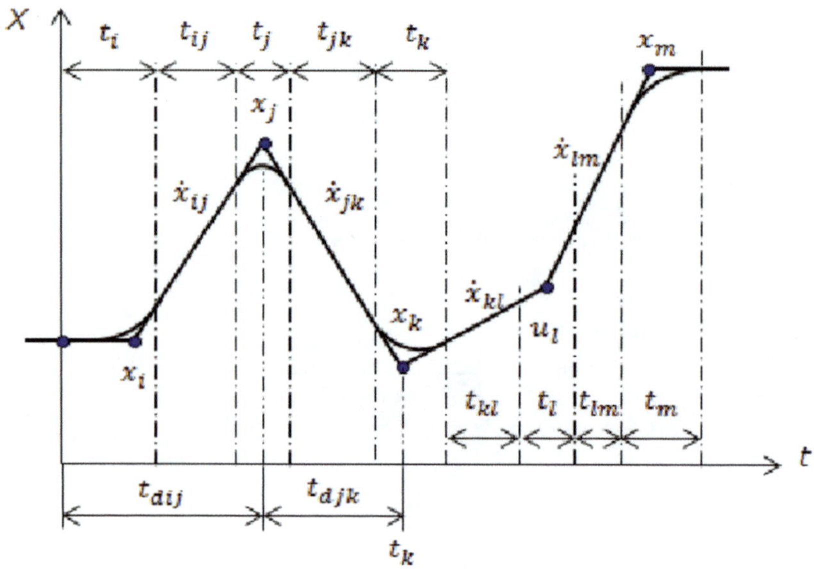

Fig. 9.11 Linear interpolation with parabolic blend

First segment

$$\ddot{x}_1 = \operatorname{sign}(x_2 - x_1)\,|\ddot{x}_1| \tag{9.98}$$

$$t_1 = t_{\mathrm{d}12} - \sqrt{t_{\mathrm{d}12}{}^2 - \frac{2\,(x_2 - x_1)}{\ddot{x}_1}} \tag{9.99}$$

$$\dot{x}_{12} = \frac{x_2 - x_1}{t_{\mathrm{d}12} - t_1/2} \tag{9.100}$$

$$t_{12} = t_{\mathrm{d}12} - t_1 - t_2/2 \tag{9.101}$$

Intermediate segment

$$\ddot{x}_{\mathrm{jk}} = \left(x_{\mathrm{k}} - x_{\mathrm{j}}\right)/t_{\mathrm{djk}} \tag{9.102}$$

$$t_{\mathrm{k}} = \left(\dot{x}_{\mathrm{kl}} - \dot{x}_{\mathrm{jk}}\right)/\ddot{x}_{\mathrm{k}} \tag{9.103}$$

$$t_{\mathrm{jk}} = t_{\mathrm{djk}} - t_{\mathrm{j}}/2 - t_{\mathrm{k}}/2 \tag{9.104}$$

Final segment

$$\ddot{x}_n = \operatorname{sign}(x_{n-1} - x_n)\,|\ddot{x}_n| \tag{9.105}$$

$$t_n = t_{d(n-1)n} - \sqrt{t_{d(n-1)n}{}^2 - \frac{2\,(x_n - x_{n-1})}{\ddot{x}_n}} \tag{9.106}$$

$$\dot{x}_{(n-1)n} = \frac{x_n - x_{n-1}}{t_{d(n-1)n} - t_n/2} \tag{9.107}$$

$$t_{(n-1)n} = t_{d(n-1)n} - t_n - t_{n-1}/2 \tag{9.108}$$

9.7 Trajectory Tracking Control

9.7.1 Second-Order System

To design a controller that makes a system behave in a designed motion, mathematical models are most often used to predict the behavior of the quantities of interest over time. Mathematical models of physical processes are the foundations of control theory. In control system design the most common mathematical models of the behavior of interest are, in the time domain, linear ordinary differential equations with constant coefficients, and in the frequency or transform domain, transfer functions obtained from time domain descriptions via Laplace transforms. Mathematical models of dynamic processes specifically how they change in response to different inputs. Solving differential equations and using Laplace transform, mathematical models provides us gain insight on how feedback control systems work and skills to design stable controller including following issues:

1. Stability and its margins of closed-loop systems.
2. Response to the command.
3. Robustness (sensor noise, disturbance, modeling error).

In this chapter, the design of reconfiguring a class of second-order dynamic systems via *Proportional Derivative* (PD) feedback is considered here. In the following, we will first consider a second-order mechanical system in some depth, and use this to introduce key ideas associated with second-order responses. Modeling and analysis methods are first introduced, followed by an overview of the classical design methods, design evaluation methods, and implementation issues. Results will be described by partial differential equations, however, tend to be more restricted and case dependent. We will develop a systematic procedure for finding controllers for simple systems. Among all possible system, PD controller is frequently employed in tracking and because PD control coincides with a generic state-feedback control of a second-order system for which the poles can be placed arbitrarily.

As shown in Fig. 9.12, the system consists of a spring and damper attached to a mass which moves laterally on a surface. The lateral position of the mass is defined as x. Mathematical models of dynamic processes are derived using physical laws

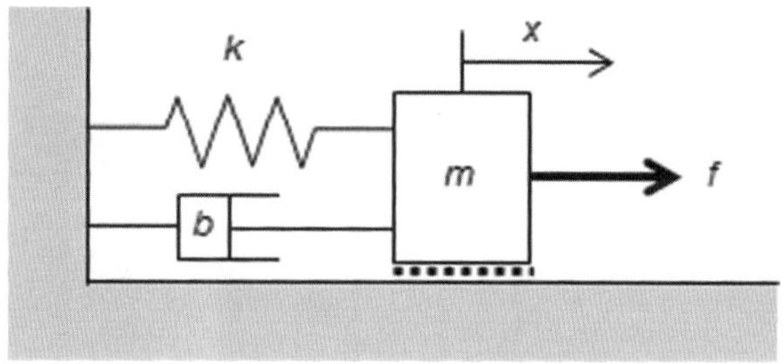

Fig. 9.12 A spring and a damper attached to a mass

such as Newton's law. According to Fig. 9.12, the equation of motion of the spring-damper-mass system is

$$m\ddot{x} + b\dot{x} + kx = f \tag{9.109}$$

where m is a mass, b is a coefficient of damper, k is a coefficient of spring and force f is applied to the system.

9.7.2 Design of Second-Order System

In the following section, we address analysis of second-order differential equation to derive stable condition from general solution of the second-order system [11, 13]. Rearranging Eq. (9.107) yields the system equation in standard form

$$M\frac{d^2x}{dt^2} + B\frac{dx}{dt} + Kx = 0. \tag{9.110}$$

In general terms, Eq. (9.108) is conveniently parameterized in terms of undamped natural frequency ω_0 and damping ratio ζ

$$\frac{d^2x}{dt^2} + 2\zeta\omega_n\frac{dx}{dt} + {\omega_0}^2x = 0 \tag{9.111}$$

where $\zeta = \frac{B}{2\sqrt{KM}}$, $\omega_n = \sqrt{\frac{K}{M}}$, with the initial conditions $x(0) = x_0$, $\dot{x}(0) = v_0$.

To solve this second-order system, we assume that $x(t)$ takes the form $x(t) = Ae^{\lambda t}$ and applying it to Eq. (9.107) yields

$$A{\lambda}^2e^{\lambda t} + 2\zeta\omega_nA\lambda e^{\lambda t} + {\omega_n}^2Ae^{\lambda t} = 0. \tag{9.112}$$

Since $Ae^{\lambda t} \neq 0$ for any λ and t, $Ae^{\lambda t}$ can be canceled and we have characteristic equation

$$\lambda^2 + 2\zeta\omega_nr + {\omega_n}^2 = 0. \tag{9.113}$$

Generally, this second-order equation has solutions are

$$\lambda_1 = -\zeta\omega_n + \omega_n\sqrt{\zeta^2 - 1} \tag{9.114}$$

and

$$\lambda_2 = -\zeta\omega_n - \omega_n\sqrt{\zeta^2 - 1}. \tag{9.115}$$

The behavior of the system depends on the two parameters, the natural frequency ω_0 and the damping ratio ζ, which are defined by mass m, spring coefficient k, damping coefficient b. In particular, the behavior of the system crucially depends on solution of Eq. (9.111) that is whether the characteristic equation has one real solution, two real solutions, or two complex conjugate solutions. Depending on the value of the damping ratio ζ, the system can be categorized into four types of responses. In this section, each types of responses are modeled and result of the step response is shown in Fig. 9.13.

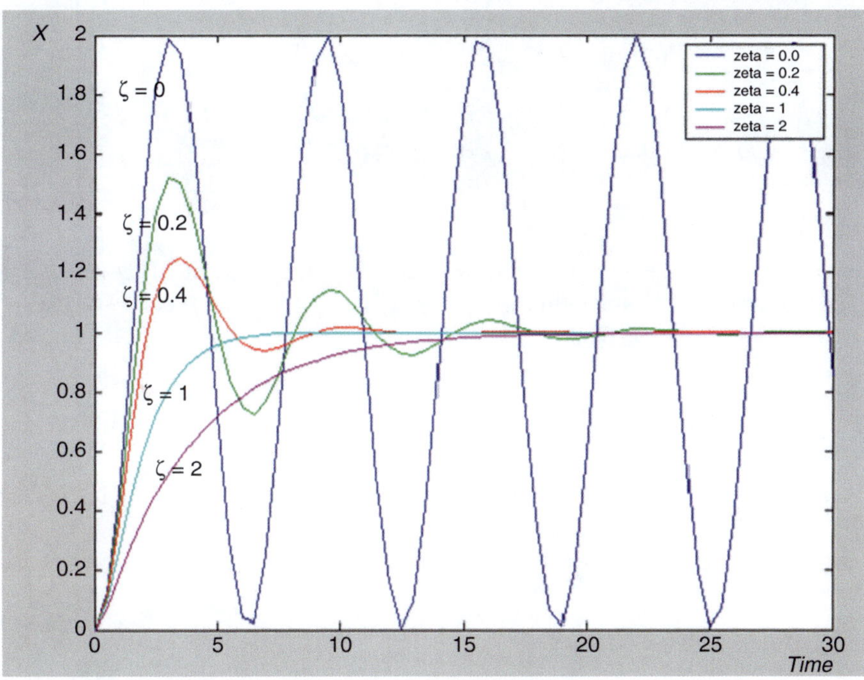

Fig. 9.13 A step response result of the second-order system

9.7.2.1 Type 1: Undamped Case

When $\zeta = 0$ and $B = 0$, the system is *undamped*. The system continues to vibrate without losing energy theoretically. On the other hand, it does not happen in real system because of any physical parameters which are not modeled in the system. In this case, the poles are complex conjugates solutions

$$\lambda_1 = j\omega_n \tag{9.116}$$

and

$$\lambda_2 = -j\omega_n. \tag{9.117}$$

The homogeneous solution takes the form

$$x(t) = c_1 e^{\lambda_1 t} + c_2 e^{\lambda_2 t} = c_1 e^{j\omega_n t} + c_2 e^{-j\omega_n t} \tag{9.118}$$

Assuming $c_1 = c_2{}^*$, and $c_1 = \alpha + j\beta$, we have

$$\begin{aligned} x(t) &= (\alpha + j\beta)\, e^{j\omega_n t} + (\alpha - j\beta)\, e^{-j\omega_n t} \\ &= \alpha \left(e^{j\omega_n t} + e^{-j\omega_n t}\right) + j\beta \left(e^{j\omega_n t} - e^{-j\omega_n t}\right) \\ &= 2\left(\alpha \cos(\omega_n t) - \beta \sin(\omega_n t)\right) \\ &= 2A \cos(\omega_n t + \varphi) \end{aligned} \tag{9.119}$$

where $A = |c_1| = \sqrt{\alpha^2 + \beta^2}$ and $\varphi = \arctan(\alpha, \beta)$. The constants α and β can be specified by the initial conditions $x(0) = x_0$ and $\dot{x}(0) = v_0$. It can be solved as $\alpha = x_0/2$ and $\beta = -v_0/\,(2\omega_n)$.

9.7.2.2 Type 2: Underdamped Case

When $0 \leq \zeta < 1$, λ is complex and the system is *underdamped*. The system oscillates at the natural damped frequency $\omega_\mathrm{d} = \omega_n\sqrt{1-\zeta^2}$, which is a function of the natural frequency ω_n and the damping ratio ζ, and gradually decrease the amplitude to zero. In this case, the poles are complex solutions

$$\lambda_1 = -\zeta\omega_n + j\omega_\mathrm{d} \tag{9.120}$$

and

$$\lambda_2 = -\zeta\omega_n - j\omega_\mathrm{d}. \tag{9.121}$$

The homogeneous solution takes the form

$$x(t) = c_1 e^{\lambda_1 t} + c_2 e^{\lambda_2 t} = c_1 e^{(-\zeta\omega_n + j\omega_d)t} + c_2 e^{(-\zeta\omega_n - j\omega_d)t} \tag{9.122}$$

Assuming $c_1 = c_2{}^*$, and $c_1 = \alpha + j\beta$, we have

$$\begin{aligned} x(t) &= (\alpha + j\beta)\, e^{(-\zeta\omega_n + j\omega_d)t} + (\alpha - j\beta)\, e^{(-\zeta\omega_n - j\omega_d)t} \\ &= e^{-\zeta\omega_n t}\left((\alpha + j\beta)\, e^{j\omega_d t} + (\alpha - j\beta)\, e^{-j\omega_d t}\right) \\ &= e^{-\zeta\omega_n t}\left(\alpha\left(e^{j\omega_d t} + e^{-j\omega_d t}\right) + j\beta\left(e^{j\omega_d t} - e^{-j\omega_d t}\right)\right) \\ &= 2e^{-\zeta\omega_n t}\left(\alpha\cos(\omega_\mathrm{d} t) - \beta\sin(\omega_\mathrm{d} t)\right) \\ &= 2Ae^{-\zeta\omega_n t}\cos(\omega_\mathrm{d} t + \varphi) \end{aligned} \tag{9.123}$$

where $A = |c_1| = \sqrt{\alpha^2 + \beta^2}$ and $\varphi = \arctan(\alpha, \beta)$. The constants α and β can be specified by the initial conditions $x(0) = x_0$ and $\dot{x}(0) = v_0$. It can be solved as $\alpha = x_0/2$ and $\beta = -(v_0 + \zeta\omega_n x_0)/(2\omega_\mathrm{d})$. The time response can be expressed in terms of $\zeta\omega_n$ and ω_d. In this case, since both $\zeta\omega_n$ and ω_d scale linearly with ω_n, the response characteristic time-scale decreases as $1/\omega_n$.

9.7.2.3 Type 3: Overdamped Case

When $\zeta > 1, \lambda$, the system is *overdamped* and two poles are at separate locations on the real axis:

$$\lambda_1 = \left(-\zeta + \sqrt{\zeta^2 - 1}\right)\omega_n \tag{9.124}$$

and

$$\lambda_2 = \left(-\zeta - \sqrt{\zeta^2 - 1}\right)\omega_n. \tag{9.125}$$

The homogeneous solution takes the form

$$x(t) = c_1 e^{\lambda_1 t} + c_2 e^{\lambda_2 t} = c_1 e^{\left(-\zeta + \sqrt{\zeta^2 - 1}\right)t} + c_2 e^{\left(-\zeta - \sqrt{\zeta^2 - 1}\right)t} \tag{9.126}$$

To estimate c_1 and c_2, we apply the initial conditions $x(0) = x_0$ and $\dot{x}(0) = v_0$ to Eq. (9.124) then we have $x(0) = c_1 + c_2 = x_0$ and $\dot{x}(0) = c_1\lambda_1 + c_2\lambda_2 = v_0$,which gives

$$c_1 = \frac{v_0 + x_0\lambda_2}{\lambda_2 - \lambda_1} = -\frac{v_0 + x_0\left(-\zeta - \sqrt{\zeta^2 - 1}\right)\omega_n}{2\omega_n\sqrt{\zeta^2 - 1}}, \tag{9.127}$$

$$c_2 = -\frac{v_0 + x_0\lambda_1}{\lambda_2 - \lambda_1} = -\frac{v_0 + x_0\left(-\zeta + \sqrt{\zeta^2 - 1}\right)\omega_n}{2\omega_n\sqrt{\zeta^2 - 1}} \tag{9.128}$$

and

$$\begin{aligned} x(t) = &-\frac{v_0 + x_0\left(-\zeta - \sqrt{\zeta^2 - 1}\right)\omega_n}{2\omega_n\sqrt{\zeta^2 - 1}} e^{\left(-\zeta + \sqrt{\zeta^2 - 1}\right)t} \\ &- \frac{v_0 + x_0\left(-\zeta + \sqrt{\zeta^2 - 1}\right)\omega_n}{2\omega_n\sqrt{\zeta^2 - 1}} e^{\left(-\zeta - \sqrt{\zeta^2 - 1}\right)t}. \end{aligned} \tag{9.129}$$

9.7.2.4 Type 4: Critically Damped Case

When $\zeta = 1$, the two poles coincide at $\lambda_1 = \lambda_2 = -\omega_n$. The system is *critically damped*. We have repeated roots and the oscillator is critically damped. The homogeneous solution takes the form

$$x(t) = c_1 e^{\lambda_1 t} + c_2 t e^{\lambda_1 t} = c_1 e^{-\omega_n t} + c_2 t e^{-\omega_n t} \tag{9.130}$$

To estimate c_1 and c_2, we apply the initial conditions $x(0) = x_0$ and $\dot{x}(0) = v_0$ to Eq. (9.128) then we have $x(0) = c_1 = x_0$ and $\dot{x}(0) = -\omega_n c_1 + c_2 = v_0$, which gives

$$c_1 = x_0, \tag{9.131}$$

$$c_2 = v_0 + \omega_n x_0 \tag{9.132}$$

and

$$x(t) = x_0 e^{-\omega_n t} + (v_0 + \omega_n x_0)\, t e^{-\omega_n t} \tag{9.133}$$

9.7.3 Trajectory Tracking Control

In the previous section, we saw that the behavior of the second-order system crucially depends on solution of Eq. (9.109) especially it depends on the damping ratio ζ. If we want to achieve a desired behavior for tracking a given trajectory, we need to specify the best parameters of the spring-damper-mass mechanical system defined by Eq. (9.107) and control gains. Generally, mechanical system has several disturbances and the system cannot be supposed to return to its initial condition such as $x = 0$. In this case, using sensors and actuators, the system can be controlled to behave as we want. Assuming that a control law of the proposed control system is to maintain the position of the block in designed position by computing the force f using the sensed position data, velocity data, position gain $\mathrm{k_p}$ and velocity gain k_v (Eq. 9.127). According to this control law, goal position tracking, that is position to position controller, is modeled

$$f = -k_p x - k_v \dot{x}. \tag{9.134}$$

By canceling f in Eqs. (9.107) and (9.132), closed-loop dynamics of the system can be represented as

$$m\ddot{x} + k_v \dot{x} + k_p x = -k_p x - k_v \dot{x}. \tag{9.135}$$

Equation (9.133) can be simplified as

$$m\ddot{x} + (b + k_v)\,\dot{x} + \left(k + k_p\right)x = 0, \tag{9.136}$$

$$m\ddot{x} + b'\dot{x} + k'x = 0. \tag{9.137}$$

where $b' = b + k_v$ and $k' = k + k_p$. Equation (9.135) is a second-order system which includes mechanical system and control system. The characteristic of the second-order system is well known framework and can be categorized into four cases such as (1) nondamped system, (2) underdamped system, (3) overdamped system, and (4) critical damped system. To have better performance of the system, we can design response of the system by gain parameters in Eq. (9.135), as we want.

9.7.4 Tracking Control for Given Trajectory

In previous section, we discussed trajectory generation from the initial position to the final position, with or without intermediate points. For given initial position, intermediate points and final position, smooth trajectory will be generated on-line. In this section, we design controller which moves the tool or the end-effector following the generated trajectory stably. Since the tool or the end-effector moves dynamically while tracking the trajectory, the controller has to compensate stable or balanced motion dynamically. As discussed in previous section, the system can be modeled as the second-order system and its behavior can be modeled accordingly. Assume that a trajectory generator is able to generate a smooth trajectory $x_{\mathrm{des}}(t)$ and $\dot{x}_{\mathrm{des}}(t)$ between the initial position and final position with the initial condition, $x(0) = x_0$ and $\dot{x}(0) = v_0$ and with several intermediate points between the initial position and final position. We define the feedback controller of trajectory tracking

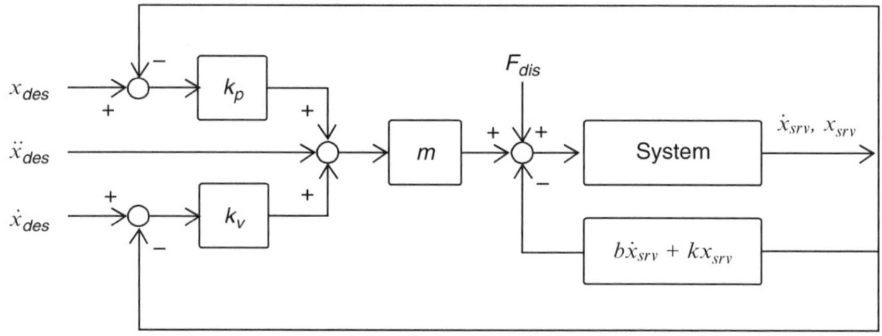

Fig. 9.14 Trajectory tracking controller

in Fig. 9.14. A control law of tracking for given trajectory has the following form

$$f = \ddot{x}_{\text{des}} + k_{\text{v}} (\dot{x}_{\text{des}} - \dot{x}_{\text{srv}}) + k_{\text{p}} (x_{\text{des}} - x_{\text{srv}}) . \tag{9.138}$$

Equation (9.136) can be modified to (9.137) using canonical equation of motion $x_{\text{srv}} = f$:

$$x_{\text{srv}} = \ddot{x}_{\text{des}} + k_{\text{v}}\dot{e} + k_{\text{p}}e. \tag{9.139}$$

or

$$\ddot{e} + k_{\text{v}}\dot{e} + k_{\text{p}}e = 0 \tag{9.140}$$

where $e = x_{\text{des}} - x_{\text{srv}}$, $\dot{e} = \dot{x}_{\text{des}} - \dot{x}_{\text{srv}}$, and $\ddot{e} = \ddot{x}_{\text{des}} - x_{\text{srv}}$. This equation captures the behavior of the system in the second-order differential equation which was discussed in previous section. Therefore, the behavior of trajectory tracking depends on the two parameters, the natural frequency and the damping ratio in the error space, which was discussed in previous section, and they can be designed according to application and desired task. If any external disturbance exists and affects to the system, Eq. (9.138) can be modified as

$$\ddot{e} + k_{\text{v}}\dot{e} + k_{\text{p}}e = f_{\text{dis}}. \tag{9.141}$$

If the external disturbance f_{dis} is constant, the system is in steady condition but the controller in Eq. (9.137) can not eliminate the steady state error caused by $e = f_{\text{dis}}/k_{\text{p}}$ since the position feedback gain k_{p} cannot be too high. To eliminate the steady state error keeping the adequate k_{p}, most commonly used approach is to add integral term to the controller, that will modifies the controller in Eq. (9.137) into

$$f' = \ddot{e} + k_{\text{v}}\dot{e} + k_{\text{p}}e + k_i \int e \, \mathrm{d}t = 0. \tag{9.142}$$

Equation (9.139) is PID control law, which is the most commonly used control system. Using PID control law, we can design the motion of the robot accordingly and the best performance in tracking motion can be designed for any given trajectory.

9.8 Dynamic Balancing with Respect to a Given Trajectory

Parallel mechanisms are in "*Statically Balanced*" when the motors do not contribute to supporting the link's weight under static conditions, for any configuration of the manipulator. Therefore, it can remain stable position for any posture without the help of motors or brakes. Generally, it can be obtained by using counterweights or springs and no computer control is involved.

On the other hand, the parallel mechanisms are in "*Dynamically Balanced*" [14] if all the dynamic reaction forces and all the dynamic reaction moments are zero and its linear momentum and angular momentum are constants at any instant. Here, the dynamic reaction forces mean the shaking forces and the dynamic reaction moments mean the shaking moments. These shaking forces and moments cause the actuation forces and moments of machines, fatigue, vibrations, noise disturbances, etc., resulting in perturbations of the motions of the moving parts. Generally, controlled system is utilized to coordinate the motions of the balancing elements. The measure advantage of dynamic balancing by control theory is that

(a) It is possible to deal with changes of the mass and inertia parameters with appropriate control schemes when the payloads or tools are variable.
(b) Dynamic balanced motion can be designed to produce adequate forces and moments by control schemes, which counteract the adverse effects of the shaking forces and moments.

To create dynamically balanced system by using control theory and realize dynamic balancing with respect to a given trajectory, our approach is

(1) To generate stable motion command by modeling dynamics of the robot and task dynamics at the tool position or the end-effector.
(2) To generate stable smooth trajectory from the initial position to final position including intermediate points.
(3) To control tracking motion stably following designed trajectory.

From Sects. 8.2–8.5, we modeled kinematics and dynamics of the Stewart Platform including physical effects such as mass/inertia, centrifugal/coriolis forces, COG of the robot in joint space and their effects in the operational space. The input torque for the system can be designed to compensate for dynamic effect of the system. Decoupled task dynamics can be applied by the Operational Space Formulation which provides the robot with higher performance in position tracking as well as in compliant motion. Therefore, advanced performance, complex behaviors and compliant posture control can be implemented for robots if torque control is applied. Measure problem which will cause the position error in Cartesian space is dynamically unbalanced situation by modeling error. In most cases, mechanical parameters are given separately from CAD but will change when assembled. Or since physical elements such as friction or damping of the mechanical system is not be modeled in most cases. Since physical model in the joint space is mapped into the operational space, modeling error in joint space will affect the motion in the Operational space. On the other hand, once we have correct model, it will compensate stable, quick, and accurate motion for any trajectory command.

In Sect. 8.6, we considered methods of trajectory generation such as cubic spline and parabolic blending, in which the path shapes are interpolated between data points in Cartesian space. For given initial point, intermediate points and final position, smooth trajectory will be generated on-line from these methods, which will guarantees stable trajectory.

In Sect. 8.7, to design a controller that makes a system behave in a designed motion, second-order system was modeled to predict the behavior of trajectory tracking motion over time. The second-order model provides us gain insight on how feedback control systems work including stability, response to the command, and robustness against sensor noise, disturbance, modeling error, etc.

Combination of precise dynamics model of the robot, on-line continuous trajectory generation and trajectory tracking controller provides the user to design stable performance with dynamically balanced control architecture for any given trajectory which is generated on-line.

References

1. Güneri, B.: Complete Dynamic Analysis of Stewart Platform Based on Workspace. LAP Lambert Academic Publishing, Saarbrücken (2011)
2. Hashimoto, K., Sugahara, Y., Sunazuka, H., Lim, H., Takanishi, A.: Human-Carrying Biped Walking Vehicle (2008)
3. Fried, G., Djouani, K., Borojeni, D., Iqbal, S.: An inverse dynamic model of the Gough-Stewart Platform. WSEAS Trans Syst **7**(2), 88–99 (2008)
4. Lee, S.-H., Song, J.-B., Choi, W.-C., Hong, D.: Position control of a Stewart platform using inverse dynamics control with approximate dynamics. Mechatronics **13**, 605–619 (2003)
5. Pedrammehr, S., Mahboubkhah, M., Khani, N.: Improved dynamic equations for the generally configured Stewart platform manipulator. J Mech Sci Technol **26**(3), 711–721 (2012)
6. Bai, X., Turner, J.D., Junkins, J.L.: Dynamic analysis and control of a Stewart Platform using a novel automatic differentiation method. In: Astrodynamics Specialist Conference and Exhibit, Keystone, CO, 21–24 August 2006
7. Bingul, Z., Karahan, O.: Dynamic Modeling and Simulation of Stewart Platform, Mechatronics Engineering. Kocaeli University, Turkey (2012)
8. Khatib, O.: A unified approach for motion and force control of robot link robots: the operational space formulation. Int J Robot Res **3**(1), 43–53 (1987)
9. Yoshikawa, T., Khatib, O.: Compliant Humanoid Robot Control by the Torque Transformer. IROS, Chicago, IL (2009)
10. De Sapio, V.: Some Approaches for Modeling and Analysis of a Parallel Mechanism with Stewart Platform Architecture. National Technical Information Service, Springfield, VA (1998). SAND98-8242 Unlimited Release, Printed

11. Craig, J.J.: Introduction to Robotics: Mechanics and Control, 3rd edn. Prentice Hall, Upper Saddle River, NJ (2004). ISBN 10:0201543613
12. Kunz, T., Stilman, M.: Turning Paths Into Trajectories Using Parabolic Blends, Technical Report, GT-GOLEM-2011-006. Georgia Institute of Technology, Atlanta, GA (2011)
13. Christiansen, D.: The Electronics Engineers' Handbook, 5th edn, pp. 19.1–19.30. McGraw-Hill, New York, NY (2005). Section 19
14. Wang, K., Luo, M.: Dynamics analysis of a three-DOF planar serial-parallel mechanism for active dynamic balancing with respect to a given trajectory. Int J Adv Robot Syst **10**(23), 1–10 (2013)
15. Aribowo, W., Terashima, K.: Cubic spline trajectory planning and vibration suppression of semiconductor wafer transfer robot arm. Int J Automat Technol **8**(2), 265–274 (2014)

Chapter 10
Dynamic Balancing and Flexible Task Execution for Dynamic Bipedal Walking Machines

Andreas Hofmann

Abstract Effective use of robots in unstructured environments requires that they have sufficient autonomy and agility to execute task-level commands with temporal constraints successfully. A challenging example of such a robot is a bipedal walking machine, particularly one of humanoid form. Key features of the human morphology include a variable base of support and a high center of mass. The high center of mass supports the ability to support a high "sensor package"; when standing erect, the head can see over obstacles. The variable base of support allows both for operation in tight spaces, by keeping the feet close together, and stability against disturbances, by keeping the feet further apart to widen the support base. The feet can also be placed in specific locations when there are constraints due to challenging terrain. Thus, the human morphology supports a range of capabilities, and is important for operating in unstructured environments as humans do. A bipedal robot with human morphology should be able to walk to a particular location within a particular time, while observing foot placement constraints, and avoiding a fall, if this is physically possible. This is a challenging problem because a biped is highly nonlinear and has limited actuation due to its limited base of support. This chapter describes a novel approach to solving this problem that incorporates three key components: (1) a robust controller that is able to use angular momentum to enhance controllability beyond the limits imposed by the support base; (2) a plan specification where task requirements are expressed in a qualitative form that provides for spatial and temporal execution flexibility; and (3) a task executive that compiles the plan into a form that makes the dynamic limitations explicit, and then executes the compiled form using the robust controller.

Keywords Bipedal • Walking • Flexible plan execution • Feedback linearization • Angular momentum

A. Hofmann (✉)
Massachusetts Institute of Technology, 77 Massachusetts Ave.,
Cambridge, MA 02139, USA
e-mail: hofma@csail.mit.edu

D. Zhang, B. Wei (eds.), *Dynamic Balancing of Mechanisms and Synthesizing of Parallel Robots*, DOI 10.1007/978-3-319-17683-3_10

10.1 Introduction

Effective use of autonomous robots in unstructured human environments requires that they have sufficient autonomy to perform useful tasks independently, have sufficient size, strength, and speed to accomplish these tasks in a timely manner, and that they operate robustly and safely in the presence of disturbances. These requirements are more challenging than the ones for today's factory robots, which are stationary, work in very restricted environments, and have very limited autonomy.

A particularly challenging example of an autonomous robot in an unstructured environment is a bipedal walking machine, as shown in Fig. 10.1. An example task for such a system is to walk to a moving soccer ball and kick it, as shown in Fig. 10.2. Stepping movement must be synchronized with ball movement so that the kick

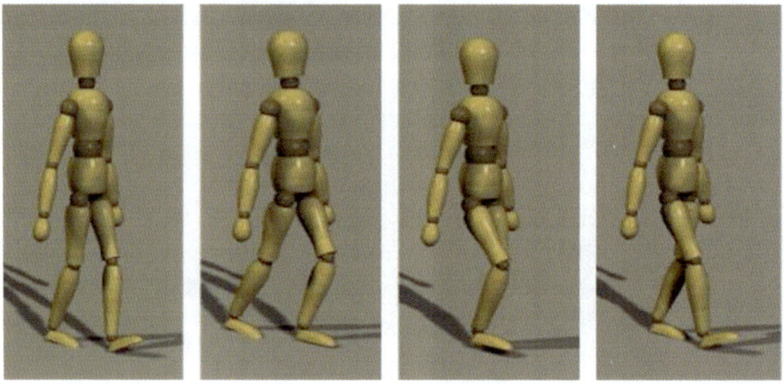

Fig. 10.1 A humanoid biped performing a walking task

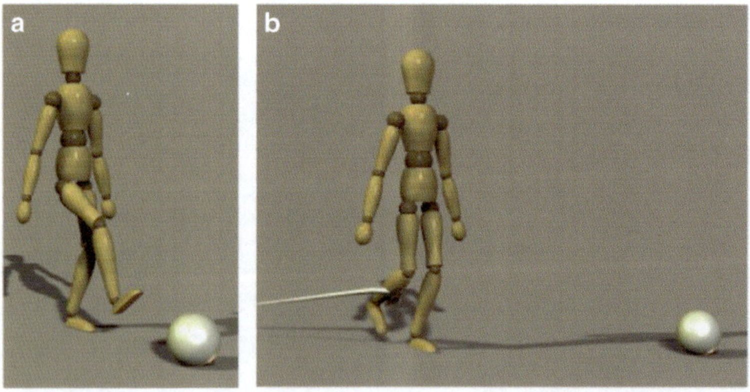

Fig. 10.2 Kicking soccer ball task, interrupted by trip disturbance. In (**a**), the goal is to kick a possibly moving soccer ball; the biped must be in an acceptable location at an acceptable time in order to perform the kick. In (**b**), the task is interrupted by a trip; the biped should try to recover, if this is physically possible

happens when the ball is close enough. More generally, such tasks require that the biped be in the right location at an acceptable time. This implies spatial and temporal constraints for such tasks. There are also important dynamic balance constraints that limit the kinds of movements the biped may make without falling down.

If the system encounters a disturbance while performing a task, it will have to compensate in some way in order to satisfy these constraints. The disturbance may cause a delay, allowing another player to kick the ball, or it may interfere with movement synchronization. For example, a trip, shown in Fig. 10.2b causes disruption of synchronization between the stepping foot, and the overall forward movement of the system's center of mass.

Another example task is walking on a constrained foot path, such as stones across a brook, or on a balance beam. As with the soccer ball example, this task has spatial, temporal, and dynamic constraints, but in this case, the spatial constraints are more stringent; the biped must reach its goal using foot placements that are precisely constrained.

Figure 10.3 shows a biped walking over blocks that constrain foot placement in a similar manner. When foot placement is constrained, the stepping pattern can't be changed arbitrarily to compensate for a disturbance. For example, if a lateral push disturbance occurs, rather than stepping the leg out to the side, other compensating techniques, such as angular movement of the body and swing leg must be used, as shown in Fig. 10.4.

In these examples, and others like them, the key challenge is to move a complex, dynamic system to the right place, at the right time, despite actuation limits, and despite disturbances. The system should be able to recover from disturbances such as slips, trips, pushes, and ground contact instability due to soft terrain, even when foot placement is constrained.

This chapter addresses the class of problems that require movement of a dynamic bipedal system according to stringent state-space and temporal requirements, despite actuation limits and disturbances. Additionally, we consider how the use of flexible link structures changes the problem and the solution. This additional

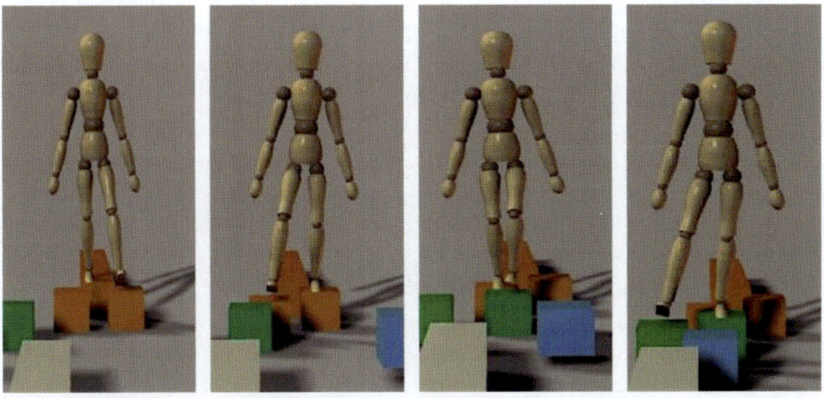

Fig. 10.3 Dynamic walking with foot placement constraints

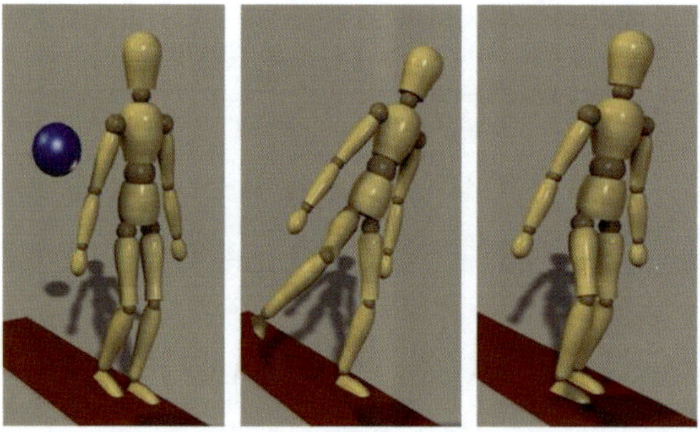

Fig. 10.4 Compensating for lateral push disturbance using angular movement of torso and swing leg

consideration is significant given the importance of using lightweight structures in mobile robots, in order to maximize energy efficiency. Use of lightweight materials for the links implies less rigidity than is the norm in industrial factory manipulators.

The remainder of this introduction describes, in more detail, the motivation for studying this class of problems, a statement of the problem being solved, and associated challenges, and an overview of the approach and innovations used to solve it.

10.1.1 Significance and Motivation

Humanoid bipedal morphologies have unique characteristics that provide significant advantages (and also some disadvantages) over quadruped or wheeled robots. Key functional features of the humanoid morphology include a variable base of support and a high center of mass. Because bipeds have only two legs, their support base is naturally constrained, allowing them to operate in environments where support base space is limited. Many human environments require this (a crowded elevator, for example). The legs can also be placed in wide stances, when space permits, enhancing stability against disturbances. The feet can also be placed in specific locations, allowing for traversal over terrain where foot placement is constrained due to obstacles. Thus, a key advantage of the humanoid morphology is that the variable base of support allows both for operation in constrained spaces, and stability against disturbances, Humanoid bipeds also have a high center of mass. This has the advantage that an elevated "sensor package" can be supported; when standing erect, the head can see over obstacles. Thus, the humanoid biped morphology supports a range of capabilities that are important for operating in unstructured environments, particularly when collaborating with humans.

Solving the problem of balance control, and task-level control of humanoid biped devices is of significant importance, in that it will be necessary for deployment of autonomous humanoid robots that can provide assistance to humans in human environments such as the home, office, construction sites, loading docks, and many others. Additionally, solution of this problem will permit the deployment of powered exoskeletons that can provide locomotion capability to disabled humans.

10.1.2 Problem Statement and Challenges

We seek to develop a robust plan execution system capable of guiding a robotic biped through a series of walking task goals, in the presence of disturbances. The system must understand commands at the task level; it must take as input a high-level specification of where it should be, and by what time, and then automatically figure out the details of how to move to accomplish these goals. It should also be able to automatically detect whether a task that it is given can be accomplished in the allotted time, and should warn the human operator when this is not the case. If a disturbance occurs during execution of the task, the system should attempt to compensate in order to avoid a fall, and should still try to complete the task on time. If this is not possible, the system should alert the user, or a higher-level control authority.

There are significant challenges to solving this problem.

First, the specification of the walking task itself should represent the true spatial and temporal constraints of the task, rather than arbitrarily setting tight, artificial constraints. For example, if the task is allowed to complete with a duration of 5–10 s, the specification should not restrict the duration any further than this. Similarly, if the goal is properly represented as a set of possible states, the specification should not restrict the goal to a single state. It is important for the specification of the task to not artificially constrain operation. This gives the control system maximum flexibility in selecting actions that maximize robustness and performance. Second, the combination of limited support base and high center of mass presents a challenge in terms of balance control in that such a system is inherently less stable (more sensitive to disturbances) than a quadrupedal or four-wheeled configuration with relatively low center of mass. The limited support base and high center of mass imply that the biped is under-actuated and has significant inertia, so future state evolution is coupled to current state through dynamics that limit acceleration. The control system must consider how current state and actions may limit achievement of future desired state. Third, a biped is a high-dimensional, highly nonlinear, tightly coupled system, so computing control actions that achieve a desired state is a challenging problem. This is complicated by the incorporation of temporal constraints, and the limits that the dynamic system imposes on temporal performance. Fourth, this type of system, because it operates in human environments, must have stringent safety requirements. Balance control is essential both for autonomous legged assistive robots and for a variety of assistive devices, including powered exoskeletons that provide locomotion to the disabled. For such systems, preventing a fall is of paramount importance. An autonomous robot that falls may damage itself, or may hurt a human in its environment. In the case of an exoskeleton, a fall implies that the human wearer of the exoskeleton has fallen. Thus, a bipedal walking machine should avoid falling, if at all physically possible, even if it encounters a significant disturbance. If a fall is inevitable, the system should recognize this sufficiently early to alert users and surrounding humans.

Addressing these challenges requires investigation of a number of questions, including:

- How should walking task goals be expressed?
- What are the fundamental requirements for achieving these goals?
- What kinds of disturbances may occur while executing walking tasks?
- What fundamental balance strategies can bipeds use?

The following discussion introduces approaches to addressing these challenges, and answering these questions.

10.1.3 Approach and Innovations

We address these challenges with three key innovative techniques.

To address the first challenge (representation of task goals), we use a specification of state-space and temporal requirements called a *Qualitative State Plan* (QSP) [9, 10]. A QSP consists of a set of *Qualitative States*, where each Qualitative State is a region of state space in which all states have a uniform property with respect to the task at hand [23]. For a biped, qualitative states are defined by foot ground contact state. The biped may be in a double-support state, where both feet are in contact with the ground, a single-support state, where either the left or the right foot is on the ground, or a jumping state, where neither foot is on the ground.

Each qualitative state may specify valid operating regions for particular state variables. Foot placement constraints are examples of such operating region constraints. Each qualitative state may also specify goal regions that particular state variables must attain. For example, it may be a requirement that the biped center of mass be in a particular region in order for the biped to kick a soccer ball. Thus, a qualitative state is hybrid in that it is defined by continuous state regions, like allowable regions for the center of mass position, as well as by discrete state, like which feet are in contact with the ground. Transitions from one qualitative state to another are defined by events. For example, the transition from double to single support is defined by a toe-off event, which is the point where the swing foot lifts off the ground. The transition from single to double support is defined by a heel-strike event, which is the point where the swing foot touches the ground after taking a step. Events represent temporal boundaries that can be restricted by temporal constraints. Thus, a QSP permits the representation of a task's true constraints, providing maximum flexibility to the control system in selecting actions to maximize robustness and performance.

To address the third challenge (nonlinearity, high dimensionality, and tight coupling), the biped system is linearized and decoupled into a set of independent, linear systems, resulting in an abstraction of the biped that is easier to control. This is accomplished through a *Dynamic Virtual Model Controller* (DVMC) [11]. The linearization and decoupling provided by this controller allows *reaction points* on the biped to be controlled directly, in a manner similar to the way that a puppeteer controls a marionette. This controller also provides a novel means of using angular momentum to enhance stability. Angular momentum balance techniques are used, for example, by tight-rope walkers in order to achieve balance on a very limited support base.

To address the second challenge (future state evolution given actuation limits), a *Task Executive*, [13, 24], capable of predicting future state is used. The Task Executive utilizes the abstraction provided by the DVMC. It implements a control policy for this abstraction by pre-compiling *Flow Tubes* that define valid operating regions for the state variables and control parameters in the abstracted biped. The Flow Tubes represent bundles of state trajectories that take into account dynamic limitations due to under-actuation, and that also satisfy plan requirements. Off-line generation of these Flow Tubes represents a pruning of infeasible trajectories, so that the on-line controller can focus on executing the plan by using only trajectories in the Flow Tubes.

The Task Executive also uses the Flow Tubes to predict whether the plan will succeed or fail. In particular, if a disturbance occurs that pushes the system into a state where no suitable Flow Tube trajectory can be found, then the Task Executive knows that the plan will fail. In this case, it alerts a higher-level control authority, such as a human user. The ability to predict failure ahead of time in this way is important since it provides some time to change the plan, or take other compensating action. For example, a soccer player chasing a ball will abort if it becomes clear that another player will get to the ball first. Similarly, if a person trips while walking and a fall is inevitable, he will put out his hands to mitigate the effects of the fall. In this way, the Task Executive and Flow Tubes also address the fourth challenge (safety of operation in human environments).

To summarize, the Task Executive interprets plan goals, as specified by an input QSP, monitors biped state, and computes control actions for the biped, as shown in Fig. 10.5. The executive computes a sequence of joint torques for the biped that results in achievement of each successive qualitative state goal in the sequence, as shown in Fig. 10.6.

The rest of this chapter provides details of this approach. We begin with a review of biped balance mechanics in Sect. 10.2. This is followed, in Sect. 10.3 by a description of the DVMC, which utilizes the balance mechanics principles. Next, Sect. 10.4 presents the QSP in detail, and Sect. 10.5 describes the Task Executive, which interprets the QSP and utilizes the abstraction provided by the DVMC. Finally, Sect. 10.6 describes experimental results, and Sect. 10.7 provides a discussion of the results and contributions.

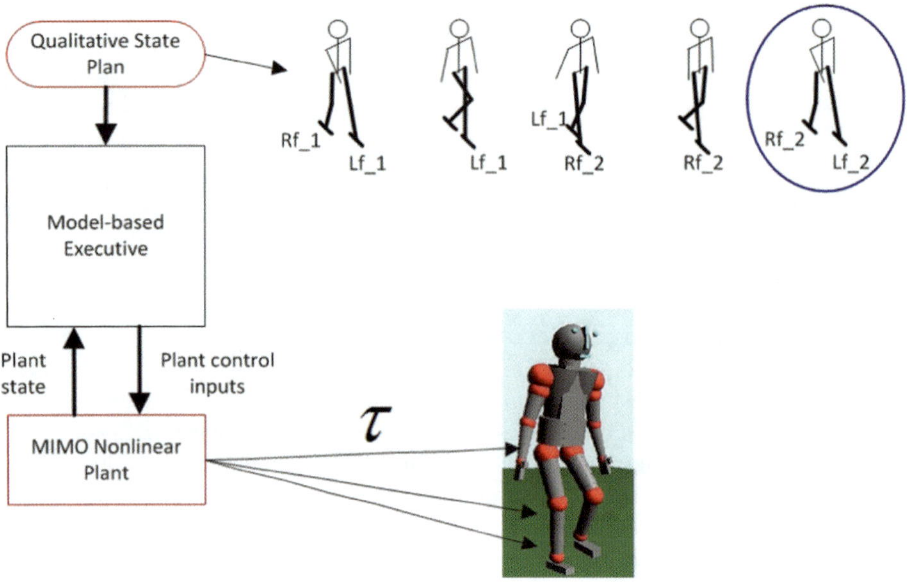

Fig. 10.5 A model-based executive computes a sequence of joint commands for the biped that results in the achievement of the successive qualitative state goals

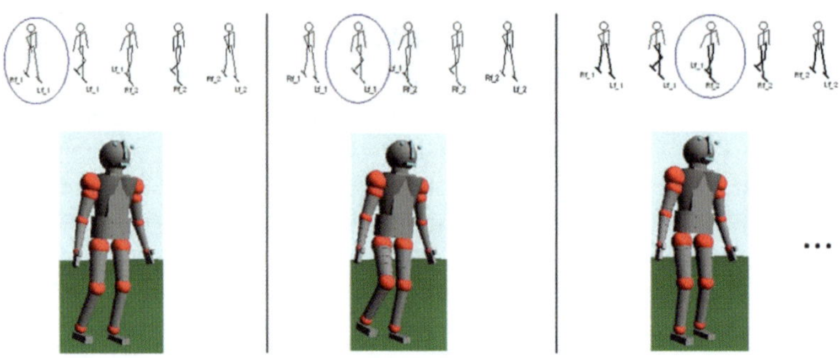

Fig. 10.6 Execution of a sequence in the qualitative state plan

10.2 Analysis of Balance Mechanics and Constraints

Balance control requires the ability to adjust the biped's linear and angular momentum. Due to conservation of momentum laws, such adjustment can only be achieved through force interaction with the environment. For a biped, this force interaction is comprised of gravity and the ground reaction force, the net force exerted by the ground against the biped. The following analysis of physical constraints and requirements for balancing leads to a simple, comprehensive model of balance control that specifies coordination of control actions that adjust the ground reaction force, and therefore, the momentum of the biped. Similar models have been used previously in a number of gait planning algorithms [12, 15, 19, 25]. These models utilize analysis of inverted pendulum dynamics [5]. A key difference in the model presented here is its ability to purposely sacrifice angular momentum control goals in order to achieve linear control goals when both cannot be met.

The model makes use of a number of physical points that summarize the system's balance state. These points are the center of mass (CM), the zero-moment point (ZMP) [21], and the centroidal-moment point (CMP) [16]. The ZMP is a point on the ground that represents the combined force interaction of all ground contact points. The CMP is the point on the ground from which the ground reaction force would have to emanate if it were to produce no torque about the CM. These points will be defined more formally in the following discussion.

A biped's *support base* [7] is defined as the smallest convex polygon that includes all points where the feet are in contact with the ground. When in single support, that is, where one foot, the stance foot, is on the ground and the other is stepping, the support base is the outline of the part of the stance foot that is in contact with the ground. When in double support, that is, where both feet are on the ground, the base of support is the smallest convex polygon that includes all points where the two feet are in contact with the ground.

The ground reaction force vector, $\mathbf{F}_{\text{gr}}$, is defined as the integral, over the base of support, of the incremental ground reaction forces emanating from each point of contact with the ground:

$$\mathbf{F}_{\text{gr}} = \iint\limits_{\text{BOS}} \mathbf{F}_{\text{gr}}(x, y)\, dxdy \tag{10.1}$$

where $\mathbf{F}_{\text{gr}}(x, y)$ is the incremental force at point x, y on the ground, and BOS refers to the base of support region.

The CM is the weighted mean of the positions of all points in the system, where the weight applied to each point is the point's mass. Thus, for a discrete distribution of masses m_i, located at positions $\mathbf{r}_i$, the position of the center of mass is given by

$$\mathbf{CM} = \frac{\sum_i m_i \mathbf{r}_i}{\sum_i m_i} \tag{10.2}$$

A bipedal mechanism consists of a set of articulated links, each of which is a rigid body with mass m_i. Each rigid body has its own CM at a point $\mathbf{r}_i$.

The CM represents the effective mass of the system, concentrated at a single point. This is valuable because it allows for simplifying the balance control problem to one of keeping the CM in the right place at the right time. Furthermore, the control dynamics of this point is expressed, simply, by Newton's law, $\mathbf{F}_{\text{gr}} = m\mathbf{a}$, where m is the total mass of the system, and $\mathbf{a}$ is the resulting acceleration of the CM.

Vukobratovic and Stepanenko defined the ZMP as the point of resulting reaction forces at the contact surface between the extremity and the ground [22]; it is the point from which the ground reaction force vector, defined by (10.1), emanates. The ZMP may be defined as the point on the ground surface about which the horizontal component of the moment of ground reaction force is zero [1, 20]. Because the base of support is defined by the convex polygon of points in contact with the ground, and because the ZMP represents the average force contribution of these points, the ZMP is always inside the biped's base of support [6].

We designate the horizontal axes as x and y, where x represents the anterior–posterior direction, and y the medio-lateral direction. We designate the vertical axis as z (positive direction is upwards). The position of the ZMP along these axes, x_{ZMP} and y_{ZMP} can be expressed in terms of CM position, force, and moment as

$$x_{\mathrm{ZMP}} = x_{\mathrm{CM}} - \frac{F_{\mathrm{gr}x}}{F_{\mathrm{gr}z}} z_{\mathrm{CM}} - \frac{\tau_y}{F_{\mathrm{gr}z}} \tag{10.3}$$

$$y_{\mathrm{ZMP}} = y_{\mathrm{CM}} - \frac{F_{\mathrm{gr}y}}{F_{\mathrm{gr}z}} z_{\mathrm{CM}} + \frac{\tau_x}{F_{\mathrm{gr}z}} \tag{10.4}$$

where x_{cm}, y_{cm}, and z_{cm} are the x, y, and z positions of the CM, $F_{\mathrm{gr}x}$, $F_{\mathrm{gr}y}$, and $F_{\mathrm{gr}z}$ are the ground reaction forces in the x, y, and z directions, and τ_x and τ_y are the CM moments about the x and y axes, respectively.

Because the gravitational force is purely vertical, $F_{\mathrm{gr}x}$ and $F_{\mathrm{gr}y}$ are the net horizontal forces on the CM. The net vertical CM force, F_z, is $F_z = F_{\mathrm{gr}z} - Mg$, where g is the gravitational acceleration and M is the total mass. The ZMP is always inside the support base [16]. If the moments in Eq. (10.2) are zero, the ground reaction force vector points directly at the CM, as shown in Fig. 10.7a.

The CMP is the point on the ground, not necessarily within the support base, from which the observed net ground reaction force vector would have to act in order to generate no torque about the CM [for τ_x and τ_y in (10.4) to be 0]. Thus, it is that point

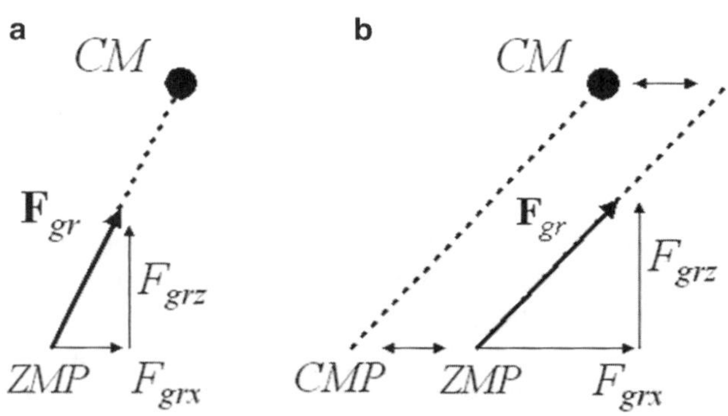

Fig. 10.7 As shown in (**a**), if there is no moment about the CM, the ground reaction force points from the ZMP to the CM position. As shown in (**b**), if there is a moment about the CM, the ZMP and CMP diverge, where the separation distance is the moment arm associated with the vertical force, $F_{\mathrm{gr}z}$ (©IEEE, 2009, reprinted with permission)

where a line parallel to the ground reaction force vector, passing through the CM, intersects with the ground, as shown in Fig. 10.7b. The CMP can be expressed as

$$(\mathbf{r}_{\mathrm{CMP}} - \mathbf{r}_{\mathrm{CM}}) \times \mathbf{F}_{\mathrm{gr}} = 0. \tag{10.5}$$

Expanding this cross product yields

$$x_{\mathrm{CMP}} = x_{\mathrm{CM}} - \frac{F_{\mathrm{gr}x}}{F_{\mathrm{gr}z}} z_{\mathrm{CM}} \tag{10.6}$$

$$y_{\mathrm{CMP}} = y_{\mathrm{CM}} - \frac{F_{\mathrm{gr}y}}{F_{\mathrm{gr}z}} z_{\mathrm{CM}} \tag{10.7}$$

Note that because $F_{\mathrm{gr}x}$ and $F_{\mathrm{gr}y}$ are the net horizontal forces on the CM, this relation can be used to compute horizontal CM force as a function of CM position, CMP point location, and vertical ground reaction force. Such horizontal forces are critical for maintaining bipedal stability since they can be applied to change CM state to desired values.

By combining Eqs. (10.4) and (10.7), we obtain a relation between ZMP and CMP:

$$x_{\mathrm{CMP}} = x_{\mathrm{ZMP}} + \frac{\tau_y}{F_{\mathrm{gr}z}} \tag{10.8}$$

$$y_{\mathrm{CMP}} = y_{\mathrm{ZMP}} - \frac{\tau_x}{F_{\mathrm{gr}z}} \tag{10.9}$$

Equation (10.9) shows that when there is no horizontal moment about the CM, the CMP and ZMP points coincide. In this case, the ground reaction force vector points directly to the CM, as shown in Fig. 10.7a. Conversely, when there is a horizontal moment about the CM, the CMP and ZMP diverge. The horizontal separation distance between these points is the moment arm for the CM moment due to vertical force, $F_{\mathrm{gr}z}$, as shown in Fig. 10.7b. Note that as the CMP and ZMP diverge, the ZMP must remain within the support base, but the CMP may leave the region of support.

The relationship between the CM and CMP indicates the specific effect that the net ground reaction force has on CM translation. Because the observed net ground reaction force always operates at the ZMP which is within the support base, whenever the net ground reaction force generates no torque about the CM, then the ZMP and CMP coincide. If the net ground reaction force generates torque, however, then the CMP and ZMP differ in location, and, in particular, the CMP may be outside the support base.

It is sometimes desirable to have the CMP and ZMP diverge as shown in Fig. 10.8b so that horizontal CM forces can be more effectively controlled. In this case the CMP can be displaced from the ZMP which reflects the increased ability

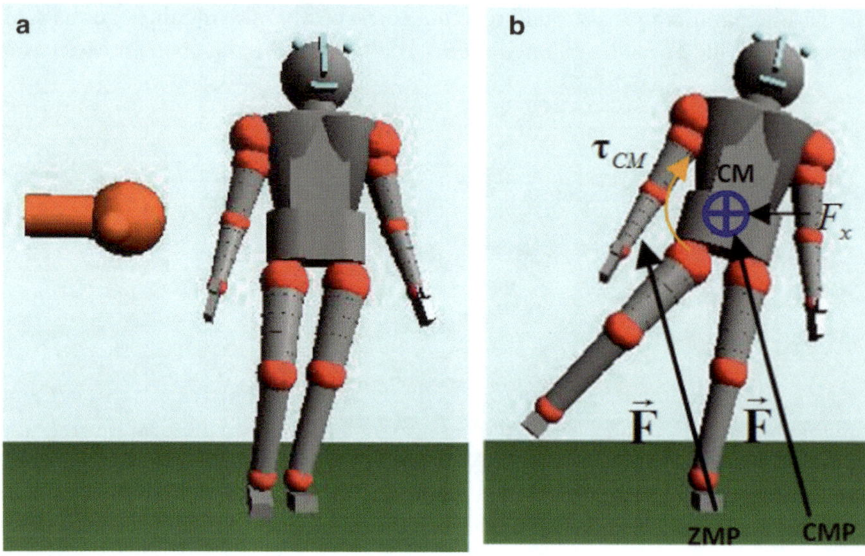

Fig. 10.8 Recovery from a lateral disturbance using CMP. In (**a**), the biped is disturbed by a lateral force. In (**b**), the use of angular momentum to recover from the disturbance is indicated by the divergence of the CMP from the ZMP

of the net ground reaction force to affect translation of the CM. The associated moment about the CM generally produces undesirable effects, such as loss of upright orientation of the upper body. In many cases, these effects are temporary, can be managed, and are well worth the overall positive effect on CM position and velocity. For example, a tightrope walker will tolerate temporary angular instability if this means that he won't fall off the tightrope.

Use of the CMP is demonstrated in Fig. 10.8, which depicts recovery from a lateral disturbance. This sequence shows an initial disturbance that pushes the biped to the right. To compensate, the system takes control actions involving rotation of the body and swing leg, that move its CMP to the right, creating a lateral compensating force to the left. Because the disturbance is significant, the CMP moves beyond the edge of the support polygon, and thus, it does not coincide with the ZMP. This compensating action corresponds to a clockwise torque about the CM, which is manifested by clockwise rotation of the torso and right leg.

The model of balance control presented here, where requirements for balance are expressed in terms of CM, ZMP, CMP, and the support base is extremely useful for planning and control, due to its simplicity. Balance control is then reduced to a problem of adjusting the base of support, adjusting the ZMP within the base of support, and, if necessary, performing motions that generate angular momentum, so that the CMP can be moved, temporarily, outside the base of support, in order to exert additional compensating force on the CM.

10.3 Dynamic Virtual Model Controller

10.3.1 Biped Model

Consider the three-dimensional humanoid biped model, shown in Fig. 10.9. The model has seven segments: two feet, two lower leg segments, two upper leg segments, and a body segment that lumps the torso, head, and arms. The leg and body segments are modeled as cylinders, whereas the feet are modeled as rectangular blocks. Segment dimensions and masses are given in Tables 10.1 and 10.2. Twelve degrees of freedom correspond to joints (six in each leg), and six degrees of freedom correspond to upper body position and orientation. Each leg is modeled with a ball-and-socket hip joint (three degrees of freedom), a pin knee joint (one degree of freedom), and a saddle-type ankle joint (two degrees of freedom). Note that although the humanoid model presented here does not include independently moving arms, the model, and the DVMC control architecture can be easily extended to include them.

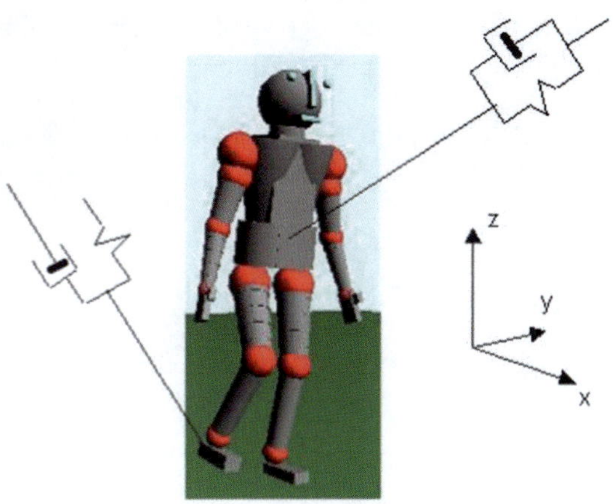

Fig. 10.9 Virtual linear spring-damper elements, attached to reaction points, allow the mechanism to be controlled as if it were a puppet. The coordinate frame is as follows: *x* is the anterior–posterior axis, *y* is the medio-lateral axis, and *z* is the vertical axis (©IEEE, 2009, reprinted with permission)

Table 10.1 Model segment masses

Model segment	Mass (kg)
Foot	1.56
Lower leg	4.48
Upper leg	10.73
Upper body	70.65

Table 10.2 Model segment dimensions

Model segment	Length (m)	Radius (m)
Upper body	0.64	0.18
Upper leg	0.46	0.08
Lower leg	0.48	0.05
Hip spacing	0.25	

Table 10.3 Outputs to be controlled

Index	Output
1	Posterior-anterior CM position
2	Medio-lateral CM position
3	Vertical CM position
4	Upper body roll angle
5	Upper body pitch angle
6	Upper body yaw angle
7	Posterior-anterior swing foot position
8	Medio-lateral swing foot position
9	Vertical swing foot position
10	Swing foot roll angle
11	Swing foot pitch angle
12	Swing foot yaw angle

For single-support case, in the global coordinate frame. For double-support, outputs 7–12 (the ones associated with the swing foot) are omitted

The outputs to be controlled are listed in Table 10.3. These outputs are values relevant to balance control and locomotion, such as CM position, upper body orientation, and stepping foot position. Thus, the purpose of the DVMC is to move the joints so that the desired motion for the outputs is achieved.

10.3.2 Controller Architecture

Desired motion behavior for the outputs is specified in a simple, straightforward way, using a linear proportional-differential (PD) law:

$$\ddot{\mathbf{y}} = \mathbf{k}_s\left(\mathbf{y}_s - \mathbf{y}\right) + \mathbf{k}_d\left(\dot{\mathbf{y}}_s - \dot{\mathbf{y}}\right), \tag{10.10}$$

where $\mathbf{y}$ is the vector of outputs to be controlled, $\mathbf{y}_s$ and $\dot{\mathbf{y}}_s$ are position and velocity setpoint vectors, and $\mathbf{k}_s$ and $\mathbf{k}_d$ are spring and damping gain vectors. Such a control law can be represented as a set of virtual spring-damper elements attached to the output points being controlled, as shown in Fig. 10.9, so that the controlled outputs move as if they were point masses attached to these spring-damper systems.

The difficulty with this is that the robot is not a linear system; the accelerations of the controlled outputs are nonlinear functions of the joint torque actuation inputs. The DVMC solves this problem by providing an abstraction of the plant, shown in Fig. 10.10, which makes it appear linear, and therefore, allows it to follow control laws in the form of Eq. (10.10).

This use of virtual elements is similar, in concept, to the one used in a virtual model controller [17]. However, unlike [17], the DVMC accounts for plant

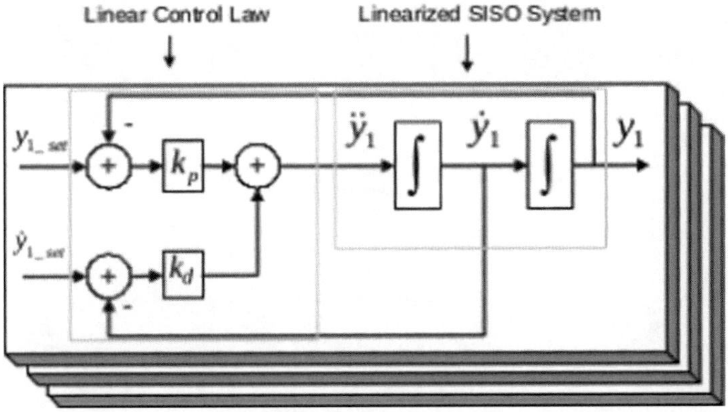

Fig. 10.10 Linear virtual element abstraction consisting of a set of SISO systems with associated linear control laws

dynamics, resulting in a linear system where controlled points move as if they were linear second order systems. Furthermore, through the use of a goal prioritization technique, the DVMC is able to generate moments about the CM in order to generate beneficial forces on the CM.

The DVMC uses a model-based input–output linearization algorithm [18] to linearize and decouple the plant. The input–output linearization approach is augmented with a slack variable relaxation technique to accommodate actuation constraints and prioritize goals. This feature is important because it is not always possible to achieve all control goals simultaneously. Actuation constraints, such as the requirement that the ZMP must remain well inside the support base in the case where foot roll is undesirable, may cause the overall system to become over-constrained, in which case some goals must be deferred. To address this problem, the controller incorporates a goal prioritization algorithm that automatically sacrifices lower-priority goals when the system becomes over-constrained in this way. For example, the system may temporarily sacrifice goals of maintaining upright posture in order to achieve CM state goals. We now describe the linearization and goal prioritization components of the controller in more detail.

10.3.3 Linear Virtual Element Abstraction

A geometric transform, **h**, is used to convert from the joint state to the workspace (output) state representation, according to

$$\mathbf{y} = \mathbf{h}(\mathbf{q}) \tag{10.11}$$

where **q** is the joint position vector, and **y** is the output vector. Thus, **h** is the kinematic transform. The controller uses a feedback linearizing transformation

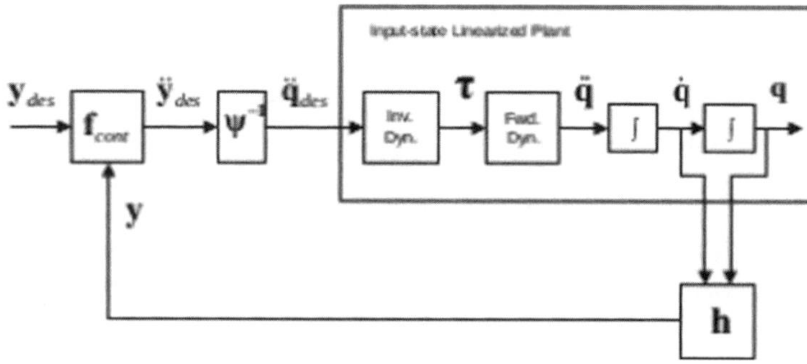

Fig. 10.11 Two-stage linearization

to convert desired workspace variable accelerations, $\ddot{\mathbf{y}}$, into corresponding joint torques. Application of these torques results in a new joint state, and associated workspace state. Use of the linearizing transformation makes the nonlinear plant appear to be a set of decoupled SISO linear second-order systems, as shown in Fig. 10.10. Each SISO system can then be controlled by a proportional-differential (PD) law, as discussed previously, resulting in a *linear virtual element abstraction*.

The linearization is accomplished using a two-stage process, as shown in Fig. 10.11. Given a desired output acceleration vector, $\ddot{\mathbf{y}}_{\text{des}}$, which is computed by the PD law, we first compute the corresponding joint acceleration vector, $\ddot{\mathbf{q}}_{\text{des}}$, using a geometric transformation. Then, we compute the joint torque vector, τ, that achieves $\ddot{\mathbf{q}}_{\text{des}}$, using an inverse dynamics transformation [3].

The inverse dynamics computation is of the form

$$\mathbf{H}(\mathbf{q})\,\ddot{\mathbf{q}} + \mathbf{C}(\mathbf{q}, \dot{\mathbf{q}}) + \mathbf{g}(\mathbf{q}) = \boldsymbol{\tau} \tag{10.12}$$

where $\mathbf{H}(\mathbf{q})$ is a matrix of inertial terms, $\mathbf{C}(\mathbf{q}, \dot{\mathbf{q}})$ is a matrix of velocity-related terms, and $\mathbf{g}(\mathbf{q})$ is a vector of gravitational terms. Hence, for a particular joint state $\left[\mathbf{q}^T, \dot{\mathbf{q}}^T\right]$, Eq. (10.12) represents a linear relation between $\boldsymbol{\tau}$ and $\ddot{\mathbf{q}}$. Note also that because $\boldsymbol{\tau}$ and $\ddot{\mathbf{q}}$ are both 12-element vectors (corresponding to the 12 actuators), $\mathbf{H}(\mathbf{q})$ is a 12×12 square matrix, so Eq. (10.12) is fully constrained.

In order to obtain a relation between $\ddot{\mathbf{y}}_{\text{des}}$ and $\ddot{\mathbf{q}}_{\text{des}}$, we differentiate Eq. (10.11) twice to obtain

$$\dot{\mathbf{y}} = \frac{\partial \mathbf{h}}{\partial \mathbf{q}}\dot{\mathbf{q}} = \mathbf{J}\dot{\mathbf{q}} \tag{10.13}$$

$$\ddot{\mathbf{y}} = \mathbf{J}\ddot{\mathbf{q}} + \dot{\mathbf{J}}\dot{\mathbf{q}} = \mathbf{J}\ddot{\mathbf{q}} + \Psi \tag{10.14}$$

where $\mathbf{J}$ is the Jacobian matrix. The matrix $\mathbf{J}$ and the vector Ψ are functions of joint state. Therefore, for a particular joint state $\left[\mathbf{q}^T, \dot{\mathbf{q}}^T\right]$, Eq. (10.14) represents a linear relation between $\ddot{\mathbf{q}}$ and $\ddot{\mathbf{y}}$. Note also that because $\mathbf{q}$ and $\mathbf{y}$ are both 12-element vectors, Eq. (10.14) represents a fully constrained system.

To achieve the linearization shown in Fig. 10.11, we combine Eqs. (10.12) and (10.14):

$$\begin{bmatrix} \mathbf{I}_{12\times12} & \mathbf{0}_{12\times12} & \mathbf{0}_{12\times12} \\ \mathbf{I}_{12\times12} & -\mathbf{J} & \mathbf{0}_{12\times12} \\ \mathbf{0}_{12\times12} & \mathbf{H} & -\mathbf{I}_{12\times12} \end{bmatrix} \begin{bmatrix} \ddot{\mathbf{y}} \\ \ddot{\mathbf{q}} \\ \boldsymbol{\tau} \end{bmatrix} = \begin{bmatrix} \ddot{\mathbf{y}}_{\text{des}} \\ \Psi \\ -\mathbf{C} \end{bmatrix} \tag{10.15}$$

Note that this is a fully constrained, linear system.

10.3.4 Multivariable Optimal Controller

The linearization of the square system represented by Eq. (10.15) is subverted if inequality constraints are introduced, and these constraints become active; the system represented by Eq. (10.15) becomes over-constrained in this case. Inequality constraints are used to represent actuation limits. An important constraint of this type is the requirement to keep the stance foot flat on the ground during single support; while balancing on one leg it is undesirable for the stance foot to roll, particularly on its lateral edge. This particular constraint is accomplished by requiring the ZMP to be inside the edge of the support envelope. Note that this constraint is distinct from the physical constraint that the ZMP not be outside the support base. If the ZMP is on the edge of the support envelope, the foot will begin to roll [16]. Hence, in order to avoid foot roll, we employ linear inequality constraints to keep the ZMP inside the edge of the support envelope.

For the humanoid model, the ZMP is given by expanding Eq. (10.4), or

$$x_{\text{ZMP}} = \frac{\sum_{i=2}^{7} m_i r_{xi}\left(\ddot{r}_{zi} + g\right) - \sum_{i=2}^{7} m_i r_{zi}\ddot{r}_{xi} - \sum_{i=2}^{7} \tau_{yi}}{\sum_{i=2}^{7} m_i\left(\ddot{r}_{zi} + g\right)} \tag{10.16}$$

$$y_{\text{ZMP}} = \frac{\sum_{i=2}^{7} m_i r_{yi}\left(\ddot{r}_{zi} + g\right) - \sum_{i=2}^{7} m_i r_{zi}\ddot{r}_{yi} + \sum_{i=2}^{7} \tau_{xi}}{\sum_{i=2}^{7} m_i\left(\ddot{r}_{zi} + g\right)} \tag{10.17}$$

$$\boldsymbol{\tau}_{xi} = \mathbf{I}_{Gi}\dot{\boldsymbol{\omega}}_{xi} \tag{10.18}$$

$$\boldsymbol{\tau}_{yi} = \mathbf{I}_{Gi}\dot{\boldsymbol{\omega}}_{yi} \tag{10.19}$$

where i is the segment index, r_{xi}, r_{yi}, and r_{zi} denote the CM position of segment i, $\mathbf{I}_{Gi}$ is the inertia matrix of segment i, and $\boldsymbol{\omega}_{xi}$ and $\boldsymbol{\omega}_{yi}$ are the angular velocities of segment i about the x and y axes, respectively. The moments, $\boldsymbol{\tau}_{xi}$ and $\boldsymbol{\tau}_{yi}$, are about the segment i CM in the x and y axes, respectively.

Equation (10.19) is transformed into a set of linear inequality constraints by replacing x_{ZMP} and y_{ZMP} with min and max terms, reflecting the bounds, so that these become constants:

$$\mathbf{H}_r \ddot{\mathbf{y}}_r \leq \mathbf{K}_r \tag{10.20}$$

where

$$\mathbf{H}_r = \begin{bmatrix} -\mathbf{m}^T \bullet \mathbf{r}_z^T & \mathbf{0}_{1\times 6} & \left(-\mathbf{x}_{\max} + \mathbf{m}^T \bullet \mathbf{r}_x^T\right) & \mathbf{0}_{1\times 6} & -\mathbf{I}_y \\ \mathbf{m}^T \bullet \mathbf{r}_z^T & \mathbf{0}_{1\times 6} & \left(\mathbf{x}_{\min} - \mathbf{m}^T \bullet \mathbf{r}_x^T\right) & \mathbf{0}_{1\times 6} & \mathbf{I}_y \\ \mathbf{0}_{1\times 6} & -\mathbf{m}^T \bullet \mathbf{r}_z^T & \left(-\mathbf{y}_{\max} + \mathbf{m}^T \bullet \mathbf{r}_y^T\right) & \mathbf{I}_x & \mathbf{0}_{1\times 6} \\ \mathbf{0}_{1\times 6} & \mathbf{m}^T \bullet \mathbf{r}_z^T & \left(\mathbf{y}_{\min} - \mathbf{m}^T \bullet \mathbf{r}_y^T\right) & -\mathbf{I}_x & \mathbf{0}_{1\times 6} \end{bmatrix} \tag{10.21}$$

$$\ddot{\mathbf{y}}_r = \left[\ddot{\mathbf{r}}_x^T \ \ddot{\mathbf{r}}_y^T \ \ddot{\mathbf{r}}_z^T \ \dot{\boldsymbol{\omega}}_x^T \ \dot{\boldsymbol{\omega}}_y^T\right]^T \tag{10.22}$$

$$\mathbf{K}_r = \begin{bmatrix} g\left(m_{\text{tot}} x_{\text{zmpmax}} - \mathbf{m}^T \bullet \mathbf{r}_x^T\right)^T \\ -g\left(m_{\text{tot}} x_{\text{zmpmin}} - \mathbf{m}^T \bullet \mathbf{r}_x^T\right)^T \\ g\left(m_{\text{tot}} y_{\text{zmpmax}} - \mathbf{m}^T \bullet \mathbf{r}_y^T\right)^T \\ -g\left(m_{\text{tot}} y_{\text{zmpmin}} - \mathbf{m}^T \bullet \mathbf{r}_y^T\right)^T \end{bmatrix} \tag{10.23}$$

$$\mathbf{x}_{\max} = x_{\text{zmpmax}} \mathbf{m}^T \tag{10.24}$$

$$\mathbf{x}_{\min} = x_{\text{zmpmin}} \mathbf{m}^T \tag{10.25}$$

$$\mathbf{y}_{\max} = y_{\text{zmpmax}} \mathbf{m}^T \tag{10.26}$$

$$\mathbf{y}_{\min} = y_{\text{zmpmin}} \mathbf{m}^T \tag{10.27}$$

where $\mathbf{m}$ is a six-element vector of masses for segments 2–7, $\mathbf{r}_x$, $\mathbf{r}_y$, and $\mathbf{r}_z$ are six-element vectors of the segments' CM x, y, and z positions, $\mathbf{I}_x$ and $\mathbf{I}_y$ are six-element vectors of the segments' inertias about the x and y axes, and x_{zmpmin}, x_{zmpmax}, y_{zmpmin}, and y_{zmpmax} are the ZMP limits. The operator $\bullet$ represents element-wise multiplication.

Now, $\mathbf{r}_x$, $\mathbf{r}_y$, $\mathbf{r}_z$, $\mathbf{I}_x$ and $\mathbf{I}_y$ are kinematic functions of $\mathbf{q}$, and $\mathbf{m}$ is a constant. Therefore, Eq. (10.20) is linear with respect to the current joint state. Furthermore, $\ddot{\mathbf{y}}_r$ is related to the joint acceleration vector through a linear function similar to Eq. (10.15):

$$\ddot{\mathbf{y}}_r = \mathbf{J}_r \ddot{\mathbf{q}} + \Psi_r \tag{10.28}$$

Therefore, if we combine the inequality constraints of Eq. (10.28) with (10.20), we have an overall system that is either fully constrained or over constrained (because $\ddot{\mathbf{y}}_r$ is fully dependent on $\ddot{\mathbf{q}}$, it does not add any flexibility). If none of the constraints in Eq. (10.20) are active, the system is fully constrained. However, if one of these constraints is active, the system becomes over constrained, and there is no feasible solution. Consequently, if the controller does not take the constraints in Eq. (10.20) into consideration, it could generate values for $\ddot{\mathbf{y}}_{\text{des}}$ that are infeasible.

One way to avoid this type of infeasibility is to use "slack" variables that provide flexibility to the overall system. Thus, the controller output, $\ddot{\mathbf{y}}_{\text{contout}}$, is given by

$$\ddot{\mathbf{y}}_{\text{des}} = \ddot{\mathbf{y}}_{\text{contout}} + \ddot{\mathbf{y}}_{\text{slack}} \tag{10.29}$$

where $\ddot{\mathbf{y}}_{\text{slack}}$ is the vector of slack variables. The goal of the overall control system is then to minimize $\ddot{\mathbf{y}}_{\text{slack}}$, taking into account the relative importance of each element.

This minimization is accomplished by formulating the control problem as a quadratic program (QP), and then using a QP optimizer to solve it. The relative importance of the slack variables is expressed in the cost function for the QP. This causes the optimizer to prioritize goals by first minimizing the slack variables for the most important outputs, and therefore, setting $\ddot{\mathbf{y}}_{\text{contout}}$ to be as close as possible to $\ddot{\mathbf{y}}_{\text{des}}$ for these outputs. For example, slack variables associated with the CM position output are given higher cost than those associated with trunk and swing leg orientation. The slack variable costs were determined empirically. Their precise value is not crucial; as long as the slack costs for CM position are higher than the slack costs for the other outputs, desired behavior is achieved.

The QP formulation is under-constrained, due to the use of the slacks. The QP optimizer explores the null space of the formulation, choosing the solution that minimizes the cost function. This cost function is of the form $\mathbf{w} \cdot \ddot{\mathbf{y}}_{\text{slack}}$ where $\mathbf{w}$ is a vector of weights reflecting the importance of minimizing the associated slack variable [11].

10.4 QSP and Problem Specification

We seek to guide a bipedal mechanism so that it accomplishes a particular motion task, such as walking at a specified speed, region, walking on a set of irregularly placed stones, or walking to a soccer ball in time to kick it. Motion tasks are specified by a QSP, which is executed by a *Model-based Executive* [9, 14]. The executive uses a *Plant Model* combined with current state estimates to generate control inputs. The flexibility provided by state and temporal constraints in the QSP allows the executive to consider multiple possible state and control input sequences, and to choose the most appropriate one given the situation. Sections 10.4.1 and 10.4 formally define the Plant Model and QSP representations and Sect. 10.4.3 defines the problem solved by the executive, based on these inputs.

10.4.1 Plant Model

We assume that the plant can be modeled as a set of subsystems whose dynamics are linear and decoupled, within specified discrete *modes*. The linearization and decoupling are provided by the DVMC, described previously, which is a component of the Model-based Executive. Thus, the Plant Model provides the Model-based Executive an abstraction of the actual system, which is easier to control. Each discrete mode, for example, single support or double support, has its own set of linearized dynamics. Such a plant model, which incorporates all discrete modes and associated continuous linearized dynamics is called a hybrid (discrete/continuous) model. We now define a plant model more formally.

Definition 1 (Plant Model). A Plant Model is a *Hybrid Concurrent Constraint Automaton* (HCCA) [8], which consists of a set of hybrid automata. Each automaton is defined by the tuple $A_a = \langle m_a, L_a, T_a \rangle$, where m_a is the discrete mode of the automaton, L_a maps each mode to a *Linearized Subsystem* that defines the continuous dynamic behavior of the mode, and T_a is a set of transition functions. A Linearized Subsystem is a tuple $\langle \mathbf{x}, \mathbf{u}, \mathbf{A}, \mathbf{B}, \mathbf{c_c} \rangle$, where $\mathbf{x} \in \Re^n$ is the state vector, $\mathbf{u} \in \Re^m$ is the input vector of the subsystem, and $\mathbf{A} \in \Re^{n \times n}, \mathbf{B} \in \Re^{m \times n}$ are matrices that represent the plant dynamics according to $\dot{\mathbf{x}} = \mathbf{Ax} + \mathbf{Bu}$. Additionally, $\mathbf{c_c}$ is a set of actuation constraints of the form $\mathbf{H_c} \begin{bmatrix} \mathbf{x} & \mathbf{u} \end{bmatrix}^T \leq \mathbf{K_c}$. Given a current mode assignment and guard condition g_a, each transition function $\tau_a (m_a, g_a)$ specifies a target mode that the automaton will transition into, if the guard is satisfied. A guard condition is associated with a Linearized Subsystem, and is represented by a set of (convex) linear algebraic equality and inequality constraints over the state vector of the Linearized Subsystem: $\mathbf{f}(\mathbf{x}) = \mathbf{c_1}$, $\mathbf{g}(\mathbf{x}) \leq \mathbf{c_2}$ where $\mathbf{c_1}$ and $\mathbf{c_2}$ are vectors of constants.

One requirement for successful plan execution is that the state trajectory satisfies the dynamics and actuation constraints of the linearized subsystems, and that it corresponds to valid mode transitions of the automata in the plant. We call such a trajectory a *Plant-feasible Trajectory*. We develop this concept by first defining a *Mode Feasible Trajectory*, for a particular mode of a particular automaton in the HCCA, and then generalizing.

Definition 2 (Mode Feasible Trajectory). Given a Plant Model, an automaton, A in the model, and a mode, M, for the automaton, imply a particular Linearized Subsystem $L = L(A, M)$. We call a state and input trajectory, $\langle \mathbf{x}(t), \mathbf{u}(t) \rangle$, a *Mode Feasible Trajectory* with respect to M if it satisfies the dynamics and actuation constraints of L, as specified in Definition 1.

We now utilize this definition as a basis for defining feasible trajectories for a mode sequence in an automaton, and for automata in a plant.

Definition 10.1 (Automaton Feasible Trajectory). Given a Plant Model, and a particular automaton, A in the model, suppose we have a sequence of trajectories

$\{T_0, T_1, \ldots, T_n\}$, where $T_i = < x_i(t), u_i(t) >, t \in [ts_i, tf_i]$. Suppose, further, that each trajectory, T_i, in the sequence is a Mode Feasible Trajectory with respect to a mode M_i in A. We call the sequence of trajectories an *Automaton Feasible Trajectory* if for every trajectory, T_i, in the sequence, the final continuous state, $x_i(tf_i)$, in the trajectory satisfies the guard condition for transition to mode M_{i+1}, and $x_i(tf_i) = x_{i+1}(ts_{i+1})$.

Definition 10.2 (Plant Feasible Trajectory). Given a Plant Model, a *Plant Feasible Trajectory* is a set of Automaton Feasible Trajectories, one for each automaton in the plant, where the start and finish times of all Automaton Feasible Trajectories in the set are the same.

10.4.2 Qualitative State Plan

In this sub-section, we begin with an informal description of a QSP, and provide an example of such a plan. We follow this with a formal definition of a QSP, and in the following sub-section, a formal definition of the problem solved by the Model-based Executive.

A QSP provides a loose, flexible specification of desired performance, in terms of state space regions and temporal ranges. This flexibility may be exploited, for example, to improve optimality or to adapt to disturbances (improve robustness).

Reaching a goal location may require the biped to take a sequence of steps. Such steps represent transitions through a sequence of qualitatively different states, defined by which feet are in contact with the ground. Thus, a stepping sequence consists of alternating between double support phases, where both feet are on the ground, and single support phases, where one foot (the stance foot) is in contact with the ground, and the other foot (the swing foot) is taking the step. These phases represent qualitatively different system states, with correspondingly different behaviors.

We formalize the concept of a *Qualitative State* as a set of constraints on state and temporal behavior. For example, a Qualitative State may contain constraints on which feet of a legged robot are on the ground, and may include constraints on the position of each foot. It may also include state constraints on quantities like center of mass, and temporal constraints specifying time ranges by which the state goals must be achieved. Thus, a qualitative state is a loose, partial specification of desired behavior for a specific maneuver, like taking a step.

For example, a plan for a biped divides the walking cycle into a sequence of Qualitative States representing single and double support gait phases. Such a plan is shown in Fig. 10.12. In this plan, the first Qualitative State represents double support with the left foot in front, the second, left single support, the third, double support with the right foot in front, and the fourth, right single support. The fifth Qualitative State repeats the first, but is one gait cycle forward.

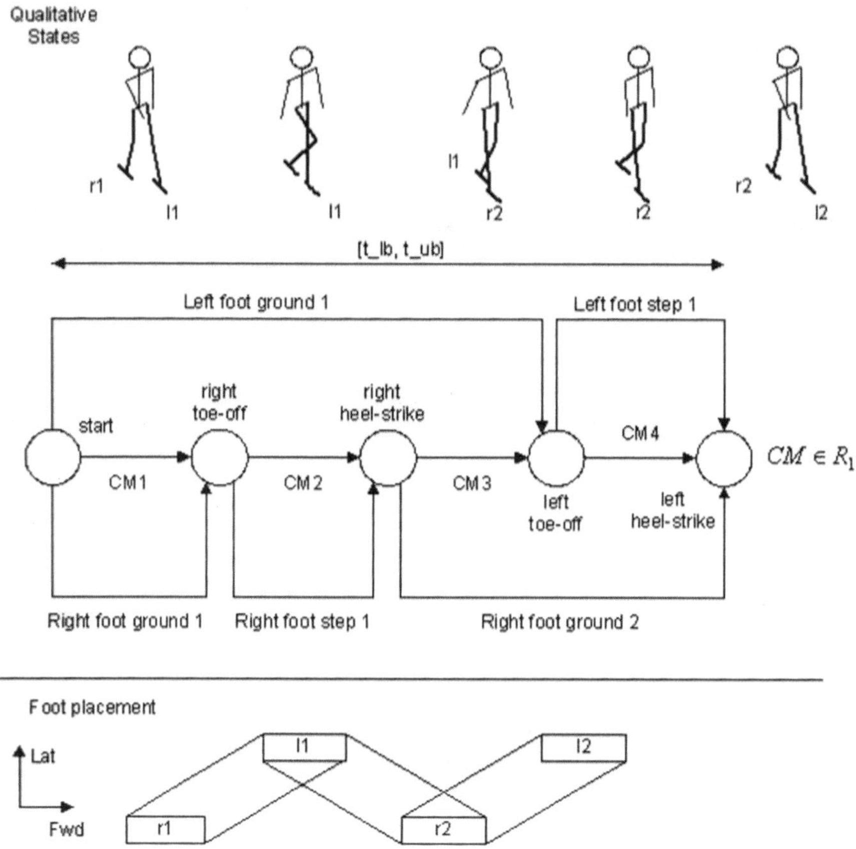

Fig. 10.12 Example QSP for walking gait cycle. *Circles* represent events, and *horizontal arrows* between events represent activities. For example, the activity "left foot ground 1" indicates that the left foot is on the ground from the event "start" to the event "left toe off". The activity "left foot step 1" indicates that the left foot is stepping (system is in right single support) from the event "left toe off" to the event "left heel strike". Similar activities for the right foot indicate when the right foot is on the ground, and when it is stepping. Activities may have associated state space constraints, such as the goal region constraint CM $\in r_1$, which specifies a goal for CM (center of mass) position and velocity. Foot placement constraints are indicated at the *bottom*; for example, rectangle r_1 represents constraints on the first right foot position on the ground, and rectangle l_1 on the first left foot position. The *lines between the rectangles* define the polygon of support when in double support

The QSP in Fig. 10.12 has a temporal constraint between the start and finish events. This constraint specifies a lower and upper bound, $[l, u]$, on the time between these events. It is a constraint on the time to complete the gait cycle, and thus, can be used to specify walking speed.

In addition to temporal constraints, QSPs include state constraints that specify valid initial, operating, and goal regions for an activity. If an initial region is specified for an activity, then the trajectory must be within this initial region, in order for the activity to begin. If an operating region is specified, then the trajectory must stay within this region for the duration of the activity. If a goal region is specified, then the trajectory must be within this region in order for the activity to end. In Fig. 10.12, the goal region constraint CM $\in r_1$ represents the requirement that the CM trajectory must be in region r_1 for the CM movement activity to finish successfully.

We now provide formal definitions for *Events* and *Activities*, and then use these components to define a QSP.

Definition 10.3 (Event). An event, e, represents a point in time. For a schedule, T, the specific time of e is given by $T(e)$.

Definition 10.4 (Activity). An activity is a tuple $\langle e_s, e_f, R_{\text{input}}, R_{op}, R_{\text{init}}, R_{\text{goal}}, s_i \rangle$, where e_s is an event representing the start of the activity, e_f is an event representing its finish, s_i is a Linearized Subsystem associated with the activity, R_{input} is a set of constraints on the inputs, $\mathbf{u}$, of s_i, R_{op} is a set of operational constraints on the state, $\mathbf{x}$, of s_i that must hold for the duration of the activity, R_{init} is a set of constraints on the state that must hold for the activity to begin, and R_{goal} is a set of constraints on the state that must hold for the activity to finish. The state constraints, R_{op}, R_{init}, and R_{goal}, are each of the form $\mathbf{Hx} \leq \mathbf{K}$, where $\mathbf{H} \in \Re^{q \times n}$ and $\mathbf{K} \in \Re^{q \times 1}$, and q is the number of linear inequalities in the set. The input constraints, R_{input}, are of the form $\mathbf{H}\left[\mathbf{x}^T\ \mathbf{u}^T\right]^T \leq \mathbf{K}$.

Definition 10.5 (QSP). A QSP is a tuple $\langle E, A, C \rangle$, where E is a set of Events, A is a set of Activities, and C is a set of externally imposed temporal constraints on the start and finish times of the activities. For example, the QSP shown in Fig. 10.12 has five Events ("start," "right toe-off," "right heel-strike," "left toe-off," "left heel-strike"), nine activities ("Left foot ground 1," "Left foot step 1," "CM1," "CM2," "CM3," "CM4," "Right foot ground 1," "Right foot step 1," "Right foot ground 2"), and one temporal constraint.

Definition 10.6 (Temporal Constraint). A temporal constraint is a tuple $\langle e_1, e_2, l, u \rangle$, where e_1 and e_2 are events, and l and u represent lower and upper bounds on the time between these events. Thus, $l \in \Re \cup \{-\infty\}$, and $u \in \Re \cup \{\infty\}$, such that $l \leq t(e_2) - t(e_1) \leq u$. In the QSP of Fig. 10.12, the temporal constraint restricts the time between the start and finish events. Events are used to represent start and finish times of an activity.

10.4.3 Plan Execution: The Problem Solved by the Model-Based Executive

Having formally defined a Plant and a QSP, we are now in a position to define the problem solved by the Model-based Executive in terms of a successful execution of a QSP. Successful execution can be expressed in terms of satisfaction of the individual activities in the QSP, and a consistent schedule, which combined, define satisfaction of a QSP.

Definition 10.7 (Schedule and Consistent Schedule). Given a QSP, Q, a *Schedule*, T, is an assignment of a specific time to each Event in Q. T is consistent with Q if it satisfies all Temporal Constraints in Q, that is, for each Temporal Constraint, $c \in C(Q)$, then a schedule assigns $t(e_1(c)) = T_1$, $t(e_2(c)) = T_2$ such that $l(c) \leq T_2 - T_1 \leq u(c)$, where e_1, e_2, l, u are the elements of c from Definition 10.8.

Definition 10.8 (Satisfaction of an Activity). Given an activity, a (Definition 10.6), with associated Linearized Subsystem s_i, a plant-feasible trajectory $\langle \mathbf{x}(t), \mathbf{t}(t) \rangle$ for s_i, and a schedule T, then a is satisfied by $\mathbf{x}(t)$ and T if the following conditions hold:

(1) $\mathbf{x}(t)$ must satisfy the initial and goal region state constraints of a. Let $t_s = T(e_s(a))$ be the start time of a under schedule T, and $t_f = T(e_f(a))$ be the finish time. Then, $\mathbf{x}(t)$ satisfies the initial and goal region constraints if $\mathbf{x}(t_s) \in R_{\text{init}}(a)$ and if $\mathbf{x}(t_f) \in R_{\text{goal}}(a)$.
(2) $\mathbf{x}(t)$ must satisfy the operating state constraints of a. That is, it must be the case that $\mathbf{x}(t) \in R_{op}(a)\, \forall t : t_s \leq t \leq t_f$.

Definition 10.9 (Satisfaction of a QSP). Given a QSP, Q, plant-feasible trajectory $\langle \mathbf{X}(t), \mathbf{U}(t) \rangle$, and a schedule, T, then Q is satisfied by $\langle \mathbf{X}(t), \mathbf{U}(t), T \rangle$ if T is consistent with Q (Definition 10.7), and $\langle \mathbf{X}(t), \mathbf{U}(t), T \rangle$ satisfies all activities in Q (Definition 10.8).

We now formally define the problem solved by the Model-Based Executive.

Definition 10.10 (Problem Solved by the Model-Based Executive). Given a QSP, Q, and a Plant Model, M, the Model-Based Executive must find a plant-feasible trajectory and schedule that satisfy the QSP (Definition 10.9), and then execute that trajectory and schedule. If no such trajectory and schedule exist, the executive will abort and indicate a plan infeasibility error. If a disturbance occurs during execution, then the Model-Based Executive must find a new plant-feasible trajectory and schedule, and continue execution. If no such trajectory and schedule exist, the executive will abort and indicate a plan infeasibility error.

The next section describes the Model-based Executive, and how it solves this problem.

10.5 Task Executive

The Task Executive (Model-based Executive) is responsible for attempting to execute a QSP, according to Definition 10.10. It does this using a two-part approach. The first part is an off-line component in which the QSP is compiled into a form that can be executed more efficiently. The second part is an on-line component that performs this execution.

10.5.1 Plan Compilation

In order to reduce runtime computational load, we construct, at compile time, a *Qualitative Control Plan* (QCP), which uses *Flow Tubes* to represent all trajectories that satisfy the QSP and the plant dynamics (see also [10]). Using the QCP, the executive achieves efficiency by selecting an appropriate trajectory, within each flow tube, that begins at the current system state. We first define the QCP, and present theorems that define conditions under which the problem is solvable by the Model-Based Executive. We then describe the algorithm for compiling a QCP, given a QSP and Plant Model.

10.5.1.1 Qualitative Control Plan

A key concept in plan feasibility for a hybrid system is *Temporal Feasibility* of individual activities in the plan. Temporal feasibility implies that the set of plant feasible trajectories includes ones that go from R_{init} to R_{goal} over the entire duration range $[l, u]$. More specifically, if a trajectory starting anywhere in the initial region, R_{init}, of the activity and ending somewhere in the goal region, R_{goal}, at any duration d such that $l \leq d \leq u$, is Plant Feasible, then the activity is temporally feasible in the duration range. This implies that the actuation limits imposed by the plant, and the operating constraints of the activity allow for actuation commands that can be used to control the linearized subsystem to the goal region from the initial region, at any duration in the duration range.

We now introduce the concept of a control policy for an activity, and use this, along with the previous definition, to define a *Temporally Controllable Activity*. A control policy maps a state, $\mathbf{x}$, associated with an activity's plant, to an actuation command $\mathbf{u}$ for the plant. A *Valid Control Policy* must provide the mapping for all states (all $\mathbf{x}$) that satisfy the operating constraints of the plant and the activity.

Definition 10.11 (Temporally Controllable Activity). Given an activity and associated plant, and given a Valid Control Policy, P, for the activity, the activity is *Temporally Controllable* by P in the duration range $[l, u]$ if the activity is Temporally Feasible in this range, and if all trajectories for the activity are consistent with (generated by) P. A trajectory is consistent with, or generated by P if for every state $x(k)$ in the trajectory, the subsequent state $x(k+1)$ results from applying P to $x(k)$.

We next use the concept of a Temporally Controllable Activity to state the conditions under which a QSP can be satisfied. Given a Plant Model and a QSP, suppose that each activity a_i in the QSP is Temporally Controllable over the duration range $[l_i, u_i]$. Let N be the Simple Temporal Network formed by combining the temporal constraints explicitly specified in the QSP, with the duration range temporal constraints $[l_i, u_i]$. If N is dispatchable [14], then a trajectory and schedule exist that satisfy the QSP (see Definition 10.11). Under these conditions, the plan is feasible with respect to the plant and control policy. This extends the notion of *dispatchability*, first introduced by Muscettola for discrete activity systems [14] , to general hybrid system.

In order to support efficient execution, the flow tube representation must allow the dispatcher to: (1) quickly determine whether a feasible state trajectory exists from the current state, and (2) if such a trajectory exists, what the control commands should be (based on a control policy) that achieve the trajectory. In order to leverage the advantages of dispatchable representations for discrete activity systems [4, 14], we require that the QCP temporal constraints be represented in minimum dispatchable form. This should include the temporal constraints specified in the QSP, and those implied by the dynamic limitations of the plant.

A variety of approaches are possible for the Flow Tube implementation. These include explicit sets (bundles) of trajectories with associated control policies, and discrete time sequences of polytope cross sections that define the feasible state space and associated control policies [2]. Regardless of the implementation, the Flow Tube representation must support in its API the following three functions.

Given a current plant state, $\mathbf{x}$, a goal region to achieve, $\mathbf{R}_g$, and an allowed duration range, $[l, u]$, the function
$[\mathbf{u}, d] = \textbf{ComputeControlAction}\left(\mathbf{x}, \mathbf{R}_g, [l, u]\right)$
must determine whether a feasible trajectory exists that will reach the goal from the current state within the allowed duration range. If so, it returns the next control action, $\mathbf{u}$, consistent with moving the state along the trajectory, along with a prediction d of the remaining duration until the goal is achieved. If no feasible trajectory exists, **ComputeControlAction** returns an error indicating that this is the case. The goal region is represented as a convex polytope: $\mathbf{R}_g = \mathbf{H}\mathbf{x} \le \mathbf{K}$, where $\mathbf{x} \in \Re^n$ is the state vector of the plant. This function is used by the Model-based Executive to generate commands to the biped, at each control step.

The function
$[l, u] = \textbf{ComputeControllableDuration}\left(\mathbf{x}, \mathbf{R}_g\right)$
computes the range of feasible (controllable) durations in which the state can be moved from the current state to the goal set. If no such duration range exists, **ComputeControllableDuration** returns an error indicating that this is the case. This function is used by the Model-based Executive to schedule activities consistently at runtime.

Given an initial region in state space, $\mathbf{R}_i$, the function
$[l, u] = \textbf{ComputeControllableDuration}\left(\mathbf{R}_i, \mathbf{R}_g\right)$
computes the range of feasible (controllable) durations in which the state can be moved from any state in the initial set to the goal set. If no such duration range exists, **ComputeControllableDuration** returns an error indicating that this is the case. This function is used by the Model-based Executive at compile time to generate QCPs that can be scheduled at runtime.

Now, consider two QSP activities, A_1 and A_2, that share the same linearized subsystem, s_i. Suppose that the finish event of A_1 is the start event of A_2. We call A_2 the *Successor Activity* of A_1. Consider feasible trajectories for A_1 and A_2 as shown in Fig. 10.13.

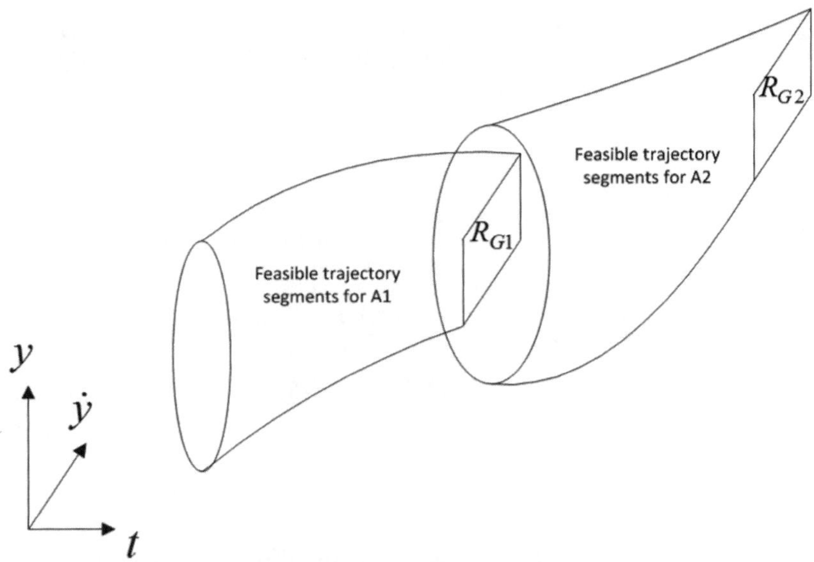

Fig. 10.13 Feasible trajectory segments for A_1 and A_2

Because A_2 is the successor of A_1, any feasible trajectory segment for A_1 must be part of a trajectory that has a feasible trajectory segment for A_2. Therefore, it is a requirement that the goal region for the flow tube for A_1 be a subset of the initial cross section of the flow tube for A_2.

We now formally define a QCP in terms of *Control Activities*, and what it means for a QCP to be executed successfully. A QCP has a structure similar to that of a QSP, but augments this with flow tube cross-sections representing feasible state trajectories and corresponding control policies. A control activity includes the information of the corresponding activity in the QSP, augmented with flow tubes specifying the activity's feasible state trajectories and corresponding control policies.

Definition 10.12. A *Control Activity* is a tuple $< A, F >$, where A is an activity (Definition 10.6), and F is a corresponding Flow Tube that implements the **ComputeControlAction** and **ComputeControllableDuration** functions.

Definition 10.13. A QCP is a triple $< E, A_c, C_t >$, where E is a set of events (Definition 10.5), C_t is a set of temporal constraints on the events (Definition 10.8), and A_c is a set of control activities (Definition 10.12). Each event is either a start event or finish event of a Control Activity.

Having specified the structure of a QCP, we now specify properties of a valid QCP. We begin with a Lemma that specifies requirements for activity succession.

Lemma 10.1. *Given a control activity,* A_2*, with predecessor activity* A_1*, for any specified duration range* $[l_{\mathrm{des}}, u_{\mathrm{des}}]$*, if*

$[l, u] = \textbf{\textit{ComputeControllableDuration}}\left(R_{\text{goal}}(A_1) \cap R_{\text{init}}(A_2), R_{\text{goal}}(A_2)\right)$ *returns an error (no duration,* $[l, u]$*), then no valid trajectories exist for the activity. Further, if a duration,* $[l, u]$ *exists, but* $[l, u] \cap [l_{\text{des}}, u_{\text{des}}]$ *is empty, then no valid trajectories exist for the activity in the desired duration range* $[l_{\text{des}}, u_{\text{des}}]$.

Note that the elimination of feasible durations for activities due to this Lemma will result in a tightening of the temporal bounds for such activities, beyond the [l, u] bounds specified in the QSP. In the extreme case, there are no feasible durations for the activity at all, leading to the following Lemma.

Lemma 10.2. *If there are no feasible durations according to Lemma 10.1 within the* $[l, u]$ *bounds specified for the control activity in the QCP, then there are no feasible trajectories for the activity, or for the QSP as a whole. Conversely, if a feasible duration exists, then a feasible trajectory exists for the activity that satisfies all initial, goal, operational, and actuation constraints.*

The tightening of temporal constraints due to Lemma 10.1 can cause a QCP to become infeasible, even if the individual activities are all feasible according to Lemma 10.2. This is expressed in the following theorem.

Theorem 10.1. *If a minimum dispatchable graph based on the temporal constraints specified in a QCP (Definitions 10.7 and 10.8), and possibly tightened according to Lemma 10.1, has a negative loop, then the QCP is infeasible.*

Proof. A minimum dispatchable graph represents a network of temporal constraints. If this graph has a negative loop, then there is an inconsistency in the temporal constraints [14]. If the minimum dispatchable graph is based on the temporal constraints explicitly specified for the QSP, as well as the additional temporal constraints implied by Lemmas 1, then a negative loop indicates an inconsistency in the overall set of temporal constraints, and the QCP is infeasible. □

A temporal inconsistency may result if the explicitly specified temporal constraints are inconsistent, or if they are inconsistent with the overall set of temporal constraints due to Lemma 10.1. The concepts presented here are useful for recognizing cases where the problem formulation, as represented by the QSP and Plant Model, is infeasible.

10.5.1.2 Plan Compilation Algorithm

The purpose of the plan compiler is to generate a QCP from a QSP. The main compilation steps are shown in Algorithm 1. The algorithm iterates over each activity in the QSP, calling ComputeFlowTubeForActivity. If the result is a flow tube with no valid trajectories, then the algorithm stops and indicates an error; the QSP is infeasible. If the flow tube is non-empty (has valid trajectories), then these trajectories are feasible in terms of plant dynamics, but not necessarily other

Algorithm 1: `CompileQSP`

```
Input: A QSP, Q, a plant model, M
Output: A QCP
foreach activity in the QSP do
    flow tube ← ComputeFlowTubeForActivity(activity);
    if !ValidTrajectories(flow tube) then
        Error (Activity goal, operational, and temporal constraints are incompatible with
        plant dynamics and actuation constraints.)
    PruneInfeasibleTrajectories (flow tube, activity, QSP);
    UpdateFeasibleDuration (flow tube, activity);
STN ← ComputeMinimumDispatchableGraph(QSP);
if STNInfeasible(STN) then
    Error (Temporal constraints, and plant dynamics and actuation constraints are
    inconsistent.)
```

aspects of the QSP. Therefore, the algorithm calls PruneInfeasibleTrajectories, which performs the intersection of initial region, predecessor goal region, and durations specified in Lemma 10.1. The consistency of temporal constraints is checked using a minimum dispatchable graph algorithm [14].

10.5.2 Plan Execution

To execute a QCP, the Dispatcher (online component of the Task Executive) must successfully execute each control activity. The dispatcher accomplishes this by, in real time, monitoring plant state, and generating plant control inputs based on the appropriate QCP control policy for the current state and time. In this way, the dispatcher indirectly schedules start and finish events so that they are consistent with the temporal constraints of the QCP. This is a key difference between the dispatcher described here, and those of discrete activity execution systems [14], in which event times are set directly by the dispatcher.

The Dispatcher performs three key functions in executing a control activity: initialization, monitoring, and transition. During initialization, the dispatcher chooses a goal duration for the control activity that is consistent with its execution window, and computes an initial control input. This control input is consistent with an optimal trajectory that will reach the activity's goal region in the chosen duration, if there are no disturbances.

After initializing an activity, the dispatcher begins monitoring its execution by obtaining an updated state estimate at each time increment, and checking whether the state is within the flow tube. If this is not the case, then a disturbance has occurred, and the dispatcher must determine the type of disturbance, and react accordingly.

As part of the monitoring function, the dispatcher also checks whether the state trajectory has achieved the activity's goal region in an acceptable time. If this is the case, it checks whether the activity's end event has occurred. This involves checking if the state trajectories of other activities whose completion must be synchronized are in their respective goal regions. If all completion conditions for a control activity are satisfied, the dispatcher switches to the transition function. If the control activity has a successor, the transition function invokes the initialization function for this new activity. As part of this transition, the dispatcher notes the time of the transition event and propagates this through the temporal constraints.

For each executing activity, the Dispatcher maintains a *target completion time*. The concept of a target completion time represents a key distinction between this system, and discrete activity execution systems [14]. This concept is needed here because activity completion is controlled indirectly, by applying control inputs. Thus, activity completion event times cannot simply be set, but rather occur as a consequence of the plant dynamics. If there are disturbances, the target completion time may have to be adjusted.

The dispatcher chooses a target activity completion time, and then selects an appropriate control policy that is predicted to complete the activity at the desired time. Because multiple activities are typically executing in parallel, achieving a desired event execution time requires synchronization of these activities. The Dispatcher uses a data structure called the *Event Horizon* to provide a mechanism for ensuring this consistency.

Definition 10.14 (Event Horizon). An Event Horizon is a set of Event Paths, each of which is a list of events and associated target execution windows. Hence, each element of an Event Path is a tuple $\langle e, l, u \rangle$ where e is the event, and l and u represent lower and upper bounds on times for the event.

The Event Paths represent events whose target execution windows have to be propagated in order to properly set target completion times for current activities in the Runtime Activity State. This propagation is a special kind of tightening of execution windows, distinct from the execution window tightening that is performed when events occur. This special propagation is necessary in order to ensure that target activity completion times are temporally consistent. This is a unique feature of our dispatcher; it is not used in discrete activity execution systems.

The algorithm for determining the event horizon involves starting at the target events of current activities, and searching back along outgoing negative arcs in the minimum dispatchable graph until executed events are reached. Consider the example QSP shown in Fig. 10.14. Activity a11 is for linearized sub-system 1, and activities a21, a22, and a23 are for linearized sub-system 2. Suppose that event ev1 has just occurred, and activities a11 and a21 are about to start executing. The dispatcher must choose target completion times for each activity. However, it cannot just choose target completion times that fit inside the event execution windows that were propagated when ev1 occurred; the Event Horizon must be taken into account. For example, suppose that the duration constraints on a11 are [3, 15], and the duration constraints on a21, a22, and a23 are [1, 5]. If event ev1 occurs at time

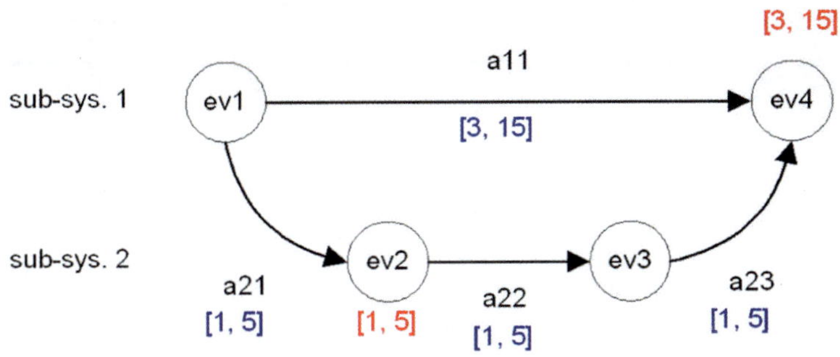

Fig. 10.14 Example event horizon. *Circles* represent events, and *horizontal arrows* between events represent activities

Fig. 10.15 Example Event Horizon. Circles represent events, and horizontal arrows between events represent activities.

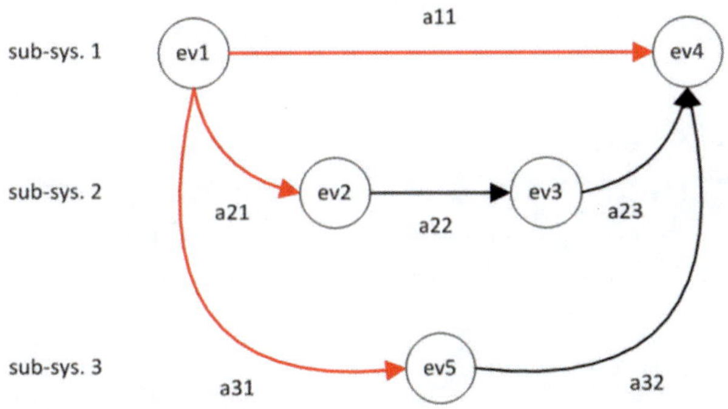

0, then the execution window for ev4 is [3, 15], and the execution window for event ev2 is [1, 5]. If the dispatcher were to base its decisions about target completion times solely on these execution windows, it could choose a target execution time of 4 for a21, and 4 for a11. This would cause a future temporal infeasibility, because a22 and a23 would have to be executed in 0 time, which violates their duration constraints.

To solve this problem, the dispatcher considers the event horizon, which, in this case, is ev1, ev2, ev3, ev4. In this case, if the dispatcher chooses duration range midpoints as target durations, then the target completion times for a21, a22, and a23 are 3, 6, and 9, respectively, and the target completion time for a11 is 9 as well.

Algorithm 2 shows the top-level dispatch loop of the executive. This algorithm is based on the one for discrete activity execution systems [14], but has some key extensions, which are highlighted.

InitializeActivities sets target execution times for the initial activities in the QCP. UpdateCurrentActivities iterates over each currently executing activity, checking if the event that has just occurred, ExecutableEvent?, is the finish event for the activity. If this is the case, it transitions the activity to the subsequent activity for the plant.

Algorithm 2: `Executive Dispatcher algorithm`

Input: A Qualitative Control Plan, QCP, and a plant model, P
Output: Actuation commands, u

```
t = 0;                                          // Initialize current time.
InitializeActivities(QCP, P);
ExecutableEvent? = StartEvent(QCP);
EnabledEvents = { ExecutableEvent? };
InitializeExecutionWindows ();
while EnabledEvents not empty do
    if ExecutableEvent? exists then
        ExecuteEvent (ExecutableEvent?) ;
        PropagateExecutionWindows (QCP, ExecutableEvent?) ;
        EnabledEvents = UpdateEnabledEvents (QCP, ExecutableEvent?) ;
        UpdateCurrentActivities(ExecutableEvent?, QCP) ;
        EventHorizon = UpdateEventHorizon(EventHorizon, ExecutableEvent?,
        QCP) ;
        UpdateTargetExecutionTimes(EventHorizon, QCP) ;
    UpdateCurrentState(P) ;
    UpdateControlInputs(P) ;
    t = t + dt ;                                // Increment current time.
    ExecutableEvent? = EventOccurred(t, EnabledEvents);
```

UpdateEventHorizon is used to update the event horizon after an event has occurred. To perform the update, the algorithm first removes any paths from the event horizon that contain the executed event. The algorithm then iterates over each currently executing activity, obtaining the target event for the activity. If the target event is not already in the event horizon, a search back from this event is started. This depth-first search proceeds back from the target event along negative out-going arcs in the minimum dispatchable graph. The search proceeds back along events that have not been executed. When an event that has been executed is encountered, the search branch stops, and search proceeds along the next branch. When there are no more branches, the path is added to the event horizon. For Fig. 10.14, this algorithm computes an event horizon $\{\{ev1, ev2, ev3, ev4\}\}$.

UpdateTargetExecutionTimes uses the event horizon to decide target completion times for activities. The algorithm first initializes target execution windows of all events in the event horizon to be identical to the execution windows computed by InitializeExecutionWindows and PropagateExecutionWindows. It then iterates over every path in the event horizon, retrieving the activity corresponding to the beginning of the path. This is always a currently executing activity. The algorithm sets the target completion time for the activity to be the midpoint of the execution window, and then propagates this decision to future events in the path. This can result in a tightening of target execution windows for these events.

The function UpdateCurrentState updates the estimate of plant state. The function UpdateControlInputs iterates over each currently executing activity, and computes new plant control inputs by accessing flow tube control policies based on

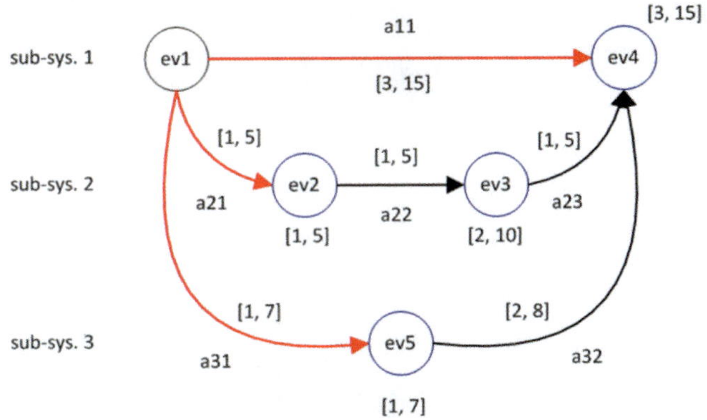

Fig. 10.16 Example Event Horizon. Circles represent events, and horizontal arrows between events represent activities.

the remaining execution time for the activity. The function EventOccurred checks whether an event has occurred, and if so, returns this as an executable event, whose occurrence must be propagated in the next iteration of the main loop in DispatchQCP.

10.6 Experiments

This section describes two types of experiments. The first type involves balancing on one leg. This type of experiment is used to evaluate the DVMC controller. The second type involves more general walking tasks. This type of experiment is used to evaluate the Model-based Executive as a whole, which the DVMC controller is a part of.

10.6.1 Single Support Leg Balance Experiments

A series of tests was performed to test the DVMC controller's ability to restore balance after a disturbance. This series of tests was performed with the humanoid model in single support. Initial conditions were such that the ground projection of the CM was outside the support polygon, and all velocities were set to zero. For such initial conditions, the CM cannot be stabilized by stance ankle torques alone without the foot rolling and the model going unstable. Simple reference trajectories consisting of single, time invariant setpoints were selected for the

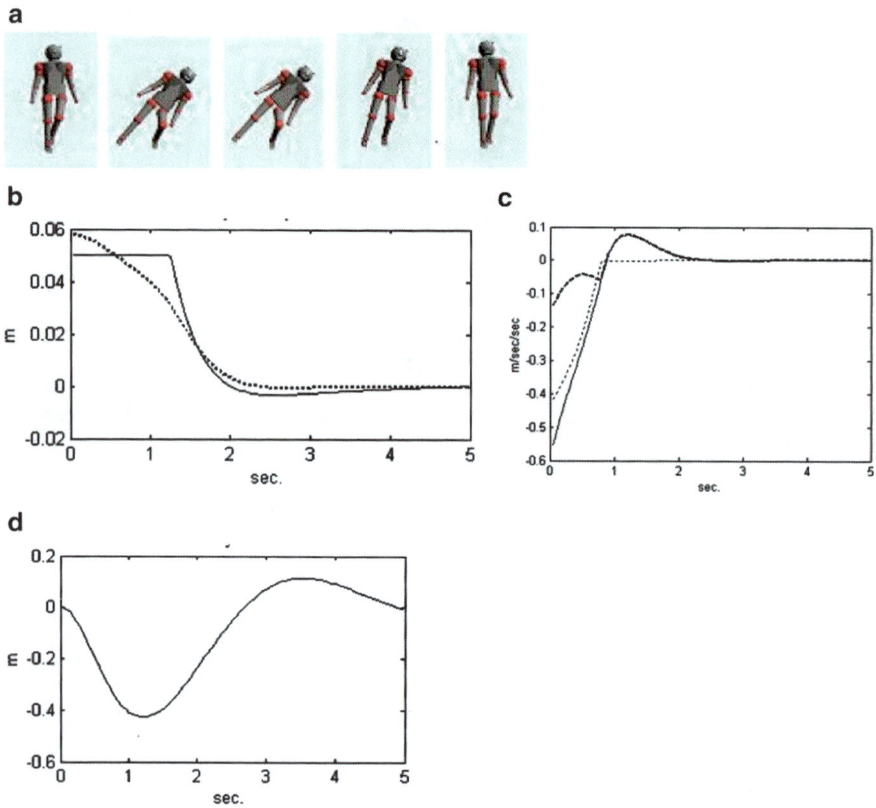

Fig. 10.17 Lateral disturbance recovery. In (**a**), several frames of the model are shown, starting from the maximally displaced CM posture (*left most image*) to the final static equilibrium posture (*right most image*). From the perspective of the model, the right leg is the swing leg and the left the stance leg. In (**b**), the lateral direction CM (*dotted line*) and the ZMP (*solid line*) are plotted versus time. In (**c**), the desired CM acceleration (*solid line*), the actual CM acceleration (*heavy dashed line*), and the slack value (*dotted line*) are plotted, showing the stabilization of the model's CM. Finally the body roll is plotted in (**d**), showing the corrective measures taken by the controller

controller. These setpoints specified the desired equilibrium positions and velocities of the model's COM and swing leg foot. Because the desired final equilibrium posture was to stand on one leg assuming a static pose, all setpoint velocities were set to zero.

Figure 10.15 shows the system's recovery from an initial displacement in the lateral (positive y) direction. From the model's perspective, the left most edge of the foot support polygon is at 0.05 m. As is shown in B, the ZMP remains within the foot support polygon, while the laterally displaced CM position begins outside the stance foot, but is brought quickly to zero by the controller. Part C shows the desired, actual, and slack values for the lateral CM acceleration in the DVMC optimal controller. Note how the slack goes to zero quickly, due to its high penalty. Part D shows the

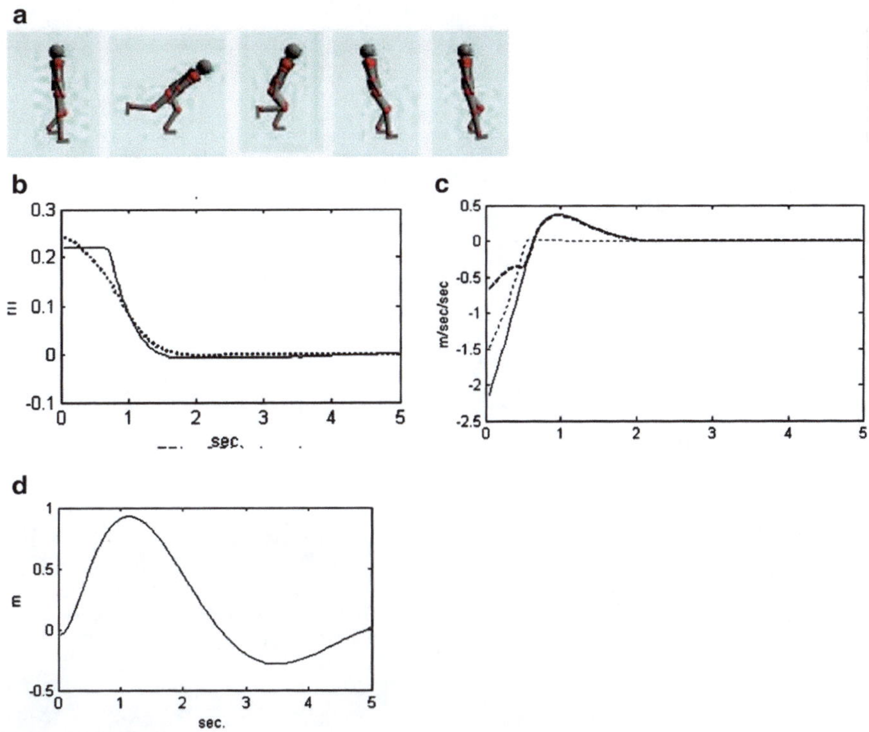

Fig. 10.18 Forward disturbance recovery. In (**a**), several frames of the model are shown, starting from the maximally displaced CM posture (*left most image*) to the final static equilibrium posture (*right most image*). From the perspective of the model, the right leg is the swing leg, left is stance leg. In (**b**), the forward direction CM (*dotted line*) and the ZMP (*solid line*) are plotted. In (**c**), the desired CM acceleration (*solid line*), the actual CM acceleration (*heavy dashed line*), and the slack value (*dotted line*) are plotted, showing the stabilization of the model's CM. Finally, in (**d**), body pitch is plotted

roll angle of the body. Because roll angle is less tightly controlled (penalty on slack variable is less than for CM position), the angle converges, but more slowly than the lateral CM position.

Figure 10.17 shows the system's recovery from a forward initial displacement. The front most edge of the foot support polygon is at 0.22 m. As is shown in B, the ZMP remains within the foot support polygon, while the forward CM position begins outside the foot, but is brought quickly to zero by the controller. Part C shows the desired, actual, and slack values for forward COM acceleration. Note how the slack goes to zero quickly, due to its high penalty. Part D shows the pitch angle of the body. Pitch converges, but more slowly than forward CM position because it is less tightly controlled (Fig. 10.18).

Figure 10.17 shows the system's recovery from a combined forward and lateral displacement,

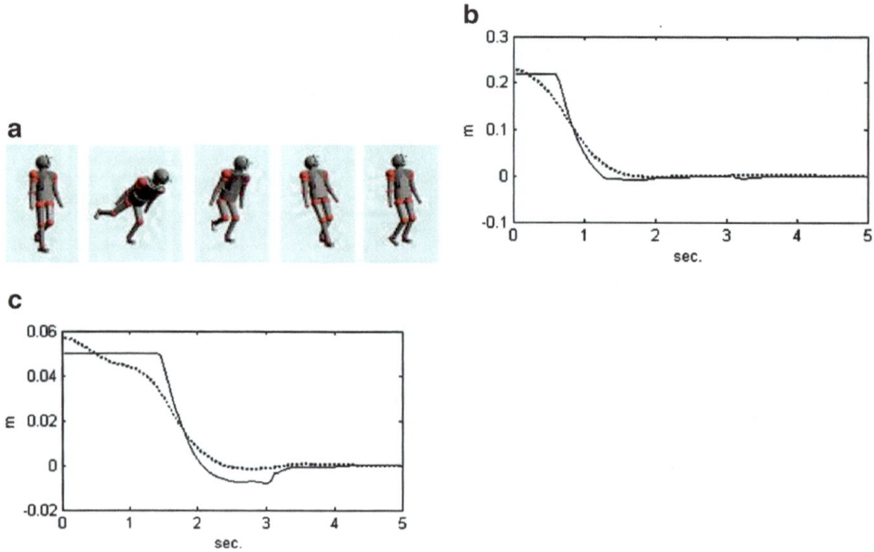

Fig. 10.19 Forward and lateral disturbance recovery. In (**a**), several frames of the model are shown, starting from the maximally displaced CM posture (*left most image*) to the equilibrium posture. In (**b**), the forward direction CM (*dotted line*) and the ZMP (*solid line*) are plotted. (**c**) shows lateral COM and ZMP

The results show that the controller makes appropriate use of non-contact limbs and stance leg ankle torques to stabilize the system. The non-contact limbs are used in two ways: to shift the ZMP, and to shift the CM. Consider, for example, the experiment shown in Fig. 10.17. From the model's perspective, the model stands on its left foot, leaning to the left (positive y direction). If the controller were to take no action, it would tip further to the left and fall down. Due to the action of the controller, the upper body leans further to the left, and the swing leg swings out to the right. Both of these actions correspond, initially, to a negative angular acceleration about the x axis.

The negative angular acceleration about the x axis allows a linear acceleration of the CM to the right (in the negative y direction) while not requiring the ZMP to shift further to the left (positive y direction). This is important since, as shown in Fig. 10.17, the ZMP begins up against the left-most edge of the foot support polygon. As the CM approaches the desired position, the ZMP moves away from the edge and towards the center of the foot support polygon. At this point, the swing leg and body are able to return to their nominal neutral positions.

The lateral acceleration of the swing leg to the right (negative y direction) is also beneficial in that it moves part of the model's mass to the right, and so, helps move the CM in the right direction. The net effect of the swing leg and body movements is an overall angular acceleration at the ankle joint that, together with the action of the stance ankle torque, moves the CM back to the center of the foot support polygon.

The extreme case of non-contact limb movement occurs when the support polygon becomes very small, as is the case for a tight-rope walker. A tight-rope walker's support polygon is very narrow, and therefore, little stance ankle torque can be exerted. Lateral forces by the foot against the tight-rope move the CM, but also create torques of the CM about the contact point. This must be countered by spin angular accelerations (angular accelerations about the CM), so that overall angular momentum is conserved. The spin angular accelerations are generated by movement of the non-contact limbs. Thus, a tight-rope walker extends his arms, and moves his arms, body, and non-contact leg to generate appropriate spin angular accelerations.

An important feature of the controller is that the coordinated behavior of the stance leg and non-contact limbs is not controlled explicitly, but rather, emerges indirectly from a high-level specification of desired behavior. This specification is given in terms of setpoints and PD gains for the CM, body orientation, and swing leg control outputs, in terms of constraints such as the one on the ZMP, and in terms of penalties for slacks and torques in the optimization cost function.

Another important feature of the controller is that, due to its extended range of operation, it can reject significant disturbances more easily than simpler controllers. This feature also means that reference trajectories for the new controller need not be as detailed as those for simpler controllers. The reference "trajectories" for the above tests were single, time invariant setpoints for CM, body orientation, and swing leg outputs. Simpler controllers require more detailed reference trajectories, with more waypoints as a function of time. This extra level of detail puts significant computational burden on the motion planning component of an integrated motion planning and control system. The motion planner has to be executed more frequently, when there are disturbances, and it must produce more detailed reference trajectories.

10.6.2 Biped Walking Tasks

We now present test results of execution of a variety of QCPs for bipedal walking with foot placement and temporal constraints, and with disturbances. Test results for nominal walking at different speeds are provided in [9].

To perform these tests, we used a high-fidelity, 20 degree-of-freedom humanoid simulation to represent the plant being controlled [9, 11]. This simulation accurately models gravity, ground reaction forces and joint torques, and the resulting link acceleration dynamics. In particular, just as with a real biped robot (or a human), this simulated humanoid will fall if inappropriate control commands are provided.

The hybrid plant model (Definition 1) has four modes: left foot single support, double support right foot in front, right single support, double support left foot in front. The mode transition guard conditions are based on toe-off and heel-strike events. The linearized plant abstraction, for each mode, is provided by a feedback linearizing controller [9, 11]. This produces linearized decoupled models for the forward, lateral, and vertical center of mass (CM) components, and stepping foot components. Thus, performing walking tasks involves synchronization of the 3

CM and the 3 stepping foot components when in single support, and the 3 CM components when in double support. For example, when the mode is left foot single support, 6 parallel activities are in the Runtime Activity State, all with right heel strike as the finish event. Each activity must be in its goal region when right heel strike occurs in order for execution to proceed successfully. The event horizon in this case is simple since the parallel activities all share the same finish event.

10.6.2.1 Irregular Foot Placement

Figure 10.20 shows dynamic walking, but with an irregular stepping pattern, which is necessary due to the blocks the biped is walking on. These blocks move slowly, so the timing of foot placement, as well as the positioning is important. At this speed, the biped can't just balance statically on each block. Instead, the moderately fast speed requires dynamic balancing and coordination of the center of mass trajectory. Figure 10.21 shows the CM trajectory and foot placements for this test. The dynamic nature is indicated by the fact that the CM trajectory barely touches the foot placement polygons, and in one case, is 0.1 m away. This indicates that the system is not statically stable in this pose, and is relying on the subsequent foot placement sequence to maintain balance.

10.6.2.2 Lateral Push Disturbances

A biped is especially sensitive to lateral push disturbances when in single support, due to the limited support base provided by one foot. In particular, a biped is most sensitive to lateral push disturbances when there are foot placement constraints, and when the push disturbance results in acceleration towards the outer edge of the stance foot.

Figure 10.22 shows recovery from a lateral push disturbance, while walking on a balance beam. The push occurs from the right side of the biped during left single support. Thus, the push results in an acceleration of the CM to the biped's left. Because foot placement is constrained by the narrowness of the balance beam, compensation by stepping is not an option. Instead, the system compensates for the disturbance by exerting a restoring torque at the ankle. This torque has a significant actuation limit; the lateral center of pressure must not get too close to the outer edge of the foot, or the foot will roll. Because the foot is relatively narrow, this presents a severe actuation limit. Additional (but also limited) compensation is accomplished through the angular movement of the torso and right leg, as shown in the third frame of the sequence [9, 11]. In particular, as shown in Fig. 10.22, the torso rotates clockwise, from the viewer's perspective, which induces a counterclockwise rotation of the stance leg, which, in turn, engenders an acceleration of the biped's CM toward the biped's right.

Due to joint acceleration limits, there is a limit to the angular acceleration that can be produced by the torso and the right leg. Therefore, recovery of lateral balance

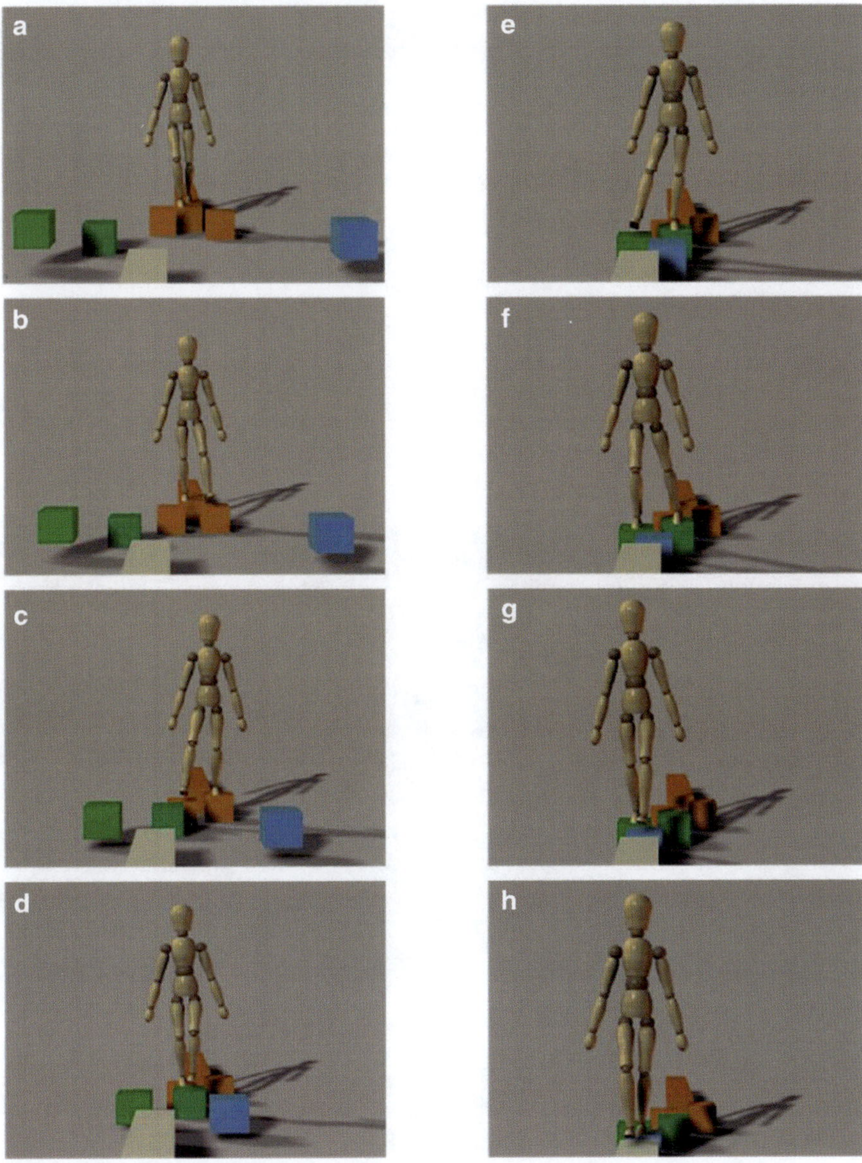

Fig. 10.20 Walking by stepping on slowly moving blocks: (**a**) biped starts on long, narrow path; (**b**) steps with left foot onto the *brown* block; (**c**) steps with right foot onto the other *brown* block; (**d**) steps with left foot onto the *green* block; (**e**, **f**) steps with right foot onto the other *green* block; (**g**) steps with left foot onto *blue* block; (**h**) finished

takes some time; the right leg is out for a significantly longer time (about 2 s) than it would be if it were just taking a normal step. This means that the forward center of

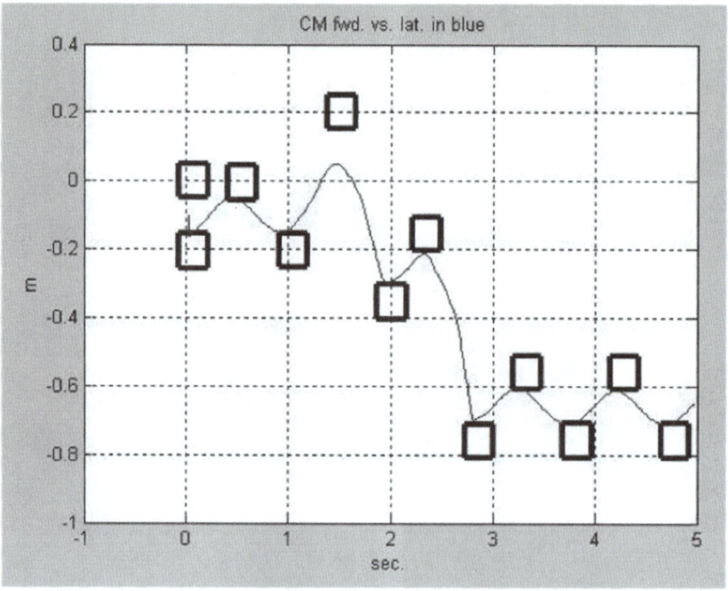

Fig. 10.21 CM trajectory and foot placements for irregular stepping task

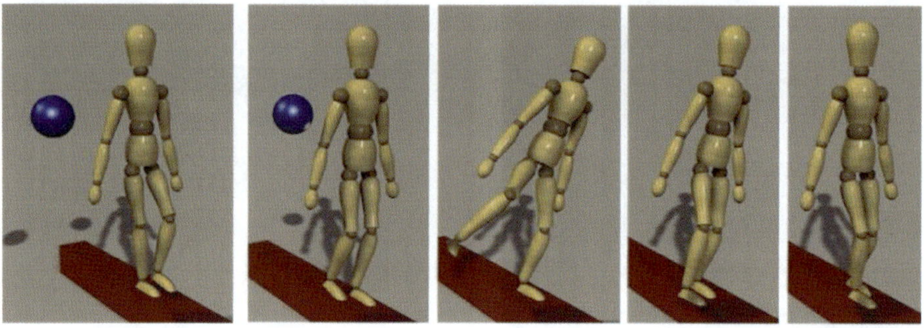

Fig. 10.22 Recovery from lateral push while walking on a balance beam

mass velocity must be reduced while the lateral compensation movement is taking place. This forward velocity reduction must be accomplished by the left (stance) foot alone. Due to support base limitations, there is a limit on the force that can be applied in this way, and therefore a limit to the negative forward acceleration that can be produced. Thus, the biped must be walking relatively slowly, in the first place, for this sort of maneuver to work at all. If this is the case, then forward movement of the CM can be slowed while the right leg is out, and then sped up again after the lateral compensation maneuver is completed. Thus, the forward CM position and the forward stepping position remain synchronized. This is one reason why people tend to walk slowly on tightropes or balance beams.

10.6.2.3 Kicking a Soccer Ball

The problem of moving a biped to kick a soccer ball (Fig. 10.2) requires synchronization of the forward and lateral components of the CM with the step movement, and with the movement of the kicking foot. Figure 10.23a, b show flow tubes and nominal trajectories for the forward and lateral CM components. The flow tubes correspond to CM movement for taking three steps before kicking a soccer ball. Figure 10.23c, d show flow tube cross sections (in the position-velocity plane) for the first activity. The intersection of the initial state with the cross sections determines the duration (temporal) controllability. If the initial state intersects all cross sections, as shown in the figure, then the executive is free to choose any duration in the range covered by the cross sections. This is important for adjusting timing of task completion when kicking a moving soccer ball. Disturbances from the nominal trajectory may restrict the controllable duration range. For example, suppose the system is subjected to a lateral push disturbance at the beginning of the task. This may cause the initial state to deviate from the nominal initial state. If the initial state intersects all cross sections for forward movement, but only a subset for lateral movement, then the overall temporal controllability is limited by the latter set of cross sections.

10.7 Discussion

10.7.1 Scientific Contributions

The techniques described in this chapter extend previously developed temporally flexible execution systems for discrete activity plans to work with hybrid discrete/continuous systems such as bipedal walking mechanisms. This is achieved by first developing a representation for temporally and spatially flexible tasks for hybrid systems, called a QSP, then developing a plan compiler that transforms the QSP to a QCP, which is easier to execute, and then developing a plan dispatcher that executes the QCP. The QCP produced by the plan compiler represents the set of feasible trajectories in an easily executable form. The dispatcher is based on ones developed for discrete activity systems [14], but extends these to allow for indirect scheduling of events through control actions applied to a dynamic system, thus allowing the approach to be used for hybrid systems.

An important attractive property of this approach is that it clearly represents the boundaries between feasible and infeasible states and times with respect to successful plan execution. This allows the system to detect that a plan will fail, sooner rather than later.

It is interesting that traditional control theory does not explicitly address the issue of temporal flexibility. Traditional control theory deals with two basic kinds of problems: stabilization, and trajectory following [18]. Stabilization is an infinite-time

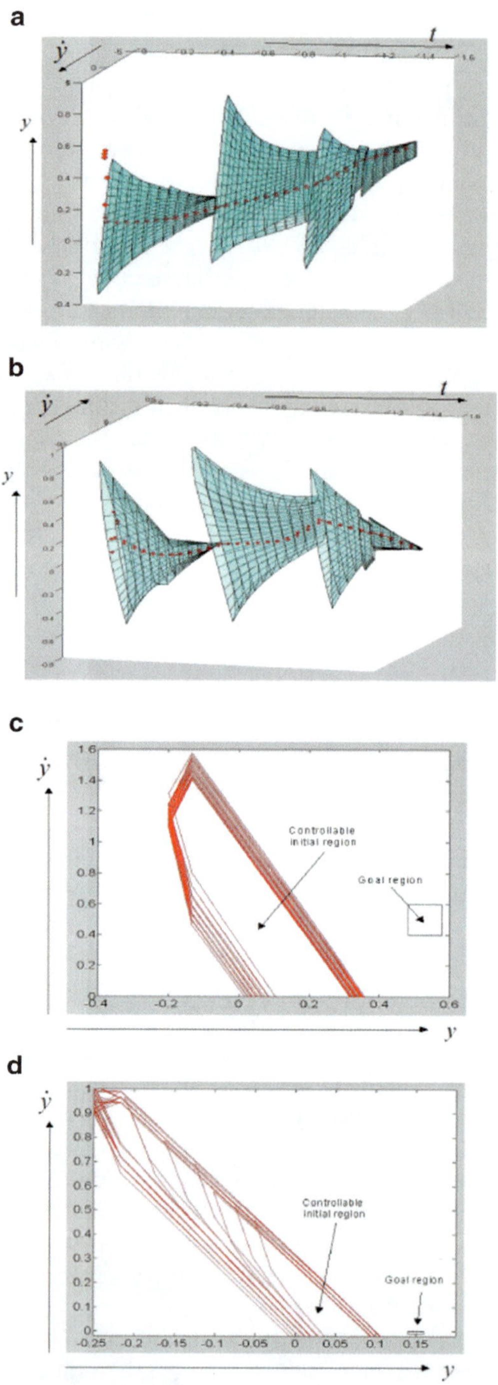

Fig. 10.23 Flow tubes for CM corresponding to taking three steps before kicking a soccer ball: (**a**) flow tubes and nominal trajectory for forward CM component; (**b**) flow tubes and nominal trajectory for lateral CM component; (**c**) flow tube cross sections corresponding to different durations, superimposed, for forward CM component, first step; (**d**) flow tube cross sections for lateral CM component, first step

concept; a system is stable if it converges to an equilibrium point at some time in the future, possibly, infinity. Thus, stabilization has infinite temporal flexibility. Reference trajectory following, on the other hand, has no temporal flexibility. If the control system tracks the reference trajectory exactly, it will reach a goal state at a specific time. We believe that this is a significant omission. Temporal flexibility exists in most task specifications, and should be taken advantage of to achieve robust and efficient plan execution.

To summarize, the system described here takes full advantage of plan specification flexibility, both temporal and spatial, in order to maximize robustness to disturbances. Key contributions are: (1) a plan specification that represents task flexibility; (2) a DVMC that decomposes a complex nonlinear system into a set of loosely coupled linear systems; (3) a plan compiler that transforms the plan, using the abstraction provided by the DVMC, into a form that can be easily executed (QCP); and (4) a model-based executive that uses the flexibility in the QCP to reject disturbances, and to detect when a disturbance is so severe that the plan will fail.

10.7.2 Applications and Impact

The approach described here is intended for inherently under-actuated systems, such as bipeds or aerial vehicles, where there are more degrees of freedom to be controlled than actuators to control them. However, even systems that are fully actuated, such as most robot manipulators, can have actuation limits that become relevant for demanding tasks. For example, moving a robot manipulator at high speeds so that it can perform tasks quickly exposes the velocity and acceleration limits of the joints. Such limits are potentially in conflict with temporal constraints imposed on tasks by the user. Thus, the techniques described here are potentially applicable to any physical system that has velocity and/or acceleration actuation limits, and temporal constraints associated with tasks it is to perform.

References

1. Arakawa, T., Fukuda, T.: Natural motion generation of biped locomotion robot using hierarchical trajectory generation method consisting of ga, ep layers. In: Proceedings of 1997 IEEE International Conference on Robotics and Automation, vol. 1, pp. 211–216. IEEE, New York (1997)
2. Bemporad, A., Morari, M., Dua, V., Pistikopoulos, E.: The explicit linear quadratic regulator for constrained systems. Automatica **38**(1), 3–20 (2002)
3. Craig, J.J.: Introduction to Robotics: Mechanics and Control. Pearson/Prentice Hall, Upper Saddle River (2005)
4. Dechter, R., Meiri, I., Pearl, J.: Temporal constraint networks. Artif. Intell. **49**(1), 61–95 (1991)
5. Formal'skii, A.: An inverted pendulum on a fixed and a moving base. J. Appl. Math. Mech. **70**(1), 56–64 (2006)

6. Goswami, A.: Postural stability of biped robots and the foot-rotation indicator (fri) point. Int. J. Robot. Res. **18**(6), 523–533 (1999)
7. Hirai, K.: Current and future perspective of honda humamoid robot. In: Proceedings of the 1997 IEEE/RSJ International Conference on Intelligent Robots and Systems (IROS'97), vol. 2, pp. 500–508. IEEE, New York (1997)
8. Hofbaur, M.W., Williams, B.C.: Hybrid estimation of complex systems. IEEE Trans. Syst. Man Cybern. Part B Cybern. **34**(5), 2178–2191 (2004)
9. Hofmann, A.: Robust execution of bipedal walking tasks from biomechanical principles. Ph.D. thesis, Massachusetts Institute of Technology (2006)
10. Hofmann, A., Williams, B.: Exploiting Spatial and temporal flexibility for plan execution of hybrid, under-actuated systems. In: AAAI 2006 (2006)
11. Hofmann, A., Massaquoi, S., Popovic, M., Herr, H.: A sliding controller for bipedal balancing using integrated movement of contact and non-contact limbs. In: Proceedings of the 2004 IEEE/RSJ International Conference on Intelligent Robots and Systems (IROS 2004), vol. 2, pp. 1952–1959. IEEE, New York (2004)
12. Kajita, S., Matsumoto, O., Saigo, M.: Real-time 3d walking pattern generation for a biped robot with telescopic legs. In: Proceedings 2001 ICRA IEEE International Conference on Robotics and Automation, vol.3, pp. 2299–2306. IEEE, New York (2001)
13. Léauté, T., Williams, B.C.: Coordinating agile systems through the model-based execution of temporal plans. In: Proceedings of the National Conference on Artificial Intelligence, vol. 20, p. 114. AAAI Press/MIT Press, Menlo Park/Cambridge (2005)
14. Muscettola, N., Morris, P., Tsamardinos, I.: Reformulating temporal plans for efficient execution. In: In Principles of Knowledge Representation and Reasoning, Citeseer (1998)
15. Nishiwaki, K., Kagami, S., Kuniyoshi, Y., Inaba, M., Inoue, H.: Online generation of humanoid walking motion based on a fast generation method of motion pattern that follows desired ZMP. In: IEEE/RSJ International Conference on Intelligent Robots and Systems, vol. 3, pp. 2684–2689. IEEE, New York (2002)
16. Popovic, M.B., Goswami, A., Herr, H.: Ground reference points in legged locomotion: definitions, biological trajectories and control implications. Int. J. Robot. Res. **24**(12), 1013–1032 (2005)
17. Pratt, J., Dilworth, P., Pratt, G.: Virtual model control of a bipedal walking robot. In: Proceedings of the 1997 IEEE International Conference on Robotics and Automation, vol. 1, pp. 193–198. IEEE, New York (1997)
18. Slotine, J.J.E., Li, W., et al.: Applied Nonlinear Control, vol. 199. Prentice Hall, New Jersey (1991)
19. Sugihara, T., Nakamura, Y., Inoue, H.: Real-time humanoid motion generation through ZMP manipulation based on inverted pendulum control. In: Proceedings of ICRA'02 IEEE International Conference on Robotics and Automation, vol. 2, pp. 1404–1409. IEEE, New York (2002)
20. Vukobratović, M., Borovac, B.: Zero-moment point—thirty five years of its life. Int. J. Humanoid Robot. **1**(01), 157–173 (2004)
21. Vukobratovic, M., Juricic, D.: Contribution to the synthesis of biped gait. IEEE Trans. Biomed. Eng. **16**(1), 1–6 (1969)
22. Vukobratović, M., Stepanenko, J.: On the stability of anthropomorphic systems. Math. Biosci. **15**(1), 1–37 (1972)
23. Williams, B.C.: The use of continuity in a qualitative physics. In: AAAI, pp. 350–354 (1984)
24. Williams, B.C., Nayak, P.P.: A reactive planner for a model-based executive. In: IJCAI, Citeseer, vol. 97, pp. 1178–1185 (1997)
25. Yokoi, K., Kanehiro, F., Kaneko, K., Fujiwara, K., Kajita, S., Hirukawa, H.: A honda humanoid robot controlled by aist software. In: Proceedings of the IEEE-RAS International Conference on Humanoid Robots, pp. 259–264 (2001)

Chapter 11
Design of Reactionless Mechanisms Based on Constrained Optimization Procedure

Himanshu Chaudhary and Kailash Chaudhary

Abstract This chapter presents an optimization technique to dynamically balance planar mechanisms by minimizing the shaking forces and shaking moments due to inertia-induced forces. Dynamically equivalent systems of point masses which represent rigid links and counterweights are useful for developing optimization technique. The point-mass parameters are explicitly identified as the design variables. The balancing problem is formulated as both single-objective and multi-objective optimization problem and solved using genetic algorithm which produces better results as compared to the conventional optimization algorithm. Also, for the multi-objective optimization problem, multiple optimal solutions are created as a *Pareto front* using the genetic algorithm. The reduction of shaking force and shaking moment is obtained by optimizing the link mass distribution and counterweight of their point masses. The inertial properties of balanced mechanism are then computed in reverse by applying dynamical equivalent conditions from the optimized design variables. The effectiveness of the methodology is shown by applying it to problems of planar four-bar, slider-crank, and Stephenson six-bar mechanisms.

Keywords Dynamic balancing • Equimomental system • Genetic algorithm • Optimization • Shaking force and shaking moment

The design of reactionless mechanisms is important in order to (1) reduce the amplitude of vibration of the frame on which the mechanism is mounted due to transmission of shaking forces and (2) smoothen highly fluctuating driving torque/force needed to obtain nearly constant drive speed. Since any vibration leads to noise, wear, fatigue, etc., in the mechanism, its reduction improves several aspects of mechanical design as well. Design a reactionless mechanism means the balancing of shaking force, shaking moment, and input-torque fluctuations together. The shaking force can be eliminated completely by attaching counterweights and/or redistributing masses of the moving links. This will increase overall mass

---

H. Chaudhary (✉) • K. Chaudhary
Department of Mechanical Engineering, Malaviya National Institute of Technology Jaipur, JLN Marg, Jaipur 302017, India
e-mail: hchaudhary.mech@mnit.ac.in

D. Zhang, B. Wei (eds.), *Dynamic Balancing of Mechanisms and Synthesizing of Parallel Robots*, DOI 10.1007/978-3-319-17683-3_11

and moment of inertia of the mechanism. As a result, shaking moment, driving torque, and reactions in the joints will increase significantly. Therefore, to design a mechanism with minimum reaction forces transmitted to the frame, it is required to reduce all the competing dynamic quantities, namely the shaking force, shaking moment, driving torque/force, and bearing reactions simultaneously. This means that design of reactionless mechanism problem can be treated as an optimization problem, whose formulation needs the following:

1. An efficient dynamic algorithm to compute the dynamic quantities
2. Identification of the design variables, and the formulation of the constraints on them that define the design space of the feasible solutions
3. An objective function to evaluate the performance of a mechanism at hand

This chapter presents a constrained optimization procedure to balance the planar mechanisms dynamically. This will minimize the shaking force and shaking moment by optimally distributing the link masses. The concept of equimomental system of point masses is used to identify the design variables and to define the constraints for the optimization problem formulation.

11.1 Equimomental Systems for Planar Motion

A study on an equimomental system of rigidly connected point masses undergoing planar motion is discussed in this section. Inertia-induced dynamic quantities, e.g., shaking force, shaking moment, and input-torque, of a mechanical system depend on the mass and inertia of its each link, and the corresponding mass center location. These inertia properties can be represented more conveniently using the dynamically equivalent system of point masses referred to as *equimomental system* [1–3].

A point mass is an idealized concept, and defined as a mass that is concentrated at a point. Two rigid systems are equimomental if their dynamic behaviors are identical; that is, they have the same mass, the same center of mass, and the same inertia tensor with respect to a common point [1]. Referring to the ith rigid link, Fig. 11.1a, of a planar mechanism, the location of its mass center, C_i, is defined by the vector, $\mathbf{d}_i$, at an angle, θ_i, from the axis $\mathrm{O}_i\mathrm{X}_i$ of the local frame, $\mathrm{O}_i\mathrm{X}_i\mathrm{Y}_i$, fixed to the link. The axis $\mathrm{O}_i\mathrm{X}_i$ is set along the vector from O_i to O_{i+1}, that is, at an angle α_i from the axis, OX, of the fixed inertial frame, OXY, Fig. 11.1a. The points, O_i and O_{i+1}, on the link are chosen as the points where the ith link is coupled to its neighboring links, whereas link's mass and the mass moment of inertia about O_i are m_i and I_i, respectively. A system of p point masses, which is equimomental to the ith link, is shown in Fig. 11.1b. The point masses, m_{ij}, for $j = 1, \ldots, p$, are fixed in the local frame, $\mathrm{O}_i\mathrm{X}_i\mathrm{Y}_i$, and their distances from the origin, O_i, are l_{ij}. The angles, θ_{ij}, are defined between the line joining the point masses from O_i, and the axis, $\mathrm{O}_i\mathrm{X}_i$. In this section, all the vectors are represented in the fixed frame, OXY, unless stated otherwise.

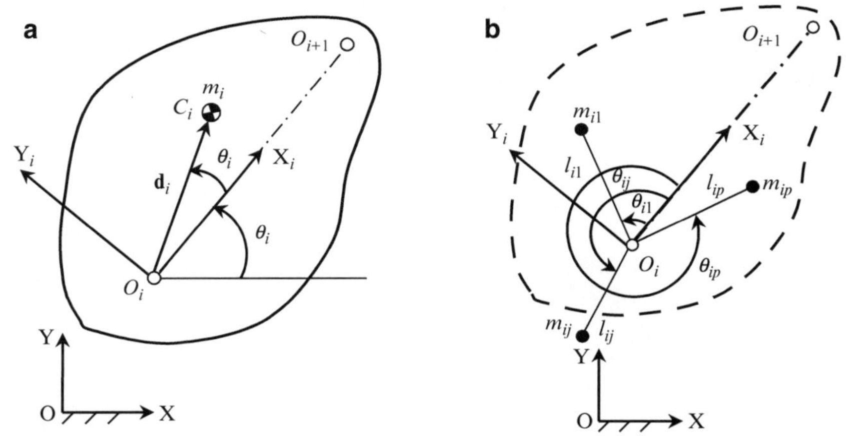

Fig. 11.1 Parameters for rigid link and its equimomental system. (**a**) The *i*th rigid link, (**b**) equimomental system of the *i*th link

If the p point masses are equimomental to the ith link, then they must satisfy the conditions of dynamic equivalence with reference to the fixed frame, OXY, given as

$$\sum_{j=1}^{p} m_{ij} = m_i \tag{11.1}$$

$$\sum_{j=1}^{p} m_{ij} l_{ij} \cos\left(\theta_{ij} + \alpha_i\right) = m_i d_i \cos\left(\theta_i + \alpha_i\right) \tag{11.2}$$

$$\sum_{j=1}^{p} m_{ij} l_{ij} \sin\left(\theta_{ij} + \alpha_i\right) = m_i d_i \sin\left(\theta_i + \alpha_i\right) \tag{11.3}$$

$$\sum_{j=1}^{p} m_{ij} l_{ij}^2 = I_i \tag{11.4}$$

The first subscript i denotes the link number, and the second one, i.e., $j = 1, \ldots, p$, represents the point masses corresponding to the ith link. Since each mass requires three parameters, (m_{ij}, l_{ij}, θ_{ij}), to identify it, a total of $3p$ parameters are necessary to completely define the equimomental system of p point masses. However, there are four constraints, namely Eqs. (11.1–11.4), that need to be satisfied. Hence, an infinite number of solutions exist for $p \geq 2$, as the resultant system of equations is underdeterminate; that is, the number of unknowns is more than the equations [2]. If $p = 1$, there is only one point mass with three unknown parameters, which cannot satisfy all the four conditions, Eqs. (11.1–11.4), unless they are consistent. This is because the resulting system of equations is overdeterminate with more equations than the number of unknowns. Typically, such system of equations does not yield any solution unless the equations are consistent. As a consequence, an equimomental system of a rigid link moving in a plane cannot be represented using one point mass, which is obvious from the fundamental knowledge of mechanics. Clearly, the minimum number of point masses is then two giving six unknown parameters, of which two need to be assigned arbitrarily. If three point masses are taken, five parameters are to be assigned arbitrarily. In general, $(3p-4)$ parameters need to be assigned arbitrarily so that the remaining four are determinate. Note here that it is not always possible to get all the point masses positive. This, however, does not hinder the process of representing the rigid link as long as the total mass and the moment of inertia about the mass center give positive values [4].

11.1.1 Two-Point-Mass Model

As explained in the previous section, an equimomental system of point masses of a rigid link moving in a plane requires at least two point masses. The representation of the link by the equimomental system of two point masses is referred to as two-point-mass model. Similarly, equimomental system of three point masses is called three-point-mass model, and so on. In this section, the conversion of a rigid link into the two-point-mass model is illustrated. Let a two-point-mass model for ith rigid link is moving in the XY plane. The polar coordinates of the point masses are (l_{ij}, θ_{ij}), for $j=1, 2$. Note that the point masses are rigidly fixed in the local frame. The system of two point masses is then equimomental to the rigid link if it satisfies the conditions given by Eqs. (11.1–11.4).

For the rigid link of given mass, its center location, and inertia, one can convert the link into an appropriate two point masses. Since the four equations are nonlinear in six unknown parameters of the two point masses, a judicious selection is required to choose for the arbitrary assigned parameters. Assuming $\theta_{i1}=0$ and $\theta_{i1}=\pi/2$ [5], the four parameters of the two point masses, namely m_{i1}, m_{i2}, l_{i1}, and l_{i2}, are determined from Eqs. (11.1–11.4) as

$$m_{i1} = \frac{\left(m_i^2\bar{x}_i^2 - m_i^2\bar{y}_i^2 + m_i I_i\right) \pm \sqrt{\left(m_i^2\bar{x}_i^2 - m_i^2\bar{y}_i^2 + m_i I_i\right)^2 - 4m_i I_i \bar{x}_i^2}}{2I_i} \tag{11.5}$$

$$m_{i2} = m_i - m_{i1} \tag{11.6}$$

$$l_{i1} = \frac{m_i\bar{x}_i}{m_{i1}} \tag{11.7}$$

$$l_{i2} = \frac{m_i\bar{y}_i}{m_{i2}} \tag{11.8}$$

where the Cartesian coordinates of the mass center of ith rigid link are $\overline{x}_i = d_i \cos\theta_i$ and $\overline{y}_i = d_i \sin\theta_i$. Hence, each point mass has two solutions. If $\theta_{i1} = 0$, i.e., the mass center of the rigid link lies on the X-axis, two sets of the solutions are as follows:

$$m_{i1} = \begin{cases} \frac{m_i^2 d_i^2}{I_i} \\ m_i \end{cases}; m_{i2} = \begin{cases} m_i - \frac{m_i^2 d_i^2}{I_i} \\ 0 \end{cases}; l_{i1} = \begin{cases} \frac{m_i d_i}{m_{i1}} \\ d_i \end{cases}; \text{ and } l_{i2} = \begin{cases} 0 \\ 0 \end{cases} \quad (11.9)$$

Thus, one can convert a given rigid link into a suitable two-point-mass model assuming any two-point-mass parameters.

11.1.2 Three-Point-Mass Model

In this section, the procedure of finding a three-point-mass model is illustrated. Consider a three-point-mass model for ith rigid link moving in the XY plane. The polar coordinates of the point masses are (l_{ij}, θ_{ij}), for $j = 1, 2$, and 3. Similar to two-point-mass model, the three-point-mass model would then be equimomental to the original rigid link if Eqs. (11.1–11.4) are satisfied.

Note that there are nine unknown parameters of point masses, namely m_{ij}, l_{ij}, and θ_{ij}, for $j = 1, 2$, and 3, in the four equimomental equations. Hence, it is important to decide which five parameters should be chosen so that the remaining four become determinate. It is advisable to choose l_{ij} and θ_{ij}, so that the dynamic equivalence conditions become linear in point masses. Assuming $l_{i2} = l_{i3} = l_{i1}$ and substituting them in Eq. (11.4) yield

$$\left(\sum_{j=1}^{3} m_{ij}\right) l_{i1}^2 = m_i k_i^2 \quad (11.10)$$

where $m_i k_i^2 = I_i^c + m_i d_i^2$, k_i being the radius of gyration about the point, O_i. Equation (11.10) gives $l_{i1} = \pm k_i$. Taking the positive value for l_{i1}, which is physically possible, Eqs. (11.1–11.3) are then written in a compact form as

$$\mathbf{Km} = \mathbf{b} \quad (11.11)$$

where the 3×3 matrix, $\mathbf{K}$, and the three vectors, $\mathbf{m}$ and $\mathbf{b}$, are as follows:

$$\mathbf{K} = \begin{bmatrix} 1 & 1 & 1 \\ k_i \cos\theta_{i1} & k_i \cos\theta_{i2} & k_i \cos\theta_{i3} \\ k_i \sin\theta_{i1} & k_i \sin\theta_{i2} & k_i \sin\theta_{i3} \end{bmatrix}; \mathbf{m} = \begin{bmatrix} m_{i1} \\ m_{i2} \\ m_{i3} \end{bmatrix}; \mathbf{b} = \begin{bmatrix} m_i \\ m_i d_i \cos\theta_i \\ m_i d_i \sin\theta_i \end{bmatrix} \quad (11.12)$$

The magnitudes of three point masses are then solved from Eq. (11.11) by assuming suitable values for θ_{ij}, $j = 1, 2$, and 3. It is clear that the solution for $\mathbf{m}$

exists if det $(\mathbf{K}) \neq 0$, i.e., $\theta_{i1} \neq \theta_{i2}$, $\theta_{i1} \neq \theta_{i3}$, and $\theta_{i2} \neq \theta_{i3}$. It means that any two point masses should not lie on the same radial line emanating from the origin, O_i. The vector **m** is obtained as

$$\mathbf{m} = \mathbf{K}^{-1}\mathbf{b} \tag{11.13}$$

where $\mathbf{K}^{-1}$ is evaluated as

$$\mathbf{K}^{-1} = \frac{k_i}{\det(\mathbf{K})}\begin{bmatrix} k_i \sin(\theta_{i3} - \theta_{i2}) & (\sin\theta_{i2} - \sin\theta_{i3}) & (\cos\theta_{i3} - \cos\theta_{i2}) \\ -k_i \sin(\theta_{i3} - \theta_{i1}) & (\sin\theta_{i3} - \sin\theta_{i1}) & (\cos\theta_{i1} - \cos\theta_{i3}) \\ -k_i \sin(\theta_{i1} - \theta_{i2}) & (\sin\theta_{i1} - \sin\theta_{i2}) & (\cos\theta_{i2} - \cos\theta_{i1}) \end{bmatrix}$$

in which $\det(\mathbf{K}) = k_i^2 [\sin(\theta_{i3} - \theta_{i2}) + \sin(\theta_{i2} - \theta_{i1}) + \sin(\theta_{i1} - \theta_{i3})]$. It is evident from the solution, Eq. (11.13), that the sum of the point masses is equal to mass of the link for any values of angles except $\theta_{i1} \neq \theta_{i2}$, $\theta_{i1} \neq \theta_{i3}$, and $\theta_{i2} \neq \theta_{i3}$. Note here that there is a possibility that some point masses are negative. It does not hinder the process of representing the rigid link as long as its mass, m, and inertia, I^c, are positive and real, as pointed out earlier. As an example, if $\theta_{i1} = 0$, $\theta_{i2} = 2\pi/3$, and $\theta_{i3} = 4\pi/3$, the point masses are calculated as

$$m_{i1} = \frac{m_i}{3}\left(1 + \frac{2d_i \cos\theta_i}{k_i}\right) \tag{11.14}$$

$$m_{i2} = \frac{m_i}{3}\left(1 - \frac{d_i \cos\theta_i}{k_i} + \frac{\sqrt{3}d_i \sin\theta_i}{k_i}\right) \tag{11.15}$$

$$m_{i3} = \frac{m_i}{3}\left(1 - \frac{d_i \cos\theta_i}{k_i} - \frac{\sqrt{3}d_i \sin\theta_i}{k_i}\right) \tag{11.16}$$

which take simpler form if the origin, O_i, coincides with the mass center of the link, C_i; that is, $d_i = 0$. Substituting $d_i = 0$ in Eqs. (11.14–11.16), one obtains $m_{i1} = m_{i2} = m_{i3} = m_i/3$. It means that the point masses of the link are distributed equally, and located on the circumference of a circle having radius k_i.

It is pointed out here that in mechanism analysis, links are often considered as one dimensional, e.g., a straight rod, in which its diameter or width and thickness are very small in comparison to the length. Considering that the mass lying along the X-axis of the local frame, the dynamical equivalence conditions, Eqs. (11.1–11.4), reduce to

$$\sum_{j=1}^{p} m_{ij} = m_i \tag{11.17}$$

$$\sum_{j=1}^{p} m_{ij}x_{ij} = m_i\bar{x}_i \tag{11.18}$$

$$\sum_{j=1}^{p} m_{ij}x_{ij}^2 = I_i^c + m_i\bar{x}_i^2 \tag{11.19}$$

It is evident from Eqs. (11.17–11.19) that a minimum of two point masses is also required to represent a one-dimensional link, introducing a total of four variables, i.e., m_{i1}, m_{i2}, x_{i1}, and x_{i2}. Specifying any one of the variables, the other three variables can be found uniquely. It is pointed out here that a common practice in the dynamics study of reciprocating engine is to replace the connecting rod by two point masses, where the masses are placed at the ends of the connecting rod. This does not provide a true equivalent system unless the three equations, Eqs. (11.17–11.19), in two unknowns, m_{i1} and m_{i2}, leading to an overdetermined system are consistent.

Using the concept of equimomental system, Sherwood and Hokey [4] presented the optimization of mass distribution in mechanisms. Hockey [6] discussed the input-torque fluctuations of mechanisms subject to external loads by means of properly distributing the link masses. Using the two-point-mass model, momentum balancing of four-bar linkages was presented in [7]. Optimum balancing of combined shaking force, shaking moment, and torque fluctuations in high-speed linkages was reported in Lee and Cheng [5] where a two-point-mass model was used. The concept can also be applied for the kinematic and dynamic analyses of mechanisms [8]. Simultaneous minimization of shaking force, shaking moment, and other quantities using the dynamical equivalent system of point masses and optimum mass distribution has been attempted in [9, 10]. However, the results do not show significant improvement in the performances.

11.2 Balancing of Planar Mechanisms

Balancing of shaking force and shaking moment in the mechanisms is important in order to obtain reactionless mechanisms. Several methods are developed to eliminate the shaking force and shaking moment in planar mechanisms. The methods to completely eliminate the shaking force are generally based on two principles: (1) making the total potential energy of a mechanism constant [11], and (2) making the total mass center of a mechanism stationary [12, 13]. Studies based on potential energy use elastic elements like springs to balance the force. On the other hand, the methods based on making total mass center stationary use mass redistribution/counterweights. Different techniques are used for tracing and making the total mass center stationary. For example, the method of *principal vectors* [14] describes the position of the mass center by a series of vectors that are directed along the links. These vectors trace the mass center of the mechanism at hand, and the conditions are derived to make the system mass center stationary. A more referred method in the literature is the method of *linearly independent vectors* [12] where the stationary condition was achieved by redistributing the link masses in such a manner that the coefficients of the time-dependent terms of the equations describing the total center of the mass trajectory vanish. Kochev [15] presented a general method using ordinary vector algebra instead of the complex number representation of the vectors [12] for full force balance of the planar linkages. One of the attractive features of a force-balanced linkage is that the shaking force vanishes, and the shaking moment reduces to a pure torque which is independent of reference point. However, only shaking force balancing is not effective in the balancing of mechanisms, as (1) it mostly increases the total mass of the mechanism, (2) it needs some arrangement like counterweights that increase the total mass, and (3) it increases the other dynamic characteristics, like shaking moment, driving torque, and bearing reactions. The influence of the complete shaking force balancing is thoughtfully investigated by Lowen et al. [16] on the bearing reactions, input-torque, and shaking moment for a family of crank-rocker four-bar linkages. This study shows that these dynamic quantities increase, and in some cases their values rise up to five times.

Several authors attempted to treat the balancing problem as a complete shaking force and shaking moment balancing. Elliot et al. [17] developed a theory to balance torque, shaking force, and shaking moment by extending the method of linearly independent vectors. Similarly, the analytical conditions are presented for complete balancing of shaking force and shaking moment in [18]. Complete moment balancing is also achieved by a cam-actuated oscillating counterweight [19], inertia counterweight [20], physical pendulum [21], geared counterweights, and inertia flywheel [22, 23]. More information on complete shaking moment balancing can be obtained in a critical review by Lowen et al. [24], Kochev [25], and Arakelian and Smith [26]. Practically, these methods not only increase the mass of the system but also increase its complexity.

An alternate way to reduce the shaking force and shaking moment along with other dynamic quantities such as input-torque and bearing reactions is to optimize all the dynamic quantities. Since shaking moment reduces to a pure torque in a force-balanced linkage, many researchers used the fact to develop their theory of shaking moment optimization. Berkof and Lowen [27] proposed an optimization method to minimize the root-mean-square (RMS) value of the shaking moment in a fully force-balanced in-line four-bar linkage whose input link rotates at a constant speed. As an extension of this method, Carson and Stephens [7] highlighted the need to consider feasibility limits of the link parameters. A different approach for the optimization of shaking moment in a force-balanced four-bar linkage is proposed by Hains [28]. Using the principle of the independence of the static balancing properties of a linkage from the axis of rotation of the counterweights, partial shaking moment balancing is suggested by Arakelian and Dahan [29]. The principle of momentum conservation is also used by Wiederich and Roth [30] to reduce the shaking moment in a fully force-balanced four-bar linkage.

11.2.1 Problem Formulation

The problem of mechanism balancing is formulated here as an optimization problem. In order to identify the design variables and the associated constraints, a set of equimomental point masses is defined for each link of a mechanism at hand. To calculate the shaking force and shaking moment dynamic equations of motion in the minimal set are derived in the parameters of the point masses. These parameters are then treated as design variables to redistribute the link masses to minimize the force transmitted to the frame.

11.2.2 Equations of Motion in Terms of Point-Mass System

The Newton-Euler (NE) equations of motion for the *i*th rigid link moving in a plane (Fig. 11.1) are given as [31]

$$\mathbf{M}_i\dot{\mathbf{t}}_i + \mathbf{C}_i\mathbf{t}_i = \mathbf{w}_i \tag{11.20}$$

where the three vectors, $\mathbf{t}_i$, $\dot{\mathbf{t}}_i$, and $\mathbf{w}_i$, are defined as the twist, twist rate, and wrench of the *i*th link with respect to the origin, O_i; that is,

$$\mathbf{t}_i = \begin{bmatrix} \omega_i \\ \mathbf{v}_i \end{bmatrix}; \dot{\mathbf{t}}_i = \begin{bmatrix} \dot{\omega}_i \\ \dot{\mathbf{v}}_i \end{bmatrix} \text{ and } \mathbf{w}_i = \begin{bmatrix} n_i \\ \mathbf{f}_i \end{bmatrix} \tag{11.21}$$

in which ω_i and $\mathbf{v}_i$ are the scalar angular velocity about the axis perpendicular to the plane of motion, and the two-vector of linear velocity of point O_i of the *i*th link, respectively. Accordingly, $\dot{\omega}_i$ and $\dot{\mathbf{v}}_i$ are the time derivatives of ω_i and $\mathbf{v}_i$, respectively. Also, the scalar, n_i, and the two-vector, $\mathbf{f}_i$, are the resultant moment about O_i and the resultant force at O_i, respectively. Moreover, the 3×3 matrices, $\mathbf{M}_i$ and $\mathbf{C}_i$, are given as

$$\mathbf{M}_i = \begin{bmatrix} I_i & -m_i\mathbf{d}_i^{\mathrm{T}}\overline{\mathbf{E}} \\ m_i\overline{\mathbf{E}}\mathbf{d}_i & m_i\mathbf{1} \end{bmatrix} \text{ and } \mathbf{C}_i = \begin{bmatrix} 0 & \mathbf{0}^{\mathrm{T}} \\ -m_i\omega_i\mathbf{d}_i & \mathbf{O} \end{bmatrix} \tag{11.22}$$

where $\mathbf{1}$ and $\mathbf{O}$ are the 2×2 identity and zero matrices, respectively, and $\mathbf{0}$ is the two-vector of zeros, and the 2×2 matrix, $\bar{\mathbf{E}}$, is defined by

$$\overline{\mathbf{E}} = \begin{bmatrix} 0 & -1 \\ 1 & 0 \end{bmatrix}$$

Upon substitution of the expressions for the scalar, I_i, and the two-vector, $m_i\mathbf{d}_i$, from Eqs. (11.1–11.4), the 3×3 matrices, $\mathbf{M}_i$ and $\mathbf{C}_i$, of Eq. (11.22) are obtained as

$$\mathbf{M}_i = \begin{bmatrix} \sum_j m_{ij} l_{ij}^2 & -\sum_j m_{ij} l_{ij} \sin\left(\theta_{ij} + \alpha_i\right) & \sum_j m_{ij} l_{ij} \cos\left(\theta_{ij} + \alpha_i\right) \\ -\sum_j m_{ij} l_{ij} \cos\left(\theta_{ij} + \alpha_i\right) & \sum_j m_{ij} & 0 \\ \sum_j m_{ij} l_{ij} \sin\left(\theta_{ij} + \alpha_i\right) & 0 & \sum_j m_{ij} \end{bmatrix}$$

$$\mathbf{C}_i = \begin{bmatrix} 0 & 0 & 0 \\ -\omega_i \sum_j m_{ij} l_{ij} \cos\left(\theta_{ij} + \alpha_i\right) & 0 & 0 \\ -\omega_i \sum_j m_{ij} l_{ij} \sin\left(\theta_{ij} + \alpha_i\right) & 0 & 0 \end{bmatrix} \tag{11.23}$$

Equations (11.20) and (11.23) are the equations of motion for the ith link in terms of its $3p$ point-mass parameters, namely m_{ij}, θ_{ij}, and l_{ij}, for $j = 1, \ldots, p$. Now, all or some of the point-mass parameters can be used as design variables based on their influence on the objective function of an optimization problem.

In some research papers, namely by Lee and Cheng [5] and Wiederrich and Roth [30], two-point-mass model was considered to represent the mass and inertia of the links. They assumed that $\theta_{i1} = 0$ and $\theta_{i2} = \pi/2$, amongst the six parameters m_{i1}, m_{i2}, θ_{i1}, θ_{i2}, l_{i1}, and l_{i2}. The remaining parameters were then considered as design variables, and used for the optimization of four-bar mechanisms. In three-point-mass model the following five parameters can be assigned arbitrarily:

$$\theta_{i1} = 0; \theta_{i2} = 2\pi/3; \theta_{i3} = 4\pi/3; \text{ and } l_{i2} = l_{i3} = l_{i1} \tag{11.24}$$

The other four parameters, namely m_{i1}, m_{i2}, m_{i3}, and l_{i1}, are then treated as the design variables for each link.

11.2.3 Definition of Shaking Force and Shaking Moment

Figure 11.2 shows n moving links in a multiloop mechanism where the fixed link, #0, is detached from the other links. The appropriate reaction forces and moments due to the fixed link are indicated on the moving links to maintain the dynamic equilibrium.

The shaking force is now defined as the reaction of the vector sum of all the inertia forces of moving links associated with the mechanism, and the shaking moment is the reaction of the resultant of the inertia moment and the moment of the inertia forces [5]. By the above definitions, the shaking force and the shaking moment with respect to O_1, transmitted to the fixed link, are given by

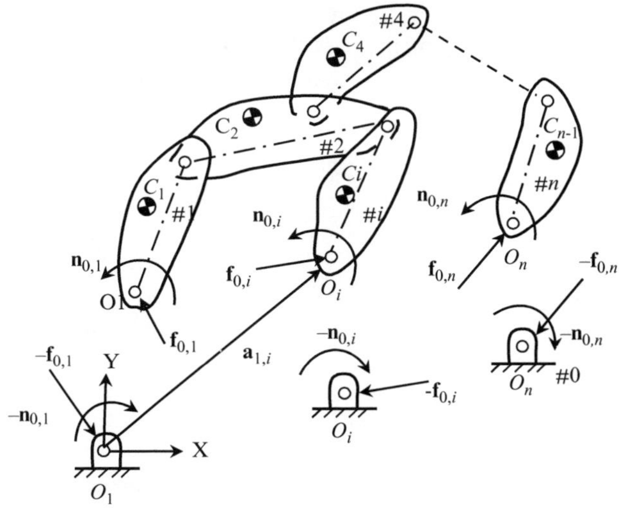

Fig. 11.2 A multiloop mechanism

$$\mathbf{f}_{\text{sh}} = -\sum_{i=1}^{n} \mathbf{f}_i^* \tag{11.25}$$

$$n_{\text{sh}} = -\sum_{i=1}^{n} \left(n_i^* - \mathbf{a}_{1,i}^{\text{T}} \overline{\mathbf{E}} \mathbf{f}_i^* \right) \tag{11.26}$$

where n_i^* and $\mathbf{f}_i^*$ are the inertia moment and the two-vector of inertia force, respectively, acting at and about the origin, O_i, of the ith link. Moreover, the two-vector, $\mathbf{a}_{1,i}$, is defined from O_1 to the origin of the ith link, as shown in Fig. 11.2. Substituting the resultant force and moment in terms of the external force and moment, and the reactions due to the adjoining joints, the force and moment balance expressions for the ith link are written as

$$\mathbf{f}_i^* = \mathbf{f}_i^{\text{e}} + \sum_{k=0}^{n} \mathbf{f}_{k,\,i} \tag{11.27}$$

$$n_i^* = n_i^{\text{e}} + \sum_{k=0}^{n} \left(n_{k,\,i} - a_{i,k}^{\text{T}} \overline{\mathbf{E}} \mathbf{f}_{k,\,i} \right) \tag{11.28}$$

where $\mathbf{f}_{k,i}$ and $n_{k,i}$ are the bearing reaction force and moment on the ith link by the kth link, respectively. Note that $\mathbf{f}_{k,i} = \mathbf{0}$ and $n_{k,i} = 0$ if kth link is not directly connected to the ith link. Furthermore, $\mathbf{f}_i^{\text{e}}$ and n_i^{e} are the external force and moment acting at and about the origin, O_i, respectively. Note that the origin for the ith link is defined at the joint where it is coupled with previous link, whereas vector, $\mathbf{a}_{i,k}$, is defined from the origin of the *i*th link to the joint where the *k*th link is connected. Upon substitution of Eqs. (11.27 and 11.28) into Eqs. (11.25 and 11.26), the shaking force and the shaking moment with respect to O_1 transmitted to the fixed link, #0, are obtained as

$$\mathbf{f}_{\mathrm{sh}} = -\sum_{j=1}^{n_{\mathrm{f}}} \mathbf{f}_{0,j} - \sum_{i=1}^{n} \mathbf{f}_i^{\mathrm{e}} \tag{11.29}$$

$$n_{\mathrm{sh}} = -\sum_{j=1}^{n_{\mathrm{f}}} \left(n_{0,j} - \mathbf{a}_{1,j}^{\mathrm{T}} \overline{\mathbf{E}} \mathbf{f}_{0,j} \right) - \sum_{i=1}^{n} \left(n_i^{\mathrm{e}} - \mathbf{a}_{1,i}^{\mathrm{T}} \overline{\mathbf{E}} \mathbf{f}_i^{\mathrm{e}} \right) \tag{11.30}$$

where $\mathbf{f}_{0,j}$ represents the reaction force on the *j*th link by the fixed link, for $j = 1, \ldots, n_{\mathrm{f}}$, n_{f} being the number of links connected to the fixed link. Hence, using Eqs. (11.29 and 11.30), the computation of the reactions at all the joints is not necessary to compute the shaking force and shaking moment. Note that the dynamic quantities, e.g., the shaking force, shaking moment, and bearing reactions, have different units and magnitudes. In order to harmonize them, the force and moment are normalized as [32]

$$\bar{f} = |\mathbf{f}| \,/\, \left(m_{\mathrm{m}}^{\mathrm{o}} a_{\mathrm{m}} \omega_{\mathrm{in}}^2 \right) \tag{11.31}$$

$$\bar{n} = n / \left(m_{\mathrm{m}}^{o} a_{\mathrm{m}}^2 \omega_{\mathrm{in}}^2 \right) \tag{11.32}$$

where a_{m} and $m_{\mathrm{m}}^{\mathrm{o}}$ are the length and mass of the reference link for the normalization, whereas ω_{in} is any input angular velocity. Superscript "o" is used for those parameters of the original mechanism, which will be changing during the optimization.

11.2.4 Optimality Criterion

There are many possible criteria by which the shaking force and shaking moment transmitted to the fixed link of the mechanism can be minimized. For example, one criterion could be based on the RMS values of the shaking force, shaking moment, and required driving torque for a given motion, and/or the combination of these. Besides the RMS values, there are other ways to specify the dynamic quantities also, namely by the maximum values, by the amplitude of the specified harmonics, or by the amplitudes at certain point during the motion cycle. Here, the RMS value is preferred over others as it gives equal emphasis on the results of every time instances, and every harmonic component. The RMS values of the normalized shaking force, $\bar{f}_{\mathrm{sh}}$, and the normalized shaking moment, $\bar{n}_{\mathrm{sh}}$, at δ discrete positions of the mechanism are defined as

$$\tilde{f}_{sh} = \sqrt{\sum \bar{f}_{sh}^2/\delta}; \text{ and } \tilde{n}_{sh} = \sqrt{\sum \bar{n}_{sh}^2/\delta} \tag{11.33}$$

where $\tilde{f}_{sh}$ and $\tilde{n}_{sh}$ are the RMS values of the normalized shaking force and the normalized shaking moment, respectively. Considering the RMS values, $\tilde{f}_{sh}$ and $\tilde{n}_{sh}$, an optimality criterion can be posed as

$$z = w_1\tilde{f}_{sh} + w_2\tilde{n}_{sh} \tag{11.34}$$

where w_1 and w_2 are the weighting factors whose values may vary depending on an application. For example, $w_1 = 1.0$ and $w_2 = 0$ if the objective is to minimize the shaking force only. The design variables and constraints depend upon whether the balancing is done through the redistribution of link masses or counterweighting the links.

11.2.4.1 Mass Redistribution Method

Consider a mechanism having n moving links, i.e., $i = 1, \ldots, n$, and each link is modeled by a system of p equimomental point masses; then the $3p$-vector of point-mass parameters for the ith link is defined as

$$\mathbf{x}_i = \left[m_{i1} \ldots m_{ip} \; l_{i1} \ldots l_{ip} \; \theta_{i1} \ldots \theta_{ip}\right]^T \tag{11.35}$$

Accordingly, the $3np$-vector of the point-mass parameters for the whole mechanism is given by

$$\mathbf{x} = \left[\mathbf{x}_i^T \ldots \mathbf{x}_n^T\right]^T \tag{11.36}$$

If three-point-mass model is used then the dimensions of the vectors, $\mathbf{x}_i$ and $\mathbf{x}$, are 9 and $9n$, respectively. If five parameters per link are assigned arbitrarily according to Eq. (11.24), the remaining four parameters, namely m_{i1}, m_{i2}, m_{i3}, and l_{i1}, per link can be treated as the *design variables* (DV). Finally the $4n$-vector, $\mathbf{x}$, of the DVs using three-point-mass model is defined as

$$\mathbf{x} = [m_{11}, m_{12}, m_{13}, l_{11}, \ldots, m_{n1}, m_{n2}, m_{n3}, l_{n1}]^T \tag{11.37}$$

The constraints on the DVs depend on the allowable minimum and maximum values of the DVs, say, mass and inertia, etc. The minimum mass, $m_{i,min}$, of the ith link and its mass distribution can be decided by the strength of its material. Furthermore, the maximum mass, $m_{i,max}$, can be taken into account according to what extent the shaking force and shaking moment are eliminated. Similarly, the limits on parameters, l_{i1}, can be determined based on the limiting values of the moment of inertia. The optimization problem is finally posed as

$$\text{Minimize } z(\mathbf{x}) = w_1\tilde{f}_{sh} + w_2\tilde{n}_{sh} \text{ (single objective) OR}$$

$$\text{Minimize } z(\mathbf{x}) = \left[\tilde{f}_{sh}, \tilde{n}_{sh}\right] \text{ (multi-objective)} \tag{11.38a}$$

$$\text{Subject to } m_{i,\min} \le m_i \le m_{i,\max} \tag{11.38b}$$

$$l_{i1,\min} \le l_{i1} \le l_{i1,\max} \tag{11.38c}$$

$$d_{i,\min} \le d_i \le d_{i,\max} \tag{11.38d}$$

$$m_i d_i^2 \le I_i \tag{11.38e}$$

for $i = 1, \ldots, n$, where $m_{i,\min}$, $m_{i,\max}$, $l_{i1,\min}$, and $l_{i1,\max}$ are the lower and upper bounds on m_i and l_{i1}, respectively, and $m_i = m_{i1} + m_{i2} + m_{i3}$. The feasibility of the mass center location and the moment of inertia of the ith link can be achieved using constraints, Eqs. (11.38d and 11.38e), where $I_i = I_i^c + m_i d_i^2$, which implies that the term $m_i d_i^2$ must be less than or equal to the moment of inertia, I_i.

11.2.4.2 Counterweighting Method

In the case of counterweight balancing, counterweights are attached to the moving links such that the shaking force and shaking moment transmitted to the frame of the mechanism are minimum. Assume that the counterweight of mass, m_i^b, with its mass center location, $(\overline{x}_i^b, \overline{y}_i^b)$, is attached to the ith link as shown in Fig. 11.3a. The equimomental system of the resulting link is shown in Fig. 11.3b, where it is assumed that the point masses of the counterweight mass, m_{ij}^b, are placed at the location of the point masses of original link, m_{ij}^o. Then the counterweight mass, its mass center location, and inertia are defined as

$$m_i^b = \sum_{j=1}^{3} m_{ij}^b \tag{11.39}$$

$$m_i^b \mathbf{d}_i^b = \sum_{j=1}^{3} m_{ij}^b l_{ij}^o \tag{11.40}$$

$$I_i^b = \sum_{j=1}^{3} m_{ij}^b \left(l_{ij}^o\right)^2 \tag{11.41}$$

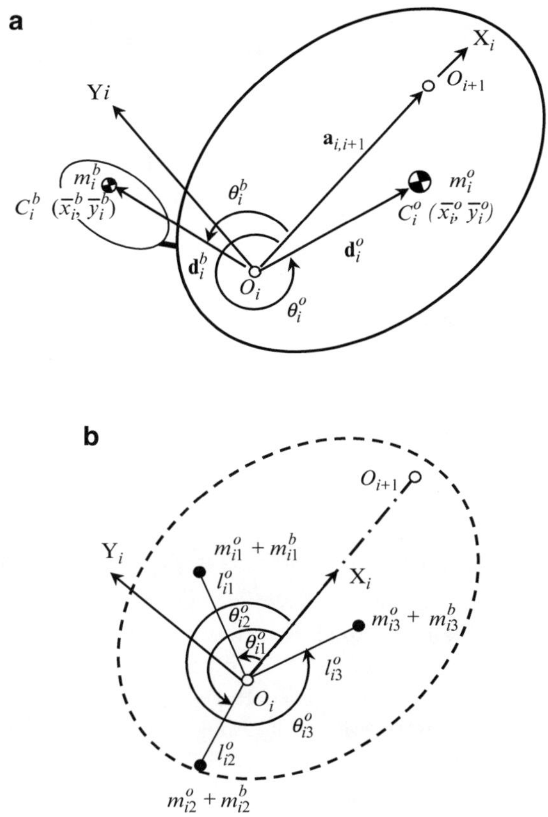

Fig. 11.3 Counterweight balancing. (**a**) Counterweight to the *i*th link, (**b**) equimomental point masses of the counterweight mass

Now, for a mechanism having n moving links, the 3n-vector of the design variables, $\mathbf{x}^{\mathrm{b}}$, is

$$\mathbf{x}^{\mathrm{b}} = \left[\mathbf{m}_1^{\mathrm{b}^{\mathrm{T}}}, \ \ldots, \ \mathbf{m}_n^{\mathrm{b}^{\mathrm{T}}}\right]^{\mathrm{T}} \tag{11.42}$$

where the three-vector, $\mathbf{m}_i^{\mathrm{b}}$, is as follows:

$$\mathbf{m}_i^{\mathrm{b}} = \left[m_{i1}^{\mathrm{b}} \ m_{i2}^{\mathrm{b}} \ m_{i3}^{\mathrm{b}}\right]^{\mathrm{T}}, \text{ for } i = 1, \ldots, n$$

Note that m_{ij}^{b} is the jth point mass of the counterweight attached to the ith link. The minimum and maximum mass of counterweight, $m_{i,\min}^{\mathrm{b}}$ and $m_{i,\max}^{\mathrm{b}}$, their locations, and the moment of inertia depend on an application. However, the counterweight balancing problem is stated to determine the mass, m_i^{b}, its mass center location, $\left(\overline{x}_i^{\mathrm{b}}, \overline{y}_i^{\mathrm{b}}\right)$, and the inertia, I_i^{b}, such that the combined effect of shaking force and shaking moment is going to be minimum; that is,

Table 11.1 Definition of normalized parameters

$a_{ij} = \left\vert \mathbf{a}_{ij} \right\vert / a_m$	Normalized distance between joints *i* and *j*
$d_i = \left\vert \mathbf{d}_i \right\vert / a_m$	Normalized distance of the mass center
$m_i = m_i / m_m^{\mathrm{o}}$	Normalized mass of the *i*th link
$I_i = I_i / \left(m_m^{\mathrm{o}} a_m^2 \right)$	Normalized moment of inertia of the *i*th link

The variables, a_m and m_m^{o}, are defined after Eq. (11.32)

$$\text{Minimize } z\left(\mathbf{x}^{\mathrm{b}}\right) = w_1 \tilde{f}_{\mathrm{sh}} + w_2 \tilde{n}_{\mathrm{sh}} \tag{11.43a}$$

$$\text{Subject to } m_{i,\min}^{\mathrm{b}} \le m_i^{\mathrm{b}} \le m_{i,\max}^{\mathrm{b}} \tag{11.43b}$$

$$d_{i,\min}^{\mathrm{b}} \le d_i^{\mathrm{b}} \le d_{i,\max}^{\mathrm{b}} \tag{11.43c}$$

$$m_i^{\mathrm{b}} \left(d_i^{\mathrm{b}}\right)^2 \le I_i^{\mathrm{b}} \tag{11.43d}$$

for $i = 1, \ldots, n$, where $d_i^{\mathrm{b}} = \sqrt{\overline{x}_i^{\mathrm{b}2} + \overline{y}_i^{\mathrm{b}2}}$. Similar to the constraints in the mass redistribution method, the mass center location and the moment of inertia of the counterweight attached to the *i*th link are constraints using inequalities of Eqs. (11.43c and 11.43d), respectively.

The optimization methodology using either the mass redistribution or counterweight methods is summarized in the following steps:

1. To harmonize the values of the link parameters, the parameters of the unbalanced mechanism are made dimensionless as explained in Table 11.1.
2. Given mass, its mass center location, and the inertia of each link: m_i, $\overline{x}_i$, $\overline{y}_i$, I_i, of the normalized unbalanced mechanism, find the set of equimomental point masses for each rigid link.
3. Define design variable for the mechanism having *n* moving links, as in Eqs. (11.37) and (11.42), for the redistribution and counterweight balancing methods, respectively.
4. Define objective function and constraints on the link masses and inertias, i.e., Eqs. (11.38a–11.38e) or (11.43a–11.43d), where the normalized shaking force and shaking moment are defined according to Eqs. (11.31) and (11.32), respectively. For the normalized mechanism operating at $\omega_{\mathrm{in}} = 1$ rad/s, the shaking force and shaking moment are the normalized shaking force and shaking moment.
5. Solve the optimization problem posed in the above step (4) using any standard optimization solver, say, the optimization toolbox of MATLAB [33]. The optimization process can be started with the parameters of the given unbalanced mechanism as the initial design vector.
6. From the optimized parameters, m_{i1}^*, m_{i2}^*, m_{i3}^*, l_{i1}^*, in redistribution method, the optimized mass, m_i^*, the location of the mass center, $(\overline{x}_i^*, \overline{y}_i^*)$, and the

inertia of each link, I_i^*, of the balanced mechanism are determined using the equimomental conditions, i.e., Eqs. (11.1–11.4). Similarly, in counterweight method the optimized total mass, m_i^{b*}, the location of the mass center ($\bar{x}_i^{b*}$, $\bar{y}_i^{b*}$), and the inertia of counterweight attached to each link, I_i^{b*}, of the balanced mechanism are determined using the equimomental conditions, i.e., Eqs. (11.39–11.41), from optimized point masses, m_{i1}^{b*}, m_{i2}^{b*}, and m_{i3}^{b*}.

7. Actual values of link masses or counterweights, their mass center location, and moments of inertia are obtained by multiplying the optimized values with the corresponding normalizing factors, namely $m_{\rm m}^{\rm o}$, $a_{\rm m}$, and $m_{\rm m}^{\rm o}a_{\rm m}^2$, respectively.

11.3 Numerical Examples

In this section, the effectiveness of the optimization methodology is shown by applying it to some planar mechanisms. The balancing problems can be framed as single-objective or multi-objective optimization problems to simultaneously minimize the shaking force and shaking moment. To solve these problems using conventional optimization algorithms, "*fmincon*" function in *Optimization Toolbox* of MATLAB is used. Alternatively, the genetic algorithm is also used as solver. Two functions "*ga*" and "*gamultiobj*" in *Genetic Algorithm and Direct Search Toolbox* of MATLAB are used for this purpose. It was observed that GA produces better results as compared to conventional optimization algorithms.

11.3.1 Planar Four-Bar Mechanism

A numerical example of standard four-bar mechanism [5, 31, 34] is solved using the methodology developed in this chapter. The minimization of inertia forces is obtained by redistributing the link masses [35]. The parameters of standard and balanced mechanisms are shown in Table 11.2 whereas the variation of shaking force, shaking moment, and driving torque for complete cycle is shown in Figs. 11.4 and 11.5.

Table 11.2 Dimensionless parameters of standard and balanced mechanism

		Standard mechanism				Balanced mechanism			
Link	Length a_i	Mass m_i	Moment of inertia $I_{i,zz}^c$	d_i	θ_i	Mass m_i	Moment of inertia $I_{i,zz}^c$	d_i	θ_i
1	1	1.0000	0.3300	0.5	0	2.0725	6.9383	0.6688	187.50
2	2	1.1597	1.0186	1.0	0	1.4979	0.5688	0.2685	309.25
3	3	1.4399	2.2880	1.5	0	1.9095	1.3829	0.3630	098.86

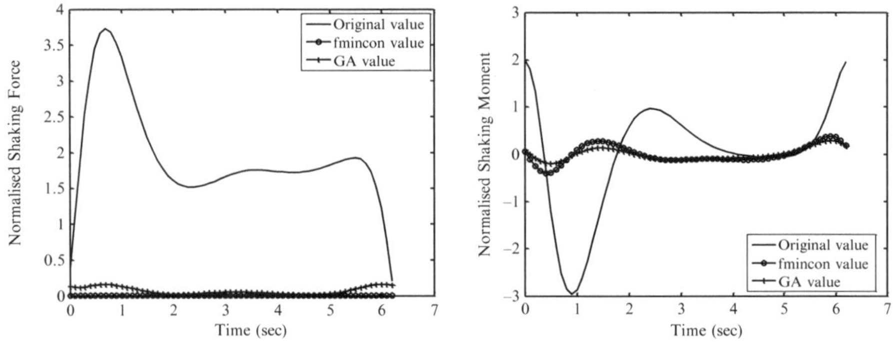

Fig. 11.4 Variations of shaking force and shaking moment for complete cycle

Fig. 11.5 Variations of driving torque for complete cycle

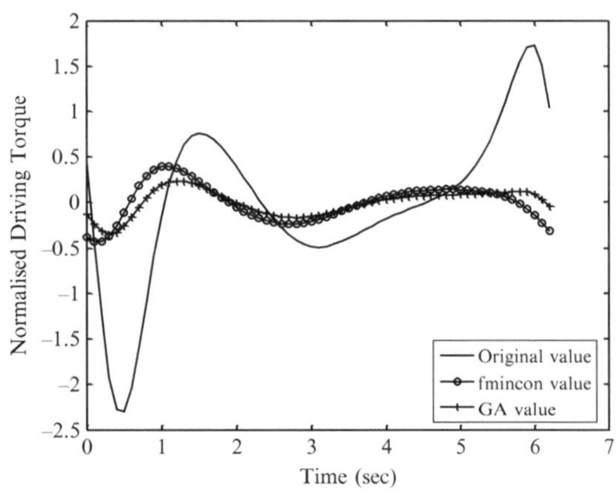

Two different approaches were used to solve this multi-objective optimization problem using GA. In a priori approach, a composite objective function is formed using equal weighting factors to both the objectives, i.e., shaking force and shaking moment as explained in Eq. (11.34). As the shaking force and the shaking moment are of different units, these quantities are made dimensionless with respect to the parameters of the driving link for adding them in a composite objective function for which the results are shown in Table 11.3. The values in the parenthesis denote the percentage increment/decrement with respect to the corresponding RMS values of the standard mechanism.

For the given problem, the genetic algorithm produced better results as compared to the results obtained using conventional optimization technique. With equal weighting to shaking force and shaking moment, about 96 %, 89 %, and 84 % reductions are achieved in shaking force, shaking moment, and driving torque, respectively.

In posterior approach, a set of optimal solutions, known as Pareto front, is found by considering both the objectives separately. Each solution in the Pareto front is

Table 11.3 RMS values of dynamic quantities of standard and optimized mechanisms

	RMS values of dimensionless dynamic quantities		
Balancing method	Shaking force	Shaking moment	Driving torque
Standard mechanism	2.0582	1.1593	0.8613
Conventional algorithm [31] $w_1 = 0.5$; $w_2 = 0.5$	3.78×10^{-6} (−100)	0.1882 (−84)	0.2051 (−76)
Genetic algorithm $w_1 = 0.5$; $w_2 = 0.5$	0.0868 (−96)	0.1233 (−89)	0.1398 (−84)

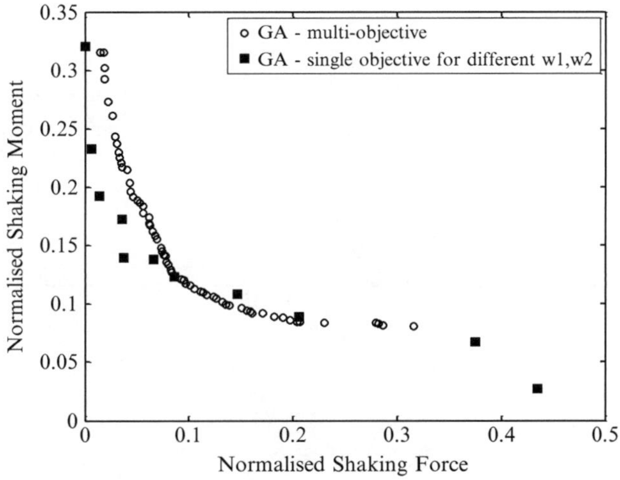

Fig. 11.6 Pareto front for four-bar mechanism problem

an optimum solution as no single solution minimizes both the objectives when compared to other solutions in the set (Fig. 11.6). The results obtained using two approaches are also compared in Fig. 11.6.

11.3.2 Planar Slider-Crank Mechanism

The optimization method presented in this chapter can be effectively used to balance the mechanisms having revolute and prismatic joints while most of the methods available in the literature are for the mechanisms with revolute joints only. A slider-crank mechanism is balanced here by optimally distributing the link masses [36] while a cam mechanism with counterweight was used to balance the same mechanism in [37].

Table 11.4 Dimensionless parameters of standard and balanced slider-crank mechanism

Link	Length a_i	Standard mechanism				Balanced mechanism			
		Mass m_i	Moment of inertia $I^c_{i,zz}$	d_i	θ_i	Mass m_i	Moment of inertia $I^c_{i,zz}$	d_i	θ_i
1	1.0000	1.0	0.1759	0.5000	0	1.5226	2.5204	1.6171	171.94
2	1.4623	1.5	0.8210	0.7329	0	1.5015	0.4222	0.1842	357.65
3	–	2.0	–	–	0	2.0011	–	0.3750	269.12

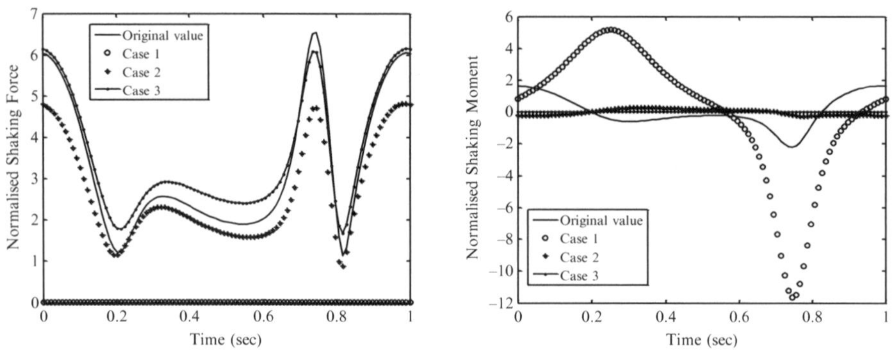

Fig. 11.7 Variations of shaking force and shaking moment for different cases

Table 11.5 RMS values of dynamic quantities of normalized standard and optimized mechanisms

Balancing method	Shaking force	Shaking moment
Standard mechanism	3.6877	1.0047
Conventional algorithm	2.9132 (−21)	0.1883 (−81)
Genetic algorithm	2.0051 (−46)	0.0105 (−99)

The problem is considered here to balance it using optimization procedure described in the chapter. The parameters of standard and balanced slider-crank mechanisms are shown in Table 11.4 whereas Fig. 11.7 shows the variation of shaking force and shaking moment over the complete cycle.

The results corresponding to different combinations of the weighting factors using conventional optimization algorithm are shown in Fig. 11.7. The case 1 is complete shaking force balancing in which the RMS value of shaking moment increases to four times that of the unbalanced mechanism. Similarly, in case 3, shaking force increases while shaking moment reduces substantially. Reductions in both the quantities occur in case 2, in which equal weights are assigned to them.

Then the same problem is solved using GA with equal weighting factors for both the quantities. The comparison of the original RMS values with the optimum RMS values of the shaking force and shaking moment obtained using conventional and genetic algorithm is presented in Table 11.5 and Fig. 11.8. The optimized link parameters are then found by using the equimomental conditions corresponding to GA solution and shown in Table 11.4.

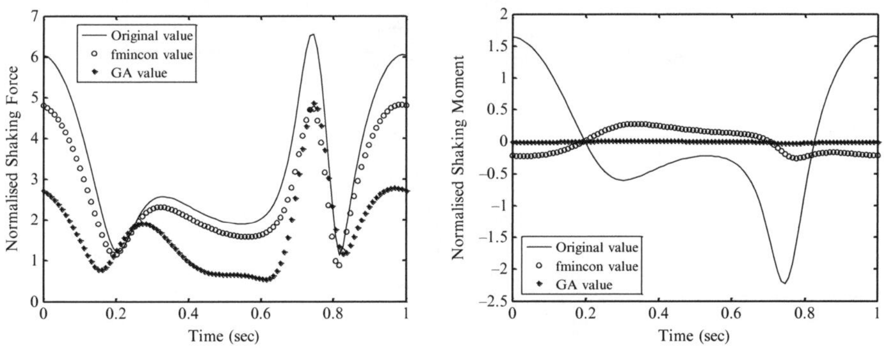

Fig. 11.8 Variations in shaking force and shaking moment for complete cycle

Note that the reductions of 21 % and 81 % in the RMS values of shaking force and shaking moment, respectively, are achieved in the conventional method. The application of the genetic algorithm results in reduction of 46 % and 99 % in the shaking force and shaking moment, respectively. The moment of inertia of slider about CG doesn't affect the values of shaking force and shaking moment and hence it is not provided in Table 11.4.

11.3.3 Planar Six-Bar Mechanism

The optimization methodology can also be used to minimize the shaking force and shaking moment in multiloop planar mechanisms. A Stephenson six-bar mechanism [12] shown in Fig. 11.9 is optimally balanced using counterweighting method [38]. First, the force balancing is achieved by optimizing the point-mass parameters of the counterweights. Next, the balancing problem is formulated as a multi-objective optimization problem which minimizes the shaking force and shaking moment simultaneously. The parameters of original unbalanced Stephenson six-bar mechanism are given in Table 11.6 whereas Fig. 11.10 shows the variation of shaking force and shaking moment over the complete cycle.

The kinematic simulation was carried out using the MotionView and MotionSolve of Altair HyperWorks 11.0 software [39]. For the problem considered, the standard and optimized values of the shaking force and shaking moment for different combinations of weighting factors using conventional optimization algorithm are presented in Table 11.7 and shown in Fig. 11.10.

For only the shaking force balancing case, results show 63.87 % reduction in the shaking force whereas 190.93 % increment in shaking moment occurred. For only the shaking moment balancing case, reduction of 39 % was found in the shaking moment while the shaking force is increased by 127 %. These two cases support the fact that the reduction in one dynamic quantity increases the other. Thus a trade-off is necessary to reduce both the shaking force and shaking moment. To reduce both

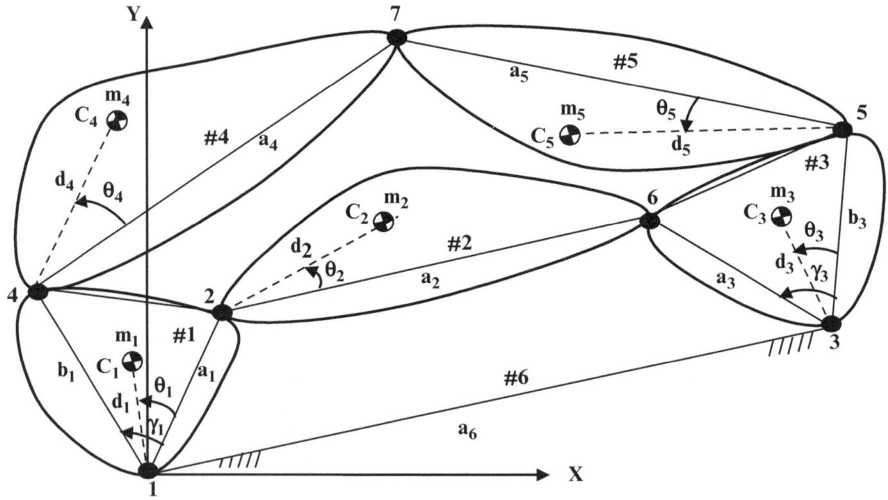

Fig. 11.9 Stephenson six-bar mechanism

Table 11.6 Parameters of unbalanced Stephenson six-bar mechanism

Link i	1	2	3	4	5	6
a_i (m)	0.0559	0.1206	0.0032	0.1397	0.0444	0.1238
b_i (m)	0.0584	–	0.0030	–	–	–
γ_i (deg)	6	–	16	–	–	–
θ_i (deg)	3	0	5	19	0	11
d_i (m)	0.0286	0.0630	0.0031	0.0836	0.0197	–
m_i (kg)	0.0608	0.0825	0.0757	0.1732	0.0395	–

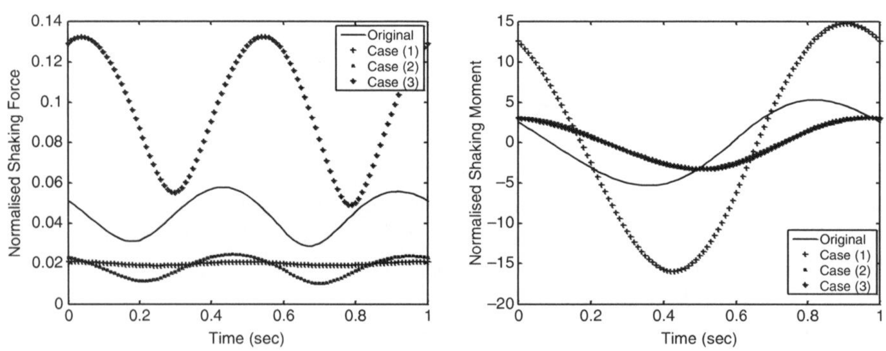

Fig. 11.10 Variations in shaking force and shaking moment for complete cycle

Table 11.7 RMS values of shaking force and shaking moment in Stephenson six-bar mechanism

	Shaking force	Shaking moment
Original value	0.0450	3.7332
Only shaking force Case (1) $w_1 = 1.0$; $w_2 = 0.0$	0.0164 (−63.87 %)	10.8610 (+190.93 %)
Both shaking force and shaking moment Case (2) $w_1 = 0.5$; $w_2 = 0.5$	0.0192 (−57.7 %)	2.2651 (−39.32 %)
Only shaking moment Case (3) $w_1 = 0.0$; $w_2 = 1.0$	0.1031 (+127.09 %)	2.2516 (−39.68 %)

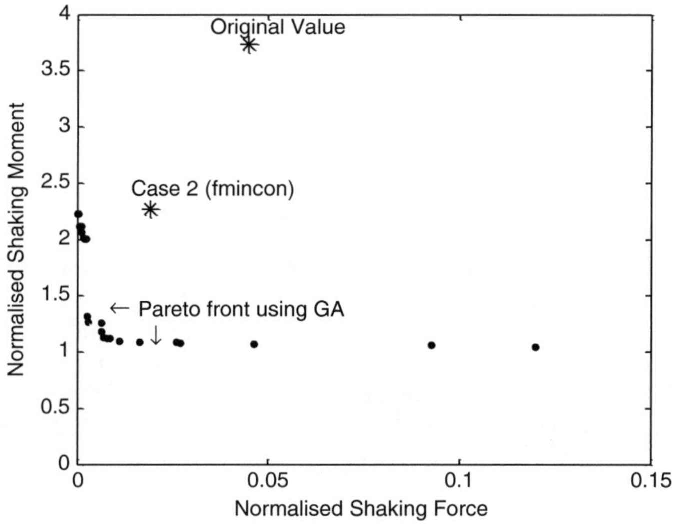

Fig. 11.11 Pareto front for six-bar mechanism problem

the shaking force and shaking moment simultaneously in the given mechanism, both the quantities are assigned equal weighting factor value, i.e., 0.5 in the objective function (Case 2). The result shows 57.7 % and 39 % reduction in the shaking force and shaking moment, respectively.

The optimum design variables obtained for case 2 are then taken as the initial population and the genetic algorithm was used to find the solution of this optimization problem. The genetic algorithm produces multiple optimal solutions (Pareto front) as shown in Fig. 11.11.

This plot shows the trade-off between the two objective functions, i.e., the shaking force and shaking moment. Thus it is advantageous to use GA for finding multiple optimal solutions without running the traditional algorithm many times. Figure 11.11 shows that the GA results are better than the results obtained using traditional optimization algorithm. The values of the shaking force and shaking moment corresponding to the best solution among available Pareto optimal solutions are given in Table 11.8.

Table 11.8 Results from GA algorithm

	Shaking force	Shaking moment
Original value	0.0450	3.7332
Optimized value	0.0069 (−84 %)	1.1260 (−69.83 %)

Table 11.9 Optimum counterweight parameters using GA algorithm

Counterweights	CW 1	CW 2	CW 3	CW 4	CW 5
Mass (kg)	0.3046	0.1124	0.0570	0.0116	0.0144
d (m)	0.0750	0.2318	0.0132	0.3336	0.2209
θ (deg)	200.55	250.13	136	144.68	214.21

The counterweight parameters for optimum design variables are calculated using the equimomental conditions and are presented in Table 11.9.

11.4 Summary

In this chapter, balancing problem of planar mechanisms is formulated as an optimization problem. The main focus of the chapter is to reduce the shaking force and shaking moment. The design variables and the constraints on them are identified by introducing the equimomental system of point masses. Using the equimomental three point masses, equations of motion are reformulated to determine the shaking force, shaking moment, and other dynamic quantities. Three planar mechanisms, namely four-bar, slider-crank, and Stephenson six-bar mechanism, are optimally balanced using the methodology given in this chapter.

Acknowledgment The following material is used "With kind permission of Springer Science + Business Media." Section 4.1 (Chapter 4) and Sects. 5.1, 5.2 (Chapter 5) including Figs. 5.1, 5.3, 5.4 and Table 5.1; pp: 87–92, 100–101, 104–105, 110–117 from book Dynamics and Balancing of Multibody Systems, Lecture notes in applied and computational mechanics, Vol. 37 by Himanshu Chaudhary, Subir Kumar Saha, published by springer-Verlag Germany, 2009 ISBN 978-3-540-78178-3.

References

1. Routh, E.J.: Treatise on the Dynamics of a System of Rigid Bodies. Elementary Part I, p. 28. Dover Publication Inc., New York (1905)
2. Wenglarz, R.A., Forarasy, A.A., Maunder, L.: Simplified dynamic models. Engineering **208**, 194–195 (1969)
3. Huang, N.C.: Equimomental system of rigidly connected equal particles. J. Guid. Control. Dyn. **16**(6), 1194–1196 (1983)
4. Sherwood, A.A., Hockey, B.A.: The optimization of mass distribution in mechanisms using dynamically similar systems. J. Mech. **4**, 243–260 (1969)

5. Lee, T.W., Cheng, C.: Optimum balancing of combined shaking force, shaking moment, and torque fluctuations in high speed linkages. Trans. ASME J. Mech. Transm. Automat. Des. **106**(2), 242–251 (1984)
6. Hockey, B.A.: The minimization of the fluctuation of input-shaft torque in plane mechanisms. Mech. Mach. Theory **7**, 335–346 (1972)
7. Carson, W.L., Stephenes, J.M.: Feasible parameter design spaces for force and root-mean-square moment balancing an in-line 4R 4-bar synthesized for kinematic criteria. Mech. Mach. Theory **13**, 649–658 (1978)
8. Attia, H.A.: A matrix formulation for the dynamic analysis of spatial mechanisms using point coordinates and velocity transformation. Acta Mech. **165**, 207–222 (2003)
9. Gill, G.S., Freudenstein, F.: Minimization of inertia-induced forces in spherical four-bar mechanisms. Part 1: The general spherical four-bar linkage. Trans. ASME J. Mech. Transm. Automat. Des. **105**, 471–477 (1983)
10. Rahman, S.: Reduction of inertia-induced forces in a generalized spatial mechanism. Ph.D. Thesis, Dept. of Mech. Eng., The New Jersey Institute of Technology. http://archives.njit.edu/vol01/etd/1990s/1996/njit-etd1996-017/njit-etd1996-017.pdf
11. Quang, P.R., Zhang, W.J.: Force balancing of robotic mechanisms based on adjustment of kinematic parameters. Trans. ASME J. Mech. Des. **127**(3), 433–440 (2005)
12. Berkof, R.S., Lowen, G.G.: A new method for completely force balancing simple linkages. Trans. ASME J. Eng. Ind. **91**(1), 21–26 (1969)
13. Fattah, A., Agarwal, S.K.: On the design of reactionless 3-DOF planar parallel mechanisms. Mech. Mach. Theory **41**, 70–82 (2006)
14. Shchepetil'nikov, V.A.: The determination of the mass centres of mechanisms in connection with the problem of mechanism balancing. J. Mech. **3**, 367–389 (1968)
15. Kochev, I.S.: A new general method for full force balancing of planar linkages. Mech. Mach. Theory **23**(6), 475–480 (1988)
16. Lowen, G.G., Tepper, F.R., Berkof, R.S.: The quantitative influence of complete force balancing on the forces and moments of certain families of four-bar linkages. Mech. Mach. Theory **9**, 299–323 (1974)
17. Elliott, J.L., Tesar, D.: The theory of torque, shaking force, and shaking moment balancing of four link mechanisms. Trans. ASME J. Eng. Ind. **99**(3), 715–722 (1977)
18. Foucault, S., Gosselin, C.M.: Synthesis, design, and prototyping of a planar three degree-of-freedom reactionless parallel mechanism. Trans. of ASME Journal of Mechanical Design **126**, 992–999 (2004)
19. Kamenskii, V.A.: On the questions of the balancing of plane linkages. J. Mech. **3**, 303–322 (1968)
20. Tricamo, S.J., Lowen, G.G.: A novel method for prescribing the maximum shaking force of a four-bar linkage with flexibility in counterweight design. Trans. ASME J. Mech. Transm. Automat. Des. **105**, 511–519 (1983)
21. Berkof, R.S.: Complete force and moment balancing of inline four-bar linkage. Mech. Mach. Theory **8**, 397–410 (1973)
22. Arakelian, V.H., Smith, M.R.: Complete shaking force and shaking moment balancing of linkages. Mech. Mach. Theory **34**, 1141–1153 (1999)
23. Arakelian, V.H., Smith, M.R.: Design of planar 3-DOF 3-RRR reactionless parallel manipulators. Mechatronics **18**, 601–606 (2008)
24. Lowen, G.G., Tepper, F.R., Berkof, R.S.: Balancing of linkages—an updates. Mech. Mach. Theory **18**(3), 213–220 (1983)
25. Kochev, I.S.: General theory of complete shaking moment balancing of planar linkages: a critical review. Mech. Mach. Theory **35**, 1501–1514 (2000)
26. Arakelian, V.H., Smith, M.R.: Shaking force and shaking moment balancing of mechanisms: a historical review with new examples. J Mech. Des. **127**, 334–339 (2005)
27. Berkof, R.S., Lowen, G.G.: Theory of shaking moment optimization of forced-balanced four-bar linkages. Trans. ASME J. Eng. Ind. **93B**(1), 53–60 (1971)

28. Hains, R.S.: Minimum RMS shaking moment or driving torque of a force-balanced linkage using feasible counterweights. Mech. Mach. Theory **16**, 185–190 (1981)
29. Arakelian, V., Dahan, M.: Partial shaking moment balancing of fully force balanced linkages. Mech. Mach. Theory **36**, 1241–1252 (2001)
30. Wiederrich, J.L., Roth, B.: Momentum balancing of four-bar linkages. Trans. ASME J. Eng. Ind. **98**(4), 1289–1295 (1976)
31. Chaudhary, H., Saha, S.K.: Balancing of four-bar linkages using maximum recursive dynamic algorithm. Mech. Mach. Theory **42**(2), 216–232 (2007)
32. Conte, F.L., George, G.R., Mayne, R.W., Sadler, J.P.: Optimum mechanism design combining kinematic and dynamic-force considerations. Trans. ASME J. Eng. Ind. **95**(2), 662–670 (1975)
33. MATLAB, R2008b: Optimization Toolbox, Version 7.7.0.471
34. Farmani, M.R.: Multiobjective optimization for force and moment balance of a four-bar mechanism using evolutionary algorithms. J. Mech. Sci. Technol. **25**(12), 2971–2977 (2011)
35. Chaudhary, K., Chaudhary, H.: Dynamic balancing of planar mechanisms using genetic algorithm. J. Mech. Sci. Technol. **28**(10), 4213–4220 (2014)
36. Chaudhary, K., Chaudhary, H.: Optimum balancing of slider-crank mechanism using equimomental system of point-masses. Procedia Technol. **14**, 35–42 (2014)
37. Arakelian, V., Briot, S.: Simultaneous inertia force/moment balancing and torque compensation of slider-crank mechanisms. Mech. Res. Commun. **37**, 265–269 (2010)
38. Chaudhary, K., Chaudhary, H.: Minimization of shaking force and shaking moment in multiloop planar mechanisms. In: Proc. of 1st International and 16th National Conference on Machines and Mechanisms (iNaCoMM2013), IIT Roorkee, 18–20 Dec 2013, pp. 346–352
39. Altair HyperWorks, Version 11

Chapter 12
Balancing of Planar Mechanisms Having Imperfect Joints Using Neural Network-Genetic Algorithm (NN-GA) Approach

Selçuk Erkaya and İbrahim Uzmay

Abstract As a result of design, manufacturing and assembly processes or a wear effect, clearances are inevitable at the joints of mechanisms. In this study, dynamic response of mechanism having revolute joints with clearance is investigated. A four-bar mechanism having two revolute joints with clearance is considered as a model mechanism. A neural network was used to model several characteristics of joint clearance. Kinematic and dynamic analyses were achieved using continuous contact mode between journal and bearing. A genetic algorithm was also used to determine the appropriate values of design variables for reducing the additional vibration effects due primarily to the joint clearance. The results show that the optimum adjusting of suitable design variables gives a certain decrease in shaking forces and their moments on the mechanism frame.

Keywords Four-bar mechanism • Joint clearance • Shaking force and moment • Neural-genetic approach • Optimisation

12.1 Introduction

One of the important factors that influence the dynamic stability and the performance of mechanisms is the joint clearance [1]. Small clearances in the kinematic joints of mechanisms are necessary for assembly and mobility. In general, in the case of dynamic analysis of mechanical systems, the kinematic joints are assumed to be perfect. Each joint is characterised as perfect adjustments, no wear or deformations, and no friction. However, the clearances always exist in the kinematic joints and they are known to be sources for impact forces. These forces not only create

S. Erkaya (✉) • İ. Uzmay
Department of Mechatronics Engineering, Engineering Faculty,
Erciyes University, 38039 Kayseri, Turkey
e-mail: serkaya@erciyes.edu.tr; iuzmay@erciyes.edu.tr

D. Zhang, B. Wei (eds.), *Dynamic Balancing of Mechanisms and Synthesizing of Parallel Robots*, DOI 10.1007/978-3-319-17683-3_12

increased vibration and noise, but also reduce system reliability, stability, life and precision. So, joint clearances play a significant role in the prediction of kinematic and dynamic behaviours of mechanisms [2].

Mathematical model of an elastic mechanical joint with clearance was formulated by Dubowsky and Freudenstein, and they derived the dynamic equation of motion [3, 4]. The developed theory was applied to the determination of the impact pair's dynamic response under the various operating conditions. Also, Dubowsky investigated the dynamic effects of joint clearance in planar mechanisms [5]. A slider-crank mechanism with one joint clearance between crank and connecting rod connection was considered, and the impact model in link connection was used to predict the dynamic force and stress amplification due to the presence of clearance. Furuhashi et al. presented a general approach for the kinematics and dynamics of mechanism having revolute joints with clearance, using the continuous contact model assumption [6–9]. The angular directions of joint clearances with respect to the position of input link were analytically performed using Lagrangian function. Different number of joints with clearance was considered in their studies, and the effects of these combinations, all joints with clearance, one, two, or three joints with clearance, etc., on the dynamic characteristics of systems were investigated and compared to each other. A literature review concerned with two-dimensional motion and impact at revolute joints with clearance was presented by Haines [10]. Bengisu et al. described the dynamic behaviour of a four-bar mechanism with a clearance at any one of its connections using a quasi-static model [11]. Theoretical and experimental studies were compared and the proposed model was applied successfully to the similar cases in literature. A procedure for dynamic analysis of a cam mechanism with bearing clearances, which mainly relied on determining clearance angles and their derivatives, was introduced to the literature by Osman et al. [12]. Rhee and Akay investigated dynamic response of a revolute joint with clearance [13]. A four-bar mechanism was implemented as an example and Lagrangian approach was used to model the motion of a rocker-arm pin at the ground connection. The effect of joint clearance in planar problems without treating the effect of lubrication was discussed by Xie et al. [14]. Bifurcation diagrams were used as a tool to investigate the dynamic behaviour of system. Jia et al. presented both theoretical and experimental study about dynamic behaviour of a slider-crank mechanism with clearance [15]. Effects of different clearance size and driving speed on the model mechanism's characteristics were investigated. Dynamic response of mechanisms and machines with revolute joint clearance was presented by Schwab et al. [16]. Mechanical system was modelled rigid or elastic bodies, and a comparison was made between several continuous contact force models and an impact model. Also, a procedure was introduced to estimate the maximum contact force during impact. Feng et al. analysed the joint forces of planar linkage with joint clearance and presented a new optimisation method, which is based on optimising mass distribution of links to decrease the change of joint forces [17]. A four-bar mechanism with joint clearances was employed as an example, and Lagrangian mechanic was used to derive the motion equation of model mechanism. Bauchau and Rodrigez investigated the effects of clearance and lubrication for revolute and spherical joints [18]. Formulation was developed within the framework of energy-preserving and decaying time integration schemes that provide unconditional stability for nonlinear, flexible multibody systems. The efficiency and accuracy of the proposed approach were verified by numerical examples. Flores et al. presented dynamic analysis of planar multibody systems with revolute joint clearances, including dry contact and lubricant effects [2]. Slider-crank mechanism with joint clearance between connecting rod and piston connection was used to discuss the assumption and the adopted procedures. Orden presented a methodology for the study of typical smooth joint clearance in multibody systems [19]. Advantage of the proposed method was proved, and some numerical applications were presented to show the stability of the proposed method especially in long-term simulations with relatively large time step size. The effects of joint clearances on mechanism path generation and transmission quality were investigated by Erkaya and Uzmay [20–22]. Four-bar and slider-crank mechanisms having joints with clearance were considered as model mechanisms, and an optimisation procedure was proposed to decrease the deviations of path generation and transmission angle. Khemili and Romdhane studied dynamic behaviour of a planar flexible slider-crank mechanism with clearance [23]. Simulation and experimental tests were carried out for this goal. For the simulation tests, model mechanism was made under the software ADAMS. A contact model based on the so-called impact function was used. An experimental set-up was designed and built to achieve some experimental validations. The results showed that in the presence of clearance, the mechanism responses were greatly influenced, and the coupler flexibility had a role of suspension for the mechanism. Flores developed a methodology for studying and quantifying the wear phenomenon in revolute clearance joints [24]. A simple model for a revolute joint in the framework of multibody system formulation was presented. Contact forces were based on a continuous contact force model, and friction effects due to the contact in the joints were also represented. A simple planar multibody mechanical system was used to perform numerical simulations.

In this study, dynamic response of mechanism having revolute joints with clearance is investigated. A four-bar (4R) mechanism whose two joints have clearances is considered as a model mechanism. In the first stage, characteristics of joint clearances, such as the magnitude of eccentricity vectors and angular directions of them, are defined using the neural network as a function of input variable, and then kinematic and dynamic analyses of the model mechanism are effected. Optimum adjusting of design variables for reducing the additional vibrations due primarily to the joint clearances on the mechanism frame is also performed using the genetic algorithm.

12.2 Four-Bar Mechanism

A planar four-bar mechanism, as shown in Fig. 12.1, is considered as an example to investigate the effects of clearances at joints between crank and coupler links, and between coupler and follower links on the dynamic response of the mechanism.

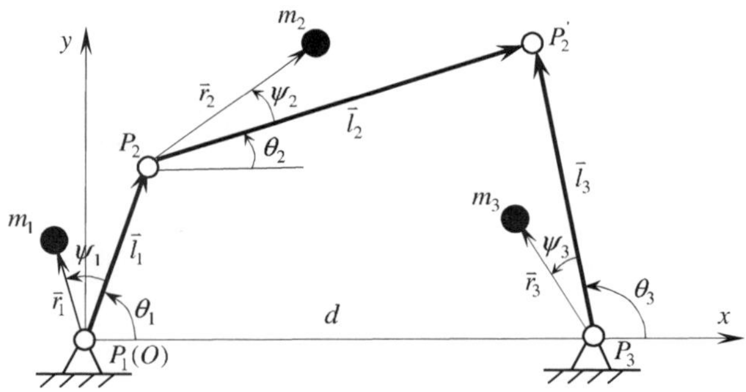

Fig. 12.1 Four-bar mechanism with clearance, schematic representation (**a**), vector representation (**b**)

Table 12.1 Parameters of the four-bar mechanism having joints with clearance

Link	L (mm)	m (kg)	I_G (kg m^2)
Fixed	800	–	–
Crank	250	1,405	1.32×10^{-2}
Coupler	750	2,866	0.157
Follower	450	2,107	5.392×10^{-2}

Link parameters of the four-bar mechanism are given in Table 12.1.

12.2.1 Modelling of Joint Clearance

The existence of clearances at the joints of mechanical systems is inevitable. A definite clearance for the joints between the links of mechanical systems is necessary to allow the relative motion of the connected links, as well as to permit the assemblage of the mechanical systems [2].

In this study, it is assumed that the adjacent mechanism links are connected to each other by revolute joints. Joint clearance, as shown in Fig. 12.2a, can be defined as the difference between journal and bearing radii, and it is modelled as a virtual, massless link.

The magnitude of the clearance vector represents the relative displacement of journal center to the bearing center. If the friction is negligible, the direction of the clearance vector coincides with the normal direction of the collision plane. When the continuous contact mode between journal and bearing at a joint is occurred, the clearance vector is equal to the difference between journal and bearing radii. But, this assumption is not valid during the whole mechanism motion due primarily to the relative penetration or free-flight mode, in which the journal moves freely inside the bearing without contact [25]. As a natural result, these phenomena change the magnitudes of the relevant clearance vectors. In the presence of clearance at a revolute joint, the two kinematic constraints lost, and two degrees of freedom

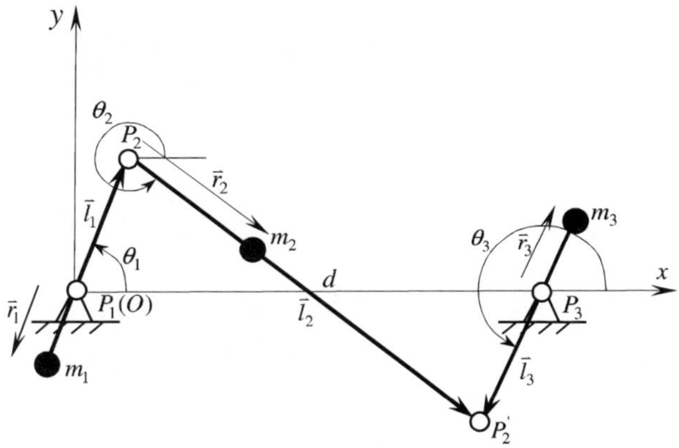

Fig. 12.2 (**a**) Equivalent clearance link, (**b**) direction of joint force

consisting of the horizontal and vertical displacement of the journal center relative to bearing center are added to the mechanism motion. These movements may lead to uncertainties in the motion of mechanism. So, additional constraints are necessary to analyse the kinematics of system.

12.2.2 Kinematic and Dynamic Analyses of the Model Mechanism Having Joints with Clearance

Kinematic analysis of the model mechanism comprises determining of displacements, velocities and accelerations of the mass centers of moving links. Loop-closure equation from the vector representation of the mechanism in Fig. 12.1b is given as

$$L_2\,\mathrm{e}^{i\theta_2} + r_2\,\mathrm{e}^{i\gamma_2} + L_3\,\mathrm{e}^{i\theta_3^c} - L_1\,\mathrm{e}^{i\theta_1} - L_4\,\mathrm{e}^{i\theta_4^c} - r_3\,\mathrm{e}^{i\gamma_3} = 0 \tag{12.1}$$

where L_i denote the lengths of corresponding links, r_j denote the clearance values, γ_j denote the angular directions of joint clearances and the superscript c denotes the "value with clearances". By separating Eq. (12.1) into its real and imaginary parts and using trigonometric relations, angular positions of coupler and follower links are given in the following form:

$$\theta_3^c = 2\tan^{-1}\left[-\frac{B}{2A} \pm \frac{1}{2A}\left(B^2 - 4AC\right)^{1/2}\right] \tag{12.2}$$

$$\theta_4^c = \cos^{-1}\left[\frac{1}{L_4}\left(L_2\cos\theta_2 + r_2\cos\gamma_2 + L_3\cos\theta_3^c - L_1\cos\theta_1 - r_3\cos\gamma_3\right)\right] \tag{12.3}$$

where A, B and C are given as follows:

$$\begin{aligned}
A &= -2r_2L_3\cos\gamma_2 - 2L_2L_3\cos\theta_2 + 2L_3r_3\cos\gamma_3 + 2L_3L_1\cos\theta_1 + 2r_3L_1\cos(\gamma_3-\theta_1)\\
&\quad + 2L_2r_2\cos(\theta_2-\gamma_2) - 2L_2L_1\cos(\theta_2-\theta_1) - L_4^2 + L_2^2 + r_2^2 + L_3^2 + r_3^2 + L_1^2\\
&\quad - 2L_2r_3\cos(\theta_2-\gamma_3) - 2r_2r_3\cos(\gamma_2-\gamma_3) - 2r_2L_1\cos(\gamma_2-\theta_1)\\
B &= 4r_2L_3\sin\gamma_2 + 4L_2L_3\sin\theta_2 - 4L_3r_3\sin\gamma_3 - 4L_3L_1\sin\theta_1\\
C &= 2r_2L_3\cos\gamma_2 + 2L_2L_3\cos\theta_2 - 2L_3r_3\cos\gamma_3 - 2L_3L_1\cos\theta_1 + 2r_3L_1\cos(\gamma_3-\theta_1)\\
&\quad + 2L_2r_2\cos(\theta_2-\gamma_2) - 2L_2L_1\cos(\theta_2-\theta_1) - L_4^2 + L_2^2 + r_2^2 + L_3^2 + r_3^2 + L_1^2\\
&\quad - 2L_2r_3\cos(\theta_2-\gamma_3) - 2r_2r_3\cos(\gamma_2-\gamma_3) - 2r_2L_1\cos(\gamma_2-\theta_1)
\end{aligned} \tag{12.4}$$

The positions of the moving links relative to the crank pivot (A_o) are given in the following form:

$$\begin{aligned}
\begin{bmatrix} x_{G_2}^c \\ y_{G_2}^c \end{bmatrix} &= K_2L_2\begin{bmatrix} \cos(\theta_2+\lambda_2) \\ \sin(\theta_2+\lambda_2) \end{bmatrix}\\
\begin{bmatrix} x_{G_3}^c \\ y_{G_3}^c \end{bmatrix} &= L_2\begin{bmatrix} \cos\theta_2 \\ \sin\theta_2 \end{bmatrix} + r_2\begin{bmatrix} \cos\gamma_2 \\ \sin\gamma_2 \end{bmatrix} + K_3L_3\begin{bmatrix} \cos\left(\theta_3^c+\lambda_3\right) \\ \sin\left(\theta_3^c+\lambda_3\right) \end{bmatrix}\\
\begin{bmatrix} x_{G_4}^c \\ y_{G_4}^c \end{bmatrix} &= L_1\begin{bmatrix} \cos\theta_1 \\ \sin\theta_1 \end{bmatrix} + K_4L_4\begin{bmatrix} \cos\left(\theta_4^c+\lambda_4\right) \\ \sin\left(\theta_4^c+\lambda_4\right) \end{bmatrix}
\end{aligned} \tag{12.5}$$

where K_2, K_3 and K_4 are used to designate the gravity centers. Time derivatives of the positions yield the mass center velocities and accelerations, respectively:

$$\begin{bmatrix} \dot{x}_{G_i}^c \\ \dot{y}_{G_i}^c \end{bmatrix} = \dot{\theta}_2\begin{bmatrix} \dfrac{\partial x_{G_i}^c}{\partial\theta_2} \\[2mm] \dfrac{\partial y_{G_i}^c}{\partial\theta_2} \end{bmatrix} + \sum_{j=2}^{3}\dot{\gamma}_j\begin{bmatrix} \dfrac{\partial x_{G_i}^c}{\partial\gamma_j} \\[2mm] \dfrac{\partial y_{G_i}^c}{\partial\gamma_j} \end{bmatrix} \tag{12.6}$$

$$\begin{bmatrix} \ddot{x}^{c}_{G_i} \\ \ddot{y}^{c}_{G_i} \end{bmatrix} = \ddot{\theta}_2 \begin{bmatrix} \dfrac{\partial x^{c}_{G_i}}{\partial \theta_2} \\ \dfrac{\partial y^{c}_{G_i}}{\partial \theta_2} \end{bmatrix} + \dot{\theta}_2^2 \begin{bmatrix} \dfrac{\partial^2 x^{c}_{G_i}}{\partial \theta_2^2} \\ \dfrac{\partial^2 y^{c}_{G_i}}{\partial \theta_2^2} \end{bmatrix} + 2\dot{\theta}_2 \sum_{j=2}^{3} \dot{\gamma}_j \begin{bmatrix} \dfrac{\partial^2 x^{c}_{G_i}}{\partial \theta_2 \partial \gamma_j} \\ \dfrac{\partial^2 y^{c}_{G_i}}{\partial \theta_2 \partial \gamma_j} \end{bmatrix} + \sum_{j=2}^{3} \ddot{\gamma}_j \begin{bmatrix} \dfrac{\partial x^{c}_{G_i}}{\partial \gamma_j} \\ \dfrac{\partial y^{c}_{G_i}}{\partial \gamma_j} \end{bmatrix}$$

$$+ \sum_{j=2}^{3} \dot{\gamma}_j^2 \begin{bmatrix} \dfrac{\partial^2 x^{c}_{G_i}}{\partial \gamma_j^2} \\ \dfrac{\partial^2 y^{c}_{G_i}}{\partial \gamma_j^2} \end{bmatrix} + \sum_{j=2}^{3} \sum_{k=2}^{3} \dot{\gamma}_j \dot{\gamma}_k \begin{bmatrix} \dfrac{\partial^2 x_{G_i}}{\partial \gamma_j \partial \gamma_k} \\ \dfrac{\partial^2 y_{G_i}}{\partial \gamma_j \partial \gamma_k} \end{bmatrix} \quad (k \neq j) \tag{12.7}$$

where $\dot{\theta}_2$ and $\ddot{\theta}_2$ denote the angular velocity and acceleration of the input link, respectively. i denotes the number of the moving links ($i = 2, 3, 4$), j denotes the number of the joint clearances and $\dot{\bullet}$ and $\ddot{\bullet}$ denote the velocity and acceleration values of corresponding parameters, respectively. Angular velocities and accelerations of coupler and follower links with clearances are given in the following form:

$$\dot{\theta}^{c}_{i} = \dot{\theta}_2 \frac{\partial \theta^{c}_{i}}{\partial \theta_2} + \sum_{j=2}^{3} \dot{\gamma}_j \frac{\partial \theta^{c}_{i}}{\partial \gamma_j} \quad (i = 3, 4) \tag{12.8}$$

$$\ddot{\theta}^{c}_{i} = \ddot{\theta}_2 \frac{\partial \theta^{c}_{i}}{\partial \theta_2} + \dot{\theta}_2^2 \frac{\partial^2 \theta^{c}_{i}}{\partial \theta_2^2} + 2\dot{\theta}_2 \sum_{j=2}^{3} \dot{\gamma}_j \frac{\partial^2 \theta^{c}_{i}}{\partial \theta_2 \partial \gamma_j} + \sum_{j=2}^{3} \ddot{\gamma}_j \frac{\partial \theta^{c}_{i}}{\partial \gamma_j} + \sum_{j=2}^{3} \dot{\gamma}_j^2 \frac{\partial^2 \theta^{c}_{i}}{\partial \gamma_j^2}$$
$$+ \sum_{j=2}^{3} \sum_{k=2}^{3} \dot{\gamma}_j \dot{\gamma}_k \frac{\partial^2 \theta^{c}_{i}}{\partial \gamma_j \partial \gamma_k} \quad (k \neq j) \tag{12.9}$$

Dynamic analysis of the model mechanism provides to define the joint forces and output torque as a function of input link's position. Assuming that the input link rotated at a constant speed as a result of a power input, dynamic force analysis was carried out considering the inertial effects of the links for determining the joint forces and output torque. When journal and bearing are in contact, a joint force occurs, and in the absence of friction, as shown in Fig. 12.2b, its direction coincides with the normal direction of the collision plane. The contact force is supposed to be a spring-damper element. If this element is linear, the approach is known as the Kelvin-Voigt model. When the relation is nonlinear, the model is generally based on the Hertz contact law [2, 26]. In the case of unlubricated joint, Hertzian contact force model is an appropriate choice [16]. This model was considered in this study, and dynamic modelling with contact-impact definition was implemented for the simulation purpose [23, 27, 28]. In the case of the dynamic analysis of mechanism, the continuous contact between journal and bearing was assumed, and the equilibrium conditions of the links, as shown in Fig. 12.3, were investigated considering the input torque and the relevant inertial effects separately.

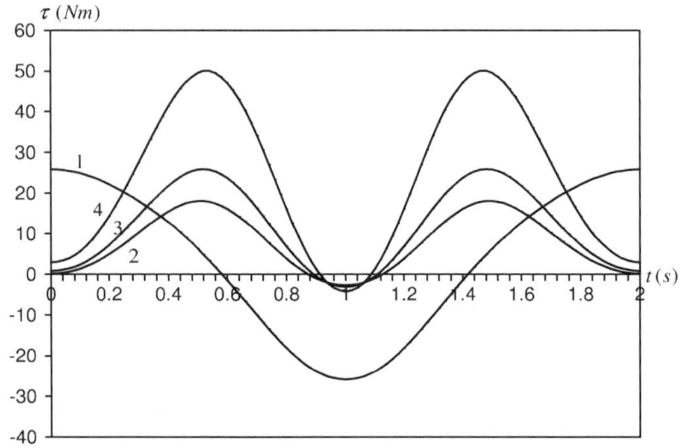

Fig. 12.3 Representation of joint forces

As well known, the inertia force is the vector sum of the joint forces acted on the mechanism links, and the inertia moment is the vector sum of the moments of those forces about any point. As seen from Fig. 12.3, equilibrium conditions of each link are given, respectively:

$$\boldsymbol{F}_{i(i+1)_x} + \boldsymbol{F}_{(i+2)(i+1)_x} + \left(-m_{(i+1)}\ddot{\boldsymbol{x}}_{G_{(i+1)}}\right) = 0 \tag{12.10}$$

$$\boldsymbol{F}_{i(i+1)_y} + \boldsymbol{F}_{(i+2)(i+1)_y} + \left(-m_{(i+1)}\ddot{\boldsymbol{y}}_{G_{(i+1)}} - m_{(i+1)}\boldsymbol{g}\right) = 0 \tag{12.11}$$

$$\sum \boldsymbol{M}_{G_{(i+1)}} - I_{G_{(i+1)}}\ddot{\boldsymbol{\theta}}_{(i+1)} = 0 \tag{12.12}$$

where $\boldsymbol{F}_{i(i+1)_x}$ denote the joint forces, exerted on link $(i+1)$ from link i. All the joint forces and output torque, obtained by superposition of the input torque and inertial effects of each link, are defined in the following form:

$$\sum \boldsymbol{F}_{i(i+1)_x} = \boldsymbol{F}^{\mathrm{I}}_{i(i+1)_x} + \boldsymbol{F}^{\mathrm{II}}_{i(i+1)_x} + \boldsymbol{F}^{\mathrm{III}}_{i(i+1)_x} + \boldsymbol{F}^{\mathrm{IV}}_{i(i+1)_x} \tag{12.13}$$

$$\sum \boldsymbol{F}_{i(i+1)_y} = \boldsymbol{F}^{\mathrm{I}}_{i(i+1)_y} + \boldsymbol{F}^{\mathrm{II}}_{i(i+1)_y} + \boldsymbol{F}^{\mathrm{III}}_{i(i+1)_y} + \boldsymbol{F}^{\mathrm{IV}}_{i(i+1)_y} \tag{12.14}$$

$$\sum \boldsymbol{T}_{\mathrm{out}} = \boldsymbol{T}^{\mathrm{I}}_{\mathrm{out}} + \boldsymbol{T}^{\mathrm{II}}_{\mathrm{out}} + \boldsymbol{T}^{\mathrm{III}}_{\mathrm{out}} + \boldsymbol{T}^{\mathrm{IV}}_{\mathrm{out}} \tag{12.15}$$

where the superscripts I, II, III and IV denote such active effects as input torque and inertial characteristics of second, third and fourth links at the related joint, respectively. The design algorithm used in this study aims at minimising the shaking forces and their moments at the main supports. Therefore, these forces and the relevant moment can be defined as

$$\sum \boldsymbol{F}_{\mathrm{sh}_x} = \sum \boldsymbol{F}_{41_x} + \sum \boldsymbol{F}_{21_x} \tag{12.16}$$

$$\sum \boldsymbol{F}_{\mathrm{sh}_y} = \sum \boldsymbol{F}_{41_y} + \sum \boldsymbol{F}_{21_y} \tag{12.17}$$

$$\boldsymbol{M}_{\mathrm{sh}} = L_1 \sin\theta_1 \boldsymbol{j} \otimes \sum \boldsymbol{F}_{41_x} \boldsymbol{i} + L_1 \cos\theta_1 \boldsymbol{i} \otimes \sum \boldsymbol{F}_{41_y} \boldsymbol{j} \tag{12.18}$$

where $\sum \boldsymbol{F}_{41}$ and $\sum \boldsymbol{F}_{21}$ denote the resultant forces at the joints of follower-frame and crank-frame, respectively. They comprise the superposed effects of input torque and inertial characteristics of second, third and fourth links.

12.3 Modelling and Optimisation Process

In this stage, a multilayered feedforward NN structure, in which the input signals propagate through the network from the input layer onwards to the output layer in a feedforward manner [29], was used for modelling the clearance vector and angular position of its direction with respect to the input position during the mechanism motion. The proposed NN structure consists of one input and output layer with one linear neuron and also eight hidden layers, which consist of 15 nonlinear neurons in each hidden layer. The neural network toolbox of MATLAB was used to develop the proposed neural model [30]. Tangent sigmoid activation function was used in nonlinear neurons as follows:

$$\mathrm{Tansig}(x) = \frac{2}{1 + \mathrm{e}^{-2x}} - 1 \tag{12.19}$$

The values of the training and testing data were normalised between 0 and 1. Levenberg–Marquardt (LM) back-propagation algorithm, which is a supervised learning method, was performed to update the weights of the neural model until a reasonable mean square error (MSE) value was achieved. In order to specify network accuracy in predicting the system outputs, designed network was tested in responses to inputs which were not used in the training step. After modelling of equivalent clearances and directions of joint clearances, genetic algorithm (GA) approach was used to solve the optimisation problem for making the dynamic performance of mechanism better, and it was performed on genetic algorithm toolbox of MATLAB. In the genetic algorithm approach, stochastic uniform was applied as selection function for choosing the next generation, and the crossover probability was adjusted as 0.8. As well known, balancing the shaking forces and their moments is very important for improving the dynamic performance of the mechanisms [31]. So, shaking force and moment fluctuations can be considered as an objective function in genetic algorithm approach. In the case of complete shaking force and moment balancing, it is aimed to close their fluctuations to the zero during the one cycle of mechanism. In this study, complete shaking force and moment balancing are not considered. It is only considered to reduce the additional effect of the joint clearance on the shaking force and moment in mechanism. The developed objective function is given as follows:

$$\begin{aligned} \text{Minimize} F(\boldsymbol{X}) = {} & W_1 \sum_{n=1}^{s} \left[\left(F^{\mathrm{d}}_{\mathrm{sh}_{x_n}} - F^{\mathrm{a}}_{\mathrm{sh}_{x_n}} \right)^2 + \left(F^{\mathrm{d}}_{\mathrm{sh}_{y_n}} - F^{\mathrm{a}}_{\mathrm{sh}_{y_n}} \right)^2 \right] \\ & + W_2 \sum_{n=1}^{s} \left(M^{\mathrm{d}}_{\mathrm{sh}_n} - M^{\mathrm{a}}_{\mathrm{sh}_n} \right)^2 \end{aligned} \tag{12.20}$$

$$\begin{aligned} \text{Subject to} \quad & g_k(\boldsymbol{X}) \le 0 \\ & x_r^{\min} \le x_r \le x_r^{\max} \\ & x_r \in \boldsymbol{X} \end{aligned}$$

where W_1 and W_2 are weighting factors, and s is the number of the considered points during the one cycle of crank. The superscripts d and a denote the desired (without clearance) and actual (with clearance) mechanisms, respectively. g_k are the constraints arising from the condition that satisfies the crank-rocker motion. The error in objective function is minimised provided that the generated solution satisfies a set of constraints. These constraints are necessary to have a functional mechanism, although they increase solution complexity. $\boldsymbol{X}$ is a vector comprising the independent design variables (x_r). $x^{\min}$ and $x^{\max}$ determine the range of each design variable. These variables consist of link lengths (L_i), structural angles (λ_i) and structural parameters (K_i). The values of the weighting factors in Eq. (12.20) are generally chosen according to the relative importance of the relevant terms [21, 32]. Sometimes the relative importance of the terms is not obvious, and to make the decision on choosing the weighting factor values is difficult. Both training of network and solving of optimisation problem using genetic algorithm were performed on a PIV processor with a CPU speed of 3.2 MHz and 1,024 Mb Ram. The neural-genetic approach for determining of unmodelled joint parameters and solving the optimisation problem is given in Fig. 12.4.

As shown, the first step of this scheme consists of neural network approach for modelling the clearance values and their directions as a function of position variable of input link, and the second step is constituted by using the genetic algorithm approach. In that step, an error-based objective function is used for making the dynamic performance of the original mechanism better by adjusting the appropriate values of design variables.

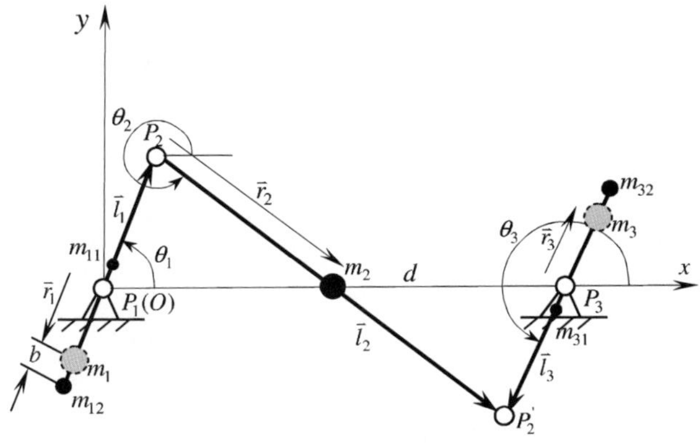

Fig. 12.4 Schematic representation of neural-genetic approach

12.4 Results

In this study, dynamic analysis of a four-bar mechanism having two revolute joints with clearance was implemented. The joint clearance, which is the difference between the journal and bearing radii at each joint, was taken to be 0.5 mm [16]. All the links in mechanism were assumed to be rigid, the running speed of the input link was considered as 600 rpm and the time step for the numerical solution was arranged as 0.00005 s. In the original mechanism, it was assumed that the mass centers of links were in the midpoint of corresponding links, and gravity effect results in a contact between journal and bearing at the joint connections in the beginning of simulation. A feedforward neural network was used to model the clearance vectors and its angular characteristics as a function of input variable. Training and testing data for neural network were obtained during the stationary phase of the simulation process [16]. Each parameter was modelled separately by using the same NN structure, and the successful training of the network weights was achieved at the different iteration numbers. Also, a genetic algorithm was used to adjust the design variables for reducing the additional vibration effects due primarily to the joint clearance. The variations of the clearance vectors with respect to the input variable are given in Fig. 12.5.

As seen from NN results in this figure, there is a good approximation; that is, the deviations between NN and simulation results for each clearance vector are minimum. The trajectories of the journal centers relative to the bearing centers for each joint with clearance are outlined in Fig. 12.6.

In this figure, solid line designates the joint clearance, which is the difference between journal and bearing radii, and the dot line denotes the journal center trajectory inside the bearing for each joint with clearance. As seen from this figure, the different types of motion between journal and bearing can be clearly observed; that is, the continuous contact or free-flight modes between journal and bearing occur. Since the contact mode is naturally affected by the previous joint clearance, the free-flight mode at joint 3 occurs at a much more rate than that of joint 2. The kinematic contribution of the previous joint clearance is superposed to the

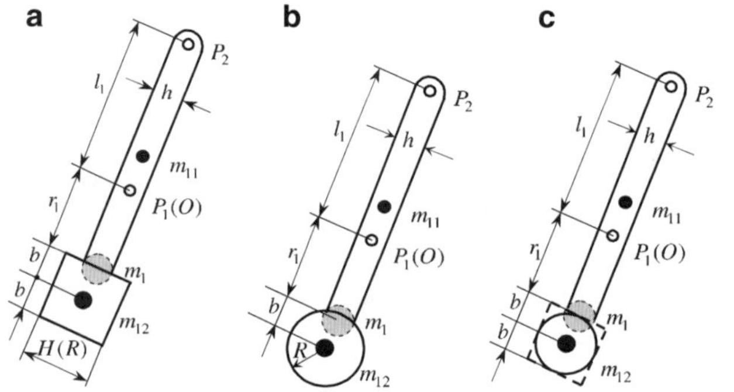

Fig. 12.5 Clearance vector variations

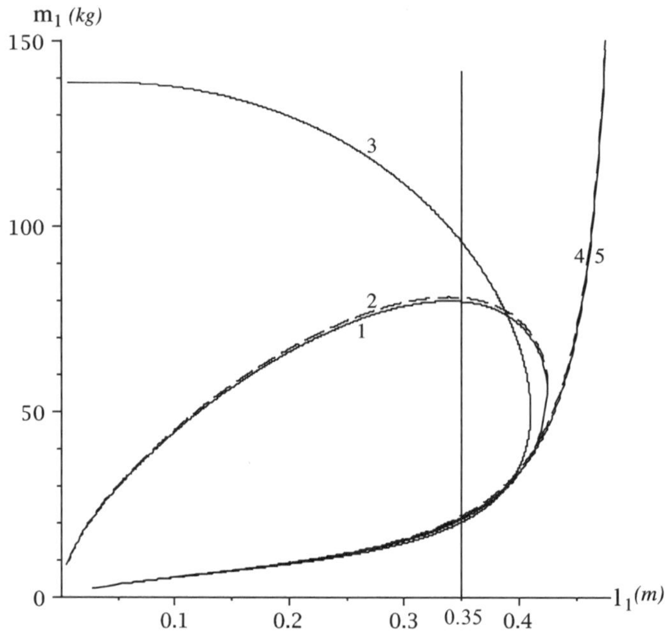

Fig. 12.6 Trajectories of the journal centers inside the bearings

kinematics of the successive joint. The obtained results are in line with the proposed results in the literature [2]. Training histories of NN for the clearance vectors and time derivatives are given in Fig. 12.7.

During the NN training, the reasonable stopping criterion was considered as 1×10^{-9} mean square of the error for modelling each joint parameter. This error is acceptable for modelling the equivalent clearances and the directions of the joint clearances as a function of input variable. For the proposed NN model, this was

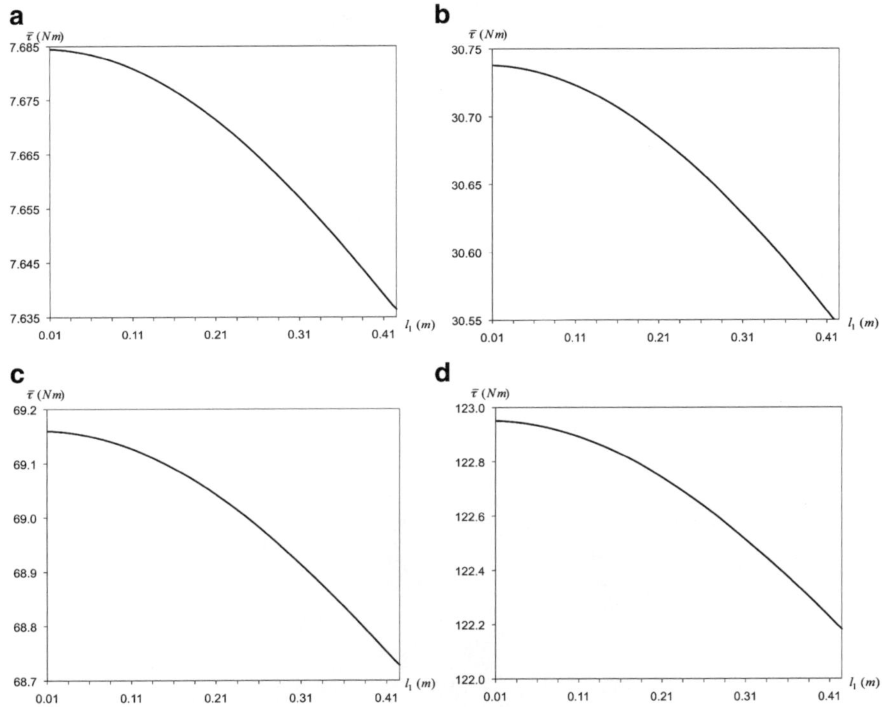

Fig. 12.7 Training histories of clearance vector

achieved at 12679, 15138, 14316, 19436, 13024, 24873, 13628 and 28486 iterations for r_2, r_3, γ_2, γ_3, $\dot{\gamma}_2$, $\dot{\gamma}_3$, $\ddot{\gamma}_2$ and $\ddot{\gamma}_3$, respectively. Also, it is concluded that these values are successful reflections of the proposed NN structure for determining of unmodelled parameters at joints with clearance. Shaking forces in *x*- and *y*-directions, transmitted to the frame, are given in Fig. 12.8.

The extreme values for the force variations are originated from higher values of the contact force acting only during a small time interval of contact. The variation of shaking moment is also given in Fig. 12.9.

The values of design parameters, obtained as a result of optimisation, are outlined in Table 12.2.

Also, the original and final values of objective function are given as 1.6×10^9 and 2.7×10^8, respectively. The reason for these high values in objective function is mainly originated from the peak values at the related force and moment statements during a small time interval. By determining the optimum design parameters using genetic algorithm, the shaking forces and their moment for actual and optimised mechanisms are given in Figs. 12.10 and 12.11, respectively.

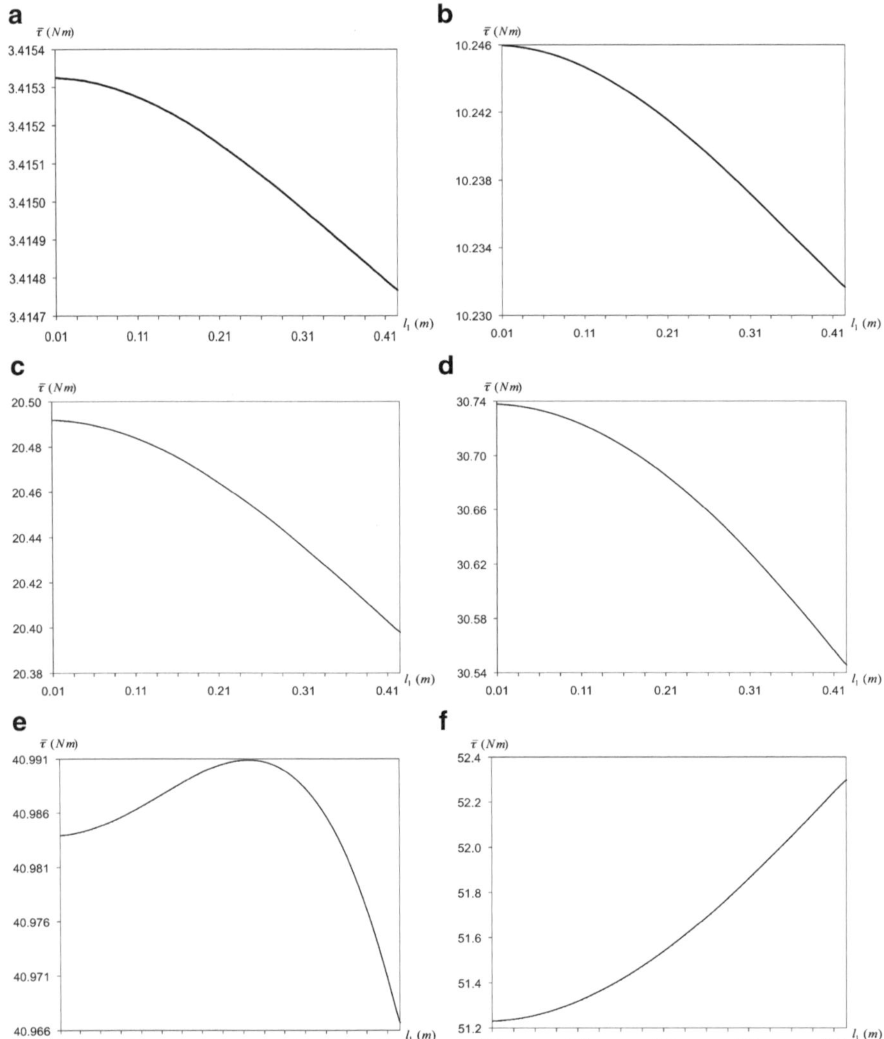

Fig. 12.8 Shaking forces

As shown in relevant figures, the extreme values of the force and moment decrease and the fluctuations of these forces and moments for the optimised mechanism are closer to zero rather than those of the actual mechanism. Appropriate design parameters, determined by using the proposed optimisation algorithm, cause the shaking forces in *x*- and *y*-directions to decrease by 69.24 % and 75.85 %, respectively. Also, the shaking moment decreases in proportion as 77.51 %. The decreases at forces being obtained are bigger than those of Feng et al.'s study [17].

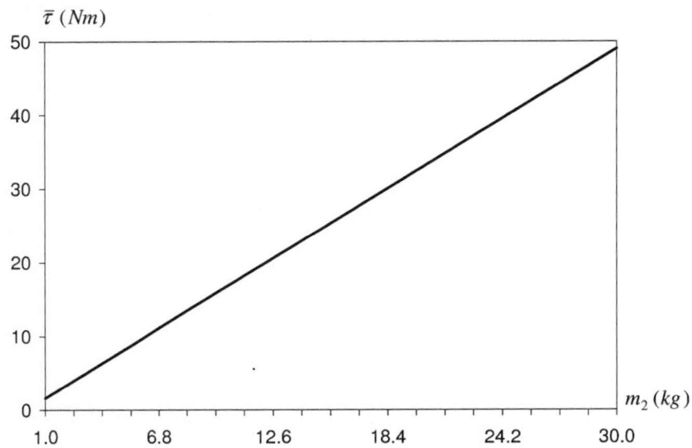

Fig. 12.9 Shaking moment

12.5 Conclusion

In this study, in order to determine the effects of joint clearance on dynamic response of mechanism, the joints with clearance at the contact points as known crank-coupler connection and coupler-follower connection were considered. The variations of the clearance vectors were studied for the stationary running phase of the mechanism. The design parameters for the joints with clearances were modelled by using a multilayered feedforward NN, and these model characteristics were used for analysing and optimising of the mechanism.

Although the kinematic and dynamic responses of the mechanism having no clearance show a periodic character, in the presence of joint clearance, these responses approximately happen as a non-periodic behaviour. Higher accelerations in the mechanism with clearances cause the inertial forces of the links to increase. The higher values of the contact force acting only during a small time interval constitute extreme values for the force and its moment variations. These effects not only cause the vibration and noise, but also reduce the system reliability, stability, life and precision. By using the proposed optimisation strategy, the undesired effects of the joint clearances such as vibration are reduced. As a practical implication, the proposed optimised values of the design variables lead to a reduction of vibration causing the dynamic performance to be worse. The proposed modelling and optimisation strategy can be adapted to the similar systems to make the mechanical working condition better.

Table 12.2 Original and optimised design variables

	W_1	W_2	L_1 (mm)	L_2 (mm)	L_3 (mm)	L_4 (mm)	K_2	K_3	K_4	λ_2 (rad)	λ_3 (rad)	λ_4 (rad)
Original	–	–	800	250	750	450	0.5	0.5	0.5	0	0	0
Optimised	0.9	0.1	799.98	269.95	740	464.69	0.472	0.176	0.166	−3.014	−0.292	2.242

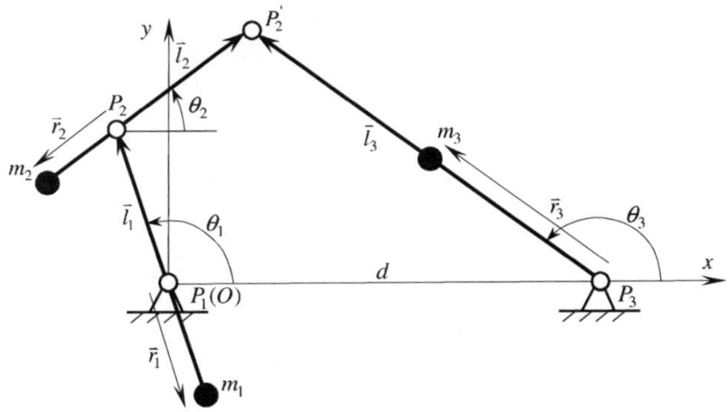

Fig. 12.10 Shaking forces in actual and optimised mechanisms

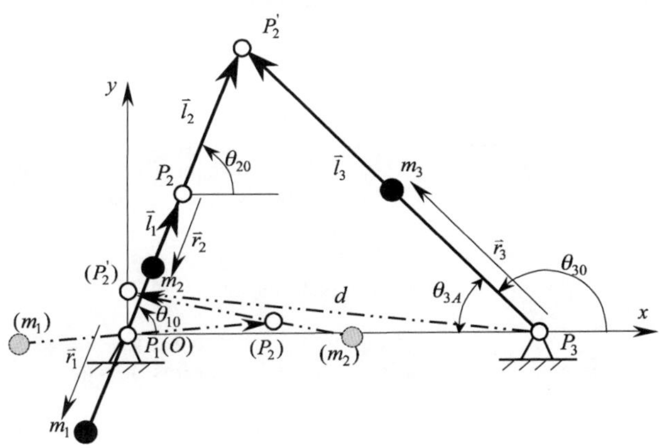

Fig. 12.11 Shaking moments in actual and optimised mechanisms

References

1. Stoenescu, E.D., Marghitu, B.D.: Dynamic analysis of a planar rigid-link mechanism with rotating slider joint and clearance. J. Sound Vib. **266**, 394–404 (2003)
2. Flores, P., Ambrosioi, J., Claro, J.P.: Dynamic analysis for planar multibody mechanical systems with lubricated joints. Multibody Syst. Dyn. **12**, 47–74 (2004)
3. Dubowsky, S., Freudenstein, F.: Dynamic analysis of mechanical systems with clearances, Part 1: Formation of dynamic model. J. Eng. Ind. Trans. ASME **93**(1), 305–309 (1971)
4. Dubowsky, S., Freudenstein, F.: Dynamic analysis of mechanical systems with clearances, Part 2: Dynamic response. J. Eng. Ind. Trans. ASME **93**(1), 310–316 (1971)
5. Dubowsky, S.: On predicting the dynamic effects of clearances in planar mechanisms. J. Eng. Ind. Trans. ASME **96**(1), 317–323 (1974)
6. Furuhashi, T., Morita, N., Matsuura, M.: Research on dynamics of four-bar linkage with clearances at turning pairs (1st Report, General theory of continuous contact model). Bull. JSME **21**, 518–523 (1978)

7. Morita, N., Furuhashi, T., Matsuura, M.: Research on dynamics of four-bar linkage with clearances at turning pairs (2nd Report, Analysis of crank-level mechanism with clearance at joint of crank and coupler using continuous contact model). Bull. JSME **21**, 1284–1291 (1978)
8. Morita, N., Furuhashi, T., Matsuura, M.: Research on dynamics of four-bar linkage with clearances at turning pairs (3rd Report, Analysis of crank-level mechanism with clearance at joint of coupler and lever using continuous contact model). Bull. JSME **21**, 1292–1298 (1978)
9. Furuhashi, T., Morita, N., Matsuura, M.: Research on dynamics of four-bar linkage with clearances at turning pairs (4th Report, Force acting at joints of crank-level mechanism). Bull. JSME **21**, 1299–1305 (1978)
10. Haines, R.S.: Survey: 2-dimensional motion and impact at revolute joints. Mech. Mach. Theory **15**, 361–370 (1980)
11. Bengisu, M.T., Hidayetoglu, T., Akay, A.: A theoretical and experimental investigation of contact loss in the clearances of a four-bar mechanism. J. Mech. Trans. Auto. Des. Trans. ASME **108**, 237–244 (1986)
12. Osman, M.O.M., Bahgat, B.M., Osman, M.: Dynamic analysis of a cam mechanism with bearing clearances. Mech. Mach. Theory **22**(4), 303–314 (1987)
13. Rhee, J., Akay, A.: Dynamic response of a revolute joint with clearance. Mech. Mach. Theory **31**(1), 121–134 (1996)
14. Xie, H., Flowers, G., Feng, L., Lawrence, C.: Steady-state dynamic behaviour of a flexible rotor with auxiliary support from a clearance bearing. J. Vib. Acoust. **121**, 78–83 (1999)
15. Jia, X., Jin, D., Ji, L., Zhang, J.: Investigation on the dynamic performance of the tripod-ball sliding joint with clearance in a crank-slider mechanism. Part 1. Theoretical and experimental results. J. Sound Vib. **252**(5), 919–933 (2002)
16. Schwab, A.L., Meijaard, J.P., Meijers, P.: A comparison of revolute joint clearance models in the dynamic analysis of rigid and elastic mechanical systems. Mech. Mach. Theory **37**, 895–913 (2002)
17. Feng, B., Morita, N., Torii, T.: A new optimization method for dynamic design of planar linkage with clearances at joints-optimizing the mass distribution of links to reduce the change of joint forces. J. Mech. Des. Trans. ASME **124**, 68–73 (2002)
18. Bauchau, O.A., Rodrigez, J.: Modelling of joints with clearance in flexible multibody systems. Int. J. Solids Struct. **39**, 41–63 (2002)
19. Orden, J.C.G.: Analysis of joint clearances in multibody systems. Multibody Syst. Dyn. **13**, 401–420 (2005)
20. Erkaya, S., Uzmay, İ.: Determining link parameters using genetic algorithm in mechanisms with joint clearance. Mech. Mach. Theory **44**(1), 222–234 (2009)
21. Erkaya, S., Uzmay, İ.: A neural-genetic (NN-GA) approach for optimising mechanisms having joints with clearance. Multibody Syst. Dyn. **20**, 69–83 (2008)
22. Erkaya, S., Uzmay, İ.: Optimization of transmission angle for slider-crank mechanism with joint clearances. Struct. Multidiscip. Opt. **37**, 493–508 (2009)
23. Khemili, I., Romdhane, L.: Dynamic analysis of a flexible slider–crank mechanism with clearance. Eur. J. Mech. A Solid **27**(5), 882–898 (2008)
24. Flores, P.: Modeling and simulation of wear in revolute clearance joints in multibody systems. Mech. Mach. Theory **44**, 1211–1222 (2009)
25. Flores, P., Ambrosio, J., Claro, J.C.P., Lankarani, H.M., Koshy, C.S.: A study on dynamics of mechanical systems including joints with clearance and lubrication. Mech. Mach. Theory **41**, 247–261 (2006)
26. Ravn, P.: A continuous analysis method for planar multibody systems with joint clearance. Multibody Syst. Dyn. **2**, 1–24 (1998)
27. Chen, M., Chen, L., Zhang, X., Bai, H., Xiao, Y.: Research and dynamic simulation of docking locks with contact-impact. Aerospace Sci. Technol. **7**, 364–372 (2003)

28. MSC.ADAMS 2005 r2: Automatic dynamic analysis of mechanical systems. MSC Software Corporation
29. Haykin, S.: Neural Networks: A Comprehensive Foundation. Macmillan College Publishing Company, New York (1994)
30. MATLAB (ver 7.0): The MathWorks Inc, 3 Apple Hill Drive, Natick, MA 01760-2098
31. Li, Z.: Sensitivity and robustness of mechanism balancing. Mech. Mach. Theory **33**(7), 1045–1054 (1998)
32. Chiou, S.T., Bai, G.J., Chang, W.K.: Optimum balancing designs of the drag-link drive of mechanical presses for precision cutting. Int. J. Mach. Tools Manuf. **38**(3), 131–141 (1998)

Chapter 13
Minimization of Shaking Force and Moment on a Four-Bar Mechanism Using Genetic Algorithm

Selçuk Erkaya

Abstract In this study, optimal balancing of a 2D articulated mechanism is investigated to minimize the shaking force and moment fluctuations. Balancing of a four-bar mechanism is formulated as an optimization problem. On the other hand, an objective function based on the subcomponents of shaking force and moment is constituted, and design variables consisting of kinematic and dynamic parameters are defined. Genetic algorithm is used to solve the optimization problem under the appropriate constraints. By using commercial simulation software, optimized values of design variables are also tested to evaluate the effectiveness of the proposed optimization process. This work provides a practical method for reducing the shaking force and moment fluctuations. The results show that both the structure of objective function and particularly the selection of weighting factors have a crucial role to obtain the optimum values of design parameters. By adjusting the value of weighting factor according to the relative sensitivity of the related term, there is a certain decrease at the shaking force and moment fluctuations. Moreover, these arrangements also decrease the initiative of mechanism designer on choosing the values of weighting factors.

Keywords Shaking force and moment • Optimal balancing • Four-bar mechanism • Genetic algorithm

13.1 Introduction

Since the dynamic performance characteristics such as shaking force, shaking moment, and input-torque, depend on the mass and inertia of each moving link, and its mass center location, it is required to optimally distribute the link masses

S. Erkaya (✉)
Department of Mechatronics Engineering, Engineering Faculty,
Erciyes University, 38039 Kayseri, Turkey
e-mail: serkaya@erciyes.edu.tr

D. Zhang, B. Wei (eds.), *Dynamic Balancing of Mechanisms and Synthesizing of Parallel Robots*, DOI 10.1007/978-3-319-17683-3_13

for dynamic balancing. Minimization both shaking force and shaking moment fluctuations is important for improving the mechanism's fatigue life by reducing vibration, noise, and wear.

Many machine designers have paid an attention to solve the balancing problems by using either classical methods or optimal approaches. Assuming that both linear and rotary inertia, Feng [1] presented a method for the complete shaking force and moment balancing of eight-bar linkages having only revolute joints. In study of Ye and Smith [2], a logical extension to the concept of mass flow was developed in which the effects of inertia moment as well as inertia force of a link were modeled by equivalent simple links. Li [3] presented sensitivity formulation of the shaking force and moment for planar articulating mechanisms. The sensitivity analysis and a robust balancing method, which was sensitive to the processing errors in manufacture, were presented. Objective function was composed of shaking force and shaking moment, and the values of weighting factor were selected as equal to each other. For reducing the shaking force and moment of mechanical presses, Chiou et al. [4] proposed optimum designs by adding disk counterweights. Two-phase optimization technique was presented for the multi-objective optimization. Arakelian and Smith [5] proposed a new solution, considering a pantograph with the crank and coupler, to the problem of complete shaking force and shaking moment balancing of linkages. By using counterweights, complete force balancing of planar linkages was presented by Tepper and Lowen [6]. Esat and Bahai [7] also showed if a linkage can be fully force balanced using the criterion of Tepper and Lowen, then it can be fully force and moment balanced using geared counter-inertias. Feng et al. [8] analyzed the joint forces of planar linkage with joint clearance and presented a new optimization method, which was based on optimizing mass distribution of links to decrease the change of joint forces. A critical review of complete shaking moment balancing was implemented in the study of Kochev [9].

Guo et al. [10] proposed a new mixed mass redistribution method to investigate the optimum dynamic design. By using genetic algorithms, optimum dynamic characteristics were obtained more efficiently than the traditional nonlinear optimization techniques. Arakelian et al. [11, 12] presented a solution of the shaking force and shaking moment balancing of planar and spatial linkages. Arakelian [13] also formulated the conditions of shaking moment balancing by using the copying properties of the pantograph linkage and the method of dynamic substitution of distributed masses by concentrated point masses. Alici and Shirinzadeh [14] presented optimum dynamic balancing of planar 2-DOF parallel manipulators. By using an objective function based on the sensitivity analysis of shaking moment with respect to the position, velocity and acceleration of the links, the dynamic balancing was formulated as an optimization problem. Chaudhary and Saha [15] proposed a method based on the maximum recursiveness of the dynamic equations to evaluate the bearing forces. Balancing problem of four-bar linkages was considered as an optimization problem, and mass distribution of linkage was embedded in the constraints to obtain the new linkage. Also, they [16] presented a general mathematical formulation of optimization problem for balancing of planar mechanisms to improve the dynamic performances. Erkaya and Uzmay [17] investigated dynamic behavior of a four-bar mechanism with joint clearances. They used an objective function based on shaking force and shaking moment. Also, they proposed a Neural-Genetic (NN-GA) approach to minimize the additional effects of joint clearances on shaking force and moment under related constraints. By using a novel and simplified approach, Ilia and Sinatra [18] studied the derivation of design equations and techniques for the dynamic balancing of a five-bar linkage. Balancing of the mechanism was formulated and solved as an optimization problem under equality constraints. Park et al. [19] studied for minimizing the moments excited in a four-stroke seven-cylinder vehicle engine and reducing the forces transmitted to the engine mounts. A computer program was developed to predict the excitation forces and moments.

Former balancing studies, which particularly consider this problem as an optimization problem, have usually chosen the values of weighting factors equal to each other. This arrangement obviously affects the results of optimization. Furthermore, the initiative of the mechanism designer has a crucial role on choosing the values of weighting factors and defining the structure of objective function. The focus of this study is to present a simple approach to constitute the structure of objective function for decreasing the shaking force and shaking moment fluctuations. On the other hand, a simple method is also proposed to reduce the initiative of the mechanism designer on choosing the values of weighting factors. An objective function based on the subcomponents of shaking force and shaking moment is constituted. Genetic algorithm is used for solving the optimization problem. Three case studies are implemented to show the effectiveness of the proposed approach. This chapter is organized as follows: Sect. 13.2.1 outlines the kinematics and dynamics of model mechanism. Optimization process is given in Sect. 13.2.2. Results and conclusions are summarized in Sects. 13.3 and 13.4, respectively.

13.2 Materials and Methods

13.2.1 Model Mechanism

A four-bar mechanism, which is frequently used in the former balancing problems, is considered as an example to investigate the effects of shaking force and shaking moment exerted in the frame (Fig. 13.1).

Kinematic analysis of the model mechanism comprises determining of displacements, velocities and accelerations of moving links. Mass center positions of the moving links relative to the crank pivot (A_o) are given in the following form,

$$\begin{bmatrix} x_{G2} \\ y_{G2} \end{bmatrix} = \overline{A_oG_2} \begin{bmatrix} \cos\ (\theta_2 + \lambda_2) \\ \sin\ (\theta_2 + \lambda_2) \end{bmatrix} \tag{13.1}$$

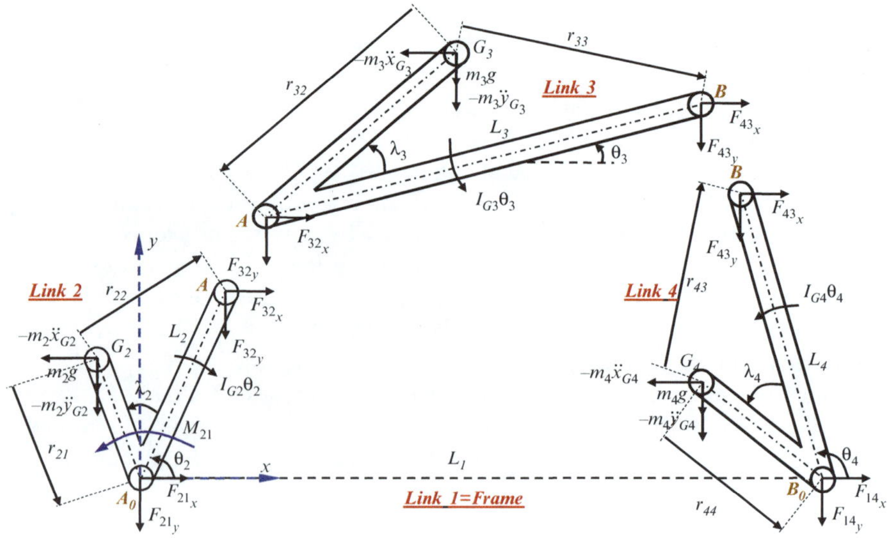

Fig. 13.1 Force representation of four-bar mechanism

$$\begin{bmatrix} x_{G3} \\ y_{G3} \end{bmatrix} = L_2 \begin{bmatrix} \cos\theta_2 \\ \sin\theta_2 \end{bmatrix} + \overline{AG_3} \begin{bmatrix} \cos\,(\theta_3+\lambda_3) \\ \sin\,(\theta_3+\lambda_3) \end{bmatrix} \tag{13.2}$$

$$\begin{bmatrix} x_{G4} \\ y_{G4} \end{bmatrix} = L_1 \begin{bmatrix} \cos\theta_1 \\ \sin\theta_1 \end{bmatrix} + \overline{B_oG_4} \begin{bmatrix} \cos\,(\theta_4+\lambda_4) \\ \sin\,(\theta_4+\lambda_4) \end{bmatrix} \tag{13.3}$$

where L_i denote the lengths of corresponding links. x_{Gi} and y_{Gi} are the displacements at the x and y directions for mass center of ith moving link, respectively. θ_3 and θ_4 define the angular positions of coupler and follower links relative to x direction, respectively.

$$\theta_3 = 2\tan^{-1}\left[-\frac{B}{2A} \pm \frac{1}{2A}\left(B^2-4AC\right)^{1/2}\right] \tag{13.4}$$

$$\theta_4 = \cos^{-1}\left[\frac{1}{L_4}\left(L_2\cos\theta_2 + L_3\cos\theta_3 - L_1\cos\theta_1\right)\right] \tag{13.5}$$

where A, B and C are given as:

$$A = 2L_3L_1\cos\theta_1 - 2L_2L_3\cos\theta_2 + L_1^2 + L_2^2 + L_3^2 - L_4^2 - 2L_2L_1\cos\,(\theta_2-\theta_1)$$

$$B = 4L_3\,(L_2\sin\theta_2 - L_1\sin\theta_1)$$

$$C = 2L_2L_3\cos\theta_2 - 2L_3L_1\cos\theta_1 + L_1^2 + L_2^2 + L_3^2 - L_4^2 - 2L_2L_1\cos\,(\theta_2-\theta_1)$$

Mass center velocities and accelerations can also be defined as the time-derivatives of Eqs. (13.1)–(13.3). Dynamic analysis of the model mechanism provides to define the joint forces and torque as a function of input link's position. Dynamic force analysis was carried out considering the inertial effects of the links for determining the joint forces and torque. Force analysis for the model mechanism is given in Eq. (13.6).

$$\begin{bmatrix} -1 & 0 & 1 & 0 & 0 & 0 & 0 & 0 & 0 \\ 0 & -1 & 0 & 1 & 0 & 0 & 0 & 0 & 0 \\ r_{21_y} & -r_{21_x} & -r_{22_y} & r_{22_x} & 0 & 0 & 0 & 0 & 1 \\ 0 & 0 & -1 & 0 & 1 & 0 & 0 & 0 & 0 \\ 0 & 0 & 0 & -1 & 0 & 1 & 0 & 0 & 0 \\ 0 & 0 & r_{32_y} & -r_{32_x} & -r_{33_y} & r_{33_x} & 0 & 0 & 0 \\ 0 & 0 & 0 & 0 & -1 & 0 & 1 & 0 & 0 \\ 0 & 0 & 0 & 0 & 0 & -1 & 0 & 1 & 0 \\ 0 & 0 & 0 & 0 & r_{43_y} & -r_{43_x} & -r_{44_y} & r_{44_x} & 0 \end{bmatrix} \begin{bmatrix} F_{21_x} \\ F_{21_y} \\ F_{32_x} \\ F_{32_y} \\ F_{43_x} \\ F_{43_y} \\ F_{14_x} \\ F_{14_y} \\ M_{21} \end{bmatrix} = \begin{bmatrix} m_2\ddot{x}_{G2} \\ m_2\ddot{y}_{G2} + m_2 g \\ I_{G2}\ddot{\theta}_2 \\ m_3\ddot{x}_{G3} \\ m_3\ddot{y}_{G3} + m_3 g \\ I_{G3}\ddot{\theta}_3 \\ m_4\ddot{x}_{G4} \\ m_4\ddot{y}_{G4} + m_4 g \\ I_{G4}\ddot{\theta}_4 \end{bmatrix} \tag{13.6}$$

Position vectors from the gravity center of link i to joint j are read from the Fig. 13.1 as:

$$\begin{bmatrix} r_{21_x} & r_{21_y} \\ r_{22_x} & r_{22_y} \\ r_{32_x} & r_{32_y} \\ r_{33_x} & r_{33_y} \\ r_{43_x} & r_{43_y} \\ r_{44_x} & r_{44_y} \end{bmatrix} = \begin{bmatrix} \overline{A_oG_2} & 0 & 0 & 0 & 0 & 0 \\ 0 & \overline{AG_2} & 0 & 0 & 0 & 0 \\ 0 & 0 & \overline{AG_3} & 0 & 0 & 0 \\ 0 & 0 & 0 & \overline{BG_3} & 0 & 0 \\ 0 & 0 & 0 & 0 & \overline{BG_4} & 0 \\ 0 & 0 & 0 & 0 & 0 & \overline{B_oG_4} \end{bmatrix} \begin{bmatrix} -\cos(\theta_2+\lambda_2) & -\sin(\theta_2+\lambda_2) \\ \cos(\theta_2-\beta_2) & \sin(\theta_2-\beta_2) \\ -\cos(\theta_3+\lambda_3) & -\sin(\theta_3+\lambda_3) \\ \cos(\theta_3-\beta_3) & \sin(\theta_3-\beta_3) \\ \cos(\theta_4-\beta_4) & \sin(\theta_4-\beta_4) \\ -\cos(\theta_4+\lambda_4) & -\sin(\theta_4+\lambda_4) \end{bmatrix} \tag{13.7}$$

where β_2, β_3, and β_4 define the acute angles for G_2AA_o, G_3BA, and G_4BB_o, respectively.

$$\beta_2 = a\cos\left(\frac{\overline{AG_2}^2 + L_2^2 - \overline{A_oG_2}^2}{2\overline{AG_2}L_2}\right)$$

$$\beta_3 = a\cos\left(\frac{\overline{BG_3}^2 + L_3^2 - \overline{AG_3}^2}{2\overline{BG_3}L_3}\right)$$

$$\beta_4 = a\cos\left(\frac{\overline{BG_4}^2 + L_4^2 - \overline{B_oG_4}^2}{2\overline{BG_4}L_4}\right)$$

According to theorem, the shaking force is considered as the reaction of the vector sum of all the inertia forces of moving links associated with the mechanism, and the shaking moment is also the reaction of the resultant of the inertia moment and the moment of the inertia forces. The design algorithm used in this study aims at minimizing the shaking force and shaking moment. Therefore, this force components and the relevant moment relative to the crank pivot can be defined as;

$$\sum \mathbf{F}_{\mathbf{sh}_x} = \mathbf{F}_{\mathbf{41}_x} + \mathbf{F}_{\mathbf{21}_x} \tag{13.8}$$

$$\sum \mathbf{F}_{\mathbf{sh}_y} = \mathbf{F}_{\mathbf{41}_y} + \mathbf{F}_{\mathbf{21}_y} \tag{13.9}$$

$$\mathbf{M}_{\mathbf{sh}} = L_1 \sin\theta_1 \mathbf{j} \otimes \mathbf{F}_{\mathbf{41}_x}\mathbf{i} + L_1 \cos\theta_1 \mathbf{i} \otimes \mathbf{F}_{\mathbf{41}_y}\mathbf{j} \tag{13.10}$$

where $\mathbf{F}_{41}$ and $\mathbf{F}_{21}$ denote the forces at the joints of follower–frame and crank–frame, respectively.

13.2.2 Optimization Process

In the optimization process, Genetic Algorithm (GA) approach was used to solve the optimization problem and it was performed on optimization toolbox of MATLAB [20]. Genetic algorithm, or any evolutionary method, differs from classical optimization methods in that there is a non-zero probability of attaining the global optimum [21]. Many Gradient-based methods, which are very efficient local optimization methods for parameter optimization, are available. However, these conventional algorithms need the gradient information of the objective function with respect to the design variables and cannot get out of local optimum points when they fall into a false peak (local optimum point). Also, they may miss a global optimum solution because they are dependent on the starting point of searching and converge on the optimum solution that is nearest to the starting point, and cannot find all the global optimum solutions [22]. Genetic algorithms, on the other hand, are simple to implement and involve evaluations of only the objective function and the use of certain genetic operators such as selection, crossover, mutation, and reproduction to explore the design space [23]. Moreover, a population of optimum points is obtained that will allow the designer to select a design that satisfies all subjective constraints as well. These characteristics make this approach well suited for finding the optimal solutions. GA operations in a typical optimization procedure are outlined in Fig. 13.2. In this study, stochastic uniform was applied as selection function for choosing the next generation, and the crossover probability was adjusted as 0.8. The solving of optimization problem using genetic algorithm was performed on a PIV processor with a CPU speed of 3.2 MHz and 1,024 Mb Ram.

In order to balance a mechanism completely, it is necessary to eliminate both the shaking force and the shaking moment. However, complete balancing of any one may result in an increased unbalance in the other one. The shaking force can be eliminated completely by attaching counterweights to the moving links of mechanism.

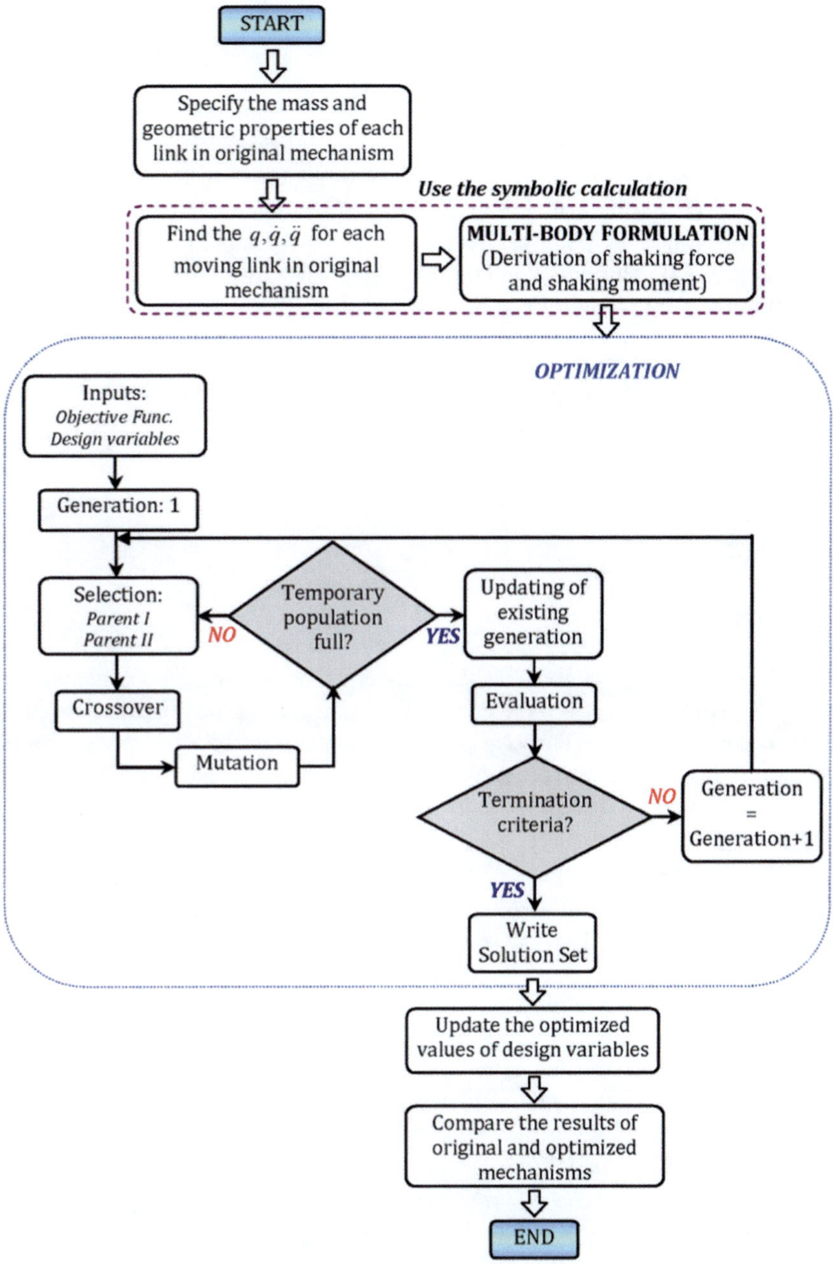

Fig. 13.2 Flowchart of design study and genetic algorithm

But, this increases overall mass and inertia of the mechanism. Also, this leads to increasing in shaking moment, required driving torque, and reactions at the joints. An alternative way to reduce the shaking force and shaking moment together with other dynamic quantities such as driving torque and bearing reactions is to optimize all the dynamic quantities. Hence, for improving the overall performance of mechanism, the balancing problem should be treated as an optimization problem [24].

In this study, objective function in the optimization process was constituted as given in Eq. (13.11) by considering the shaking force and shaking moment [3, 15–17]. This function comprises each subcomponent of shaking force and shaking moment;

$$\begin{aligned} \text{Minimize} \quad & F(\mathbf{X}) = \sum_{n=1}^{s} \left[W_1 \left(F_{21x_n}\right) + W_2 \left(F_{21y_n}\right) + W_3 \left(F_{41x_n}\right) \right. \\ & \left. + W_4 \left(F_{41y_n}\right) + W_5 \left(M_{\mathrm{sh}_n}\right) \right] \\ \text{Subject to} \quad & g_k(\mathbf{X}) \leq 0 \\ & x_r^{\min} \leq x_r \leq x_r^{\max} \\ & x_r \in \mathbf{X} \end{aligned} \tag{13.11}$$

where W_h are weighting factors, s is the number of the considered points during the one cycle of crank link. g_k are the constraints arising from the condition satisfying the crank–rocker motion. The objective function minimizes the related shaking force and shaking moment provided that the generated solution satisfies a set of constraints. These constraints are necessary to have a functional mechanism, although they increase solution complexity. $\mathbf{X}$ is a vector comprising the 16 independent design variables (x_r). These variables are given as:

$$\mathbf{X} = \begin{bmatrix} L_i & \lambda_i & m_i & I_{G_i} & r_{21} & r_{32} & r_{44} \end{bmatrix}^{\mathrm{T}} \tag{13.12}$$

where L_i denote the link lengths as L_1, L_2, L_3, and L_4. λ_i consists of structural angles of moving links as λ_2, λ_3, and λ_4. m_i and I_{Gi} are the masses and inertial moments of moving links, respectively, that is, m_2, m_3, m_4, I_{G2}, I_{G3}, and I_{G4}. r_{21}, r_{32}, and r_{44} are the position vectors of crank, coupler, and follower links, respectively. $x^{\min}$ and $x^{\max}$ are lower and upper bounds of design variables. These bounds have to be arranged by considering the working space of the mechanism. For the verification of the proposed approach, lower bounds of link lengths were arranged as $L_i - 0.1 \times L_i$. Similarly, upper bounds of link lengths were arranged as $L_i + 0.1 \times L_i$. Also, lower and upper bounds for λ_i were considered 0–360°, respectively. Lower and upper bounds for m_i, I_{Gi}, r_{21}, r_{32}, and r_{44} were arranged by considering the link geometries. Depth, thickness, and length of each moving link were used for these definitions. The weighting factor's value has an important effect on the optimum adjusting of design variables. Since the selecting criterion is not obvious, it is always difficult to make the decision on choosing the values of weighting factors [4, 25]. In general, initiative of mechanism designer has a crucial role upon the definition of these values. Each weighting factor must satisfy the condition;

$$0 \leq W_h \leq 1 \quad \text{and} \quad \sum_{h=1}^{5} W_h = 1$$

In this study, by using the total value of the shaking force at the main support, the relative importance of each subcomponent inside the shaking force was calculated. These calculated values were considered as weighting factors. In the proposed optimization process, the values of the weighting factors were adjusted as 0.40, 0.24, 0.16, 0.1, and 0.1 for W_1, W_2, W_3, W_4, and W_5, respectively.

13.3 Results

In the present study, a theoretical model was used to investigate the effects of shaking force and moment. The operation speed of the mechanism was constant and it was adjusted as 300 rpm. Assuming that an objective function based on the subcomponents of shaking force and shaking moment, genetic algorithm was used to solve the optimization problem. Design variables which consisted of kinematic and dynamic parameters of the mechanism were also defined. Three case studies were implemented. Proposed structure of objective function and values of weighting factors, which were defined in Sect. 13.3, were considered in the first case. By using the different values of weighting factors, second and third cases were also performed. Dimensions and inertial parameters of the original (unbalanced) and optimized (balanced) mechanisms for three case studies are given in Table 13.1.

By using the optimized values of each case study, dynamic analysis of the mechanism was performed to obtain the force and moment results. The convergence history for Case I is given in Fig. 13.3. The algorithm shows good convergence. After 111 generations, the best individual fitness stays as 5711.4096 and the average fitness occurs as 5711.6292.

Figure 13.4 gives the crank–frame and follower–frame joint forces, which are also the subcomponent of shaking force.

After the optimization, there is a certain decrease at the force values. x and y components of the crank–frame joint force decrease by 95.52 % and 77.18 %, respectively. The decreasing ratios for the maximum values are observed as 97.51 % and 95.97 % for x and y components of the crank–frame joint force, respectively. In the case of minimum values, the decreasing ratios occur as 96.23 % and 71.56 % for x and y components of that force, respectively. For the case of follower–frame joint, the force components for x and y directions decrease by 84.69 % and 74.95 %, respectively. Maximum values for x and y components of the follower–frame joint force are reduced as 93.45 % and 89.81 %, respectively. The decreasing ratios for the minimum values are also obtained as 79.07 % and 80.96 % for x and y components of that force, respectively.

Table 13.1 Original and optimized parameters of four-bar mechanism

			Optimized values		
Parameter	Description	Original value	Case I	Case II	Case III
L_1 (mm)	Length of fixed link	600	570	570	570.2
L_2 (mm)	Length of crank link	100	95	95	95
r_{21} (mm)	Position vector of crank link	50	66.7	66.7	72.54
m_2 (kg)	Mass of crank link	0.360	2.027	3.470	1.755
I_{G2} (kg m^2)	Inertial moment of crank link	4.13×10^{-4}	42.30×10^{-4}	98.28×10^{-4}	48.93×10^{-4}
λ_2 (rad)	Structural angle of crank link	0	3.0332	3.065	3.032
L_3 (mm)	Length of coupler link	400	420	420	420
r_{32} (mm)	Position vector of coupler link	200	77.5	88.73	87.98
m_3 (kg)	Mass of coupler link	1.296	1.264	2.06	1.23
I_{G3} (kg m^2)	Inertial moment of coupler link	1.87×10^{-2}	4.87×10^{-2}	9.96×10^{-2}	4.43×10^{-2}
λ_3 (rad)	Structural angle of coupler link	0	0.1275	0.417	0.1619
L_4 (mm)	Length of follower link	320	329.8	313.9	330
r_{44} (mm)	Position vector of follower link	160	100.4	128	97
m_4 (kg)	Mass of follower link	1.046	0.866	1.425	1.22
I_{G4} (kg m^2)	Inertial moment of follower link	9.85×10^{-3}	14.30×10^{-3}	16×10^{-3}	15×10^{-3}
λ_4 (rad)	Structural angle of follower link	0	0.0002	0.0013	0.0023

Case I: $W_1 = 0.40$, $W_2 = 0.24$, $W_3 = 0.16$, $W_4 = 0.1$, $W_5 = 0.1$; Case II: $W_1 = 0.45$, $W_2 = 0.27$, $W_3 = 0.18$, $W_4 = 0.1$, $W_5 = 0$; Case III: $W_1 = 0.2$, $W_2 = 0.2$, $W_3 = 0.2$, $W_4 = 0.2$, $W_5 = 0.2$

As a natural result of the optimization, shaking force and shaking moment at the optimized mechanism are more close to zero than that of the original mechanism. As shown in Fig. 13.5a, b, shaking force decreases by 90.96 % and 77.54 % for *x* and *y* directions, respectively.

During the one period of the crank link, the maximum values of this force components decrease 91.11 % and 97.78 % for *x* and *y* directions, respectively. In the case of minimum values, the decreasing ratios occur as 85.36 % and 76.25 % for *x* and *y* directions, respectively. As seen from Fig. 13.5c, shaking moment decreases by 76.21 % as well. This ratio is better than that of literature [3]. The decreasing ratios for the maximum and minimum values occur as 90.32 % and 81.91 %, respectively.

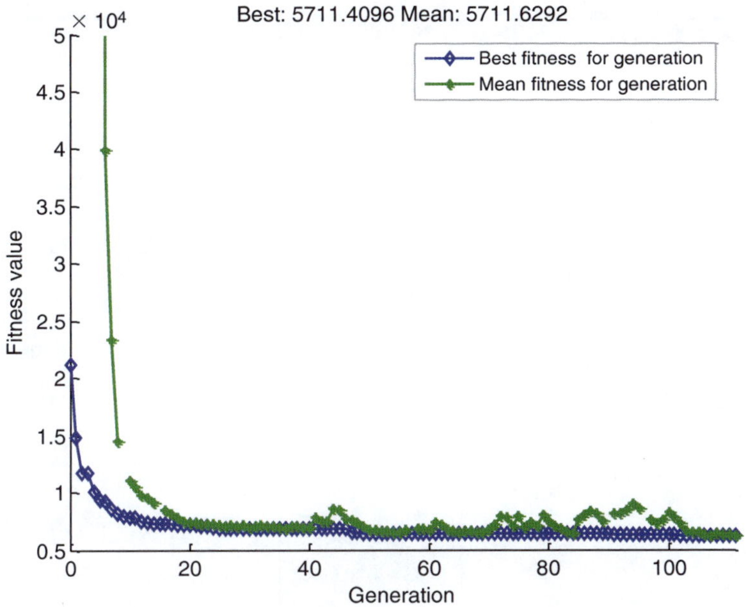

Fig. 13.3 Convergence history of GA evolution for Case I

After the optimum adjusting of design variables, driving torque decreases by 73.46 %. Decreasing ratios for the maximum and minimum values are read from Fig. 13.5d as 76.76 % and 82.48 %, respectively. The commercial simulation software is also used to model the mechanism and to test the optimized values of design variables [26]. Simulation results for original and optimized values of forces and moments are given in appendix. Force and moment results of Case II and III are also given in Figs. 13.6 and 13.7.

By using the different values of weighting factors, decreasing ratios at the total values of forces and moments are outlined in Table 13.2 to evaluate the results of three case studies.

Case I shows the decreasing ratios for the proposed structure of the objective function and the values of weighting factors in Sect. 13.3. Contrary to Case I, objective function of Case II is constituted by using only subcomponents of shaking force, that is, shaking moment is eliminated due to the values of the fifth weighting factor ($W_5 = 0$). The evaluations of Case II with respect to Case I show that the objective function should comprise both shaking force and shaking moment while their dimensions do not match. When the values of weighting factors are selected equals to each other as in Case III [3, 14], the decreasing ratios are smaller than that of Case I. So the weighting factor's value has to be defined by considering the relative sensitivity of the related term.

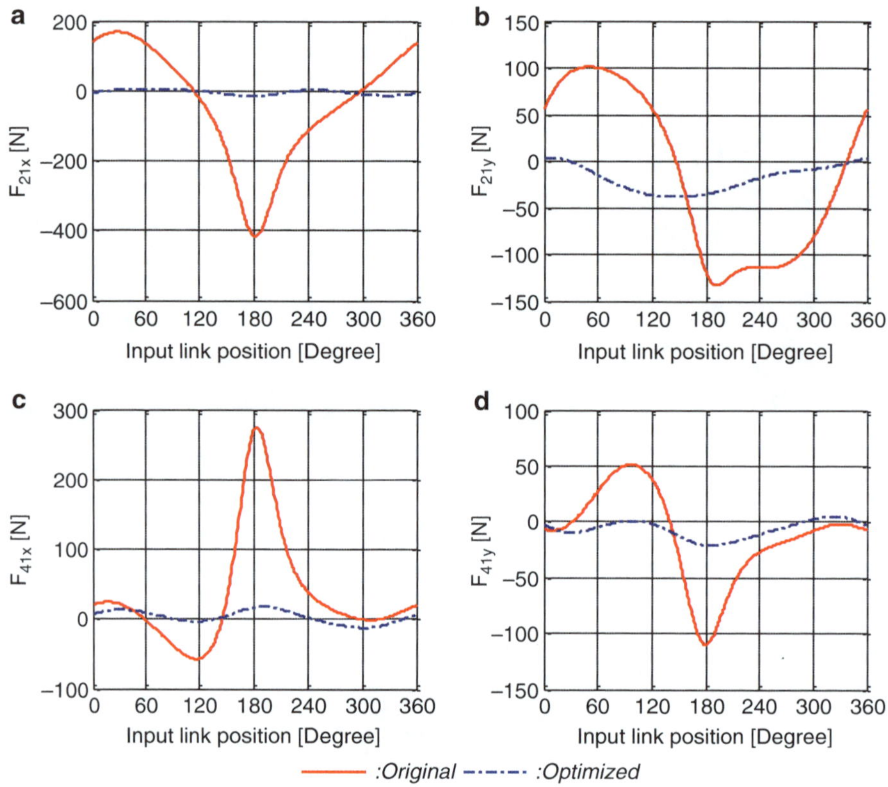

Fig. 13.4 Original and optimized values of joint forces for Case I; (**a**) and (**b**) Crank–frame joint, (**c**) and (**d**) Follower–frame joint

In addition to three case studies, if the objective function only consists of F_{shx}, F_{shy}, and M_{sh}, that is, not comprising their subcomponents, the obtained decreasing ratios are worse than that of the proposed Case I. So, this proves that the proposed structure of the objective function is very effective for the optimum balancing.

13.4 Conclusions and Discussions

The focus of this study is to minimize the shaking force and moment fluctuations at the planar mechanism. This phenomenon is considered as an optimization problem. In addition to the similar studies in literature, the subcomponents of shaking force and shaking moment are considered together to constitute the objective function. Also, relative importance of the force component inside the total shaking force is evaluated to define the value of related weighting factor. Therefore, it is possible to reduce the negative reflection on the optimization process arising from mechanism designer's initiative.

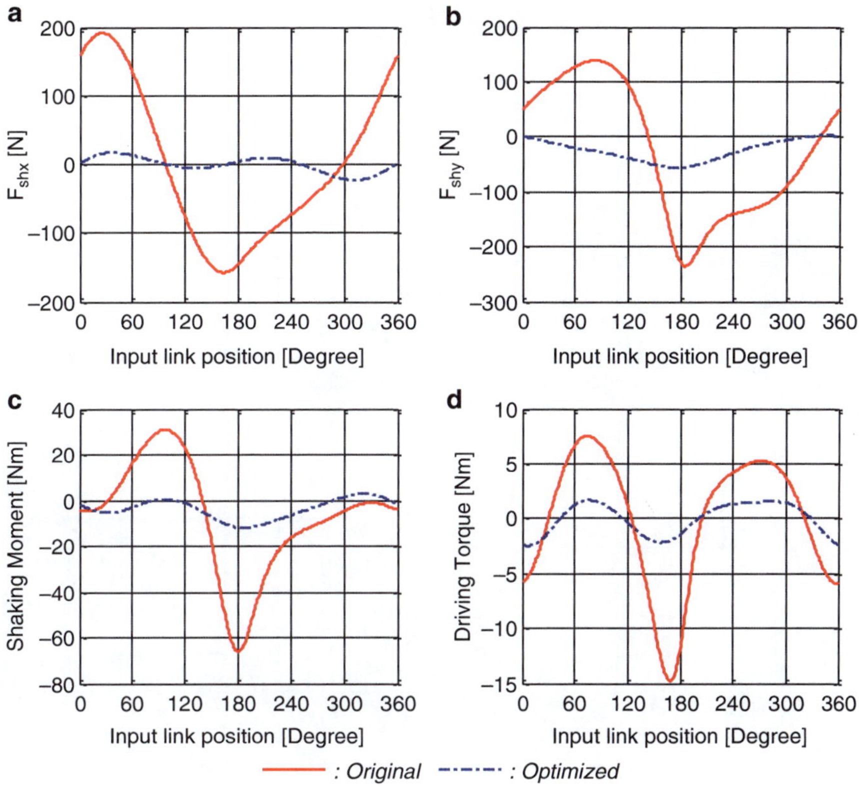

Fig. 13.5 Original and optimized values for Case I; (**a**) and (**b**) Shaking force components, (**c**) Shaking moment, (**d**) Driving torque

Three case studies indicate that both the structure of the objective function and the value of the weighting factor have a crucial role to minimize the shaking force and moment fluctuations. Objective function should comprise both shaking force and shaking moment while their dimensions do not match. Although the objective functions have the same structure, definition of the weighting factors' values is very important for the optimization process. By using the shaking force and moment in objective function, and evaluating the values of weighting factors according to their relative importance of the related forces, Case I gives the better results for solving the present optimization problem than that of the other cases. The obtained results show that the proposed structure of the objective function and the values of weighting factors are very effective to decrease the force and moment fluctuations, and power consumption for driving torque. Due to the flexibility of the proposed approach, mechanism designer can individually decrease each subcomponent of force, and this approach can also be applied to other planar and spatial mechanisms.

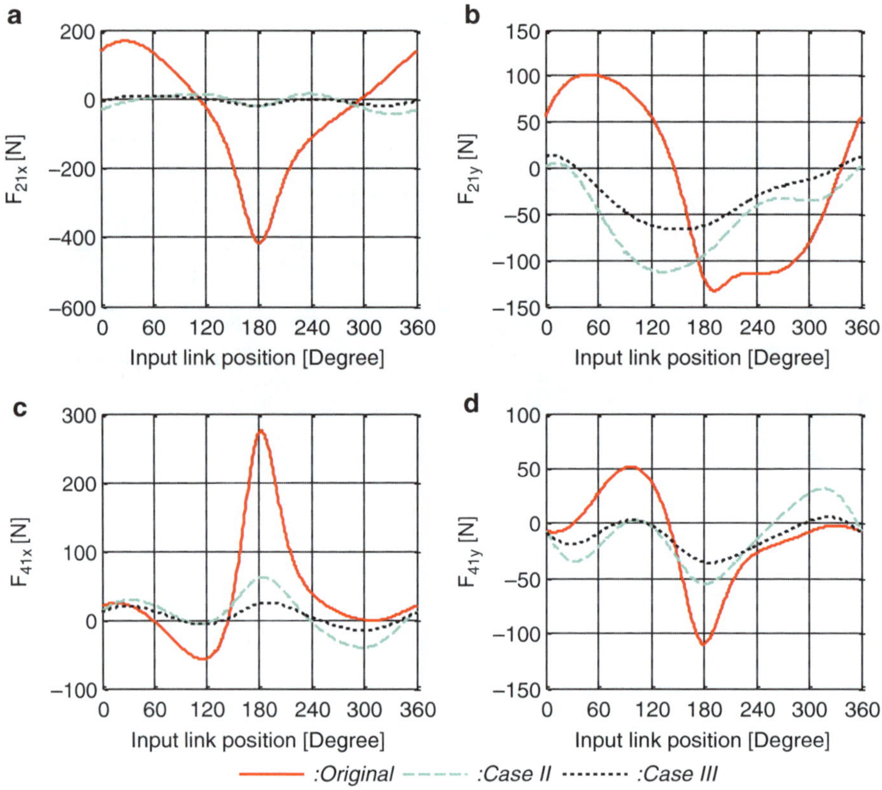

Fig. 13.6 Original and optimized values of joint forces for Case II and III; (**a**) and (**b**) Crank–frame joint, (**c**) and (**d**) Follower–frame joint

Appendix

Simulation results of force and moment characteristics for Case I are given in Figs. 13.8 and 13.9.

Simulation result of bearing vibrations for Case I is outlined in Fig. 13.10.

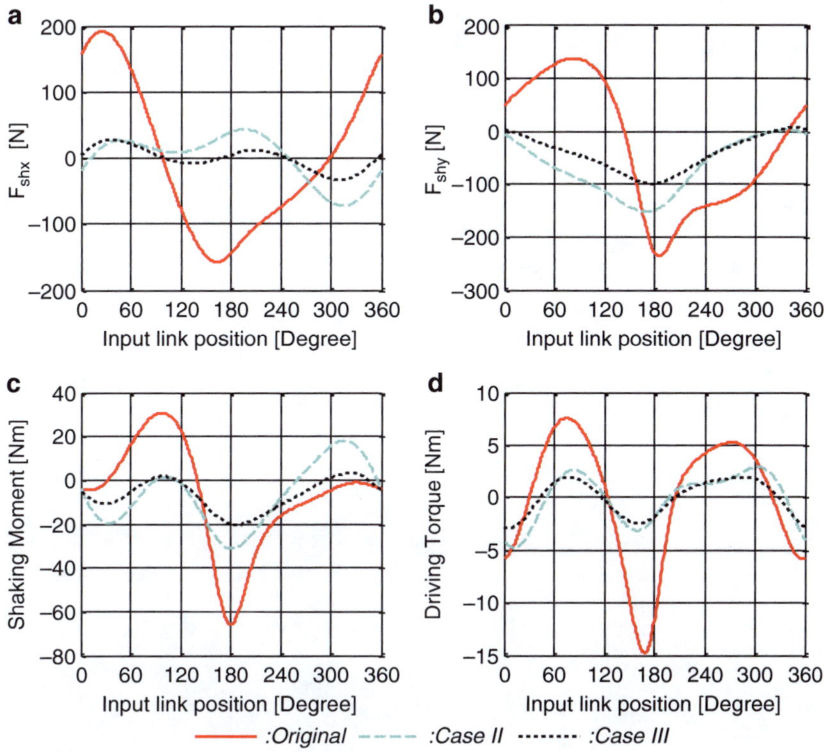

Fig. 13.7 Original and optimized values for Case II and III; (**a**) and (**b**) Shaking force components, (**c**) Shaking moment, (**d**) Driving torque

Table 13.2 Decreasing ratios for three case studies

	Decreasing ratio (%)		
	Case I	Case II	Case III
F_{21x}	95.52	88.04	93.50
F_{21y}	77.18	31.66	59.10
F_{41x}	84.69	51.48	78.28
F_{41y}	74.95	21.59	56.58
F_{shx}	90.96	69.35	86.30
F_{shy}	77.54	37.61	61.54
M_{sh}	76.21	25.51	58.73
M_{21}	73.46	57.65	70.49

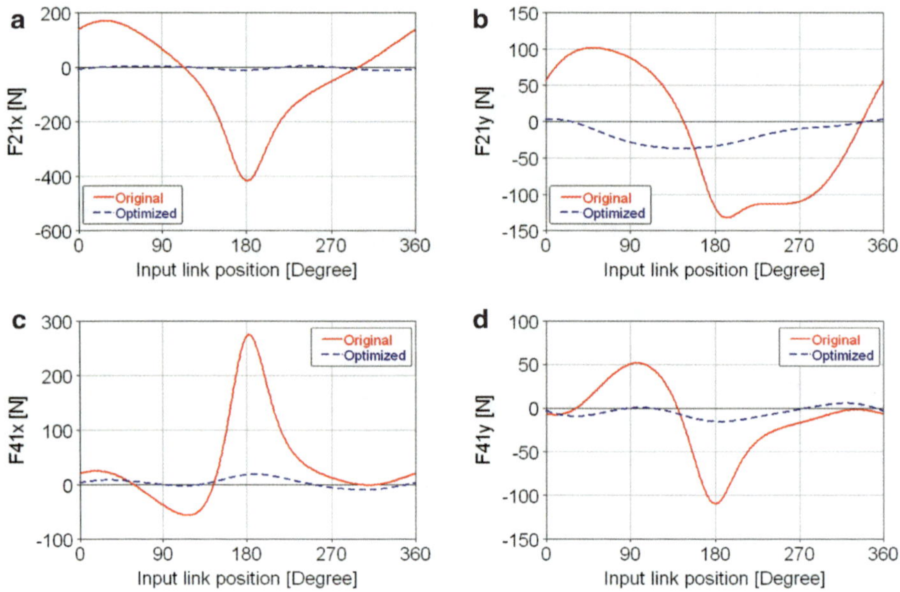

Fig. 13.8 Simulation results for Case I; (**a**) and (**b**) Crank–frame joint force, (**c**) and (**d**) Follower–frame joint force

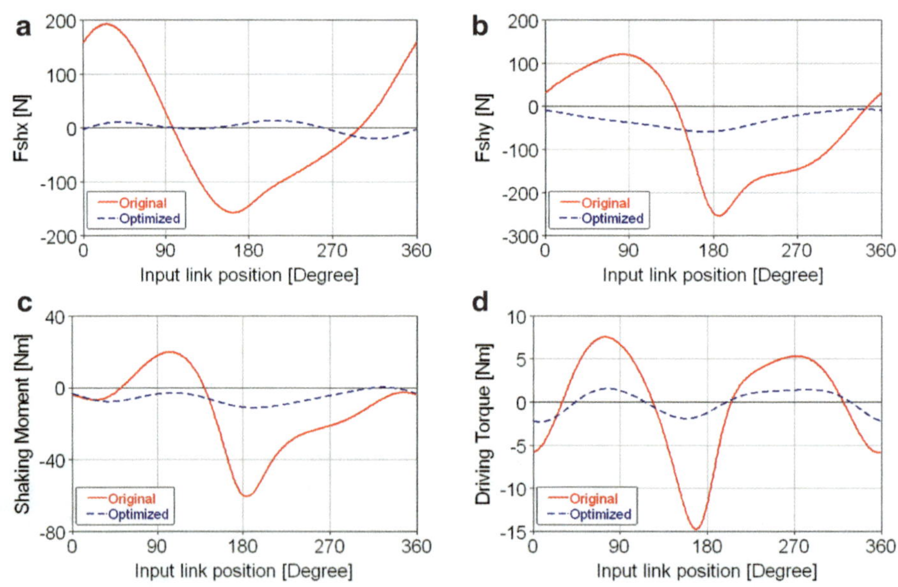

Fig. 13.9 Simulation results for Case I; (**a**) and (**b**) Shaking force components, (**c**) Shaking moment, (**d**) Driving torque

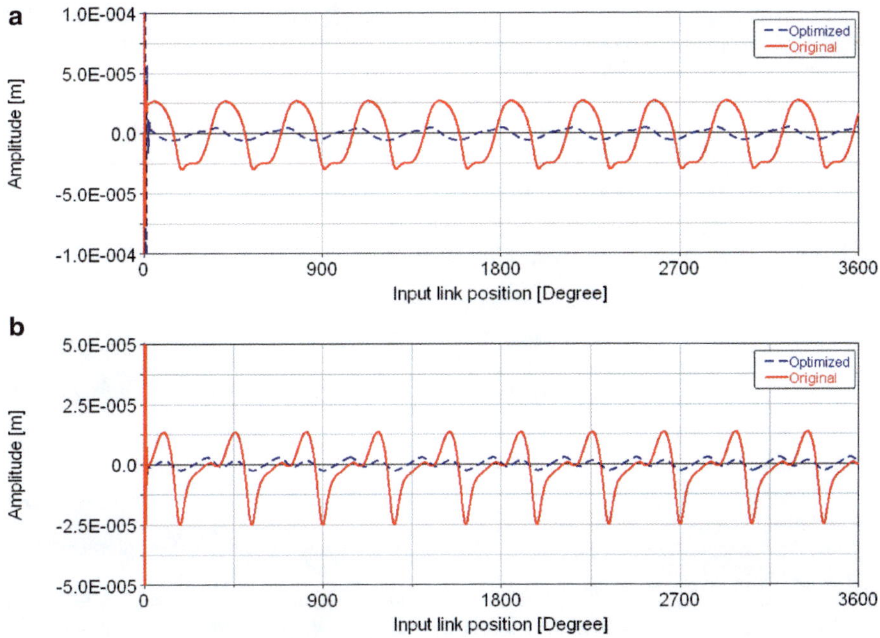

Fig. 13.10 Bearing vibrations in vertical direction of original and optimized mechanisms for Case I; (**a**) Left bearing, (**b**) Right bearing

References

1. Feng, G.: Complete shaking force and shaking moment balancing of 17 types of eight-bar linkages only with revolute pairs. Mech. Mach. Theory **26**(2), 197–206 (1991)
2. Ye, Z., Smith, M.R.: Complete balancing of planar linkages by an equivalence method. Mech. Mach. Theory **29**(5), 701–712 (1994)
3. Li, Z.: Sensitivity and robustness of mechanism balancing. Mech. Mach. Theory **33**(7), 1045–1054 (1998)
4. Chiou, S.T., Bai, G.J., Chang, W.K.: Optimum balancing designs of the drag-link drive of mechanical presses for precision cutting. Int. J. Mach. Tools Manuf. **38**(3), 131–141 (1998)
5. Arakelian, V.H., Smith, M.R.: Complete shaking force and shaking moment balancing of linkages. Mech. Mach. Theory **34**, 1141–1153 (1999)
6. Tepper, F.R., Lowen, G.G.: General theorems concerning full force balancing of planar linkage by internal mass redistribution. J. Manuf. Sci. Eng. **94**, 789–796 (1972)
7. Esat, I., Bahai, H.: A theory of complete force and moment balancing of planer linkage mechanisms. Mech. Mach. Theory **34**, 903–922 (1999)
8. Feng, B., Morita, N., Torii, T.: A new optimization method for dynamic design of planar linkage with clearances at joints-optimizing the mass distribution of links to reduce the change of joint forces. J. Mech. Des. **124**, 68–73 (2002)
9. Kochev, I.S.: General theory of complete shaking moment balancing of planar linkages: a critical review. Mech. Mach. Theory **35**, 1501–1514 (2000)
10. Guo, G., Morita, N., Torii, T.: Optimum dynamic design of planar linkage using genetic algorithms. JSME Int. J. C **43**(2), 372–377 (2000)

11. Arakelian, V., Dahan, M.: Partial shaking moment balancing of fully force balanced linkages. Mech. Mach. Theory **36**, 1241–1252 (2001)
12. Arakelian, V.H., Smith, M.R.: Shaking force and shaking moment balancing of mechanisms: a historical review with new examples. J. Mech. Des. **127**(2), 334–339 (2005)
13. Arakelian, V.H.: Shaking moment cancellation of self-balanced slider-crank mechanical systems by means of optimum mass redistribution. Mech. Res. Commun. **33**, 846–850 (2006)
14. Alici, G., Shirinzadeh, B.: Optimum dynamic balancing of planar parallel manipulators based on sensitivity analysis. Mech. Mach. Theory **41**(12), 1520–1532 (2006)
15. Chaudhary, H., Saha, S.K.: Balancing of four-bar linkages using maximum recursive dynamic algorithm. Mech. Mach. Theory **42**, 216–232 (2007)
16. Chaudhary, H., Saha, S.K.: Balancing of shaking forces and shaking moments for planar mechanisms using the equimomental systems. Mech. Mach. Theory **43**, 310–334 (2008)
17. Erkaya, S., Uzmay, İ.: Investigation on effect of joint clearance on dynamics of four-bar mechanism. Nonlinear Dyn. **58**, 179–198 (2009)
18. Ilia, D., Sinatra, R.: A novel formulation of the dynamic balancing of five-bar linkages with applications to link optimization. Multibody Syst. Dyn. **21**, 193–211 (2009)
19. Park, J.G., Jeong, W.B., Seo, Y.S., Yoo, W.S.: Optimization of crank angles to reduce excitation forces and moments in engines. J. Mech. Sci. Technol. **21**(2), 272–281 (2007)
20. MATLAB (Symbolic Math Toolbox and Optimization Toolbox). The MathWorks Inc, 3 Apple Hill Drive, Natick, MA 01760-2098
21. Kunjur, A., Krishnamurty, S.: Genetic algorithms in mechanism synthesis. J. Appl. Mech. Rob. **4**, 18–24 (1997)
22. Yang, B.S., Choi, S.P., Kim, Y.C.: Vibration reduction optimum design of a steam-turbine rotor-bearing system using a hybrid genetic algorithm. Struct. Multidiscip. Optim. **30**, 43–53 (2005)
23. Goldberg, D.E.: Genetic Algorithms in Search, Optimization and Machine Learning. Addison-Wesley, Boston, MA (1989)
24. Chaudhary, H., Saha, S.K.: Dynamics and Balancing of Multibody Systems. Lecture Notes in Applied and Computational Mechanics, vol. 37. Springer, Berlin (2009)
25. Erkaya, S., Uzmay, İ.: A neural-genetic (NN-GA) approach for optimising mechanisms having joints with clearance. Multibody Syst. Dyn. **20**, 69–83 (2008)
26. MSC.ADAMS 2005 r2: Automatic dynamic analysis of mechanical systems. MSC Software Corporation

Chapter 14
Optimal Balancing of the Robotic Manipulators

A. Nikoobin and M. Moradi

Abstract The balancing of robotic systems is an important issue, because it allows for significant reduction of torques. However, the literature review shows that the balancing of robotic systems is performed without considering the traveling trajectory. Although in static balancing the gravity effects on the actuators are removed, and in complete balancing the Coriolis, centripetal, gravitational, and cross-inertia terms are eliminated, it does not mean that the required torque to move the manipulator from one point to another point is minimum. In this chapter, "optimal balancing" is presented for the open-chain robotic system based on the indirect solution of open-loop optimal control problem. Indeed, optimal balancing is an optimal trajectory planning problem in which states, controls, and all the unknown parameters associated with the counterweight masses or springs must be determined simultaneously to minimize the given performance index for a predefined point-to-point task. For this purpose, on the base of the fundamental theorem of calculus of variations, the necessary conditions for optimality are derived which lead to the optimality conditions associated with the Pontryagin's minimum principle and an additional condition associated with the constant parameters. In this chapter, after presenting the formulation of the optimal balancing and static balancing, the obtained optimality conditions are developed for a two-link manipulator in details. Finally the efficiency of the suggested approach is illustrated by simulation for a two-link manipulator and a PUMA-like robot. The obtained results show that the proposed method has dominant superiority over the previous methods such as static balancing or complete balancing.

Keywords Optimal balancing • Static balancing • Serial robot manipulators • Trajectory planning • Optimal control • Pontryagin minimum principle • Counterweight • Spring

---

A. Nikoobin (✉)
Faculty of Mechanical Engineering, Semnan University, P.O.B. 3513119111, Semnan, Iran
e-mail: anikoobin@semnan.ac.ir

M. Moradi
Department of Mechanical Engineering, Amirkabir University of Technology, Tehran, Iran
e-mail: m_moradi@aut.ac.ir

D. Zhang, B. Wei (eds.), *Dynamic Balancing of Mechanisms and Synthesizing of Parallel Robots*, DOI 10.1007/978-3-319-17683-3_14

14.1 Introduction

Achieving the optimal performance of robot manipulators in repeating tasks have been attracted many attentions in the recent years. In an optimal task, minimum consumed energy, minimum torque, or minimum time can be considered. Often the used manipulators in an assembly or manufacturing line are fixed so for a new product, the end effector of the manipulators and their predefined trajectories can be changed. Since changing the robot and its structure is a hard task or often impossible, beside the trajectory planning, one efficient way to increase the robot performance is balancing. Balancing introduces some simple modifications in the architecture of the original mechanism, which may simplifies its dynamic model and, as a result, its control as well. Besides the control simplification, balancing can also provide reduction of the driving torques. Basically balancing can be categorized into the two types: active and passive balancing. In the active balancing an external electric, pneumatic or hydraulic force is applied to the system [1], while in the passive balancing, compensation inertia [2] or springs [3] are used. Since the additional actuators are not required in the passive balancing, it is more economical and simpler than active one.

Two methods studied in the literature for the passive balancing are using counterweights and/or springs. The balancing by masses is due to added counterweights or link's mass redistribution. Counterweight balancing is simple and has some advantageous, but it increases inertia of manipulators. In case of balancing by springs, changes in mass and inertia parameters of robot mechanism are insignificant because of negligible weight of springs in comparison to link. Unlike the counterweight balancing which is straight forward and almost simple, spring balancing can be performed in different forms as [4]: Balancing by springs jointed directly with links, balancing by using a cable and pulley arrangement, balancing by using an auxiliary linkage, balancing by using a cam mechanism and balancing by using gear train.

Arakelian et al. [3] has reviewed different types of spring balancing mechanisms. Kolarski et al. [5] compared dynamical behavior of the unbalanced, spring and counterweight balanced PUMA robot configuration. Banala et al. [6] presented spring balancing for 2-DOF spatial manipulator and 3-DOF spatial manipulator. They described the theory of gravity balanced spatial robotic manipulators through a hybrid strategy which uses springs in addition to identification of the center of mass using auxiliary parallelograms. Regardless of the balancing implementation method by mass or spring, the passive balancing approaches are classified into the four types:

- Static balancing: A machine is said to be static balanced if its potential energy is constant for all possible configurations [5].
- Dynamic balancing: It has one step more than static balancing, and it is reducing the reaction forces and moments on the base and among actuators, for all situations. Thus, the dynamic balanced robot will lightly transfer some reactions to its adjacent actuators and environments when it is operated [7].

- Complete balancing: brings some modifications to unbalanced mechanisms in such away to obtain static balancing and complete decoupling of dynamic equations [2].
- Optimal balancing: Selection of the balancing parameters by consideration of the system trajectory. According to the structure of the trajectory, again it can be divided to two subcategories as Closed Loop optimal balancing and Open Loop optimal balancing [8–10].

In the Closed Loop optimal balancing the controller is considered for the system. In this field, Saravanan et al. performed optimum balancing for an industrial robot, while a rectangular path must be tracked [8]. Ravichandran et al. by considering a nonlinear PD controller for a 2-DOF robot manipulator applied optimal balancing in order to determine the controller parameters and counterweight values [9]. This optimal balancing approach can be considered as a more general problem called Integrated Plant and Controller design problem. Design for control approach (DFC), in which structure and parameter of the both machine body and control algorithm are designed to fulfill the specific task of a five bar close chain robot by Cheng et al. [11, 12]. To this end, a PD plus robust term control algorithm in the DFC approach have been proposed to obtain the desired performance in terms of change of the tasks [11].

In the Open Loop optimal balancing, the trajectory and the balancing unknowns are simultaneously designed. Generally speaking, this open loop optimal balancing is the trajectory planning problem with some unknown parameters. Optimal trajectory planning of the manipulators is based on the optimization of an objective function while the dynamic equations of motion as well as bounds on joint positions, velocities and torques must be taken into account [13]. The idea is presented by Nikoobin and Moradi [10] and it have been tested for the counterweight balanced robot manipulators in the point to point motion. In fact, an optimal trajectory planning problem is outlined in which states, controls, and the values of counterweights are determined simultaneously in order to minimize the given performance index for a predefined point-to-point task. At this point, two strategies are well-known for path optimization: indirect methods [14] and direct methods [15]. These techniques are used in many articles and they have own benefits and weaknesses [13].

Although the optimal counterweight balancing method in comparison with previous methods such as unbalancing [16], static balancing [5], or complete balancing has significant superiorities to optimize the given performance index, it suffers from increasing the inertia. In order to overcome this drawback, in this chapter optimal balancing is developed for the spring balanced robot manipulators and it is shown that spring balancing is more practical and efficient than counterweight balancing. Then this approach has been applied on the spring balancing of robotic manipulators [4]. The chapter is organized as follows: general formulation of optimal balancing and static balancing is presented in Sects. 14.2 and 14.3, respectively. Then in Sect. 14.4 using the obtained general formulation, modeling and optimality conditions are derived for a two-link manipulator in details. Finally in order to verify the method, simulation results for a two-link manipulator and a PUMA-like robot are presented in Sect. 14.5.

14.2 The Optimal Balancing Formulation

Optimal balancing is defined as: simultaneously achieving of the unknown parameters and the trajectory of the system using the solution of the optimal control. The optimal control problem for the dynamic system involving parameters can be stated as follows [10]: Find the parameter vector **b** and continuous admissible control history $\mathbf{u}:[t_0,t_f] \rightarrow \Omega \subseteq \mathbb{R}^m$ generating the corresponding state trajectory **x**: $[t_0,t_f] \rightarrow \mathbb{R}^n$ which defined as continuous state vector as

$$\mathbf{x} = \left[\mathbf{x}_1^{\mathrm{T}}\ \mathbf{x}_2^{\mathrm{T}}\right]^{\mathrm{T}} = \left[\mathbf{q}^{\mathrm{T}}\ \dot{\mathbf{q}}^{\mathrm{T}}\right]^{\mathrm{T}}, \tag{14.1}$$

which minimize the objective function

$$J = \phi\left(\mathbf{x}_f, \mathbf{b}\right) + \int_{t_0}^{t_f} L\left(\mathbf{x}, \mathbf{u}, \mathbf{b}\right) dt, \tag{14.2}$$

subject to the system dynamics

$$\mathbf{M}\left(\mathbf{q}, \mathbf{b}\right) \ddot{\mathbf{q}} + \mathbf{C}\left(\mathbf{q}, \dot{\mathbf{q}}, \mathbf{b}\right) + \mathbf{G}\left(\mathbf{q}, \mathbf{b}\right) = \mathbf{u}, \tag{14.3}$$

where $\mathbf{M} \in \mathbb{R}^{n \times n}$ is the mass matrix, $\mathbf{C} \in \mathbb{R}^n$ is the coupling matrix, $\mathbf{G} \in \mathbb{R}^n$ is the gravity dependent terms, $\mathbf{q} \in \mathbb{R}^n$ is the position vector of the manipulator, and $\mathbf{b} \in \mathbb{R}^r$ is the design parameters vector. Generally, the vector **b** contains all the unknown constant parameters which their optimal value must be obtained during the problem solution. The vector $\mathbf{u} \in \mathbb{R}^m$ is the control vector, Ω is an admissible subset of $\mathbb{R}^m$, t_0 and t_{f} are the initial and the final time and $\mathbf{x}_{\mathrm{f}}$ is predefined final state. ϕ and L are scalar continuously differentiable functions in which ϕ is the final state penalty term and L is the integrand of the cost function. ϕ and L can be selected to obtain different optimal control problem such as minimum time, terminal control, minimum effort, tracking problem, or regulator problem. Dynamic equation (14.3) can be rewritten in state space form as

$$\begin{bmatrix} \dot{\mathbf{x}}_1 \\ \dot{\mathbf{x}}_2 \end{bmatrix} = \begin{bmatrix} \mathbf{x}_2 \\ \mathbf{M}^{-1}\left(\mathbf{x}_1, \mathbf{b}\right)\left[\mathbf{u} - \mathbf{C}\left(\mathbf{x}_1, \mathbf{x}_2, \mathbf{b}\right) - \mathbf{G}\left(\mathbf{x}_1, \mathbf{b}\right)\right] \end{bmatrix}, \tag{14.4}$$

with the given initial condition

$$\mathbf{x}\left(t_0\right) = \mathbf{x}_0, \tag{14.5}$$

and the prescribed final conditions

$$\mathbf{x}\,(\mathrm{t_f}) = \mathbf{x}_f. \tag{14.6}$$

where $\mathbf{x} \in \mathbb{R}^n$ is the state vector and t_0 is the initial time. Constant state vector can be appended for considering the parameters as $\dot{\mathbf{b}} = \mathbf{0}$. Then by defining the Hamiltonian function as

$$H = L + \boldsymbol{\lambda}^{\mathrm{T}}\mathbf{f} + \boldsymbol{\mu}^{\mathrm{T}}\dot{\mathbf{b}}, \tag{14.7}$$

Theorem 1: Optimality conditions to minimize the objective function (14.4), subject to dynamic equation and boundary conditions (14.5) and (14.6) can be derived as follows:

$$\begin{aligned}
&\mathrm{ODE}: \dot{\mathbf{x}} = \nabla_{\boldsymbol{\lambda}} H, \quad \dot{\boldsymbol{\lambda}} = -\nabla_{\mathbf{x}} H, \quad \dot{\boldsymbol{\mu}} = -\nabla_{\mathbf{b}} H\\
&\text{Algebraic Equation}: \nabla_{\mathbf{u}} H = 0\\
&\mathrm{BCs}: \mathbf{x}\,(t_0) = \mathbf{x}_0, \quad \mathbf{x}\,(\mathrm{t_f}) = \mathbf{x}_f, \quad \boldsymbol{\lambda}\,(\mathrm{t_f}) = -\nabla_{\mathbf{x}_f}\phi, \quad \boldsymbol{\mu}(0) = 0, \quad \boldsymbol{\mu}\,(\mathrm{t_f}) = \nabla_{\mathbf{b}}\phi.
\end{aligned} \tag{14.8}$$

With the realization that the parameters behave like states, for the optimal control **u*** the Legendre-Clebsch condition,

$$\nabla^2 H\left(\mathbf{x}^*, \mathbf{u}^*, \boldsymbol{\lambda}^*, \mathbf{b}^*\right) \geq 0, \tag{14.9}$$

must be satisfied.

Proof: Adjoining the constraints (Dynamic equations and constancy of the parameters) to the performance index by Lagrange multipliers λ leads to the augmented performance index [10]

$$J' = \phi\left(\mathbf{x}_f, \mathbf{b}\right) + \int_{t_0}^{t_f} \left[H\left(\mathbf{x}, \mathbf{u}, \lambda\right) - \lambda^{\mathrm{T}}\dot{\mathbf{x}}\right] dt. \tag{14.10}$$

According to the fundamental theorem of calculus of variations, if $\mathbf{x}^*$ is an extremal of functional J', the variation of J' must vanish on $\mathbf{x}'$, that is $\delta J'(\mathbf{x}^*, \delta\mathbf{x}) = 0$ for all $\delta\mathbf{x}$. Then, taking the variation of J' and performing the integration by parts gives

$$\begin{aligned}
\delta J' = &\left(\nabla_{\mathbf{x}_f}\phi - \lambda_f^{\mathrm{T}}\right)\delta\mathbf{x}_f + \left(\nabla_{\mathbf{b}}\phi + \int_{t_0}^{t_f} \nabla_{\mathbf{b}} H\, dt\right)\delta\mathbf{b} + \lambda_0 \delta\mathbf{x}_0\\
&+ \int_{t_0}^{t_f} \left[\left(\nabla_{\mathbf{x}} H + \dot{\lambda}^{\mathrm{T}}\right)\delta\mathbf{x} + \nabla_{\mathbf{u}} H \delta\mathbf{u} + \left(\nabla_{\lambda} H - \dot{\mathbf{x}}^{\mathrm{T}}\right)\delta\lambda\right] dt
\end{aligned} \tag{14.11}$$

If the initial and the final states are predefined then $\delta\mathbf{x}_0$ and $\delta\mathbf{x}_f$ are zero and if not, the corresponding multipliers should be considered zero. It means, for free initial it should consider $\boldsymbol{\lambda}_0 = \mathbf{0}$ and for free final state it forms $\lambda_f^{\mathrm{T}} = \nabla_{\mathbf{x}_f}\phi$. Since the parameter vector **b** and control **u** are independent, consider the class of admissible comparison paths where there is no change in the parameter ($\delta\mathbf{b} = \mathbf{0}$). The problem reduces to the standard fixed final time problem for which

$$\dot{\mathbf{x}} = \nabla_{\lambda} H, \quad \dot{\boldsymbol{\lambda}} = -\nabla_{\mathbf{x}} H. \tag{14.12}$$

Then, for admissible comparison paths where $\delta\mathbf{b} \neq \mathbf{0}$, the first variation yields the condition

$$\nabla_{\mathbf{b}}\phi + \int_{t_0}^{t_f} \nabla_{\mathbf{b}} H \, dt = 0. \tag{14.13}$$

Now by selecting of the multiplier function $\boldsymbol{\mu}$ as

$$\dot{\boldsymbol{\mu}} = -\nabla_{\mathbf{b}} H, \quad \boldsymbol{\mu}(0) = 0, \tag{14.14}$$

Eq. (14.13) becomes

$$\boldsymbol{\mu}\,(\mathrm{t_f}) = \nabla_{\mathbf{b}}\phi. \tag{14.15}$$

Also with the realization that the parameters behave like states, the Weiestrass condition,

$$H\left(x, u^{*}, \lambda, b\right) \geq H\left(x, u, \lambda, b\right) \text{ for all u}^{*} \tag{14.16}$$

must be satisfied as must the Legendre–Clebsch condition to have minimum in the Hamiltonian,

$$H_{uu}\left(x, u, \lambda, b\right) \geq 0 \tag{14.17}$$

As it can be seen, eliminating the parameter vector **b** in the above equations leads to the well-known optimality conditions obtained from Pontryagin's minimum principle.

14.3 Static Balancing

For statically balanced robotic systems, the weight of the links does not exert any force at the actuators for any configuration. In the other word, it removes the gravitational effects in mechanical systems. Another appropriate and practical meaning of this concept can be stated as to be constant potential energy of the manipulator. This can be applied by establishing the additional mechanical elements into the system, such as counterweights or springs to make potential energy constant. The use of counterweights has some advantages along with disadvantages that serve to limit its usefulness. For instance, this is undesirable to provide an extra mass on robot where minimum weight is an important criterion. Also, adding the counterweights increases the moment of inertia of the manipulator. Here the general theory for static balancing of manipulators based energy method is stated. The static balancing using energy index can be stated as

$$\frac{\partial P}{\partial q_i} = 0, \quad i = 1, \ldots, n, \tag{14.18}$$

where P is the total potential function of manipulator and q_i is the position of link i as generalized coordination of the system. Consequently for systems consisting of springs, the unknown parameters can be founded by using these n equations. In some cases, these equations have no solution and this means such systems are not completely balanceable. If complete static balancing occurs, then the gravitational forces become entirely eliminated as

$$\mathbf{G}(\mathbf{q}) = 0. \tag{14.19}$$

So, system dynamic described in Eq. (14.3) is reduced to

$$\widehat{\mathbf{u}} = \widehat{\mathbf{M}}\ddot{\mathbf{q}} + \widehat{\mathbf{C}}(\mathbf{q}, \dot{\mathbf{q}}), \tag{14.20}$$

where $\widehat{\mathbf{M}}$ is inertia matrix and $\hat{\mathbf{C}}$ is the vector of centripetal and Coriolis forces of the static balanced manipulator.

Theorem 2: Static balancing is a special case of the optimal balancing with an infinite horizon performance index.

Proof: In order to show the relation between the static and optimal balancing, the scaled time $\tau \in [0\ 1]$ is defined to represent the time as

$$t = \mathrm{t_f}\tau. \tag{14.21}$$

Using this time-scaling the derivatives of position vector becomes

$$\dot{\mathbf{q}} = \frac{d\mathbf{q}}{dt} = \frac{d\mathbf{q}}{t_f\, d\tau} = \frac{\mathbf{q}'}{t_f}, \qquad \ddot{\mathbf{q}} = \frac{d^2\mathbf{q}}{dt^2} = \frac{\mathbf{q}''}{t_f^2}, \tag{14.22}$$

so Eq. (14.3) can be rewritten as follows:

$$\mathbf{M}(\mathbf{q}, \mathbf{b})\frac{\mathbf{q}''}{t_f^2} + \mathbf{C}(\mathbf{q}, \mathbf{b})\frac{\mathbf{q}''}{t_f} + \mathbf{G}(\mathbf{q}, \mathbf{b}) = \mathbf{u}. \tag{14.23}$$

The effort-optimal pay-off functional now selected as

$$J = \int_0^1 \|\mathbf{u}\|^2 d\tau. \tag{14.24}$$

By substituting the Eq. (14.23) into Eq. (14.24) one can write

$$J = \int_0^1 \left\| \mathbf{M}(\mathbf{q}, \mathbf{b}) \frac{\mathbf{q}''}{t_f^2} + \mathbf{C}(\mathbf{q}, \mathbf{b}) \frac{\mathbf{q}''}{t_f} + \mathbf{G}(\mathbf{q}, \mathbf{b}) \right\|^2 d\tau. \tag{14.25}$$

By approaching $t_f \to \infty$ the all terms in Eq. (14.25) will be vanished except the gravitational one. Therefore,

$$\lim_{t_f \to \infty} J = \int_0^1 \|\mathbf{G}(\mathbf{q}_s, \mathbf{b}_s)\|^2 d\tau \tag{14.26}$$

where $\mathbf{q}_s$ denotes optimal static trajectory, $\mathbf{b}_s$ denotes static balanced vector. The minimum solution of this functional is defined as static balancing of the robotic manipulator in terms of optimal balancing.

As usual, static balancing is considered as solution of the vector equation

$$\mathbf{G}(\mathbf{q}_s, \mathbf{b}_s) \equiv \mathbf{0}. \tag{14.27}$$

It is obvious this leads to global minimum of the Eq. (14.26). Thus by approaching the final time to infinity, the optimal balancing leads to static balancing.

14.3.1 Counterweight Balancing

Counterweight static balancing is typically achieved by pioneering additional counterweights or reshaping of the links. This method has advantages along with disadvantages that serve to limit its usefulness. For example, in space robots this is undesirable to provide an extra mass on robot where minimum weight is an important criterion. Also, adding the balancing masses increases the moment of inertia of the manipulator. In contrast, the required torque to remain the manipulator at rest decreases considerably. The necessary number of counterweight for balancing argued before by others, e.g., Kamenski [17].

14.3.2 Spring Balancing

The general spring balancing schematic is presented in Fig. 14.1 for open chain robots, and it has been used for structure in static and optimal balancing. In this figure, k_i denotes the spring between bodies whereas the ground body is inertial and considered as 0th body.

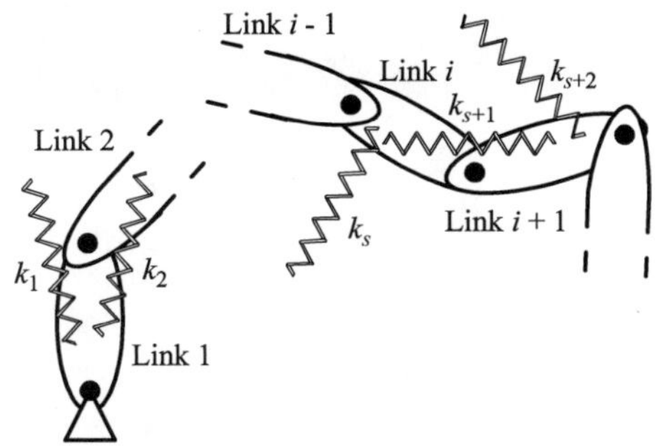

Fig. 14.1 The general representation of open-chain robot manipulator including springs

The existence of spring in the manipulator will change the potential energy of the manipulator. For the general system shown in Fig. 14.1 the potential energy can be computed as

$$P = \sum_{j=1}^{n_m} P_{j\mathrm{G}} + \sum_{j=1}^{n_k} P_{j\mathrm{E}} = \sum_{j=1}^{n_m} \left(\frac{1}{2} k_j (l'_j - l_j)^2 \right) + \sum_{j=1}^{n_k} (m_j g h_j) \tag{14.28}$$

where P_{G} is gravitational potential function and P_{E} is elastic potential of system. l denotes initial length of spring, l' denotes deflected length of spring, m denotes mass of link, h denotes height of center of gravity for link, g denotes gravitational acceleration, n_k denotes number of springs and n_m denotes number of masses in the system. Substituting Eq. (14.28) into Eq. (14.18) yields

$$-\sum_{j=1}^{n_m} \left(k_j \left(l'_j - l_j \right) \frac{\partial l_j}{\partial q_i} \right) + \sum_{j=1}^{n_k} \left(m_j g \frac{\partial h_j}{\partial q_i} \right) = 0; \quad i = 1, \ldots, n, \tag{14.29}$$

where represents n nonlinear equations with n unknown parameters, which depend on the choice of springs structure, may has one, many or no solution. If there is a solution for Eq. (14.29), n unknown parameters are obtained which eliminates the gravitational torques.

14.4 Optimal Balancing of Two-Link Manipulator

14.4.1 Modeling of Two-Link manipulator

Here three different conditions are considered: unbalanced, statically balanced and optimally balanced manipulator. Dynamic equations of all these cases can be presented in the general form. Using the structure presented in Fig. 14.2 a two-link manipulator can be statically balanced.

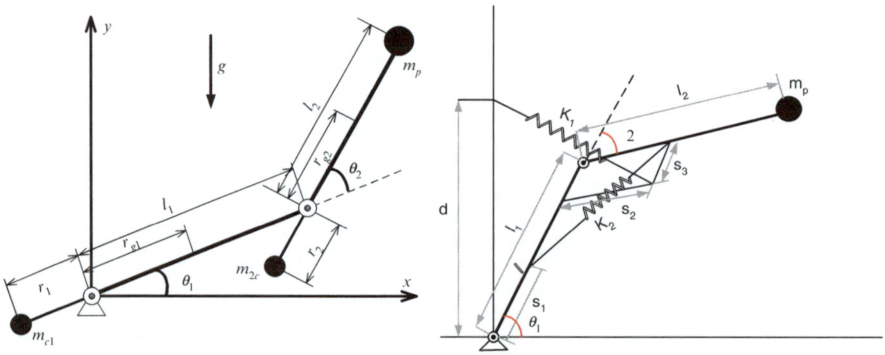

Fig. 14.2 Two-link manipulator with counterweigh balancing (*left*) and balancing springs (*right*)

14.4.1.1 Counterweight Balanced

The dynamic equations for such general two-link manipulator with supposing of arbitrary value of counterweights can be described as follows:

$$\begin{bmatrix} M_{11} & M_{12} \\ M_{12} & M_{22} \end{bmatrix} \begin{bmatrix} \ddot{\theta}_1 \\ \ddot{\theta}_2 \end{bmatrix} + \begin{bmatrix} C_1 \\ C_2 \end{bmatrix} + \begin{bmatrix} G_1 \\ G_2 \end{bmatrix} = \begin{bmatrix} \tau_1 \\ \tau_2 \end{bmatrix} \tag{14.30}$$

where

$$\begin{aligned}
M_{11} &= m_1 r_{g1}^2 + m_2 \left(l_1^2 + 2 l_1 r_{g2} \cos\theta_2 + r_{g2}^2\right) \\
&\quad + m_{\mathrm{p}} \left(l_1^2 + 2 l_1 l_2 \cos\theta_2 + l_2^2\right) + m_{c1} r_1^2 \\
&\quad + m_{c2} \left(l_1^2 - 2 l_1 l_2 \cos\theta_2 + r_2^2\right) + I_1 + I_2 \\
M_{12} &= m_2 \left(l_1 r_{g2} \cos\theta_2 + r_{g2}^2\right) + m_{\mathrm{p}} \left(l_1 l_2 \cos\theta_2 + l_2^2\right) \\
&\quad + m_{c2} \left(-l_1 r_2 \cos\theta_2 + r_2^2\right) + I_2 \\
M_{22} &= m_2 r_{g2}^2 + m_{\mathrm{p}} l_2^2 + m_{\mathrm{c2}} r_2^2 + I_2 \\
C_1 &= l_1 \sin\theta_2 \left(-m_2 r_{g2} - m_{\mathrm{p}} l_2 + r_2 m_{\mathrm{c2}}\right) \left(2\dot{\theta}_1 \dot{\theta}_2 + \dot{\theta}_2^2\right) \\
C_2 &= l_1 \sin\theta_2 \left(m_{\mathrm{c2}} r_2 - m_2 r_{g2} - m_{\mathrm{p}} l_2\right) \dot{\theta}_1^2 \\
G_1 &= \left(m_1 r_{g1} + m_2 l_1 + m_p l_1 - m_{\mathrm{c1}} r_1 + m_{\mathrm{c2}} l_1\right) g \cos\theta_1 \\
&\quad + \left(m_2 r_{g2} + m_{\mathrm{p}} l_2 - m_{\mathrm{c2}} r_2\right) g \cos(\theta_1 + \theta_2) \\
G_2 &= \left(m_{\mathrm{c2}} r_2 - m_2 r_{g2} - m_{\mathrm{p}} l_2\right) g \cos(\theta_1 + \theta_2)
\end{aligned} \tag{14.31}$$

For link i ($i = 1$, 2), denotes the mass, l_i denotes the length, I_i denotes the mass moment of inertia, m_{p} denotes the payload mass, r_{gi} denotes the distance from joint i to the center of mass of the link i, m_{ci} denotes the counterweights attached to link i, and r_i denotes the distance from the joint to the counterweights. In unbalanced case (namely normal case), counterweights of manipulator are zero ($m_{c1} = m_{c2} = 0$) which called normal case in this article. Now, in order to achieve the static balancing, m_{c1} and m_{c2} must be obtained in such a way that the gravity effects in Eq. (14.31) vanish. It implies that $G_1 = G_2 = 0$, so by defining the counterweights as follows:

$$\begin{aligned} m_{c2} &= \left(m_2 r_{g2} + m_{\mathrm{p}} l_2\right)/r_2 \\ m_{c1} &= \left(m_1 r_{g1} + m_2 l_1 + m_{\mathrm{p}} l_1 + m_{c2} l_1\right)/r_1 \end{aligned} \tag{14.32}$$

the dynamic parameters in Eq. (14.30) are changed to

$$\begin{bmatrix} M_{11} & M_{12} \\ M_{12} & M_{22} \end{bmatrix} \begin{bmatrix} \ddot{\theta}_1 \\ \ddot{\theta}_2 \end{bmatrix} = \begin{bmatrix} \tau_1 \\ \tau_2 \end{bmatrix}, \tag{14.33}$$

where the value of inertia matrix is

$$\begin{aligned} M_{11} &= m_1\left(r_{g1}^2 + r_g r_1\right) \\ &+ m_2\left(l_1^2 + r_{g2}^2 + l_1 r_{g2}\frac{r_1 + l_1}{r_2} + l_1 r_1 + r_{g2} r_2\right) \\ &+ m_{\mathrm{p}}\left(l_1^2 + l_2^2 + l_1 r_1 + l_2 r_2 + l_1 l_2 \frac{r_1 + l_1}{r_2}\right) + I_1 \\ M_{12} &= M_{21} = m_2 r_{g2}^2 + m_{\mathrm{p}} l_2^2 + m_2 r_{g2} r_2 + m_{\mathrm{p}} l_2 r_2 \\ M_{22} &= m_2 r_{g2}^2 + m_{\mathrm{p}} l_2^2 + m_2 r_{g2} r_2 + m_{\mathrm{p}} l_2 r_2 + I_2 \end{aligned}$$

All required parameters of the robot manipulator are given in Table 14.1. The counterweights of static counterweight balanced manipulator are calculated to be: $m_{c1} = 17$ kg, $m_{c2} = 5$ kg.

Table 14.1 Parameters of two-link manipulator [10]

Parameters	Values	Unit
Mass	$m_1 = m_2 = 1$	kg
Payload mass	$m_{\mathrm{p}} = 2$	kg
Length of links	$L_1 = L_2 = 1$	m
Moment of inertia	$I_1 = I_2 = 1/12$	kg m^2
Length of adjacent links	$r_1 = r_2 = 0.5$	m
Length of parallelogram side	$s_2 = 0.5$	m

14.4.1.2 Spring Balanced

In deriving the dynamic equations the zero free springs are supposed. For zero free length springs the potential energy using Eq. (14.28) can be written as follows:

$$\begin{aligned} P = &-m_1 g r_{g1} \sin\theta_1 - m_2 g r_{g2} \sin(\theta_1+\theta_2) \\ &+ m_{s1} g \left[\left(l_1 - \frac{1}{2}s_1\right)\sin\theta_1 + \frac{1}{2}s_2\cos(\theta_1+\theta_2)\right] \\ &- m_{s2} g\,[s_1\sin\theta_1 + s_2\sin(\theta_1+\theta_2)] \\ &+ m_p\,[l_1\sin\theta_1 + l_2\sin(\theta_1+\theta_2)] + \frac{1}{2}k_1x_1^2 + \frac{1}{2}k_2x_2^2, \end{aligned} \tag{14.34}$$

where m_{s1} and m_{s2} are mass of fractional mechanism, m_1 and m_2 are mass of links, m_p is payload mass, l_1 and l_2 are length of links, r_{gi} denotes the distance from joint i to the center of mass of the link i, k_1 and k_2 are stiffness of springs, x_{01} and x_{02} are initial length of springs, d, s_1, s_2, and s_3 are the connecting position of springs as shown in Fig. 14.2. x_1 and x_2 are instantaneous length of springs which are functions of θ_1 and θ_2 as follows:

$$\begin{aligned} x_1^2 &= 2(l_1 - s_3)\,(s_2\cos\theta_2 - d\sin\theta_1) - 2ds_2\sin(\theta_1+\theta_2) + d^2 + s_2^2 + (l_1 - s_3)^2 \\ x_2^2 &= s_1^2 + (l_1 - s_2)^2 + 2s_2(l_1 - s_1)\cos\theta_2. \end{aligned} \tag{14.35}$$

For convenience the parameters α and β are defined as follows:

$$\alpha = m_2 r_{g2} + l_2 m_p, \qquad \beta = m_1 r_{g1} + (m_2 + m_p)\,l_1 \tag{14.36}$$

The dynamic equations for such general two-link manipulator can be described as follows:

$$\begin{bmatrix} M_{11} & M_{12} \\ M_{12} & M_{22} \end{bmatrix}\begin{bmatrix} \ddot\theta_1 \\ \ddot\theta_2 \end{bmatrix} + \begin{bmatrix} C_1 \\ C_2 \end{bmatrix} + \begin{bmatrix} G_1 \\ G_2 \end{bmatrix} = \begin{bmatrix} \tau_1 \\ \tau_2 \end{bmatrix}. \tag{14.37}$$

where

$$\begin{aligned} M_{22} &= m_2 r_{g2}^2 + m_p l_2^2 + I_2 \\ M_{11} &= m_1 r_{g1}^2 + (m_2 + m_p)\,l_1^2 + I_1 + 2l_1\alpha\cos\theta_2 + M_{22} \\ M_{12} &= l_1\alpha\cos\theta_2 + M_{22} \end{aligned}$$

$$
\begin{aligned}
C_1 &= -l_1 \alpha \sin \theta_2 \left(2\dot{\theta}_1 + \dot{\theta}_2\right) \dot{\theta}_2 \\
C_2 &= l_1 \alpha \dot{\theta}_1^2 \sin \theta_2 \\
G_1 &= \cos \theta_1 \left(k_1 d \left(s_3 - l_1\right) + \beta g\right) + \cos\left(\theta_1 + \theta_2\right)\left(-k_1 d s_2 + \alpha g\right) \\
G_2 &= \left(s_2 \left(k_2 \left(s_1 - l_1\right) - k_1 \left(l_1 + s_3\right)\right)\right) \sin \theta_2 + \left(\alpha g - k_1 d s_2\right) \cos\left(\theta_1 + \theta_2\right)
\end{aligned} \tag{14.38}
$$

where I_i denotes the mass moment of inertia of links.

Again in the unbalanced case, spring parameters of manipulator are zero ($k_1 = k_2 = 0$) which called normal case. Now, in order to achieve the static balancing, s_1, d and s_3 must be obtained in such a way that the gravity effects in Eq. (14.38) vanish. Here k_1, k_2 and s_2 are supposed to be known parameters. Static balancing implies that $G_1 = G_2 = 0$, so by defining the spring parameters as follows:

$$
d = \frac{g\alpha}{k_1}, \qquad s_1 = l_1 + \frac{k_1}{k_2}\frac{\beta}{\alpha} s_2, \qquad s_3 = l_1 - \frac{\beta}{\alpha} s_2, \tag{14.39}
$$

static balancing is applied and then, the dynamic parameters in Eq. (14.38) becomes

$$
\begin{aligned}
M_{22} &= m_2 r_{g2}^2 + m_p l_2^2 + I_2 \\
M_{11} &= m_1 r_{g1}^2 + \left(m_2 + m_p\right) l_1^2 + I_1 + 2 l_1 \alpha \cos \theta_2 + M_{22} \\
M_{12} &= l_1 \alpha \cos \theta_2 + M_{22} \\
C_1 &= -l_1 \alpha \sin \theta_2 \left(2\dot{\theta}_1 + \dot{\theta}_2\right) \dot{\theta}_2 \\
C_2 &= l_1 \alpha \dot{\theta}_1^2 \sin \theta_2 \\
G_1 &= G_2 = 0.
\end{aligned} \tag{14.40}
$$

A two-link manipulator at vertical plan is considered as shown in Fig. 14.2. All required parameters of the robot manipulator are given in Table 14.1. The static balancing results can be seen in the Table 14.2.

Table 14.2 Manipulator parameters for static and optimal balanced cases

Parameter	Static	Optimal
Ground joint of spring, d (m)	0.4905	1.3
Length of parallelogram sides, s_1, s_2, s_3 (m)	1.7, 0.5, 0.3	0.4, 0.5, 0.2
First spring stiffness, k_1 (N m^{-1})	100	82.02
Second spring's stiffness, k_2 (N m^{-1})	100	0

14.4.2 Optimality Conditions for Normal and Static Balanced Case

In this section, using the general formulation mentioned in Sect. 14.2, optimality conditions are derived for the considered two-link manipulator at unbalanced and static balanced cases. For the unbalanced case, all the parameters associated with the springs are supposed to be zero, and finding the optimal trajectory between the two given points of the end-effector is considered. For the static balanced case, at first unknown parameters are obtained using Eq. (14.39), then the optimal trajectory for the given performance index will be achieved. Consequently in both unbalanced and statically balanced cases, the unknown parameters do not appear in trajectory optimization procedure. The initial position of the end-effector in XY plane at $t = 0$ is $P_0 = (x_0, y_0)$ and the final position at $t = t_f$ is $P_f = (x_f, y_f)$. The initial and final velocity is considered to be zero. So by solving the inverse kinematic equations, one can write the boundary conditions as follows:

$$\begin{aligned} &\theta_1(0) = \theta_{10}, \quad \theta_2(0) = \theta_{20}, \quad \theta_1(\mathrm{t_f}) = \theta_{1\mathrm{f}}, \quad \theta_2(\mathrm{t_f}) = \theta_{2\mathrm{f}} \\ &\dot{\theta}_1(0) = \dot{\theta}_2(0) = \dot{\theta}_1(\mathrm{t_f}) = \dot{\theta}_2(\mathrm{t_f}) = 0. \end{aligned} \tag{14.41}$$

At the first step, using Eq. (14.1) by defining the continuous state vector as follows:

$$X_1 = \begin{bmatrix} \theta_1(t) \\ \theta_2(t) \end{bmatrix} = \begin{bmatrix} x_1(t) \\ x_2(t) \end{bmatrix}, \qquad X_2 = \begin{bmatrix} \dot{\theta}_1(t) \\ \dot{\theta}_2(t) \end{bmatrix} = \begin{bmatrix} x_3(t) \\ x_4(t) \end{bmatrix}, \tag{14.42}$$

the state space form of Eq. (14.37), using Eq. (14.4) becomes

$$\begin{aligned} \dot{x}_1 &= x_3 \\ \dot{x}_2 &= x_4 \\ \dot{x}_3 &= \frac{M_{22}(X_1)\,(\tau_1 - C_1(X_1, X_2) - G_1(X_1)) - M_{12}(X_1)\,(\tau_2 - C_2(X_1, X_2) - G_2(X_1))}{M_{11}(X_1)\,M_{22}(X_1) - M_{12}(X_1)\,M_{21}(X_1)} \\ \dot{x}_4 &= \frac{M_{11}(X_1)\,(\tau_2 - C_2(X_1, X_2) - G_2(X_1)) - M_{21}(X_1)\,(\tau_1 - C_1(X_1, X_2) - G_1(X_1))}{M_{11}(X_1)\,M_{22}(X_1) - M_{12}(X_1)\,M_{21}(X_1)}. \end{aligned} \tag{14.43}$$

where M_{ij}, C_i, and G_i $(i, j = 1, 2)$ are substituted from Eqs. (14.38) and (14.40) for normal case and static balanced case, respectively. For unbalanced case unknown parameters are considered to be zero in Eq. (14.38).

Now according to Eq. (14.7), by considering the performance index as minimum control effort is defined as follows:

$$J = \int_{t_0}^{t_f} \left(\tau_1^2 + \tau_2^2\right) dt, \tag{14.44}$$

and the costate vector as $\lambda = \begin{bmatrix} x_5 & x_6 & x_7 & x_8 \end{bmatrix}$, the Hamiltonian function becomes

$$H = \tau_1^2 + \tau_2^2 + x_5\dot{x}_1 + x_6\dot{x}_2 + x_7\dot{x}_3 + x_8\dot{x}_4, \tag{14.45}$$

where $\dot{x}_i,\ i = 1, \ldots, 4$ can be substituted from Eq. (14.43). Then by substituting Eq. (14.43) into Eq. (14.45), and differentiating the Hamiltonian function with respect to the states, according to Eq. (14.8), the costate equations are obtained as follows:

$$\begin{aligned}
\dot{x}_5 &= -\frac{\partial H}{\partial x_1} = -\frac{\partial}{\partial x_1}\left(\frac{(-C_1 - G_1)(x_7 M_{22} - x_8 M_{12}) + (-C_2 - G_2)(-x_7 M_{12} + x_8 M_{11})}{M_{11}M_{22} - M_{12}M_{21}}\right) \\
\dot{x}_6 &= -\frac{\partial H}{\partial x_2} = -\frac{\partial}{\partial x_2}\left(\frac{(-C_1 - G_1)(x_7 M_{22} - x_8 M_{12}) + (-C_2 - G_2)(-x_7 M_{12} + x_8 M_{11})}{M_{11}M_{22} - M_{12}M_{21}}\right) \\
\dot{x}_7 &= -\frac{\partial H}{\partial x_3} = -x_5 - \frac{\partial}{\partial x_3}\left(\frac{(-C_1)(x_7 M_{22} - x_8 M_{12}) + (-C_2)(-x_7 M_{12} + x_8 M_{11})}{M_{11}M_{22} - M_{12}M_{21}}\right) \\
\dot{x}_8 &= -\frac{\partial H}{\partial x_4} = -x_6 - \frac{\partial}{\partial x_4}\left(\frac{(-C_1)(x_7 M_{22} - x_8 M_{12}) + (-C_2)(-x_7 M_{12} + x_8 M_{11})}{M_{11}M_{22} - M_{12}M_{21}}\right)
\end{aligned} \tag{14.46}$$

After that the control values can be obtained by solving the following equations

$$\frac{\partial H}{\partial \tau_1} = 0, \qquad \frac{\partial H}{\partial \tau_2} = 0. \tag{14.47}$$

So by substituting the Hamiltonian function from Eq. (14.45) into Eq. (14.47), the optimal control laws become

$$\begin{aligned}
\tau_1 &= \frac{0.5}{M_{11}M_{22} - M_{12} - M_{21}}\left(-x_7 M_{22} + x_8 M_{21}\right) \\
\tau_2 &= \frac{0.5}{M_{11}M_{22} - M_{12} - M_{21}}\left(-x_7 M_{22} + x_8 M_{11}\right).
\end{aligned} \tag{14.48}$$

Finally by substituting Eq. (14.48) into Eqs. (14.43) and (14.46), eight nonlinear ordinary differential equations will be obtained which with eight boundary conditions given in Eq. (14.41), construct a two point boundary value problem. This problem can be solved using the bvp4c command in MATLAB®.

14.4.2.1 Optimal Counterweight Balancing

Unlike the static and adaptive balanced cases in which the counterweights are dependent on manipulator parameters as shown by Eq. (14.32), in the optimal balanced case the values of counterweights depends on the dynamic equations, the performance index, and the boundary conditions. Therefore, the counterweights and the optimal trajectory are obtained simultaneously in such a way that the given performance index is minimized. Dynamic equations, costate equations, and optimal control law are the same as obtained in Eqs. (14.43), (14.46), and (14.48), respectively. The dynamic parameters M_{ij}, C_i and G_i $(i, j = 1, 2)$ in Eqs. (14.43), (14.46), and (14.48), can be substituted form Eq. (14.31). Here, in all equations m_{c1} and m_{c2} are considered to be unknown parameters.

Now using Eq. (14.14), by defining the two new state variables x_9 and x_{10}, the optimality conditions associated with the parameters are given by

$$\dot{x}_9 = -\frac{\partial H}{\partial m_{c1}}, \qquad \dot{x}_{10} = -\frac{\partial H}{\partial m_{c2}} \tag{14.49}$$

where according to Eqs. (14.14) and (14.15) the associated boundary conditions become

$$x_9(0) = x_{10}(0) = x_9\,(t_f) = x_{10}\,(t_f) = 0. \tag{14.50}$$

At last, by substituting Eq. (14.48) into Eqs. (14.43), (14.46), and (14.51), ten nonlinear ordinary differential equations with respect to state $\left[x_1\ x_2\ x_3\ x_4\right]$, costate $\left[x_5\ x_6\ x_7\ x_8\right]$, new states $\left[x_9\ x_{10}\right]$ and unknown parameters $\left[m_{c1}\ m_{c2}\right]$ will be achieved. These ten equations with 12 boundary conditions given in Eqs. (14.41) and (14.52) construct a two point boundary value problem and by solving it using bvp4c command in MATLAB, all the states and unknown parameters can be obtained.

14.4.2.2 Optimal Spring Balancing

Unlike the static balanced case in which the unknown parameters are dependent on manipulator parameters as Eq. (14.39), in optimal balanced case the values of unknowns are dependent on dynamic equations, performance index, and boundary conditions according to Eq. (14.8). Therefore, the unknown parameters and optimal trajectory are obtained simultaneously in such a way that the given performance index is minimized. In this case, optimal control problem involving parameters which its optimality conditions are given in Eq. (14.8) must be considered. All of dynamic equations, costate equations, and optimal control law are the same as unbalanced case obtained in the last section. For convenience the optimization process selection of unknown parameters is divided into two steps. At the first step, k_1, k_2, and s_2 are considered to be known parameters and the optimal value of s_1, d, and s_3 are obtained, on the other hand the parameter vector in Eq. (14.3) is considered to be $\mathbf{b} = [\ s_1\ d\ s_3]^T$. At the second step, the obtained values for s_1, d, and s_3 at the first step are rounded, and unknown parameters vector is considered to be $\mathbf{b} = [\ k_1\ k_2]^T$. In the first step, by defining the three new state variables as x_9, x_{10}, and x_{11}, the optimality conditions associated with the parameters become

$$\dot{x}_9 = -\frac{\partial H}{\partial s_1}, \qquad \dot{x}_{10} = -\frac{\partial H}{\partial d}, \qquad \dot{x}_{11} = -\frac{\partial H}{\partial s_3}, \tag{14.51}$$

where according to Eq. (14.8) the associated boundary conditions become

$$x_{9,10,11}(0) = x_{9,10,11}\,(\mathrm{t_f}) = 0. \tag{14.52}$$

For the second step, by defining of two new state variables as x_9 and x_{10}, one can write the optimality conditions as

$$\dot{x}_9 = -\frac{\partial H}{\partial k_1}, \qquad \dot{x}_{10} = -\frac{\partial H}{\partial k_2}, \tag{14.53}$$

where according to Eq. (14.8) the associated boundary conditions become

$$x_{9,10}(0) = x_{9,10}\,(\mathrm{t_f}) = 0. \tag{14.54}$$

At last, by substituting Eq. (14.48) into Eqs. (14.43), (14.46), and (14.51), 11 nonlinear ordinary differential equations in terms of the state $[x_1\ x_2\ x_3\ x_4]$, costate $[x_5\ x_6\ x_7\ x_8]$, new states $[x_9\ x_{10}\ x_{11}]$, and unknown parameters $(s_1,\ d,\ s_3)$ will be achieved. These 11 equations with 14 boundary conditions given in Eqs. (14.41) and (14.52), construct a two point boundary value problem which by solving it all the states and unknown parameters can be obtained.

14.4.3 Simulation Results

In these simulations, the five different methods are compared. Normal case means unbalanced form of manipulator, counterweight static balanced means static balanced of manipulator using mass, counterweight optimal means optimal balanced of manipulator using mass, zero free spring-static means static balanced with spring, and zero free spring-optimal means optimal balanced of manipulator using spring.

The initial position of the end-effector in *XZ* plane at $t = 0$ is $p_0 = (1,\ 0)\,m$ and the final position at $t = 1$ s is $p_f = (0,\ 1.73)\,m$. The initial and final velocities are zero. From the inverse kinematic equations, the boundary condition can be expressed as

$$\begin{aligned}&\theta_1(0) = 60^\circ, \quad \theta_2(0) = 120^\circ, \quad \theta_1\,(\mathrm{t_f}) = 120^\circ, \quad \theta_2\,(\mathrm{t_f}) = -60^\circ\\ &\dot{\theta}_1(0) = \dot{\theta}_2(0) = \dot{\theta}_1\,(\mathrm{t_f}) = \dot{\theta}_2\,(\mathrm{t_f}) = 0.\end{aligned} \tag{14.55}$$

The results of simulations for the normal case, counterweight static balanced case and counterweight optimal balanced case are the same as reported in [10]. For counterweight optimal balanced case, at first the values of parameters are obtained using Eq. (14.39). Then the corresponding boundary value problem derived in Sect. 14.4.2.1 is solved to obtain the states and controls. For spring optimal balanced case the corresponding boundary value problem derived in Sect. 14.4.2.2 is solved to obtain states, controls, and unknown parameters. The manipulator parameters for static and optimal balanced cases are given in Table 14.2. In optimal case, since the second spring's stiffness k_2 is zero, the value of s_1 is unimportant and it is probable to eliminate the second spring in practice.

The obtained optimal trajectories between the initial and final points for the five cases are shown in Fig. 14.3. Figure 14.4 shows the obtained torque of motors. The angular position and angular velocity of links are illustrated in Fig. 14.5. The second column of Table 14.3 shows the values of performance index defined in Eq. (14.44) for five considered cases. The third and forth column represent the improvement relative to the normal case in state of amplification or reduction. As it reported in [10] and can be shown in Table 14.3, the performance index for counterweight-optimal balanced case is less than normal case and counterweight-static balanced case. While the performance index for spring-optimal balanced case is less than the all other cases. In the following figures, readers should notice optimal balancing decrement of input torques, and its effect on the trajectory (path and velocity profile of joints).

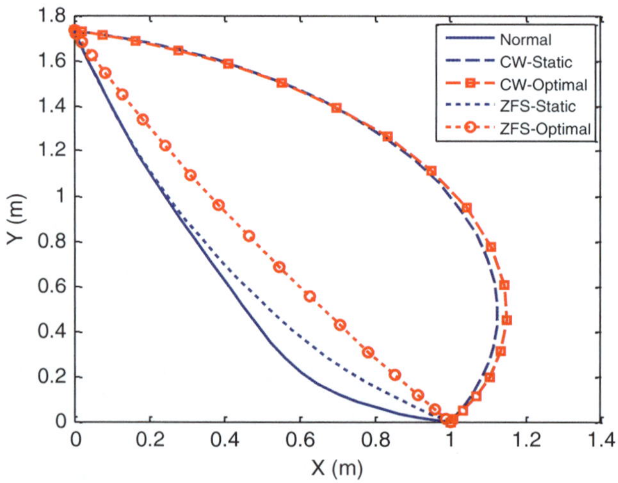

Fig. 14.3 Optimal trajectories for different cases

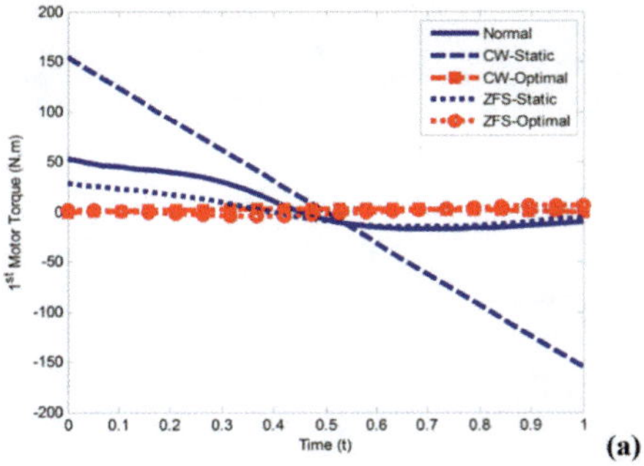

Fig. 14.4 Input torques of motor 1 and 2 (effect of optimal spring and mass balancing)

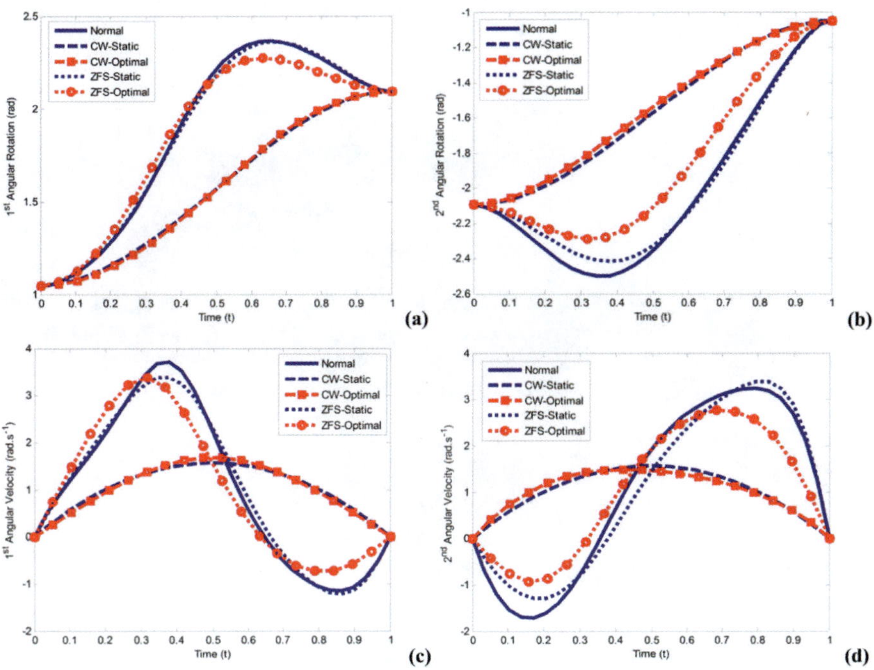

Fig. 14.5 Angular position and angular velocity of links (The effect of optimal balancing on velocity)

14.5 PUMA-Like Robot

14.5.1 Modeling

A spatial three-jointed PUMA robot is considered as shown in Fig. 14.6. DH parameters and links parameters are given in Tables 14.4 and 14.5. In this robot the first spring is connected between the base and the parallel fractional mechanism. Second spring is connected between second link and third link as shown in Fig. 14.6.

Table 14.3 Comparison of performance indexes

Case	Pay-off $(\mathrm{Nm})^2$ s	Reduction (times)	Amplification (times)
Normal	1,090	1	1
Counterweight static balanced	5,770	–	5.29
Counterweight optimal balanced	564	1.93	–
Spring static balanced	361	3.02	–
Spring optimal balanced	52	20.96	–

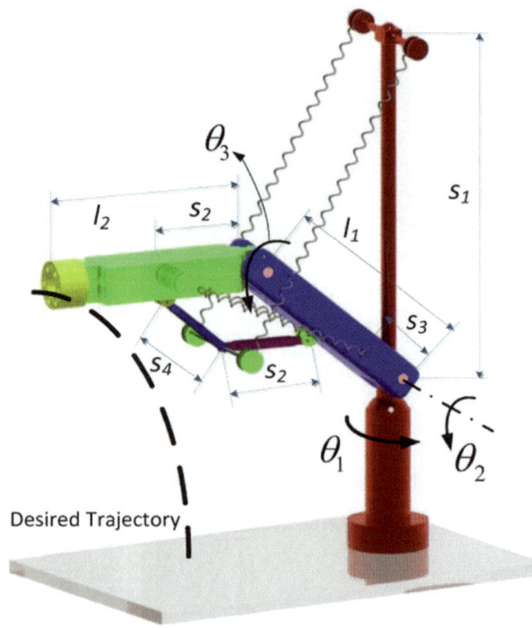

Fig. 14.6 PUMA-like robot with additional springs and parallelogram mechanism

Table 14.4 Denavit–Hartenberg parameters for a PUMA-Like robot

Link	θ_i (rad)	α (rad)	a_i (m)	d_i (m)
1	q_1	$\pi/2$	0	0.4
2	q_2	0	0.5	0
3	q_3	0	0.5	0

For obtaining the dynamic equations, the Lagrangian formulation is used. Total Lagrangian for this robot can be written as follows:

$$L_t = L + L_{sp}, \tag{14.56}$$

where L_t is total Lagrangian, L is Lagrangian of robot, and L_{sp} is additional Lagrangian due to springs. The additional Lagrangian can be stated as

Table 14.5 Link parameters and inertia properties [4]

Link	Length (m)	Mass (kg)	Moment of inertia (kg m^2)			Link center of mass (m)
1	0.4	12	0	0	0	0
			0	0.2	0	−0.2
			0	0	0	0
2	0.5	10	0	0	0	−0.25
			0	0.2	0	0
			0	0	0.2	0
3	0.5	5	0	0	0	−0.25
			0	0.1	0	0
			0	0	0.1	0

$$L_{sp} = K - U = 0 - U = -\sum_{i=1}^{2} \frac{1}{2} k_i x_i^2, \tag{14.57}$$

where k is stiffness of spring and x_i is deformed length of springs.

14.5.2 Static Balancing

The static balancing is considered by two counterweights and its application of the balancing equation with the manipulator's parameters given in the Table 14.5 leads to the following counterweight's masses as

$$\begin{aligned} m_{c2} r_2 &= 0.5 \times m_p + 1.25 \\ m_{c1} r_1 &= 0.5 \times \left(m_p + m_{c2} + 5\right) + 2.5. \end{aligned} \tag{14.1}$$

These can be rearranged as

$$m_{c2} = \frac{m_p + 2.5}{2r_2}, \quad m_{c1} = \frac{(2r_2 + 1)\, m_p + 20r_2 + 2.5}{2r_1}, \tag{14.2}$$

Therefore for any selection of r_1 and r_2 and payload m_p, there are corresponding m_{c1} and m_{c2}. If the value of r_1 and r_2 considered as 0.25 and 0.125 correspondingly, two methods for static balancing are possible: with payload 2 kg and without payload value. In the first method, $m_{c1} = 40$ kg and $m_{c2} = 10$ kg. These counterweights are considered for comparison, and maybe their application is impossible.

14.5.3 Point-to-Point Motion

After deriving the dynamic equations for this robot, using Eq. (14.8) the optimality condition can be obtained as the same way presented for the two-link manipulator. The boundary conditions are considered as follows:

$$\theta_1(0) = 17°, \quad \theta_1(t_f) = 29.22°, \quad \theta_2(0) = 29°, \quad \theta_2(t_f) = -24°$$
$$\theta_3(0) = 11.45°, \quad \theta_3(t_f) = 32.23°,$$
$$\dot{\theta}_1(0) = \dot{\theta}_1(t_f) = \dot{\theta}_2(0) = \dot{\theta}_2(t_f) = \dot{\theta}_3(0) = \dot{\theta}_3(t_f) = 0. \tag{14.58}$$

For this robot simulations are performed for two cases: normal case and optimal balanced case. For the normal case all the parameters dealing with the springs are considered to be zero. For optimal balanced case, at first the stiffness of springs, k_1 and k_2 are considered to be known and the values of distance between joints and spring connection points are determined. In the next step, by considering the rounded position values, the optimal value of stiffness are obtained. Optimal values of parameters are listed in Table 14.6.

The obtained optimal controls are shown in Fig. 14.7. The angular position and angular velocity of links are illustrated in Fig. 14.8. The optimal trajectories for normal and optimal balanced cases are given in Figs. 14.9 and 14.10 respectively. Performance index for normal case is found to be 8.81 Nm^2 and for spring optimal balanced case is found to be 7.08 Nm^2 that this value is 20 % less than normal case.

Table 14.6 Optimal values of parameters for PUMA-like robot in point-to-point motion

Parameter	Value (unit)
First spring stiffness, k_1	1.23 (N m^{-1})
Second spring stiffness, k_2	0 (N m^{-1})
Ground joint of Spring, s_1	1.3 (m)
Length of second spring application point, s_2	0.186 (m)

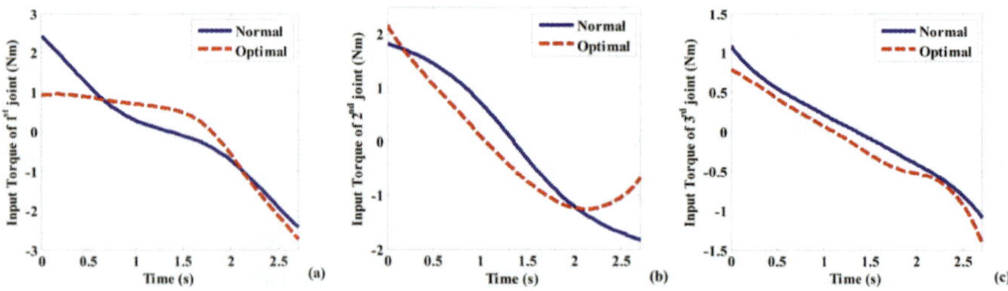

Fig. 14.7 Input torques of motors

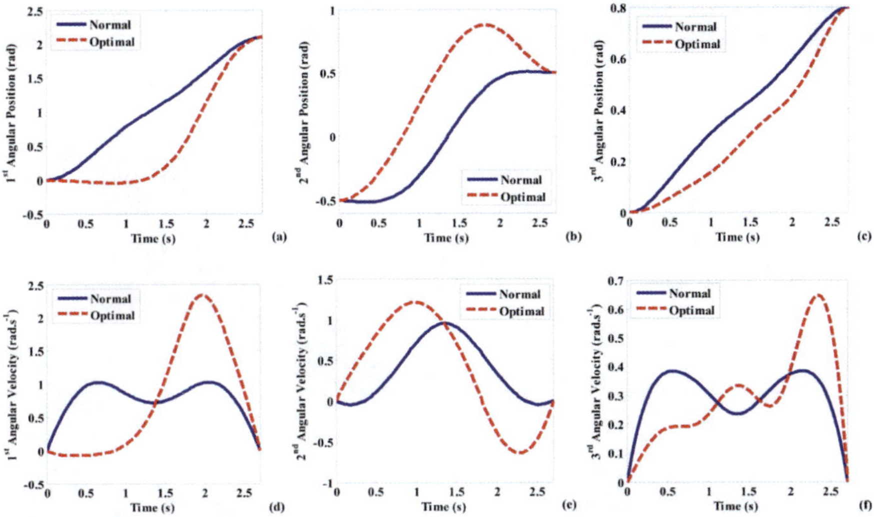

Fig. 14.8 Angular position and velocity of links

Fig. 14.9 Optimal trajectories for normal case

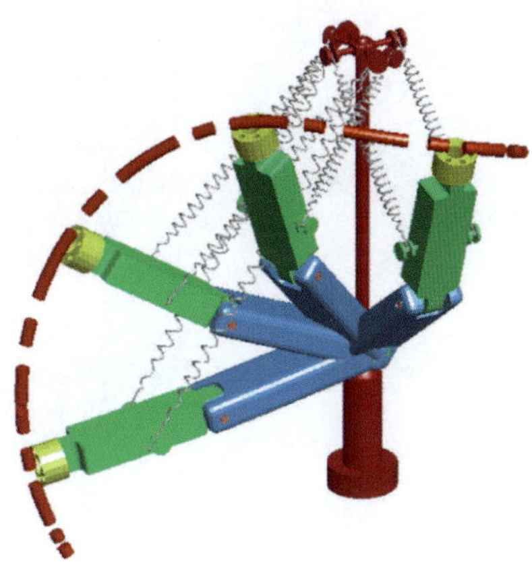

14.5.4 Specified Path Tracking

For motion in predefined path, the performance index is considered as follows:

$$J = \int_{t_0}^{t_f} \left(w \left(\left(x_{\mathrm{end}} - x_{\mathrm{p}} \right)^2 + \left(y_{\mathrm{end}} - y_{\mathrm{p}} \right)^2 + \left(z_{\mathrm{end}} - z_{\mathrm{p}} \right)^2 \right) + \tau_1^2 + \tau_2^2 + \tau_3^2 \right) dt, \tag{14.59}$$

where $x_{\mathrm{end}}, y_{\mathrm{end}}, z_{\mathrm{end}}$ are end effector coordinate and $x_{\mathrm{p}}, y_{\mathrm{p}}, z_{\mathrm{p}}$ are path coordinate. The trajectory is a quadrant that start point is $p_0 = (0.3, 0.8, 0.3)$ and final point is $p_{\mathrm{f}} = (3.0, 0.3, 0.8)$. The equation of path is defined as following:

Fig. 14.10 Optimal trajectories for optimal cases

Table 14.7 Optimal values of parameters for PUMA-like robot

Parameter	Value (unit)
First spring stiffness, k_1	3.5 (N m^{-1})
Second spring stiffness, k_2	4 (N m^{-1})
Length of first spring connection point, s_1, s_2	1.3, 0 (m)
Length of second spring connection point, s_2	0, 0.5 (m)

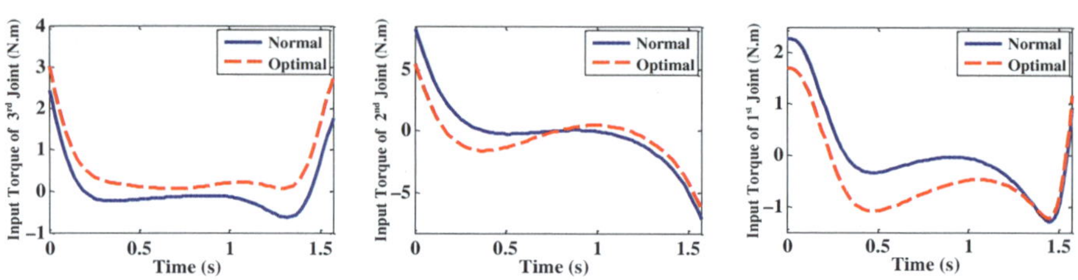

Fig. 14.11 Input torques of motors

$$\begin{cases} x_p = 0.3 \\ y_p = 0.3 + 0.5\cos(t) \\ z_p = 0.3 + 0.5\sin(t) \end{cases} \tag{14.60}$$

After deriving the dynamic equations, using Eq. (14.8) the optimality condition can be obtained. The obtained optimal value of the parameters, are given in Table 14.7.

The obtained optimal controls are shown in Fig. 14.11. The angular position and angular velocity of links are illustrated in Fig. 14.12. The optimal trajectories for normal and optimal balanced cases are given in Fig. 14.13. Performance index for normal case is found to be 10.6Nm^2 and for spring optimal balanced case is found to be 6.903 Nm^2 that this value is 34.9 % less than normal case.

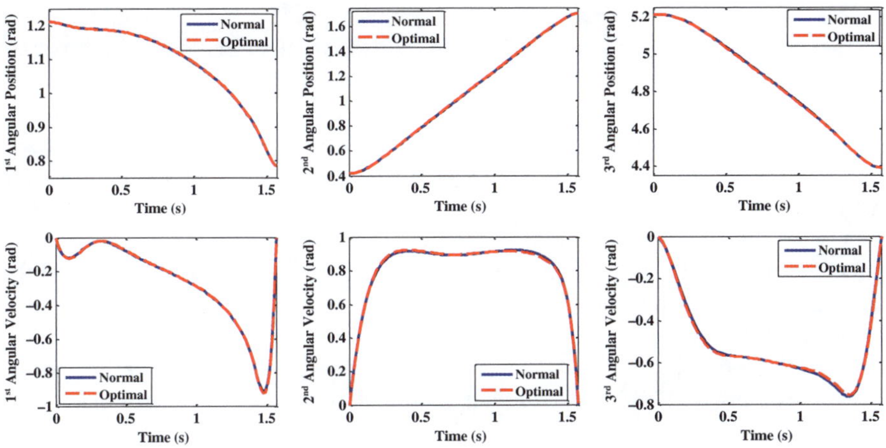

Fig. 14.12 Angular position and velocity of links (should be as same as each other)

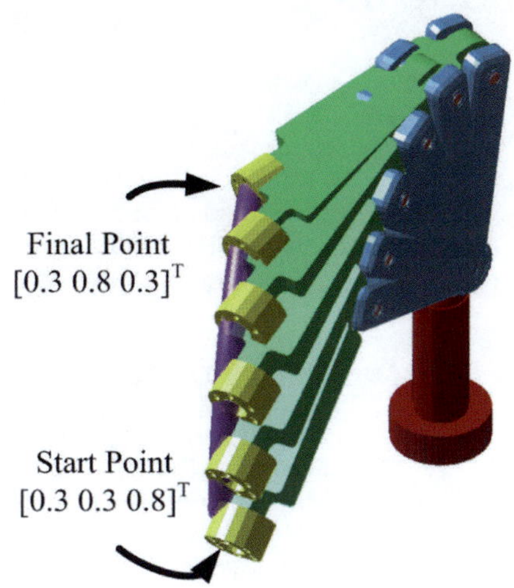

Fig. 14.13 The trajectory of the end effector

14.6 Conclusion

In this chapter, the optimal balancing for robotic system based on the indirect solution of optimal control problem is formulated and then verified by simulation. The method uses the well-known Pontryagin's minimum principle. The obtained equations lead to a standard form of a two-point boundary value problem which can be solved by computer programs such as MATLABs *bvp4c* command or Fortan's *twpbvp* code.

The efficiency of the proposed method is investigated through computer simulations for a two-link manipulator and the PUMA-like manipulator. The obtained results show that, although the performance index for the static balanced manipulator has been reduced 66.8 % with respect to unbalanced case, by applying the proposed method this reduction reaches to 95 % by using spring balancing. It is also shown that the performance index for spring balancing is very less than the performance index for the counterweight balancing. This result is expected, because in the counterweight balancing the moment of inertia is increased due to the added masses. Finally, simulation is performed for a PUMA-like robot and the capability of the method to solve the complicated problem is shown. For this case study, performance index for optimal balanced case is obtained 20 % less than the unbalanced case. Also, effect of the predefined trajectory is examined for the optimal spring balancing for PUMA-like robot. By simulation, the performance index for the optimal balanced case is obtained 34.9 % less than the unbalanced case.

References

1. Park, J., Haan, J., Park, F.C.: Convex optimization algorithms for active balancing of humanoid robots. IEEE Trans. Robotics **23**(4), 817–822 (2007)
2. Moradi, M., Nikoobin, A., Azadi, S.: Adaptive decoupling for open chain planar robots. Sci. Iran. Trans. B Mech. Eng. **17**(5B), 376–386 (2010)
3. Arakelian, V., Ghazaryan, S.: Improvement of balancing accuracy of robotic systems: Application to leg orthosis for rehabilitation devices. Mech. Mach. Theory **43**, 565–575 (2008)
4. Nikoobin, M., Moradi, A.E.: Optimal spring balancing of robot manipulators in point-to-point motion. Robotica **31**(4), 611–621 (2013)
5. Kolarski, M., Vukobratovic, M., Borovac, B.: Dynamic analysis of balanced robot mechanisms. Mech. Mach. Theory **29**(3), 427–454 (1994)
6. Banala, S.K., Agrawal, S.K., Fattah, A., Krishnamoorthy, V., Hsu, W.L., Scholz, J., Rudolph, K.: Gravity-balancing leg orthosis and its performance evaluation. IEEE Trans. Robotics **22**(6), 1228–1239 (2006)
7. Kochev, I.S.: General theory of complete shaking moment balancing of planar linkages: a critical review. Mech. Mach. Theory **35**, 1501–1514 (2000)
8. Saravanan, R., Ramabalan, S., Babu, P.D.: Optimum static balancing of an industrial robot mechanism. Eng. Appl. Artif. Intel. **21**(6), 824–834 (2008)
9. Ravichandran, T., Wang, D., Heppler, G.: Simultaneous plant-controller design optimization of a two-link planar manipulator. Mechatronics **16**, 233–242 (2006)
10. Nikoobin, A., Moradi, M.: Optimal balancing of robot manipulators in point-to-point motions. Robotica **29**(2), 233–244 (2011)
11. Cheng, L., Lin, Y., Hou, Z.G., Tan, M., Huang, J., Zhang, W.J.: Integrated design of machine body and control algorithm for improving the robustness of a closed-chain five-bar machine. IEEE/ASME Trans. Mechatron. **17**(3), 587–591 (2012)
12. Cheng, L., Hou, Z.G., Tan, M., Zhang, W.J.: Tracking control of a closed-chain five-bar robot with two degrees of freedom by integration of approximation-based approach and mechanical design. IEEE Trans. Syst. Man Cybern. B Cybernetics **42**(5), 1470–1479 (2012)
13. Chettibi, T., Lehtihet, H.E., Haddad, M., Hanchi, S.: Minimum cost trajectory planning for industrial robots. Eur. J. Mech. A/Solids **23**(4), 703–715 (2004)
14. Callies, R., Rentrop, P.: Optimal control of rigid-link manipulators by indirect methods. GAMM-Mitteilungen **31**(1), 27–58 (2008)

15. Betts, J.T.: Survey of numerical methods for trajectory optimization. J. Guid. Cont. Dyn. **21**(2), 193–207 (1998)
16. Korayem, M.H., Nikoobin, A.: Formulation and numerical solution of robot manipulators in point-to-point motion with maximum load carrying capacity. Sci. Iran. J. **16**(1), 101–109 (2009)
17. Kamenskii, V.A.: On the problem of the number of counterweights in the balancing of plane linkages. J. Mech. **4**, 323–333 (1969)

Chapter 15
Dynamics and Control of Planar, Translational, and Spherical Parallel Manipulators

Victor Glazunov and Sergey Kheylo

Abstract This chapter focuses on a study of the dynamics and control of planar, translational, and spherical parallel manipulators. These mechanisms are the most often used in different applications. The synthesis of these mechanisms is carried out by means of screw groups. The dynamics and control are considered by means the constraint equations. The control algorithms based on dynamical model without linearization use the concept of inverse dynamic problems.

Keywords Parallel manipulator • Planar • Translational and spherical mechanisms • Dynamics • Control • Singularity

15.1 Introduction

It is well known that the closed-loops of parallel manipulators cause high stiffness and payload capacity [1–9]. The parallel manipulators can have high velocities, because the actuators are situated on the base and the mass of the links are lights. As an example, one could consider robot Delta, which is well known. That is why investigations of dynamic properties represent a big challenge for scientists.

As a rule, the control problem is solved basing on the kinematic equations taking into account no dynamical properties and mutual influence between the actuators. Therefore, the control based on dynamical model represents a challenge too.

The most applicable of them are planar, translational, and spherical parallel mechanisms that are used as machine tools, simulators, surgery robots, measure equipment, etc. For structural synthesis of these mechanisms the theory of screws or Lie groups can be used [10, 11]. The structure and kinematics of planar, translational, and spherical mechanisms in many publications is investigated [12–18].

V. Glazunov (✉)
Institute of Machines Sciences of RAS, Moscow, Russia
e-mail: vaglznv@mail.ru

S. Kheylo
Moscow State University of Design and Technology, Moscow, Russia
e-mail: sheilo@yandex.ru

D. Zhang, B. Wei (eds.), *Dynamic Balancing of Mechanisms and Synthesizing of Parallel Robots*, DOI 10.1007/978-3-319-17683-3_15

On the basis of consideration of the planar, translational, and spherical mechanisms many decoupled manipulators were synthesized [19–21]. But the dynamics and control of them are considered not so often [22–25]. Therefore, this chapter is devoted to dynamical analysis and control of these manipulators. The approach used for control of the robots is based on the algorithm suggested in [26].

15.2 Description of Planar, Translational, and Spherical Mechanisms by Means Screw Groups

Let us consider the closed screw groups corresponding to planar, translational, and spherical mechanisms. These groups include all the screw products of their main members. We use the simplest representation of the main screws (twists) of these groups by Plücker coordinates.

The three-member closed screw groups are:

- Three-member screw group which can be represented by Plücker coordinates of the main members $\boldsymbol{\Omega}_1$ $(0, 0, 0, 1, 0, 0)$, $\boldsymbol{\Omega}_2$ $(0, 0, 0, 0, 1, 0)$, and $\boldsymbol{\Omega}_3$ $(0, 0, 1\ 0, 0, p)$. This group corresponds to one screw kinematic pair and two prismatic kinematic pairs whose axes are perpendicular to each other. If $p = 0$ then one kinematic pair gives rotation and these pairs express planar mechanisms.
- Three-member screw group which can be represented by Plücker coordinates of the main members $\boldsymbol{\Omega}_1$ $(0, 0, 0, 1, 0, 0)$, $\boldsymbol{\Omega}_2$ $(0, 0, 0, 0, 1, 0)$, and $\boldsymbol{\Omega}_3$ $(0, 0, 0, 0, 0, 1)$. This group corresponds to three prismatic kinematic pairs whose axes are not coplanar, in particular perpendicular to each other. These pairs express translational mechanisms.
- Three-member screw group which can be represented by Plücker coordinates of the main members $\boldsymbol{\Omega}_1$ $(1, 0, 0, 0, 0, 0)$, $\boldsymbol{\Omega}_2$ $(0, 1, 0, 0, 0, 0)$, and $\boldsymbol{\Omega}_3$ $(0, 0, 1, 0, 0, 0)$. This group corresponds to three rotation kinematic pairs whose axes intersect at one point. These pairs express spherical mechanisms.

Note that all the screw products of the main screws of these groups are members of the same group. If all the motions of a rigid body are described by one of these groups, then after any finite displacement of this body the screw group corresponding to all its motions will be the same as before motion. It means that a rigid body can be connected to the base by any number of kinematic chains corresponding to one of the closed screw groups and the degree of freedom will be determined by the number of the main members of this group.

Let us consider parallel manipulators corresponding to three-member screw groups. We use the notification (Fig. 15.1): (a) actuated prismatic pair (linear actuator), (b) actuated rotation pair (rotating actuator), (c) twist of zero pitch, (d) twist of infinite pitch, (e) wrench of zero pitch, (f) wrench of infinite pitch.

Firstly, let us consider the three-member screw group corresponding to a planar parallel mechanism (Fig. 15.2a). Each kinematic chain can consist of one rotation

Fig. 15.1 Twists and wrenches

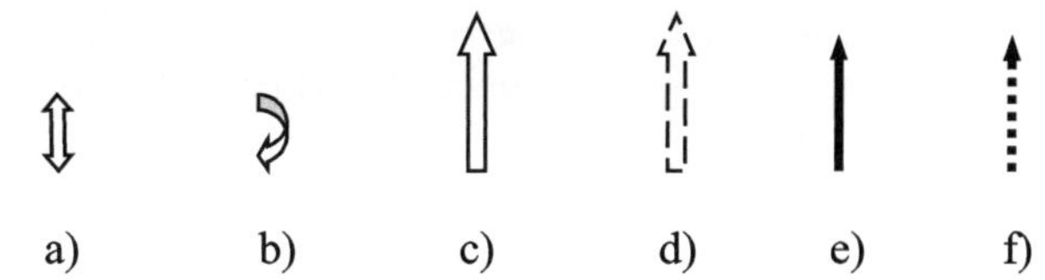

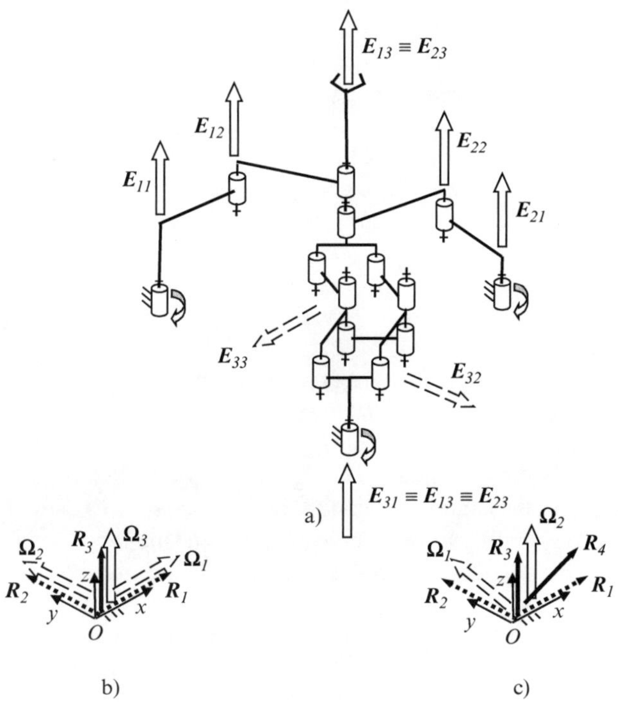

Fig. 15.2 Planar parallel mechanism

kinematic pair and two prismatic kinematic pairs (the axis of the rotation pair is perpendicular to the axes of the prismatic pairs), or of two rotation kinematic pairs and one prismatic kinematic pair (the axes of the rotation pairs are parallel to each other and are perpendicular to the axis of the prismatic pair), or of three rotation kinematic pairs with parallel axes. In our mechanism two kinematic chains consist of three rotation kinematic pairs (one of them is actuated and situated on the base) and one kinematic chains consists of one actuated rotation kinematic pair situated on the base (rotating actuator) and two prismatic kinematic pairs represented as four-bar parallelograms. The unit screws of the axes of these kinematic pairs have coordinates: $\boldsymbol{E_{11}}$ $(0, 0, 1, 0, 0, 0)$, $\boldsymbol{E_{12}}$ $(0, 0, 1, e_{12x}, e_{12y}, 0)$, $\boldsymbol{E_{13}}$ $(0, 0, 1, e_{13x}, e_{13y}, 0)$, $\boldsymbol{E_{21}}$ $(0, 0, 1, 0, 0, 0)$, $\boldsymbol{E_{22}}$ $(0, 0, 1, e_{22x}, e_{22y}, 0)$, $\boldsymbol{E_{23}}$ $(0, 0, 1, e_{23x}, e_{23y}, 0)$, $\boldsymbol{E_{31}}$ $(0, 0, 1, 0, 0, 0)$, $\boldsymbol{E_{32}}$ $(0, 0, 0, e_{32x}, e_{32y}, 0)$, $\boldsymbol{E_{33}}$ $(0, 0, 0, e_{33x}, e_{33y}, 0)$.

The screws $\boldsymbol{E}_{32}$ and $\boldsymbol{E}_{33}$ are of infinite pitch. All other screws are of zero pitch. All three kinematic chains impose the same constraints, so that one can insert other similar chains between the base and moving platform and the degree of freedom will remain equal to three. The wrenches of the constraints imposed by kinematic chains have coordinates (Fig. 15.2b): $\boldsymbol{R}_1$ $(0, 0, 0, 1, 0, 0)$, $\boldsymbol{R}_2$ $(0, 0, 0, 0, 1, 0)$, $\boldsymbol{R}_3$ $(0, 0, 1, 0, 0, 0)$. All the twists of motions of the platform can be represented by the twists reciprocal to the wrenches of the imposed constraints (Fig. 15.2b): $\boldsymbol{\Omega}_1$ $(0, 0, 0, 1, 0, 0)$, $\boldsymbol{\Omega}_2$ $(0, 0, 0, 0, 1, 0)$, $\boldsymbol{\Omega}_3$ $(0, 0, 1, 0, 0, 0)$. The twists $\boldsymbol{\Omega}_1$ and $\boldsymbol{\Omega}_2$ are of infinite pitch, while the twist $\boldsymbol{\Omega}_3$ is of zero pitch.

In this mechanism singularities corresponding to loss of one degree of freedom exist if three screws $\boldsymbol{E}_{i1}$, $\boldsymbol{E}_{i2}$, and $\boldsymbol{E}_{i3}$ $(i = 1, 2, 3)$ are linearly dependent which is possible if three screws $\boldsymbol{E}_{i1}$, $\boldsymbol{E}_{i2}$, and $\boldsymbol{E}_{i3}$ $(i = 1, 2)$ are situated in the same plane or if two screws $\boldsymbol{E}_{32}$, $\boldsymbol{E}_{33}$ are parallel. In particular if $\boldsymbol{E}_{32} = \boldsymbol{E}_{33}$ (Fig. 15.2c) then there exist four wrenches of constraints imposed by the kinematic chains: $\boldsymbol{R}_1$ $(0, 0, 0, 1, 0, 0)$, $\boldsymbol{R}_2$ $(0, 0, 0, 0, 1, 0)$, $\boldsymbol{R}_3$ $(0, 0, 1, 0, 0, 0)$, $\boldsymbol{R}_4$ $(r_{4x}, r_{4y}, 0, 0, 0, 0)$ and only two twists of motion of the platform reciprocal to these wrenches $\boldsymbol{\Omega}_1$ $(0, 0, 0, v_{1x}, v_{1y}, 0)$ and $\boldsymbol{\Omega}_2$ $(0, 0, 1, 0, 0, 0)$. Note that $\boldsymbol{R}_4$ is perpendicular to $\boldsymbol{E}_{32}$ and $\boldsymbol{E}_{33}$, and $\boldsymbol{\Omega}_1$ is parallel to them.

If the actuators are fixed, then there exist six wrenches imposed by the kinematic chains: $\boldsymbol{R}_1$ $(0, 0, 0, 1, 0, 0)$, $\boldsymbol{R}_2$ $(0, 0, 0, 0, 1, 0)$, $\boldsymbol{R}_3$ $(0, 0, 1, 0, 0, 0)$, $\boldsymbol{R}_4$ $(r_{4x}, r_{4y}, 0, 0, 0, 1)$, $\boldsymbol{R}_5$ $(r_{5x}, r_{5y}, 0, 0, 0, 1)$, and $\boldsymbol{R}_6$ $(0, 0, 0, 0, 0, 1)$. The wrenches $\boldsymbol{R}_4$ and $\boldsymbol{R}_5$, are of zero pitch, they are situated along the axes of the links connecting passive rotation pairs of the first and the second kinematic chains. $\boldsymbol{R}_6$ is of infinite pitch. Singularities corresponding to non-controlled infinitesimal motions of the moving platform exist if the wrenches $\boldsymbol{R}_1$, $\boldsymbol{R}_2$, $\boldsymbol{R}_3$, $\boldsymbol{R}_4$, $\boldsymbol{R}_5$, $\boldsymbol{R}_6$ are linearly dependent which is possible if the wrenches $\boldsymbol{R}_4$, and $\boldsymbol{R}_5$ coincide. In this case the twist of infinite pitch $\boldsymbol{\Omega}$ $(0, 0, 0, v_x, v_y, 0)$ exists which is perpendicular to the axes of the wrenches $\boldsymbol{R}_4$ and $\boldsymbol{R}_5$ and therefore reciprocal to all the wrenches $\boldsymbol{R}_1$, $\boldsymbol{R}_2$, $\boldsymbol{R}_3$, $\boldsymbol{R}_4$, $\boldsymbol{R}_5$, $\boldsymbol{R}_6$.

Note that singularities exist corresponding both to loss of one degree of freedom and to non-controlled infinitesimal motion of the moving platform. By this any three screws $\boldsymbol{E}_{i1}$, $\boldsymbol{E}_{i2}$, $\boldsymbol{E}_{i3}$ $(i = 1, 2, 3)$ and the wrenches $\boldsymbol{R}_1$, $\boldsymbol{R}_2$, $\boldsymbol{R}_3$, $\boldsymbol{R}_4$, $\boldsymbol{R}_5$, $\boldsymbol{R}_6$ are linearly dependent.

Now, we consider the three-member screw group corresponding to a translating parallel mechanism (Fig. 15.3a). Each kinematic chain consists of one actuated prismatic pair (linear actuator) situated on the base and two prismatic kinematic pairs represented as four-bar parallelograms. The unit screws of the axes of these kinematic pairs have coordinates: $\boldsymbol{E}_{11}$ $(0, 0, 0, 1, 0, 0)$, $\boldsymbol{E}_{12}$ $(0, 0, 0, 0, e_{12y}, e_{12z})$, $\boldsymbol{E}_{13}$ $(0, 0, 0, 0, e_{13y}, e_{13z})$, $\boldsymbol{E}_{21}$ $(0, 0, 0, 0, 1, 0)$, $\boldsymbol{E}_{22}$ $(0, 0, 0, e_{22x}, 0, e_{22z})$, $\boldsymbol{E}_{23}$ $(0, 0, 0, e_{23x}, 0, e_{23z})$, $\boldsymbol{E}_{31}$ $(0, 0, 0, 0, 0, 1)$, $\boldsymbol{E}_{32}$ $(0, 0, 0, e_{32x}, e_{32y}, 0)$, $\boldsymbol{E}_{33}$ $(0, 0, 0, e_{33x}, e_{33y}, 0)$.

All the screws are of infinite pitch. This mechanism is isotropic so that each actuator corresponds to one Cartesian coordinate x, y or z. All three kinematic chains impose the same constraints, so that one can insert other similar chains between the base and moving platform and the degree of freedom will remain equal to three. The wrenches of the constraints imposed by kinematic chains have coordinates

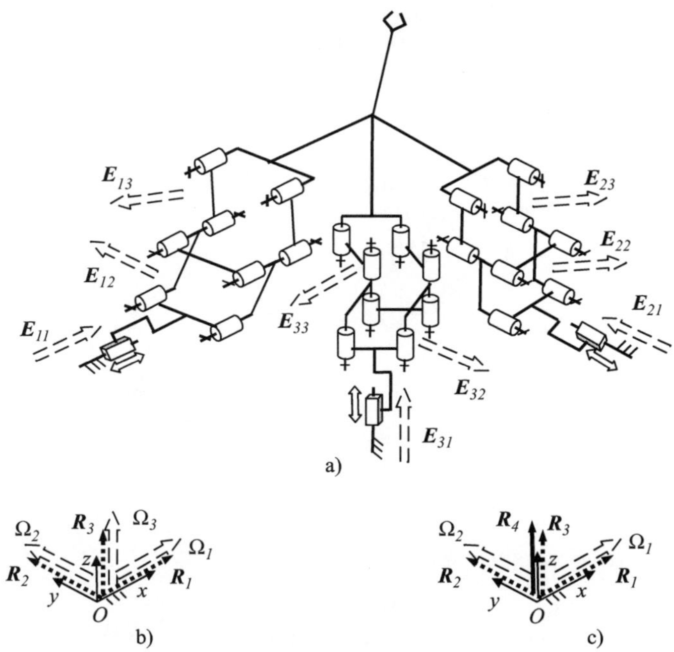

Fig. 15.3 Translational parallel mechanism

(Fig. 15.2b): $\boldsymbol{R_1}$ *(0, 0, 0, 1, 0, 0)*, $\boldsymbol{R_2}$ *(0, 0, 0, 0, 1, 0)*, $\boldsymbol{R_3}$ *(0, 0, 0, 0, 0, 1)*. All the twists of motions of the platform can be represented by the twists reciprocal to the wrenches of the imposed constraints (Fig. 15.3b): $\boldsymbol{\Omega_1}$ *(0, 0, 0, 1, 0, 0)*, $\boldsymbol{\Omega_2}$ *(0, 0, 0, 0, 1, 0)*, $\boldsymbol{\Omega_3}$ *(0, 0, 0, 0, 0, 1)*. All three twists are of infinite pitch.

In this mechanism, singularities corresponding to loss of one degree of freedom exist if three screws $\boldsymbol{E_{i1}}$, $\boldsymbol{E_{i2}}$ and $\boldsymbol{E_{i3}}$ ($i = 1, 2, 3$) are linearly dependent which is possible if any two screws $\boldsymbol{E_{i2}}$, $\boldsymbol{E_{i3}}$ are parallel. In particular, if $\boldsymbol{E_{22}}$ *(0, 0, 0, 1, 0, 0)* = $\boldsymbol{E_{23}}$ *(0, 0, 0, 1, 0, 0)* (Fig. 15.3c) then there exist four wrenches of constraints imposed by kinematic chains: $\boldsymbol{R_1}$ *(0, 0, 0, 1, 0, 0)*, $\boldsymbol{R_2}$ *(0, 0, 0, 0, 1, 0)*, $\boldsymbol{R_3}$ *(0, 0, 0, 0, 0, 1)*, and $\boldsymbol{R_4}$ *(0, 0, 1, 0, 0, 0)* and only two twists of motion of the platform reciprocal to these wrenches $\boldsymbol{\Omega_1}$ *(0, 0, 0, 1, 0, 0)* and $\boldsymbol{\Omega_2}$ *(0, 0, 0, 0, 1, 0)*. If the parallelograms in each kinematic chain are replaced by general prismatic kinematic pairs then this singularity does not exist.

Now let us consider the three-member screw group corresponding to a spherical parallel mechanism (Fig. 15.4a). Each kinematic chain consists of one actuated rotation pair (rotating actuator) situated on the base and two passive rotation kinematic pairs. The unit screws of the axes of these kinematic pairs have coordinates (note that the origin of the coordinate system is the point O in which the axes of all the pairs intersect): $\boldsymbol{E_{11}}$ *(1, 0, 0, 0, 0, 0)*, $\boldsymbol{E_{12}}$ $(e_{12x}, e_{12y}, e_{12z}, 0, 0, 0)$, $\boldsymbol{E_{13}}$ $(e_{13x}, e_{13y}, e_{13z}, 0, 0, 0)$, $\boldsymbol{E_{21}}$ *(0, 1, 0, 0, 0, 0)*, $\boldsymbol{E_{22}}$ $(e_{22x}, e_{22y}, e_{22z}, 0, 0, 0)$, $\boldsymbol{E_{23}}$ $(e_{23x}, e_{23y}, e_{23z}, 0, 0, 0)$, $\boldsymbol{E_{31}}$ *(0, 0, 1, 0, 0, 0)*, $\boldsymbol{E_{32}}$ $(e_{32x}, e_{32y}, e_{32z}, 0, 0, 0)$, $\boldsymbol{E_{33}}$ $(e_{33x}, e_{33y}, e_{33z}, 0, 0, 0)$.

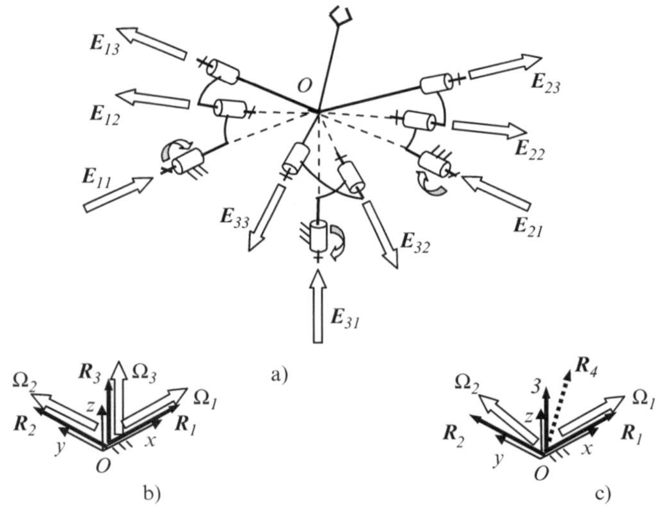

Fig. 15.4 Spherical parallel mechanism

All the screws are of zero pitch. All three kinematic chains impose the same constraints, so that one can insert other similar chains between the base and moving platform and the degree of freedom will remain equal to three. The wrenches of the constraints imposed by kinematic chains have coordinates (Fig. 15.4b): $\boldsymbol{R_1}$ $(1, 0, 0, 0, 0, 0)$, $\boldsymbol{R_2}$ $(0, 1, 0, 0, 0, 0)$, $\boldsymbol{R_3}$ $(0, 0, 1, 0, 0, 0)$, these wrenches are of zero pitch. All the twists of motions of the platform can be represented by the twists reciprocal to the wrenches of the imposed constraints (Fig. 15.4b): $\boldsymbol{\Omega_1}$ $(1, 0, 0, 0, 0, 0)$, $\boldsymbol{\Omega_2}$ $(0, 1, 0, 0, 0, 0)$, $\boldsymbol{\Omega_3}$ $(0, 0, 1, 0, 0, 0)$. All three twists are of zero pitch.

In this mechanism singularities expressed by loss of one degree of freedom exist if any three screws $\boldsymbol{E_{i1}}$, $\boldsymbol{E_{i2}}$, $\boldsymbol{E_{i3}}$ $(i = 1, 2, 3)$ are linearly dependent which is possible if they are coplanar (they are situated in the same plane). In particular if the unit screws $\boldsymbol{E_{11}}$ $(1, 0, 0, 0, 0, 0)$, $\boldsymbol{E_{12}}$ $(e_{12x}, e_{12y}, e_{12z}, 0, 0, 0)$, $\boldsymbol{E_{13}}$ $(e_{13x}, e_{13y}, e_{13z}, 0, 0, 0)$ are coplanar (Fig. 15.4c) then there exist four wrenches of constraints imposed by kinematic chains: $\boldsymbol{R_1}$ $(1, 0, 0, 0, 0, 0)$, $\boldsymbol{R_2}$ $(0, 1, 0, 0, 0, 0)$, $\boldsymbol{R_3}$ $(0, 0, 1, 0, 0, 0)$, and $\boldsymbol{R_4}$ $(0, 0, 0, 0, r_{4y}, r_{4z})$ and only two twists of motion of the platform reciprocal to these wrenches $\boldsymbol{\Omega_1}$ $(1, 0, 0, 0, 0, 0)$ and $\boldsymbol{\Omega_2}$ $(\omega_{2x}, \omega_{2y}, \omega_{2z}, 0, 0, 0)$, and these twists are of zero pitch. The wrench $\boldsymbol{R_4}$ is of infinite pitch, it is perpendicular to the axes $\boldsymbol{E_{11}}$, $\boldsymbol{E_{12}}$, $\boldsymbol{E_{13}}$.

If the actuators are fixed then there exist six wrenches imposed by kinematic chains: $\boldsymbol{R_1}$ $(1, 0, 0, 0, 0, 0)$, $\boldsymbol{R_2}$ $(0, 1, 0, 0, 0, 0)$, $\boldsymbol{R_3}$ $(0, 0, 1, 0, 0, 0)$, $\boldsymbol{R_4}$ $(0, 0, 0, r_{4x}, r_{4y}, r_{4z})$, $\boldsymbol{R_5}$ $(0, 0, 0, r_{5x}, r_{5y}, r_{5z})$, and $\boldsymbol{R_6}$ $(0, 0, 0, r_{6x}, r_{6y}, r_{6z})$. The wrenches $\boldsymbol{R_4}$, $\boldsymbol{R_5}$, $\boldsymbol{R_6}$ are of infinite pitch. Singularities corresponding to non-controlled infinitesimal motion of the moving platform (end-effector) exist if the wrenches $\boldsymbol{R_1}$, $\boldsymbol{R_2}$, $\boldsymbol{R_3}$, $\boldsymbol{R_4}$, $\boldsymbol{R_5}$, $\boldsymbol{R_6}$ are linearly dependent which is possible if the wrenches $\boldsymbol{R_4}$, $\boldsymbol{R_5}$, $\boldsymbol{R_6}$ are coplanar. In this case the twist of zero pitch $\boldsymbol{\Omega}$ $(\omega_x, \omega_y, \omega_z, 0, 0, 0)$ exists which is perpendicular to the axes of the wrenches $\boldsymbol{R_4}$, $\boldsymbol{R_5}$, $\boldsymbol{R_6}$ and therefore reciprocal to all the wrenches $\boldsymbol{R_1}$, $\boldsymbol{R_2}$, $\boldsymbol{R_3}$, $\boldsymbol{R_4}$, $\boldsymbol{R_5}$, $\boldsymbol{R_6}$.

Moreover singularities exist corresponding both to loss of one degree of freedom and to non-controlled motion of the moving platform. By this any three screws $\boldsymbol{E}_{i1}$, $\boldsymbol{E}_{i2}$, $\boldsymbol{E}_{i3}$ ($i = 1, 2, 3$) and the wrenches $\boldsymbol{R}_1$, $\boldsymbol{R}_2$, $\boldsymbol{R}_3$, $\boldsymbol{R}_4$, $\boldsymbol{R}_5$, $\boldsymbol{R}_6$ are linearly dependent.

15.3 Planar Parallel Manipulators

Let us consider dynamics and control of planar parallel manipulators. We begin our investigation from the mechanism with two degrees of freedom.

15.3.1 Control of 2-DOF Planar Robot While Intersect a Singularity

Let us consider the device (Fig. 15.5), made for the laser installation [9], in which the fixed laser ray is deflected with the help of a two-mirror set. The device presents a planar five-bar mechanism. The input links *DB* and *CE* are connected with the rotating actuators fixed at the base. The other two links *AB* and *AC* are connected with each other. These links transfer the kinematic chain *AF* which dials with the laser situated on the base. Kinematic chain *AF* comprises the sliding pair placed between the *A* and *F* points. Along the revolving pair *F* there is a laser optical axis. The ray is deflected by two mirrors, placed in points *F* and *A* at angle of 45 degrees to the mechanism plane and to the *AF* line. Extra actuator could be placed in the kinematic pair *B*.

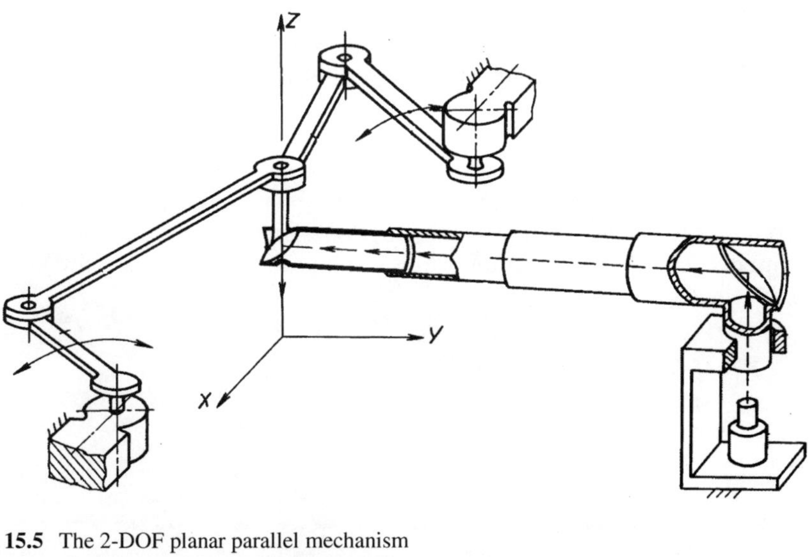

Fig. 15.5 The 2-DOF planar parallel mechanism

To control the output link motion, the model is obtained capable to solve kinematics and dynamics tasks. We got the following expression for actuator moments M_{n1}, M_{n2}:

$$\begin{cases} M_{n1} = J_{11} \cdot \dfrac{d\omega_1}{dt} + \dfrac{1}{2}\omega_1^2 \cdot \dfrac{\partial J_{11}}{\partial q_1} + \omega_1 + \omega_2 \cdot \dfrac{\partial J_{11}}{\partial q_2} \\ \quad + \omega_2^2 \cdot \left(\dfrac{\partial J_{12}}{\partial q_2} - \dfrac{1}{2} \cdot \dfrac{\partial J_{22}}{\partial q_1} \right) + J_{12} \cdot \dfrac{d\omega_2}{dt}, \\ \quad \ldots\ldots\ldots \end{cases} \tag{15.1}$$

where J_{11}, J_{12}, J_{22} are variable inertia moments, q_1, q_2 are generalized coordinates, and ω_1, ω_2 are generalized velocities.

We add the following expression to electrical circuit:

$$\begin{aligned} &L_1 \cdot \tfrac{dI_1}{dt} + R_1 \cdot I_1 + k_{w1} \cdot N_1 \cdot \dot{q}_1 = U_1, \\ &I_1 \cdot k_{m1} = M_1, \end{aligned} \tag{15.2}$$

Here L is inductance, I—the strength of the current, R—electric resistance, M—actuator moment, k_w—constant parameter, k_m—parameter binding the current and the moment, N—reduction gear transmission ratio. (For the second actuator there are analogous equations.)

Having done the transformation, we get the equation system in vector form:

$$\begin{aligned} &\mathbf{A}\,(\overline{q}) \cdot \ddot{\overline{\mathbf{q}}} + \mathbf{B}\left(\overline{q}, \dot{\overline{q}}\right) \cdot \dot{\overline{\mathbf{q}}} = \overline{\mathbf{M}}, \\ &\overline{\mathbf{U}} = \tfrac{\overline{\mathbf{L}}}{\overline{\mathbf{k}}_m \cdot \overline{\mathbf{N}}} \cdot \dot{\overline{\mathbf{M}}} + \tfrac{\overline{\mathbf{R}}}{\overline{\mathbf{k}}_m \cdot \overline{\mathbf{N}}} \overline{\mathbf{M}} + \overline{\mathbf{k}}_w \cdot \overline{\mathbf{N}} \cdot \dot{\overline{\mathbf{q}}}. \end{aligned} \tag{15.3}$$

Here A and B are matrices depending on generalized coordinates and velocities. We use the following optimal control algorithm [26] for mechanism control, using differential equation analysis, which defines the fault at every generalized coordinate.

The program trajectory is presented as a function, $q_p(t)$, $t \in [t_0, T]$. Approximation of this trajectory is by splines obtained. It is expected that we construct the algorithm of control, which transfers the system from the initial stage to the set neighborhood above mentioned trajectory in finite time, minimizing functional J built in the deviation $\Delta(t)$:

$$\Delta(t) = q_p(t) - q(t) \tag{15.4}$$

$$J = \int_{t_0}^{T} \left(\Delta^2 + k_1 \cdot \dot{\Delta}^2 + k_2 \cdot \ddot{\Delta}^2\right) \mathrm{d}t \tag{15.5}$$

where J, Δ, k_1, k_2 present functional, fault, and weight coefficients. Let us write:

$$\gamma_{i1} \cdot \dot{\Delta}_{i0}^2 + \gamma_{i0} \cdot \left(2 \cdot \dot{\Delta}_{i0} \cdot \Delta_{i0} + \gamma_{i1} \cdot {\Delta_{i0}}^2\right) = C_i\left(\Delta_{i0}, \dot{\Delta}_{i0}\right) \tag{15.6}$$

where $i = 1, 2$; γ_{i0}, γ_{i1} are coefficients, then

$$\int_{t_0}^{T} \left(\ddot{\Delta}_i + \gamma_{i1} \cdot \dot{\Delta}_i + \gamma_{i0} \cdot \Delta_i\right)^2 \mathrm{d}t + C_i\left(\cdot\Delta_{i0}, \dot{\Delta}_{i0}\right)$$

$$= \int_{t_0}^{T} \left\{\ddot{\Delta}_i^2 + \dot{\Delta}_i^2 \cdot \left({\gamma_{i1}}^2 - 2 \cdot \gamma_{i0}\right) + {\gamma_{i0}}^2 \cdot {\Delta_i}^2\right\} \mathrm{d}t \tag{15.7}$$

We use forward approach depend on accuracy of model. Indeed, the more accurate the model, the more accurate the control.

Thus, functional J minimum is in trajectories when the following conditions are realized:

$$\ddot{\Delta}_i + \gamma_{i1} \cdot \dot{\Delta}_i + \gamma_{i0} \cdot \Delta_i = 0 \tag{15.8}$$

where $i = 1, 2$, $\gamma_{i0} = \sqrt{k_{i2}}$, $\quad \gamma_{i1} = \sqrt{k_{i1} + 2 \cdot \gamma_{i0}}$, $\Delta_i(t) = q_{\mathrm{p}i}(t) - q_i(t)$.

We have the formula for the generalized acceleration vector as this:

$$\ddot{\overline{\mathbf{q}}} = \ddot{\overline{\mathbf{q}}}_{\mathrm{p}} + \overline{\boldsymbol{\gamma}}_1 \cdot \left(\dot{\overline{\mathbf{q}}}_{\mathrm{p}} - \dot{\overline{\mathbf{q}}}\right) + \overline{\boldsymbol{\gamma}}_0 \cdot \left(\overline{\mathbf{q}}_{\mathrm{p}} - \overline{\mathbf{q}}\right) \tag{15.9}$$

From this, we can calculate the generalized forces.

This algorithm can be applied for parallel manipulators. Actually, while approaching singular configurations the system of equations becomes degenerate, and required generalized forces become too large. In this case an additional actuator should be used, which has to be taken into consideration in the control algorithm.

In parallel manipulator the singular configuration is determined by links AB and AC forming one line. This singular zone is one-dimensional, as in the case of constant relative position of the abovementioned links the manipulator has only one degree of freedom.

We introduce the criterion of singular configurations: the overrun of the generalized moment's marginal tolerance value. It is necessary that the moment surpass the nominal value not more than two times. On reaching such configuration there should be a load transfer with taking extra actuator into account.

The singular configuration would not be a singular one if the actuators are situated in E and B pairs (Fig. 15.5). The algorithm could be realized by this way: at initial position, the two main actuators E and D are in operation. When the torque of one of them reaches the surpass nominal value, the other extra actuator (the point B) is put into operation.

The parallel manipulator motion was considered for search of feasibilities for the algorithm mentioned. The A point moves towards ellipsis, and the kinematical chain passes nearby singular configuration. At first, the calculations were made without possible limitations on the actuator moment. It was necessary to test how the algorithm copes with the task of the manipulator optimal control.

Trajectory is ellipse (Fig. 15.6a). It is reasonable that the moments in actuators reach unfeasibly high values in the case (Fig. 15.6b), though the fault in generalized coordinate and generalized velocity tends to zero rather quickly.

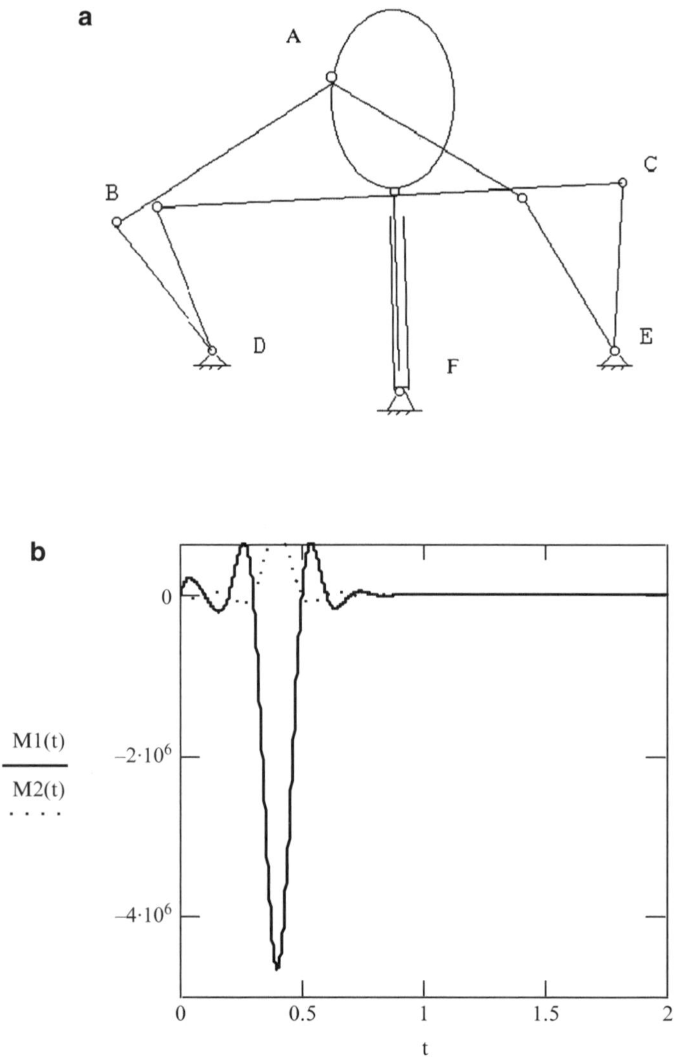

Fig. 15.6 (**a**) Trajectory of end-effector, (**b**) the actuator moments $M1$, $M2$ [N × m] (t[s]) without extra actuator

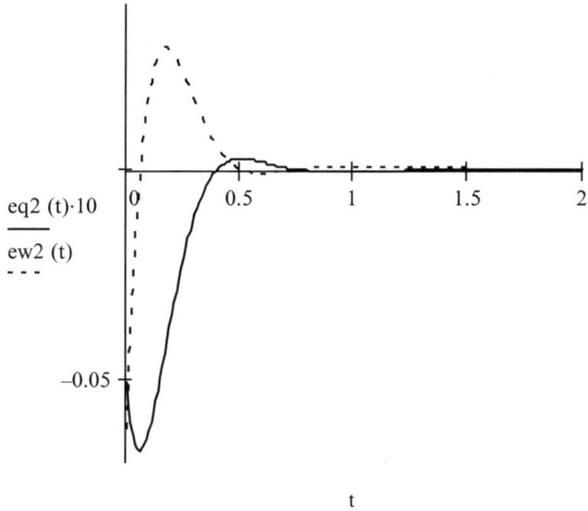

Fig. 15.7 Fault alteration of the second generalized coordinate eq2[rad] (t[s]) and velocity ew2[rad/s] (t[s]) without extra actuator

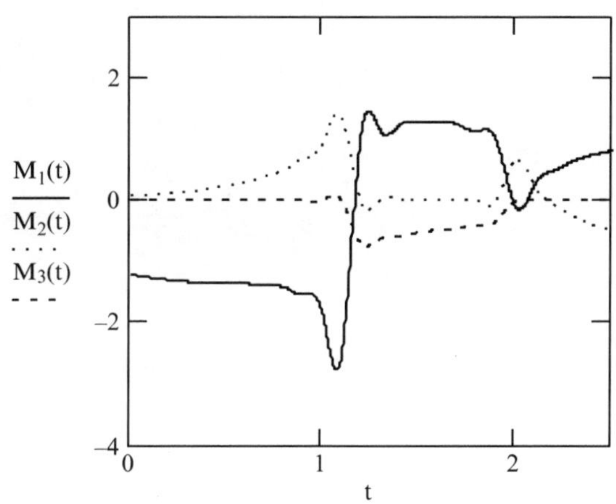

Fig. 15.8 The actuator moments M_1, M_2, M_3 [N × m] (t[s]) meaning with extra actuator

Pattern (Fig. 15.7) presents the fault, connected with the movement of the second actuator located in *E*-pair. In order to escape unfeasibly wide actuator moment values, we used extra actuator *B*. The actuator moments underwent sharp changes (Fig. 15.8).

After getting away from singular configuration there was the extra actuator shutdown and putting the main actuator into operation. These conditions influence the fault value of generalized coordinates (Fig. 15.9).

At whole, the algorithm is characterized by the acceptable index on stability under fault minimization by motion in trajectory.

Thus, the criteria of singular configuration could be the admissible moment of actuator. Extra actuator allows escaping uncontrolled motions, as well as wide moments in actuators.

Fig. 15.9 Fault alteration $eq_{1,2} \times 10^2$ [rad] (t[s]) according to generalized coordinates with extra actuator

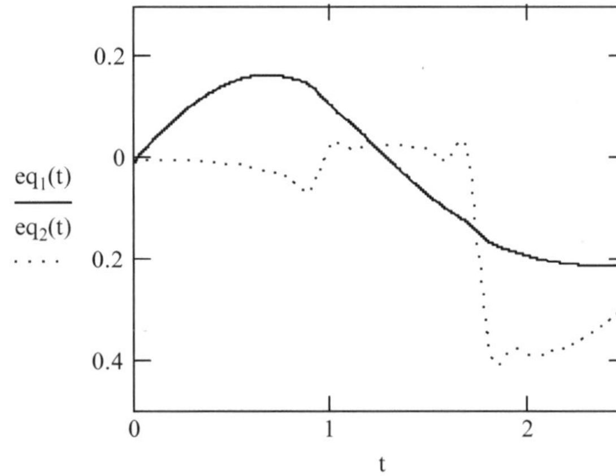

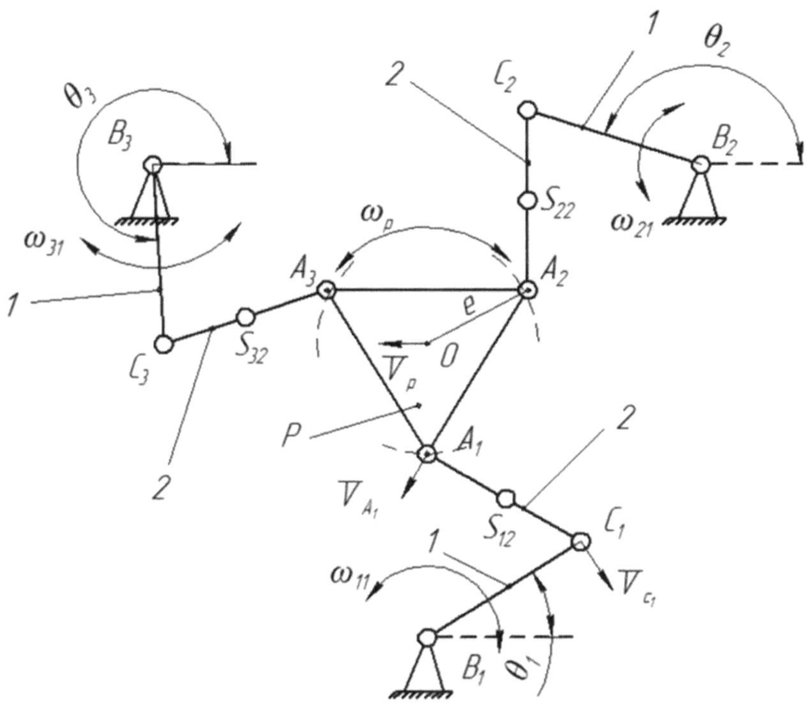

Fig. 15.10 Planar parallel robot

15.3.2 *Oscillations of Planar Parallel Manipulator*

Let us consider oscillations of the planar parallel robot (Fig. 15.10). Let the generalized forces be proportional to the angles of rotations in the actuators. The equations of dynamics by taking into account the mass of the output link are φ:

$$m\ddot{x} = -F_x;$$
$$m\ddot{y} = -F_y;$$
$$J\ddot{\varphi} = -M_z, \tag{15.10}$$

Here m is the mass of the output link; $\ddot{x}$, $\ddot{y}$ are the accelerations of the center of the mass; J is the inertia moment; $\ddot{\varphi}$ is angular acceleration; F_x, F_y are the forces acting to the output link; M_z is the torque.

Let us assume masses of other links are equal to 0. $A_1A_2A_3$ is the output link.

The generalized forces are:

$$M_i = -c \cdot \Delta q \tag{15.11}$$

Here c is the stiffness of the actuators and Δqi is the angles of rotations.

We use the principle of virtual displacements:

$$m\ddot{x}\delta x + M_1\frac{\partial q_1}{\partial x}\delta x + M_2\frac{\partial q_2}{\partial x}\delta x + M_3\frac{\partial q_3}{\partial x}\delta x = 0$$
$$m\ddot{y}\delta x + M_1\frac{\partial q_1}{\partial x}\delta y + M_2\frac{\partial q_2}{\partial y}\delta y + M_3\frac{\partial q_3}{\partial y}\delta y = 0$$
$$J\ddot{\varphi}\delta\varphi + M_1\frac{\partial q_1}{\partial\varphi}\delta\varphi + M_2\frac{\partial q_2}{\partial\varphi}\delta\varphi + M_3\frac{\partial q_3}{\partial\varphi}\delta\varphi = 0 \tag{15.12}$$

Here δx, δy, $\delta\varphi$ are the virtual displacements; M_1, M_2, M_3 are the torques in the actuators.

After transformations we have:

$$\ddot{x} = \frac{M_1\frac{\partial q_1}{\partial x} + M_2\frac{\partial q_2}{\partial x} + M_3\frac{\partial q_3}{\partial x}}{m};$$
$$\ddot{y} = \frac{M_1\frac{\partial q_1}{\partial y} + M_2\frac{\partial q_2}{\partial y} + M_3\frac{\partial q_3}{\partial y}}{m}$$
$$\ddot{z} = \frac{M_1\frac{\partial q_1}{\partial\varphi} + M_2\frac{\partial q_2}{\partial\varphi} + M_3\frac{\partial q_3}{\partial\varphi}}{J} \tag{15.13}$$

These equations contain the partial derivatives of the constraints:

$$\frac{\partial q_i}{\partial x} = -\frac{\partial F_i/\partial x}{\partial F_i/\partial q_i}; \quad \frac{\partial q_i}{\partial y} = -\frac{\partial F_i/\partial y}{\partial F_i/\partial q_i}; \quad \frac{\partial q_i}{\partial\varphi} = -\frac{\partial F_i/\partial\varphi}{\partial F_i/\partial q_i} \tag{15.14}$$

Let us consider the configuration of the manipulator where $q_1 = 0.34$ rad; $q_2 = 2.43$ rad; $q_3 = -1.75$ rad; $x = 0$; $y = 0$; $\varphi = 0$.

The partial derivatives are:

$$\frac{\partial q_1}{\partial x} = \frac{\partial F_1/\partial x}{\partial F_1/\partial q_1} = 1; \quad \frac{\partial q_1}{\partial y} = \frac{\partial F_1/\partial y}{\partial F_1/\partial q_1} = -0.353; \quad \frac{\partial q_1}{\partial \varphi} = \frac{\partial F_1/\partial \varphi}{\partial F_1/\partial q_1} = 1$$

$$\frac{\partial q_2}{\partial x} = \frac{\partial F_2/\partial x}{\partial F_2/\partial q_2} = -0.153; \quad \frac{\partial q_2}{\partial y} = \frac{\partial F_2/\partial y}{\partial F_2/\partial q_2} = 1.042; \quad \frac{\partial q_2}{\partial \varphi} = \frac{\partial F_2/\partial \varphi}{\partial F_2/\partial q_2} = 1$$

$$\frac{\partial q_3}{\partial x} = \frac{\partial F_3/\partial x}{\partial F_3/\partial q_3} = -0.806; \quad \frac{\partial q_3}{\partial y} = \frac{\partial F_3/\partial y}{\partial F_3/\partial q_3} = -0.686; \quad \frac{\partial q_3}{\partial \varphi} = \frac{\partial F_3/\partial \varphi}{\partial F_3/\partial q_3} = 1\,.$$

In initial position $\varphi = 0.1$ rad. After numerical solution of the equations above we have oscillations. We can see that the oscillations are nonlinear with three degree of freedom (Fig. 15.11).

We use the approach based on the principle virtual displacements. This approach is used not only for initial configuration but also for any consequent one. To our opinion, there exists a goal to limit the magnitudes of the oscillations by means the constraints without any damping.

15.3.3 Some Singular Positions of a Planar Manipulator

It is well known that singularities are very important from the point of view controllability. That is why we consider some singularities existing in a prototype of a planar manipulator. One of singularities corresponds to the case when three wrenches acting to the output link from kinematic chains intersect in one point (Fig. 15.12). The matrix of the Plücker coordinates of the constraint wrenches is:

$$(\mathbf{R}) = \begin{pmatrix} 7.8 & 6.3 & 0 \\ -9.4 & 3.6 & 0 \\ 1.6 & -9.9 & 0 \end{pmatrix}$$

Other singularity corresponds to the case when three wrenches acting to the output link from kinematic chains are parallel (Fig. 15.13). The matrix of the Plücker coordinates of the constraint wrenches is:

$$(\mathbf{R}) = \begin{pmatrix} 9.5 & -3.1 & -82 \\ -9.5 & 3.1 & -3 \\ 9.5 & -3.1 & -13.5 \end{pmatrix}$$

All the twists of one of the kinematic chains can be situated in one plane (Fig. 15.14). In this case one degree of freedom is lost. The matrix of the Plücker coordinates of the twists of one kinematic chain can be written as: .

$$(\mathbf{E})_2 = \begin{pmatrix} 1 & -1.1 & 3.1 \\ 1 & 1.6 & -6.5 \\ 1 & 4.3 & -16.1 \end{pmatrix}$$

The third line is a linear combination of the first two lines.

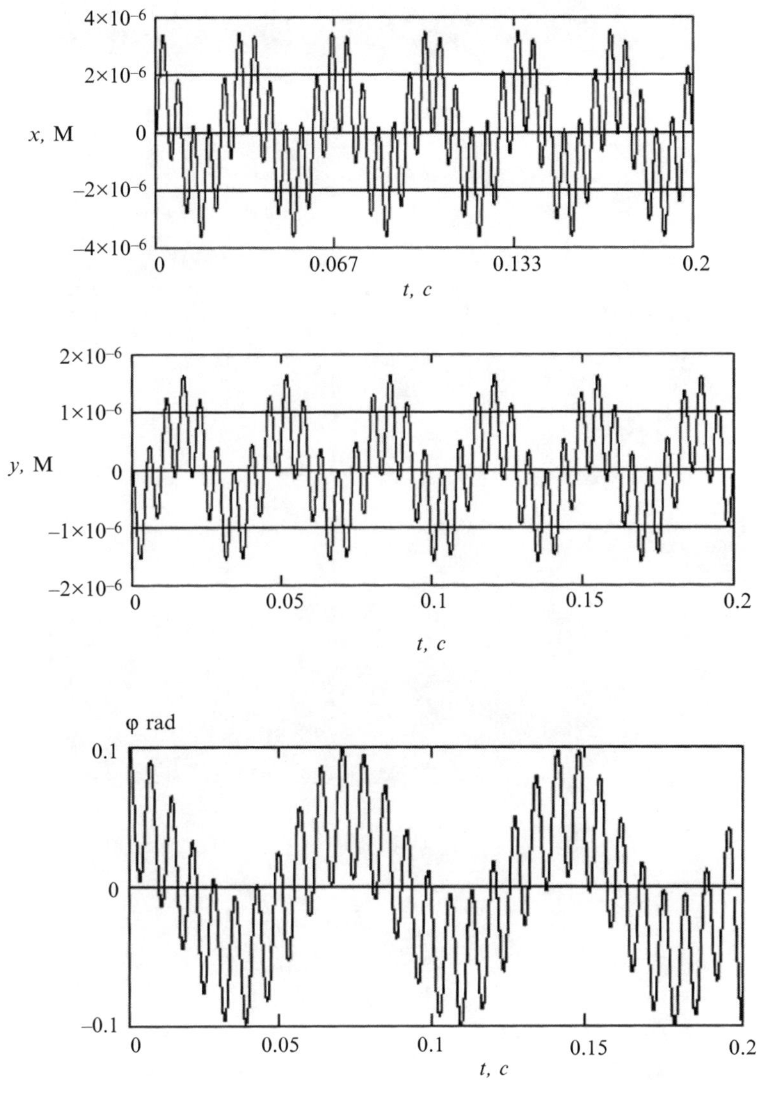

Fig. 15.11 The oscillations of the output link

Two of the constraint wrenches can coincide and the third kinematic chain can be reason to lost one degree of freedom (Fig. 15.15). The matrices of the Plücker coordinates of the wrenches and twists can be written as:

$$(\mathbf{E})_2 = \begin{pmatrix} 1 & -5.7 & 1.3 \\ 1 & -0.7 & -7.4 \\ 1 & 4.3 & -16.1 \end{pmatrix}$$

Fig. 15.12 The singular configuration corresponding to the intersection of the constraint wrenches

Fig. 15.13 The singular configuration corresponding to the parallelism of the constraint wrenches

$$(\mathbf{R}) = \begin{pmatrix} 4.3 & 9 & 66 \\ -9.6 & -3 & 45 \\ -4.3 & -9 & -66 \end{pmatrix}$$

Both the matrices vanish.

Fig. 15.14 One degree of freedom is lost

Fig. 15.15 One degree of freedom and controllability are lost

15.4 Translational Parallel Manipulators

Translational parallel manipulators with three degrees of freedom have attracted much attention from researchers. The most famous translational parallel manipulator is the Clavel's Delta robot [13]. This mechanism has three R-R-Pa-R legs and provides pure translational motion to its moving platform in three dimensions. This mechanism is widely used in packaging and pick-and-place operations because of its phenomenal speed capability and low inertia. In this chapter 3-DOF translational mechanism 3PPaPa is presented.

15.4.1 The Structure of the Manipulator

The proposed translational mechanism is shown in Fig. 15.16.

Each leg of the mechanism has two parallelograms. The first P-joint in each leg is actuated.

The constraint equations represent a relationship between Cartesian coordinates x_0, y_0, z_0 and generalized coordinates q_1, q_2, q_3 and can be represented by the following system:

$$\begin{cases} F_1 = q_1 - x_0 - l_3 - l_2 \cos\left(-\arcsin\dfrac{y_0}{l_2}\right) - l_1 \cos\left(\arcsin\dfrac{z_0}{l_1}\right) = 0; \\ F_2 = q_2 - y_0 - l_3 - l_2 \cos\left(-\arcsin\dfrac{z_0}{l_2}\right) - l_1 \cos\left(\arcsin\dfrac{x_0}{l_1}\right) = 0; \\ F_3 = q_3 - z_0 - l_3 - l_2 \cos\left(-\arcsin\dfrac{x_0}{l_2}\right) - l_1 \cos\left(\arcsin\dfrac{y_0}{l_1}\right) = 0. \end{cases} \quad (15.15)$$

These conditions are not static force balance equations.

15.4.2 Dynamical Analysis

Differentiating these expressions with respect to t, we obtain a system of equations of velocities of the input and output links:

$$\frac{\partial F_i}{\partial x}\dot{x} + \frac{\partial F_i}{\partial y}\dot{y} + \frac{\partial F_i}{\partial z}\dot{z} + \frac{\partial F_i}{\partial q_i}\dot{q}_i = 0, \qquad i = 1, 2, 3 \quad (15.16)$$

Fig. 15.16 (**a**) Kinematic scheme of translational manipulator, (**b**) displacement in kinematic chains

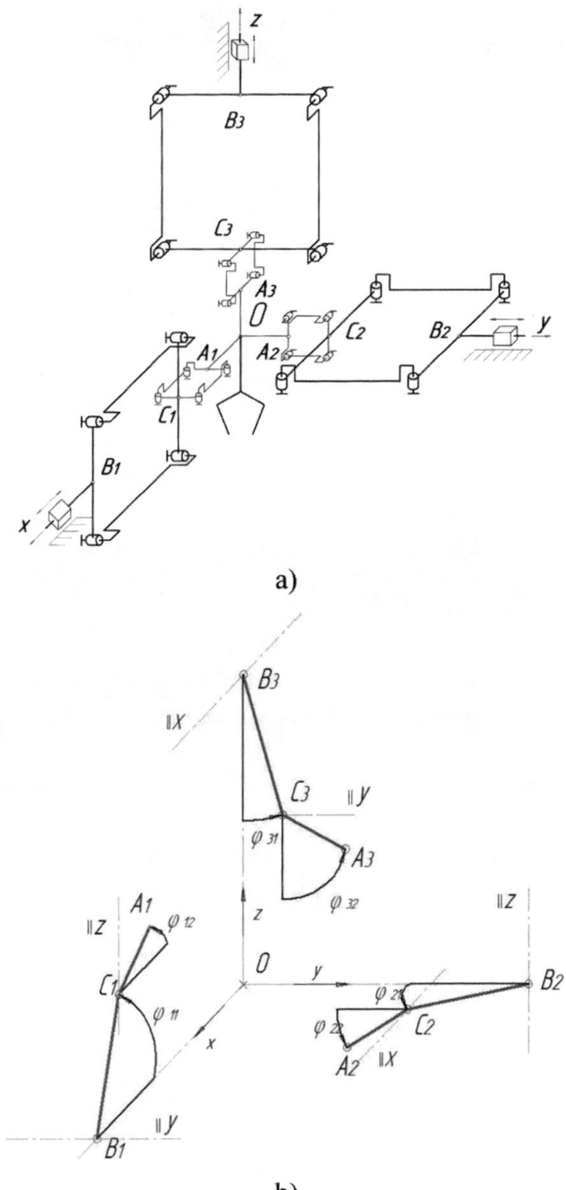

Differentiating the equations again with respect to t, we obtain equations of accelerations:

$$\left(\frac{\partial^2 F_i}{\partial x^2}\dot{x}+\frac{\partial^2 F_i}{\partial x\partial y}\dot{y}+\frac{\partial^2 F_i}{\partial x\partial z}\dot{z}+\frac{\partial^2 F_i}{\partial x\partial q_i}\dot{q}_i\right)\dot{x}+\frac{\partial F_i}{\partial x}\ddot{x}$$
$$+\left(\frac{\partial^2 F_i}{\partial x\partial y}\dot{x}+\frac{\partial^2 F_i}{\partial y^2}\dot{y}+\frac{\partial^2 F_i}{\partial y\partial z}\dot{z}+\frac{\partial F_i}{\partial y\partial q_i}\dot{q}_i\right)\dot{y}+\frac{\partial F_i}{\partial y}\ddot{y}$$
$$+\left(\frac{\partial^2 F_{\mathrm{i}}}{\partial x\partial z}\dot{x}+\frac{\partial^2 F_i}{\partial y\partial z}\dot{y}+\frac{\partial^2 F_i}{\partial z^2}\dot{z}+\frac{\partial^2 F_i}{\partial z\partial q_1}\dot{q}_i\right)\dot{z}+\frac{\partial F_i}{\partial z}\ddot{z}$$
$$+\left(\frac{\partial^2 F_i}{\partial x\partial q_i}\dot{x}+\frac{\partial^2 F_i}{\partial y\partial q_i}\dot{y}+\frac{\partial^2 F_i}{\partial z\partial q_i}\dot{z}+\frac{\partial^2 F_i}{\partial {q^2}_i}\dot{q}_i\right)\dot{q}+\frac{\partial F_i}{\partial q_i}\ddot{q}_i \tag{15.17}$$

Let us assume the equation of motion: $x = 1 - \cos(t)$; $y = 2 - 2\cdot\cos(t)$; $z = 3 - 3\cdot\cos(t)$. Solution of the equation is represented graphically (Fig. 15.17).

15.4.3 Oscillations

The equations of motion of a translational mechanism with three degrees of freedom are:

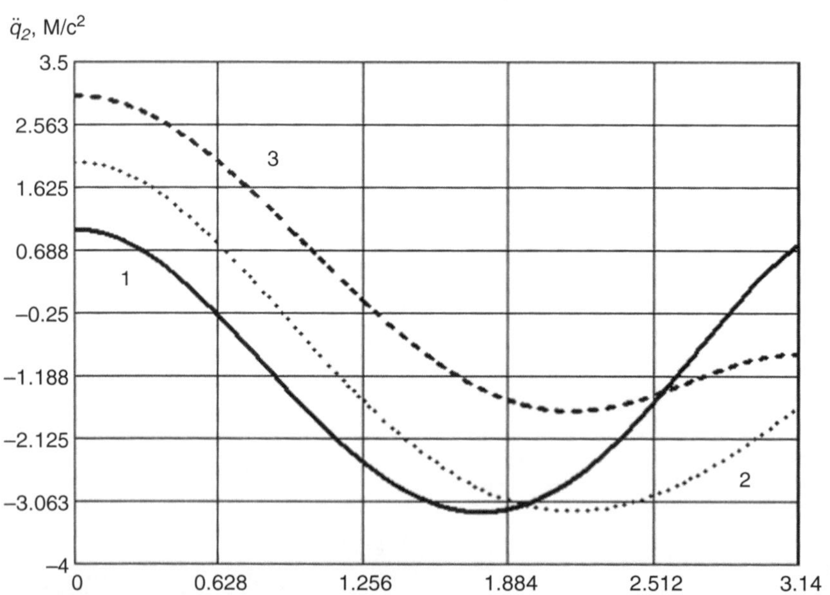

Fig. 15.17 Acceleration graph of input links. 1—$\ddot{q}_1$, 2—$\ddot{q}_2$, 3—$\ddot{q}_3$

$$m\ddot{x}\delta x + P_1\frac{\partial q_1}{\partial x}\delta x + P_2\frac{\partial q_2}{\partial x}\delta x + P_3\frac{\partial q_3}{\partial x}\delta x + m_1\ddot{q}_1\frac{\partial q_1}{\partial x}\delta x$$
$$+ m_2\ddot{q}_2\frac{\partial q_2}{\partial x}\delta x + m_3\ddot{q}_3\frac{\partial q_3}{\partial x}\delta x = 0$$
$$m\ddot{y}\delta y + P_1\frac{\partial q_1}{\partial y}\delta y + P_2\frac{\partial q_2}{\partial y}\delta y + P_3\frac{\partial q_3}{\partial y}\delta y + m_1\ddot{q}_1\frac{\partial q_1}{\partial y}\delta y$$
$$+ m_2\ddot{q}_2\frac{\partial q_2}{\partial y}\delta y + m_3\ddot{q}_3\frac{\partial q_3}{\partial y}\delta y = 0$$
$$m\ddot{z}\delta y + P_1\frac{\partial q_1}{\partial z}\delta z + P_2\frac{\partial q_2}{\partial z}\delta z + P_3\frac{\partial q_3}{\partial z}\delta z + m_1\ddot{q}_1\frac{\partial q_1}{\partial z}\delta z$$
$$+ m_2\ddot{q}_2\frac{\partial q_2}{\partial z}\delta z + m_3\ddot{q}_3\frac{\partial q_3}{\partial z}\delta z + mg\delta z = 0 \tag{15.18}$$

Here m is the mass of the output link, P_1, P_2, P_3 are torques in the drives, $\frac{\partial q_i}{\partial x}$, $\frac{\partial q_i}{\partial y}$, $\frac{\partial q_i}{\partial \varphi}$ are variable factors.

The variable factors are situated by the generalized forces P:

$$\frac{\partial x}{\partial q_i} = -\frac{\partial F_i}{\partial q_i} \Big/ \frac{\partial F_i}{\partial x}, \quad \frac{\partial y}{\partial q_i} = -\frac{\partial F_i}{\partial q_i} \Big/ \frac{\partial F_i}{\partial y}, \quad \frac{\partial z}{\partial q_i} = -\frac{\partial F_i}{\partial q_i} \Big/ \frac{\partial F_i}{\partial z}.$$

Using numerical integration, we find coordinate change of output link for different initial conditions: $x = 0.25$; $y = 0.15$; $z = 0.01$ m (Fig. 15.18).

This is the result of tracking the motion reference trajectory.

15.4.4 Control of the Manipulator

The desired laws of the coordinates of the mobile platform are described by equations $x_T(t)$, $y_T(t)$, $z_T(t)$. The task is to minimize the errors $\Delta_1(t) = x_T(t) - x(t)$, $\Delta_2(t) = y_T(t) - y(t)$, $\Delta_3(t) = z_T(t) - z(t)$. Here $x(t)$, $y(t)$, $z(t)$ are the actual coordinates of the mobile platform. Equations of errors are:

$$\ddot{\Delta}x + K_D\dot{\Delta}x + K_P\Delta x = 0;$$
$$\ddot{\Delta}y + K_D\dot{\Delta}y + K_P\Delta y = 0;$$
$$\ddot{\Delta}z + K_D\dot{\Delta}z + K_P\Delta z = 0. \tag{15.19}$$

Here K_D, K_P are the feedback coefficients. The value feedback gains coefficients are determined according the theory of robotic control [26].

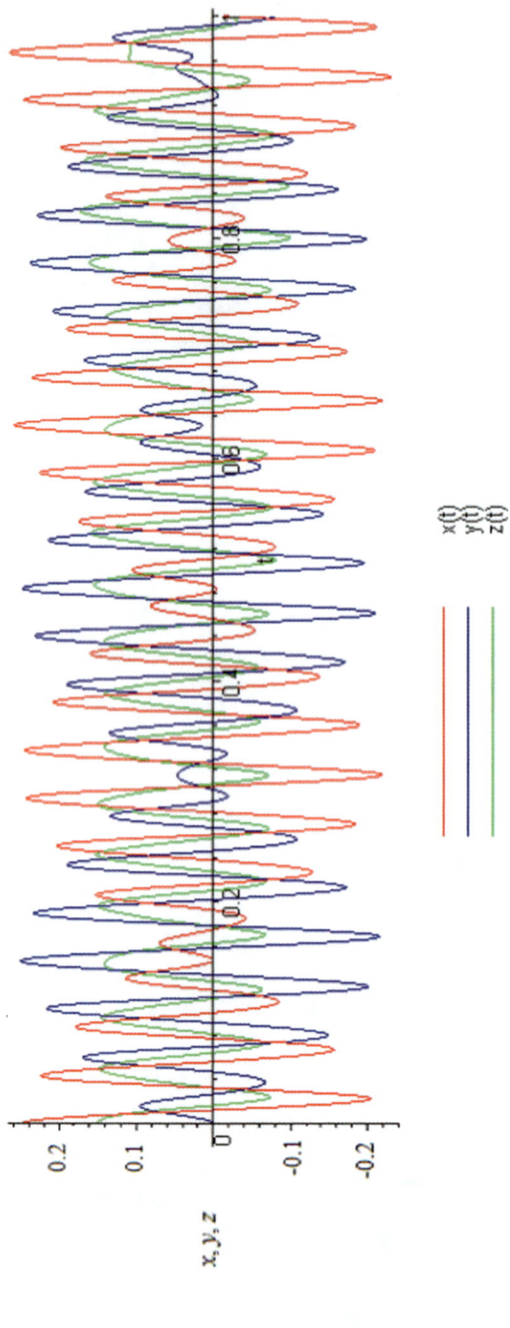

Fig. 15.18 The low of coordinates

According to the constraint equations the actual accelerations become:

$$\ddot{x} = \ddot{x}_T + K_D \cdot (\dot{x}_T - \dot{x}) + K_P \cdot (x_T - x)$$
$$\ddot{y} = \ddot{y}_T + K_D \cdot (\dot{y}_T - \dot{y}) + K_P \cdot (y_T - y)$$
$$\ddot{z} = \ddot{z}_T + K_D \cdot (\dot{z}_T - \dot{z}) + K_P \cdot (z_T - z)\,; \tag{15.20}$$

The laws of errors are described by oscillatory second-order systems, by which control time is minimized. Let us consider an example: the desired laws of the coordinates of the moving platform are $x_T(t) = 0.1 \cdot \mathbf{sin}\,(\boldsymbol{\omega t})$, $y_T(t) = 0.12 \cdot \mathbf{sin}\,(\omega t)$, $z_T(t) = 0.15 \cdot \mathbf{sin}\,(\omega t)$.

The result of the simulation is presented in Figs. 15.19 and 15.20. The maximal errors are about 5×10^{-3} mm.

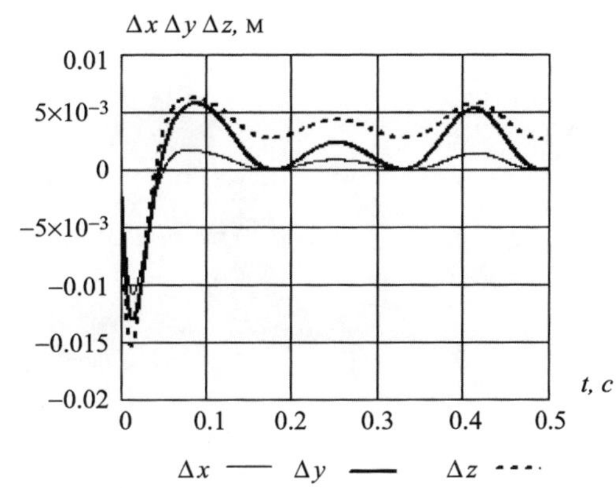

Fig. 15.19 The result of simulations

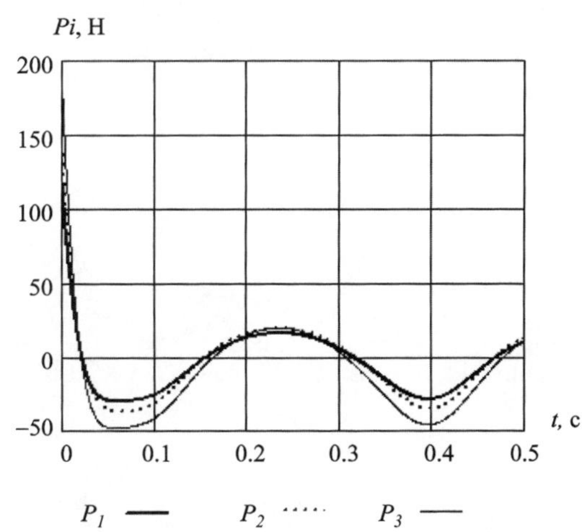

Fig. 15.20 The simulation result of driving torques

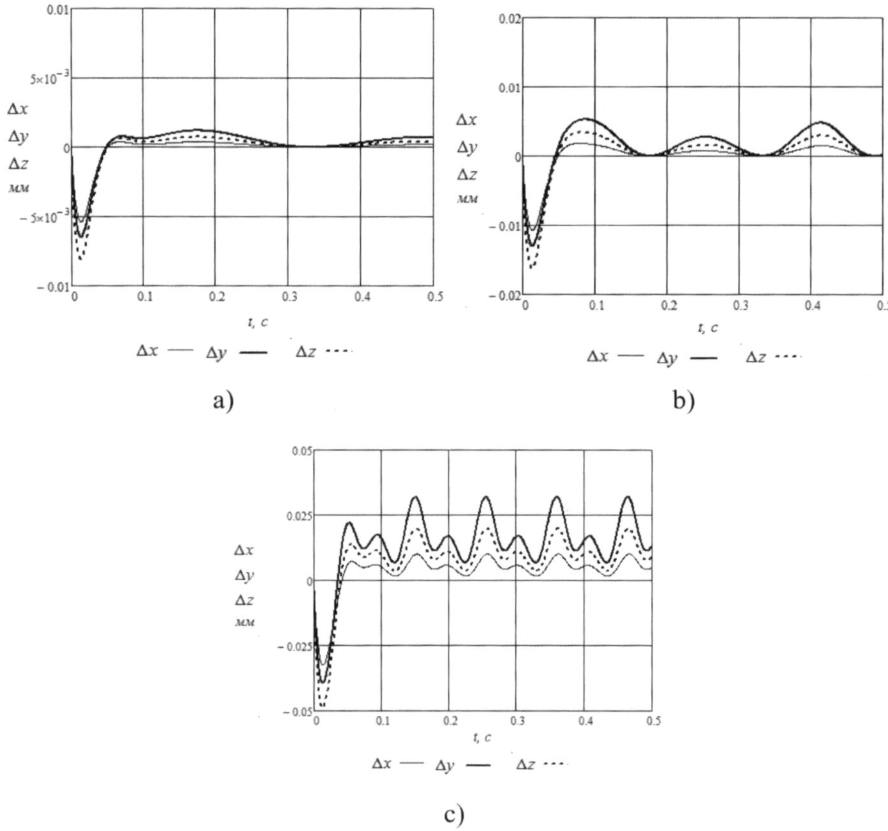

Fig. 15.21 The lows of the faults. (**a**) $\omega = 10$ rad/s, (**b**) $\omega = 20$ rad/s, (**c**) $\omega = 40$ rad/s

Let us consider how the frequency influences on the faults. Let the feedback coefficients be equal to $\gamma_0 = 7,200$, $\gamma_1 = 120$, the own frequency is $\omega_0 = \sqrt{\gamma_0 - \frac{\gamma_1^2}{4}} = 60$ rad/s. By the frequencies $\omega = 10$ rad/s, $\omega = 20$ rad/s, $\omega = 40$ rad/s the faults the faults are different values (Fig. 15.21).

Let us consider how the frequency influences on the faults. Let the feedback coefficients be equal to $\gamma_0 = 7,200$, $\gamma_1 = 120$, the own frequency is $\omega_0 = \sqrt{\gamma_0 - \frac{\gamma_1^2}{4}} = 60$ rad/s. By the frequencies $\omega = 10$ rad/s, $\omega = 20$ rad/s, $\omega = 40$ rad/s the faults the faults are different values (Fig. 15.21).

Thus, the frequency influences on the fault.

15.5 Kinematics, Dynamics, Control and Accuracy of Spherical Parallel Robot

Here we consider kinematics, dynamics, control and accuracy of spherical parallel mechanisms.

15.5.1 Mechanism Structure

In the considered spherical mechanism with five kinematic pairs in each chain (Fig. 15.22), the input link is connected to the engine. The output link is a platform that revolves around three axes intersecting at a point O. The output coordinates are angles of rotation of the platform α, β, γ around the axes, whose relative positions are described by a fiction kinematic chain (Fig. 15.23a) corresponding to the angles between the coordinate axes (Fig. 15.23b). The generalized coordinates are angles φ_{11}, φ_{21}, φ_{31}. Each of the three kinematic chains has three joints with intersecting axes.

The spherical manipulator under consideration is built basing on the principle of selecting the degree of freedom in the kinematic chain. Each input link in the

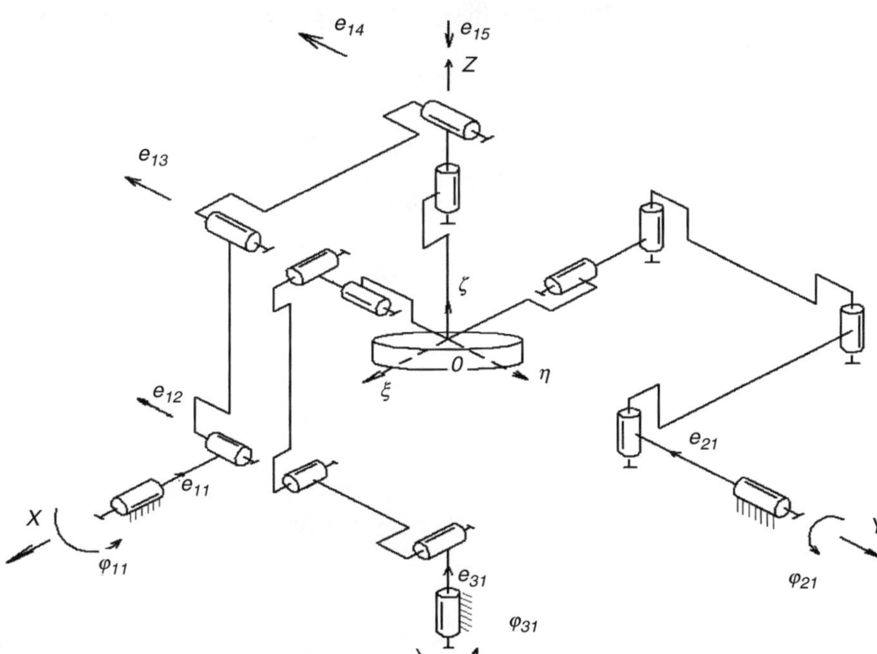

Fig. 15.22 Spherical mechanism with five kinematic pairs in each chain

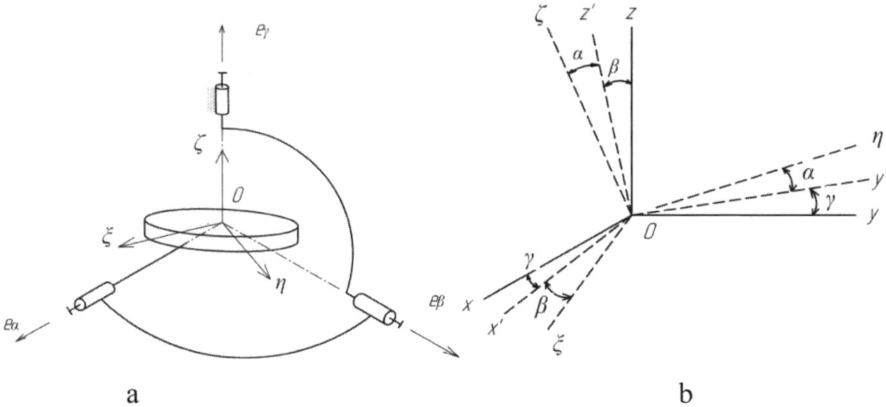

Fig. 15.23 The angles α, β, γ: (**a**) fiction kinematic chain, (**b**) coordinate systems

chain is connected to the engine. Output link is a platform that rotates around three axes intersecting at one point O. Output coordinates are angles of the rotation α, β, γ around the axes x, y, z respectively. Generalized coordinates are angles φ_{11}, φ_{21}, φ_{31}—the rotation angles of input units of the first, second and third kinematic chains (for the problem solution of accelerating the input angles are designated q_1, q_2, q_3). Each of three kinematic chains has two hinges with intersecting axes and three hinges with parallel axes. Solution of the kinematic problem of such a manipulator are well known.

Let us associate the output link of the mechanism with a moving coordinate system ξ, η, ζ, whose axes are situated along the main inertia axes of this link. Therefore, for the orientation angles ($\alpha = \beta = \gamma = 0$) the directions of the axes ξ, η, and ζ coincide with directions of the axes x, y, and z, respectively. The constraint equations are derived from geometry of mechanism, transfer matrixes absolute and moving coordinate systems.

The constraint equations for a spherical mechanism with three kinematic chains can be represented by the following system:

$$F_1 = \mathrm{tg}\varphi_{11} + \frac{\cos\alpha \cdot \sin\gamma \cdot \sin\beta - \cos\gamma \cdot \sin\alpha}{\cos\alpha \cdot \cos\beta} = 0$$

$$F_2 = \frac{\sin\beta}{\cos\gamma \cdot \cos\beta} - \mathrm{tg}\varphi_{21} = 0$$

$$F_3 = \mathrm{tg}\varphi_{31} + \frac{\cos\gamma \cdot \sin\alpha \cdot \sin\beta - \cos\alpha \cdot \sin\gamma}{\cos\alpha \cdot \cos\gamma + \sin\alpha \cdot \sin\beta \cdot \sin\gamma} = 0 \qquad (15.21)$$

Three rotating pairs with parallel axes can be replaced by one rotating kinematic pair (Fig. 15.24).

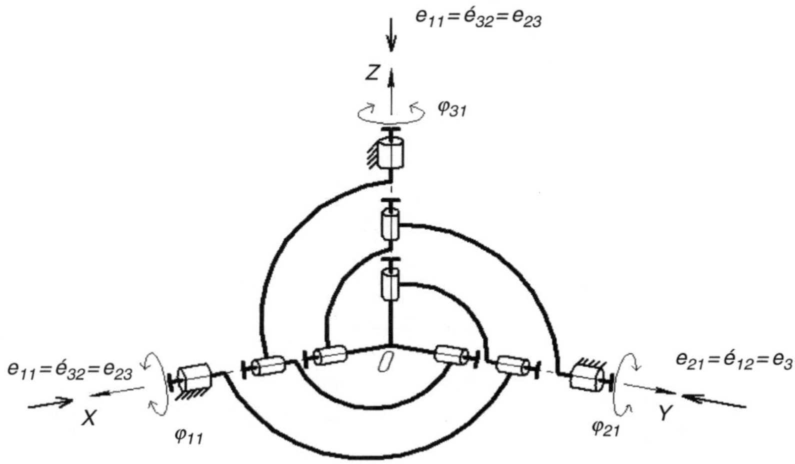

Fig. 15.24 Spherical mechanism with three kinematic pairs in each chain

15.5.2 Dynamical Analysis

Differentiating the constraint equations with respect to t, we obtain a system of equations of velocities of the input and output links:

$$\frac{\partial F_i}{\partial \alpha}\dot{\alpha} + \frac{\partial F_i}{\partial \beta}\dot{\beta} + \frac{\partial F_i}{\partial \gamma}\dot{\gamma} + \frac{\partial F_i}{\partial \varphi_{i1}}\dot{\varphi}_{i1} = 0$$
$$\frac{\partial F_2}{\partial \alpha}\dot{\alpha} + \frac{\partial F_2}{\partial \beta}\dot{\beta} + \frac{\partial F_2}{\partial \gamma}\dot{\gamma} + \frac{\partial F_2}{\partial \varphi_{21}}\dot{\varphi}_{21} = 0$$
$$\frac{\partial F_3}{\partial \alpha}\dot{\alpha} + \frac{\partial F_3}{\partial \beta}\dot{\beta} + \frac{\partial F_3}{\partial \gamma}\dot{\gamma} + \frac{\partial F_3}{\partial \varphi_{31}}\dot{\varphi}_{31} = 0 \qquad (15.22)$$

Differentiating these equations again with respect to t, we obtain equations of accelerations:

$$\frac{\partial F_1}{\partial \varphi_{11}}\ddot{\varphi}_{11} = \frac{\partial^2 F_1}{\partial \alpha^2}\dot{\alpha}^2 + 2\frac{\partial^2 F_1}{\partial \alpha \partial \beta}\dot{\alpha}\dot{\beta} + 2\frac{\partial^2 F_1}{\partial \alpha \partial \gamma}\dot{\alpha}\dot{\gamma} + \frac{\partial^2 F_1}{\partial \beta^2}\dot{\beta}^2 + 2\frac{\partial^2 F_1}{\partial \beta \partial \gamma}\dot{\beta}\dot{\gamma} + \frac{\partial^2 F_1}{\partial \gamma^2}\dot{\gamma}^2$$
$$+ \frac{\partial^2 F_1}{\partial \phi_{11}^2}\dot{\varphi}_{11} + \frac{\partial F_1}{\partial \alpha}\ddot{\alpha} + \frac{\partial F_1}{\partial \beta}\ddot{\beta} + \frac{\partial F_1}{\partial \gamma}\ddot{\gamma};$$

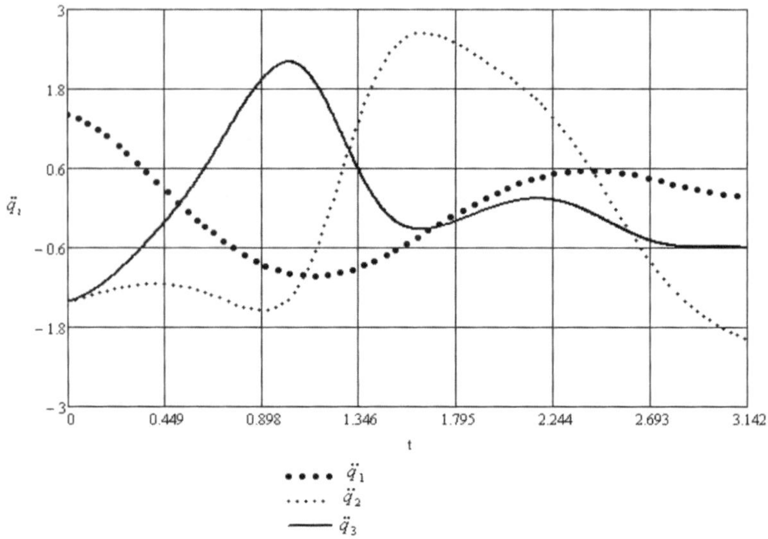

Fig. 15.25 Acceleration graph of input links

$$\frac{\partial F_2}{\partial \varphi_{21}}\ddot{\varphi}_{21} = \frac{\partial^2 F_2}{\partial \beta^2}\dot{\beta}^2 + 2\frac{\partial^2 F_2}{\partial \beta \partial \gamma}\dot{\beta}\dot{\gamma} + \frac{\partial^2 F_2}{\partial \gamma^2}\dot{\gamma}^2 + \frac{\partial^2 F_2}{\partial \phi_{21}^2}\dot{\varphi}_{21} + \frac{\partial F_2}{\partial \beta}\ddot{\beta} + \frac{\partial F_2}{\partial \gamma}\ddot{\gamma}$$

$$\frac{\partial F_3}{\partial \varphi_{31}}\ddot{\varphi}_{31} = \frac{\partial^2 F_3}{\partial \alpha^2}\dot{\alpha}^2 + 2\frac{\partial^2 F_3}{\partial \alpha \partial \beta}\dot{\alpha}\dot{\beta} + 2\frac{\partial^2 F_3}{\partial \alpha \partial \gamma}\dot{\alpha}\dot{\gamma} + \frac{\partial^2 F_3}{\partial \beta^2}\dot{\beta}^2 + 2\frac{\partial^2 F_3}{\partial \beta \partial \gamma}\dot{\beta}\dot{\gamma} + \frac{\partial^2 F_3}{\partial \gamma^2}\dot{\gamma}^2$$
$$+ \frac{\partial^2 F_3}{\partial \phi_{31}^2}\dot{\varphi}_{31} + \frac{\partial F_3}{\partial \alpha}\ddot{\alpha} + \frac{\partial F_3}{\partial \beta}\ddot{\beta} + \frac{\partial F_3}{\partial \gamma}\ddot{\gamma}. \tag{15.23}$$

Let us assume the equation of motion as:

$\alpha = 0.706 \cdot t^2 - 0.15 \cdot t^3$; $\beta = 0.706 \cdot t^2 - 0.15 \cdot t^3$; $\gamma = 0.706 \cdot t^2 - 0.15 \cdot t^3$

Solution of the equation is represented graphically in Fig. 15.25.

The equation of motion of a spherical mechanism with three degrees of freedom has the following form (mass of the arc-shaped links is ignored).

15.5.3 Oscillations

Nonlinearity of mechanical system is conditional on its geometry and influence of actuators. To solve the problem of motion of a spherical mechanism, we introduce a mobile system of coordinates ξ, η, ζ. Then the equation of motion of a spherical mechanism with three degrees of freedom is:

$$\begin{cases} J_\xi \cdot \ddot{\varphi}_\xi = M_1 \cdot \dfrac{\partial \varphi_{11}}{\partial \varphi_\xi} + M_2 \cdot \dfrac{\partial \varphi_{21}}{\partial \varphi_\xi} + M_3 \cdot \dfrac{\partial \varphi_{31}}{\partial \varphi_\xi} + \dot{\varphi}_\eta \cdot \dot{\varphi}_\zeta \cdot \left(J_\zeta - J_\eta\right) \\ J_\eta \cdot \ddot{\varphi}_\eta = M_1 \cdot \dfrac{\partial \varphi_{11}}{\partial \varphi_\eta} + M_2 \cdot \dfrac{\partial \varphi_{21}}{\partial \varphi_\eta} + M_3 \cdot \dfrac{\partial \varphi_{31}}{\partial \varphi_\eta} + \dot{\varphi}_\xi \cdot \dot{\varphi}_\zeta \cdot \left(J_\xi - J_\zeta\right) \\ J_\zeta \cdot \ddot{\varphi}_\zeta = M_1 \cdot \dfrac{\partial \varphi_{11}}{\partial \varphi_\zeta} + M_2 \cdot \dfrac{\partial \varphi_{21}}{\partial \varphi_\zeta} + M_3 \cdot \dfrac{\partial \varphi_{31}}{\partial \varphi_\zeta} + \dot{\varphi}_\xi \cdot \dot{\varphi}_\eta \cdot \left(J_\eta - J_\xi\right) \end{cases} \tag{15.24}$$

Here $J_\xi = J_\eta = \frac{1}{12} \cdot m \cdot r^2$, $J_\zeta = \frac{1}{2} \cdot m \cdot r^2$ are moments of inertia relative to the axes ξ, η, ζ, m is the mass of the output link, r is the radius of the platform of output link, M_1, M_2, M_3 are torques in the drives, $\frac{\partial \varphi_{ij}}{\partial \varphi_\xi}$—variable factors, $\ddot{\varphi}_i$, $\dot{\varphi}_i$—respectively the projections of acceleration and speed on the ith axis.

Variable factors are determined from the equation of direct problem of speeds by method screw calculus. Since the wrench interacts the unit vectors of the non-driving pairs, then the relative moments mom($\mathbf{R}_i$, Ω_{i2}) = 0, mom($\mathbf{R}_i$, Ω_{i3}) = 0. So mom($\mathbf{R}_i$, Ω_i) = mom($\mathbf{R}_i$, Ω_{i1}). Substituting the coordinate values of the wrenches and twists, we obtain the equations of relative moments:

$$\begin{aligned} \operatorname{mom}\left(\mathbf{R}_i, \boldsymbol{\Omega}_i\right) &= \omega_x r_{ix}^0 + \omega_y r_{iy}^0 + \omega_z r_{iz}^0 \\ \operatorname{mom}\left(\mathbf{R}, \boldsymbol{\Omega}_{i1}\right) &= \omega_{i1}\left(x_{i1} r_{ix}^0 + y_{i1} r_{iy}^0 + z_{i1} r_{i1z}^0\right) \end{aligned} \tag{15.25}$$

Here (x_{i1}, y_{i1}, z_{i1}) are the Plücker coordinates of the unit vectors e_{i1}, $\mathbf{r}_i^0$ is the moment part of the wrench with coordinates $r_{1x}^0, r_{1y}^0, r_{1z}^0$.

The system of equations for three kinematic chains can be written as following forms:

$$\begin{aligned} \omega_x r_{1x}^0 + \omega_y r_{1y}^0 + \omega_z r_{1z}^0 &= \omega_{11}\left(x_{11} r_{1x}^0 + y_{11} r_{1y}^0 + z_{11} r_{1z}^0\right); \\ \omega_x r_{2x}^0 + \omega_y r_{2y}^0 + \omega_z r_{2z}^0 &= \omega_{21}\left(x_{21} r_{2x}^0 + y_{21} r_{2y}^0 + z_{21} r_{2z}^0\right); \\ \omega_x r_{3x}^0 + \omega_y r_{3y}^0 + \omega_z r_{3z}^0 &= \omega_{31}\left(x_{31} r_{3x}^0 + y_{31} r_{3y}^0 + z_{31} r_{3z}^0\right) \end{aligned} \tag{15.26}$$

Here $\omega_\xi, \omega_\eta, \omega_\zeta$ are the angular velocities of the output link around axes ξ, η, ζ, $r_{i\xi}^0, r_{i\eta}^0, r_{i\zeta}^0$ are the coordinates of moment part of the ith wrench.

Then, the variable factors will be defined as following:

$$\frac{\partial\varphi_{11}}{\partial\varphi_\xi}=\frac{\omega_{11}}{\omega_\xi}=\frac{r^0_{1\xi}}{r^0_{1x}};\quad \frac{\partial\varphi_{11}}{\partial\varphi_\eta}=\frac{\omega_{11}}{\omega_\eta}=\frac{r^0_{1\eta}}{r^0_{1x}};\quad \frac{\partial\varphi_{11}}{\partial\varphi_\zeta}=\frac{\omega_{11}}{\omega_\zeta}=\frac{r^0_{1\zeta}}{r^0_{1x}}=0$$
$$\frac{\partial\varphi_{21}}{\partial\varphi_\xi}=\frac{\omega_{21}}{\omega_\xi}=\frac{r^0_{2\xi}}{r^0_{2y}}=0;\quad \frac{\partial\varphi_{21}}{\partial\varphi_\eta}=\frac{\omega_{21}}{\omega_\eta}=\frac{r^0_{2\eta}}{r^0_{2y}};\quad \frac{\partial\varphi_{21}}{\partial\varphi_\zeta}=\frac{\omega_{21}}{\omega_\zeta}=\frac{r^0_{2\zeta}}{r^0_{2y}}$$
$$\frac{\partial\varphi_{31}}{\partial\varphi_\xi}=\frac{\omega_{31}}{\omega_\xi}=\frac{r^0_{3\xi}}{r^0_{3z}};\quad \frac{\partial\varphi_{31}}{\partial\varphi_\eta}=\frac{\omega_{31}}{\omega_\eta}=\frac{r^0_{3\eta}}{r^0_{3z}}=0;\quad \frac{\partial\varphi_{31}}{\partial\varphi_\zeta}=\frac{\omega_{31}}{\omega_\zeta}=\frac{r^0_{3\zeta}}{r^0_{3z}} \tag{15.27}$$

The relationship between velocities and velocity projections on the axes ξ, η, ζ can be written as following form:

$$\begin{aligned}\omega_\xi &= \dot\alpha\cdot\alpha_\xi+\dot\beta\cdot\beta_\xi+\dot\gamma\cdot\gamma_\xi\\ \omega_\eta &= \dot\alpha\cdot\alpha_\eta+\dot\beta\cdot\beta_\eta+\dot\gamma\cdot\gamma_\eta\\ \omega_\zeta &= \dot\alpha\cdot\alpha_\zeta+\dot\beta\cdot\beta_\zeta+\dot\gamma\cdot\gamma_\zeta\end{aligned} \tag{15.28}$$

Here $\begin{pmatrix}\alpha_\xi\\ \alpha_\eta\\ \alpha_\zeta\end{pmatrix}$ are the projections of vector α on axes ξ, η, ζ; $\begin{pmatrix}\beta_\xi\\ \beta_\eta\\ \beta_\zeta\end{pmatrix}$ are the projections of vector β on axes ξ, η, ζ, equal to the product of the rotation matrix by an angle α around the axis ξ

$$\begin{pmatrix}\beta_\alpha\\ \beta_\eta\\ \beta_\eta\end{pmatrix}=\begin{pmatrix}1&0&0\\0&\cos\alpha&\sin\alpha\\0&-\sin\alpha&\cos\alpha\end{pmatrix}\begin{pmatrix}0\\1\\0\end{pmatrix}=\begin{pmatrix}0\\ \cos\alpha\\ -\sin\alpha\end{pmatrix}$$

$\begin{pmatrix}\alpha_\gamma\\ \beta_\gamma\\ \gamma_\gamma\end{pmatrix}$ are the projections of vector γ on axes ξ, η, ζ, equal to the product of the rotation matrix to the coordinates of the output link in the initial position.

$$\begin{pmatrix}\xi_\gamma\\ \eta_\gamma\\ \zeta_\gamma\end{pmatrix}=\begin{pmatrix}1&0&0\\0&\cos\alpha&\sin\alpha\\0&-\sin\alpha&\cos\alpha\end{pmatrix}\cdot\begin{pmatrix}\cos\beta&0&-\sin\beta\\0&1&0\\ \sin\beta&0&\cos\beta\end{pmatrix}\cdot\begin{pmatrix}\cos\gamma&\sin\gamma&0\\-\sin\gamma&\cos\gamma&0\\0&0&1\end{pmatrix}\begin{pmatrix}0\\0\\1\end{pmatrix}$$
$$=\begin{pmatrix}-\sin\beta\\ \cos\beta\cdot\sin\alpha\\ \cos\beta\cdot\cos\alpha\end{pmatrix}$$

The coordinates of the unit vectors of the second (x_{12}, y_{12}, z_{12}) and the third (x_{13}, y_{13}, z_{13}) pairs of the first chain are calculated:

$$\begin{pmatrix} x_{12} \\ y_{12} \\ z_{12} \end{pmatrix} = \begin{pmatrix} 1 & 0 & 0 \\ 0 & \cos\varphi_{11} & -\sin\varphi_{11} \\ 0 & \sin\varphi_{11} & \cos\varphi_{11} \end{pmatrix} \cdot \begin{pmatrix} 0 \\ 1 \\ 0 \end{pmatrix} = \begin{pmatrix} 0 \\ \cos\varphi_{11} \\ \sin\varphi_{11} \end{pmatrix}$$

$$\begin{pmatrix} x_{13} \\ y_{13} \\ z_{13} \end{pmatrix} = (A) \cdot \begin{pmatrix} 0 \\ 0 \\ 1 \end{pmatrix} = \begin{pmatrix} \sin\alpha \cdot \sin\gamma + \cos\alpha \cdot \cos\gamma \cdot \sin\beta \\ \cos\gamma \cdot \sin\alpha \cdot \sin\beta - \cos\alpha \cdot \sin\gamma \\ \cos\beta \cdot \cos\gamma \end{pmatrix}$$

Here A is the matrix which has a form: $A = A_3A_2A_1$, where A_1 is the matrix of rotation around the axis x; A_2 is the matrix of rotation around the axis y; A_3 is the matrix of rotation around the axis z.

For the second chain, the coordinate of the unit vectors of the second and third pairs are:

$$\begin{pmatrix} x_{22} \\ y_{22} \\ z_{22} \end{pmatrix} = \begin{pmatrix} \cos\varphi_{21} & 0 & \sin\varphi_{21} \\ 0 & 1 & 0 \\ -\sin\varphi_{21} & 0 & \cos\varphi_{21} \end{pmatrix} \cdot \begin{pmatrix} 0 \\ 0 \\ 1 \end{pmatrix} = \begin{pmatrix} \sin\varphi_{21} \\ 0 \\ \sin\varphi_{21} \end{pmatrix}$$

$$\begin{pmatrix} x_{13} \\ y_{13} \\ z_{13} \end{pmatrix} = (A) \cdot \begin{pmatrix} 1 \\ 0 \\ 0 \end{pmatrix} = \begin{pmatrix} \cos\alpha \cdot \cos\beta \\ \cos\beta \cdot \sin\alpha \\ -\sin\beta \end{pmatrix}$$

For the third chain, the coordinates of the unit vectors of the second and third pairs are:

$$\begin{pmatrix} x_{32} \\ y_{32} \\ z_{32} \end{pmatrix} = \begin{pmatrix} \cos\varphi_{31} & -\sin\varphi_{31} & 0 \\ \sin\varphi_{31} & \cos\varphi_{31} & 0 \\ 0 & 0 & 1 \end{pmatrix} \cdot \begin{pmatrix} 1 \\ 0 \\ 0 \end{pmatrix} = \begin{pmatrix} \cos\varphi_{31} \\ \sin\varphi_{31} \\ 0 \end{pmatrix}$$

$$\begin{pmatrix} x_{33} \\ y_{33} \\ z_{33} \end{pmatrix} = (A) \cdot \begin{pmatrix} 0 \\ 1 \\ 0 \end{pmatrix} = \begin{pmatrix} \cos\alpha \cdot \sin\beta \cdot \sin\gamma - \cos\gamma \cdot \sin\alpha \\ \cos\alpha \cdot \cos\gamma + \sin\alpha \cdot \sin\beta \cdot \sin\gamma \\ -\cos\beta \cdot \sin\gamma \end{pmatrix}$$

The coordinates of the unit vectors of the first, second, and third chains in the moving coordinate system is defined by the matrix A^{-1}, which is the inverse of matrix A.

$$\begin{pmatrix} e_{\xi 12} \\ e_{\eta 12} \\ e_{\zeta 12} \end{pmatrix} = (A)^{-1} \cdot \begin{pmatrix} x_{12} \\ y_{12} \\ z_{12} \end{pmatrix}, \begin{pmatrix} e_{\xi 22} \\ e_{\eta 22} \\ e_{\zeta 22} \end{pmatrix} = (A)^{-1} \cdot \begin{pmatrix} x_{2} \\ y_{22} \\ z_{22} \end{pmatrix}, \begin{pmatrix} e_{\xi 32} \\ e_{\eta 32} \\ e_{\zeta 32} \end{pmatrix} = (A)^{-1} \cdot \begin{pmatrix} x_{32} \\ y_{32} \\ z_{32} \end{pmatrix}$$

The coordinates of the unit vectors of the third pairs of the first, second, and third chains are:

$$\begin{pmatrix} e_{\xi 13} \\ e_{\eta 13} \\ e_{\zeta 13} \end{pmatrix} = \begin{pmatrix} 0 \\ 0 \\ 1 \end{pmatrix}, \begin{pmatrix} e_{\xi 23} \\ e_{\eta 23} \\ e_{\zeta 23} \end{pmatrix} = \begin{pmatrix} 1 \\ 0 \\ 0 \end{pmatrix}, \begin{pmatrix} e_{\xi 33} \\ e_{\eta 33} \\ e_{\zeta 33} \end{pmatrix} = \begin{pmatrix} 0 \\ 1 \\ 0 \end{pmatrix}$$

Let us express $\dot{\alpha}, \dot{\beta}, \dot{\gamma}$ via $\omega_\xi, \omega_\eta, \omega_\zeta$:

$$\dot{\alpha} = \frac{\omega_\xi \cdot \cos\beta + \omega_\eta \cdot \sin\alpha \cdot \sin\beta + \omega_\zeta \cdot \sin\beta \cdot \cos\alpha}{\cos\beta}$$

$$\dot{\beta} = \omega_\eta \cdot \cos\alpha - \omega_\zeta \cdot \sin\alpha$$

$$\dot{\gamma} = \frac{\omega_\eta \cdot \sin\alpha + \omega_\zeta \cdot \cos\alpha}{\cos\beta} \tag{15.29}$$

Moment part of the wrench in the mobile system of coordinates is equal:

$$\begin{pmatrix} r_{1x}^0 \\ r_{1y}^0 \\ r_{1z}^0 \end{pmatrix} = \begin{pmatrix} \cos\varphi_{12} \\ -\sin\varphi_{11} \cdot \sin\varphi_{12} \\ \cos\varphi_{11} \cdot \sin\varphi_{12} \end{pmatrix}, \quad \begin{pmatrix} r_{2x}^0 \\ r_{2y}^0 \\ r_{2z}^0 \end{pmatrix} = \begin{pmatrix} -\cos\varphi_{21} \cdot \sin\varphi_{22} \\ \cos\varphi_{22} \\ \sin\varphi_{21} \cdot \sin\varphi_{22} \end{pmatrix},$$

$$\begin{pmatrix} r_{3x}^0 \\ r_{3y}^0 \\ r_{3z}^0 \end{pmatrix} = \begin{pmatrix} \sin\varphi_{31} \cdot \sin\varphi_{32} \\ \cos\varphi_{31} \cdot \sin\varphi_{32} \\ \cos\varphi_{32} \end{pmatrix}$$

The dependence of rotation angles of input φ_{i1} and intermediate pairs φ_{i2} are determined by solving the problem of the position:

$$\mathrm{tg}\varphi_{11} = -\frac{\sin\gamma \cdot \cos\alpha \cdot \sin\beta - \cos\gamma \cdot \sin\alpha}{\cos\beta \cdot \cos\alpha};$$

$$\mathrm{tg}\varphi_{21} = \frac{\sin\beta}{\cos\gamma \cdot \cos\beta};$$

$$\mathrm{tg}\varphi_{31} = -\frac{\cos\gamma \cdot \sin\beta \cdot \sin\alpha - \cos\alpha \cdot \sin\gamma}{\cos\alpha \cdot \cos\gamma + \sin\alpha \cdot \sin\beta \cdot \sin\gamma};$$

$$\sin\varphi_{12} = \sin\gamma \cdot \sin\alpha + \cos\alpha \cdot \cos\gamma \cdot \sin\beta;$$

$$\sin\varphi_{22} = \cos\beta \cdot \sin\gamma; \qquad \sin\varphi_{32} = \cos\beta \cdot \sin\alpha \tag{15.30}$$

The torques in the drives are equal:

$$\begin{cases} M_1 = -c_1 \cdot \varphi_{11} \\ M_2 = -c_2 \cdot \varphi_{21} \\ M_3 = -c_3 \cdot \varphi_{31} \end{cases} \tag{15.31}$$

Here c_i is the drive rigidity.

Spherical mechanism gets in balance with the following angles: $\varphi_{11} = 0$, $\varphi_{21} = 0$, $\varphi_{11} = 0$, $\varphi_{31} = 0$, $\alpha = \beta = \gamma = 0$. Using numerical integration, we will find coordinate change of output link for different initial conditions: $\alpha = 0.1$ rad; $\beta = 0.05$ rad; $\gamma = 0.325$ rad (Fig. 15.26).

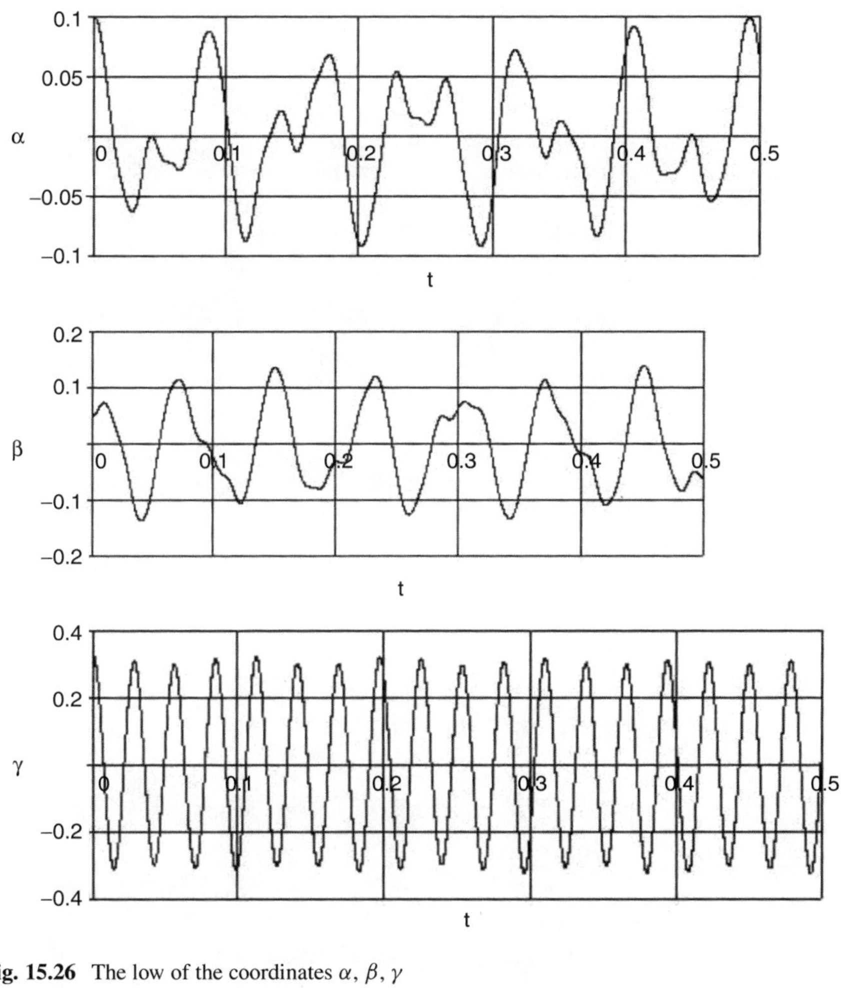

Fig. 15.26 The low of the coordinates α, β, γ

15.5.4 Control of the Manipulator

Control of parallel manipulators is one of the most important problems. There exist different approaches to solve this problem. The applied algorithm is based on the inverse problems of dynamics [26].

The desired laws of the coordinates of the mobile platform are described by equations $\alpha_T(t)$, $\beta_T(t)$, $\gamma_T(t)$. The task is to minimize the errors $\Delta\alpha = \alpha_T(t) - \alpha(t)$, $\Delta\beta = \beta_T(t) - \beta(t)$, $\Delta\gamma = \gamma_T(t) - \gamma(t)$, where $\alpha(t)$, $\beta(t)$, $\gamma(t)$ are the actual coordinates of the mobile platform. Equations of errors are:

$$\begin{aligned}
&\ddot{\Delta}\alpha + K_D\dot{\Delta}\alpha + K_P\Delta\alpha = 0;\\
&\ddot{\Delta}\beta + K_D\dot{\Delta}\beta + K_P\Delta\beta = 0;\\
&\ddot{\Delta}\gamma + K_D\dot{\Delta}\gamma + K_P\Delta\gamma = 0.
\end{aligned} \tag{15.32}$$

Here K_D, K_P are the feedback coefficients. The value feedback gains coefficients are determined according the theory robotic control [26].

According to the equations above the accelerations become:

$$\begin{aligned}
\ddot{\alpha} &= \ddot{\alpha}_T + \gamma_1 \cdot (\dot{\alpha}_T - \dot{\alpha}) + \gamma_0 \cdot (\alpha_T - \alpha);\\
\ddot{\beta} &= \ddot{\beta}_T + \gamma_1 \cdot \left(\dot{\beta}_T - \dot{\beta}\right) + \gamma_0 \cdot (\beta_T - \beta);\\
\ddot{\gamma} &= \ddot{\gamma}_T + \gamma_1 \cdot (\dot{\gamma}_T - \dot{\gamma}) + \gamma_0 \cdot (\gamma_T - \gamma).
\end{aligned} \tag{15.33}$$

The laws of errors is described as oscillatory second-order systems, by which control time is minimized.

Let us consider an example: the desired laws of the coordinates of the moving platform are $\alpha_T(t) = 0.1{\cdot}\sin(\omega{\cdot}t)$, $\beta_T(t) = 0.11{\cdot}\sin(\omega{\cdot}t)$, $\gamma_T(t) = 0.12{\cdot}\sin(\omega{\cdot}t)$. Moments of inertia are $J_\xi = J_\eta = 0.0012$ kg m^2, $J_\zeta = 0.002$ kg m^2, $K_D = 120$, $K_P = 7{,}200$. Result of the simulation is presented in Figs. 15.27 and 15.28. The maximal errors are about 6×10^{-3} rad.

15.5.5 Accuracy of the Manipulator

The full differential of input–output equation can be written:

$$\frac{\partial F_i}{\partial \alpha}\delta\alpha + \frac{\partial F_i}{\partial \beta}\delta\beta + \frac{\partial F_i}{\partial \gamma}\delta\gamma + \frac{\partial F_i}{\partial \theta_{i1}}\delta\theta_{i1} + \frac{\partial F_i}{\partial \theta_{i2}}\delta\theta_{i2} + \frac{\partial F_i}{\partial \varphi_{i1}}\delta\varphi_{i1} = 0 \tag{15.34}$$

Here θ_{i1}, θ_{i2} are the angels between the axes of kinematic pairs.

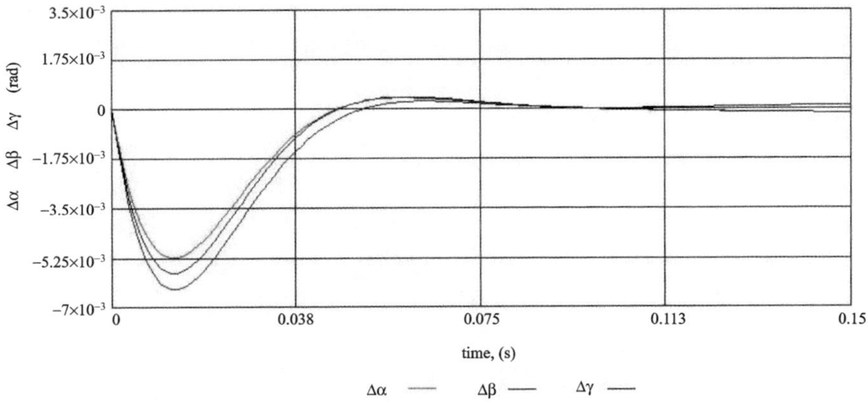

Fig. 15.27 The simulation result of displacement

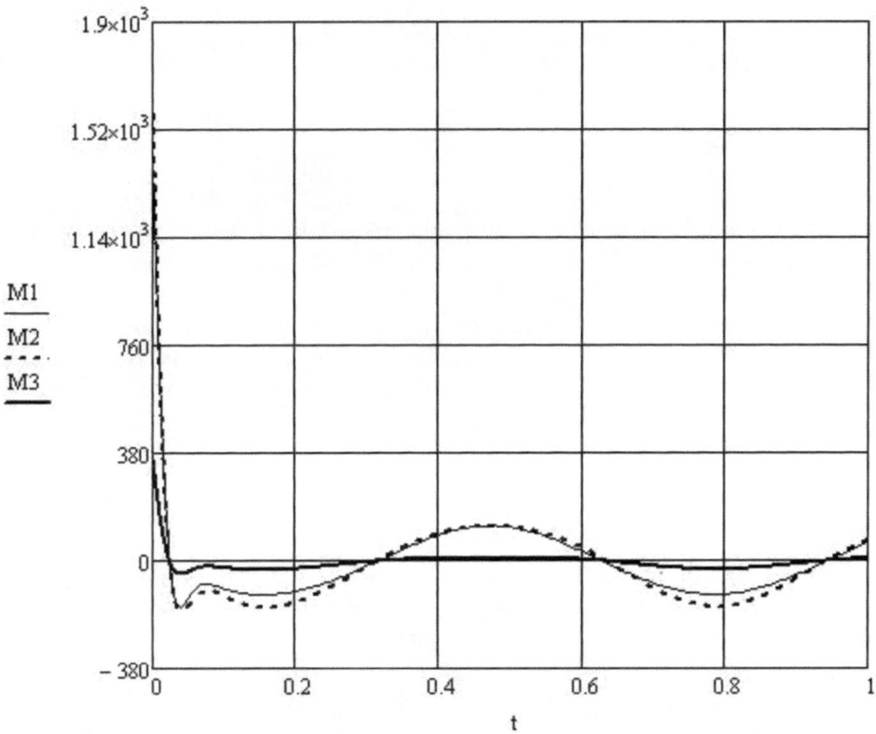

Fig. 15.28 The simulation result of driving torques

According to the linear theory of accuracy, we assume increment in the actuators are equal to zero: $\delta\varphi_{11} = \delta\varphi_{21} = \delta\varphi_{31} = 0$. Therefore, the system of equations can be represented as:

$$\begin{aligned}
\frac{\partial F_1}{\partial \alpha}\delta\alpha + \frac{\partial F_1}{\partial \beta}\delta\beta + \frac{\partial F_1}{\partial \gamma}\delta\gamma &= -\left(\frac{\partial F_1}{\partial \theta_{12}}\delta\theta_{12} + \frac{\partial F_1}{\partial \theta_{11}}\delta\theta_{11}\right);\\
\frac{\partial F_2}{\partial \alpha}\delta\alpha + \frac{\partial F_2}{\partial \beta}\delta\beta + \frac{\partial F_2}{\partial \gamma}\delta\gamma &= -\left(\frac{\partial F_2}{\partial \theta_{22}}\delta\theta_{22} + \frac{\partial F_2}{\partial \theta_{21}}\delta\theta_{21}\right)\\
\frac{\partial F_3}{\partial \alpha}\delta\alpha + \frac{\partial F_3}{\partial \beta}\delta\beta + \frac{\partial F_3}{\partial \gamma}\delta\gamma &= -\left(\frac{\partial F_3}{\partial \theta_{32}}\delta\theta_{32} + \frac{\partial F_3}{\partial \theta_{31}}\delta\theta_{31}\right).
\end{aligned} \tag{15.35}$$

Thus, knowing the deviations of angles between the axes of kinematic pairs we can determine the deviations of the position of the output link.

For example, if errors are: $\theta_{11} = 0.01$ rad, $\theta_{12} = 0.005$ rad, $\theta_{31} = 0.01$ rad, $\theta_{i2} = 0.0075$ rad, $\theta_{12} = 0.005$ rad, $\theta_{22} = 0.005$ rad, $\theta_{23} = 0.005$ rad, then the deviations of the position of the output link are $\delta\alpha = 0.012$ rad, $\delta\beta = 0.017$ rad, $\delta\gamma = 0.005$ rad.

15.5.6 Prototyping of the Manipulator

We considered the mechanism where all the axes of the kinematic pairs intersect at a single point. One of versions of spherical mechanisms with five kinematic pairs in any chain is prototyped. One of the versions of spherical mechanism with five kinematic pairs in two kinematic chains and three kinematic pairs in one kinematic chain is prototyped (Fig. 15.29). This prototype is used to prove the kinematic properties of the manipulator. In future we plan to investigate the dynamical properties of the prototype.

15.6 Conclusion

Therefore, in this chapter the dynamical properties of planar, translational, and spherical parallel manipulators are considered.

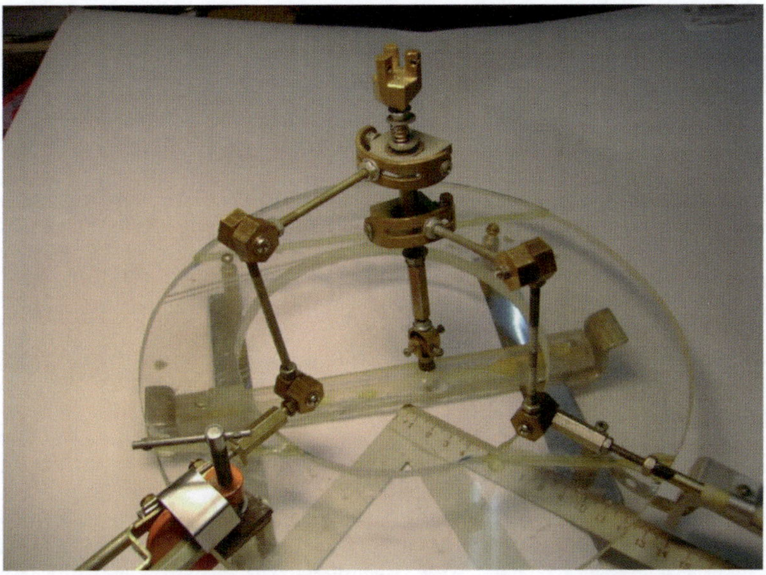

Fig. 15.29 Prototype of spherical mechanism with three kinematic chains

References

1. Hunt, K.: Structural kinematics of in-parallel-actuated robot arms. ASME J. Mech. Transm Autom Des **105**(4), 705–712 (1983)
2. Merlet, J.-P.: Parallel Robots, 2nd edn. Springer, Netherlands (2006)
3. Kong, X., Gosselin, C.: Type Synthesis of Parallel Mechanisms. Springer, Berlin (2007)
4. Ceccarelli, M.: Fundamentals of Mechanics of Robotic Manipulations. Kluwer Academic Publishers, Boston (2004)
5. Tsai, L.: Robot Analysis: The Mechanics of Serial and Parallel Manipulators. Wiley, New York (1999)
6. Glazunov, V.A., Koliskor, A.S., Krainev, A.F., Model, B.I.: Classification principles and analysis methods for parallel-structure spatial mechanisms. J. Mach. Manuf. Reliabil. **1**, 30–37 (1990)
7. Glazunov, V.: Principles of the construction and analysis of spatial parallel structure mechanisms. J. Mach. Manuf. Reliabil. **1**, 10–15 (1995)
8. Gogu, G.: Structural Synthesis of Parallel Robots, Part 1: Methodology. Springer, Dordrecht (2008)
9. Gogu, G.: Structural Synthesis of Parallel Robots, Part 2: Translational Topologies with Two and Three Degrees of Freedom. Springer, Dordrecht (2009)
10. Dimentberg, F.M.: The Screw Calculus and Its Applications in Mechanics. Nauka, Moscow (1965) (English translation: AD680993, Clearinghouse for Federal Technical and Scientific Information, Virginia)
11. Herve, J.: The Lie Group of Rigid Body Displacements, a Fundamental Tool for Mechanism Design. Mech. Mach. Theory **34**(8), 719–730 (1991)
12. Gogu, G.: Structural Synthesis of Fully-isotropic Translational Parallel robots via Theory of Linear Transformations. Eur. J. Mech. A Solids **23**, 1021–1039 (2004)

13. R. Clavel: Device for displacing and positioning an element in space. Patent WO 87/03528, 1987
14. Gosselin, C.M., Angeles, J.: The optimum design of spherical three degree of freedom parallel manipulator. Trans. ASME J. Mech. Des. **111**, 202–207 (1989)
15. Huda, S., Takeda Y.: Dimension synthesis of 3-URU pure rotation parallel mechanism with respect to singularity and workspace. In: 12th IFToMM World Congress, Becasson, 2007, pp. 235–242
16. Liu, X.J., Jin, Z.L., Gao, F.: Optimum design of 3-DOF spherical parallel mechanism with respect to the conditioning and stiffness indices. Mech. Mach. Theory **35**(9), 257–1267 (2000)
17. Bonev, I.A., Zlatanov, D., Gosselin, C.M.: Singularity analysis of 3-DOF planar parallel mechanisms via Screw Theory. Trans. ASME J. Mech. Des. **125**, 573–581 (2003)
18. Carricato, M., Parenti-Castelli, V.: On the topological and geometrical synthesis and classification of translational parallel mechanisms. In: Pr. of the XI World Congress in Mechanism and Machine Science. Tianjin. China, 2004, pp. 1624–1628
19. Mianovski, K.: Singularity analysis of parallel manipulator POLMAN 3×2 with six degrees of freedom. In: 12th IFToMM World Congress, Besançon, France,18–21 June 2007
20. Yan Jin, I-Ming Chen, Guilin Yang: Mobility and singularity analysis of a selectively actuated parallel mechanism. Theory and practice of robots and manipulators (ROMANSY), Proceedings of XV CISM-IFToMM Symposium, Montreal, 2004
21. Glazunov, V.: Design of decoupled parallel manipulators by means of the theory of screws. Mech. Mach. Theory **45**, 239–250 (2010)
22. Glazunov, V.A., Bykov, R.E., Esina, M.G.: Control of the parallel-structure mechanisms when passing through the singular positions. J. Mach. Manuf. Reliabil. **2**, 73–79 (2004)
23. Kheylo, S., Glazunov, V., Nguyen Minh Thanh: Aceleration and nonlinear oscillations of parallel spherical mechanism. In: Proceedings of the 3rd IFToMM International Symposium on Robotics and Mechatronics, Singapore. ISBN: 978-981-07-7744-9. 2–4 October 2013, pp. 504–512
24. Kheylo, S., Glazunov, V., Kulemkin, Y., Ephros, V.: Analysis of accelerations and nonlinear oscillations of spherical mechanism of parallel structure. J. Mach. Manuf. Reliabil. **4**, 184–191 (2013)
25. Kheylo, S., Glazunov, V.: Kinematics, dynamics, control and accuracy of spherical parallel robot. Advances on theory and practice of robots and manipulators. In: Proceedings of ROMANSY 2014 XX CISM-IFToMM Symposium on Theory and Practice of Robots and Manipulators. Springer Cham Heidelberg New York, Dordrecht, London, 2014, pp. 133–140
26. Craig, J.J.: Introduction to Robotics: Mechanics and Control, 2nd edn. Addisson-Wesley, Reading, MA (1989)

Chapter 16
Dynamic Modelling and Control of Balanced Parallel Mechanisms

Renato Maia Matarazzo Orsino, André Garnier Coutinho, and Tarcisio Antonio Hess Coelho

Abstract Balancing is an important issue related to the design of mechanical systems in general, and also parallel mechanisms, in particular. In fact, the performance of parallel mechanisms associated with specific applications depends on the choice of the balancing method, namely, either static or dynamic, either passive or active, whether it is valid for a given trajectory or even for any motion. The main contribution of this work is to highlight the importance of the dynamic modelling process in order to achieve the compensation conditions associated with the chosen balancing technique. Due to the fact that parallel mechanisms have highly complex structures, the use of dynamic formalisms that employ redundant generalized coordinates, in association with the successive coupling of additional balancing elements to the original system model, can bring remarkable benefits. Additionally, this book chapter also discusses the impact of the dynamic model, developed in accordance with the methodology shown here, for the control strategy of parallel mechanisms. Finally, the simulation results demonstrate how effective is the presented methodology for the planar 5-bar with revolute joints (5R).

Keywords Dynamic modelling • Balancing • Parallel mechanisms • Control

Nomenclature

$a, b, \dots$	Scalars, components of column-matrices or indexes
$A, B, \dots$	Scalars or components of matrices
$\mathbb{a}, \mathbb{b}, \dots$	Column-matrices
$\mathbb{A}, \mathbb{B}, \dots$	Matrices

R.M.M. Orsino
Department of Mechanical Engineering, Escola Politecnica, University of Sao Paulo, Brazil
e-mail: renato.orsino@gmail.com

A.G. Coutinho • T.A.H. Coelho (✉)
Department of Mechatronics and Mechanical Systems Engineering, Escola Politecnica, University of Sao Paulo, Brazil
e-mail: andre.garnier.coutinho@usp.br; tarchess@usp.br

D. Zhang, B. Wei (eds.), *Dynamic Balancing of Mechanisms and Synthesizing of Parallel Robots*, DOI 10.1007/978-3-319-17683-3_16

$\boldsymbol{a}, \boldsymbol{b}, \ldots$	Vectors
$\boldsymbol{A}, \boldsymbol{B}, \ldots$	Tensors
$\mathtt{a}, \mathtt{b}, \ldots$	Points
$\mathtt{A}, \mathtt{B}, \ldots$	Coordinate systems
$\mathcal{A}, \mathcal{B}, \ldots$	Rigid bodies or reference frames
$\mathscr{A}, \mathscr{B}, \ldots$	Sets or multibody mechanical systems[1]
$a_{n,l}$	Arbitrary physical parameter
$\mathbb{A}_n$	Jacobian matrix of kinematic invariants ($\mathbb{c}_n$) with respect to quasi-accelerations ($\dot{\mathbb{p}}_n$)
$\boldsymbol{a}_{\mathtt{p}\mid\mathcal{E}}$	Acceleration of point $\mathtt{p}$ measured relatively to reference frame $\mathcal{E}$
$\mathbb{c}_n$	Kinematic invariants (constraints) column-matrix
$\mathscr{C}^s$	Differentiability class
$\mathbb{d}_n$	Dynamic invariants column-matrix
d	Differential operator
$f_{n,j}$	Generalized force
$\mathbb{f}_n$	Generalized forces column-matrix
$\boldsymbol{f}_{\mathcal{B}}$	Resultant force acting on body $\mathcal{B}$ (excluding constraint forces)
$g_{n,j}$	Generalized gyroscopic inertia force
$\mathbb{g}_n$	Generalized gyroscopic inertia forces column-matrix
I_s	Arbitrary moment of inertia
$\mathbf{I}_{\mathcal{B}\mid\mathtt{p}}$	Inertia tensor of rigid body $\mathcal{B}$ relative to point $\mathtt{p}$
$\mathscr{I}_x(\mathcal{S}_n)$	Set of indexes of variables $x_{n,r}$ defined in the model of system $\mathcal{S}_n$, i.e., $\mathscr{I}_x(\mathcal{S}_n) = \{r \mid x_{n,r} \in \mathcal{S}_n\}$
m_s	Arbitrary mass
$\mathbb{M}_n$	Generalized inertia matrix
$\boldsymbol{m}_{\mathcal{B}\mid\mathtt{p}}$	Resultant moment (torque) acting on body $\mathcal{B}$ measured relatively to pole $\mathtt{p}$ (excluding constraint moments)
$\mathcal{N}$	Inertial reference frame
$p_{n,j}$	Quasi-velocity
$\dot{p}_{n,j}$	Quasi-acceleration
$\mathbb{p}_n$	Quasi-velocities column-matrix
$\dot{\mathbb{p}}_n$	Quasi-accelerations column-matrix
$q_{n,i}$	Generalized coordinate
$\mathbb{q}_n$	Generalized coordinates column-matrix
$\boldsymbol{r}_{\mathtt{p}_2\mid\mathtt{p}_1}$	Position of point $\mathtt{p}_2$ relative to point $\mathtt{p}_1$
t	Time
$u_{n,k}$	Control input
$\boldsymbol{v}_{\mathtt{p}\mid\mathcal{E}}$	Velocity of point $\mathtt{p}$ measured relatively to reference frame $\mathcal{E}$

[1] A multibody mechanical system will be conceived as a set whose elements are material bodies, joints, actuators, energy storage, dissipation and transformation elements and a mathematical model (which includes physical parameters, model variables and constitutive, constraint and dynamic equations).

$w_{n,j}$	Arbitrary term of generalized force or generalized gyroscopic inertia force linear or bilinear with respect to quasi-velocities
$\mathbb{w}_n$	Column-matrix whose entries are $w_{n,j}$
$z_{n,j}$	Arbitrary term of generalized force or generalized gyroscopic inertia force independent of quasi-velocities
$\mathbb{z}_n$	Column-matrix whose entries are $z_{n,j}$
δ	Variation operator
$\nu_x(\mathcal{S}_n)$	Number of elements of the set $\mathscr{I}_x(\mathcal{S}_n)$
$\nu^{\#}(\mathcal{S}_n)$	Number of degrees of freedom of the mechanical system $\mathcal{S}_n$
$\boldsymbol{\omega}_{\mathcal{B}\mid\mathcal{E}}$	Angular velocity of rigid body or reference frame $\mathcal{B}$ measured relatively to reference frame $\mathcal{E}$
$\mathbb{0}$	Null column-matrix or null matrix
$\mathbf{0}$	Null vector or null tensor
$\mathbb{1}$	Identity matrix
$\mathbf{1}$	Identity tensor
$[\cdot]^{\mathsf{T}}$	Matrix transposition
$[\cdot]^{-1}$	Matrix inversion
$\lVert \cdot \rVert_\infty$	Matrix infinity norm

16.1 Introduction and Literature Review

Balancing is an important issue related to the design of mechanical systems in general, and also parallel mechanisms, in particular. In fact, the performance of parallel mechanisms associated with specific applications depends on the choice of the balancing method, namely, either static or dynamic, either passive or active, whether it is valid for a given trajectory or even for any motion.

In a statically balanced mechanism, the potential energy is invariant. Hence, the actuator torques/forces are null at any configuration [14]. On the other hand, in a dynamically balanced mechanism, the shaking forces and shaking moments at its frame are reduced or even eliminated. As a consequence, the mechanism components are less susceptible to vibration, wear, and fatigue [13], improving its life.

Passive balancing means that the original mechanism architecture is modified by using some techniques. The most common techniques are the addition of counterweights [4, 6, 9–15], the use of counter-rotating inertias [4, 6, 9, 13], and the redistribution of masses [2, 3]. Alternatively, other works propose the addition of extra links [1, 8, 10]. However, for high speed manipulators, these techniques might cause the increase of the actuator torques and the size of links and joints. Hence, in order to avoid such undesirable effects, some authors [1–3, 7, 14, 15] recommend the use of elastic springs attached to the mechanism.

For the active balancing [4–6, 9, 11, 16], more complex modifications are needed to implement it. For instance, the counterweights position in the moving links might be altered according to the end-effector load or the given trajectory. Hence, additional actuators/sensors and control are usually employed to reach satisfactory performance levels. Arakelian and Smith [4] employ a computer control of the motion of an inertia flywheel connected to the mechanism.

Moreover, Coelho et al. [6] and Moradi et al. [9] use the adaptive balancing to achieve the decoupling of dynamic equations for open loop kinematic chain mechanisms. Consequently, this action simplifies the control of manipulators due to the fact that the actuators can be controlled independently.

The main contribution of this work is to highlight the importance of the dynamic modelling process (see Fig. 16.1) in order to achieve the compensation conditions associated with the chosen balancing technique. Due to the fact that parallel mechanisms have highly complex structures, the use of dynamic formalisms that employ redundant generalized coordinates, in association with the successive coupling of additional balancing elements to the original system model, can bring remarkable benefits. Additionally, this book chapter also discusses the impact of the dynamic model, developed in accordance with the methodology shown here, for the control strategy of parallel mechanisms.

This chapter is organized as follows. Section 16.2 treats the theoretical background for the dynamic modelling and control, the adaptive balancing techniques are described in Sects. 16.3, and 16.4 shows a case study for the planar 5-bar mechanism with revolute joints. Finally, the conclusions are drawn in Sect. 16.5.

16.2 Theoretical Background for the Dynamic Modelling and Control

16.2.1 Basic Structure of a Dynamic Model

Let $\mathcal{S}_n$ be a multibody mechanical system with a finite number $\nu^{\#}(\mathcal{S}_n)$ of degrees of freedom. Consider the definition of a set of $\nu_q(\mathcal{S}_n)$ generalized coordinates $q_{n,i}$, $i \in \mathscr{I}_q(\mathcal{S}_n)$, for the description of every possible configuration of such system. A set of coordinates is useful for such description if and only if an arbitrary assignment of values for these coordinates correspond either to a finite number of non-neighboring configurations or to no configuration. In the first case, the values assigned to the coordinates are called *compatible*. It is important to notice that descriptions in which compatible values of coordinates correspond to more than one configuration can be avoided by adding new coordinates to the original set, so that compatible values in this latter set of variables uniquely describe any possible configuration of the system.

Consider a particular compatible set of values for the generalized coordinates and suppose that some of these values are varied infinitesimally. The maximum number of independent variations still leading to compatible values corresponds to the local number of degrees of freedom associated with the respective configurations.

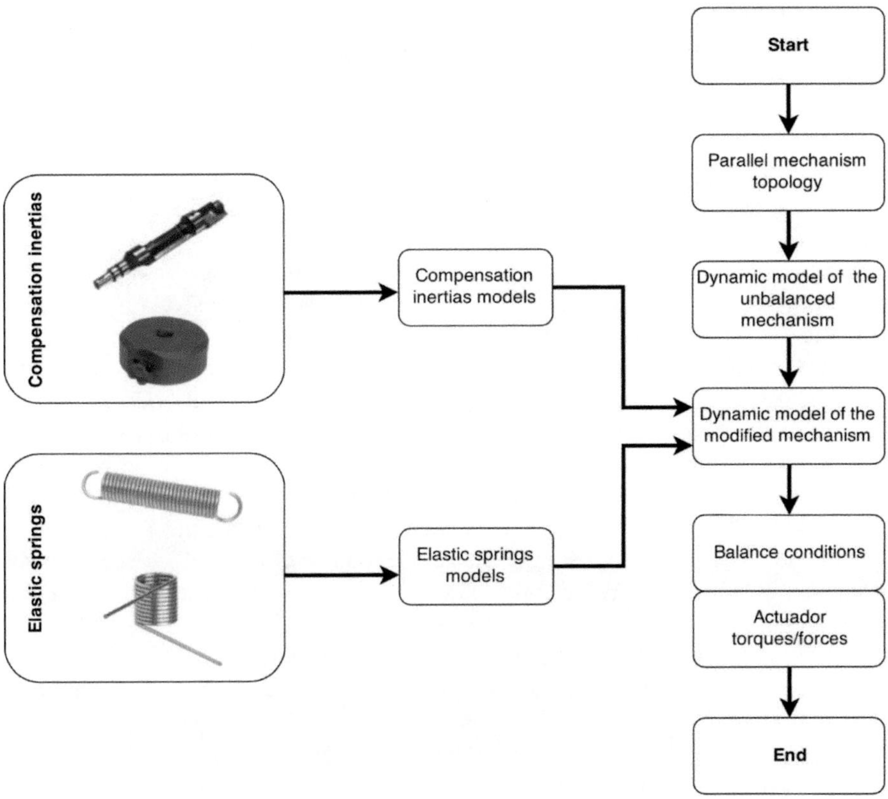

Fig. 16.1 Dynamic modelling process

The maximum local number of degrees of freedom among all the possible configurations of the system is the number of degrees of freedom of $\mathcal{S}_n$ (which, by hypothesis, is finite). A particular set of generalized coordinates is called *redundant* if the number of coordinates is greater than the number of degrees of freedom of $\mathcal{S}_n$, i.e., if $\nu_q(\mathcal{S}_n) > \nu^{\#}(\mathcal{S}_n)$. In such cases there must be as much independent invariants as the excess of generalized coordinates over the number of degrees of freedom of the system, i.e., $\nu_q(\mathcal{S}_n) - \nu^{\#}(\mathcal{S}_n)$ independent invariants, imposing conditions that limit the maximum number of independent variations (and, consequently, time derivatives) of the generalized coordinates but still lead to compatible values for these variables. Denoting by $\mathbb{q}_n$ the column-matrix whose entries are the $q_{n,i}$, these invariants can be expressed by the so-called *constraint equations*, once they represent the kinematic constraints of system $\mathcal{S}_n$, which can be presented in the following form:

$$\psi_{n,r}(t, \mathbb{q}_n, \dot{\mathbb{q}}_n) = 0 \qquad \text{for} \qquad r \in \{1, 2, \ldots, \nu_q(\mathcal{S}_n) - \nu^{\#}(\mathcal{S}_n)\} \tag{16.1}$$

If $\psi_{n,r}(t, \mathbb{q}_n, \mathrm{d}\mathbb{q}_n)$ is an exact differential, the corresponding constraint is classified as *holonomic*. In these cases, there is a function $h_{n,r}(t, \mathbb{q}_n)$ such that $\mathrm{d}h_{n,r} = \psi_{n,r}(t, \mathbb{q}_n, \mathrm{d}\mathbb{q}_n) = 0$, i.e., $h_{n,r}(t, \mathbb{q}_n)$ is an invariant representing a constraint which depends exclusively on the instantaneous configurations of the system, but not on its instantaneous motion. When $\psi_{n,r}(t, \mathbb{q}_n, \mathrm{d}\mathbb{q}_n)$ is an inexact differential, the corresponding constraint is called *nonholonomic*. Actually, the class of nonholonomic constraints include constraints that cannot even be expressed by equations. These latter cases, however, are out of the scope of this book chapter.

Now, consider that, in order to replace time derivatives of generalized coordinates in the description of instantaneous motions of system $\mathcal{S}_n$, a set of $\nu_p(\mathcal{S}_n)$ quasi-velocities $p_{n,j}$, $j \in \mathscr{I}_p(\mathcal{S}_n)$, is defined. The most common example of the use of quasi-velocities occurs in Newton–Euler equations: angular velocity components are typically used to describe instantaneous rotations of a rigid body instead of time derivatives of the generalized coordinates associated with the instantaneous orientations of this body (which can be Euler angles, quaternion components, etc.). However, time integrals of the components of angular velocity, in general, cannot be used to describe instantaneous configurations of a rigid body. That is, although it may be convenient to use a particular quasi-velocity in the description of instantaneous motions of a mechanical system, it is not required that its time integral have any physical sense.

Once quasi-velocities replace time derivatives of generalized coordinates, the following set of independent equations defining such transformation of variables must be provided:

$$\phi_{n,s}(t, \mathbb{q}_n, \dot{\mathbb{q}}_n, \mathbb{p}_n) = 0 \qquad \text{for} \qquad s \in \{1, 2, \ldots, \max\left(\nu_q(\mathcal{S}_n),\ \nu_p(\mathcal{S}_n)\right)\} \tag{16.2}$$

Suppose that all the $\phi_{n,s}(t, \mathbb{q}_n, \dot{\mathbb{q}}_n, \mathbb{p}_n)$ are class $\mathscr{C}^1$ functions and that for all states $(t, \mathbb{q}_n, \dot{\mathbb{q}}_n, \mathbb{p}_n)$ satisfying both systems of Eqs. (16.1) and (16.2) the conditions of the Theorem of Implicit Functions are satisfied, so that it is possible to use (some of the) Eq. (16.2) to obtain the following solutions (valid in some neighborhood of such states):

$$\dot{q}_{n,i} = \dot{q}^*_{n,i}(t, \mathbb{q}_n, \mathbb{p}_n) \qquad \forall\, i \in \mathscr{I}_q(\mathcal{S}_n) \tag{16.3}$$

$$p_{n,j} = p^*_{n,j}(t, \mathbb{q}_n, \dot{\mathbb{q}}_n) \qquad \forall\, j \in \mathscr{I}_p(\mathcal{S}_n) \tag{16.4}$$

Now, consider the following system of equations:

$$\begin{cases} \psi_{n,r}(t, \mathbb{q}_n, \dot{\mathbb{q}}^*_n(t, \mathbb{q}_n, \mathbb{p}_n)) = 0 & \text{for} \quad r \in \{1, 2, \ldots, \nu_q(\mathcal{S}_n) - \nu^{\#}(\mathcal{S}_n)\} \\ \phi_{n,s}(t, \mathbb{q}_n, \dot{\mathbb{q}}^*_n(t, \mathbb{q}_n, \mathbb{p}_n), \mathbb{p}_n) = 0 & \text{for} \quad s \in \{1, 2, \ldots, \max\left(\nu_q(\mathcal{S}_n),\ \nu_p(\mathcal{S}_n)\right)\} \end{cases} \tag{16.5}$$

Once the expressions of $\dot{q}^*_{n,i}(t, \mathbb{q}_n, \mathbb{p}_n)$ are obtained from local solutions of (some of the) Eq. (16.2), at least $\nu_q(\mathcal{S}_n)$ equations must be eliminated from the system (16.5) in order to obtain a system of independent equations. Indeed, for any possible state, system (16.5) will not have more than $\nu_c(\mathcal{S}_n) = \nu_p(\mathcal{S}_n) - \nu^{\#}(\mathcal{S}_n)$

independent equations, and when it has less than $\nu_c(\mathcal{S}_n)$ independent equations, the corresponding state is called *singular*. It is clear that the time derivative of the independent equations in system (16.5) can be expressed in the following form (replacing $\dot{q}_{n,i} = \dot{q}^*_{n,i}(t, \mathbb{q}_n, \mathbb{p}_n)$, when necessary):

$$c_{n,r}(t, \mathbb{q}_n, \mathbb{p}_n, \dot{\mathbb{p}}_n) = \sum_j A_{n,rj}(t, \mathbb{q}_n, \mathbb{p}_n)\, \dot{p}_{n,j} + b_{n,r}(t, \mathbb{q}_n, \mathbb{p}_n)$$
$$= 0 \quad \text{for} \quad r \in \{1, 2, \ldots, \nu_c(\mathcal{S}_n)\} \tag{16.6}$$

It is also possible to express these equations in the following matrix form:

$$\mathbb{c}_n(t, \mathbb{q}_n, \mathbb{p}_n, \dot{\mathbb{p}}_n) = \mathbb{A}_n(t, \mathbb{q}_n, \mathbb{p}_n)\, \dot{\mathbb{p}}_n + \mathbb{b}_n(t, \mathbb{q}_n, \mathbb{p}_n) = \mathbb{0} \tag{16.7}$$

In order to determine completely all the quasi-accelerations $\dot{p}_{n,j}$ associated with a given state of the mechanical system, it is necessary to form a system of equations in which (16.7) is taken along with the dynamic equations of the model.

Consider that $\mathcal{S}_n$ is a multiple rigid-body mechanical system, composed of a set of constrained rigid bodies $\mathcal{B}_{n,s}$, $s \in \mathscr{I}_{\mathcal{B}}(\mathcal{S}_n)$ and a set of actuators providing control inputs $u_{n,k}$, $k \in \mathscr{I}_u(\mathcal{S}_n)$. The following notation is adopted:

- $\mathrm{b}^\star_{n,s}$ represents the centre of mass of $\mathcal{B}_{n,s}$;
- $\boldsymbol{v}_{\mathrm{b}^\star_{n,s} | \mathcal{N}}$ and $\boldsymbol{a}_{\mathrm{b}^\star_{n,s} | \mathcal{N}}$ are, respectively, the velocity and the acceleration of $\mathrm{b}^\star_{n,s}$ measured relatively to an inertial reference frame $\mathcal{N}$;
- $\boldsymbol{\omega}_{\mathcal{B}_{n,s} | \mathcal{N}}$ and $\dot{\boldsymbol{\omega}}_{\mathcal{B}_{n,s} | \mathcal{N}}$ are respectively the angular velocity and the angular acceleration of $\mathcal{B}_{n,s}$ measured relatively to an inertial reference frame $\mathcal{N}$;
- $m_{\mathcal{B}_{n,s}}$ and $I_{\mathcal{B}_{n,s} | \mathrm{b}^\star_{n,s}}$ are, respectively, the mass of $\mathcal{B}_{n,s}$ and its inertia tensor relative to point $\mathrm{b}^\star_{n,s}$;
- $\boldsymbol{f}_{\mathcal{B}_{n,s}}$ and $\boldsymbol{m}_{\mathcal{B}_{n,s} | \mathrm{b}^\star_{n,s}}$ are, respectively, the resultant force and resultant moment measured relatively to pole $\mathrm{b}^\star_{n,s}$, including control inputs and excluding constraint effects, acting on body $\mathcal{B}_{n,s}$.

Applying the Principle of Virtual Power to $\mathcal{S}_n$, it can be stated that:

$$\sum_s \Big(\delta \boldsymbol{v}_{\mathrm{b}^\star_{n,s} | \mathcal{N}} \cdot \Big(\boldsymbol{f}_{\mathcal{B}_{n,s}} - m_{\mathcal{B}_{n,s}}\, \boldsymbol{a}_{\mathrm{b}^\star_{n,s} | \mathcal{N}} \Big)$$
$$+ \delta \boldsymbol{\omega}_{\mathcal{B}_{n,s} | \mathcal{N}} \cdot \Big(\boldsymbol{m}_{\mathcal{B}_{n,s} | \mathrm{b}^\star_{n,s}} - I_{\mathcal{B}_{n,s} | \mathrm{b}^\star_{n,s}} \cdot \dot{\boldsymbol{\omega}}_{\mathcal{B}_{n,s} | \mathcal{N}}$$
$$- \boldsymbol{\omega}_{\mathcal{B}_{n,s} | \mathcal{N}} \times (I_{\mathcal{B}_{n,s} | \mathrm{b}^\star_{n,s}} \cdot \boldsymbol{\omega}_{\mathcal{B}_{n,s} | \mathcal{N}}) \Big) \Big) = 0 \tag{16.8}$$

Consider that at a given time t', all the values of the generalized coordinates $q_{n,i}(t')$ and of the quasi-velocities $p_{n,j}(t')$ are known, implying that $\delta \mathbb{q}_n(t') = \mathbb{0}$ and $\delta \mathbb{p}_n(t') = \mathbb{0}$. Thus, given the expressions of $\boldsymbol{v}_{\mathrm{b}^\star_{n,s} | \mathcal{N}}$ and $\boldsymbol{\omega}_{\mathcal{B}_{n,s} | \mathcal{N}}$ in terms of $q_{n,i}$ and $p_{n,j}$, i.e., $\boldsymbol{v}_{\mathrm{b}^\star_{n,s} | \mathcal{N}} = \boldsymbol{v}^*_{\mathrm{b}^\star_{n,s} | \mathcal{N}}(t, \mathbb{q}_n, \mathbb{p}_n)$ and $\boldsymbol{\omega}_{\mathcal{B}_{n,s} | \mathcal{N}} = \boldsymbol{\omega}^*_{\mathcal{B}_{n,s} | \mathcal{N}}(t, \mathbb{q}_n, \mathbb{p}_n)$,

$\delta \boldsymbol{v}_{\mathrm{b}^\star_{n,s}|\mathcal{N}}(t') = \mathbf{0}$ and $\delta \boldsymbol{\omega}_{\mathcal{B}_{n,s}|\mathcal{N}}(t') = \mathbf{0}$. Consider the expansion in Taylor series of $\delta \boldsymbol{v}_{\mathrm{b}^\star_{n,s}|\mathcal{N}}(t'+\tau)$ and $\delta \boldsymbol{\omega}_{\mathcal{B}_{n,s}|\mathcal{N}}(t'+\tau)$:

$$\delta \boldsymbol{v}_{\mathrm{b}^\star_{n,s}|\mathcal{N}}(t'+\tau) = \delta \boldsymbol{v}_{\mathrm{b}^\star_{n,s}|\mathcal{N}}(t') + \tau\, \delta \boldsymbol{a}_{\mathrm{b}^\star_{n,s}|\mathcal{N}}(t') + O(\tau^2)$$

$$= \tau \sum_j \frac{\partial \boldsymbol{v}_{\mathrm{b}^\star_{n,s}|\mathcal{N}}}{\partial p_{n,j}}(t')\, \delta \dot{p}_{n,j}(t') + O(\tau^2) \tag{16.9}$$

$$\delta \boldsymbol{\omega}_{\mathcal{B}_{\backslash,\int}|\mathcal{N}}(t'+\tau) = \delta \boldsymbol{\omega}_{\mathcal{B}_{\backslash,\int}|\mathcal{N}}(t') + \tau\, \delta \dot{\boldsymbol{\omega}}_{\mathcal{B}_{\backslash,\int}|\mathcal{N}}(t') + O(\tau^2)$$

$$= \tau \sum_j \frac{\partial \boldsymbol{\omega}_{\mathcal{B}_{\backslash,\int}|\mathcal{N}}}{\partial p_{n,j}}(t')\, \delta \dot{p}_{n,j}(t') + O(\tau^2) \tag{16.10}$$

At time instant $t' + \tau$, Eq. (16.8) can be rewritten in the following form, with all the first member evaluated at t':

$$\sum_j \delta \dot{p}_{n,j}\left(f_{n,j}(t, \mathbb{q}_n, \mathbb{p}_n, \mathbb{u}_n) + g_{n,j}(t, \mathbb{q}_n, \mathbb{p}_n) - \sum_r M_{n,jr}(t, \mathbb{q}_n, \mathbb{p}_n)\, \dot{p}_{n,r} \right) = \frac{O(\tau^2)}{\tau} \tag{16.11}$$

In Eq. (16.11), $f_{n,j}$ is the *j-th generalized force of system* $\mathcal{S}_n$, $g_{n,j}$ is the *j-th generalized gyroscopic inertia force of system* $\mathcal{S}_n$, and $M_{n,jr}$ is the *generalized inertia associated with* $\dot{p}_{n,j}$ *and* $\dot{p}_{n,r}$. Their expressions satisfy the following identities:

$$f_{n,j}(t, \mathbb{q}_n, \mathbb{p}_n, \mathbb{u}_n) = \sum_s \left(\frac{\partial \boldsymbol{v}_{\mathrm{b}^\star_{n,s}|\mathcal{N}}}{\partial p_{n,j}} \cdot \boldsymbol{f}_{\mathcal{B}_{n,s}} + \frac{\partial \boldsymbol{\omega}_{\mathcal{B}_{n,s}|\mathcal{N}}}{\partial p_{n,j}} \cdot \boldsymbol{m}_{\mathcal{B}_{n,s}|\mathrm{b}^\star_{n,s}} \right) \tag{16.12}$$

$$\begin{aligned} g_{n,j}(t, \mathbb{q}_n, \mathbb{p}_n) - \sum_r M_{n,jr}(t, \mathbb{q}_n, \mathbb{p}_n)\, \dot{p}_{n,r} = -\sum_s \Bigg(& \frac{\partial \boldsymbol{v}_{\mathrm{b}^\star_{n,s}|\mathcal{N}}}{\partial p_{n,j}} \cdot \left(m_{\mathcal{B}_{n,s}}\, \boldsymbol{a}_{\mathrm{b}^\star_{n,s}|\mathcal{N}} \right) \\ & + \frac{\partial \boldsymbol{\omega}_{\mathcal{B}_{n,s}|\mathcal{N}}}{\partial p_{n,j}} \cdot \Big(I_{\mathcal{B}_{n,s}|\mathrm{b}^\star_{n,s}} \cdot \dot{\boldsymbol{\omega}}_{\mathcal{B}_{n,s}|\mathcal{N}} + \boldsymbol{\omega}_{\mathcal{B}_{n,s}|\mathcal{N}} \\ & \times \left(I_{\mathcal{B}_{n,s}|\mathrm{b}^\star_{n,s}} \cdot \boldsymbol{\omega}_{\mathcal{B}_{n,s}|\mathcal{N}} \right) \Big) \Bigg) \end{aligned} \tag{16.13}$$

Taking the limit $\tau \to 0$ in Eq. (16.11), using matrix notation, it can be stated that, at time instant t':

$$\delta \dot{\mathbb{p}}_n^{\mathsf{T}} \left(\mathbb{f}_n + \mathbb{g}_n - \mathbb{M}_n\, \dot{\mathbb{p}}_n \right) = 0 \tag{16.14}$$

Also, at time instant t', Eq. (16.7) implies that:

$$\mathbb{A}_n\, \delta \dot{\mathbb{p}}_n = \mathbb{0} \tag{16.15}$$

Obviously, if it is possible to find a matrix $\mathbb{C}_n$ such that $\mathbb{A}_n \mathbb{C}_n = \mathbb{0}$, then $\delta\dot{\mathbb{p}}_n = \mathbb{C}_n \mathbb{\pi}_n$ will satisfy the condition imposed by Eq. (16.15) for any column-matrix $\mathbb{\pi}_n$ (with adequate dimensions). If, at time instant t', $\mathcal{S}_n$ is in a non-singular state, then Eq. (16.15) imposes $\nu_c(\mathcal{S}_n) = \nu_p(\mathcal{S}_n) - \nu^{\#}(\mathcal{S}_n)$ conditions for $\nu_p(\mathcal{S}_n)$ variations $\delta\dot{p}_{n,j}$. Thus, the most general solution for Eq. (16.15) is to express all the $\nu_p(\mathcal{S}_n)$ variations $\delta\dot{p}_{n,j}$ in terms of $\nu^{\#}(\mathcal{S}_n)$ arbitrary parameters. This can be achieved by finding a full rank, $\nu_p(\mathcal{S}_n)$ by $\nu^{\#}(\mathcal{S}_n)$ matrix $\mathbb{C}_n$ such that $\mathbb{A}_n \mathbb{C}_n = \mathbb{0}$, i.e., matrix $\mathbb{C}_n$ must be an *orthogonal complement* of matrix $\mathbb{A}_n$. Therefore, replacing $\delta\dot{\mathbb{p}}_n = \mathbb{C}_n \mathbb{\pi}_n$ in Eq. (16.14):

$$\mathbb{\pi}_n^{\mathsf{T}} \mathbb{C}_n^{\mathsf{T}} (\mathbb{f}_n + \mathbb{g}_n - \mathbb{M}_n \dot{\mathbb{p}}_n) = 0 \tag{16.16}$$

Considering that $\mathbb{C}_n$ is full rank and that $\mathbb{\pi}_n$ is column-matrix with $\nu^{\#}(\mathcal{S}_n)$ entries, the only possible way to ensure that this equation is satisfied is to state that:

$$\mathbb{d}_n = \mathbb{C}_n^{\mathsf{T}} (\mathbb{f}_n + \mathbb{g}_n - \mathbb{M}_n \dot{\mathbb{p}}_n) = \mathbb{0} \tag{16.17}$$

The system composed by Eqs. (16.7) and (16.17) completely determines, for a given time t' when all the state variables $q_{n,i}(t')$ and $p_{n,j}(t')$ are known, all the associated quasi-accelerations $\dot{p}_{n,j}(t')$, given the values of the control inputs $u_{n,k}(t')$. Thus, Eqs. (16.7) and (16.17) are a dynamic model of the multibody mechanical system $\mathcal{S}_n$. Finally, it is worth mentioning that, although the derivation of Eq. (16.17) was performed for a multiple rigid-body mechanical system, every multibody mechanical system with a finite number of degrees of freedom can be expressed by a system of equations in the form of (16.7) and (16.17).

16.2.2 Dynamic Coupling of Subsystems

Let $\mathscr{M}$ be a multibody mechanical system which can be interpreted as an assemble of constrained subsystems $\mathscr{S}_n$, $n \in \mathscr{I}_{\mathscr{S}}(\mathscr{M})$. Consider that the dynamic model of each of the subsystems $\mathcal{S}_n$, when not constrained to the others, is given by the following pair of matrix equations:

$$\mathbb{c}_n = \mathbb{A}_n \dot{\mathbb{p}}_n + \mathbb{b}_n = \mathbb{0} \tag{16.18}$$

$$\mathbb{d}_n = \mathbb{C}_n^{\mathsf{T}} (\mathbb{f}_n + \mathbb{g}_n - \mathbb{M}_n \dot{\mathbb{p}}_n) = \mathbb{0} \tag{16.19}$$

The system of Eq. (16.18), $\mathbb{c}_n = \mathbb{0}$, represents the (*internal*) kinematic constraints of the subsystems $\mathcal{S}_n$, while the system of Eq. (16.19), $\mathbb{d}_n = \mathbb{0}$, describes its dynamics when no *external* kinematic constraints are present. Together, Eqs. (16.18) and (16.19) are called the *free subsystem $\mathcal{S}_n$ model*, which can be expressed in terms of the following variables and parameters:

- Generalized coordinates $q_{n,i}$ and quasivelocities $p_{n,j}$ describing every admissible state of $\mathcal{S}_n$ considering the absence of external kinematic constraints;
- Control inputs $u_{n,k}$ describing the effect of external actuations on $\mathcal{S}_n$;
- Fixed and adjustable physical parameters $a_{n,l}$ of $\mathcal{S}_n$.

When the subsystems are assembled to a single multibody mechanical system $\mathscr{M}$, additional kinematical constraints are imposed. Consider that these further constraints can be expressed by the following matrix equation:

$$\hat{\mathbb{c}} = \sum_n \hat{\mathbb{A}}_n \dot{\mathbb{p}}_n + \hat{\mathbb{b}} = \mathbb{0} \tag{16.20}$$

The satisfaction of all the kinematic constraints expressed by (16.18) and (16.20) in a particular (admissible) state of the system $\mathscr{M}$ implies that:

$$\mathbb{A}_n \, \delta\dot{\mathbb{p}}_n = \mathbb{0} \qquad \forall\, n \in \mathscr{I}_{\mathcal{S}}(\mathscr{M}) \tag{16.21}$$

$$\sum_n \hat{\mathbb{A}}_n \, \delta\dot{\mathbb{p}}_n = \mathbb{0} \tag{16.22}$$

As shown in the previous section, a solution for each of Eq. (16.21) is already known. Once $\mathbb{A}_n\mathbb{C}_n = \mathbb{0}$, $\forall\, n \in \mathscr{I}_{\mathcal{S}}(\mathscr{M})$, for every column-matrix $\mathbb{\pi}_n$ with as much elements as the number of degrees of freedom of the free subsystem $\mathcal{S}_n$ model, Eq. (16.21) are identically satisfied for:

$$\delta\dot{\mathbb{p}}_n = \mathbb{C}_n \, \mathbb{\pi}_n \qquad \forall\, n \in \mathscr{I}_{\mathcal{S}}(\mathscr{M}) \tag{16.23}$$

Replacing (16.23) in (16.22):

$$\sum_n \hat{\mathbb{A}}_n \, \mathbb{C}_n \, \mathbb{\pi}_n = \mathbb{0} \tag{16.24}$$

Thus, suppose that it is possible to define matrices $\hat{\mathbb{C}}_n$, $\forall\, n \in \mathscr{I}_{\mathcal{S}}(\mathscr{M})$, such that for every column-matrix $\mathbb{\pi}$ with as much elements as the number of degrees of freedom of the whole system $\mathscr{M}$, Eq. (16.24) is identically satisfied for:

$$\mathbb{\pi}_n = \hat{\mathbb{C}}_n \, \mathbb{\pi} \qquad \text{or} \qquad \delta\dot{\mathbb{p}}_n = \mathbb{C}_n \, \hat{\mathbb{C}}_n \, \mathbb{\pi} \qquad \forall\, n \in \mathscr{I}_{\mathcal{S}}(\mathscr{M}) \tag{16.25}$$

For the sake of simplicity in notation, suppose that $\mathscr{I}_S(\mathscr{M}) = \{1, 2, \ldots, \nu_S\}$. In this case, Eqs. (16.21) and (16.22) can be put in the following matrix form:

$$\begin{bmatrix} \mathbb{A}_1 & \mathbb{0} & \ldots & \mathbb{0} \\ \mathbb{0} & \mathbb{A}_2 & \ldots & \mathbb{0} \\ \vdots & \vdots & \ddots & \vdots \\ \mathbb{0} & \mathbb{0} & \ldots & \mathbb{A}_{\nu_S} \\ \hat{\mathbb{A}}_1 & \hat{\mathbb{A}}_2 & \ldots & \hat{\mathbb{A}}_{\nu_S} \end{bmatrix} \begin{bmatrix} \delta\dot{\mathbb{p}}_1 \\ \delta\dot{\mathbb{p}}_2 \\ \vdots \\ \delta\dot{\mathbb{p}}_{\nu_S} \end{bmatrix} = \mathbb{0} \tag{16.26}$$

Using Eq. (16.25) and considering that the choice of $\mathbb{\pi}$ is arbitrary, it can be stated that:

$$\begin{bmatrix} \mathbb{A}_1 & \mathbb{0} & \ldots & \mathbb{0} \\ \mathbb{0} & \mathbb{A}_2 & \ldots & \mathbb{0} \\ \vdots & \vdots & \ddots & \vdots \\ \mathbb{0} & \mathbb{0} & \ldots & \mathbb{A}_{\nu_S} \\ \hat{\mathbb{A}}_1 & \hat{\mathbb{A}}_2 & \ldots & \hat{\mathbb{A}}_{\nu_S} \end{bmatrix} \begin{bmatrix} \mathbb{C}_1 & \mathbb{0} & \ldots & \mathbb{0} \\ \mathbb{0} & \mathbb{C}_2 & \ldots & \mathbb{0} \\ \vdots & \vdots & \ddots & \vdots \\ \mathbb{0} & \mathbb{0} & \ldots & \mathbb{C}_{\nu_S} \end{bmatrix} \begin{bmatrix} \hat{\mathbb{C}}_1 \\ \hat{\mathbb{C}}_2 \\ \vdots \\ \hat{\mathbb{C}}_{\nu_S} \end{bmatrix} = \mathbb{0} \tag{16.27}$$

Once $\mathbb{A}_n\mathbb{C}_n = \mathbb{0}$, $\forall\, n \in \mathscr{I}_S(\mathscr{M})$, the satisfaction of Eq. (16.27) is ensured if and only if:

$$\begin{bmatrix} \hat{\mathbb{A}}_1\mathbb{C}_1 & \hat{\mathbb{A}}_2\mathbb{C}_2 & \ldots & \hat{\mathbb{A}}_{\nu_S}\mathbb{C}_{\nu_S} \end{bmatrix} \begin{bmatrix} \hat{\mathbb{C}}_1 \\ \hat{\mathbb{C}}_2 \\ \vdots \\ \hat{\mathbb{C}}_{\nu_S} \end{bmatrix} = \mathbb{0} \tag{16.28}$$

Adopting the notation:

$$\hat{\mathbb{A}} = \begin{bmatrix} \hat{\mathbb{A}}_1\mathbb{C}_1 & \hat{\mathbb{A}}_2\mathbb{C}_2 & \ldots & \hat{\mathbb{A}}_{\nu_S}\mathbb{C}_{\nu_S} \end{bmatrix} \qquad \text{and} \qquad \hat{\mathbb{C}} = \begin{bmatrix} \hat{\mathbb{C}}_1 \\ \hat{\mathbb{C}}_2 \\ \vdots \\ \hat{\mathbb{C}}_{\nu_S} \end{bmatrix}$$

equation (16.28) can be rewritten in the following form:

$$\hat{\mathbb{A}}\,\hat{\mathbb{C}} = \mathbb{0} \tag{16.29}$$

As each matrix $\mathbb{C}_n$ is an orthogonal complement of the respective $\mathbb{A}_n$, so is matrix $\hat{\mathbb{C}}$ an orthogonal complement of $\hat{\mathbb{A}}$. Thus, Eq. (16.29) proves that the same algorithm that is used to obtain matrices $\mathbb{C}_n$ can be used to obtain $\hat{\mathbb{C}}$ and, consequently, $\hat{\mathbb{C}}_1$, $\hat{\mathbb{C}}_2, \ldots, \hat{\mathbb{C}}_{\nu_S}$.

Applying the Principle of Virtual Power for the system $\mathscr{M}$, it can be stated that:

$$\sum_n \delta\dot{\mathbb{p}}_n^{\mathsf{T}}(\mathbb{f}_n + \mathbb{g}_n - \mathbb{M}_n\,\dot{\mathbb{p}}_n) = 0 \tag{16.30}$$

Using Eq. (16.25):

$$\mathbb{\pi}^{\mathsf{T}}\left(\sum_n \hat{\mathbb{C}}_n^{\mathsf{T}}\mathbb{C}_n^{\mathsf{T}}(\mathbb{f}_n + \mathbb{g}_n - \mathbb{M}_n\,\dot{\mathbb{p}}_n)\right) = \mathbb{\pi}^{\mathsf{T}}\left(\sum_n \hat{\mathbb{C}}_n^{\mathsf{T}}\,\mathbb{d}_n\right) = 0 \tag{16.31}$$

Again, once the choice of $\mathbb{\pi}$ is arbitrary, Eq. (16.31) implies that:

$$\mathbb{d} = \sum_n \hat{\mathbb{C}}_n^{\mathsf{T}}\,\mathbb{d}_n = \mathbb{0} \tag{16.32}$$

Adopting the notation:

$$\mathbb{c} = \begin{bmatrix} \mathbb{c}_1 \\ \mathbb{c}_2 \\ \vdots \\ \mathbb{c}_{\nu_S} \\ \hat{\mathbb{c}} \end{bmatrix} \qquad \text{and} \qquad \hat{\mathbb{d}} = \begin{bmatrix} \mathbb{d}_1 \\ \mathbb{d}_2 \\ \vdots \\ \mathbb{d}_{\nu_S} \end{bmatrix}$$

the mathematical model of the system $\mathscr{M}$ is given by the following pair of matrix equations:

$$\mathbb{c} = \mathbb{0} \tag{16.33}$$

$$\mathbb{d} = \hat{\mathbb{C}}^{\mathsf{T}}\,\hat{\mathbb{d}} = \mathbb{0} \tag{16.34}$$

This approach for obtaining mathematical models of multibody mechanical systems based on already known free subsystem models is a basis for the development of a methodology for adaptive balancing of mechanical systems.

16.2.3 Introduction to Sliding Modes Control

In this subsection, a brief introduction to the sliding modes control will be done. The theme will be explored only to perform second order systems control, without parametric uncertainties, to not escape the scope of the chapter.

Consider a dynamical system given by the following differential equation:

$$\ddot{x} = u \tag{16.35}$$

A curve in the error phase plan, called sliding surface, can be defined:

$$s(e, \dot{e}) = -(\dot{e} + \lambda e) = 0,\ \lambda > 0 \tag{16.36}$$

Being $e = x^{\diamond} - x$ the error signal and $x^{\diamond}$ reference signal. Note that if the system is on the sliding surface, then:

$$\dot{e} + \lambda e = 0 \Rightarrow e(t) = c\,\mathrm{e}^{-\lambda t} \tag{16.37}$$

Thus, the error drops exponentially to zero, with time constant $1/\lambda$.

To find a control law that brings the system to the sliding surface, define:

$$s = -(\dot{e} + \lambda e)$$

Differentiating with respect to time:

$$\dot{s} = -(\ddot{e} + \lambda \dot{e}) = \ddot{x} - \ddot{x}^{\diamond} - \lambda \dot{e} \tag{16.38}$$

Substituting (16.35) into (16.38):

$$\dot{s} = u - \ddot{x}^{\diamond} - \lambda \dot{e} \tag{16.39}$$

Using the following control law:

$$u = \ddot{x}^{\diamond} + \lambda \dot{e} - k\,\mathrm{sign}(s),\ k > 0 \tag{16.40}$$

it can be stated that:

$$\dot{s} = -k\,\mathrm{sign}(s) \tag{16.41}$$

Suppose that the system starts at $s(0) = s_0 > 0$. Solving the ODE for $s > 0$:

$$\begin{aligned} \dot{s} = -k &\Rightarrow s = -kt + c \\ s(0) = s_0 &\Rightarrow c = s_0 \\ \therefore s = s_0 &- kt,\ s > 0 \end{aligned}$$

According to the solution found, when $t \to t_s = |s_0|/k$, $s \to 0$. Solving the ODE for $s(t_s) = 0$:

$$\begin{aligned} \dot{s} = 0 &\Rightarrow s = c \\ s(t_s) = 0 &\Rightarrow c = 0 \end{aligned}$$

Therefore, the solution of the ODE for $s(0) = s_0 > 0$ is given by:

$$s(t) = \begin{cases} s_0 - kt, & t < t_s \\ 0, & t \geq t_s \end{cases} \tag{16.42}$$

An analogous result is found solving the ODE for $s(0) = s_0 < 0$:

$$s(t) = \begin{cases} s_0 + kt, & t < t_s \\ 0, & t \geq t_s \end{cases} \tag{16.43}$$

Thus, it can be concluded that the ODE (16.41) converges to $s = 0$, regardless of the initial condition. Therefore, the control law (16.40) makes the system represented by (16.35) follow the reference signal, because the error signal converges to zero.

16.2.4 Extended Sliding Modes Control Techniques

As seen in Sects. 16.2.1 and 16.2.2, it is very convenient to use redundant coordinates to perform parallel mechanism dynamic modelling. The aim of this subsection is to propose a control law for systems described by redundant coordinates.

Let $\mathscr{M}$ be a multibody mechanical system whose mathematical model is given by Eqs. (16.7) and (16.17). For the sake of brevity, no system index n will be used in this subsection. Suppose that each $\mathbb{f}$ is an affine function of the control inputs u_k in which the coefficients of the u_k may depend on the instantaneous configuration of the system. Suppose additionally that all the $\mathbb{A}$ (of each subsystem) are independent of the quasi-velocities p_j and all the $\mathbb{M}$, $\mathbb{g}$, $\mathbb{A}$ and $\mathbb{b}$ are independent of the u_k. Under these conditions, matrices $\mathbb{C}$ will not depend on any quasi-velocity, $\mathbb{d}$ can be expressed as an affine function of the control inputs and $\mathbb{c}$ is independent of them. Considering that the number of control inputs in $\mathscr{M}$ is exactly equal to the number of degrees of freedom of $\mathscr{M}$, there exists a particular matrix $\mathbb{C}(t, \mathbb{q})$ such that:

$$\mathbb{u} = \mathbb{C}^{\mathsf{T}}(t, \mathbb{q})\Big(\mathbb{M}(t, \mathbb{q})\,\dot{\mathbb{p}} + \mathbb{w}(t, \mathbb{q}, \mathbb{p}) + \mathbb{z}(t, \mathbb{q})\Big) \tag{16.44}$$

In Eq. (16.44), $\mathbb{w}$ is a column-vector representing terms of generalized force or generalized gyroscopic inertia force which are linear or bilinear with respect to quasi-velocities and $\mathbb{z}$ stems from terms that are independent of these variables.

From the control perspective, it is convenient to work with mathematical models in which $\mathbb{p} = \dot{\mathbb{q}}$, in order to have a position feedback control. Thus, based on

Eqs. (16.44) and (16.18), consider that the mathematical model of $\mathscr{M}$ is given by the following equations:

$$\begin{cases} \mathbb{C}^{\mathsf{T}}(t,\mathbb{q})\Big(\mathbb{M}(t,\mathbb{q})\,\ddot{\mathbb{q}} + \mathbb{w}(t,\mathbb{q},\dot{\mathbb{q}}) + \mathbb{z}(t,\mathbb{q})\Big) = \mathbb{u} \\ \mathbb{A}(t,\mathbb{q})\,\ddot{\mathbb{q}} + \mathbb{b}(t,\mathbb{q},\dot{\mathbb{q}}) = \mathbb{0} \end{cases} \tag{16.45}$$

Rewriting in a compact matrix form:

$$\begin{bmatrix} \mathbb{C}^{\mathsf{T}}\mathbb{M} \\ \mathbb{A} \end{bmatrix} \ddot{\mathbb{q}} = \begin{bmatrix} \mathbb{u} - \mathbb{C}^{\mathsf{T}}(\mathbb{w} + \mathbb{z}) \\ -\mathbb{b} \end{bmatrix} \tag{16.46}$$

The desired control law should satisfy, in closed loop, the condition $\ddot{\mathbb{q}} = \mathbb{v}$, with $\mathbb{v}$ being a control input column-matrix. Thus, the following control law should be used:

$$\mathbb{u} = \mathbb{C}^{\mathsf{T}}(\mathbb{M}\,\mathbb{v} + \mathbb{w} + \mathbb{z}) \tag{16.47}$$

Once $\ddot{\mathbb{q}} = \mathbb{v}$, and $\ddot{\mathbb{q}}$ has to satisfy constraint equations, $\mathbb{v}$ must respect the same restrictions, i.e.:

$$\mathbb{A}\,\mathbb{v} + \mathbb{b} = \mathbb{0} \tag{16.48}$$

Applying the control law (16.47) and the restrictions (16.48) in (16.46):

$$\begin{bmatrix} \mathbb{C}^{\mathsf{T}}\mathbb{M} \\ \mathbb{A} \end{bmatrix} \ddot{\mathbb{q}} = \begin{bmatrix} \mathbb{C}^{\mathsf{T}}(\mathbb{M}\,\mathbb{v} + \mathbb{w} + \mathbb{z}) - \mathbb{C}^{\mathsf{T}}(\mathbb{w} + \mathbb{z}) \\ \mathbb{A}\,\mathbb{v} \end{bmatrix} = \begin{bmatrix} \mathbb{C}^{\mathsf{T}}\mathbb{M}\,\mathbb{v} \\ \mathbb{A}\,\mathbb{v} \end{bmatrix} = \begin{bmatrix} \mathbb{C}^{\mathsf{T}}\mathbb{M} \\ \mathbb{A} \end{bmatrix} \mathbb{v}$$

Once the matrix $\begin{bmatrix} \mathbb{C}^{\mathsf{T}}\mathbb{M} \\ \mathbb{A} \end{bmatrix}$ is non-singular:

$$\ddot{\mathbb{q}} = \mathbb{v} \tag{16.49}$$

Let $\mathbb{v}'$ be given by the sliding modes control law:

$$\mathbb{v}' = \ddot{\mathbb{q}}^{\diamond} + \lambda\dot{\mathbb{e}} + k\,\mathrm{sign}(\dot{\mathbb{e}} + \lambda\mathbb{e}) \tag{16.50}$$

Being $\mathbb{e} = \mathbb{q}^{\diamond} - \mathbb{q}$ the error signal and $\mathbb{q}^{\diamond}$ the reference signal. If there were no restrictions, it could be stated that $\mathbb{v} = \mathbb{v}'$, and:

$$\ddot{\mathbb{q}} = \mathbb{v} \quad \Rightarrow \quad \ddot{\mathbb{e}} + \lambda\dot{\mathbb{e}} + k\,\mathrm{sign}(\dot{\mathbb{e}} + \lambda\mathbb{e}) = \mathbb{0} \quad \Leftrightarrow \quad \dot{\mathbb{s}} = -k\,\mathrm{sign}(\mathbb{s})$$

This would ensure that $\mathbb{e} \to 0$ when $t \to \infty$ for any initial condition, as seen in the last subsection.

Once $\mathbb{v}$ cannot be freely set as $\mathbb{v}'$, the following optimization problem is proposed:

$$\min_{\mathbb{v}} \quad (\mathbb{v}-\mathbb{v}')^{\mathsf{T}}\,\mathbb{M}\,(\mathbb{v}-\mathbb{v}') \qquad \text{s.t.} \qquad \mathbb{A}\,\mathbb{v}+\mathbb{b}=\mathbb{0} \tag{16.51}$$

As $\mathbb{M}$ is positive-semidefinite, then $(\mathbb{v}-\mathbb{v}')^{\mathsf{T}}\,\mathbb{M}\,(\mathbb{v}-\mathbb{v}') \geq 0$ for any $\mathbb{v}$.

Applying the method of Lagrange undetermined multipliers, it can be stated that this optimization problem is equivalent to minimize with respect to $\mathbb{v}$ and $\mathbb{\lambda}$ the following functional:

$$L = (\mathbb{v}-\mathbb{v}')^{\mathsf{T}}\,\mathbb{M}\,(\mathbb{v}-\mathbb{v}') + (\mathbb{A}\mathbb{v}+\mathbb{b})^{\mathsf{T}}\,\mathbb{\lambda} \tag{16.52}$$

To solve the problem, the Lagrangian function must be stationary. Thus:

$$\begin{aligned} \delta L = 0 \quad &\Rightarrow \quad \delta\mathbb{v}^{\mathsf{T}}\mathbb{M}(\mathbb{v}-\mathbb{v}') + (\mathbb{v}-\mathbb{v}')^{\mathsf{T}}\mathbb{M}\delta\mathbb{v} + (\mathbb{A}\delta\mathbb{v})^{\mathsf{T}}\mathbb{\lambda} + (\mathbb{A}\mathbb{v}+\mathbb{b})^{\mathsf{T}}\delta\mathbb{\lambda} = 0 \\ &\Rightarrow \quad \delta\mathbb{v}^{\mathsf{T}}\left((\mathbb{M}+\mathbb{M}^{\mathsf{T}})(\mathbb{v}-\mathbb{v}') + \mathbb{A}^{\mathsf{T}}\mathbb{\lambda}\right) + \delta\mathbb{\lambda}^{\mathsf{T}}(\mathbb{A}\mathbb{v}+\mathbb{b}) = 0 \end{aligned} \tag{16.53}$$

Once $\mathbb{M}$ is symmetric and $\delta\mathbb{v}$ and $\delta\mathbb{\lambda}$ are arbitrary:

$$\begin{cases} 2\,\mathbb{M}\,(\mathbb{v}-\mathbb{v}') + \mathbb{A}^{\mathsf{T}}\mathbb{\lambda} = \mathbb{0} \\ \mathbb{A}\,\mathbb{v}+\mathbb{b} = \mathbb{0} \end{cases} \tag{16.54}$$

Considering that $\mathbb{C}$ is an orthogonal complement of $\mathbb{A}$, pre-multiplying the first equation of (16.54) by $\mathbb{C}^{\mathsf{T}}$ leads to:

$$2\,\mathbb{C}^{\mathsf{T}}\,\mathbb{M}(\mathbb{v}-\mathbb{v}') + \mathbb{C}^{\mathsf{T}}\,\mathbb{A}^{\mathsf{T}}\mathbb{\lambda} = \mathbb{0} \quad \Rightarrow \quad \mathbb{C}^{\mathsf{T}}\,\mathbb{M}(\mathbb{v}-\mathbb{v}') = \mathbb{0} \quad \therefore \quad \mathbb{C}^{\mathsf{T}}\,\mathbb{M}\,\mathbb{v} = \mathbb{C}^{\mathsf{T}}\,\mathbb{M}\,\mathbb{v}' \tag{16.55}$$

Thus, the control law that makes the closed loop system (Fig. 16.2) as close as possible of $\ddot{\mathbb{q}} = \mathbb{v}'$, according to the optimization criterion adopted, is:

$$\mathbb{u} = \mathbb{C}^{\mathsf{T}}(\mathbb{M}\,\mathbb{v}' + \mathbb{w} + \mathbb{z}) \tag{16.56}$$

16.3 Adaptive Balancing Techniques

16.3.1 Definitions

Let $\mathscr{M}$ be a multibody mechanical system whose mathematical model is given by Eqs. (16.7) and (16.17). For the sake of brevity, no system index n will be used in this subsection. Consider that the number of control inputs in $\mathscr{M}$ is exactly equal to the number of degrees of freedom of $\mathscr{M}$ and that $\mathbb{M}$, $\mathbb{g}$, $\mathbb{f}$, $\mathbb{A}$, and $\mathbb{b}$ satisfy the

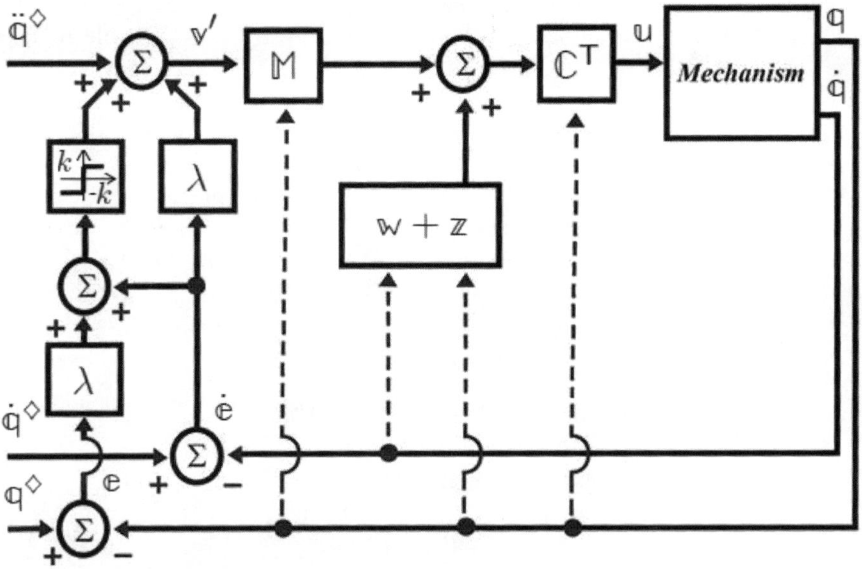

Fig. 16.2 Extended sliding modes control loop

same hypotheses which led to Eq. (16.44). It can be stated that there are functions $M'_{kr}(t, \mathbb{q})$, $w'_k(t, \mathbb{q}, \mathbb{p})$ and $z'_k(t, \mathbb{q})$ such that, for $k \in \{1, \ldots, \nu^{\#}(\mathscr{M})\}$:

$$u_k = \sum_r M'_{kr}(t, \mathbb{q})\, \dot{p}_r + w'_k(t, \mathbb{q}, \mathbb{p}) + z'_k(t, \mathbb{q}) \tag{16.57}$$

Noting that each $w'_k(t, \mathbb{q}, \mathbb{p})$ can be expressed as a sum of a bilinear and a linear function on the quasi-velocities, there are functions $D'_{krs}(t, \mathbb{q})$ and $B'_{kr}(t, \mathbb{q})$ such that:

$$w'_k(t, \mathbb{q}, \mathbb{p}) = \sum_r \sum_s D'_{krs}(t, \mathbb{q})\, p_r p_s + \sum_r B'_{kr}(t, \mathbb{q})\, p_r \tag{16.58}$$

Adaptive balancing types are defined based on Eq. (16.57).

Definition 3.1. Let $\mathscr{M}$ be a $\nu^{\#}(\mathscr{M})$ degrees of freedom multibody mechanical system with $\nu^{\#}(\mathscr{M})$ control inputs. Suppose that, except from singular states, it is possible to express the dynamic equations of the model of $\mathscr{M}$ in the form (16.57). The following types of adaptive balancing can be defined:

- Static adaptive balancing: when, by a particular choice of the values of the adjustable physical parameters a_l, denoted by a'_l, it is possible to make all the $z'_k(t, \mathbb{q}) = 0$ for every configuration $(t, \mathbb{q})$ corresponding to a nonsingular state; in such cases, in every static equilibrium state of $\mathscr{M}$, all the control inputs will be null.

- Gyroscopic adaptive balancing: when, by a particular choice of the values of the adjustable physical parameters a_l, denoted by a'_l, it is possible to make all the $z'_k(t, \mathbb{q}) = 0$, all the $D'_{krs}(t, \mathbb{q}) = 0$, and all the $B'_{kr}(t, \mathbb{q}) = 0$ for every configuration $(t, \mathbb{q})$ corresponding to a nonsingular state; in such cases, the control inputs will be proportional to the quasi-accelerations of the system, independent of the values of the quasi-velocities.
- Decoupling adaptive balancing: when, by a particular choice of the values of the adjustable physical parameters a_l, denoted by a'_l, and by some reordering of indices, it is possible to make all the $z'_k(t, \mathbb{q}) = 0$, all the $D'_{krs}(t, \mathbb{q}) = 0$, all the $B'_{kr}(t, \mathbb{q}) = 0$ and $M'_{kr}(t, \mathbb{q}) = 0$ whenever $k \neq r$, for every configuration $(t, \mathbb{q})$ corresponding to a nonsingular state; in such cases, each control input will be proportional to a particular quasi-acceleration.

The adjective "fully" will be used whenever a particular choice of adjustable parameters leads to the following equations, for $k \in \{1, \ldots, \nu^{\#}(\mathcal{M})\}$:

$$u_k = \sum_r M''_{kr} \dot{p}_r + \sum_r \sum_s D''_{krs} p_r p_s + \sum_r \sum_s B''_{kr} p_r \tag{16.59}$$

with M''_{kr}, D''_{krs}, and B''_{kr} independent of the configuration $(t, \mathbb{q})$ of the system. In such cases, when some $D''_{krs} \neq 0$ or some $B''_{kr} \neq 0$, it corresponds to a fully static adaptive balancing; when all the $D''_{krs} = 0$ and all the $B''_{kr} = 0$, it corresponds to a fully gyroscopic adaptive balancing and when all the $D''_{krs} = 0$, all the $B''_{kr} = 0$ and $M''_{kr} = 0$ whenever $k \neq r$, it corresponds to a fully decoupling adaptive balancing.

16.3.2 Adaptive Balancing of Serial and Parallel Mechanisms

Let $\mathcal{U}$ be an "unbalanced" serial or parallel mechanism. A serial mechanism can be interpreted as an assemble of two subsystems: an open loop kinematic chain (representing the actual mechanism) and a payload. On the other hand, a parallel mechanism will be conceived as an assemble of the following subsystems: open loop kinematic chains (representing the active and passive chains of the actual mechanism) and an end-effector rigidly attached to a payload.

Let $\mathcal{S}_n$ denote mechanical systems with adjustable parameters that are assembled to $\mathcal{U}$, in order to obtain a multibody mechanical system $\mathcal{M}$ that can be balanced according to some of the strategies shown in Definition 3.1.

Consider the results presented in the following propositions.

Proposition 3.2. *The dynamic coupling of any static adaptive balanced subsystems is a static adaptive balanced multibody mechanical system.*

Proof. Consider some static adaptive balanced subsystems $\mathcal{S}_n$. For each of these subsystems there is a particular matrix $\mathbb{C}_n$ such that:

$$\mathbb{d}_n = \mathbb{u}_n - \mathbb{M}'_n(t, \mathbb{q}_n)\,\dot{\mathbb{p}}_n - \mathbb{w}'_n(t, \mathbb{q}_n, \mathbb{p}_n) \tag{16.60}$$

In this equation, $\mathbb{w}'_n(t, \mathbb{q}_n, \mathbb{p}_n)$ denotes the column-matrix whose entries are $w'_{n,k}(t, \mathbb{q}_n, \mathbb{p}_n)$.

Consider that these subsystems are assembled. The dynamic equations of the multibody mechanical system $\mathcal{M}$ thus obtained is given by

$$\mathbb{d} = \sum_n \hat{\mathbb{C}}_n^{\mathsf{T}} \left(\mathbb{u}_n - \mathbb{M}'_n(t, \mathbb{q}_n)\,\dot{\mathbb{p}}_n - \mathbb{w}'_n(t, \mathbb{q}_n, \mathbb{p}_n)\right) = \mathbb{0} \tag{16.61}$$

Supposing that all but $\nu^{\#}(\mathcal{M})$ of the $u_{n,k}$ are identically zero in the system $\mathcal{M}$, Eq. (16.61) can be manipulated to obtain explicit equations for the remaining $\nu^{\#}(\mathcal{M})$. It is evident that such equations will have the same form as Eq. (16.57) with all the $z'_k(t, \mathbb{q}) = 0$. This concludes the proof.

Analogously, it can also be proved that:

Proposition 3.3. *The dynamic coupling of any gyroscopic adaptive balanced subsystems is a gyroscopic adaptive balanced multibody mechanical system.*

Considering the results presented on Propositions 3.2 and 3.3 the following strategy for balancing $\mathcal{U}$ can be stated:

- If $\mathcal{U}$ is a serial mechanism, attach to it a subsystem $\mathcal{S}_n$ with adjustable parameters to each of its links (or to pairs of links) and verify which values of parameters have to be chosen in order to achieve a desired balancing. If there are not enough parameters, some of the subsystems $\mathcal{S}_n$ can be substituted or others can be added.
- If $\mathcal{U}$ is a parallel mechanism, consider it as an assemble of open loop kinematic chains and an end-effector rigidly attached to a payload. Consider each open loop kinematic chain as a serial mechanism and attach to its end-effector part of the inertia of the end-effector and the payload of the original mechanism. (This inertia redistribution has to be able to reproduce adequately the inertial effects of the end-effector and payload on each chain of the mechanism.) The next step is to balance each of these serial mechanisms according to the strategy previously presented, and obtain the model of the balanced parallel mechanism by recoupling the models of the balanced versions of these subsystems (according to the same constraints already existing in the original mechanism).

16.4 Case Study: Adaptive Balancing of a 5-Bar Mechanism

16.4.1 Preliminaries

Consider a planar two degrees of freedom 5-bar mechanism with revolute joints, also called 5R, which can be conceived as two RR mechanisms constrained by a revolute joint. Thus, considering the strategy presented on Sect. 16.3.2, the first step to balance a 5R mechanism is to balance a RR mechanism.

Consider the RR mechanism presented in Fig. 16.3. It is constituted by two articulated rigid arms $\mathcal{B}_{\mathrm{RR},1}$ and $\mathcal{B}_{\mathrm{RR},2}$ ($m_{\mathrm{RR},\mathcal{B}_s}$ denotes the mass of $\mathcal{B}_{\mathrm{RR},s}$, $\mathrm{b}^{\star}_{\mathrm{RR},s}$ denotes the centre of mass of $\mathcal{B}_{\mathrm{RR},s}$, and $I_{\mathrm{RR},\mathcal{B}_s}$ denotes moments of inertia of $\mathcal{B}_{\mathrm{RR},s}$ relative to an axis passing through $\mathrm{b}^{\star}_{\mathrm{RR},s}$ perpendicularly to its plane of motion, $s \in \{1, 2\}$). Points $\mathrm{p}_{\mathrm{RR},1}$ and $\mathrm{p}_{\mathrm{RR},2}$ are the geometric centres of the revolute joints and $\mathrm{p}_{\mathrm{RR},3}$ represents the geometric centre of the end-effector of the mechanism (which is rigidly attached to its second arm). The main geometric parameters of this system are: the distance between points $\mathrm{p}_{\mathrm{RR},1}$ and $\mathrm{p}_{\mathrm{RR},2}$, denoted by $a_{\mathrm{RR},1}$ and the distance between points $\mathrm{p}_{\mathrm{RR},2}$ and $\mathrm{p}_{\mathrm{RR},3}$, denoted by $a_{\mathrm{RR},2}$. Two additional adimensional parameters $\gamma_{\mathrm{RR},1}$ and $\gamma_{\mathrm{RR},2}$ are defined to express the ratio of the distance between $\mathrm{p}_{\mathrm{RR},1}$ and $\mathrm{b}^{\star}_{\mathrm{RR},1}$ to $a_{\mathrm{RR},1}$ and the ratio of the distance between $\mathrm{p}_{\mathrm{RR},2}$ and $\mathrm{b}^{\star}_{\mathrm{RR},2}$ to $a_{\mathrm{RR},2}$, respectively.

In order to model this mechanism, consider a coordinate system $\mathtt{N}$ fixed to an inertial reference frame $\mathcal{N}$: the origin of $\mathtt{N}$ will be denoted by n_0 and its orthonormal vector basis is given by the unit vectors $\hat{\boldsymbol{n}}_1$ (horizontal direction in the plane of the motions of the arms), $\hat{\boldsymbol{n}}_2$ (vertical) and $\hat{\boldsymbol{n}}_3 = \hat{\boldsymbol{n}}_1 \times \hat{\boldsymbol{n}}_2$. The following generalized coordinates are defined:

- $q_{\mathrm{RR},\mathcal{R}_1}$ and $q_{\mathrm{RR},\mathcal{R}_2}$ (see Fig. 16.3) describing the relative motions associated with the revolute joints: $q_{\mathrm{RR},\mathcal{R}_1}$ is the angle between the horizontal and the line joining $\mathrm{p}_{\mathrm{RR},1}$, and $\mathrm{p}_{\mathrm{RR},2}$ and $q_{\mathrm{RR},\mathcal{R}_2}$ is the angle between this latter line and the line joining $\mathrm{p}_{\mathrm{RR},2}$ and $\mathrm{p}_{\mathrm{RR},3}$.
- $q_{\mathrm{RR},\mathrm{p}_r,1}$ and $q_{\mathrm{RR},\mathrm{p}_r,2}$, for $r \in \{1, 2, 3\}$, such that $\boldsymbol{r}_{\mathrm{p}_{\mathrm{RR},r}|\mathrm{n}_0} = q_{\mathrm{RR},\mathrm{p}_r,1}\,\hat{\boldsymbol{n}}_1 + q_{\mathrm{RR},\mathrm{p}_r,2}\,\hat{\boldsymbol{n}}_2$.

Fig. 16.3 RR mechanism

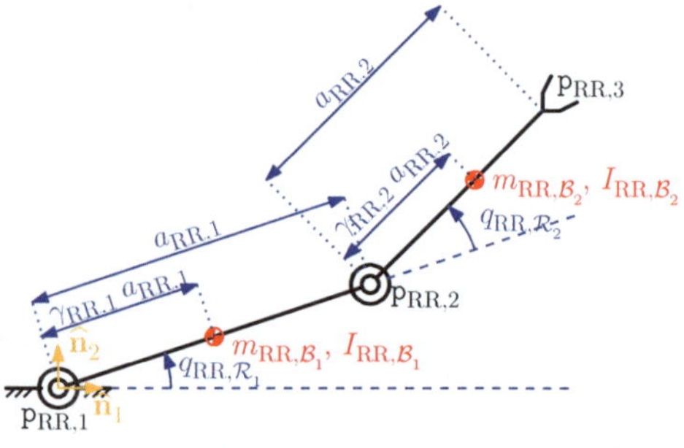

The following quasi-velocities are defined:

- $p_{\mathrm{RR},\mathcal{R}_1} = \dot{q}_{\mathrm{RR},\mathcal{R}_1}$ and $p_{\mathrm{RR},\mathcal{R}_2} = \dot{q}_{\mathrm{RR},\mathcal{R}_2}$.
- $p_{\mathrm{RR},\mathcal{B}_s,1}, p_{\mathrm{RR},\mathcal{B}_s,2}$ and $p_{\mathrm{RR},\mathcal{B}_s,3}$, for $s \in \{1,2\}$, such that $\boldsymbol{v}_{\mathrm{b}^{\star}_{\mathrm{RR},s}|\mathcal{N}} = p_{\mathrm{RR},\mathcal{B}_s,1}\,\hat{\boldsymbol{n}}_1 + p_{\mathrm{RR},\mathcal{B}_s,2}\,\hat{\boldsymbol{n}}_2$ and $\boldsymbol{\omega}_{\mathcal{B}_{\mathrm{RR},s}|\mathcal{N}} = p_{\mathrm{RR},\mathcal{B}_s,3}\,\hat{\boldsymbol{n}}_3$.

The following column-matrices can also be defined:

$$\mathbb{q}_{\mathrm{RR}} = \left[\, q_{\mathrm{RR},\mathcal{R}_1}\; q_{\mathrm{RR},\mathcal{R}_2}\; q_{\mathrm{RR,p_2},1}\; q_{\mathrm{RR,p_2},2}\; q_{\mathrm{RR,p_3},1}\; q_{\mathrm{RR,p_3},2} \,\right]^{\mathsf{T}} \tag{16.62}$$

$$\mathbb{p}_{\mathrm{RR}} = \left[\, p_{\mathrm{RR},\mathcal{R}_1}\; p_{\mathrm{RR},\mathcal{R}_2}\; p_{\mathrm{RR},\mathcal{B}_1,1}\; p_{\mathrm{RR},\mathcal{B}_1,2}\; p_{\mathrm{RR},\mathcal{B}_1,3}\; p_{\mathrm{RR},\mathcal{B}_2,1}\; p_{\mathrm{RR},\mathcal{B}_2,2}\; p_{\mathrm{RR},\mathcal{B}_2,3} \,\right]^{\mathsf{T}} \tag{16.63}$$

The transformation of variables relating the time derivatives of generalized coordinates to the quasi-velocities of the model can be given by the following identities:

$$\begin{cases} p_{\mathrm{RR},\mathcal{R}_1} = \dot{q}_{\mathrm{RR},\mathcal{R}_1} \\ p_{\mathrm{RR},\mathcal{R}_2} = \dot{q}_{\mathrm{RR},\mathcal{R}_2} \\ \boldsymbol{v}_{\mathrm{b}^{\star}_{\mathrm{RR},1}|\mathcal{N}} = (1-\gamma_{\mathrm{RR},1})\dot{\boldsymbol{r}}_{\mathrm{p_{RR,1}}|\mathrm{n_0}} + \gamma_{\mathrm{RR},1}\,\dot{\boldsymbol{r}}_{\mathrm{p_{RR,2}}|\mathrm{n_0}} \\ \boldsymbol{v}_{\mathrm{b}^{\star}_{\mathrm{RR},2}|\mathcal{N}} = (1-\gamma_{\mathrm{RR},2})\dot{\boldsymbol{r}}_{\mathrm{p_{RR,2}}|\mathrm{n_0}} + \gamma_{\mathrm{RR},2}\,\dot{\boldsymbol{r}}_{\mathrm{p_{RR,3}}|\mathrm{n_0}} \end{cases} \tag{16.64}$$

Solving these equations for the time derivatives of generalized coordinates, it can be stated that:

$$\begin{cases} \dot{q}^{*}_{\mathrm{RR},\mathcal{R}_1} = p_{\mathrm{RR},\mathcal{R}_1} \\ \dot{q}^{*}_{\mathrm{RR},\mathcal{R}_2} = p_{\mathrm{RR},\mathcal{R}_2} \\ \dot{q}^{*}_{\mathrm{RR,p_1},1} = 0 \\ \dot{q}^{*}_{\mathrm{RR,p_1},2} = 0 \\ \dot{q}^{*}_{\mathrm{RR,p_2},1} = \dfrac{p_{\mathrm{RR},\mathcal{B}_1,1}}{\gamma_{\mathrm{RR},\mathcal{B}_1}} \\ \dot{q}^{*}_{\mathrm{RR,p_2},2} = \dfrac{p_{\mathrm{RR},\mathcal{B}_1,2}}{\gamma_{\mathrm{RR},\mathcal{B}_1}} \\ \dot{q}^{*}_{\mathrm{RR,p_3},1} = \dfrac{p_{\mathrm{RR},\mathcal{B}_2,1}}{\gamma_{\mathrm{RR},\mathcal{B}_2}} - \dfrac{(1-\gamma_{\mathrm{RR},\mathcal{B}_2})\,p_{\mathrm{RR},\mathcal{B}_1,1}}{\gamma_{\mathrm{RR},\mathcal{B}_1}\gamma_{\mathrm{RR},\mathcal{B}_2}} \\ \dot{q}^{*}_{\mathrm{RR,p_3},2} = \dfrac{p_{\mathrm{RR},\mathcal{B}_2,2}}{\gamma_{\mathrm{RR},\mathcal{B}_2}} - \dfrac{(1-\gamma_{\mathrm{RR},\mathcal{B}_2})\,p_{\mathrm{RR},\mathcal{B}_1,2}}{\gamma_{\mathrm{RR},\mathcal{B}_1}\gamma_{\mathrm{RR},\mathcal{B}_2}} \end{cases} \tag{16.65}$$

The holonomic constraint equations for this model can be given by the following identities:

$$\begin{cases} \dot{\boldsymbol{r}}_{\mathrm{p_{RR,2}|p_{RR,1}}} = \boldsymbol{\omega}_{\mathcal{B}_{\mathrm{RR,1}}|\mathcal{N}} \times \boldsymbol{r}_{\mathrm{p_{RR,2}|p_{RR,1}}} \\ \dot{\boldsymbol{r}}_{\mathrm{p_{RR,3}|p_{RR,2}}} = \boldsymbol{\omega}_{\mathcal{B}_{\mathrm{RR,2}}|\mathcal{N}} \times \boldsymbol{r}_{\mathrm{p_{RR,3}|p_{RR,2}}} \\ \boldsymbol{\omega}_{\mathcal{B}_{\mathrm{RR,1}}|\mathcal{N}} = p_{\mathrm{RR},\mathcal{R}_1}\,\hat{\boldsymbol{n}}_3 \\ \boldsymbol{\omega}_{\mathcal{B}_{\mathrm{RR,2}}|\mathcal{N}} = \left(p_{\mathrm{RR},\mathcal{R}_1} + p_{\mathrm{RR},\mathcal{R}_2}\right)\hat{\boldsymbol{n}}_3 \end{cases} \tag{16.66}$$

This leads to the following constraint equations:

$$\begin{cases} \psi_{\mathrm{RR},1} = \gamma_{\mathrm{RR},\mathcal{B}_1}\, p_{\mathrm{RR},\mathcal{B}_1,3}\left(q_{\mathrm{RR,p_2,2}} - q_{\mathrm{RR,p_1,2}}\right) + p_{\mathrm{RR},\mathcal{B}_1,1} = 0 \\ \psi_{\mathrm{RR},2} = \gamma_{\mathrm{RR},\mathcal{B}_1}\, p_{\mathrm{RR},\mathcal{B}_1,3}\left(q_{\mathrm{RR,p_1,1}} - q_{\mathrm{RR,p_2,1}}\right) + p_{\mathrm{RR},\mathcal{B}_1,2} = 0 \\ \psi_{\mathrm{RR},3} = \gamma_{\mathrm{RR},\mathcal{B}_1}\left(\gamma_{\mathrm{RR},\mathcal{B}_2}\, p_{\mathrm{RR},\mathcal{B}_2,3}\left(q_{\mathrm{RR,p_3,2}} - q_{\mathrm{RR,p_2,2}}\right) + p_{\mathrm{RR},\mathcal{B}_2,1}\right) - p_{\mathrm{RR},\mathcal{B}_1,1} = 0 \\ \psi_{\mathrm{RR},4} = \gamma_{\mathrm{RR},\mathcal{B}_1}\left(\gamma_{\mathrm{RR},\mathcal{B}_2}\, p_{\mathrm{RR},\mathcal{B}_2,3}\left(q_{\mathrm{RR,p_2,1}} - q_{\mathrm{RR,p_3,1}}\right) + p_{\mathrm{RR},\mathcal{B}_2,2}\right) - p_{\mathrm{RR},\mathcal{B}_1,2} = 0 \\ \psi_{\mathrm{RR},5} = p_{\mathrm{RR},\mathcal{R}_1} - p_{\mathrm{RR},\mathcal{B}_1,3} = 0 \\ \psi_{\mathrm{RR},6} = p_{\mathrm{RR},\mathcal{R}_1} + p_{\mathrm{RR},\mathcal{R}_2} - p_{\mathrm{RR},\mathcal{B}_2,3} = 0 \end{cases} \tag{16.67}$$

Calculating the time derivatives of the invariants $\psi_{\mathrm{RR},s}$, it can be stated that:

$$\mathbb{A}_{\mathrm{RR}} = \begin{bmatrix} 0 & 0 & 1 & 0 & \gamma_{\mathrm{RR},\mathcal{B}_1}\left(q_{\mathrm{RR,p_2,2}} - q_{\mathrm{RR,p_1,2}}\right) & 0 & 0 & 0 \\ 0 & 0 & 0 & 1 & \gamma_{\mathrm{RR},\mathcal{B}_1}\left(q_{\mathrm{RR,p_1,1}} - q_{\mathrm{RR,p_2,1}}\right) & 0 & 0 & 0 \\ 0 & 0 & -1 & 0 & 0 & \gamma_{\mathrm{RR},\mathcal{B}_1} & 0 & \gamma_{\mathrm{RR},\mathcal{B}_1}\gamma_{\mathrm{RR},\mathcal{B}_2}\left(q_{\mathrm{RR,p_3,2}} - q_{\mathrm{RR,p_2,2}}\right) \\ 0 & 0 & 0 & -1 & 0 & 0 & \gamma_{\mathrm{RR},\mathcal{B}_1} & \gamma_{\mathrm{RR},\mathcal{B}_1}\gamma_{\mathrm{RR},\mathcal{B}_2}\left(q_{\mathrm{RR,p_2,1}} - q_{\mathrm{RR,p_3,1}}\right) \\ 1 & 0 & 0 & 0 & -1 & 0 & 0 & 0 \\ 1 & 1 & 0 & 0 & 0 & 0 & 0 & -1 \end{bmatrix} \tag{16.68}$$

A full rank matrix $\mathbb{C}_{\mathrm{RR}}$ satisfying the condition $\mathbb{A}_{\mathrm{RR}}\,\mathbb{C}_{\mathrm{RR}} = \mathbb{0}$ is the following:

$$\mathbb{C}_{\mathrm{RR}} = \begin{bmatrix} 1 & 0 \\ 0 & 1 \\ \gamma_{\mathrm{RR},\mathcal{B}_1}\left(q_{\mathrm{RR,p_1,2}} - q_{\mathrm{RR,p_2,2}}\right) & 0 \\ \gamma_{\mathrm{RR},\mathcal{B}_1}\left(q_{\mathrm{RR,p_2,1}} - q_{\mathrm{RR,p_1,1}}\right) & 0 \\ 1 & 0 \\ q_{\mathrm{RR,p_1,2}} + \left(\gamma_{\mathrm{RR},\mathcal{B}_2} - 1\right) q_{\mathrm{RR,p_2,2}} - \gamma_{\mathrm{RR},\mathcal{B}_2} q_{\mathrm{RR,p_3,2}} & \gamma_{\mathrm{RR},\mathcal{B}_2}\left(q_{\mathrm{RR,p_2,2}} - q_{\mathrm{RR,p_3,2}}\right) \\ q_{\mathrm{RR,p_1,1}} + \left(\gamma_{\mathrm{RR},\mathcal{B}_2} - 1\right) q_{\mathrm{RR,p_2,1}} - \gamma_{\mathrm{RR},\mathcal{B}_2} q_{\mathrm{RR,p_3,1}} & \gamma_{\mathrm{RR},\mathcal{B}_2}\left(q_{\mathrm{RR,p_2,1}} - q_{\mathrm{RR,p_3,1}}\right) \\ 1 & 1 \end{bmatrix} \tag{16.69}$$

Suppose that there are actuators in both revolute joints, providing control torques $u_{\text{RR},1}$ and $u_{\text{RR},2}$ in the joints $\mathcal{R}_{\text{RR},1}$ and $\mathcal{R}_{\text{RR},2}$, respectively. Suppose also that the gravitational acceleration is given by $-g\hat{\boldsymbol{n}}_2$. Using Eqs. (16.12) and (16.13) it can also be stated that:

$$\mathbb{M}_{\text{RR}} = \begin{bmatrix} 0 & 0 & 0 & 0 & 0 & 0 & 0 & 0 \\ 0 & 0 & 0 & 0 & 0 & 0 & 0 & 0 \\ 0 & 0 & m_{\text{RR},\mathcal{B}_1} & 0 & 0 & 0 & 0 & 0 \\ 0 & 0 & 0 & m_{\text{RR},\mathcal{B}_1} & 0 & 0 & 0 & 0 \\ 0 & 0 & 0 & 0 & I_{\text{RR},\mathcal{B}_1} & 0 & 0 & 0 \\ 0 & 0 & 0 & 0 & 0 & m_{\text{RR},\mathcal{B}_2} & 0 & 0 \\ 0 & 0 & 0 & 0 & 0 & 0 & m_{\text{RR},\mathcal{B}_2} & 0 \\ 0 & 0 & 0 & 0 & 0 & 0 & 0 & I_{\text{RR},\mathcal{B}_2} \end{bmatrix} \tag{16.70}$$

$$\mathbb{g}_{\text{RR}} = \mathbb{0} \tag{16.71}$$

$$\mathbb{f}_{\text{RR}} = \begin{bmatrix} u_{\text{RR},1} \\ u_{\text{RR},2} \\ 0 \\ -g\, m_{\text{RR},\mathcal{B}_1} \\ 0 \\ 0 \\ -g\, m_{\text{RR},\mathcal{B}_2} \\ 0 \end{bmatrix} \tag{16.72}$$

Now, consider the mathematical modelling of the balancing masses (BM). When unconstrained to the RR mechanism, they can be conceived as punctual masses that can freely execute a planar motion. Each balancing mass will be considered as a particle $\mathcal{B}_{\text{BM},1}$ whose mass is equal to $m_{\text{BM},\mathcal{B}_1}$. The point in the plane representing $\mathcal{B}_{\text{BM},1}$ will be denoted by $\text{p}_{\text{BM},1}$. Suppose also that $\boldsymbol{r}_{\text{p}_{\text{BM},1}|\text{n}_0} = q_{\text{BM},\text{p}_1,1}\,\hat{\boldsymbol{n}}_1 + q_{\text{BM},\text{p}_1,2}\,\hat{\boldsymbol{n}}_2$ and $\boldsymbol{v}_{\text{p}_{\text{BM},1}|\mathcal{N}} = \dot{\boldsymbol{r}}_{\text{p}_{\text{BM},1}|\text{n}_0} = p_{\text{BM},\text{p}_1,1}\,\hat{\boldsymbol{n}}_1 + p_{\text{BM},\text{p}_1,2}\,\hat{\boldsymbol{n}}_2$. The dynamical model of system BM is defined by the following matrices:

$$\mathbb{C}_{\text{BM}} = \mathbb{1} \tag{16.73}$$

$$\mathbb{M}_{\text{BM}} = \begin{bmatrix} m_{\text{BM},\mathcal{B}_1} & 0 \\ 0 & m_{\text{BM},\mathcal{B}_1} \end{bmatrix} \tag{16.74}$$

$$\mathbb{g}_{\text{BM}} = \mathbb{0} \tag{16.75}$$

$$\mathbb{f}_{\text{BM}} = \begin{bmatrix} 0 \\ -g\, m_{\text{BM},\mathcal{B}_1} \end{bmatrix} \tag{16.76}$$

In this description, the system BM has no (internal) kinematic constraints.

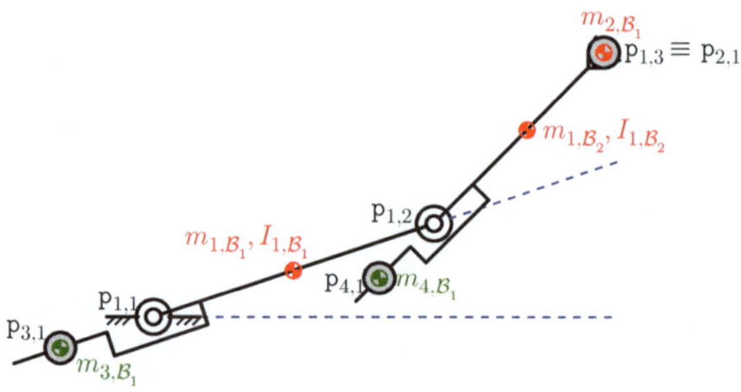

Fig. 16.4 Adaptively balanced RR mechanism

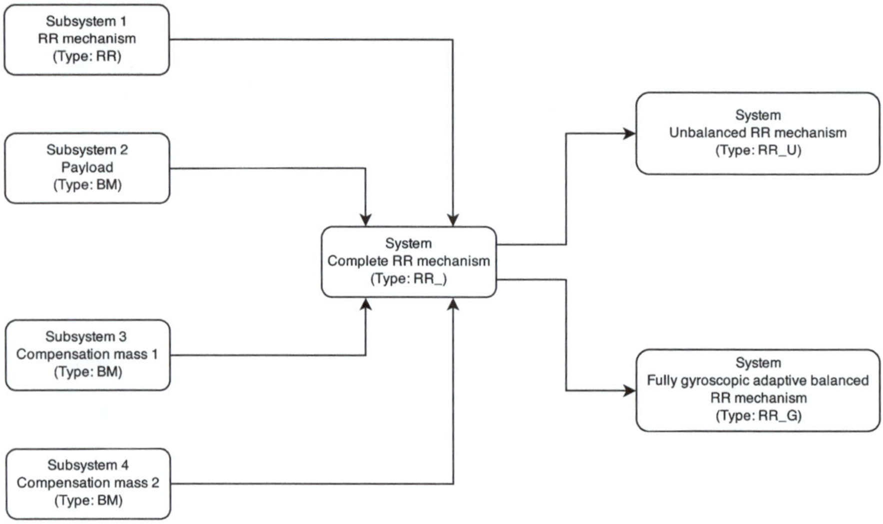

Fig. 16.5 Dynamic modelling of the complete RR mechanism

16.4.2 Adaptive Balancing of a RR Mechanism

Consider a multibody mechanical system denoted by RR_ that is constituted by a RR mechanism, two balancing masses (one attached to each of the arms of the RR mechanism) and a payload, as represented in Fig. 16.4.

The modelling process leading to the dynamical equations of system RR_ is synthesized in the diagram of Fig. 16.5. Four subsystems will be considered: subsystem 1 will be a RR mechanism whose model is given by Eqs. (16.69)–(16.72); subsystems 2 is a payload rigidly attached to the point $p_{1,3}$ of subsystem 1 and will be modelled identically to a balancing mass (BM), whose model is given by Eqs. (16.73)–(16.76); subsystems 3 and 4 are balancing masses (BM) rigidly attached to the arms $\mathcal{B}_{1,1}$ and $\mathcal{B}_{2,1}$ of subsystem 1, satisfying the relations $\boldsymbol{r}_{p_{3,1}|p_{1,1}} = -\gamma_{3,\mathcal{B}_1}\, \boldsymbol{r}_{p_{1,2}|p_{1,1}}$ and $\boldsymbol{r}_{p_{4,1}|p_{1,2}} = -\gamma_{4,\mathcal{B}_1}\, \boldsymbol{r}_{p_{1,3}|p_{1,2}}$.

The generalized coordinates and quasi-velocities of the model of system RR_ are the following:

$$\begin{cases} \mathbb{q}_{\mathrm{RR\_}} = \left[\mathbb{q}_1^{\mathsf{T}}\ \mathbb{q}_2^{\mathsf{T}}\ \mathbb{q}_3^{\mathsf{T}}\ \mathbb{q}_4^{\mathsf{T}}\right]^{\mathsf{T}} \\ \mathbb{q}_1 = \left[q_{1,\mathcal{R}_1}\ q_{1,\mathcal{R}_2}\ q_{1,\mathrm{p}_1,1}\ q_{1,\mathrm{p}_1,2}\ q_{1,\mathrm{p}_2,1}\ q_{1,\mathrm{p}_2,2}\ q_{1,\mathrm{p}_3,1}\ q_{1,\mathrm{p}_3,2}\right]^{\mathsf{T}} \\ \mathbb{q}_2 = \left[q_{2,\mathrm{p}_1,1}\ q_{2,\mathrm{p}_1,2}\right]^{\mathsf{T}} \\ \mathbb{q}_3 = \left[q_{3,\mathrm{p}_1,1}\ q_{3,\mathrm{p}_1,2}\right]^{\mathsf{T}} \\ \mathbb{q}_4 = \left[q_{4,\mathrm{p}_1,1}\ q_{4,\mathrm{p}_1,2}\right]^{\mathsf{T}} \end{cases} \tag{16.77}$$

$$\begin{cases} \mathbb{p}_{\mathrm{RR\_}} = \left[\mathbb{p}_1^{\mathsf{T}}\ \mathbb{p}_2^{\mathsf{T}}\ \mathbb{p}_3^{\mathsf{T}}\ \mathbb{p}_4^{\mathsf{T}}\right]^{\mathsf{T}} \\ \mathbb{p}_1 = \left[p_{1,\mathcal{R}_1}\ p_{1,\mathcal{R}_2}\ p_{1,\mathcal{B}_1,1}\ p_{1,\mathcal{B}_1,2}\ p_{1,\mathcal{B}_1,3}\ p_{1,\mathcal{B}_2,1}\ p_{1,\mathcal{B}_2,2}\ p_{1,\mathcal{B}_2,3}\right]^{\mathsf{T}} \\ \mathbb{p}_2 = \left[p_{2,\mathcal{B}_1,1}\ p_{2,\mathcal{B}_1,2}\right]^{\mathsf{T}} \\ \mathbb{p}_3 = \left[p_{3,\mathcal{B}_1,1}\ p_{3,\mathcal{B}_1,2}\right]^{\mathsf{T}} \\ \mathbb{p}_4 = \left[p_{4,\mathcal{B}_1,1}\ p_{4,\mathcal{B}_1,2}\right]^{\mathsf{T}} \end{cases} \tag{16.78}$$

Thus, the holonomic constraint equations describing how the payload and balancing masses are constrained to the RR mechanism can be expressed in the following form:

$$\begin{cases} \hat{\psi}_{\mathrm{RR\_},1} = (\gamma_{1,\mathcal{B}_2} - 1)\, p_{1,\mathcal{B}_1,1} + \gamma_{1,\mathcal{B}_1}\, (p_{1,\mathcal{B}_2,1} - \gamma_{1,\mathcal{B}_2}\, p_{2,\mathcal{B}_1,1}) = 0 \\ \hat{\psi}_{\mathrm{RR\_},2} = (\gamma_{1,\mathcal{B}_2} - 1)\, p_{1,\mathcal{B}_1,2} + \gamma_{1,\mathcal{B}_1}\, (p_{1,\mathcal{B}_2,2} - \gamma_{1,\mathcal{B}_2}\, p_{2,\mathcal{B}_1,2}) = 0 \\ \hat{\psi}_{\mathrm{RR\_},3} = \gamma_{3,\mathcal{B}_1}\, p_{1,\mathcal{B}_1,1} + \gamma_{1,\mathcal{B}_1}\, p_{3,\mathcal{B}_1,1} = 0 \\ \hat{\psi}_{\mathrm{RR\_},4} = \gamma_{3,\mathcal{B}_1}\, p_{1,\mathcal{B}_1,2} + \gamma_{1,\mathcal{B}_1}\, p_{3,\mathcal{B}_1,2} = 0 \\ \hat{\psi}_{\mathrm{RR\_},5} = \gamma_{4,\mathcal{B}_1}\, (\gamma_{1,\mathcal{B}_1}\, p_{1,\mathcal{B}_2,1} - p_{1,\mathcal{B}_1,1}) + \gamma_{1,\mathcal{B}_2}\, (\gamma_{1,\mathcal{B}_1}\, p_{4,\mathcal{B}_1,1} - p_{1,\mathcal{B}_1,1}) = 0 \\ \hat{\psi}_{\mathrm{RR\_},6} = \gamma_{4,\mathcal{B}_1}\, (\gamma_{1,\mathcal{B}_1}\, p_{1,\mathcal{B}_2,2} - p_{1,\mathcal{B}_1,2}) + \gamma_{1,\mathcal{B}_2}\, (\gamma_{1,\mathcal{B}_1}\, p_{4,\mathcal{B}_1,2} - p_{1,\mathcal{B}_1,2}) = 0 \end{cases} \tag{16.79}$$

The first time derivative of these equations corresponds to the external kinematic constraint equations of system RR_, $\hat{\mathbb{c}}_{\mathrm{RR\_}} = \mathbb{0}$.

Following the procedure described in Sect. 16.2.2, a matrix $\hat{\mathbb{C}}_{\mathrm{RR\_}}$ satisfying the condition $\hat{\mathbb{A}}_{\mathrm{RR\_}}\ \hat{\mathbb{C}}_{\mathrm{RR\_}} = \mathbb{0}$ is given by:

$$\hat{\mathbb{C}}_{RR\_} = \begin{bmatrix} 1 & 0 \\ 0 & 1 \\ q_{1,p_1,2} - q_{1,p_3,2} & q_{1,p_2,2} - q_{1,p_3,2} \\ q_{1,p_3,1} - q_{1,p_1,1} & q_{1,p_3,1} - q_{1,p_2,1} \\ 1 & 1 \\ \gamma_{3,\mathcal{B}_1}\left(q_{1,p_2,2} - q_{1,p_1,2}\right) & 0 \\ \gamma_{3,\mathcal{B}_1}\left(q_{1,p_1,1} - q_{1,p_2,1}\right) & 0 \\ q_{1,p_1,2} - \left(\gamma_{4,\mathcal{B}_1} + 1\right) q_{1,p_2,2} + \gamma_{4,\mathcal{B}_1} q_{1,p_3,2} & \gamma_{4,\mathcal{B}_1}\left(q_{1,p_3,2} - q_{1,p_2,2}\right) \\ q_{1,p_1,1} - \left(\gamma_{4,\mathcal{B}_1} + 1\right) q_{1,p_2,1} + \gamma_{4,\mathcal{B}_1} q_{1,p_3,1} & \gamma_{4,\mathcal{B}_1}\left(q_{1,p_3,1} - q_{1,p_2,1}\right) \end{bmatrix} \tag{16.80}$$

Therefore, the dynamical model of system RR_ corresponds to the equation $\mathbb{d}_{RR\_} = \hat{\mathbb{C}}_{RR\_}^{\mathsf{T}}\ \hat{\mathbb{d}}_{RR\_} = \mathbb{0}$.

It is remarkable, however, that such a mathematical model has a high number of variables among generalized coordinates and quasi-velocities. Many of them are useful only to simplify the modelling process, being unnecessary to keep then in the model once the dynamical equations have already been obtained. In order to eliminate some of the variables of the model, consider the following identities obtained from mathematical manipulation of Eqs. (16.67) and (16.79):

$$\begin{cases} p_{1,\mathcal{B}_1,1} = \gamma_{1,\mathcal{B}_1} p_{1,\mathcal{R}_1}\left(q_{1,p_1,2} - q_{1,p_2,2}\right) \\ p_{1,\mathcal{B}_1,2} = \gamma_{1,\mathcal{B}_1} p_{1,\mathcal{R}_1}\left(q_{1,p_2,1} - q_{1,p_1,1}\right) \\ p_{1,\mathcal{B}_1,3} = p_{1,\mathcal{R}_1} \\ p_{1,\mathcal{B}_2,1} = q_{1,p_1,2} p_{1,\mathcal{R}_1} + q_{1,p_2,2}\left(\left(\gamma_{1,\mathcal{B}_2} - 1\right) p_{1,\mathcal{R}_1} + \gamma_{1,\mathcal{B}_2} p_{1,\mathcal{R}_2}\right) - \gamma_{1,\mathcal{B}_2}\left(p_{1,\mathcal{R}_1} + p_{1,\mathcal{R}_2}\right) q_{1,p_3,2} \\ p_{1,\mathcal{B}_2,2} = -q_{1,p_1,1} p_{1,\mathcal{R}_1} - q_{1,p_2,1}\left(\left(\gamma_{1,\mathcal{B}_2} - 1\right) p_{1,\mathcal{R}_1} + \gamma_{1,\mathcal{B}_2} p_{1,\mathcal{R}_2}\right) + \gamma_{1,\mathcal{B}_2}\left(p_{1,\mathcal{R}_1} + p_{1,\mathcal{R}_2}\right) q_{1,p_3,1} \\ p_{1,\mathcal{B}_2,3} = p_{1,\mathcal{R}_1} + p_{1,\mathcal{R}_2} \\ p_{2,\mathcal{B}_1,1} = q_{1,p_1,2} p_{1,\mathcal{R}_1} + p_{1,\mathcal{R}_2} q_{1,p_2,2} - \left(p_{1,\mathcal{R}_1} + p_{1,\mathcal{R}_2}\right) q_{1,p_3,2} \\ p_{2,\mathcal{B}_1,2} = -q_{1,p_1,1} p_{1,\mathcal{R}_1} - p_{1,\mathcal{R}_2} q_{1,p_2,1} + \left(p_{1,\mathcal{R}_1} + p_{1,\mathcal{R}_2}\right) q_{1,p_3,1} \\ p_{3,\mathcal{B}_1,1} = \gamma_{3,\mathcal{B}_1} p_{1,\mathcal{R}_1}\left(q_{1,p_2,2} - q_{1,p_1,2}\right) \\ p_{3,\mathcal{B}_1,2} = \gamma_{3,\mathcal{B}_1} p_{1,\mathcal{R}_1}\left(q_{1,p_1,1} - q_{1,p_2,1}\right) \\ p_{4,\mathcal{B}_1,1} = q_{1,p_1,2} p_{1,\mathcal{R}_1} - q_{1,p_2,2}\left(\left(\gamma_{4,\mathcal{B}_1} + 1\right) p_{1,\mathcal{R}_1} + \gamma_{4,\mathcal{B}_1} p_{1,\mathcal{R}_2}\right) + \gamma_{4,\mathcal{B}_1}\left(p_{1,\mathcal{R}_1} + p_{1,\mathcal{R}_2}\right) q_{1,p_3,2} \\ p_{4,\mathcal{B}_1,2} = -q_{1,p_1,1} p_{1,\mathcal{R}_1} + q_{1,p_2,1}\left(\left(\gamma_{4,\mathcal{B}_1} + 1\right) p_{1,\mathcal{R}_1} + \gamma_{4,\mathcal{B}_1} p_{1,\mathcal{R}_2}\right) - \gamma_{4,\mathcal{B}_1}\left(p_{1,\mathcal{R}_1} + p_{1,\mathcal{R}_2}\right) q_{1,p_3,1} \end{cases} \tag{16.81}$$

Consider also the following identities that can be easily obtained from the geometry of the RR mechanism:

$$\begin{cases} q_{1,\mathrm{p}_2,1} = q_{1,\mathrm{p}_1,1} + a_{1,1} \cos(q_{1,\mathcal{R}_1}) \\ q_{1,\mathrm{p}_2,2} = q_{1,\mathrm{p}_1,2} + a_{1,1} \sin(q_{1,\mathcal{R}_1}) \\ q_{1,\mathrm{p}_3,1} = q_{1,\mathrm{p}_1,1} + a_{1,1} \cos(q_{1,\mathcal{R}_1}) + a_{1,2} \cos(q_{1,\mathcal{R}_1} + q_{1,\mathcal{R}_2}) \\ q_{1,\mathrm{p}_3,2} = q_{1,\mathrm{p}_1,2} + a_{1,1} \sin(q_{1,\mathcal{R}_1}) + a_{1,2} \sin(q_{1,\mathcal{R}_1} + q_{1,\mathcal{R}_2}) \end{cases} \tag{16.82}$$

Making the following definitions:

$$\begin{aligned} q_{\mathrm{RR\_},\mathcal{R}_1} &= q_{1,\mathcal{R}_1} & p_{\mathrm{RR\_},\mathcal{R}_1} &= p_{1,\mathcal{R}_1} \\ q_{\mathrm{RR\_},\mathcal{R}_2} &= q_{1,\mathcal{R}_2} & p_{\mathrm{RR\_},\mathcal{R}_2} &= p_{1,\mathcal{R}_2} \\ q_{\mathrm{RR\_},\mathrm{p}_1,1} &= q_{1,\mathrm{p}_1,1} & u_{\mathrm{RR\_},1} &= u_{1,1} \\ q_{\mathrm{RR\_},\mathrm{p}_1,2} &= q_{1,\mathrm{p}_1,2} & u_{\mathrm{RR\_},2} &= u_{1,2} \\ q_{\mathrm{RR\_},\mathrm{p}_3,1} &= q_{1,\mathrm{p}_3,1} \\ q_{\mathrm{RR\_},\mathrm{p}_3,2} &= q_{1,\mathrm{p}_3,2} \end{aligned}$$

the new variables of the RR_ model are given by:

$$\mathbb{q}_{\mathrm{RR\_}} = \left[q_{\mathrm{RR\_},\mathcal{R}_1}\ q_{\mathrm{RR\_},\mathcal{R}_2}\ q_{\mathrm{RR\_},\mathrm{p}_1,1}\ q_{\mathrm{RR\_},\mathrm{p}_1,2}\ q_{\mathrm{RR\_},\mathrm{p}_3,1}\ q_{\mathrm{RR\_},\mathrm{p}_3,2} \right]^{\mathsf{T}} \tag{16.83}$$

$$\mathbb{p}_{\mathrm{RR\_}} = \left[p_{\mathrm{RR\_},\mathcal{R}_1}\ p_{\mathrm{RR\_},\mathcal{R}_2} \right]^{\mathsf{T}} \tag{16.84}$$

Define the following parameters:

$$\begin{cases} I_{\mathrm{RR\_\ U},1} = a_{1,2}^2 \big(m_{1,\mathcal{B}_2}\, \gamma_{1,\mathcal{B}_2}^2 + m_{2,\mathcal{B}_1} + m_{4,\mathcal{B}_1}\, \gamma_{4,\mathcal{B}_1}^2 \big) + I_{1,\mathcal{B}_2} \\ I_{\mathrm{RR\_\ U},2} = a_{1,1}^2 \big(m_{1,\mathcal{B}_1}\, \gamma_{1,\mathcal{B}_1}^2 + m_{3,\mathcal{B}_1}\, \gamma_{3,\mathcal{B}_1}^2 + m_{1,\mathcal{B}_2} + m_{2,\mathcal{B}_1} + m_{4,\mathcal{B}_1} \big) + I_{1,\mathcal{B}_1} \\ I_{\mathrm{RR\_\ U},3} = a_{1,1}\, a_{1,2} \big(m_{1,\mathcal{B}_2}\, \gamma_{1,\mathcal{B}_2} - m_{4,\mathcal{B}_1}\, \gamma_{4,\mathcal{B}_1} + m_{2,\mathcal{B}_1} \big) \\ \mu_{\mathrm{RR\_\ U},1} = g\, a_{1,1} \big(m_{1,\mathcal{B}_1}\, \gamma_{1,\mathcal{B}_1} - m_{3,\mathcal{B}_1}\, \gamma_{3,\mathcal{B}_1} + m_{1,\mathcal{B}_2} + m_{2,\mathcal{B}_1} + m_{4,\mathcal{B}_1} \big) \\ \mu_{\mathrm{RR\_\ U},2} = g\, a_{1,2} \big(m_{1,\mathcal{B}_2}\, \gamma_{1,\mathcal{B}_2} - m_{4,\mathcal{B}_1}\, \gamma_{4,\mathcal{B}_1} + m_{2,\mathcal{B}_1} \big) \end{cases} \tag{16.85}$$

If the adjustable physical parameters $m_{3,\mathcal{B}_1}$, $\gamma_{3,\mathcal{B}_1}$, $m_{4,\mathcal{B}_1}$, and $\gamma_{4,\mathcal{B}_1}$ are chosen in such a way that at least one of the parameters among $I_{\mathrm{RR\_\ U},3}$, $\mu_{\mathrm{RR\_\ U},1}$ and $\mu_{\mathrm{RR\_\ U},2}$ are not null, model RR_ represents an unbalanced RR mechanism and will be denoted as RR_ U. Its mathematical model is given by:

$$\mathbb{C}_{\mathrm{RR\_\ U}} = \mathbb{1} \tag{16.86}$$

$$\mathbb{M}_{\mathrm{RR\_\ U}} =$$

$$\begin{bmatrix} I_{RR\_U,1}+I_{RR\_U,2}+2\cos\left(q_{RR\_U,\mathcal{R}_2}\right)I_{RR\_U,3} & I_{RR\_U,1}+\cos\left(q_{RR\_U,\mathcal{R}_2}\right)I_{RR\_U,3} \\ I_{RR\_U,1}+\cos\left(q_{RR\_U,\mathcal{R}_2}\right)I_{RR\_U,3} & I_{RR\_U,1} \end{bmatrix} \tag{16.87}$$

$$\mathbb{g}_{RR\_U} = \begin{bmatrix} I_{RR\_U,3}\,p^2_{RR\_U,\mathcal{R}_2}\sin\left(q_{RR\_U,\mathcal{R}_2}\right)+2I_{RR\_U,3}\,p_{RR\_U,\mathcal{R}_1}\,p_{RR\_U,\mathcal{R}_2}\sin\left(q_{RR\_U,\mathcal{R}_2}\right) \\ -I_{RR\_U,3}\,p^2_{RR\_U,\mathcal{R}_1}\sin\left(q_{RR\_U,\mathcal{R}_2}\right) \end{bmatrix} \tag{16.88}$$

$$\mathbb{f}_{RR\_U}=\begin{bmatrix} u_{RR\_U,1}-\left(\mu_{RR\_U,1}\cos\left(q_{RR\_U,\mathcal{R}_1}\right)+\mu_{RR\_U,2}\cos\left(q_{RR\_U,\mathcal{R}_1}+q_{RR\_U,\mathcal{R}_2}\right)\right) \\ u_{RR\_U,2}-\mu_{RR\_U,2}\cos\left(q_{RR\_U,\mathcal{R}_1}+q_{RR\_U,\mathcal{R}_2}\right) \end{bmatrix} \tag{16.89}$$

On the other hand, if parameters $m_{3,\mathcal{B}_1}$, $\gamma_{3,\mathcal{B}_1}$, $m_{4,\mathcal{B}_1}$, and $\gamma_{4,\mathcal{B}_1}$ are chosen to satisfy the following relations:

$$\begin{cases} m_{1,\mathcal{B}_2}\,\gamma_{1,\mathcal{B}_2} - m_{4,\mathcal{B}_1}\,\gamma_{4,\mathcal{B}_1} + m_{2,\mathcal{B}_1} = 0 \\ m_{1,\mathcal{B}_1}\,\gamma_{1,\mathcal{B}_1} - m_{3,\mathcal{B}_1}\,\gamma_{3,\mathcal{B}_1} + m_{1,\mathcal{B}_2} + m_{2,\mathcal{B}_1} + m_{4,\mathcal{B}_1} = 0 \end{cases} \tag{16.90}$$

model RR_ represents a fully gyroscopic adaptive balanced RR mechanism and will be denoted as RR_ G. Defining the following parameters:

$$\begin{cases} I_{RR\_G,1} = a^2_{1,2}\left(m_{1,\mathcal{B}_2}\,\gamma^2_{1,\mathcal{B}_2} + m_{4,\mathcal{B}_1}\,\gamma^2_{4,\mathcal{B}_1} + m_{2,\mathcal{B}_1}\right) + I_{1,\mathcal{B}_2} \\ I_{RR\_G,2} = a^2_{1,1}\left(m_{1,\mathcal{B}_1}\,\gamma^2_{1,\mathcal{B}_1} + m_{3,\mathcal{B}_1}\,\gamma^2_{3,\mathcal{B}_1} + m_{1,\mathcal{B}_2} + m_{2,\mathcal{B}_1} + m_{4,\mathcal{B}_1}\right) + I_{1,\mathcal{B}_1} \end{cases} \tag{16.91}$$

it can be stated that RR_ G model is given by:

$$\mathbb{C}_{RR\_G} = \mathbb{1} \tag{16.92}$$

$$\mathbb{M}_{RR\_G} = \begin{bmatrix} I_{RR\_G,1} + I_{RR\_G,2} & I_{RR\_G,1} \\ I_{RR\_G,1} & I_{RR\_G,1} \end{bmatrix} \tag{16.93}$$

$$\mathbb{g}_{RR\_G} = \mathbb{0} \tag{16.94}$$

$$\mathbb{f}_{RR\_G} = \begin{bmatrix} u_{RR\_G,1} \\ u_{RR\_G,2} \end{bmatrix} \tag{16.95}$$

16.4.3 Adaptive Balancing of a 5R Mechanism

Consider a mechanical system denoted by 5R_ that represents a 5-bar mechanism with revolute joints, transporting a payload (punctual mass) and with balancing masses attached to each of its movable links as shown in Fig. 16.6.

The strategy to model this mechanical system, synthesized in the diagram of Fig. 16.7, is to consider that it can be conceived as two RR_ subsystems, denoted by the indexes 1 and 2, constrained by a revolute joint in their end effectors. These RR_ subsystems, however, do not have actuators in their second joints. This fact does not represent a limitation on the strategy, once these active joints can be modelled as passive by imposing an identically null torque input. Basically, it will be considered that $u_{1,2} = 0$ and $u_{2,2} = 0$. Also, consider the notation $u_{5R\_,1} = u_{1,1}$ and $u_{5R\_,2} = u_{2,1}$, such that $\mathbb{u}_{5R\_} = [u_{5R\_,1}\ u_{5R\_,2}]^{\mathsf{T}}$.

In addition to the generalized coordinates from each RR_ subsystem, define a third set of coordinates, represented by a column-matrix $\mathbb{q}_3$, with a single angular coordinate $q_{3,\mathcal{R}_3}$ that represents the angle between the longitudinal axes of bodies $\mathcal{B}_{1,2}$ and $\mathcal{B}_{2,2}$ (see Fig. 16.6). Thus, the generalized coordinates of the model 5R_ are given by:

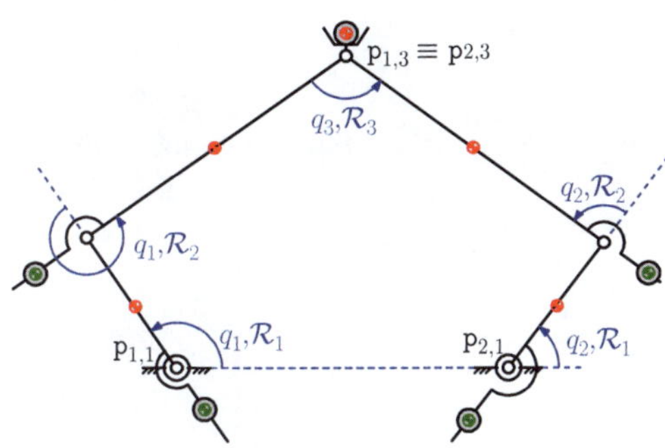

Fig. 16.6 5R mechanism

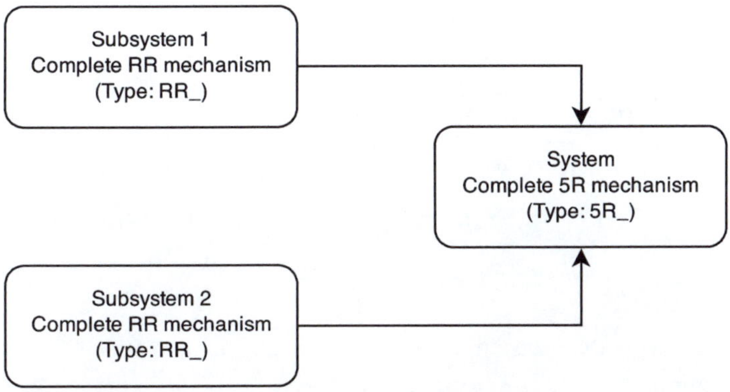

Fig. 16.7 Dynamic modelling of the complete 5R mechanism

$$\begin{cases} \mathbb{q}_{5R\_} = \left[\mathbb{q}_1^{\mathsf{T}}\ \mathbb{q}_2^{\mathsf{T}}\ \mathbb{q}_3^{\mathsf{T}}\right]^{\mathsf{T}} \\ \mathbb{q}_1 = \left[q_{1,\mathcal{R}_1}\ q_{1,\mathcal{R}_2}\ q_{1,\mathrm{p}_1,1}\ q_{1,\mathrm{p}_1,2}\ q_{1,\mathrm{p}_3,1}\ q_{1,\mathrm{p}_3,2}\right]^{\mathsf{T}} \\ \mathbb{q}_2 = \left[q_{2,\mathcal{R}_1}\ q_{2,\mathcal{R}_2}\ q_{2,\mathrm{p}_1,1}\ q_{2,\mathrm{p}_1,2}\ q_{2,\mathrm{p}_3,1}\ q_{2,\mathrm{p}_3,2}\right]^{\mathsf{T}} \\ \mathbb{q}_3 = \left[q_{3,\mathcal{R}_3}\right]^{\mathsf{T}} \end{cases} \tag{16.96}$$

The (internal) holonomic constraints of subsystems 1 and 2 and the external holonomic constraints between them can be described by the following invariants:

$$\begin{cases} h_{1,1} = q_{1,\mathrm{p}_3,1} - \cos(q_{1,\mathcal{R}_1})\, a_{1,1} - \cos(q_{1,\mathcal{R}_1} + q_{1,\mathcal{R}_2})\, a_{1,2} - q_{1,\mathrm{p}_1,1} = 0 \\ h_{1,2} = q_{1,\mathrm{p}_3,2} - \sin(q_{1,\mathcal{R}_1})\, a_{1,1} - \sin(q_{1,\mathcal{R}_1} + q_{1,\mathcal{R}_2})\, a_{1,2} - q_{1,\mathrm{p}_1,2} = 0 \\ h_{2,1} = q_{2,\mathrm{p}_3,1} - \cos(q_{2,\mathcal{R}_1})\, a_{2,1} - \cos(q_{2,\mathcal{R}_1} + q_{2,\mathcal{R}_2})\, a_{2,2} - q_{2,\mathrm{p}_1,1} = 0 \\ h_{2,2} = q_{2,\mathrm{p}_3,2} - \sin(q_{2,\mathcal{R}_1})\, a_{2,1} - \sin(q_{2,\mathcal{R}_1} + q_{2,\mathcal{R}_2})\, a_{2,2} - q_{2,\mathrm{p}_1,2} = 0 \\ \hat{h}_{5R\_,1} = q_{1,\mathrm{p}_3,1} - q_{2,\mathrm{p}_3,1} = 0 \\ \hat{h}_{5R\_,2} = q_{1,\mathrm{p}_3,2} - q_{2,\mathrm{p}_3,2} = 0 \\ \hat{h}_{5R\_,3} = q_{1,\mathcal{R}_1} + q_{1,\mathcal{R}_2} - q_{2,\mathcal{R}_1} - q_{2,\mathcal{R}_2} + q_{3,\mathcal{R}_3} - 2\pi = 0 \end{cases} \tag{16.97}$$

Also, the quasi-velocities of the model 5R_ are given by:

$$\begin{cases} \mathbb{p}_{5R\_} = \left[\mathbb{p}_1^{\mathsf{T}}\ \mathbb{p}_2^{\mathsf{T}}\right]^{\mathsf{T}} \\ \mathbb{p}_1 = \left[p_{1,\mathcal{R}_1}\ p_{1,\mathcal{R}_2}\right]^{\mathsf{T}} \\ \mathbb{p}_2 = \left[p_{2,\mathcal{R}_1}\ p_{2,\mathcal{R}_2}\right]^{\mathsf{T}} \end{cases} \tag{16.98}$$

Due to the constraints between subsystems 1 and 2 the following conditions must be satisfied:

$$\begin{cases} \hat{\psi}_{5R\_,1} = \sin(q_{1,\mathcal{R}_1})\, a_{1,1}\, p_{1,\mathcal{R}_1} + \sin(q_{1,\mathcal{R}_1} + q_{1,\mathcal{R}_2})\, a_{1,2}\, (p_{1,\mathcal{R}_1} + p_{1,\mathcal{R}_2}) \\ \qquad - \sin(q_{2,\mathcal{R}_1})\, a_{2,1}\, p_{2,\mathcal{R}_1} - \sin(q_{2,\mathcal{R}_1} + q_{2,\mathcal{R}_2})\, a_{2,2}\, (p_{2,\mathcal{R}_1} + p_{2,\mathcal{R}_2}) \\ \hat{\psi}_{5R\_,2} = \cos(q_{1,\mathcal{R}_1})\, a_{1,1}\, p_{1,\mathcal{R}_1} + \cos(q_{1,\mathcal{R}_1} + q_{1,\mathcal{R}_2})\, a_{1,2}\, (p_{1,\mathcal{R}_1} + p_{1,\mathcal{R}_2}) \\ \qquad - \cos(q_{2,\mathcal{R}_1})\, a_{2,1}\, p_{2,\mathcal{R}_1} - \cos(q_{2,\mathcal{R}_1} + q_{2,\mathcal{R}_2})\, a_{2,2}\, (p_{2,\mathcal{R}_1} + p_{2,\mathcal{R}_2}) \end{cases} \tag{16.99}$$

Again following the procedure described in Sect. 16.2.2, a matrix $\hat{\mathbb{C}}_{5R\_}$ satisfying the condition $\hat{\mathbb{A}}_{5R\_}\ \hat{\mathbb{C}}_{5R\_} = \mathbb{0}$ is given by:

$$\hat{\mathbb{C}}_{5R\_} = \begin{bmatrix} 1 & 0 \\ -\dfrac{\sin(q_{1,\mathcal{R}_2}+q_{3,\mathcal{R}_3})\,a_{1,1}}{\sin(q_{3,\mathcal{R}_3})\,a_{1,2}} - 1 & \dfrac{\sin(q_{2,\mathcal{R}_2})\,a_{2,1}}{\sin(q_{3,\mathcal{R}_3})\,a_{1,2}} \\ 0 & 1 \\ -\dfrac{\sin(q_{1,\mathcal{R}_2})\,a_{1,1}}{\sin(q_{3,\mathcal{R}_3})\,a_{2,2}} & \dfrac{\sin(q_{2,\mathcal{R}_2}-q_{3,\mathcal{R}_3})\,a_{2,1}}{\sin(q_{3,\mathcal{R}_3})\,a_{2,2}} - 1 \end{bmatrix} \tag{16.100}$$

When subsystems 1 and 2 are fully gyroscopic adaptive balanced RR mechanisms (model RR_ G), the 5R_ model obtained corresponds to a gyroscopic adaptive balanced 5R mechanism, and will be denoted by 5R_ G. The dynamical equations of model 5R_ G can be written in the following matrix form:

$$\mathbb{u}_{5R\_G} = \mathbb{M}'_{5R\_G}\,\dot{\mathbb{p}}_{5R\_G} \tag{16.101}$$

with matrix $\mathbb{M}'_{5R\_G}$ given by:

$$\mathbb{M}'_{5R\_G} = \begin{bmatrix} I_{1,2} - \dfrac{\sin(q_{1,\mathcal{R}_2}+q_{3,\mathcal{R}_3})\,a_{1,1}\,I_{1,1}}{\sin(q_{3,\mathcal{R}_3})\,a_{1,2}} & \dfrac{\sin(q_{2,\mathcal{R}_2})\,a_{2,1}\,I_{1,1}}{\sin(q_{3,\mathcal{R}_3})\,a_{1,2}} \\ -\dfrac{\sin(q_{1,\mathcal{R}_2}+q_{3,\mathcal{R}_3})\,a_{1,1}\,I_{1,1}}{\sin(q_{3,\mathcal{R}_3})\,a_{1,2}} & \dfrac{\sin(q_{2,\mathcal{R}_2})\,a_{2,1}\,I_{1,1}}{\sin(q_{3,\mathcal{R}_3})\,a_{1,2}} \\ -\dfrac{\sin(q_{1,\mathcal{R}_2})\,a_{1,1}\,I_{2,1}}{\sin(q_{3,\mathcal{R}_3})\,a_{2,2}} & \dfrac{\sin(q_{2,\mathcal{R}_2}-q_{3,\mathcal{R}_3})\,a_{2,1}\,I_{2,1}}{\sin(q_{3,\mathcal{R}_3})\,a_{2,2}} + I_{2,2} \\ -\dfrac{\sin(q_{1,\mathcal{R}_2})\,a_{1,1}\,I_{2,1}}{\sin(q_{3,\mathcal{R}_3})\,a_{2,2}} & \dfrac{\sin(q_{2,\mathcal{R}_2}-q_{3,\mathcal{R}_3})\,a_{2,1}\,I_{2,1}}{\sin(q_{3,\mathcal{R}_3})\,a_{2,2}} \end{bmatrix}^{\mathsf{T}} \tag{16.102}$$

On the other hand, when subsystems 1 and 2 are unbalanced RR mechanisms (model RR_ U), so will be the corresponding 5R mechanism, whose model will be denoted by 5R_ U. In this case, the dynamical equations can be written in the following form:

$$\begin{aligned} u_{5R\_U,1} &= \left(\frac{a_{1,2}(I_{1,2}+\cos(q_{1,\mathcal{R}_2})I_{1,3}) - \csc(q_{3,\mathcal{R}_3})\sin(q_{1,\mathcal{R}_2}+q_{3,\mathcal{R}_3})a_{1,1}(I_{1,1}+\cos(q_{1,\mathcal{R}_2})I_{1,3})}{a_{1,2}}\right)\dot{p}_{1,\mathcal{R}_1} \\ &+ \left(\cos(q_{1,\mathcal{R}_2})I_{1,3} - \frac{\csc(q_{3,\mathcal{R}_3})\sin(q_{1,\mathcal{R}_2}+q_{3,\mathcal{R}_3})a_{1,1}I_{1,1}}{a_{1,2}}\right)\dot{p}_{1,\mathcal{R}_2} \\ &+ \left(-\frac{\csc(q_{3,\mathcal{R}_3})\sin(q_{1,\mathcal{R}_2})a_{1,1}(I_{2,1}+\cos(q_{2,\mathcal{R}_2})I_{2,3})}{a_{2,2}}\right)\dot{p}_{2,\mathcal{R}_1} \\ &+ \left(-\frac{\csc(q_{3,\mathcal{R}_3})\sin(q_{1,\mathcal{R}_2})a_{1,1}I_{2,1}}{a_{2,2}}\right)\dot{p}_{2,\mathcal{R}_2} \end{aligned}$$

$$
\begin{aligned}
&+\frac{1}{a_{1,2}\,a_{2,2}}\Big(a_{1,2}\big(a_{2,2}\big(\mu_{1,1}\cos\left(q_{1,\mathcal{R}_1}\right)-I_{1,3}\left(p_{1,\mathcal{R}_1}+p_{1,\mathcal{R}_2}\right)^2\sin\left(q_{1,\mathcal{R}_2}\right)\big)\\
&-a_{1,1}\sin\left(q_{1,\mathcal{R}_2}\right)\csc\left(q_{3,\mathcal{R}_3}\right)\left(I_{2,3}p_{2,\mathcal{R}_1}^2\sin\left(q_{2,\mathcal{R}_2}\right)+\mu_{2,2}\cos\left(q_{2,\mathcal{R}_1}+q_{2,\mathcal{R}_2}\right)\right)\big)\\
&-a_{1,1}a_{2,2}\sin\left(q_{3,\mathcal{R}_3}+q_{1,\mathcal{R}_2}\right)\csc\left(q_{3,\mathcal{R}_3}\right)\big(I_{1,3}p_{1,\mathcal{R}_1}^2\sin\left(q_{1,\mathcal{R}_2}\right)\\
&+\mu_{1,2}\cos\left(q_{1,\mathcal{R}_1}+q_{1,\mathcal{R}_2}\right)\big)\Big) \qquad (16.103)
\end{aligned}
$$

$$
\begin{aligned}
&u_{5\mathrm{R\_\ U},2}\\
&=\left(\frac{\csc\left(q_{3,\mathcal{R}_3}\right)\sin\left(q_{2,\mathcal{R}_2}\right)a_{2,1}\left(I_{1,1}+\cos\left(q_{1,\mathcal{R}_2}\right)I_{1,3}\right)}{a_{1,2}}\right)\dot{p}_{1,\mathcal{R}_1}\\
&+\left(\frac{\csc\left(q_{3,\mathcal{R}_3}\right)\sin\left(q_{2,\mathcal{R}_2}\right)a_{2,1}I_{1,1}}{a_{1,2}}\right)\dot{p}_{1,\mathcal{R}_2}\\
&+\left(\frac{\csc\left(q_{3,\mathcal{R}_3}\right)\sin\left(q_{2,\mathcal{R}_2}-q_{3,\mathcal{R}_3}\right)a_{2,1}\left(I_{2,1}+\cos\left(q_{2,\mathcal{R}_2}\right)I_{2,3}\right)+a_{2,2}\left(I_{2,2}+\cos\left(q_{2,\mathcal{R}_2}\right)I_{2,3}\right)}{a_{2,2}}\right)\dot{p}_{2,\mathcal{R}_1}\\
&+\left(\frac{\csc\left(q_{3,\mathcal{R}_3}\right)\sin\left(q_{2,\mathcal{R}_2}-q_{3,\mathcal{R}_3}\right)a_{2,1}I_{2,1}}{a_{2,2}}+\cos\left(q_{2,\mathcal{R}_2}\right)I_{2,3}\right)\dot{p}_{2,\mathcal{R}_2}\\
&+\frac{1}{a_{1,2}a_{2,2}}\Big(a_{2,1}\csc\left(q_{3,\mathcal{R}_3}\right)\big(a_{1,2}\sin\left(q_{2,\mathcal{R}_2}-q_{3,\mathcal{R}_3}\right)\big(I_{2,3}p_{2,\mathcal{R}_1}^2\sin\left(q_{2,\mathcal{R}_2}\right)\\
&+\mu_{2,2}\cos\left(q_{2,\mathcal{R}_1}+q_{2,\mathcal{R}_2}\right)\big)+a_{2,2}\sin\left(q_{2,\mathcal{R}_2}\right)\big(I_{1,3}p_{1,\mathcal{R}_1}^2\sin\left(q_{1,\mathcal{R}_2}\right)\\
&+a_{1,2}a_{2,2}\big(\mu_{2,1}\cos\left(q_{2,\mathcal{R}_1}\right)-I_{2,3}\left(p_{2,\mathcal{R}_1}+p_{2,\mathcal{R}_2}\right)^2\sin\left(q_{2,\mathcal{R}_2}\right)\big)\\
&+\mu_{1,2}\cos\left(q_{1,\mathcal{R}_1}+q_{1,\mathcal{R}_2}\right)\big)\big)\Big) \qquad (16.104)
\end{aligned}
$$

Comparing the equations of models 5R_ G and 5R_ U it is remarkable the simplicity of the first one for calculating the control torques based on the current generalized coordinates, quasi-velocities, and quasi-accelerations. Indeed, the following subsections will present a discussion, based on simulations, on the advantages of controlling a gyroscopic adaptive balanced 5R mechanism (over its unbalanced counterpart).

16.4.4 Inverse Dynamics and Control Simulations

In this subsection, given the control law presented in Sect. 16.2.4 and the dynamic model presented in Sect. 16.4.3, the inverse dynamics simulation and the closed loop control system simulation of the RR_ G and RR_ U models will be performed. Thus, it is necessary to define values for the parameters of the models, initial conditions of the systems, reference trajectories, and the controller parameters.

Thus, define:

- Physical fixed parameters of the RR_ models used to obtain the 5R_ model:

  — $a_{1,1} = 0.1\,\text{m}$
  — $a_{1,2} = 0.2\,\text{m}$
  — $m_{1,\mathcal{B}_1} = 0.2\,\text{kg}$
  — $m_{1,\mathcal{B}_2} = 0.4\,\text{kg}$
  — $m_{2,\mathcal{B}_1} = 1.5\,\text{kg}$
  — $I_{1,\mathcal{B}_1} = 67 \cdot 10^{-5}\,\text{kg m}^2$
  — $I_{1,\mathcal{B}_2} = 134 \cdot 10^{-5}\,\text{kg m}^2$
  — $\gamma_{1,\mathcal{B}_1} = \gamma_{1,\mathcal{B}_2} = 0.5$

- Physical adjustable parameters of the RR_ G model:

  — $\gamma_{3,\mathcal{B}_1} = 3.0$
  — $\gamma_{4,\mathcal{B}_1} = 1.7$
  — $m_{3,\mathcal{B}_1} = m_{4,\mathcal{B}_1} = 1.0\,kg$

- Physical adjustable parameters of the RR_ U model:

  — $m_{3,\mathcal{B}_1} = m_{4,\mathcal{B}_1} = 0$

- Positioning of the bases of the RR_ models to obtain the 5R_ model:

  — $q_{1,\text{p}_1,1} = -0.1\,\text{m}$
  — $q_{1,\text{p}_1,2} = 0$
  — $q_{2,\text{p}_1,1} = 0.1\,\text{m}$
  — $q_{2,\text{p}_1,2} = 0$

- Controller parameters:

  — $\lambda = 40$
  — $k = 10$

- Position initial conditions of the 5R_ model:

$$\begin{cases} q_{1,\text{p}_3,1}(0) = q_{2,\text{p}_3,1}(0) = 0 \\ q_{1,\text{p}_3,2}(0) = q_{2,\text{p}_3,2}(0) = 0.02\,\text{m} \\ q_{3,\mathcal{R}_3}(0) = 173.282^\circ \\ q_{1,\mathcal{R}_1}(0) = 175.249^\circ \\ q_{1,\mathcal{R}_2}(0) = 188.11^\circ \\ q_{2,\mathcal{R}_1}(0) = 4.75078^\circ \\ q_{2,\mathcal{R}_2}(0) = 171.89^\circ \end{cases}$$

- Reference trajectory 1:

$$\begin{cases} q^{\diamond}_{1,\text{p}_3,1}(t) = q^{\diamond}_{2,\text{p}_3,1}(t) = 0 \\ q^{\diamond}_{1,\text{p}_3,2}(t) = q^{\diamond}_{2,\text{p}_3,2}(t) = 0.02 + 0.22\left(\frac{t}{5} - \frac{1}{2\pi}\sin\left(\frac{2\pi t}{5}\right)\right) \end{cases}$$

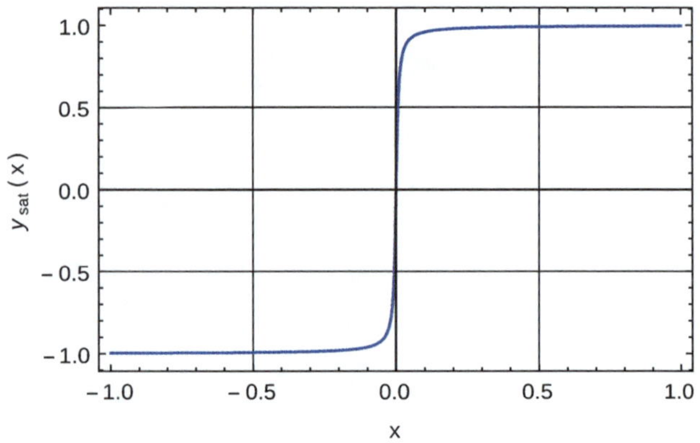

Fig. 16.8 Saturation function

- Reference trajectory 2:

$$\begin{cases} q^{\diamond}_{1,\mathrm{p}_3,1}(t) = q^{\diamond}_{2,\mathrm{p}_3,1}(t) = 0.005\sin(7t) \\ q^{\diamond}_{1,\mathrm{p}_3,2}(t) = q^{\diamond}_{2,\mathrm{p}_3,2}(t) = 0.14 - 0.12\cos(7t) \end{cases}$$

Only some coordinates of the reference trajectory were made explicit, because, defining these, the other can be found numerically or analytically by the position constraints presented in Eq. (16.97),

For the closed loop system simulation, the function $y(x) = \mathrm{sign}(x)$ has been replaced by the function $y_{\mathrm{sat}}(x) = \frac{2}{\pi}\arctan(159.1x)$, which presents the following properties (Fig. 16.8): $y_{\mathrm{sat}}(0.2) = -y_{\mathrm{sat}}(-0.2) = 0.98$ and $y_{\mathrm{sat}}(\infty) = -y_{\mathrm{sat}}(-\infty) = 1$. Its use makes the numerical simulation much more efficient, avoids control inputs chattering, and ensures a negligible error in steady state for this application.

In the inverse dynamics simulations, the actuator efforts necessary for the mechanism to follow the reference trajectory are calculated, ignoring the velocity initial conditions that will be defined.

(A) Trajectory 1 simulations:
In the closed loop system simulation, the following velocity initial conditions are used:

$$\begin{cases} \dot{q}_{1,\mathrm{p}_3,1}(0) = \dot{q}_{2,\mathrm{p}_3,1}(0) = 0 \\ \dot{q}_{1,\mathrm{p}_3,2}(0) = \dot{q}_{2,\mathrm{p}_3,2}(0) = 1\,\mathrm{m/s} \\ \dot{q}_{3,\mathcal{R}_3}(0) = -5.871\,\mathrm{rad/s} \\ \dot{q}_{1,\mathcal{R}_1}(0) = -4.153\,\mathrm{rad/s} \\ \dot{q}_{1,\mathcal{R}_2}(0) = 7.089\,\mathrm{rad/s} \\ \dot{q}_{2,\mathcal{R}_1}(0) = 4.153\,\mathrm{rad/s} \\ \dot{q}_{2,\mathcal{R}_2}(0) = -7.089\,\mathrm{rad/s} \end{cases}$$

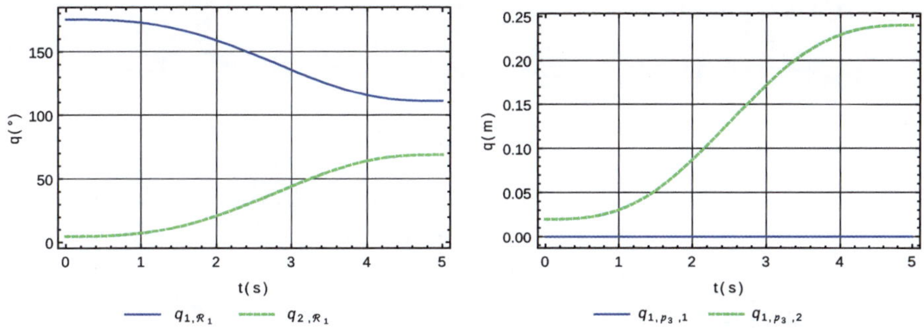

Fig. 16.9 Reference trajectory 1

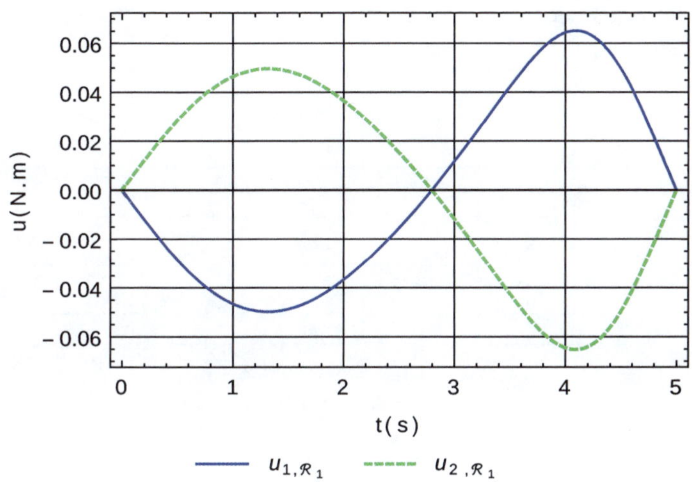

Fig. 16.10 Inverse dynamics simulation

This means that there is a non-zero velocity error for $t = 0$ and makes the error dynamics analysis more interesting.

Here are the graphs of the reference trajectory 1 (Fig. 16.9) for some coordinates:

(A.1) Balanced mechanism
Simulation of the efforts applied by the actuators (Figs. 16.10–16.12):
Error signal dynamics (Fig. 16.13):

(A.2) Unbalanced mechanism
Simulation of the efforts applied by the actuators (Figs. 16.14–16.15):
Error dynamics (Fig. 16.16):

[(B)] It is possible to observe that, most of the time, the efforts made by the actuators are significantly higher in the unbalanced mechanism than in the balanced one. This result is expected, because the reference trajectory 1 is a slow movement, in which the gravitational effects prevail over the inertial effects.

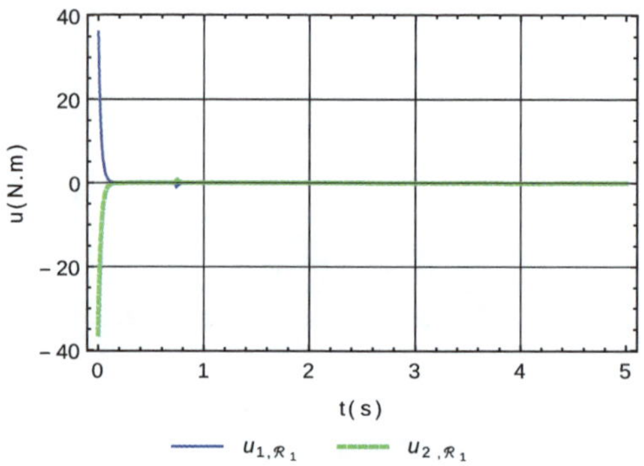

Fig. 16.11 Control inputs

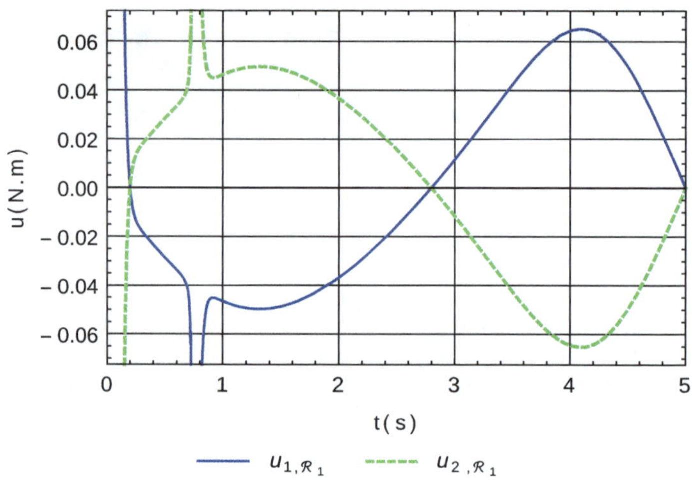

Fig. 16.12 Control inputs (zoom)

(B) Trajectory 2 simulation:
In the closed loop system simulation, the velocity initial conditions are null. Here are the graphs of the reference trajectory 2 (Fig. 16.17) for some coordinates:

(B.1) Balanced mechanism
Simulation of the efforts applied by the actuators (Figs. 16.18–16.19):
Error signal dynamics (Fig.16.20):

(B.2) Unbalanced mechanism
Simulation of the efforts applied by the actuators (Figs. 16.21–16.22):
Error signal dynamics (Fig.16.23):

It is possible to observe that, most of the time, the efforts made by the actuators are significantly higher in the balanced mechanism than in the unbalanced one. This result is expected, because the reference trajectory is a reasonably fast movement, in which the inertial effects prevail over the gravitational effects.

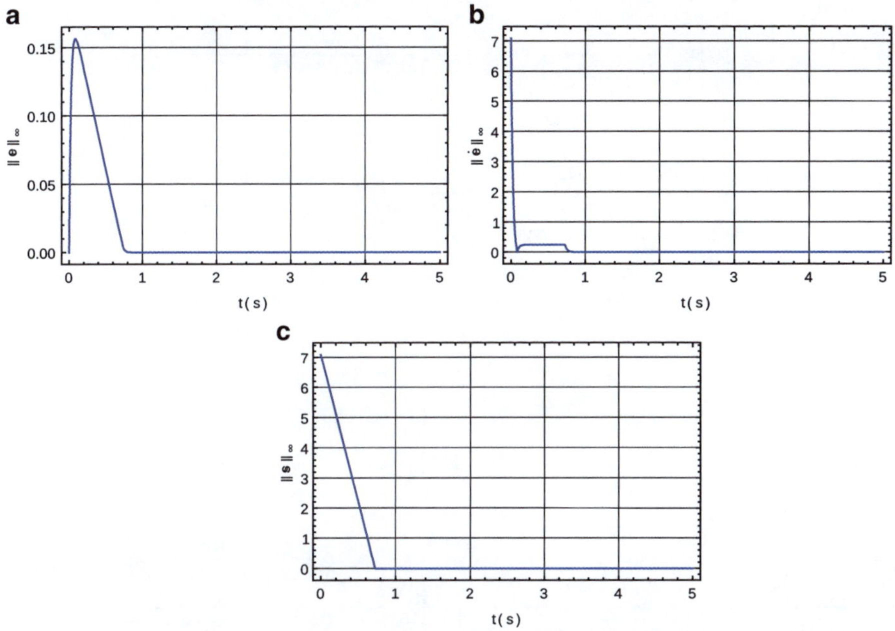

Fig. 16.13 Error dynamics. (a) Position error norm; (b) velocity error norm; (c) $\| \mathbb{s} \|_\infty = \| \dot{\mathbb{e}} + \lambda \mathbb{e} \|_\infty$

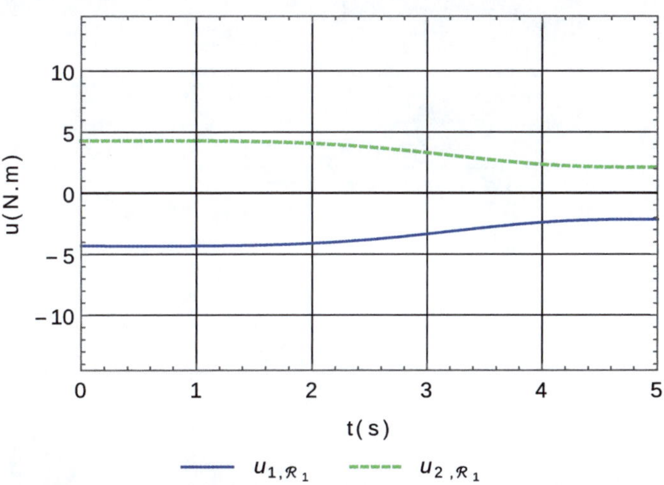

Fig. 16.14 Inverse dynamics simulation

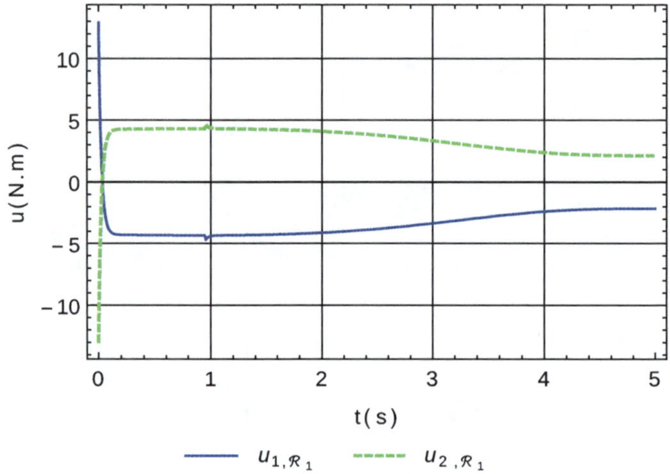

Fig. 16.15 Control inputs

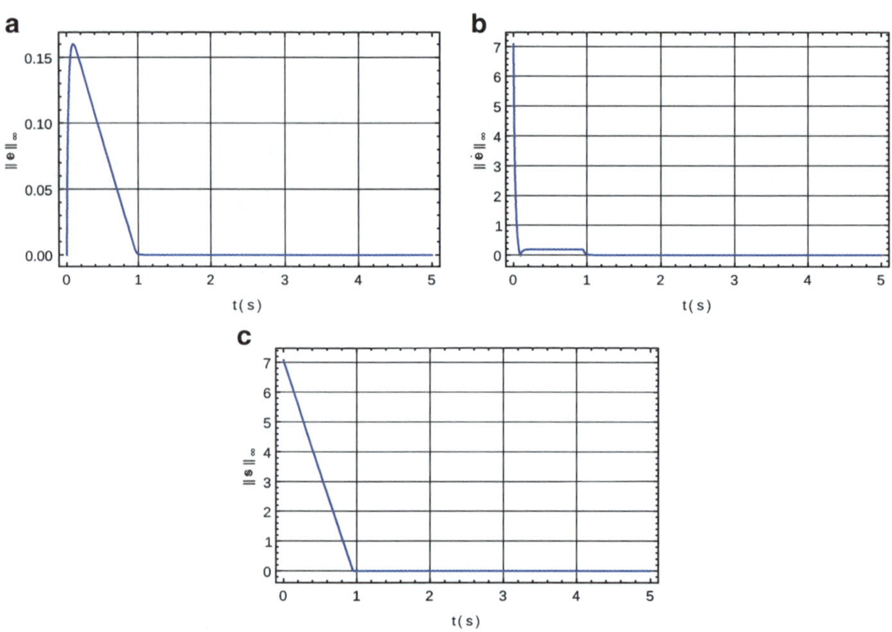

Fig. 16.16 Error dynamics. (**a**) Position error norm; (**b**) velocity error norm; (**c**) $\| \mathbb{s} \|_\infty = \| \dot{\mathbb{e}} + \lambda \mathbb{e} \|_\infty$

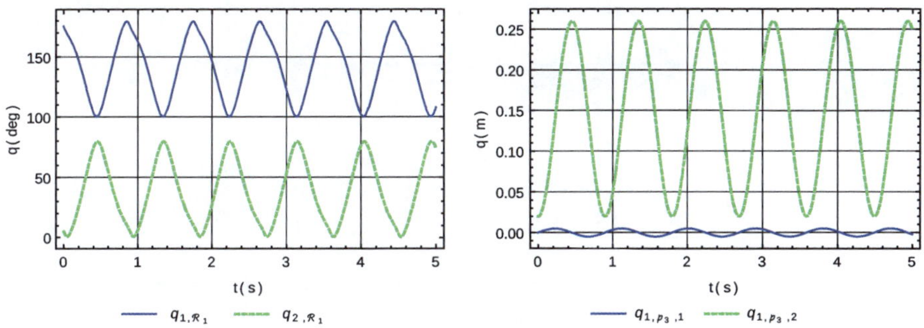

Fig. 16.17 Reference trajectory 2

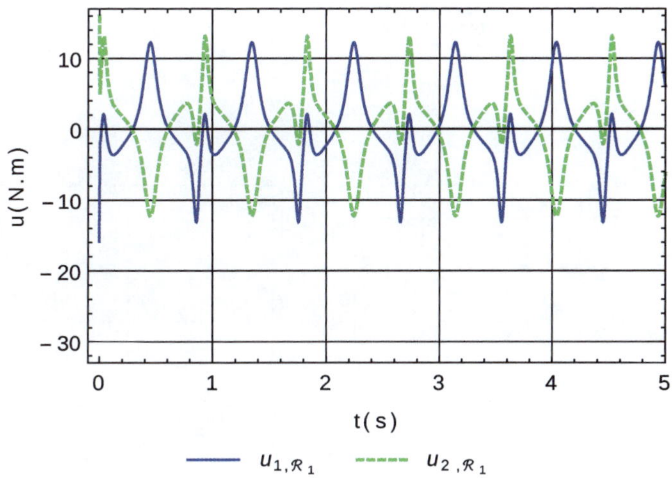

Fig. 16.18 Inverse dynamics simulation

16.5 Conclusions

This work dealt with the dynamic modelling and control of balanced parallel mechanisms. The dynamic modelling process constitutes an important issue considering the structural complexity of parallel mechanisms. Therefore, this book chapter described a dynamic formalism capable to deal with redundant generalized coordinates in association with the successive coupling of additional balancing elements to the original system model. This represents *a posterior* procedure because the analyst can successively include compensation inertias during the modelling without the need of rewriting the dynamic equations. Then, not only the compensation conditions can be derived but also the desired input torques for the

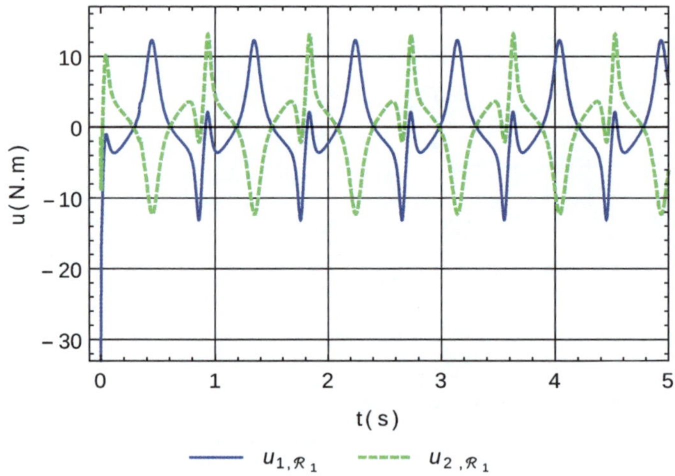

Fig. 16.19 Control inputs

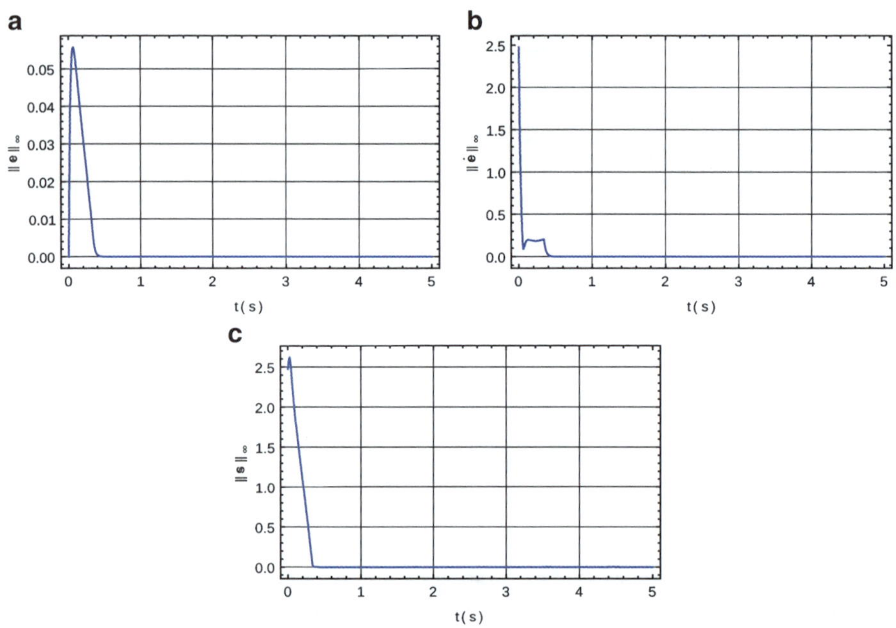

Fig. 16.20 Error dynamics. (**a**) Position error norm. (**b**) Velocity error norm. (**c**) $\| \mathbb{s} \|_\infty = \| \dot{\mathbb{e}} + \lambda \mathbb{e} \|_\infty$

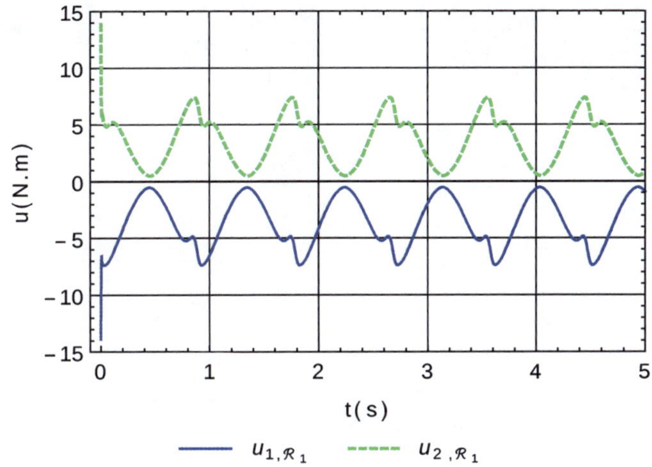

Fig. 16.21 Inverse dynamics simulation

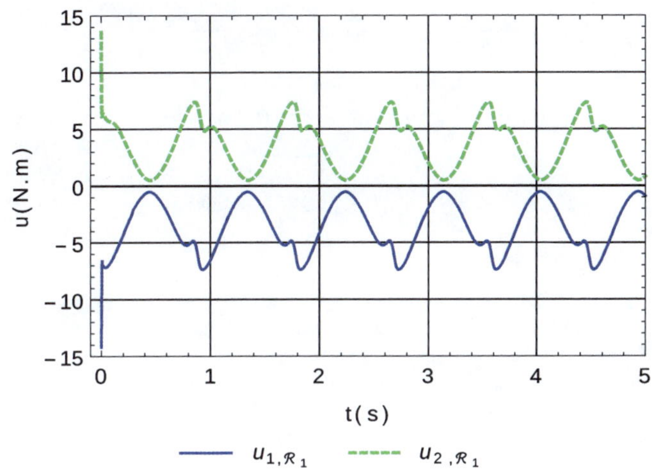

Fig. 16.22 Control inputs

motion control of the parallel mechanism. This work discussed the advantages of the dynamic model, developed in accordance with the methodology shown here, for the sliding modes control. Finally, the simulation results have demonstrated how effective is the presented methodology for the planar 5-bar mechanism with revolute joints.

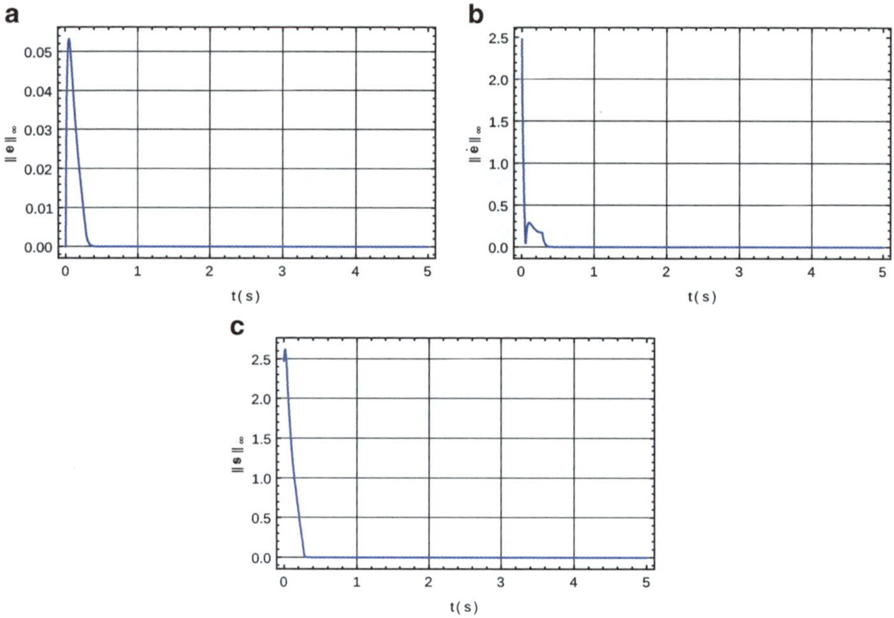

Fig. 16.23 Error dynamics. (**a**) Position error norm; (**b**) velocity error norm; (**c**) $\| \mathbb{s} \|_\infty = \| \dot{\mathbb{e}} + \lambda \mathbb{e} \|_\infty$

References

1. Agrawal, S.K., Fattah, A.: Gravity-balancing of spatial robotic manipulators. Mech. Mach. Theory **39**, 1331–1344 (2004)
2. Alici, G., Shirinzadeh, B.: Optimum force balancing with mass distribution and a single elastic element for a five-bar parallel manipulator. In: Proceedings of the IEEE International Conference on Robotics and Automation, pp. 3666–3671, Taipei, 14–19 September 2003
3. Alici, G., Shirinzadeh, B.: Optimum dynamic balancing of planar parallel manipulators based on sensitivity analysis. Mech. Mach. Theory **41**, 1520–1532 (2006)
4. Arakelian, V.H., Smith, M.R.: Design of planar 3-dof 3-RRR reactionless parallel manipulators. Mechatronics **18**, 601–606 (2008)
5. Briot, S., Arakelian, V., Le Baron, J.-P.: Shaking force minimization of high-speed robots via centre of mass acceleration control. Mech. Mach. Theory **57**, 1–12 (2012)
6. Coelho, T.A.H., Yong, L., Alves, V.F.A.: Decoupling of dynamic equations by means of adaptive balancing of 2-dof open-loop mechanisms. Mech. Mach. Theory **39**, 871–881 (2004)
7. Dehkordi, M.B., Frisoli, A., Sotgiu, E., Bergamasco, M.: Modelling and experimental evaluation of a static balancing technique for a new horizontally mounted 3-UPU parallel mechanism. Int. J. Adv. Robot. Syst. **9**, 193–205 (2012)
8. Gosselin, C.M., Vollmer, F., Côté, G., Wu, Y.: Synthesis and design of reactionless three-degree-of-freedom parallel mechanisms. IEEE Trans. Robot. Autom. **20**(2), 191–199 (2004)
9. Moradi, M., Nikoobin, A., Azadi, S.: Adaptive decoupling for open chain planar robots. Trans. B Mech. Eng. **17**(5), 376–386 (2010)
10. Russo, A., Sinatra, R., Xi, F.: Static balancing of parallel robots. Mech. Mach. Theory **40**, 191–202 (2005)

11. Seo, J.-T., Woo, J.H., Lim, H., Chung, J., Kim, W.K., Yi, B.-J.: Design of an antagonistically counter-balancing parallel mechanism. In: IEEE/RSJ International Conference on Intelligent Robots and Systems (IROS), pp. 2882–2887, Tokyo, 3–7 November 2013
12. Van der Wijk, V.: Shaking moment balancing of mechanisms with principal vectors and moments. Front. Mech. Eng. **8**(1), 10–16 (2013)
13. Wu, Y., Gosselin, C.M.: Design of reactionless 3-dof and 6-dof parallel manipulators using parallelepiped mechanisms. IEEE Trans. Robot. **21**(5), 821–833 (2005)
14. Wang, J., Gosselin, C.M.: Static balancing of spatial four-degree-of-freedom parallel mechanisms. Mech. Mach. Theory **35**, 563–592 (2000)
15. Wang, J., Gosselin, C.M.: Static balancing of spatial three-degree-of-freedom parallel mechanisms. Mech. Mach. Theory **34**, 437–452 (1999)
16. Wang, K., Luo, M., Mei, T., Zhao, J., Cao, Y.: Dynamics analysis of a three-DOF planar serial-parallel mechanism for active dynamic balancing with respect to a given trajectory. Int. J. Adv. Robot. Syst. **10**, 23–33 (2013)

Chapter 17
Controlled Biped Balanced Locomotion and Climbing

Benjamin Kenwright

Abstract This chapter describes the control principles necessary for an articulated biped model to accomplish balanced locomotion during walking and climbing. We explain the synthesizes mechanism for coordinated control of lower-body joints (i.e., ankle, hip, and knee). A humanoid biped can have a large number of degrees of freedom (DOF) that make it challenging to create physically correct, plausible and efficient motions. While we are able to define the physical principles of unintelligent models (e.g., multi-rigid body systems), the area of actively controlling a virtual character to mimic real-world creatures is an ongoing area of research. We focus on the control strategy and stability factors during continuous motion for the performing of essential rudimentary tasks (i.e., walking and climbing). We use a multi-level feedback mechanism to generated motion trajectories for the different actions, such as, stepping and walking. For example, the support leg is controlled through active forces (i.e., actuated joint feedback) based upon the control strategy to create a targeted set of parabolic trajectories for the action (e.g., stepping or climbing). The parabolic trajectories control the articulated skeleton while taking into account environmental influences (e.g., terrain height and balance information); with control parameters, such as leg-length, centre-of-mass (COM) location, and step-length being fed-back into the control mechanism.

Keywords Control • Trajectory generation • Balanced • Locomotion • Walking • Climbing • Biped • Stability • Control architecture • Jacobian

---

B. Kenwright (✉)
Edinburgh Napier University, Edinburgh, UK
e-mail: b.kenwright@napier.ac.uk
http://www.napier.ac.uk

D. Zhang, B. Wei (eds.), *Dynamic Balancing of Mechanisms and Synthesizing of Parallel Robots*, DOI 10.1007/978-3-319-17683-3_17

17.1 Introduction

17.1.1 Preface

Synthesizing controlled walking and climbing biped motions is challenging and important. We study and analyse real-world human motion for popular actions, such as, walking, so that we can better understand certain principles (e.g., balance logic, stability, control reasoning, and navigation). This chapter focuses on the physical simulation of articulated character motions using a virtual environment, i.e., an interconnected set of rigid body limbs represent the biped skeleton, which are controlled through joint actuator torques. Determining what torques and when to apply them to accomplish specific motions is the challenging control tasks. We use real-world masses and dimensions (e.g., size and weight of an average human) for our model. The joint actuator torques are kept within reasonable limits during controlled motions, such as leg stepping transitions, to emulate real-world constraints. We try and make the control strategies as straightforward and graceful as possible. We demonstrate the biomechanical simulation and investigation into the creation of controlled humanoid locomotion and climbing (e.g., navigating complex terrain and posture control). We present the fundamental concepts to inspire the development of better biomechanical animation systems that are controllable and self-driven.

17.1.2 Inspiration

Humans are an inspiration to us all. Humans are capable of an infinite array of motions. Even the simplest tasks, such as stepping and jumping, are difficult to synthesize. Capturing the life-like qualities while remaining physically correct is challenging and important. Over the past few decades, robotics research has introduced ever interesting and novel approaches [1–3] which have fed into other fields (e.g., graphics [4, 5] and biomechanics [6]). For example, Raibert et al. [1], developed numerous walking models (i.e., skipping and running) based on energy preserving considerations, while Kazerooni et al. [7] exploited a weight-force ration to influence the control of lower body using pneumatic actuators. While we focus on the active joint control for steering and navigating the lower body of an articulated skeleton to accomplished coordinated balanced stepping motions, there are other interesting sub-fields that explore related areas, such as the inverted pendulum model and its numerous flavours (e.g., spring loaded inverted pendulum) for creating responsive stepping data (i.e., location of the centre-of-mass (COM) and pin-point foot location) [8, 9]. We must be reminded, as Ryave and Schenkein explains [10], that we take walking for granted, forgetting the navigational challenges involved in traversing complex terrain, such as steps, whilst avoiding collisions with objects and keeping a controlled pace.

17.2 Overview

17.2.1 Lower Body

Each leg is represented by four limbs (i.e., pelvis, foot, lower-leg, and upper-leg) and is controlled by local actuators as shown in Fig. 17.1. We model the degrees of freedom (DOF) based upon human anatomy (e.g., the knee has one degree of freedom) giving five DOF in total for each leg (i.e., ankles and hips have 2-DOF while the knees have 1-DOF). The low-dimensional skeleton has a reduced number of DOF to simplify the problem ambiguity while allowing enough flexibility to navigate complex terrain (e.g., see Fig. 17.2).

17.2.2 Joint Torques

The joint torque actuation is derived based on a Jacobian transpose methodology (i.e., virtual work displacement [11, 12]). Whereby, we analyse the current model's parameters (e.g., foot position, COM, momentum, and desired target) to create specific motion trajectories to feed into the joint controller to produce the necessary

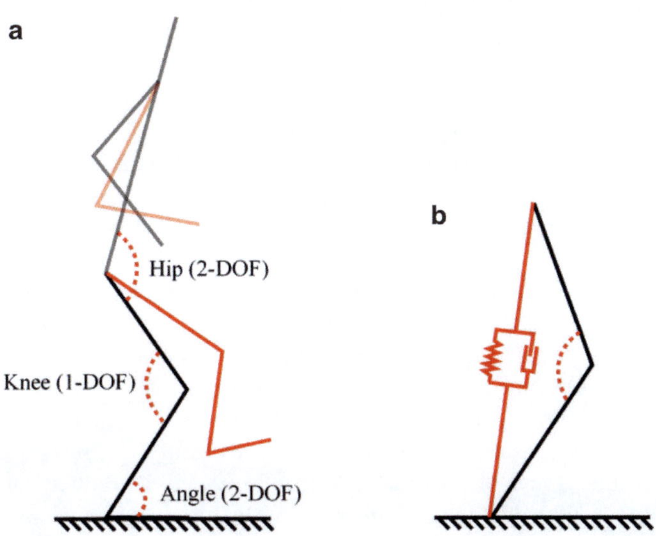

Fig. 17.1 Lower body structure—(**a**) general interconnected rigid body model for an articulated biped stepping controller; (**b**) hip and the centre-of-pressure form a low-dimensional model composed of the ground contact point and pelvis. The hip and ground contact point form a spring-damper analogy (i.e., force between the hip and ground can be converted to a torque for the knee joint)

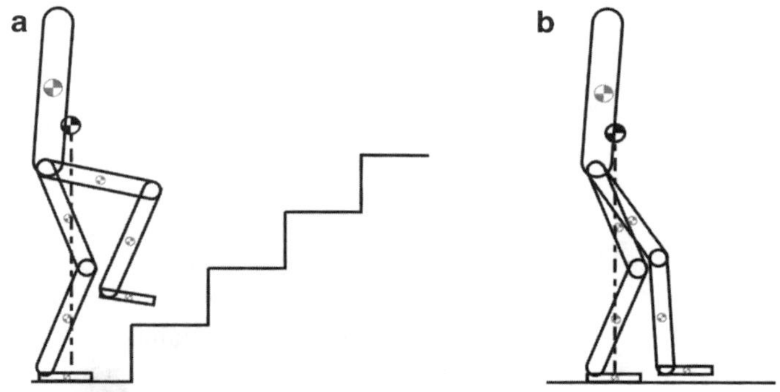

Fig. 17.2 Statically balanced stepping—(**a**) climbing a set of steps, and (**b**) walking on flat terrain. The mass of the individual limbs sum up to provide the overall centre-of-mass, which is essential for statically balanced stepping transitions

control torques. For example, the ankle and hip joints have the ability to steer, while the knee joint is limited to a single DOF (i.e., height control). We are able to produce both roll and pitch motions with the hip and ankle joint. In particular, the combined joints are highly coupled and need to work in unison to actuate the articulated model to perform directed effortless motions. This will be explained in detail in later sections as we expand upon the kinematic details of the model.

17.2.3 Balance

For models with telescopic-legs and pin-point feet (i.e., feet with no support area), the model must constantly keep stepping to remain upright and balanced. However, we focus on a model with feet, which provide a support area to create controlled statically balanced motions. Whereby, as long as the overall COM remains above the support region the biped does not need to worry about falling over. We maintain balance by manipulating the ground reaction forces (GRFs) (i.e., manipulating the centre-of-pressure COP) of the stance foot (or feet) during stepping transitions to achieve stable, controlled locomotion. While we focus on statically balanced motions (i.e., movements with low momentum where the overall COM remains above the support region), it should be noted that if we shift the centre-of-pressure around the support area (i.e., foot or feet), we can alter the body's momentum (e.g., moving the centre-of-pressure towards the toes will reduce forward body momentum while shifting it towards the back of the foot helps accelerate the body).

Clarification: keeping the overall mechanism's COM above the support region does not guarantee the stability for dynamic situations—e.g., acceleration and deceleration of the overall mass causes momentum, which would cause instability issues, such as unbalancing the mechanism. Hence, we must stress low momentum situations (i.e., motions with minimum amount of energy, such as, walking and stepping). The principle gives a fundamental grounding that can be embellished with other techniques, such as the inverted pendulum, to form a computationally efficient and algorithmically uncomplicated concept.

17.2.4 Dynamic Balance

Dynamic balancing is not required to return to a statically balanced state at any point during motion. Where "dynamic balancing" is sometimes referred to as "actively balancing", since during "dynamic" movement the control system must constantly take actions to keep the body from falling over. In effect, dynamic balancing is achieved by shifting the body into a state of a continuous controlled fall.

17.2.5 Quasi-dynamic Balance

Quasi-dynamic control solutions, as we propose in this chapter, attempt to solve dynamic problems using static system approximations. They provide extra flexibility over basic static balancing solution with the ability to break the rule of always being continuously statically balanced. However, they can require long static periods of time to recover from balance disturbances. The solutions are not "truly" physically correct.

17.3 Static and Dynamic Walking

If the walking motion is done at slow speeds it is referred to as "static walking", while at the speed of a typical human walk or faster it is referred to as "dynamic walking" [13] as shown in Fig. 17.3.

17.3.1 Static Walking

Static walking is known as *slow walking*. During foot support transitions the COM is always within the foot support area. That is, while the next foot is being placed at a new location the COM remains with the support foot region. Only once the new foot has been placed does the COM move towards the newly placed foot (staying within the foot support region of both feet). The dynamics of the body does not help the stability since the COM remains within the foot support area.

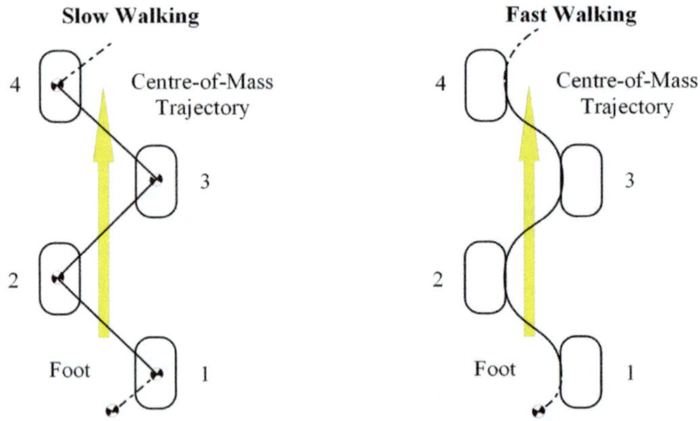

Fig. 17.3 Static and dynamic walking—during slow walking, the body's centre-of-mass (COM) always remains within the foot support area, while when the body walks at a smooth fast rate the COM is not always within the foot support area

17.3.2 Dynamics Walking

Dynamic walking is known as "fast walking". During step transitions the COM is **not** inside the foot support; however, the zero moment point (ZMP) must be inside the foot support region [14, 15].

17.4 Motion Kinematics

Environmental sensors (e.g., foot position, COM, and target direction), are used to generate joint trajectories that control the articulated model. At any moment, we can extract the character model's status, such as the current and desired joint angles. We feedback into the algorithm to derive control articulations for the desired action. A number of interesting approaches based upon this principle have already been proposed, such as Kajita et al. [16] and Muscato et al. [17]. The geometrical structure is fixed and enables us to formulate a Jacobian matrix. The Jacobian control matrix is inverted to derive control parameters.

$$\dot{x} = J\dot{q} \tag{17.1}$$

where J is the Jacobian matrix based on the link transforms, $\dot{q}$ are the joint rate of change, and $\dot{x}$ are the Cartesian links rate of change. Once the Jacobian matrix is formulated it can be inverted to feedback kinematic information [18]. Similarly, a Jacobian matrix is formulated for the knee structure.

$$\dot{q} = J^{-1}\dot{x} \tag{17.2}$$

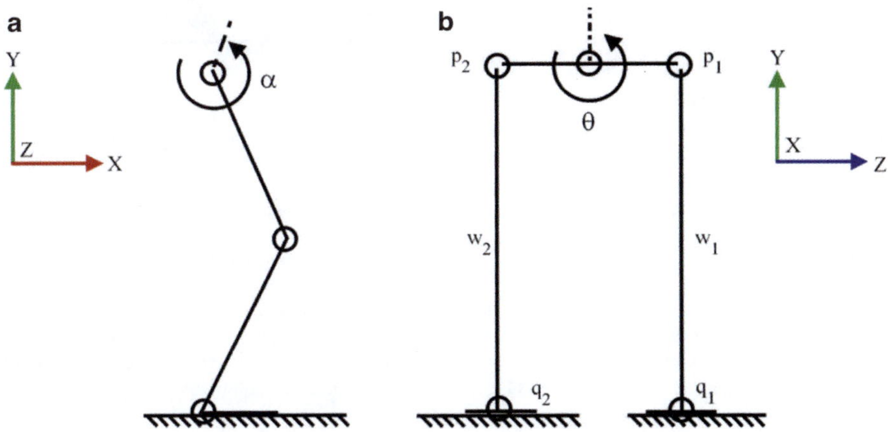

Fig. 17.4 Parameters—simplified articulated geometric lower body structure (i.e., hip and ankle) joint for deriving a set of control equations

with q representing the knee position with reference to the ankle joint shown in Fig. 17.4. The kinematic analysis leads to the derivation of the relationship between forces and joint torques that are fed to the control actuators. The virtual work principle is applied to the Jacobian transpose [11] to give the relationship between Cartesian forces and joint torques:

$$\tau = J^{\mathrm{T}} F \tag{17.3}$$

where F are the Cartesian forces in the link coordinate frame, τ are the static joint torques, and J defines the Jacobian matrix. The Jacobian matrix is calculated iteratively each frame using the current skeleton's configuration [19] (i.e., the linked manipulator formation).

17.5 Control Architecture

We explain an interconnected framework to manage the control of the biped movements at various stages (e.g., stepping transitions). We use a state machine logic to decide on the current and next state of action. A feedback loop continuously monitors the status of the model and controls the generated trajectories (e.g., interpolating end-effectors, such as the feet, between the current and target location over a specified duration). The foot transitions are defined using a set of parabolic trajectories to create arc-like swinging-leg motion and can be formulated using a

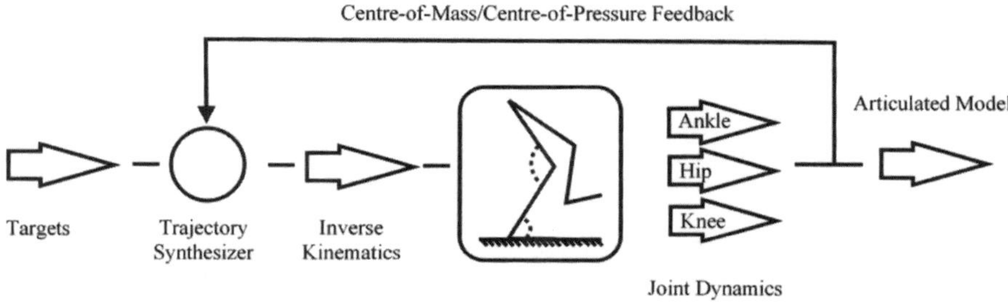

Fig. 17.5 Control framework—the interconnected framework demonstrate how the model functions to create the final motions. The trajectory information (e.g., step-height and size) are fed into the trajectory synthesizer in conjunction with feedback from the current articulated model to formulate the desired skeleton pose. The desired pose is used to calculate the necessary joint torques to feed the articulated skeleton and create the final motion

set of geometric equations [20, 21]. As shown below in Fig. 17.5, the framework is decomposed into manageable components (e.g., inverse kinematics and joint dynamics).

The low-dimensional model of the hip and ground contact point forms a fundamental underpinning of the biped stepping mechanism. The Jacobian matrix is calculated according to the posture of the biped and the centre-of-pressure position of the foot. The external virtual forces, i.e., F_{ext}, provide a feedback vector for the support leg during step transitions. The external forces are integrated into the model so that we can compensate through posture and foot reaction forces to accomplish the desired motion.

17.6 Simulation Considerations

There are a variety of open source and commercial physics-based simulation packages available for constructing articulated rigid body skeletons (e.g., collisions, revolute joints, and rigid body mechanics). A popular and well-known dynamic simulator is the Open Dynamics Engine [22, 23]. For the low-dimensional model presented in this chapter, the computational cost and memory overhead should be minimal enabling simulations to be run in virtually real-time. However, the concept is scalable and can be applied to more complex avatars with a greater number of DOF, but introduces more ambiguity, singularities, and the possibility of producing unnatural looking motions (e.g., Monty Python's famous 'ministry of silly walks' sketch).

17.7 Conclusion

This chapter explained an uncomplicated approach for creating active joint torques to synthesize autonomous lower-body motions that remain stable during continuous locomotion (e.g., compared to penalty-based methods, such as angular joint springs). The algorithm is derived and implemented without artistic intervention (e.g., emotion or style), hence the walking patterns lacked human personality. Future work would be the extrapolation of behavioural walking parameters from real-world motion capture data for injection and control into the algorithm (e.g., mixing pre-recorded trajectory motion patterns).

References

1. Raibert, M.H., et al.: Legged Robots That Balance, vol. 3. MIT press, Cambridge (1986)
2. Niiyama, R., Nishikawa, S., Kuniyoshi, Y.: Athlete robot with applied human muscle activation patterns for bipedal running. In: 2010 10th IEEE-RAS International Conference on Humanoid Robots (Humanoids), pp. 498–503. IEEE, New York (2010)
3. Gienger, M., Loffler, K., Pfeiffer, F.: Towards the design of a biped jogging robot. In: Proceedings of 2001 ICRA IEEE International Conference on Robotics and Automation, vol. 4, pp. 4140–4145. IEEE, New York (2001)
4. Abe, Y., Popović, J.: Interactive animation of dynamic manipulation. In: Proceedings of the 2006 ACM SIGGRAPH/Eurographics Symposium on Computer Animation, pp. 195–204. Eurographics Association (2006)
5. Fang, A.C., Pollard, N.S.: Efficient synthesis of physically valid human motion. In: ACM Transactions on Graphics (TOG), vol. 22, pp. 417–426. ACM, New York (2003)
6. Rodriguez, I., Boulic, R., Meziat, D.: A joint-level model of fatigue for the postural control of virtual humans. In: Proceedings of the 5th International Conference on Human and Computer (HC02), no. VRLAB-CONF-2007-040 (2002)
7. Kazerooni, H., Racine, J.-L., Huang, L., Steger, R.: On the control of the berkeley lower extremity exoskeleton (bleex). In: Proceedings of the 2005 IEEE International Conference on Robotics and Automation, 2005 (ICRA 2005), pp. 4353–4360 (2005)
8. Kenwright, B.: Responsive biped character stepping: When push comes to shove. In: 2012 International Conference on Cyberworlds (CW), pp. 151–156. IEEE, New York (2012)
9. Kenwright, B., Davison, R., Morgan, G.: Dynamic balancing and walking for real-time 3d characters. In: Motion in Games, pp. 63–73. Springer, New York (2011)
10. Ryave, A.L., Schenkein, J.N.: Notes on the art of walking. Ethnomethodology, 265–274 (1974)
11. Pratt, J., Chew, C.-M., Torres, A., Dilworth, P., Pratt, G.: Virtual model control: An intuitive approach for bipedal locomotion. Int. J. Robot. Res. **20**(2), 129–143 (2001)
12. Kenwright, B.: Joint-torque control of character motions. Technical Report, 2013
13. Honda, M.C.I.: Inc. corporate affairs and communications american honda motor co. technical information. Technical Report, January 2003
14. Saavedra, M.R.A.: Stable locomotion of humanoid robots based on mass concentrated model. Ph.D. Thesis, Department of System and Automation Engineering, Universidad Carlos III De Madrid (October 2008)
15. Vukobratović, M., Borovac, B.: Zero-moment point' thirty five years of its life. Int. J. Humanoid Robot. **1**(01), 157–173 (2004)

16. Kajita, S., Matsumoto, O., Saigo, M.: Real-time 3d walking pattern generation for a biped robot with telescopic legs. In: Proceedings of 2001 ICRA IEEE International Conference on Robotics and Automation, vol. 3, pp. 2299–2306. IEEE, New York (2001)
17. Muscato, G., Spampinato, G.: Kinematical model and control architecture for a human inspired five dof robotic leg. Mechatronics **17**(1), 45–63 (2007)
18. Sciavicco, L., Siciliano, B.: Modelling and Control of Robot Manipulators. Springer, Berlin (2000)
19. Kenwright, B.: Synthesizing balancing character motions. In: VRIPHYS, pp. 87–96 (2012)
20. Saibene, F., Minetti, A.E.: Biomechanical and physiological aspects of legged locomotion in humans. Eur. J. Appl. Phys. **88**(4–5), 297–316 (2003)
21. Shih, C.-L., Gruver, W.A., Lee, T.-T.: Inverse kinematics and inverse dynamics for control of a biped walking machine. J. Robot. Syst. **10**(4), 531–555 (1993)
22. Smith, R., et al.: Open dynamics engine (2005)
23. Kenwright, B., Morgan, G.: Practical introduction to rigid body linear complementary problem (lcp) constraint solvers. In: Algorithmic and Architectural Gaming Design, pp. 159–205 (2012)

Chapter 18
Dynamic Balancing of Mobile Robots in Simulation and Real Environments

Adrian Boeing and Thomas Bräunl

Abstract Transferring an evolved control system from a simulated environment to the physical world poses a number of challenges. One of the most challenging control tasks is to generate a stable walking gait for a bipedal robot. We describe a method using a combination of repetitive splines and genetic algorithms to evolve a simulated control system for a humanoid robot, which is subsequently transferred to a real robot hardware. Multiple dynamic simulation systems can be simultaneously employed to provide a valid range of simulation variance. This will result in a much smaller reality gap and ultimately in a more robust control algorithm for the real robot.

Keywords Mobile robot • Biped walking • Dynamic balance • Robot simulation • Genetic algorithms • Spline encoding

18.1 Introduction

Historically robots have been limited by simple control and mechanical designs. Increasingly complex problems are being solved by increasingly complex robotic systems interacting with varied environments. Automated design techniques can assist robot designers providing an optimized solution to a problem, thereby alleviating some of the difficulties in understanding the ever-increasing robotic system complexity [1].

Computer simulations provide a number of advantages during system design and experimentation including decreased cost, rapid prototyping, increased ability to observe and alter the experiment, and ability to enable more sophisticated automated design process [2]. The key disadvantage of simulation-based approaches to physical experimentation is in creating an adequate mathematical model that accurately describes the physical reality of the system [3].

A. Boeing (✉) • T. Bräunl
The University of Western Australia, Crawley, WA, Australia
e-mail: adrian.boeing@uwa.edu.au; thomas.braunl@uwa.edu.au

D. Zhang, B. Wei (eds.), *Dynamic Balancing of Mechanisms and Synthesizing of Parallel Robots*, DOI 10.1007/978-3-319-17683-3_18

Transferring a control system designed in a simulated environment to the physical world poses a number of challenges. The difficulty in accomplishing such a task increases as the complexity of the system that is being simulated and controlled increases. It is generally accepted that the cause of the failures in the transfer of control systems from a simulation to the real world lies in the imperfections of the simulated representation of the real world [4].

Despite imperfect simulations, several evolutionary simulation experiments for a restricted class of robot and environments have been successfully validated on real robots. However, Nolfi and Folreano [2] state:

> It is not clear if current simulation methods can scale up to significantly more complex cases.

Thus, there is a need for an approach that will allow the simulation of complex robots and environments that produces a control system that is capable of successfully transferring from a simulated environment to the real world.

18.2 The Reality Gap

Brooks [5] states that one of the fundamental reasons for avoiding robot simulations is that there is a great danger that the simulations will not match the real world. This difference between the simulated and real world is sometimes referred to as the reality gap. Despite this, there are a number of examples of successful evolution of controllers using simulations.

A number of approaches have been proposed for "crossing the reality gap" [6]. Nolfi and Folreano [2] outlined the problems associated with crossing the reality gap for traditional simulation systems and identified the modeling of the sensor and actuator behavior as well as the robot body and environment characteristics as the key difficulties.

An early approach to solving the difficulty in modeling the robot behavior was recording extensive data sets for each sensor [7]. There have been claims that the empirical measurements, whilst quite extensive, were still too coarse [4]. Nevertheless, individuals evolved in simulations based on this technique continue to perform satisfactorily when transferred to the real environment [2].

Despite this, the approach is not very applicable to complex robots due to the exponentially increased number of situations that must be sampled for complex interactions. To alleviate this problem mathematical models of the sensor and actuator behavior based on known engineering concepts can be constructed instead. To reduce the problems associated with the uncertainty in sensor readings and actuator commands, noise can be introduced into the simulation at all levels.

Jakobi [6] proposed an alternative method named "minimal simulation." Jakobi et al. [4] demonstrated that if the noise model is significantly different from the real system then the controller is less likely to work when transferred to the real world. Furthermore it was noted that controllers evolved in simulations would come to depend on particular aspects only available in the simulation and hence fail in reality.

The minimal simulation approach attempted to reduce the differences between simulation and reality by only simulating the aspects of the robot and its environment that were critical to the success of the control system. These critical aspects (also known as "base-set" aspects) are reliably simulated and the aspects deemed to be non-critical (or "implementation-related" aspects) are varied for each trial to be unreliable. As a result, controllers only evolve to depend on the reliable aspects of the system.

Jakobi successfully applied the minimal simulation method to a number of robots, including an octopod robot. This approach was also successfully applied to a quadruped robot by Hornby et al. [8].

Whilst the minimal simulation approach allowed for a range of robot controllers to cross the reality gap, it does require the designer to assess the problem task to precisely and accurately identify the reliable, valid behavior of the system and build a custom simulation system that will only allow those behaviors. Nolfi and Folreano [2] illustrate cases where problems that are decomposed by Jakobi into base-set and implementation-related aspects eliminate the opportunity for some evolvable solutions. This indicates the difficulty of correctly identifying valid base-set features for any robotics problem, including relatively simple robots and robot tasks, such as two-wheeled mobile robot maze navigation. Furthermore, in the case of complex robot-environment interactions such as an underwater vehicle or developing a dynamic gait for a bipedal robot the base-set features may be difficult to identify. As a result, there may be very few implementation-related aspects removing the advantages of minimal simulation over a more traditional high-fidelity simulation-based approach.

Zagal et al. [9] proposed a hybrid simulation and real-world architecture named "back to reality" (BTR). The key feature of the approach that minimizes the effect of the reality gap is the coevolution of the simulation model with reality. BTR consists of three learning algorithms: one for evolving the simulated controller, another for evolving the physical robot controller, and finally a learning algorithm for modifying the simulation model to better fit the real-world data [10]. For the evolution of the robot (in both simulation and reality) the experimenter provides a fitness function indicating the ability of the controller to achieve the desired task. The simulation model is evolved based on the average fitness from both simulation and reality, relative to just the real robot fitness value. In this way, the discrepancies between the simulator and the real world are slowly minimized until a controller can successfully cross the reality gap.

The BTR method was successfully applied for evolving a ball kicking behavior for a quadruped robot. The evolved control system was successfully transferred to the real robot hardware, making this one of the few approaches to successfully cross the reality gap for a complex robotic system.

Another hybrid approach is the estimation-exploration algorithm (EEA) [11]. The EEA defines operators for the input space as well as a similarity metric for the output space. Given an approximate model of the target system, the EEA enters an exploration phase, in which the tests are evolved to determine the best test for the system. The next phase is the estimation phase, in which the models are evolved that best explain the correlation between the inputs and outputs of the system. These two phases are repeated until it converges to a solution. This approach was applied to a four-legged robot, and the EEA was able to evolve a simulation model representing the real robot.

Hybrid simulation and hardware evolution approaches have been demonstrated to successfully cross the reality gap for complex robotic systems and tasks. However, this approach requires the fully constructed robot hardware. This may not be available when evaluating a robot design or it may not be practical or too expensive to construct the appropriate physical test harness.

18.3 Evolution with Multiple Simulators

The process for the traditional simulation approach begins with the construction of an accurate model of the robot dynamics, the environment, and empirically based sensor and actuator models. For an evolutionary controller design, the process then is:

1. Initialize a set of potential controller designs.
2. Evaluate each design in the simulator.
3. Assign a fitness value indicating how well the design solves the desired task.
4. Use an evolutionary algorithm to generate a new set of controller designs.
5. Return to Step 2, unless the task is solved.

Our approach extends this process by evaluating the design not just for a single simulator, but rather on multiple simulators. Typically this would require reconstructing the robot and environment model each time for each simulation system. In order to remove this demand a simulation abstraction system is required that can transform a single system representation to a valid representation for various simulators and provide a single programming interface [28, 29].

Having multiple simulators alters the evolutionary design process:

1. Initialize a set of potential controller designs.
2. Evaluate each design in a set of each of the simulators.
3. Use statistical methods to assign a fitness value indicating how well the design solves the desired task across all simulators.
4. Use an evolutionary algorithm to generate a new set of controller designs.
5. Return to Step 2, unless the task is solved within a confidence interval, for all of the simulators.

In this way, it is hypothesized that the evaluation of each controller by multiple simulators will provide a valid source of noise to the robot dynamics, similar to that provided by adding a source of noise to the sensor and actuator models. A second hypothesis is that the transfer of an evolved control system from one simulator to a different simulator will simulate the process of transferring a control system from one simulator to the real world. This should cause the evolved controller to be robust enough to provide acceptable behavior when transferred to the real world.

There are a number of factors that influence the characteristics of a dynamics simulation. These range from the simulation paradigm, collision detection, and response to the type of numerical integrator, to whether air resistance is considered. As a result each dynamic simulation package will provide quite different results despite stimulating the exact same system.

Boeing and Bräunl [12] published an extensive comparison of various simple test cases and found significant discrepancies between the performance of each simulator, primarily due to the design choices implemented in the dynamic simulation packages.

18.4 Biped Locomotion

Evolving control systems for robot locomotion and balancing is becoming a standard approach for the generation of improved or newer control systems for robots [6, 7, 13–15]. There have been a number of successful demonstrations of legged robot control being transferred from a simulated environment to a physical environment. Satisfactory results for quadruped, hexapod, and octopod robots have been obtained [6, 8, 14]; however results for bipedal robots have not been generally satisfactory [16], often resulting in shuffling movements rather than walking motions. This chapter presents a simple control system and describes a method to overcome some of the difficulties encountered when making the transition from a simulated world into the real world.

If a physically accurate simulation model can be constructed for a robot, then a number of advantages for robot development present themselves. Physical simulation of the robot allows a robot designer to prototype and visualize a robot's design without requiring physical construction. The simulation also simplifies the task of evolving a control system for the robot. Thus, the workload on the designer is reduced and robot walking motions can be tested and optimized at an early design stage allowing the designer to modify the design, if necessary.

The target hardware for the controller is a small humanoid robot called *Andy Droid* (see Fig. 18.1) [17]. Cost and weight were important design considerations in Andy's development. As a result, Andy stands approximately 350 mm tall, and weighs around 1,400 g. Andy has 10 degrees of freedom in his legs, and each joint is powered by a Hi-Tec 945 MG servo with links made from 3 mm thick aluminum plate. These connections result in a substantial amount of inherent flexibility.

Andy can be equipped with a number of sensors, including a color camera, PSDs, inclinometers, gyroscopes, and pressure sensors. The pressure sensors are permanently mounted as Andy's feet, which are constructed from three metal "toes." Each toe has two strain gauges that are used to produce a voltage in proportion to the applied force.

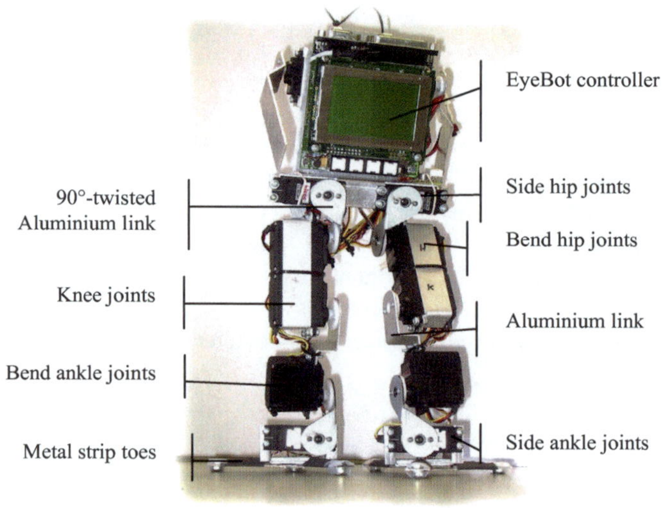

Fig. 18.1 Andy Droid humanoid robot

The biped's processing requirements are provided by an EyeBot MK3 controller with an LCD display, user input buttons, parallel and serial ports, as well as digital and analog inputs as well as digital outputs [17].

18.5 Simulation

18.5.1 Physics Simulation

There are a number of advantages in using a simulated environment to evolve robot controllers. Simulations eliminate the risk of damaging robot hardware and other hardware-related concerns, such as battery power and temperature effects. Simulation also provides the added convenience and freedom to manipulate any force or environment variable to suit the situation. This aids greatly in general experimentation, and in resetting the robot to an identical initial position for each evolution trial. Simulations typically execute faster than control programs executed on robot hardware, and information from the motions is easier to extract. These factors make simulations an attractive option.

The simulation tool used to simulate the robot was the freely available DynaMechs library [18]. DynaMechs is an efficient rigid-body dynamic simulation library that is based on the Articulated Body Algorithm (ABA) developed by Featherstone [19]. The robot's structure is defined using multiple chains, starting with a mobile base with each link described in terms of the previous link using modified Denavit-Hartenberg parameters [20].

The ABA is based on the observation that the accelerations of bodies in a rigid-body system are always linear functions of the applied forces. Initially the velocities of each joint are calculated by working from the base link to the terminal links. The Articulated Body Inertia (ABI) matrix can then be calculated by traversing back from the terminal links to the base link.

Extensions to the simulation package were made to include the fitness evaluation functions required to operate the genetic algorithm, as well as the simulated servo and sensor models. The simulated control system was also added to the model.

18.5.2 Robot Model

A schematic of the robot legs is illustrated in Fig. 18.2. The robot model required by DynaMechs was constructed using the RobotBuilder [21] package. For each link, DynaMechs requires information including its relative position and orientation, mass, center of gravity, and inertia matrix.

RobotModeler (part of the RobotBuilder package) allows the use of primitive shapes such as cubes and spheres to approximate the physical shape of each link and subsequently allows calculation of the inertia matrix of each link. The center of gravity was estimated using a similar method.

To model the inherent flexibility in Andy's toes, an extra joint was added to each toe. The flexibility in Andy's toes is a result of the steel springs that are used for pressure sensing. The flexibility was replicated using a rotational joint with very

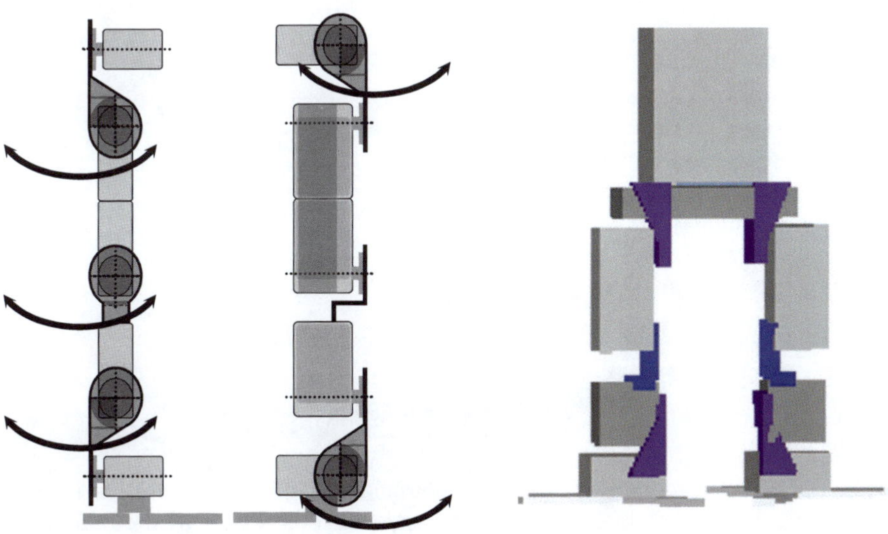

Fig. 18.2 Leg schematics and simulator model

small joint limits and a large friction value. Once the joint moves outside its limits, a spring restoring force is applied, mimicking the memory effect of the steel spring.

An accurate model of the torque produced by the servomotors is critical if the walking gaits produced by the mechanical simulator are to provide a reasonable approximation to the gaits produced in the real world. A servomotor comprises essentially two components, a control system, and a DC motor. The DC motor torque can be mathematically modeled using the standard armature-controlled DC motor model [22], represented by (Eq. 18.1). The control system for a servo is generally a proportional-integral-derivative (PID) controller. The controller can be mathematically simplified by ignoring the integral and derivative terms, since the proportional term dominates its behavior. Incorporating the DC motor model into the servo P-controller and using $\theta_\varepsilon = \theta_{\text{output}} - \theta_{\text{input}}$ for the error signal give (Eq. 18.2).

Considering the case where the armature is stationary ($\omega_{\text{n}} = 0$) and the maximum supply voltage is applied to the armature ($V_{\text{a}} = 4.8$ V) allows us to determine:

Equation 1: Armature-Controlled DC Motor Torque Equation [22]:

$$T_A(t) = K_T \frac{V_a(t) - K_b \omega_n(t)}{R_a} \tag{18.1}$$

Equation 2: Torque Equation for Servo Control System [23]:

$$T_A = \frac{N \cdot K_T}{R_a} \left(K_a K_\varepsilon \theta_\varepsilon(t) - K_b \omega_n(t) \right) \tag{18.2}$$

where:

N—gear reduction
R_a—armature resistance (Ω)
K_T—motor torque constant (Nm/A)
K_a—power amplifier gain
K_b—proportional gain for error signal
K_ε—proportional gain for error signal
ω_n—angular velocity of motor (rad/s)
V_a—applied armature voltage (V)
θ_ε—angular error signal ($\theta_{\text{output}} - \theta_{\text{input}}$)

Equation 3: *Stall Torque Test*:

$$\frac{N \cdot K_T}{R_A} = \frac{T_{A,stall}}{V_{a,max}} = 0.180\ \text{N.m/V} \tag{18.3}$$

When the motor is at top speed the applied armature voltage equals the back EMF, i.e., $V_{\text{a}}(t) = K_{\text{e}} \times w_{n,\max}(t)$. From the servo specifications we know the maximum angular velocity, allowing us to solve the motor back EMF constant:

Equation 4: *Maximum Speed Test*:

$$K_e = \frac{V_{a,\max}}{\omega_{n,\max}} = 0.733 \text{ V s/rad} \tag{18.4}$$

Equation 5: *Servo Torque Equation*:

$$T_A = 0.180\left(V_a(t) - 0.733 \cdot \omega_n(t)\right) \tag{18.5}$$

The proportional component of the controller was modeled with (Eq. 18.6). The model assumes that the maximum supply voltage is applied to the motor until it gets within a tolerance of the desired angle. The voltage applied to the motor is then linearly decreased until the servo reaches its final destination.

Equation 6: *Armature Voltage, P-Controller Model*:

$$V_a = \begin{cases} K_\varepsilon \theta_\varepsilon(t), & K_\varepsilon \theta_\varepsilon(t) < V_{a,max} \\ V_{a,max}, & otherwise \end{cases} \tag{18.6}$$

This model performed adequately for large movements; however, it was found that for small angle movements, where the maximum armature voltage was not achieved, the servo model was not accurate since the full stall torque is not applied. To overcome this, it is assumed that the maximum supply voltage is always applied to the armature. This is a reasonable assumption since the slowing down of the servo has only a minor effect on its time response.

The deadband specification of the servo was used to decide when the servo model had reached its target angle. Once the servo is decreed to have reached its destination, a torque is no longer applied to the joint. This is shown in (Eq. 18.7).

***Equation 7: Servo Deadband Model*:**

$$T = \begin{cases} 0, & \left|\theta_{current} - \theta_{target}\right| < \frac{\theta_{deadband}}{2} \\ T_A, & otherwise \end{cases} \tag{18.7}$$

18.6 Control System

A large variety of different control strategies have been devised for balancing and locomotion of biped walking robots. One of the most popular methods is the zero-moment-point (ZMP) method as introduced by [24] and described in [25], [26], or [27]. ZMP is a method of maintaining dynamic balance, so in some sense it can be seen as the dynamic equivalent of the static center of mass (CM). For static balance the CM has to stay within the robot's support area, which is the convex hull of all its contact points (feet) with the ground. The ZMP method uses pressure sensors or strain gauges in a legged robot's feet, whose data feed into the kinematic model of the robot, considering its current body attitude and weight distribution, in order to calculate the ZMP. If the ZMP falls outside a robot's support area, the robot is dynamically unstable and will fall over.

We are not following any of these methods. Instead we will be using genetic algorithms (GA) to evolve suitable control parameters for a spline-based set of motion trajectories for each of the robot's joints. This can be done with or without (simulated and real) sensors for the robot. A gait generated without sensors will be much simpler to evolve; however, it will not be as stable as a gait with sensors, as this will not allow the robot to react to any changes in the walking terrain, such as slopes or obstacles, or any other small disturbances.

In the following we represent any robot movement through a set of spline curves, one for each of the robot's joints. A spline-based control system is then responsible for manipulating all of the robot's servo inputs [13]. The spline controller comprises a set of connected Hermite splines. Each spline can be defined by a variable number of control points allowing variable degrees of freedom. The function used to interpolate the control points, given starting point p_1, ending point p_2, tangent values t_1 and t_2, and interpolation point s, is shown below:

***Equation 8: Hermite Splines*:**

$$f(s) = h_1 \cdot p_1 + h_2 \cdot p_2 + h_3 \cdot t_1 + h_4 \cdot t_2 \tag{18.8}$$

where:

$$h_1 = 2s3 - 2s2 + 1$$

$$h_2 = -2s3 + 3s2$$

$$h_3 = s3 - 2s2 + s$$

$$h_4 = s3 - s2$$

Three connected splines are combined to form the overall control structure for one servo. The three splines are responsible for three different phases of the robot's walk. The initial phase of the walk is considered to be responsible for moving the robot from a stationary position into the walking motion. The servo inputs for this phase are represented by the start spline. The repeated motions that sustain the walk correspond to the cyclic spline. And finally, the end spline is used to move the robot safely back to a stationary position.

The advantages of a Hermite spline controller are the following:

- It can be represented by a compact chromosome which aids the genetic algorithm convergence speed.
- Relatively computationally inexpensive and hence can execute comfortably on the EyeBot platform.
- Joint positions and velocities are always continuous.

The spline controller used to control Andy's movements contains eight control points for each cycle and has a cycle time of 3 s. The end section of the spline controller is discarded, and the start spline contains only two control points. Each spline has an associated hardcoded offset value and an amplitude modifier. This allows the masking of the effects of motor wear and enables some compensation between the small differences between the individual servos mounted on the biped robot.

18.7 Genetic Algorithms

18.7.1 Encoding

Success of evolved spline control systems has already been demonstrated for various legged robot configurations for a wide range of activities from walking to jumping [13]. The spline controller can be directly encoded with each joint's control point parameters encoded using 8-bit fixed-point values.

Typically, all of the control point's parameters are encoded (position in time and output value, and tangent) to enable the complete description of the spline. However, to enforce the approximate solution required to overcome the reality gap, the control point locations are limited. Each control point is forced to be an equal distance from all other control points, and the tangents are forced to be zero (see Figs. 18.3 and 18.4). Thus only the servo input value at any specific time position is evolved.

Limiting the controller's output in this way stops the GA from overoptimizing the controller so as to take advantage of specific features only present in the simulation. Essentially it is this aspect that enables the robot to cross the reality gap. This approach also requires less evolution time as the resultant controller chromosome is more compact.

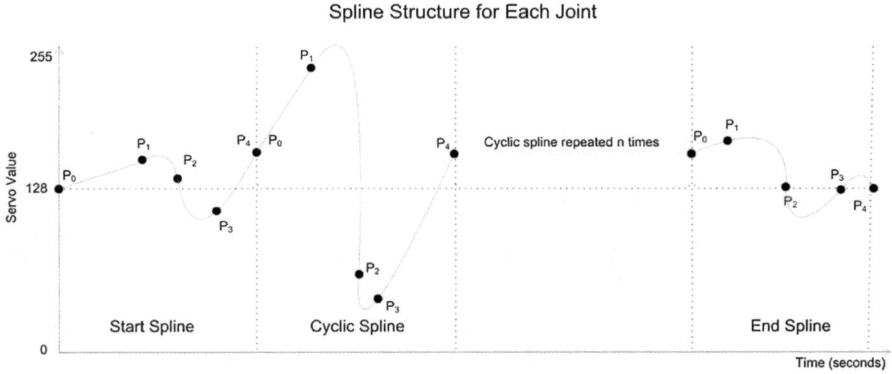

Fig. 18.3 Spline controller

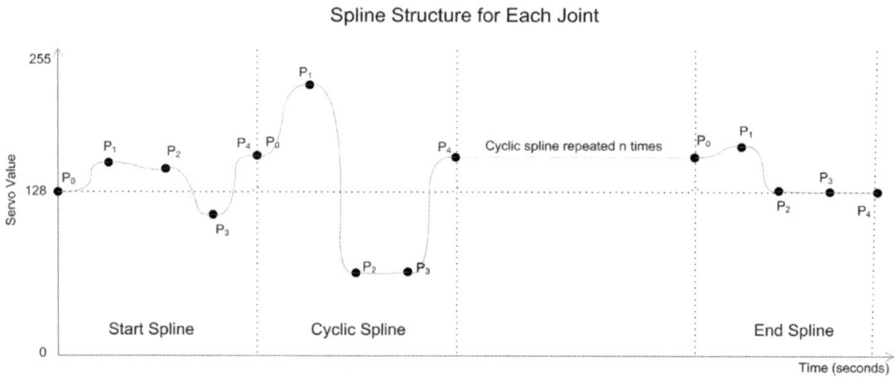

Fig. 18.4 Limited spline controller

Table 18.1 GA parameters

Operator name	Selection chance (%)
Bitwise mutate	10
Bitwise crossover	30
Byte-wise mutate	25
Byte-wise average	30
Byte-wise creep	5

18.7.2 Genetic Algorithm Configuration

The genetic algorithm employed a simple fitness proportionate selection scheme. The operators implemented were a bitwise mutate, a bitwise crossover, a byte-wise mutate, average, and creep. The byte-wise mutate replaced a randomly selected byte with a randomly generated byte value. The byte-wise creep randomly incremented or decremented a byte by 1, and the average operator generated a new chromosome from the byte-wise average of each byte in the two parent chromosomes. The parameter configurations are given in Table 18.1.

In order to evaluate the appropriateness of each gait a fitness function is employed, which returns information to the genetic algorithm about the performance of each gait. To evaluate the fitness of each robot, the function takes into consideration the forward distance of the walk, and the average velocity at which the robot's center is lowering [16].

***Equation 9: Fitness Function*:**

$$fitness = 5 \times forwards\_distance - 50 \times ave\_vel\_lowering \tag{18.9}$$

Noise and robustness for fitness functions are discussed in [30] to improve convergence. (Ref. to [30]) In order to decrease the evolution time, a terminating condition was included to the fitness function. Termination would occur if the torso (main reference point of the robot) touched the ground, i.e., the robot fell over.

18.8 Results

18.8.1 Servo Model

The simulated servo model was verified against the physical servo responses by attaching an inclinometer to the servos and recording the data generated during movement. Figures 18.5 and 18.6 show the outputs of both the simulated servos and the actual servos.

Figure 18.5 shows a close approximation between the software simulation and Andy for the ankle joint. The ankle joint has only a small amount of mass, mainly the feet, as a load. However, in Fig. 18.6, a larger discrepancy can be seen between the simulated model and the physical response. This is due to the extra load on these joints. The hip joint, in particular, has the entire leg as load. The observed overshoot in these cases is due to several factors. The first factor is due to the overshoot of the PID controller in the servo motor. The second factor is due to the flexibility of the plastic shafts of the servos. Another factor is the reaction torque of the servo inducing vibrations in the robot that tend to affect the inclinometer readings.

18.8.2 Simulation Results

The parameters of the control system can be extracted from the gene pool in a device-independent format. This allows the parameters to be either transparently imported into the simulated program via a file or downloaded to Andy's control program over a serial connection. Figures 18.7 and 18.8 illustrate the same evolved control program executing on the real and simulated Andy Droid robot.

Figure 18.7 depicts the simulated robot movements. The robot achieves locomotion by initially pressing downward with its left toes, causing the robot to tilt to its

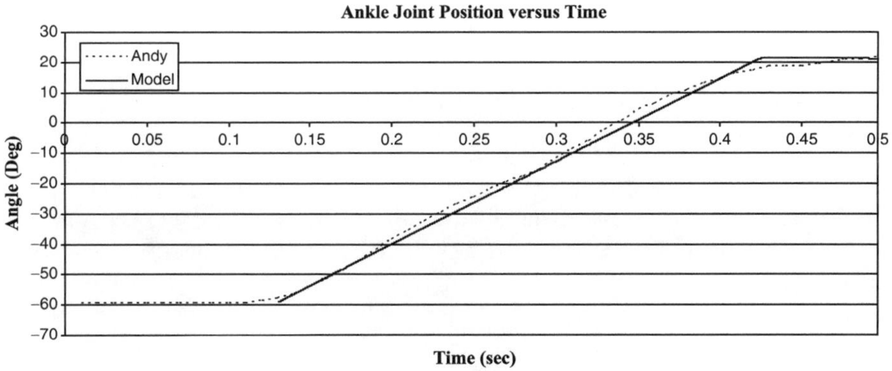

Fig. 18.5 Ankle joint response

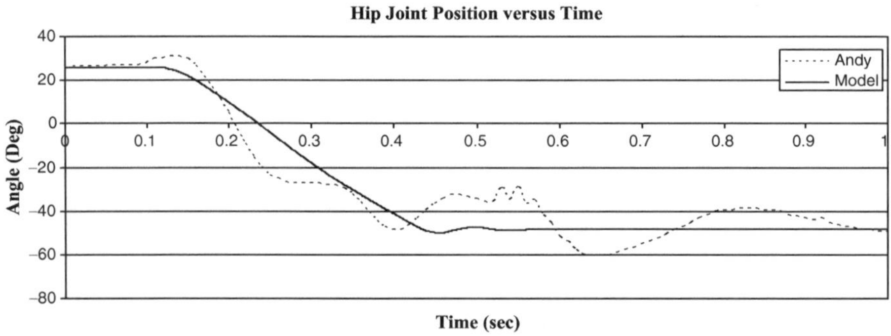

Fig. 18.6 Hip joint response

right (Fig. 7.2). The robot then drags its left foot along the ground in front of it. The robot then presses downwards with its right toes and lifts its right foot off the ground and places it in front of it. This cycle repeats itself to produce a slow forward walk.

18.8.3 Real Robot Results

The figures show a close mapping between the simulated robot and the physical robot. One significant discrepancy between the simulated and physical walks is illustrated in Figs. 7.4 and 8.4. This difference between the simulator and Andy is most likely due to worn motors, whose behavior has changed over time.

Whilst the method did result in transferable walking patterns between the simulated robot and the physical robot, the resulting locomotion still performed worse than a good manually designed gait. Furthermore, whilst almost all transferred walks allowed Andy to sustain balance and motion for continuous walking cycles, a number of factors. A number of factors hampered the performance of Andy's gait, including battery power, servo jitter, flexibility in the plastic joints, and motor wear.

18.9 Conclusions

Most problems observed in this system are believed to arise from the incomplete servo model and the shortcomings of the robot hardware. The evolved walking patterns can potentially be improved by relaxing the restrictions on the evolved controller. The servo model could be improved if the internal parameters of the servo PID controller were fully known. This can be achieved by additional in-depth testing of the servomotor. Alternatively, the real robot hardware could be redesigned to operate with DC motors with separately designed and modeled PID controllers.

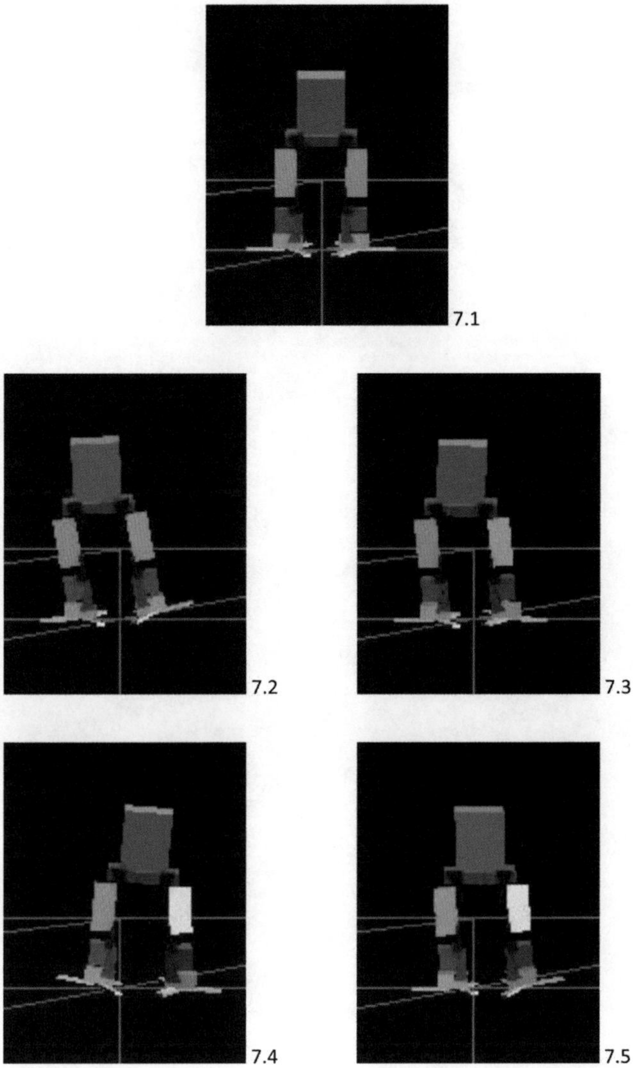

Fig. 18.7 Simulated walk

Further investigations into the encoding of the spline controller could reveal an optimal configuration enabling further optimizations of the controller, yet still limiting its complexity such as to avoid the problems related with the reality gap. Incorporating sensor feedback into the controller could also increase the robustness of the generated gait. This can then assist the controller in bridging the reality gap.

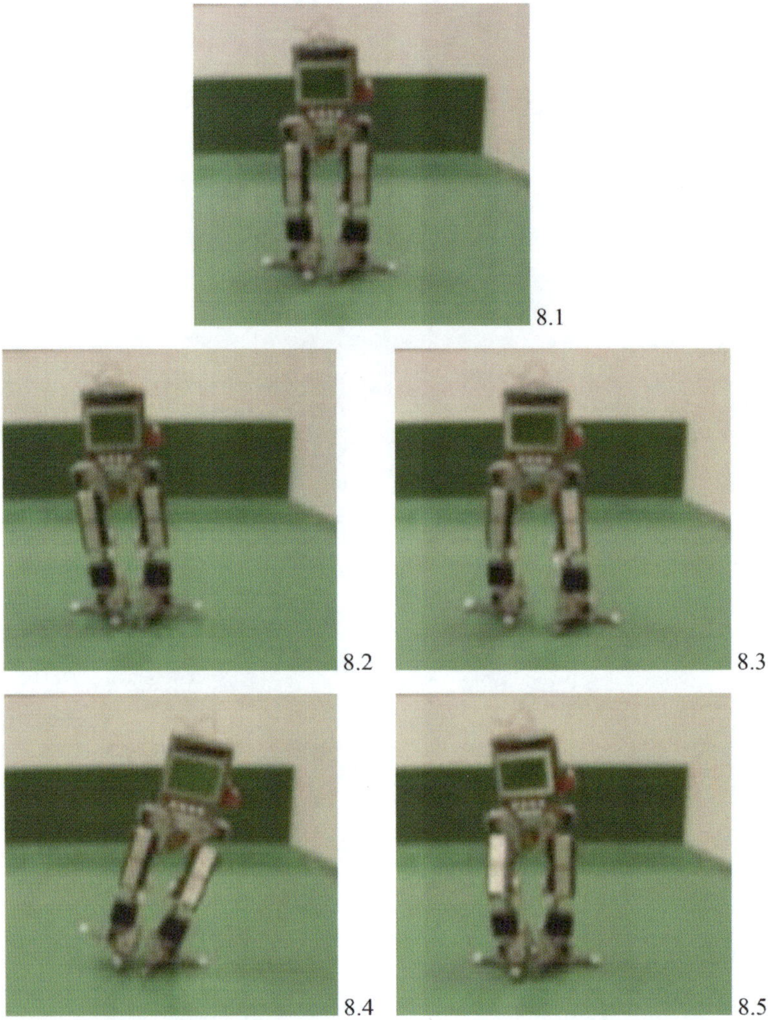

Fig. 18.8 Andy Droid robot walking

Our system has successfully managed to evolve controllers in a simulated environment that were then able to be transferred to and successfully drive to the physical biped robot. However, more research is required to generate more general solutions and to further increase robustness of the generated robot controllers.

Acknowledgements We would like to thank J. Zimmermann and S. Hanham for their contributions in software development to this project.

References

1. Leger, C.: Darwin 2k: An Evolutionary Approach to Automated Design for Robotics. Kluwer Academic Publishers, Norwell, MA (2000)
2. Nolfi, S., Folreano, D.: Evolutionary Robotics. The MIT Press, Cambridge, MA (2000)
3. Law, A.M., Kelton, D.: Simulation Modelling and Analysis, 3rd edn. McGraw-Hill, New York (2000)
4. Jacobi, N., Husbands, P., Harvey, I.: Noise and the reality gap: the use of simulation in evolutionary robotics. Proceedings of the Third European Conference on Advances in Artificial Life **929**, 704–720 (1995)
5. Brooks, R.A.: Artificial life and real robots. In: Toward a Practice of Autonomous Systems: Proceedings of the First European Conference on Artificial Life, 1992, pp. 3–10
6. Jakobi, N.: Minimal simulations for evolutionary robotics. PhD Thesis, University of Sussex, Sussex (1998)
7. Miglino, O., Lund, H.H., Nolfi, S.: Evolving mobile robots in simulated and real environments. Artif. Life **2**(4), 417–434 (1995)
8. Hornby, G.S, Takamura, S., Yokono, J., Hanagata, O., Yamamoto, T., Fujita, M.: Evolving robust gaits with AIBO. In: IEEE International Conference on Robotics and Automation, 2000, pp. 3040–3045
9. Zagal, C., Ruiz-del-Solar, J., Vallejos, P.: Back to reality: crossing the reality gap in evolutionary robotics. In: Proceedings of the 5th IFAC/EURON Symposium on Intelligent Autonomous Vehicles, 2004
10. Zagal, C., Ruiz-Del-Solar, J.: Combining simulation and reality in evolutionary robotics. J. Intell. Robot. Syst. **50**(1), 19–39 (2007)
11. Bongard, J., Lipson, H.: Nonlinear system identification using coevolution of models and tests. IEEE Trans. Evol. Comput. **9**(4), 361–384 (2005)
12. Boeing, A., Bräunl, T.: Evaluation of real-time physics simulation systems. In: Proceedings of the 5th International Conference on Computer graphics and Interactive Techniques in Australia and Southeast Asia, 2007, pp. 281–288
13. Boeing, A., Bräunl, T.: Evolving splines: an alternative locomotion controller for a bipedal robot. In: Proceedings of the Seventh International Conference on Control Automation, Robotics and Vision (ICARCV 2002), 2002
14. Ziegler J, Banzhaf, W.: Evolution of robot leg movements in a physical simulation. In: Proceedings of the Fourth International Conference on Climbing and Walking Robots, CLAWAR, 2001
15. Ziegler, J., Barnholt, J., Bush, J., Banzhaf, W.: Automatic evolution of control programs for a small humanoid walking robot. In: 5th International Conference on Climbing and Walking Robots (CLAWAR), 2002
16. A. Boeing: Design of a physics abstraction layer for improving the validity of evolved robot control simulations. Ph.D Dissertation, School of Electrical, Electronic and Computer Engineering, The University of Western Australia, WA (2009)
17. Bräunl, T.: Embedded robotics, 3rd edn. Springer-Verlag, Berlin (2008)
18. McMillan: DynaMechs: A Multibody Dynamic Simulation Library. 6 November 2003, available from: http://dynamechs.sourceforge.net/
19. Featherstone, R.: The calculation of robot dynamics using articulated-body inertias. Int. J. Robotics Res. **2**(1), 13–30 (1983)
20. McMillan, S., Orin, D.E., McGhee, R.B.: Efficient dynamic simulation of an underwater vehicle with a robotic manipulator. IEEE Trans. Syst. Man Cybernet. **25**(8), 1194–1206 (1995)
21. Rodenbaugh, S., Orin, D.E.: RobotBuilder, 6 November 2003, available from: http://www.eleceng.ohiostate.edu/~orin/RobotBuilder/RobotBuilder.html
22. Dorf, R.C., Bishop, R.H.: Modern control systems. Prentice-Hall, Upper Saddle River (2001)
23. Landee, R., Davis, D., Albrecht, A.: Electronics Designers' Handbook. McGraw-Hill, New York (1977)

24. Vukobratovic, M., Borovac, B.: Zero-moment point—thirty five years of its life. Int. J. Hum. Robot. **1**(1), 157–173 (2004)
25. Kim, J.-Y., Park, I.-W., Oh, J.-H.: Walking control algorithm of biped humanoid robot on uneven and inclined floor. J. Intell. Robot. Syst. **48**, 457–484 (2007)
26. Shih, C.-L., Lee, W.-Y., Wu, C.-P.: Planning and control of stable walking for a 3D bipedal robot. Int. J. Adv. Robot. Syst. **9**(47), 10 (2012)
27. Huy, T.D., Cuong, N., Phuong, N.: Control of biped robot with stable walking. Am. J. Eng. Res. **2**(11), 129–150 (2013)
28. Boeing, A., Bräunl, T.: SubSim: an autonomous underwater vehicle simulation package. Proceedings of the 3rd International Symposium on Autonomous Minirobots for Research and Edutainment (AMiRE 2005), 2006, pp. 33–38
29. Bräunl, T., Boeing, A., Gonzalez, L., Koestler, A., Nguyen, M.: Design, modelling and simulation of an autonomous underwater vehicle. Int. J. Vehicle Auton. Syst. **4**(2–4), 106–121 (2006)
30. Jin, Y., Branke, J.: Evolutionary optimization in uncertain environments—a survey. IEEE Trans. Evol. Comput. **9**(3), 303–317 (2005)

Chapter 19
Balancing Conditions of Planar and Spatial Mechanisms in the Algebraic Form

Nguyen Van Khang and Nguyen Phong Dien

Abstract This chapter deals with an approach to formulate balancing conditions for the shaking force and shaking moment of planar mechanisms and spatial mechanisms. In the Mechanism Theory, every Mechanism has p moving members and a non-moving frame. According to tradition, a planar 8R-eightbar mechanism is a multibody system with 7 moving bodies.

Keywords Mechanism • Balancing condition • Shaking force • Shaking moment

19.1 Introduction

Dynamic balancing of mechanisms is a classical problem of machine dynamics [1–9]. Dynamic balancing of the moving links brings about a reduction of the variable dynamic loads on the mechanism frame. In effect, this minimizes the noise and wear, and improves the dynamic performance of the mechanism [3, 4]. The main objective of mass balancing is to completely eliminate or partially reduce the resultant inertia force (shaking force) and the resultant inertia moment with respect to the ground link (shaking moment) caused by all moving links of a mechanism. Although different methods and solutions have been proposed and reported, the balancing theory continues to develop and new approaches are regularly being published. Summaries of much of the past work are given in refs. [2–4]. Recently, the terminology "reactionless mechanism" has usually been used in design and dynamic synthesis of mechanisms, e.g., [19, 20]. A mechanism is said to be reactionless or dynamically balanced if the shaking force and the shaking moment are completely eliminated for any arbitrary motion of the mechanism.

N. Van Khang (✉) • N.P. Dien
Department of Applied Mechanics, Hanoi University of Science and Technology, 1. Dai Co Viet Road, Hanoi, Vietnam
e-mail: khang.nguyenvan2@hust.edu.vn; dien.nguyenphong@hust.edu.vn

D. Zhang, B. Wei (eds.), *Dynamic Balancing of Mechanisms and Synthesizing of Parallel Robots*, DOI 10.1007/978-3-319-17683-3_19

In other words, no dynamic reaction forces and no dynamic reaction moments are transmitted to the base during the motion.

In our opinion, the problem of shaking force and shaking moment balancing consists of two aspects. The first is to find all feasible design solutions (mass redistribution, using counterweights or adding supplementary members as cams, gears, parallelogram chains, planetary gears, etc.) in order to compensate the shaking force and shaking moment. For this purpose different approaches and solutions have been developed and reported. Berkof [11] presented a review of the methods based on the different movements of the counterweights for the shaking force balancing. Feng [30] used the concept of inertia counterweight proposed by Berkof [13] to carry out the dynamic balancing of a number of single degree freedom mechanisms. The publications by Lowen et al. [7] and Kochev [16] provide a critical review of the methods employing additional members for complete shaking moment balancing. Arakelian and Smith [9] investigated the dynamic balancing of single degree of freedom mechanisms by using the pantograph copying properties. A number of other solutions for the complete shaking force and shaking moment balancing can be found in the studies presented by Kochev [17], Wu and Gosselin [21], Dresig et al. [14, 15], Arakelian [26–28], and Moore [32].

The second aspect is related to the formulation of balancing conditions which are usually expressed in terms of the design variables (such as masses, moments of inertia, and geometrical parameters of the links) of the mechanism. There are several convenient ways to formulate balancing conditions of the shaking force. For instance, the *method of linearly independent vectors* was proposed by Berkof and Lowen [11] and later successfully employed by Kaufman and Sandor [12], Feng [31] to obtain full force balancing conditions for linkages, the *equivalence method* was proposed by Ye and Smith [18]. The *method of principal vectors* was used by Shchepetilnikov [10] to investigate the static balancing conditions of mechanisms. Because the shaking force is related to the first derivative of the total linear momentum with respect to time, the *linear momentum method* can also be used to establish balancing conditions of the shaking force [16, 31]. Conversely, research on efficient methods for deriving balancing conditions of the shaking moment has been less productive due to the complexity of the problem. It is well known that the shaking moment of a mechanism is related to the first derivative of the total angular momentum with respect to time. This relationship leads to an approach for the formulation of balancing conditions of the shaking moment, known as the *angular momentum method*. This method was used by several authors such as Kochev [9, 10], Feng [31], and Nguyen [22–25]. Arakelian and Dahan [27] formulated the moment balancing conditions of a multi-link planar mechanism by minimizing the root-mean-square value of the resultant inertia moment. Another recent approach to derive balancing conditions of planar multi-loop mechanisms using the equivalent method is investigated by Chaudhary and Saha [6, 29].

In contrast to the rapid progress in balancing theory of planar mechanisms, the development on the balancing theory of spatial mechanisms is still limited. Balancing methods of planar mechanisms cannot be directly applicable to spatial mechanisms since kinematic and dynamic properties of spatial mechanisms are much more complicated. The literature on this respect therefore is little [22, 24, 33–41]. One of the problems of the complete shaking force and shaking moment balancing of the mechanism consist of the deriving the so-called balancing conditions. These balancing conditions will be used to determine the size and location of counterweights or supplementary links which must be added to the initial mechanism, in order to eliminate the shaking force and the shaking moment.

Using the methods of multibody dynamics, this chapter deals with an approach to derive balancing conditions in the algebraic form for the shaking force and shaking moment of planar and spatial multi-loop mechanisms. The developed methods are suitable for the application of the widely accessible computer algebra systems such as MAPLE®. In the examples, the conditions for complete shaking force and shaking moment balancing of a planar multi-loop, multi-DOF mechanism and a spatial one-DOF mechanism are given.

19.2 Balancing Theory of Constrained Multibody Systems

We consider a multibody system with holonomic and rheonomic constraints as a set of p linked rigid bodies in a closed loop structure shown in Fig. 19.1.

The shaking force $\vec{F}^*$ and the shaking moment $\vec{M}_O^*$ referred to a fixed point O of the considered system, which are caused by all moving bodies, can be expressed in the form [1, 2, 22, 24]

$$\vec{F}^* = -\frac{^{R_O}\mathrm{d}}{\mathrm{d}t}\sum_{i=1}^{p} m_i \vec{v}_{Si}, \tag{19.1}$$

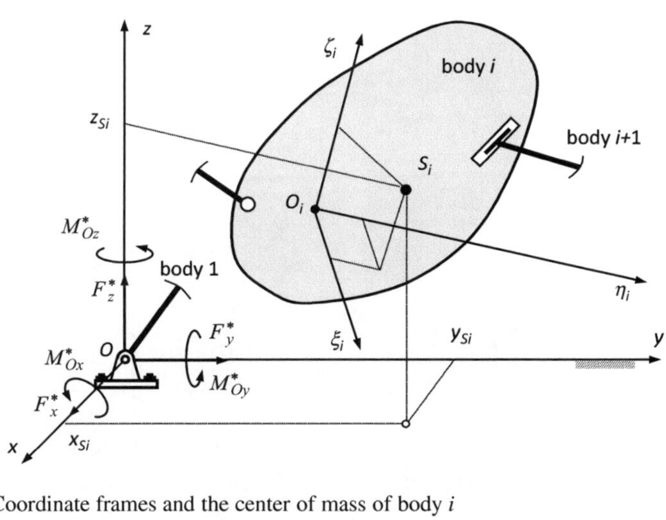

Fig. 19.1 Coordinate frames and the center of mass of body i

$$\overrightarrow{M}_{\mathrm{O}}^{*} = -\frac{^{R_O}d}{\mathrm{d}t}\sum_{i=1}^{p}\left(\overrightarrow{\overrightarrow{I}}_{Si}\cdot\overrightarrow{\omega}_{i} + \overrightarrow{r}_{Si}\times m_{i}\overrightarrow{v}_{Si}\right). \tag{19.2}$$

In Eqs. (19.1) and (19.2) the following symbols are used:

m_i mass of body i.
p number of bodies.
$\overrightarrow{r}_{Si}$ position vector of center of mass S_i of body i in the fixed coordinate frame $R_0\{x, y, z\}$.
$\overrightarrow{v}_{Si}$ velocity vector of center of mass S_i in the coordinate frame R_0.
$\overrightarrow{\overrightarrow{I}}_{Si}$ mass inertia tensor of body i referred to S_i.
$\overrightarrow{\omega}_i$ angular velocity of body i with respect to the coordinate frame R_0.

The multibody system is completely balanced if the shaking force and the shaking moment vanish at every position [1, 2]

$$\overrightarrow{F}^{*} = 0, \quad \overrightarrow{M}_{\mathrm{O}}^{*} = 0. \tag{19.3}$$

It follows that

$$\frac{^{R_O}d}{\mathrm{d}t}\sum_{i=1}^{p} m_i\overrightarrow{v}_{Si} = 0, \tag{19.4}$$

$$\frac{^{R_O}d}{\mathrm{d}t}\sum_{i=1}^{p}\left(\overrightarrow{\overrightarrow{I}}_{Si}\cdot\overrightarrow{\omega}_{i} + \overrightarrow{r}_{Si}\times m\overrightarrow{v}_{Si}\right) = 0. \tag{19.5}$$

Equations (19.4) and (19.5) can be rewritten in the matrix form as follows:

$$\frac{d}{\mathrm{d}t}\sum_{i=1}^{p} m_i\mathbf{v}_{Si} = 0, \tag{19.6}$$

$$\frac{d}{\mathrm{d}t}\left[\sum_{i=1}^{p}(\mathbf{I}_{Si}\boldsymbol{\omega}_i + m_i\tilde{\mathbf{r}}_{Si}\mathbf{v}_{Si})\right] = 0, \tag{19.7}$$

where

$$\mathbf{r}_{Si} = \begin{bmatrix} x_{Si} \\ y_{Si} \\ z_{Si} \end{bmatrix}, \quad \tilde{\mathbf{r}}_{Si} = \begin{bmatrix} 0 & -z_{Si} & y_{Si} \\ z_{Si} & 0 & -x_{Si} \\ -y_{Si} & x_{Si} & 0 \end{bmatrix}. \tag{19.8}$$

For a f-DOF stationary multibody system described by n generalized coordinates q_1, q_2, ..., q_n and $n \geq f$, position vector $\mathbf{r}_{Si}$ can be expressed in form of generalized coordinates

$$\mathbf{r}_{Si} = \mathbf{r}_{Si}\left(q_1, q_2, \ldots, q_n\right), \ (i = 1, 2, \ldots, p). \tag{19.9}$$

Differentiating Eq. (19.9) with respect to time in the coordinate frame R_0 yields

$$\mathbf{v}_{Si} = \frac{\mathrm{d}\mathbf{r}_{Si}}{\mathrm{d}t} = \frac{\partial \mathbf{r}_{Si}}{\partial \mathbf{q}}\dot{\mathbf{q}} = \mathbf{J}_{Ti}\left(\mathbf{q}\right)\dot{\mathbf{q}}, \tag{19.10}$$

where $\mathbf{J}_{Ti}(\mathbf{q})$ is the translation Jacobi matrix

$$\mathbf{J}_{Ti}\left(\mathbf{q}\right) = \frac{\partial \mathbf{r}_{Si}}{\partial \mathbf{q}} = \begin{bmatrix} \frac{\partial x_{Si}}{\partial q_1} & \frac{\partial x_{Si}}{\partial q_2} & \cdots & \frac{\partial x_{Si}}{\partial q_n} \\ \frac{\partial y_{Si}}{\partial q_1} & \frac{\partial y_{Si}}{\partial q_2} & \cdots & \frac{\partial y_{Si}}{\partial q_n} \\ \frac{\partial z_{Si}}{\partial q_1} & \frac{\partial z_{Si}}{\partial q_2} & \cdots & \frac{\partial z_{Si}}{\partial q_n} \end{bmatrix}, \tag{19.11}$$

and $\mathbf{q} = [q_1, q_2, \ldots, q_n]^T$. By introducing φ_i as the rotation vector of body i, the angular velocity $\boldsymbol{\omega}_i$ is defined by

$$\boldsymbol{\omega}_i = \frac{\mathrm{d}\boldsymbol{\varphi}_i}{\mathrm{d}t} = \frac{\partial \boldsymbol{\varphi}_i}{\partial \mathbf{q}}\dot{\mathbf{q}} = \mathbf{J}_{Ri}\left(\mathbf{q}\right)\dot{\mathbf{q}}, \tag{19.12}$$

where $\mathbf{J}_{Ri}(\mathbf{q})$ denotes the rotation Jacobi matrix

$$\mathbf{J}_{Ri} = \frac{\partial \boldsymbol{\varphi}_i}{\partial \mathbf{q}} = \frac{\partial \boldsymbol{\omega}_i}{\partial \dot{\mathbf{q}}} = \begin{bmatrix} \frac{\partial \omega_{ix}}{\partial \dot{q}_1} & \frac{\partial \omega_{ix}}{\partial \dot{q}_2} & \cdots & \frac{\partial \omega_{ix}}{\partial \dot{q}_n} \\ \frac{\partial \omega_{iy}}{\partial \dot{q}_1} & \frac{\partial \omega_{iy}}{\partial \dot{q}_2} & \cdots & \frac{\partial \omega_{iy}}{\partial \dot{q}_n} \\ \frac{\partial \omega_{iz}}{\partial \dot{q}_1} & \frac{\partial \omega_{iz}}{\partial \dot{q}_2} & \cdots & \frac{\partial \omega_{iz}}{\partial \dot{q}_n} \end{bmatrix}. \tag{19.13}$$

Substitution of Eq. (19.10) into Eq. (19.6) yields

$$\frac{d}{\mathrm{d}t}\left\{\left[\sum_{i=1}^{p} m_i \mathbf{J}_{Ti}\left(\mathbf{q}\right)\right]\dot{\mathbf{q}}\right\} = 0. \tag{19.14}$$

Substituting Eqs. (19.10) and (19.12) into Eq. (19.7), one obtains

$$\frac{d}{\mathrm{d}t}\left\{\left[\sum_{i=1}^{p}\left(\mathbf{I}_{Si}\mathbf{J}_{Ri}\left(\mathbf{q}\right) + m_i\tilde{\mathbf{r}}_{Si}\mathbf{J}_{Ti}\left(\mathbf{q}\right)\right]\dot{\mathbf{q}}\right\} = 0. \tag{19.15}$$

Note that the inertia matrix $\mathbf{I}_{Si}$ with respect to the fixed frame R_0 can be written in term of the matrix $\mathbf{I}_{Si}^{(i)}$ using the formula

$$\mathbf{I}_{Si} = \mathbf{A}_i \mathbf{I}_{Si}^{(i)} \mathbf{A}_i^T, \tag{19.16}$$

where $\mathbf{A}_i$ denotes the direction cosine matrix of body i referred to the fixed frame R_0, $\mathbf{I}_{Si}^{(i)}$ is the matrix of the mass inertia tensor relative to the axes of the body-fixed coordinate system $R_i\{\xi_i, \eta_i, \zeta_i\}$ (see Fig. 19.1).

It follows that

$$\mathbf{I}_{Si}\mathbf{J}_{Ri}(\mathbf{q}) = \mathbf{A}_i \mathbf{I}_{Si}^{(i)} \mathbf{A}_i^T \mathbf{J}_{Ri}(\mathbf{q}). \tag{19.17}$$

Since $\mathbf{A}_i = \mathbf{A}_i(\mathbf{q})$ and $\frac{\partial \mathbf{A}_i}{\partial \dot{\mathbf{q}}} = 0$, it follows

$$\mathbf{J}_{Ri}(\mathbf{q}) = \frac{\partial \boldsymbol{\omega}_i}{\partial \dot{\mathbf{q}}} = \frac{\partial \left(\mathbf{A}_i \boldsymbol{\omega}_i^{(i)}\right)}{\partial \dot{\mathbf{q}}} = \mathbf{A}_i(\mathbf{q}) \frac{\partial \boldsymbol{\omega}_i^{(i)}}{\partial \dot{\mathbf{q}}} = \mathbf{A}_i \mathbf{J}_{Ri}^{(i)}, \tag{19.18}$$

where matrix $\mathbf{J}_{Ri}^{(i)}(\mathbf{q})$ is defined by

$$\mathbf{J}_{Ri}^{(i)}(\mathbf{q}) = \frac{\partial \boldsymbol{\omega}_i^{(i)}}{\partial \dot{\mathbf{q}}}. \tag{19.19}$$

Substitution of Eqs. (19.18) and (19.19) into Eq. (19.15) yields

$$\frac{d}{\mathrm{d}t}\left\{\left[\sum_{i=1}^{p} \mathbf{A}_i \mathbf{I}_{Si}^{(i)} \mathbf{J}_{Ri}^{(i)}(\mathbf{q}) + m_i \tilde{\mathbf{r}}_{Si} \mathbf{J}_{Ti}(\mathbf{q})\right] \dot{\mathbf{q}}\right\} = 0. \tag{19.20}$$

It follows from Eqs. (19.14) and (19.20) the general balancing conditions of a multibody system

$$\sum_{i=1}^{p} m_i \mathbf{J}_{Ti}(\mathbf{q}) = 0, \tag{19.21}$$

$$\sum_{i=1}^{p} \left[\mathbf{A}_i \mathbf{I}_{Si}^{(i)} \mathbf{J}_{Ri}^{(i)}(\mathbf{q}) + m_i \tilde{\mathbf{r}}_{Si} \mathbf{J}_{Ti}(\mathbf{q})\right] = 0. \tag{19.22}$$

19.3 Balancing Conditions of Planar Mechanisms

19.3.1 Theory and Procedure for Deriving Dynamic Balancing Conditions

19.3.1.1 General Balancing Conditions

We consider an arbitrary link of a multi DOF planar mechanism as depicted in Fig. 19.2. The mechanism consists of a set of p moving links in a closed loop structure with revolute joints. Parameters x_{Si}, y_{Si} are the coordinates of the center of mass S_i of link i in the ground-fixed coordinate frame {Oxy}, φ_i is the rotation angle, ξ_{Si}, η_{Si} are coordinates of S_i in the link-fixed coordinate frame $\{O_i\xi_i\eta_i\}$.

From Eqs. (19.1) and (19.2) the shaking force and the shaking moment transmitted to the base from all moving links can be expressed in the form [15]

$$F_x^* = -\frac{d}{\mathrm{d}t}\left(\sum_{i=1}^{p} m_i\dot{x}_{Si}\right),\ F_y^* = -\frac{d}{\mathrm{d}t}\left(\sum_{i=1}^{p} m_i\dot{y}_{Si}\right) \tag{19.23}$$

$$M_\mathrm{O}^* = \frac{d}{\mathrm{d}t}\left\{\sum_{i=1}^{p}\left[m_i\left(x_{Si}\dot{y}_{Si} - y_{Si}\dot{x}_{Si}\right) + J_{Si}\dot{\varphi}_i\right]\right\} \tag{19.24}$$

where m_i denotes the mass and I_{Si} the moment of inertia of the link about the axis passing through S_i and perpendicular to the plane of motion.

The planar mechanism can then be completely balanced if the shaking force and the shaking moment vanish. This yields the following sufficient conditions

$$\sum_{i=1}^{p} m_i\dot{\mathbf{r}}_i = 0, \tag{19.25}$$

$$\sum_{i=1}^{p}\left[m_i\left(x_{Si}\dot{y}_{Si} - y_{Si}\dot{x}_{Si}\right) + I_{Si}\dot{\varphi}_i\right] = 0, \tag{19.26}$$

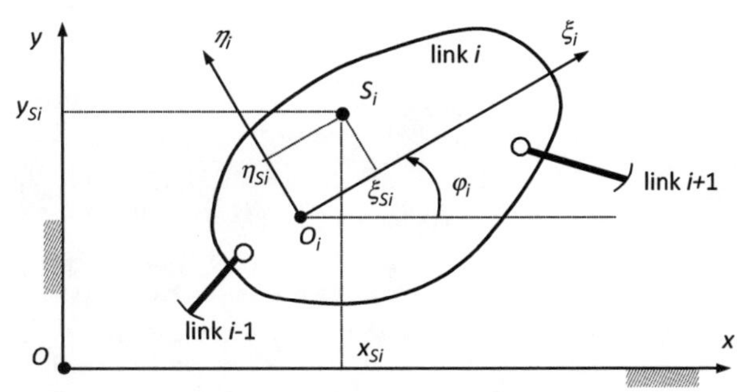

Fig. 19.2 Definition of parameters and coordinates

where $\mathbf{r}_i = [x_{Si}, y_{Si}]^T$ and $\dot{\mathbf{r}}_i = [\dot{x}_{Si}, \dot{y}_{Si}]^T$. Based on the general condition (19.25) for the shaking force balancing, there are some ways to derive the balancing conditions in form of algebraic expressions of parameters m_i, I_{Si}, ξ_{Si} and η_{Si} as mentioned in the previous section. Conversely, it is more difficult to formulate the dynamic balancing conditions of the shaking moment due to the presence of the term $I_{Si}\dot{\varphi}_i$ in Eq. (19.26).

19.3.1.2 Generalized Coordinates of the Second Type

Since the considered mechanism has only revolute joints, rotation angles φ_i $(i = 1, 2, \ldots p)$ can be chosen as generalized coordinates which describe the motion of particular links. Angle φ_i is known as "the generalized coordinates of the first type." Now we introduce vector $\mathbf{u}$

$$\mathbf{u} = \left[\cos\varphi_1, \sin\varphi_1, \ldots, \cos\varphi_p, \sin\varphi_p\right]^T, \tag{19.27}$$

where elements u_k $(k = 1, 2, \ldots, 2p)$ are trigonometric functions of φ_i. Logically, elements u_k are called "the generalized coordinates of the second type." As can be seen later, vector $\mathbf{u}$ can be used as the basis for developing a systematic procedure for deriving balancing conditions of the shaking force and moment.

19.3.1.3 Procedure to Derive Balancing Conditions of the Shaking Force

Generally, the position vector of the center of mass S_i can always be expressed in term of vector $\mathbf{u}$ as

$$\mathbf{r}_i = \mathbf{e}_i^u + \mathbf{C}_i\mathbf{u}, \tag{19.28}$$

where $i = 1, 2, \ldots, p$ and $\mathbf{e}_i^u$ is a vector of constants. The elements of matrix $\mathbf{C}_i$ $(2 \times 2p)$ are geometrical parameters and independent of $\mathbf{u}$. Similarly, the loop equations of the mechanism can be expressed in the compact matrix form

$$\mathbf{f}(\mathbf{u}) = \mathbf{d}. \tag{19.29}$$

In the cases of planar mechanisms articulated by revolute joints, Eq. (19.7) can be rewritten in linear form with vector $\mathbf{u}$

$$\mathbf{D}\,\mathbf{u} = \mathbf{d}, \tag{19.30}$$

where the elements of matrix $\mathbf{D}$ are geometrical parameters of the mechanism and independent of $\mathbf{u}$, vector $\mathbf{d}$ is constant. It follows that Eq. (19.28) can then be rewritten in term of a minimal set of elements u_k of **u**. The following partitioning of **u** from Eq. (19.30)

$$\mathbf{u} = \begin{bmatrix} \mathbf{v} \\ \mathbf{w} \end{bmatrix}, \tag{19.31}$$

leads to the following relationship

$$\mathbf{D}^{v}\mathbf{v} + \mathbf{D}^{w}\mathbf{w} = \mathbf{d}. \tag{19.32}$$

where vector **v** consists of elements from this set, and the dimension of vector **w** is equal to the number of the loop equations. Matrix $\mathbf{D}^{w}$ is chosen so that it is a square and nonsingular matrix. When vectors **v** and **w** are assigned, an easily way to obtain matrices $\mathbf{D}^{v}$ and $\mathbf{D}^{w}$ is by taking the partial derivatives

$$\mathbf{D}^{v} = \frac{\partial \mathbf{f}}{\partial \mathbf{v}}, \ \mathbf{D}^{w} = \frac{\partial \mathbf{f}}{\partial \mathbf{w}}. \tag{19.33}$$

From Eq. (19.32) we find

$$\mathbf{w} = \left(\mathbf{D}^{w}\right)^{-1}\left(\mathbf{d} - \mathbf{D}^{v}\mathbf{v}\right) = \mathbf{b} - \mathbf{G}\mathbf{v}, \tag{19.34}$$

where

$$\mathbf{G} = (\mathbf{D}^{w})^{-1}\mathbf{D}^{v}, \ \mathbf{b} = (\mathbf{D}^{w})^{-1}\mathbf{d}. \tag{19.35}$$

Differentiating Eq. (19.34) with respect to time yields

$$\dot{\mathbf{w}} = -\mathbf{G}\dot{\mathbf{v}}. \tag{19.36}$$

Using Eq. (19.31) one can rewrite Eq. (19.28) in the following form

$$\mathbf{r}_i = \mathbf{e}_i^{u} + \mathbf{C}_i^{v}\mathbf{v} + \mathbf{C}_i^{w}\mathbf{w}, \tag{19.37}$$

where matrices $\mathbf{C}_i^{v}$, $\mathbf{C}_i^{w}$ are given by

$$\mathbf{C}_i^{v} = \frac{\partial \mathbf{r}_i}{\partial \mathbf{v}}, \ \mathbf{C}_i^{w} = \frac{\partial \mathbf{r}_i}{\partial \mathbf{w}}, \tag{19.38}$$

and the vector of constant parameters $\mathbf{e}_i^{u}$ is the remaining term from Eq. (19.15).

Substitution of Eq. (19.34) into Eq. (19.37) yields

$$\mathbf{r}_i = \mathbf{e}_i^{u} + \mathbf{C}_i^{w}\mathbf{b} + \left(\mathbf{C}_i^{v} - \mathbf{C}_i^{w}\mathbf{G}\right)\mathbf{v}. \tag{19.39}$$

This can be written as

$$\mathbf{r}_i = \mathbf{e}_i + \mathbf{B}_i \mathbf{v}, \tag{19.40}$$

where

$$\mathbf{e}_i = \mathbf{e}_i^u + \mathbf{C}_i^w \mathbf{b}, \tag{19.41}$$

$$\mathbf{B}_i = \mathbf{C}_i^v - \mathbf{C}_i^w \mathbf{G}. \tag{19.42}$$

Note that the elements of vector $\mathbf{e}_i$ and matrix $\mathbf{B}_i$ are geometrical parameters of the mechanism and independent of $\mathbf{v}$. Differentiating Eq. (19.40) with respect to time yields

$$\dot{\mathbf{r}}_i = \mathbf{B}_i \dot{\mathbf{v}}. \tag{19.43}$$

Substituting Eq. (19.43) into Eq. (19.25) leads to

$$\sum_{i=1}^{p} m_i \mathbf{B}_i \dot{\mathbf{v}} = 0. \tag{19.44}$$

As a result, the balancing conditions for the shaking force reduce to the algebraic form

$$\sum_{i=1}^{p} m_i \mathbf{B}_i = \mathbf{0}. \tag{19.45}$$

If the mechanism has p moving links and r loop equations, then vector $\mathbf{w}$ contains r elements whereas matrix $\mathbf{B}_i$ has the dimension of $2 \times (2p - r)$. From Eq. (19.45) we obtain $2\,(2p - r)$ balancing conditions in form of algebraic expressions of inertia and geometrical parameters.

19.3.1.4 Procedure to Derive Balancing Conditions of the Shaking Moment

The general balancing condition of the shaking moment according to Eq. (19.4) contains two terms. The first term is

$$h_1 = \sum_{i=1}^{p} m_i\,(x_{Si}\dot{y}_{Si} - y_{Si}\dot{x}_{Si}). \tag{19.46}$$

We note that

$$x_{Si}\dot{y}_{Si} - y_{Si}\dot{x}_{Si} = \mathbf{r}_i^T \mathbf{I}^* \,\dot{\mathbf{r}}_i, \tag{19.47}$$

where $\mathbf{I}^* = \begin{bmatrix} 0 & 1 \\ -1 & 0 \end{bmatrix}$. With the use of this relationship, Eq. (19.46) leads to

$$h_1 = \sum_{i=1}^{p} m_i \mathbf{r}_i^T \mathbf{I}^* \, \dot{\mathbf{r}}_i. \tag{19.48}$$

Substitution of Eqs. (19.40) and (19.43) into Eq. (19.48) yields

$$\begin{aligned} h_1 &= \sum_{i=1}^{p} m_i (\mathbf{e}_i + \mathbf{B}_i \mathbf{v})^T \mathbf{I}^* \, \mathbf{B}_i \dot{\mathbf{v}} \\ &= \mathbf{v}^T \left(\sum_{i=1}^{p} m_i \mathbf{B}_i^T \mathbf{I}^* \, \mathbf{B}_i \right) \dot{\mathbf{v}} + \left(\sum_{i=1}^{p} m_i \mathbf{e}_i^T \mathbf{I}^* \, \mathbf{B}_i \right) \dot{\mathbf{v}} \\ &= \mathbf{v}^T \mathbf{S}_1 \dot{\mathbf{v}} + \mathbf{k}_1^T \dot{\mathbf{v}}, \end{aligned} \tag{19.49}$$

where

$$\mathbf{S}_1 = \sum_{i=1}^{p} m_i \mathbf{B}_i^T \mathbf{I}^* \, \mathbf{B}_i, \quad \mathbf{k}_1^T = \sum_{i=1}^{p} m_i \mathbf{e}_i^T \mathbf{I}^* \, \mathbf{B}_i. \tag{19.50}$$

Now we consider the second term of Eq. (19.26)

$$h_2 = \sum_{i=1}^{p} I_{Si} \dot{\varphi}_i. \tag{19.51}$$

One can verify that

$$\dot{\varphi}_i = u_1^{(i)} \dot{u}_2^{(i)} - u_2^{(i)} \dot{u}_1^{(i)}, \tag{19.52}$$

where $u_1^{(i)} = \cos\varphi_i$ and $u_2^{(i)} = \sin\varphi_i$. Equation (19.52) can also rewritten in the matrix form as

$$\dot{\varphi}_i = \begin{bmatrix} u_1^{(i)} \\ u_2^{(i)} \end{bmatrix}^T \begin{bmatrix} 0 & 1 \\ -1 & 0 \end{bmatrix} \begin{bmatrix} \dot{u}_1^{(i)} \\ \dot{u}_2^{(i)} \end{bmatrix}. \tag{19.53}$$

Substitution of Eq. (19.53) into Eq. (19.51) yields

$$h_2 = \sum_{i=1}^{p} \begin{bmatrix} u_1^{(i)} \\ u_2^{(i)} \end{bmatrix}^T \begin{bmatrix} 0 & I_{Si} \\ -I_{Si} & 0 \end{bmatrix} \begin{bmatrix} \dot{u}_1^{(i)} \\ \dot{u}_2^{(i)} \end{bmatrix} = \mathbf{u}^T \mathbf{H} \dot{\mathbf{u}}, \tag{19.54}$$

where $\mathbf{H}$ is a $2p \times 2p$ matrix defined by

$$\mathbf{H} = \begin{bmatrix} 0 & I_{S1} & 0 & 0 & \dots & 0 & 0 \\ -I_{S1} & 0 & 0 & 0 & \dots & 0 & 0 \\ 0 & 0 & 0 & I_{S2} & \dots & 0 & 0 \\ 0 & 0 & -I_{S2} & 0 & \dots & 0 & 0 \\ \vdots & \vdots & \vdots & \vdots & \ddots & \vdots & \vdots \\ 0 & 0 & 0 & 0 & \dots & 0 & I_{Sp} \\ 0 & 0 & 0 & 0 & \dots & -I_{Sp} & 0 \end{bmatrix} \tag{19.55}$$

Matrix $\mathbf{H}$ can be partitioned in four sub-matrices corresponding to vectors $\mathbf{v}$ und $\mathbf{w}$ as follows:

$$\mathbf{H} = \left[\frac{\mathbf{H}_1 \; \mathbf{H}_2}{\mathbf{H}_3 \; \mathbf{H}_4}, \right] \tag{19.56}$$

where $\mathbf{H}_1$ is a $(2p-r) \times (2p-r)$ matrix, $\mathbf{H}_2$ is a $(2p-r) \times r$ matrix of zero, $\mathbf{H}_3$ is a $r \times (2p-r)$ matrix of zeros and $\mathbf{H}_4$ a $r \times r$ matrix. Then Eq. (19.54) takes the form

$$h_2 = \begin{bmatrix} \mathbf{V}^T & \mathbf{W}^T \end{bmatrix} \begin{bmatrix} \mathbf{H}_1 & \mathbf{0} \\ \mathbf{0} & \mathbf{H}_4 \end{bmatrix} \begin{bmatrix} \dot{\mathbf{V}} \\ \dot{\mathbf{W}} \end{bmatrix} = \mathbf{V}^T \mathbf{H}_1 \dot{\mathbf{V}} + W^T \mathbf{H}_4 \dot{\mathbf{W}}. \tag{19.57}$$

Substitution of Eqs. (19.34) and (19.35) into Eq. (19.57) yields

$$\begin{aligned} h_2 &= \mathbf{v}^T \mathbf{H}_1 \dot{\mathbf{v}} + (\mathbf{b} - \mathbf{G}\mathbf{v})^T \mathbf{H}_4 (-\mathbf{G}\dot{\mathbf{v}}) \\ &= \mathbf{v}^T \left(\mathbf{H}_1 + \mathbf{G}^T \mathbf{H}_4 \mathbf{G} \right) \dot{\mathbf{v}} - \left(\mathbf{b}^T \mathbf{H}_4 \mathbf{G} \right) \dot{\mathbf{v}} \\ &= \mathbf{v}^T \mathbf{S}_2 \dot{\mathbf{v}} + \mathbf{k}_2^T \dot{\mathbf{v}}. \end{aligned} \tag{19.58}$$

where matrix $\mathbf{S}_2$ and vector $\mathbf{k}_2$ are defined by

$$\mathbf{S}_2 = \mathbf{H}_1 + \mathbf{G}^T \mathbf{H}_4 \mathbf{G}, \tag{19.59}$$

$$\mathbf{k}_2^T = -\mathbf{b}^T \mathbf{H}_4 \mathbf{G}. \tag{19.60}$$

Using Eqs. (19.49) and (19.58), the general balancing condition of the shaking moment can be written in the matrix form as

$$\mathbf{v}^T (\mathbf{S}_1 + \mathbf{S}_2) \dot{\mathbf{v}} + \left(\mathbf{k}_1^T + \mathbf{k}_2^T \right) \dot{\mathbf{v}} = \mathbf{0}. \tag{19.61}$$

Finally, the balancing conditions for the shaking moment reduce to the algebraic form

$$\mathbf{S}_1 + \mathbf{S}_2 = \mathbf{0}, \quad \mathbf{k}_1 + \mathbf{k}_2 = \mathbf{0}. \tag{19.62}$$

where matrices $\mathbf{S}_1$ and $\mathbf{S}_2$ have the dimension of $(2p - r) \times (2p - r)$ and $\mathbf{k}_1$, $\mathbf{k}_2$ are vectors of $2p - r$ elements. With the use of Eq. (19.40) we obtain a set of balancing conditions for the shaking moment in term of inertia and geometrical parameters of the mechanism, such as m_i, ξ_{Si}, η_{Si} and I_{Si}. In summary, the following steps are required to realize the proposed procedure:

- Formulating r loop equations and p position vectors of the mass centers of moving links according to Eqs. (19.28) and (19.30).
- Selecting the elements of vector $\mathbf{w}$ from elements of vector $\mathbf{u}$ based on the following rule: The number of elements in $\mathbf{w}$ is equal to r, and matrix $\mathbf{D}^w$ must be a square and nonsingular matrix.
- Calculating matrices $\mathbf{D}^v$ and $\mathbf{D}^w$ using Eq. (19.33), $\mathbf{C}_i^v$, $\mathbf{C}_i^w$ using Eq. (19.38), Matrix $\mathbf{G}$ and vector $\mathbf{b}$ using Eq. (19.35), matrices $\mathbf{B}_i$ and vectors $\mathbf{e}_i$ $(i = 1, 2, \ldots, p)$ using Eqs. (19.41) and (19.42).
- Substituting the expressions of matrices $\mathbf{B}_i$ into Eq. (19.45) to obtain the balancing conditions for the shaking force.
- Determining the elements of matrices $\mathbf{H}_1$ and $\mathbf{H}_4$ according to Eqs. (19.55) and (19.56).
- Calculating matrix $\mathbf{S}_1$ and vector $\mathbf{k}_1$ using Eq. (19.50), matrix $\mathbf{S}_2$ using Eq. (19.59) and vector $\mathbf{k}_2$ using Eq. (19.60).
- Substituting the expressions of $\mathbf{S}_1$,$\mathbf{S}_2$, $\mathbf{k}_1$ and $\mathbf{k}_2$ into Eq. (19.62) to get the balancing conditions for the shaking moment.

19.3.2 Application Example

A planar 8R-eightbar mechanism depicted in Fig. 19.3 is a multi degrees-of-freedom and multi-loop planar mechanism with seven moving links, where links 1, 4, and 6 are the driving links.

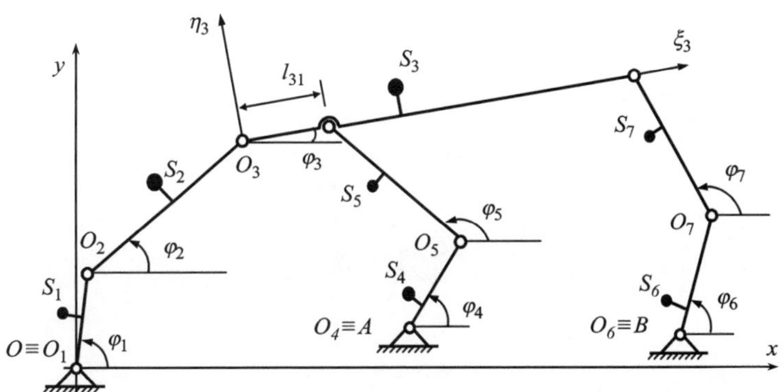

Fig. 19.3 Kinematic diagram of a planar 8R-eightbar mechanism

19.3.2.1 Formulation of Loop Equations

As shown in Fig. 19.2, the origin of the ground-fixed coordinate frame coincides with joint O of pivot link 1, and O_i denotes the origin of the link-fixed coordinate frame of link i. The loop equations of the mechanism can be written in the form

$$\begin{aligned} & l_1 \cos\varphi_1 + l_2 \cos\varphi_2 + l_{31} \cos\varphi_3 - l_4 \cos\varphi_4 - l_5 \cos\varphi_5 = x_A \\ & l_1 \sin\varphi_1 + l_2 \sin\varphi_2 + l_{31} \sin\varphi_3 - l_4 \sin\varphi_4 - l_5 \sin\varphi_5 = y_A \\ & l_1 \cos\varphi_1 + l_2 \cos\varphi_2 + l_3 \cos\varphi_3 - l_6 \cos\varphi_6 - l_7 \cos\varphi_7 = x_B \\ & l_1 \sin\varphi_1 + l_2 \sin\varphi_2 + l_3 \sin\varphi_3 - l_6 \sin\varphi_6 - l_7 \sin\varphi_7 = y_B \end{aligned} \tag{19.63}$$

where l_i denotes the length of link i, x_A, y_A and x_B, y_B are coordinates of the fixed points A and B in the fixed coordinate frame $\{Oxy\}$ respectively. According to Eq. (19.27), vector $\mathbf{u}$ is given by

$$\mathbf{u} = [\cos\varphi_1,\ \sin\varphi_1,\ \cos\varphi_2,\ \sin\varphi_2, \ldots,\ \cos\varphi_7,\ \sin\varphi_7]^T \tag{19.64}$$

According to Eq. (19.30), vector $\mathbf{d}$ are then determined from Eq. (19.63)

$$\mathbf{d} = [x_A,\ y_A,\ x_B,\ y_B]^T \tag{19.65}$$

Vector $\mathbf{w}$ and $\mathbf{v}$ is selected from the original vector $\mathbf{u}$ as follows:

$$\mathbf{w} = [\cos\varphi_4,\ \sin\varphi_4,\ \cos\varphi_6,\ \sin\varphi_6]^T \tag{19.66}$$

$$\mathbf{v} = [\cos\varphi_1,\ \ldots,\ \sin\varphi_3,\ \cos\varphi_5,\ \sin\varphi_5,\ \cos\varphi_7,\ \sin\varphi_7]^T \tag{19.67}$$

Note that there are other possibilities to choose the elements of $\mathbf{w}$ in order to obtain a nonsingular matrix $\mathbf{D}^w$. With vectors $\mathbf{v}$ and $\mathbf{w}$ given by Eqs. (19.66) and (19.67), matrices $\mathbf{D}^v$ and $\mathbf{D}^w$ are calculated from Eq. (19.63) by using Eq. (19.33)

$$\mathbf{D}^v = \begin{bmatrix} l_1 & 0 & l_2 & 0 & l_{31} & 0 & -l_5 & 0 & 0 & 0 \\ 0 & l_1 & 0 & l_2 & 0 & l_{31} & 0 & -l_5 & 0 & 0 \\ l_1 & 0 & l_2 & 0 & l_3 & 0 & 0 & 0 & -l_7 & 0 \\ 0 & l_1 & 0 & l_2 & 0 & l_3 & 0 & 0 & 0 & -l_7 \end{bmatrix} \tag{19.68}$$

$$\mathbf{D}^w = \begin{bmatrix} -l_4 & 0 & 0 & 0 \\ 0 & -l_4 & 0 & 0 \\ 0 & 0 & -l_6 & 0 \\ 0 & 0 & 0 & -l_6 \end{bmatrix}. \tag{19.69}$$

From Eq. (19.69) we get

$$(\mathbf{D}^w)^{-1} = \begin{bmatrix} -1/l_4 & 0 & 0 & 0 \\ 0 & -1/l_4 & 0 & 0 \\ 0 & 0 & -1/l_6 & 0 \\ 0 & 0 & 0 & -1/l_6 \end{bmatrix} \tag{19.70}$$

19.3.2.2 Balancing Conditions of the Shaking Force

Matrix $\mathbf{G}$ and vector $\mathbf{b}$ are calculated using the obtained matrices $\mathbf{D}^v$, $\mathbf{D}^w$ and vector $\mathbf{d}$ as follows:

$$\mathbf{G} = \begin{bmatrix} -\frac{l_1}{l_4} & 0 & -\frac{l_2}{l_4} & 0 & -\frac{l_{31}}{l_4} & 0 & \frac{l_5}{l_4} & 0 & 0 & 0 \\ 0 & -\frac{l_1}{l_4} & 0 & -\frac{l_2}{l_4} & 0 & -\frac{l_{31}}{l_4} & 0 & \frac{l_5}{l_4} & 0 & 0 \\ -\frac{l_1}{l_6} & 0 & -\frac{l_2}{l_6} & 0 & -\frac{l_3}{l_6} & 0 & 0 & 0 & \frac{l_7}{l_6} & 0 \\ 0 & -\frac{l_1}{l_6} & 0 & -\frac{l_2}{l_6} & 0 & -\frac{l_3}{l_6} & 0 & 0 & 0 & \frac{l_7}{l_6} \end{bmatrix}, \quad \mathbf{b} = \begin{bmatrix} -\frac{x_A}{l_4} & -\frac{y_A}{l_4} & -\frac{x_B}{l_6} & -\frac{y_B}{l_6} \end{bmatrix}^T. \tag{19.71}$$

Now we can determine matrices $\mathbf{C}_i^v$, $\mathbf{C}_i^w$ and vector $\mathbf{e}_i^u$ related to vector $\mathbf{r}_i$ using Eq. (19.28). For example, for $i = 1$:

$$\mathbf{r}_1 = \begin{bmatrix} \xi_{S1}\cos\varphi_1 - \eta_{S1}\sin\varphi_1 \\ \xi_{S1}\sin\varphi_1 + \eta_{S1}\cos\varphi_1 \end{bmatrix}, \mathbf{C}_1^v = \begin{bmatrix} \xi_{S1} & -\eta_{S1} & 0\,0\,0\,0\,0\,0\,0\,0 \\ \eta_{S1} & \xi_{S1} & 0\,0\,0\,0\,0\,0\,0\,0 \end{bmatrix},$$

$$\mathbf{C}_1^w = \begin{bmatrix} 0\,0\,0\,0 \\ 0\,0\,0\,0 \end{bmatrix} \quad \mathbf{e}_1^u = \begin{bmatrix} 0 \\ 0 \end{bmatrix}.$$

For $i = 7$ we get

$$\mathbf{r}_7 = \begin{bmatrix} x_B + l_6\cos\varphi_6 + \xi_{S7}\cos\varphi_7 - \eta_{S7}\sin\varphi_7 \\ y_B + l_6\sin\varphi_6 + \xi_{S7}\sin\varphi_7 + \eta_{S7}\cos\varphi_7 \end{bmatrix},$$

$$\mathbf{C}_7^v = \begin{bmatrix} 0\,0\,0\,0\,0\,0\,0\,0 & \xi_{S7} & -\eta_{S7} \\ 0\,0\,0\,0\,0\,0\,0\,0 & \eta_{S7} & \xi_{S7} \end{bmatrix}, \; \mathbf{C}_7^w = \begin{bmatrix} 0\,0\,l_6\,0 \\ 0\,0\,0\,l_6 \end{bmatrix}, \; \mathbf{e}_7^u = \begin{bmatrix} x_B \\ y_B \end{bmatrix}.$$

Then, matrices $\mathbf{B}_i$ ($i = 1, 2, \ldots, 7$) are calculated using Eq. (19.42). Finally, by substituting matrices $\mathbf{B}_i$ into Eq. (19.45), we find balancing conditions of the shaking force as follows:

$$m_1\frac{\xi_{S1}}{l_1} + m_2 + m_3 + m_4\frac{\xi_{S4}}{l_4} + m_5 + m_6\frac{\xi_{S6}}{l_6} + m_7 = 0 \tag{19.72}$$

$$m_2\frac{\xi_{S2}}{l_2} + m_3 + m_4\frac{\xi_{S4}}{l_4} + m_5 + m_6\frac{\xi_{S6}}{l_6} + m_7 = 0 \tag{19.73}$$

$$m_3\frac{\xi_{S3}}{l_3} + m_4\frac{l_{31}}{l_3}\frac{\xi_{S4}}{l_4} + m_5\frac{l_{31}}{l_3} + m_6\frac{\xi_{S6}}{l_6} + m_7 = 0 \tag{19.74}$$

$$m_4\frac{\xi_{S4}}{l_4} + m_5\left(1 - \frac{\xi_{S5}}{l_5}\right) = 0 \tag{19.75}$$

$$m_6\frac{\xi_{S6}}{l_6} + m_7\left(1 - \frac{\xi_{S7}}{l_7}\right) = 0 \tag{19.76}$$

$$m_1\frac{\eta_{S1}}{l_1} + m_4\frac{\eta_{S4}}{l_4} + m_6\frac{\eta_{S6}}{l_6} = 0 \tag{19.77}$$

$$m_2\frac{\eta_{S2}}{l_2} + m_4\frac{\eta_{S4}}{l_4} + m_6\frac{\eta_{S6}}{l_6} = 0 \tag{19.78}$$

$$m_3\frac{\eta_{S3}}{l_3} + m_4\frac{l_{31}}{l_3}\frac{\eta_{S4}}{l_4} + m_6\frac{\eta_{S6}}{l_6} = 0 \tag{19.79}$$

$$m_4\frac{\eta_{S4}}{l_4} - m_5\frac{\eta_{S5}}{l_5} = 0 \tag{19.80}$$

$$m_6\frac{\eta_{S6}}{l_6} - m_7\frac{\eta_{S7}}{l_7} = 0. \tag{19.81}$$

19.3.2.3 Balancing Conditions of the Shaking Moment

Since matrices $\mathbf{B}_i$ ($i = 1, 2, \ldots, 7$) are known and vectors of constants $\mathbf{e}_i$ are given by Eq. (19.41), matrix $\mathbf{S}_1$ and vector $\mathbf{k}_1$ can be easily calculated using Eq. (19.50). Matrix $\mathbf{H}$ takes the same form as Eq. (19.55) for $p = 7$. By partitioning of matrix $\mathbf{H}$ related to Eq. (19.56), we obtain sub-matrices $\mathbf{H}_1$ and $\mathbf{H}_4$. Then, matrix $\mathbf{S}_2$ and vector $\mathbf{k}_2$ are calculated using Eqs. (19.59) and (19.60). By substituting the obtained expressions of $\mathbf{S}_1$,$\mathbf{S}_2$, $\mathbf{k}_1$,$\mathbf{k}_2$ into Eq. (19.62), the balancing conditions of the shaking moment are then derived in the following form

$$m_1\lambda_1^2 + \frac{I_{S1}}{l_1^2} + m_2 + m_3 + m_4\lambda_4^2 + \frac{I_{S4}}{l_4^2} + m_5 + m_6\lambda_6^2 + \frac{I_{S6}}{l_6^2} + m_7 = 0 \tag{19.82}$$

$$m_2\lambda_2 + m_3 + m_4\lambda_4^2 + \frac{I_{S4}}{l_4^2} + m_5 + m_6\lambda_6^2 + \frac{I_{S6}}{l_6^2} + m_7 = 0 \tag{19.83}$$

$$m_3\lambda_3 + \frac{l_{31}}{l_3}\left(m_4\lambda_4^2 + \frac{I_{S4}}{l_4^2} + m_5\right) + m_6\lambda_6^2 + \frac{I_{S6}}{l_6^2} + m_7 = 0 \tag{19.84}$$

$$m_4\lambda_4^2 + \frac{I_{S4}}{l_4^2} + m_5\left(1 - \lambda_5\right) = 0 \tag{19.85}$$

$$m_6\lambda_6^2 + \frac{I_{S6}}{l_6^2} + m_7(1-\lambda_7) = 0 \tag{19.86}$$

$$m_2\lambda_2^2 + \frac{I_{S2}}{l_2^2} + m_3 + m_4\lambda_4^2 + \frac{I_{S4}}{l_4^2} + m_5 + m_6\lambda_6^2 + \frac{I_{S6}}{l_6^2} + m_7 = 0 \tag{19.87}$$

$$m_3\lambda_3^2 + \frac{I_{S3}}{l_3^2} + \frac{l_{31}^2}{l_3^2}\left(m_4\lambda_4^2 + \frac{I_{S4}}{l_4^2} + m_5\right) + m_6\lambda_6^2 + \frac{I_{S6}}{l_6^2} + m_7 = 0 \tag{19.88}$$

$$m_4\lambda_4^2 + \frac{I_{S4}}{l_4^2} + m_5\lambda_5^2 + \frac{I_{S5}}{l_5^2} + m_5(1-2\lambda_5) = 0 \tag{19.89}$$

$$m_6\lambda_6^2 + \frac{I_{S6}}{l_6^2} + m_7\lambda_7^2 + \frac{I_{S7}}{l_7^2} + m_7(1-2\lambda_7) = 0 \tag{19.90}$$

$$m_4\lambda_4^2 + \frac{I_{S4}}{l_4^2} - m_4\lambda_4 = 0 \tag{19.91}$$

$$m_6\lambda_6^2 + \frac{I_{S6}}{l_6^2} - m_6\lambda_6 = 0 \tag{19.92}$$

$$\eta_{S1} = \eta_{S2} = \eta_{S3} = \eta_{S4} = \eta_{S5} = \eta_{S6} = \eta_{S7} = 0 \tag{19.93}$$

where $\lambda_i = \frac{\xi_{Si}}{l_i}$ for $i = 1, 2, \ldots, 7$.

In the case that S_i is positioned along the link line, that is, $\eta_{Si} = 0$ for $i = 1, 2, \ldots, 7$, the balancing conditions for the shaking force and shaking moment of the 8R-eightbar mechanism, Eqs. (19.72)–(19.93), are reduced into the following set of equations

$$m_1\lambda_1 + m_2(1-\lambda_2) = 0, \tag{19.94}$$

$$m_2\lambda_2 + m_3 + m_5\lambda_5 + m_7\lambda_7 = 0, \tag{19.95}$$

$$m_3\lambda_3 + m_5\lambda_5\frac{l_{31}}{l_3} + m_7\lambda_7 = 0, \tag{19.96}$$

$$m_4\lambda_4 + m_5(1-\lambda_5) = 0, \tag{19.97}$$

$$m_6\lambda_6 + m_7(1-\lambda_7) = 0, \tag{19.98}$$

$$m_i\lambda_i^2 + \frac{I_{Si}}{l_i^2} - m_i\lambda_i = 0 \text{ for } i = 1, 2, 4, 5, 6, 7, \tag{19.99}$$

Table 19.1 Initial parameters of the 8R-eightbar mechanism

Link i	l_i (m)	ξ_{Si}^0 (m)	η_{Si}^0 (m)	m_i^0 (kg)	λ_i^0
1	0.08	0.04	0.01	2.4	0.5
2	0.20	0.07	0.025	3.5	0.35
3	0.35	0.15	0.035	3.6	0.428
4	0.12	0.05	0.015	2.2	0.4167
5	0.15	0.08	0.01	2.4	0.5333
6	0.12	0.06	0.02	2.0	0.5
7	0.15	0.1	0.02	2.7	0.6667

$$m_3\lambda_3^2 + \frac{I_{S3}}{l_3^2} - m_3\lambda_3 - \frac{l_{31}}{l_3} m_5\lambda_5 \left(1 - \frac{l_{31}}{l_3}\right) = 0, \tag{19.100}$$

where Eqs. (19.94)–(19.98) are the balancing conditions of the shaking force and Eqs. (19.99)–(19.100) are the balancing conditions of the shaking moment of the fully force balanced mechanism.

19.3.2.4 Numerical Study

A numerical calculation is implemented to verify the correctness of the obtained balancing conditions. The geometry and mass distribution parameters of the links are given in Table 19.1, where m_i^0, ξ_{Si}^0, η_{Si}^0 and $\lambda_i^0 = \xi_{Si}^0/l_i$ denote the initial parameters. The other geometry parameters are: $x_A = 0.17$ (m), $x_B = 0.3$ (m), $y_A = y_B = 0.03$ (m)and $l_{31} = 0.07$ (m).

Upon assuming that parameter $\eta_{Si} = 0$ for $i = 1, 2, \ldots, 7$, the remaining five conditions (19.94)–(19.98) contain a set of 14 variables m_i and λ_i. We can establish a balancing scheme with counterweights by keeping the parameters of links 3 and 5, i.e., $m_3\lambda_3 = m_3^0\lambda_3^0$, $m_5\lambda_5 = m_5^0\lambda_5^0$, and solving parameters of the other links from these conditions as follows:

$$m_1\lambda_1 = -m_3^0\left(1 - \lambda_3^0\right) - m_5^0\lambda_5^0\left(1 - \frac{l_{31}}{l_3}\right) - m_2,$$

$$m_2\lambda_2 = -m_3^0\left(1 - \lambda_3^0\right) - m_5^0\lambda_5^0\left(1 - \frac{l_{31}}{l_3}\right),$$

$$m_4\lambda_4 = -m_5^0\left(1 - \lambda_5^0\right),\ m_6\lambda_6 = -\left(m_3^0\lambda_3^0 + m_5^0\lambda_5^0\frac{l_{31}}{l_3}\right) - m_7,$$

$$m_7\lambda_7 = -\left(m_3^0\lambda_3^0 + m_5^0\lambda_5^0\frac{l_{31}}{l_3}\right).$$

It follows that parameters λ_1, λ_2, λ_4, λ_6 and λ_7 will take negative values since $0 < \lambda_3^0 < 1$ and $0 < \lambda_5^0 < 1$. As a result, the centers of mass S_1, S_2, S_4, S_6 and S_7 must be positioned at the other side of joints O_1, O_2, O_4, O_6 and O_7 respectively.

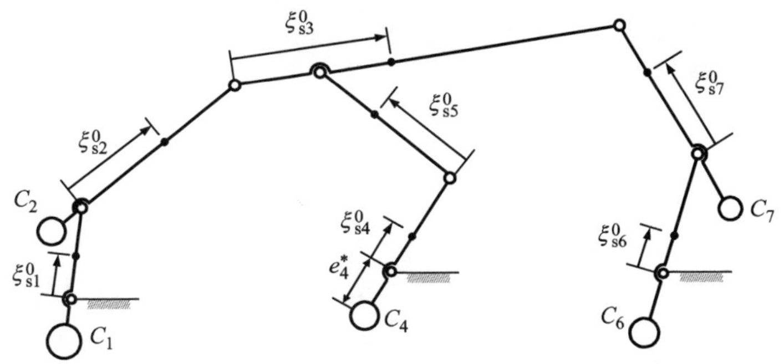

Fig. 19.4 A balancing scheme using five counterweights for the full force balancing

Table 19.2 Parameters of the force balanced the 8R-eightbar mechanism with counterweights

				Counterweights	
Link i	ξ_{Si} (m)	η_{Si} (m)	m_i (kg) (with counterweight)	e_i^* (m)	m_i^* (kg)
1	−0.0891	0.0	10.4	0.128	8.0
2	−0.0725	0.0	8.5	0.172	5.0
3	0.15	0.0	3.6	0.0	0.0
4	−0.0187	0.0	7.2	0.049	5.0
5	0.080	0.0	2.4	0.0	0.0
6	−0.1425	0.0	8.0	0.210	6.0
7	−0.0350	0.0	7.7	0.108	5.0

For this purpose, a balancing scheme with five counterweights attached to the corresponding links as shown in Fig. 19.4 is suggested. Using the same way, we can establish other force balancing schema by assigning the parameters of two arbitrary links and calculating parameters of the other links from Eqs. (19.94)–(19.98).

The mass m_i^* and the distance e_i^* of the counterweight C_i attached to link i can then be easily calculated by applying the relationship $m_i\lambda_i = m_i^0\lambda_i^0 - m_i^*\frac{e_i^*}{l_i}$. The mass distribution parameters of the links and counterweights of the fully force balanced mechanism are given in Table 19.2.

Figure 19.5 shows two components of the shaking force produced by the initial mechanism and the force balanced mechanism. The numerical results verified that the shaking force is completely eliminated during the motion of the force balanced mechanism.

In the next step, the moment balancing conditions, Eqs. (19.99) and (19.100), will be taken into account for canceling the shaking moment of the fully force balanced mechanism. The moments of inertia of links of the full force balanced mechanism are as follows: $I^0_{S1} = 0.2$, $I^0_{S2} = 0.35$, $I^0_{S3} = 0.11$, $I^0_{S4} = 0.09$, $I^0_{S5} = 0.08$, $I^0_{S6} = 0.41$, $I^0_{S7} = 0.15$ (kg m^2). A number of balancing schema using additional members were applied to balance the shaking moment at any rotating

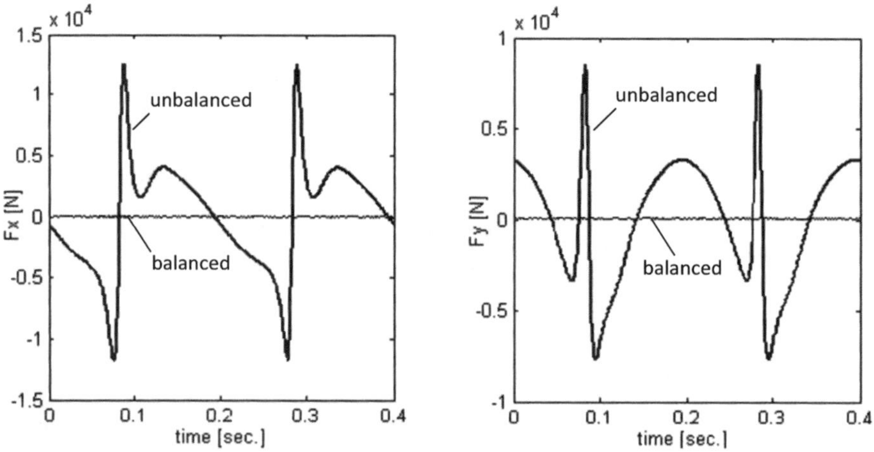

Fig. 19.5 Shaking forces of the unbalanced mechanism and the fully force balanced mechanism. (Rotating speeds of the cranks 1, 4 and 6 are assumed to be the same value of 300 rpm)

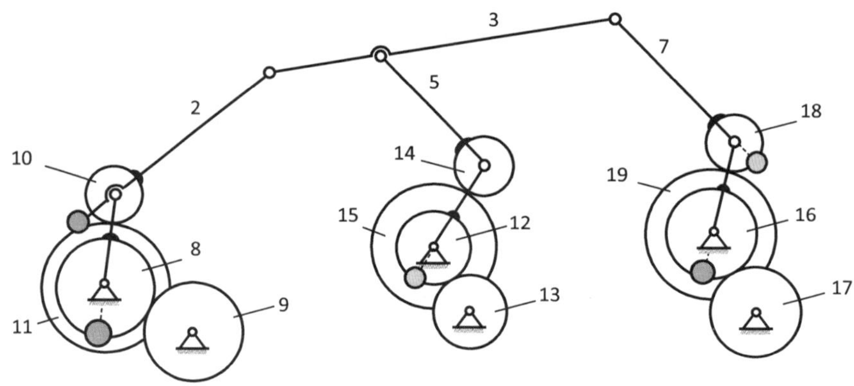

Fig. 19.6 A balancing scheme of the shaking moment of the fully force balanced mechanism

speed of the driving links, e.g., [12, 14, 15]. A well-known balancing scheme with counter-rotating balancers in Fig. 19.6 is used to verify the correctness of the conditions of moment balancing. The required moment of inertia of link 3 is calculated using Eq. (19.100) with the parameters given in Table 19.2, that yields $I_{S3} = 0.133\ \left(\text{kg m}^2\right)$. This value can be attained by mass redistribution for link 3.

As shown in Fig. 19.6, gears 11, 15, and 19 are mounted on the rotation axis of the input cranks 1, 4, and 6, respectively. They mesh with planetary gears 10, 14, and 18 mounted on links 2, 5 and 7 respectively. Using this balancing scheme, the additional balancing moments will be produced to balance correspondingly the inertia moments of all links. In other words, the shaking moment can be balanced, while the shaking force is still fully balanced. For brevity, the transmission ratios of the gear-pairs of the considered balancing scheme are chosen as follows:

$$\frac{r_8}{r_9} = \frac{r_{12}}{r_{13}} = \frac{r_{16}}{r_{17}} = 1, \quad \frac{r_{10}}{r_{11}} = \frac{r_{14}}{r_{15}} = \frac{r_{18}}{r_{19}} = \frac{1}{2}.$$

According to Fig. 19.6, the kinematic relationship of gear-pair 10–11 is

$$r_{10}\dot{\varphi}_{10} + r_{11}\dot{\varphi}_{11} - (r_{10} + r_{11})\,\dot{\varphi}_1 = 0, \quad \dot{\varphi}_{10} = \dot{\varphi}_2, \tag{19.101}$$

where r_i is the rolling circle radius of ith gear. Using Eq. (19.101) we obtain

$$I_{S11}\varphi_{11} = I_{S11}\left(\frac{r_{10}}{r_{11}} + 1\right)\varphi_1 - I_{S11}\frac{r_{10}}{r_{11}}\varphi_2, \tag{19.102}$$

$$I_{S10}\dot{\varphi}_{10} = I_{S10}\dot{\varphi}_2, \tag{19.103}$$

where I_{Si} is the moment of inertia of ith gear. Using Eqs. (19.102), (19.103) and the balancing condition (19.99) we obtain the following balancing condition for link 2 with the additional planetary gear

$$m_2\lambda_2^2 + \frac{1}{l_2^2}\left(I_{S2} + I_{S10} - \frac{r_{10}}{r_{11}}J_{S11}\right) - m_2\lambda_2 = 0. \tag{19.104}$$

By the same way, the balancing conditions with the additional gears for links 5 and 7 can be formulated as follows:

$$m_5\lambda_5^2 + \frac{1}{l_5^2}\left(I_{S5} + J_{S14} - \frac{r_{14}}{r_{15}}J_{S15}\right) - m_5\lambda_5 = 0, \tag{19.105}$$

$$m_7\lambda_7^2 + \frac{1}{l_7^2}\left(I_{S7} + J_{S18} - \frac{r_{18}}{r_{19}}J_{S19}\right) - m_7\lambda_7 = 0. \tag{19.106}$$

The moment of inertia of gear-pairs 10–11, 14–15, and 18–19 can then be chosen in order to satisfy Eqs. (19.104)–(19.106). Using Eq. (19.99), the moment of inertia of the gear-pairs 8–9, 12–13, and 16–17 can be determined by the similar way. Their values are given in Table 19.3.

Figure 19.7 shows the numerical results for the shaking moment of the fully moment balanced mechanism, where the input speeds of cranks 1, 4 and 6 are $\dot{\varphi}_1 =$

Table 19.3 Moments of inertia of the gears

Gear i	8	9	10	11	12	13	14	15	16	17	18	19
I_{Si} (kg m^2)	0.02	1.0	0.05	0.42	0.05	0.51	0.05	0.23	0.1	1.56	0.05	0.5

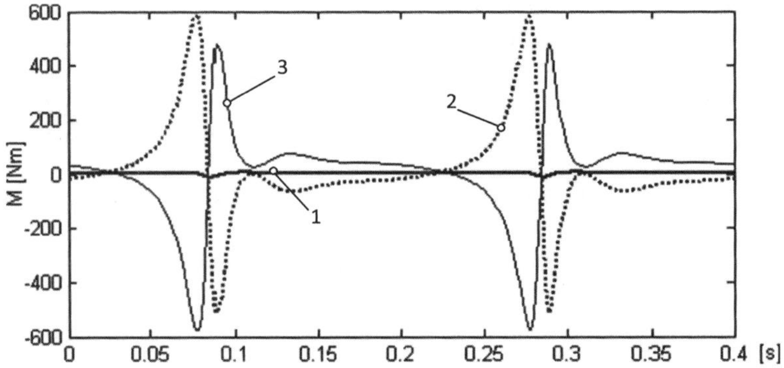

Fig. 19.7 The shaking moment (*curve 1*) of the fully moment balanced mechanism as a sum of the first term (*curve 2*) and the second term (*curve 3*)

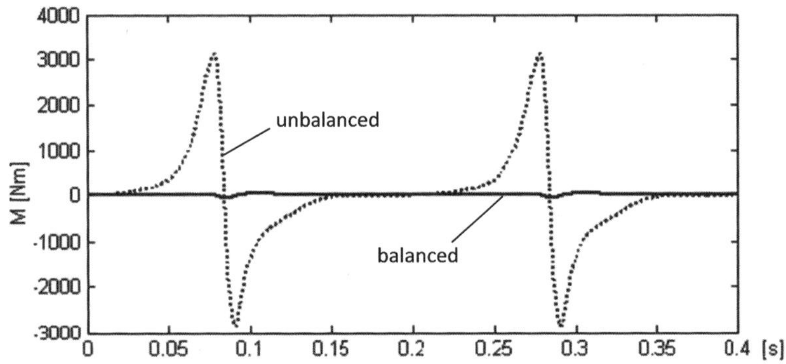

Fig. 19.8 Shaking moments of the unbalanced and the fully moment balanced mechanism

$\dot{\varphi}_4 = \dot{\varphi}_6 = 31.4$ (rad/s). The results shown in Fig. 19.8 demonstrated that the shaking moment of the 8R-eightbar mechanism is eliminated after balancing.

19.4 Balancing Conditions of Spatial One-DOF Mechanisms

19.4.1 Theory and Procedure for Deriving Balancing Conditions

19.4.1.1 The General Balancing Conditions of Spatial One-DOF Mechanisms

This section presents a method to algebraically derive the balancing conditions for shaking force and shaking moment of spatial one-degree-of freedom mechanisms. Let q be the independent generalized coordinate which describes the position of the mechanism. According to Eqs. (19.10) and (19.12), the velocity $\mathbf{v}_{Si}$ and the angular velocity $\boldsymbol{\omega}_i$ are given by

$$\mathbf{v}_{Si} = \mathbf{J}_{Ti}(q)\dot{q}(t), \tag{19.107}$$

$$\boldsymbol{\omega}_i = \mathbf{J}_{Ri}(q)\dot{q}(t), \tag{19.108}$$

where $\mathbf{J}_{Ti}(q)$ and $\mathbf{J}_{Ri}(q)$ are 3×1 Jacobian matrices and can be written in the form

$$\mathbf{J}_{Ti} = \left[x'_{Si}\ y'_{Si}\ z'_{Si}\right]^T,\ \mathbf{J}_{Ri} = \left[s'_{ix}\ s'_{iy}\ s'_{iz}\right]^T, \tag{19.109}$$

where the prime represents the derivative with respect to the generalized coordinate q and s_{ix}, s_{iy}, s_{iz} are three components of rotational vector φ_i for link i (see Sect. 19.2).

We recall that the inertia matrix $\mathbf{I}_{Si}$ is defined with respect to the fixed coordinate frame $\{Oxyz\}$ as shown in Fig. 19.9. The elements of matrix $\mathbf{I}_{Si}$ are time dependent

$$\mathbf{I}_{Si} = \begin{bmatrix} I_{ixx} & I_{ixy} & I_{ixz} \\ I_{iyx} & I_{iyy} & I_{iyz} \\ I_{izx} & I_{izy} & I_{izz} \end{bmatrix}. \tag{19.110}$$

Using Eqs. (19.107) and (19.108), Eqs. (19.6) and (19.7) take the following form

$$\frac{d}{dt}\left\{\left[\sum_{i=1}^{p} m_i\, \mathbf{J}_{Ti}(q)\right]\dot{q}\right\} = \mathbf{0}, \tag{19.111}$$

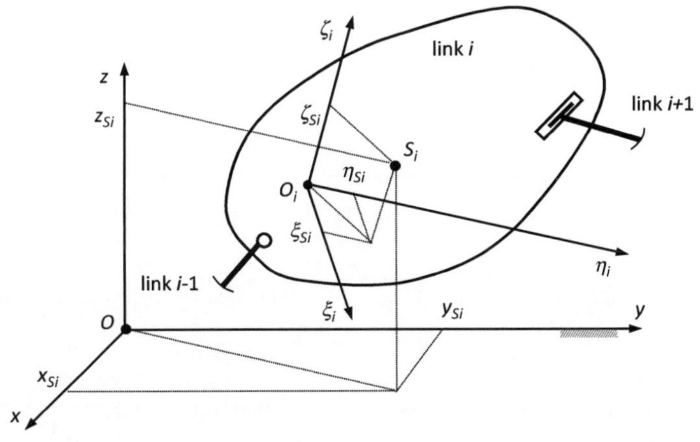

Fig. 19.9 Definition of coordinates

$$\frac{d}{\mathrm{d}t}\left\{\left[\sum_{i=1}^{p}\mathbf{I}_{Si}\boldsymbol{J}_{Ri}(q)+m_i\tilde{\mathbf{r}}_{Si}\,\mathbf{J}_{Ti}(q)\right]\dot{q}\right\}=\mathbf{0}. \tag{19.112}$$

The use of Eqs. (19.109)–(19.112) yields

$$\left(\ddot{q}+\dot{q}^2\frac{d}{dq}\right)\sum_{i=1}^{n}m_i x'_{Si}=0, \tag{19.113}$$

$$\left(\ddot{q}+\dot{q}^2\frac{d}{\mathrm{d}q}\right)\sum_{i=1}^{n}m_i y'_{Si}=0, \tag{19.114}$$

$$\left(\ddot{q}+\dot{q}^2\frac{d}{\mathrm{d}q}\right)\sum_{i=1}^{n}m_i z'_{Si}=0 \tag{19.115}$$

$$\left(\ddot{q}+\dot{q}^2\frac{d}{\mathrm{d}q}\right)\sum_{i=1}^{n}\left[m_i\left(y_{Si}z'_{Si}-z_{Si}y'_{Si}\right)+I_{ixx}s'_{ix}+I_{ixy}s'_{iy}+I_{ixz}s'_{iz}\right]=0, \tag{19.116}$$

$$\left(\ddot{q}+\dot{q}^2\frac{d}{\mathrm{d}q}\right)\sum_{i=1}^{n}\left[m_i\left(z_{Si}x'_{Si}-x_{Si}z'_{Si}\right)+I_{iyx}s'_{ix}+I_{iyy}s'_{iy}+I_{iyz}s'_{iz}\right]=0, \tag{19.117}$$

$$\left(\ddot{q}+\dot{q}^2\frac{d}{\mathrm{d}q}\right)\sum_{i=1}^{n}\left[m_i\left(x_{Si}y'_{Si}-y_{Si}x'_{Si}\right)+I_{izx}s'_{ix}+I_{izy}s'_{iy}+I_{izz}s'_{iz}\right]=0. \tag{19.118}$$

This yields the general conditions for complete balancing of spatial mechanisms

$$\sum_{i=1}^{n}m_i x'_{Si}=0,\ \sum_{i=1}^{n}m_i y'_{Si}=0,\ \sum_{i=1}^{n}m_i z'_{Si}=0, \tag{19.119}$$

$$\sum_{i=1}^{n}\left[m_i\left(y_{Si}z'_{Si}-z_{Si}y'_{Si}\right)+I_{ixx}s'_{ix}+I_{ixy}s'_{iy}+I_{ixz}s'_{iz}\right]=0 \tag{19.120}$$

$$\sum_{i=1}^{n}\left[m_i\left(z_{Si}x'_{Si}-x_{Si}z'_{Si}\right)+I_{iyx}s'_{ix}+I_{iyy}s'_{iy}+I_{iyz}s'_{iz}\right]=0 \tag{19.121}$$

$$\sum_{i=1}^{n}\left[m_i\left(x_{Si}y'_{Si}-y_{Si}x'_{Si}\right)+I_{izx}s'_{ix}+I_{izy}s'_{iy}+I_{izz}s'_{iz}\right]=0 \tag{19.122}$$

19.4.1.2 Algebraic Balancing Conditions of the Shaking Force

The position vector $\mathbf{r}_{Si}$ with respect to the fixed coordinate frame is given by

$$\mathbf{r}_{Si}=\mathbf{r}_{Oi}+\mathbf{A}_i\mathbf{r}_{Si}^{(i)}, \tag{19.123}$$

where $\mathbf{r}_{Oi}$ is position vector of origin O_i in the fixed coordinate frame $\{Oxyz\}$ and $\mathbf{r}_i^{(i)}$ is position vector of S_i in the moving coordinate frame $\{O_i\xi_i\eta_i\zeta_i\}$ shown in Fig. 19.9.

$$\mathbf{r}_{Si}^{(i)} = \left[\xi_{Si}\ \eta_{Si}\ \zeta_{Si}\right]^T. \tag{19.124}$$

The coordinates of the center of mass S_i, $\mathbf{r}_{Si} = \left[x_{Si}\ y_{Si}\ z_{Si}\right]^T$, can be rewritten as [14, 15]

$$x_{Si} = e_{xi}^* + \mathbf{a}_i^T\mathbf{z},\ y_{Si} = e_{yi}^* + \mathbf{b}_i^T\mathbf{z},\ z_{Si} = e_{zi}^* + \mathbf{c}_i^T\mathbf{z}, \quad i = 1,\ 2, .., n \tag{19.125}$$

where the vectors $\mathbf{a}_i$, $\mathbf{b}_i$ and $\mathbf{c}_i$ consist of components which are independent of q, the elements of vector $\mathbf{z}$ are functions of the generalized coordinates which describe the motion of particular links, e_{xi}^*, e_{yi}^* and e_{zi}^* are constant values.

Analog to Eq. (19.125), the loop equations of the mechanism may be written in the matrix form

$$\mathbf{D}\,\mathbf{z} = \mathbf{f},\ \ \mathbf{D} = [\mathbf{D}_{\mathrm{I}},\ \mathbf{D}_{\mathrm{II}}]\ . \tag{19.126}$$

Here the matrix $\mathbf{D}$ and the vector $\mathbf{f}$ include the components which are geometrical parameters and independent of q. A partitioning of vector $\mathbf{z}$ from Eq. (19.126)

$$\mathbf{z} = \begin{bmatrix} \mathbf{v} \\ \mathbf{w} \end{bmatrix}, \tag{19.127}$$

leads to the following relation

$$\mathbf{D}_{\mathrm{I}}\mathbf{v} + \mathbf{D}_{\mathrm{II}}\mathbf{w} = \mathbf{f}. \tag{19.128}$$

The matrix $\mathbf{D}_{\mathrm{II}}$ is chosen so that it must be a square matrix and nonsingular. The dimension of vector $\mathbf{w}$ and the number of the loop equations are equal. By solving Eq. (19.128) with the vector of variables $\mathbf{w}$, we get

$$\mathbf{w} = \mathbf{D}_{\mathrm{II}}^{-1}\,(\,\mathbf{f} – \mathbf{D}_{\mathrm{I}}\mathbf{v})\,. \tag{19.129}$$

Using Eqs. (19.127) and (19.129), the coordinates of the center of mass S_i and their derivatives can be expressed in terms of the reduced vector of variables $\mathbf{v}$ as

$$x_{Si} = e_{xi} + \mathbf{g}_i^T\mathbf{v},\ y_{Si} = e_{yi} + \mathbf{h}_i^T\mathbf{v},\ z_{Si} = e_{yi} + \mathbf{k}_i^T\mathbf{v} \tag{19.130}$$

$$x'_{Si} = \mathbf{g}_i^T\frac{\mathrm{d}\mathbf{v}}{\mathrm{d}q}\,,\ y'_{Si} = \mathbf{h}_i^T\frac{\mathrm{d}\mathbf{v}}{\mathrm{d}q},\ z'_{Si} = \mathbf{k}_i^T\frac{\mathrm{d}\mathbf{v}}{\mathrm{d}q}, \tag{19.131}$$

where

$$\mathbf{g}_i^T = \mathbf{a}_{i\,\mathrm{I}}^T - \mathbf{a}_{i\,\mathrm{II}}^T\,\mathbf{D}_{\mathrm{II}}^{-1}\,\mathbf{D}_{\mathrm{I}},\;\; \mathbf{h}_i^T = \mathbf{b}_{i\,\mathrm{I}}^T - \mathbf{b}_{i\,\mathrm{II}}^T\,\mathbf{D}_{\mathrm{II}}^{-1}\,\mathbf{D}_{\mathrm{I}},\;\; \mathbf{k}_i^T = \mathbf{c}_{i\,\mathrm{I}}^T - \mathbf{c}_{i\,\mathrm{II}}^T\,\mathbf{D}_{\mathrm{II}}^{-1}\,\mathbf{D}_{\mathrm{I}}$$
$$e_{xi} = e_{xi}^* + \mathbf{a}_{i\,\mathrm{II}}^T\,\mathbf{D}_{\mathrm{II}}^{-1}\,\mathbf{f},\;\; e_{yi} = e_{yi}^* + \mathbf{b}_{i\,\mathrm{II}}^T\,\mathbf{D}_{\mathrm{II}}^{-1}\,\mathbf{f},\;\; e_{zi} = e_{zi}^* + \mathbf{c}_{i\,\mathrm{II}}^T\,\mathbf{D}_{\mathrm{II}}^{-1}\,\mathbf{f}, \tag{19.132}$$

where vectors $\mathbf{a}_{i\mathrm{I}}, \mathbf{a}_{i\mathrm{II}}, \mathbf{b}_{i\mathrm{I}}, \mathbf{b}_{i\mathrm{II}}, \mathbf{c}_{i\mathrm{I}}, \mathbf{c}_{i\mathrm{II}}$ include elements which are independent of q. Substituting Eq. (19.131) into balancing conditions (19.119), we obtain

$$\left(\sum_{i=1}^{n} m_i \mathbf{g}_i^T\right)\frac{\mathrm{d}\mathbf{v}}{\mathrm{d}q} = 0\,,\;\; \left(\sum_{i=1}^{n} m_i \mathbf{h}_i^T\right)\frac{\mathrm{d}\mathbf{v}}{\mathrm{d}q} = 0,\;\; \left(\sum_{i=1}^{n} m_i \mathbf{k}_i^T\right)\frac{\mathrm{d}\mathbf{v}}{\mathrm{d}q} = 0\,. \tag{19.133}$$

Finally, the algebraic balancing conditions for shaking force take the compact matrix form

$$\sum_{i=1}^{n} m_i \mathbf{g}_i^T = \mathbf{0},\; \sum_{i=1}^{n} m_i \mathbf{h}_i^T = \mathbf{0},\; \sum_{i=1}^{n} m_i \mathbf{k}_i^T = \mathbf{0}. \tag{19.134}$$

19.4.1.3 Algebraic Balancing Conditions of the Shaking Moment

To extract the conditions for the shaking moment balancing, some additional transformations are required. The substitution of Eqs. (19.130) and (19.131) into Eqs. (19.120)–(19.122) yields

$$\mathbf{u}_1^T\frac{\mathrm{d}\mathbf{v}}{\mathrm{d}q} + \mathbf{v}^T\mathbf{S}_1\frac{\mathrm{d}\mathbf{v}}{\mathrm{d}q} + \sum_{i=1}^{n}\left(I_{ixx}s'_{ix} + I_{ixy}s'_{iy} + I_{ixz}s'_{iz}\right) = 0, \tag{19.135}$$

$$\mathbf{u}_2^T\frac{\mathrm{d}\mathbf{v}}{\mathrm{d}q} + \mathbf{v}^T\mathbf{S}_2\frac{\mathrm{d}\mathbf{v}}{\mathrm{d}q} + \sum_{i=1}^{n}\left(I_{iyx}s'_{ix} + I_{iyy}s'_{iy} + I_{iyz}s'_{iz}\right) = 0, \tag{19.136}$$

$$\mathbf{u}_3^T\frac{\mathrm{d}\mathbf{v}}{\mathrm{d}q} + \mathbf{v}^T\mathbf{S}_3\frac{\mathrm{d}\mathbf{v}}{\mathrm{d}q} + \sum_{i=1}^{n}\left(I_{izx}s'_{ix} + I_{izy}s'_{iy} + I_{izz}s'_{iz}\right) = 0, \tag{19.137}$$

where

$$\mathbf{u}_1^T = \sum_{i=1}^{n} m_i\left(e_{yi}\mathbf{k}_i^T - e_{zi}\mathbf{h}_i^T\right),\;\; \mathbf{u}_2^T = \sum_{i=1}^{n} m_i\left(e_{zi}\mathbf{g}_i^T - e_{xi}\mathbf{k}_i^T\right),$$
$$\mathbf{u}_3^T = \sum_{i=1}^{n} m_i\left(e_{xi}\mathbf{h}_i^T - e_{yi}\mathbf{g}_i^T\right), \tag{19.138}$$

and skew-symmetric matrices

$$\mathbf{S}_1 = \sum_{i=1}^{n} m_i \left(\mathbf{h}_i\mathbf{k}_i^T - \mathbf{k}_i\mathbf{h}_i^T\right), \quad \mathbf{S}_2 = \sum_{i=1}^{n} m_i \left(\mathbf{k}_i\mathbf{g}_i^T - \mathbf{g}_i\mathbf{k}_i^T\right),$$

$$\mathbf{S}_3 = \sum_{i=1}^{n} m_i \left(\mathbf{g}_i\mathbf{h}_i^T - \mathbf{h}_i\mathbf{g}_i^T\right). \tag{19.139}$$

Analog to Eq. (19.125), the elements of the rotational vector φ_i can be rewritten as [14, 15]

$$s_{ix} = s_{ix}^* + \mathbf{r}_{1i}^T\mathbf{z}, \; s_{iy} = s_{iy}^* + \mathbf{r}_{2i}^T\mathbf{z}, \; s_{iz} = s_{iz}^* + \mathbf{r}_{3i}^T\mathbf{z}, \tag{19.140}$$

where the vectors $\mathbf{r}_{1i}$, $\mathbf{r}_{2i}$ and $\mathbf{r}_{3i}$include components which are independent of q, the values s_{ix}^*, s_{iy}^* and s_{iz}^* are constant. The corresponding derivatives are given by

$$s'_{ix} = \mathbf{r}_{1i}^T\frac{\mathrm{d}\mathbf{z}}{\mathrm{d}q}, \; s'_{iy} = \mathbf{r}_{2i}^T\frac{\mathrm{d}\mathbf{z}}{\mathrm{d}q}, \; s'_{iz} = \mathbf{r}_{3i}^T\frac{\mathrm{d}\mathbf{z}}{\mathrm{d}q}. \tag{19.141}$$

With the vector of variables $\mathbf{z}$, the elements of the inertia matrix $\mathbf{I}_i$ may be rewritten in the matrix form as

$$\begin{aligned} I_{ixx} &= \mathbf{z}^T\mathbf{d}_{ixx}, \; I_{ixy} = \mathbf{z}^T\mathbf{d}_{ixy}, \; I_{ixz} = \mathbf{z}^T\mathbf{d}_{ixz}, \\ I_{iyx} &= \mathbf{z}^T\mathbf{d}_{iyx}, \; I_{iyy} = \mathbf{z}^T\mathbf{d}_{iyy}, \; I_{iyz} = \mathbf{z}^T\mathbf{d}_{iyz}, \\ I_{izx} &= \mathbf{z}^T\mathbf{d}_{izx}, \; I_{izy} = \mathbf{z}^T\mathbf{d}_{izy}, \; I_{izz} = \mathbf{z}^T\mathbf{d}_{izz}, \end{aligned} \tag{19.142}$$

where all elements in the vectors $\mathbf{d}_{ixx}, \mathbf{d}_{ixy}, \mathbf{d}_{ixz}, \mathbf{d}_{iyx}, \mathbf{d}_{iyy}, \mathbf{d}_{iyz}, \mathbf{d}_{izx}, \mathbf{d}_{izy}, \mathbf{d}_{izz}$ are independent of the generalized coordinate q. By using Eqs. (19.141), (19.142) and introducing the new matrices

$$\begin{aligned} \mathbf{H}_1 &= \mathbf{d}_{ixx}\mathbf{r}_{1i}^T + \mathbf{d}_{ixy}\mathbf{r}_{2i}^T + \mathbf{d}_{ixz}\mathbf{r}_{3i}^T, \\ \mathbf{H}_2 &= \mathbf{d}_{iyx}\mathbf{r}_{1i}^T + \mathbf{d}_{iyy}\mathbf{r}_{2i}^T + \mathbf{d}_{iyz}\mathbf{r}_{3i}^T, \\ \mathbf{H}_3 &= \mathbf{d}_{izx}\mathbf{r}_{1i}^T + \mathbf{d}_{izy}\mathbf{r}_{2i}^T + \mathbf{d}_{izz}\mathbf{r}_{3i}^T, \end{aligned} \tag{19.143}$$

the third term in Eqs. (19.135)–(19.137) may be expressed in the matrix form as

$$\sum_{i=1}^{n} \left(I_{ixx}s'_{ix} + I_{ixy}s'_{iy} + I_{ixz}s'_{iz}\right) = \mathbf{z}^T\mathbf{H}_1\frac{\mathrm{d}\mathbf{z}}{\mathrm{d}q}, \tag{19.144}$$

$$\sum_{i=1}^{n} \left(I_{iyx}s'_{ix} + I_{iyy}s'_{iy} + I_{iyz}s'_{iz}\right) = \mathbf{z}^T\mathbf{H}_2\frac{\mathrm{d}\mathbf{z}}{\mathrm{d}q}, \tag{19.145}$$

$$\sum_{i=1}^{n} \left(I_{izx}s'_{ix} + I_{izy}s'_{iy} + I_{izz}s'_{iz}\right) = \mathbf{z}^T\mathbf{H}_3\frac{\mathrm{d}\mathbf{z}}{\mathrm{d}q}. \tag{19.146}$$

The matrix $\mathbf{H}_j$ can be partitioned in four sub-matrices corresponding to the vector of variables $\mathbf{v}$ und $\mathbf{w}$ in Eq. (19.128) as follows:

$$\mathbf{H}_j = \begin{bmatrix} \mathbf{H}_{j1} & \mathbf{H}_{j2} \\ \mathbf{H}_{j3} & \mathbf{H}_{j4} \end{bmatrix}, \; j = 1,\ 2,\ 3. \tag{19.147}$$

By using Eqs. (19.129) and (19.147), the following relation is found from Eqs. (19.144)–(19.146)

$$\begin{aligned} \mathbf{z}^T \mathbf{H}_j \tfrac{\mathrm{d}\mathbf{z}}{dq} &= \begin{bmatrix} \mathbf{v} \\ \mathbf{w} \end{bmatrix}^T \begin{bmatrix} \mathbf{H}_{j1} & \mathbf{H}_{j2} \\ \mathbf{H}_{j3} & \mathbf{H}_{j4} \end{bmatrix} \begin{bmatrix} \tfrac{\mathrm{d}\mathbf{v}}{dq} \\ \tfrac{\mathrm{d}\mathbf{w}}{dq} \end{bmatrix} \\ &= \mathbf{v}^T \left[\mathbf{H}_{j1} + \left(\mathbf{D}_{\mathrm{II}}^{-1}\mathbf{D}_{\mathrm{I}}\right)^{\mathrm{T}} \left(\mathbf{H}_{j4}\mathbf{D}_{\mathrm{II}}^{-1}\mathbf{D}_{\mathrm{I}} - \mathbf{H}_{j3}\right) - \mathbf{H}_{j2}\mathbf{D}_{\mathrm{II}}^{-1}\mathbf{D}_{\mathrm{I}} \right] \tfrac{\mathrm{d}\mathbf{v}}{dq} \\ &\quad + \left(\mathbf{D}_{\mathrm{II}}^{-1}\mathbf{f}\right)^{\mathrm{T}} \left(\mathbf{H}_{j3} - \mathbf{H}_{j4}\mathbf{D}_{\mathrm{II}}^{-1}\mathbf{D}_{\mathrm{I}}\right) \tfrac{\mathrm{d}\mathbf{v}}{dq}. \end{aligned} \tag{19.148}$$

By introducing the vector $\mathbf{u}_j^*$

$$\left(\mathbf{u}_j^*\right)^T = \left(\mathbf{D}_{\mathrm{II}}^{-1}\mathbf{f}\right)^{\mathrm{T}} \left(\mathbf{H}_{j3} - \mathbf{H}_{j4}\mathbf{D}_{\mathrm{II}}^{-1}\mathbf{D}_{\mathrm{I}}\right), j = 1,\ 2,\ 3 \tag{19.149}$$

and the matrices $\mathbf{S}_j^*$ ($j = 1, 2, 3$)

$$\mathbf{S}_j^* = \mathbf{H}_{j1} + \left(\mathbf{D}_{\mathrm{II}}^{-1}\mathbf{D}_{\mathrm{I}}\right)^T \left(\mathbf{H}_{j4}\mathbf{D}_{\mathrm{II}}^{-1}\mathbf{D}_{\mathrm{I}} - \mathbf{H}_{j3}\right) - \mathbf{H}_{j2}\mathbf{D}_{\mathrm{II}}^{-1}\mathbf{D}_{\mathrm{I}}, \tag{19.150}$$

Equations (19.135)–(19.137) take the compact form

$$\left(\mathbf{u}_j + \mathbf{u}_j^*\right)^T \frac{\mathrm{d}\mathbf{v}}{dq} + \mathbf{v}^T \left(\mathbf{S}_j + \mathbf{S}_j^*\right) \frac{\mathrm{d}\mathbf{v}}{dq} = 0, j = 1,\ 2,\ 3. \tag{19.151}$$

Finally, the following algebraic balancing conditions for shaking moment are found from Eq. (19.151)

$$\mathbf{u}_j + \mathbf{u}_j^* = \mathbf{0}, \quad \mathbf{S}_j + \mathbf{S}_j^* = \mathbf{0}, j = 1,\ 2,\ 3. \tag{19.152}$$

Equations (19.134) and (19.152) can be used to derive the dynamic balancing conditions in form of algebraic expressions for spatial one-DOF mechanisms.

19.4.2 Application Example

In the following example we introduce the application of the balancing theory described above to a spatial slider crank mechanism shown in Fig. 19.10. The configuration of the mechanism is also prescribed by rotation angles ϕ, β and γ.

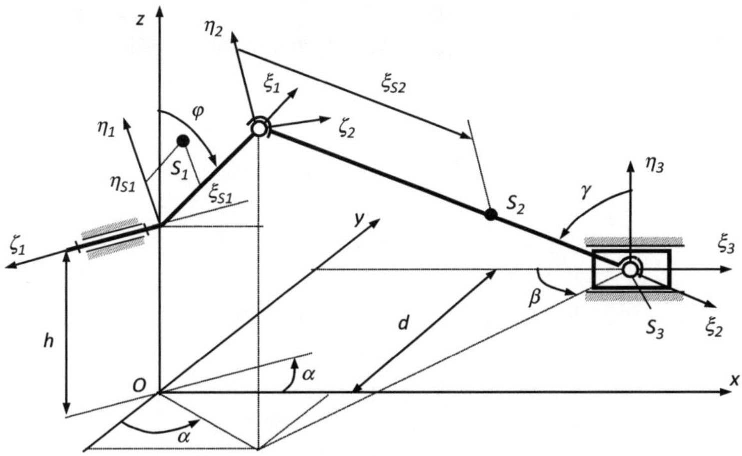

Fig. 19.10 A spatial slider crank mechanism

The angle φ is chosen as the independent generalized coordinates $q = \varphi$. The loop equations of the mechanism can be expressed in the form

$$\begin{aligned} h + l_1 \cos\phi - l_2 \cos\gamma &= 0, \\ -l_1 \sin\phi \cos\alpha + l_2 \sin\gamma \sin\beta - d &= 0, \end{aligned} \tag{19.153}$$

where l_i denotes the length of link i.

The direction cosine matrix $\mathbf{A}_i$ of link i referred to the fixed coordinate frame $\{Oxyz\}$ are given by

$$\mathbf{A}_1 = \begin{bmatrix} \sin\varphi \ \sin\alpha & -\cos\varphi \sin\alpha & -\cos\alpha \\ \sin\varphi \cos\alpha & \cos\varphi \cos\alpha & -\sin\alpha \\ \cos\varphi & \sin\varphi & 0 \end{bmatrix} \tag{19.154}$$

$$\mathbf{A}_2 = \begin{bmatrix} \sin\gamma \cos\beta & \cos\gamma \cos\beta & \sin\beta \\ \sin\gamma \sin\beta & \cos\gamma \sin\beta & -\cos\beta \\ -\cos\gamma & \sin\gamma & 0 \end{bmatrix} \tag{19.155}$$

According to the elements of matrices $\mathbf{A}_1$,$\mathbf{A}_2$, we choose the vector $\mathbf{z}$ with the following form

$$\begin{aligned} z = &\Big[\cos\varphi, \cos\beta, \cos\gamma, \sin\varphi, \sin\beta, \sin\gamma, \sin\gamma\cos\beta, \\ &\sin\gamma\sin\beta, \cos\gamma\cos\beta, \cos\gamma\sin\beta, 1\Big]^T \\ = &[z_1, z_2, z_3, z_4, z_5, z_6, z_7, z_8, z_9, z_{10}, z_{11}]^T \end{aligned} \tag{19.156}$$

For brevity, we assume that the center of mass S_2 of link 2 is positioned along the link line, the center of mass S_1 is positioned in the plane of axes ξ_2 and η_2. Then, $\eta_{S2} = 0,\ \zeta_{S1} = \zeta_{S2} = 0$. The coordinates of the center of mass S_i ($i = 1, 2, 3$) are expressed in term of the vector $\mathbf{z}$ as

$$\begin{aligned}
x_{S1} &= \left[-\eta_{S1}\sin\alpha\ 0\ 0\ \xi_{S1}\sin\alpha\ 0\ 0\ 0\ 0\ 0\ 0\ 0\right]\mathbf{z} \\
y_{S1} &= \left[\eta_{S1}\cos\alpha\ 0\ 0\ -\xi_{S1}\cos\alpha\ 0\ 0\ 0\ 0\ 0\ 0\ 0\right]\mathbf{z} \\
z_{S1} &= h + \left[\xi_{S1}\ 0\ 0\ \eta_{S1}\ 0\ 0\ 0\ 0\ 0\ 0\ 0\right]\mathbf{z} \\
x_{S2} &= \left[0\ 0\ 0\ l_1\sin\alpha\ 0\ 0\ \xi_{S2}\ 0\ 0\ 0\ 0\right]\mathbf{z} \\
y_{S2} &= \left[0\ 0\ 0\ -l_1\cos\alpha\ 0\ 0\ 0\ \xi_{S2}\ 0\ 0\ 0\right]\mathbf{z} \\
z_{S2} &= h + \left[l_1\ 0\ -\xi_{S2}\ 0\ 0\ 0\ 0\ 0\ 0\ 0\ 0\right]\mathbf{z} \\
x_{S3} &= \left[0\ 0\ 0\ l_1\sin\alpha\ 0\ 0\ l_2\ 0\ 0\ 0\ 0\right]\mathbf{z} \\
y_{S3} &= \left[0\ 0\ 0\ -l_1\cos\alpha\ 0\ 0\ 0\ l_2\ 0\ 0\ 0\right]\mathbf{z} \\
z_{S3} &= h + \left[l_1\ 0\ -l_2\ 0\ 0\ 0\ 0\ 0\ 0\ 0\ 0\right]\mathbf{z}
\end{aligned} \tag{19.157}$$

It can be shown that the loop equations in Eq. (19.153) have the form

$$\begin{bmatrix} l_1 & 0 & -l_2 & 0 & 0\ 0\ 0 & 0 & 0\ 0\ 0 \\ 0 & 0 & 0 & l_1\cos\alpha & 0\ 0\ 0 & -l_2 & 0\ 0\ 0 \end{bmatrix}\mathbf{z} = -\begin{bmatrix} d \\ h \end{bmatrix} \tag{19.158}$$

The reduced vector of variables $\mathbf{v}$ and the vector of eliminated variables $\mathbf{w}$ are selected from the original vector $\mathbf{z}$ as follows:

$$\mathbf{v} = [\cos\phi,\ \cos\beta,\ \sin\phi,\ \sin\beta,\ \sin\gamma,\ \ \sin\gamma\cos\beta,\ \cos\gamma\cos\beta,\ \cos\gamma\sin\beta,\ 1]^T \tag{19.159}$$

$$\mathbf{w} = [\cos\gamma,\ \sin\gamma\sin\beta]^T \tag{19.160}$$

The matrices $\mathbf{D}_{\mathrm{I}}$, $\mathbf{D}_{\mathrm{II}}$ and $\mathbf{D}_{\mathrm{II}}^{-1}$ in Eq. (19.128) are given by

$$\mathrm{D}_{\mathrm{I}} = \begin{bmatrix} l_1 & 0 & 0 & 0\ 0\ 0 & 0\ 0\ 0 \\ 0 & 0 & l_1\cos\alpha & 0\ 0\ 0 & 0\ 0\ 0 \end{bmatrix},\ \mathrm{D}_{\mathrm{II}} = \begin{bmatrix} -l_2 & 0 \\ 0 & -l_2 \end{bmatrix},\ D_{\mathrm{II}}^{-1} = -\frac{1}{l_2}\begin{bmatrix} 1 & 0 \\ 0 & 1 \end{bmatrix} \tag{19.161}$$

19.4.2.1 Conditions of the Shaking Force Balancing

With the known coordinates of the center of masses from Eq. (19.157) and matrices $\mathbf{D}_{\mathrm{I}}$, $\mathbf{D}_{\mathrm{II}}^{-1}$ from Eq. (19.161), vectors $\mathbf{g}_i$, $\mathbf{h}_i$, and $\mathbf{k}_i$ can be determined according to Eq. (19.132) without any difficulty. Then, by substituting all these results into Eq. (19.134), we get the following conditions for the complete shaking force balancing

$$
\begin{aligned}
&-m_1\eta_{S1}\sin\alpha = 0\\
&(m_1\xi_{S1} + m_2 l_1 + m_3 l_1)\sin\ \alpha = 0\\
&m_2\xi_{S2} + m_3 l_2 = 0\\
&m_1\eta_{S1}\cos\ \alpha = 0\\
&m_1\xi_{S1} + m_2 l_1\left(1 - \tfrac{\xi_{S2}}{l_2}\right) = 0\\
&m_1\xi_{S1}\cos\ \alpha + m_2 l_1\left(1 - \tfrac{\xi_{S2}}{l_2}\right)\cos\ \alpha = 0
\end{aligned}
\tag{19.162}
$$

By simplifying the expressions in Eq. (19.162), the balancing conditions for shaking force of the mechanism are reduced into three equations $f_1 = 0$, $f_2 = 0$, $f_3 = 0$, in which

$$
f_1 = \eta_{S1},\ f_2 = m_1\xi_{S1} + m_2 l_1 + m_3 l_1,\ f_3 = m_2\xi_{S2} + m_3 l_2. \tag{19.163}
$$

These conditions may be satisfied by internal mass redistribution or adding counterweights mounted on the links as shown in Fig. 19.11.

A simple numerical simulation is implemented in order to verify the correctness of these conditions for the static balancing. Parameters of the initial mechanism are given as follows: $m_1 = 7.0$ (kg), $m_2 = 12.5$ (kg), $m_3 = 10.5$ (kg), $l_1 = 0.1$ (m), $l_2 = 0.3$ (m), $h = 0.1$ (m), $d = 0.15$ (m), $\xi_{S1} = 0.01$ (m), $\eta_{S1} = 0.02$ (m), $\xi_{S2} = 0.05$ (m), $\eta_{S2} = 0$ (m). Using the conditions according to Eq. (19.163) we can determine the size and the location of the counterweights (see also Fig. 19.11): $m_1^*\xi_{S1}^* = 4.77$ (kg m), $m_2^*\xi_{S2}^* = 3.78$ (kg m). Figure 19.12 simultaneously shows three components of the shaking force produced by the unbalanced mechanism and the full force balanced mechanism. The results verified that the shaking force of the force balanced mechanism is completely eliminated and there is no forces transmitted to the base during the motion of the mechanism.

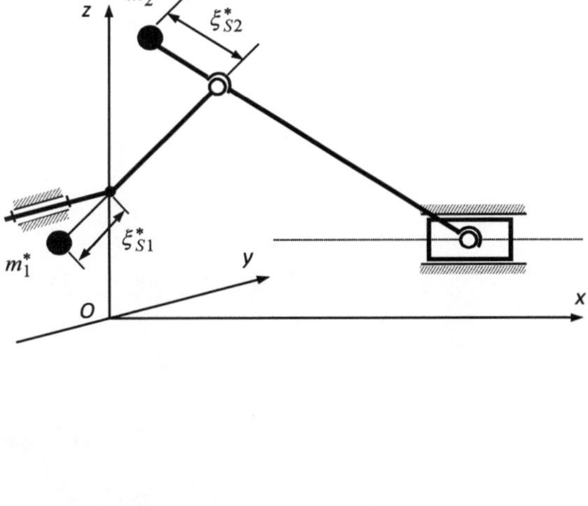

Fig. 19.11 Full force balanced mechanism with counterweights

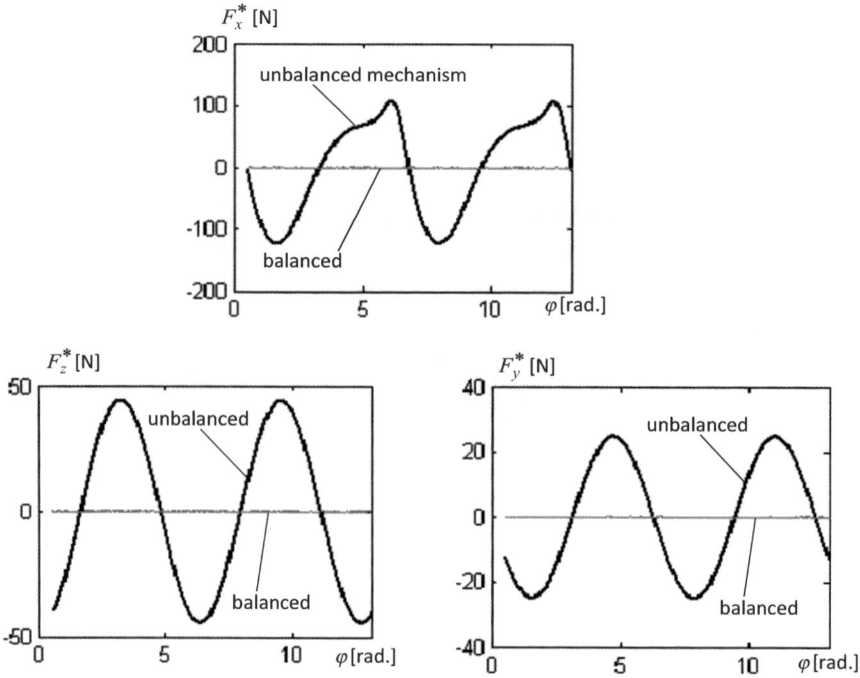

Fig. 19.12 Comparing shaking forces between the unbalanced mechanism and the full force balanced mechanism

19.4.2.2 Conditions of the Shaking Moment Balancing

To derive the conditions for the shaking moment balancing, the angular velocities of the links with respect to the fixed coordinate frame must be determined. Based on theory of multibody kinematics, these angular velocities can be calculated from the known matrices of the direction cosines $\mathbf{A}_i$ as follows:

$$\tilde{\boldsymbol{\omega}}_i = \dot{\mathbf{A}}_i \mathbf{A}_i^T, i = 1,\ 2,\ 3. \tag{19.164}$$

Using Eq. (19.164), we get

$$\boldsymbol{\omega}_1 = \begin{bmatrix} w_{1x} \\ w_{1y} \\ w_{1z} \end{bmatrix} = \begin{bmatrix} \cos\alpha \\ \sin\alpha \\ 0 \end{bmatrix} \dot{\varphi},\ \boldsymbol{\omega}_2 = \begin{bmatrix} \omega_{2x} \\ \omega_{2y} \\ \omega_{2z} \end{bmatrix} = \begin{bmatrix} \gamma' \sin\beta \\ \gamma' \cos\beta \\ \beta' \end{bmatrix} \dot{\varphi},$$

$$\boldsymbol{\omega}_2 = \begin{bmatrix} \omega_{3x} \\ \omega_{3y} \\ \omega_{3z} \end{bmatrix} = \begin{bmatrix} 0 \\ 0 \\ 0 \end{bmatrix} \dot{\varphi} \tag{19.165}$$

where $\gamma' = \frac{d\gamma}{d\varphi},\ \beta' = \frac{d\beta}{d\varphi}$.

Upon assuming that axes ξ_i, η_i, ζ_i of the link-fixed coordinate frame are principal axes. The inertia matrix $\mathbf{I}_{Si}^{(i)}$ of link i about the center of mass S_i, referred to these principal axes, can be written in the simplified form

$$\mathbf{I}_{Si}^{(i)} = \begin{bmatrix} I_{i\xi\xi} & 0 & 0 \\ 0 & I_{i\eta\eta} & 0 \\ 0 & 0 & I_{i\zeta\zeta} \end{bmatrix}, i = 1,\ 2,\ 3 \tag{19.166}$$

By comparing elements of $\boldsymbol{\omega}_i$ in Eq. (19.165) with elements of $\mathbf{J}_{Ri}$ in Eq. (19.109), we obtain

$$\begin{aligned} s'_{1x} &= \cos\alpha, & s'_{2x} &= \gamma' \sin\beta, & s'_{3x} &= 0, \\ s'_{1y} &= \sin\alpha, & s'_{2y} &= -\gamma' \cos\beta, & s'_{3y} &= 0, \\ s'_{1z} &= 0, & s'_{2z} &= \beta'. & s'_{3z} &= 0. \end{aligned} \tag{19.167}$$

So, the expressions in the left-hand side of Eqs. (19.144)–(19.146) can be established. With vector $\mathbf{z}$ according to Eq. (19.156) we find that

$$\begin{aligned} z_1 z'_4 - z_4 z'_1 &= 1, & z_2 z'_5 - z_5 z'_2 &= \beta', \\ z_{11} z'_1 &= -\sin\phi, & z_{11} z'_4 &= \cos\phi, \end{aligned} \tag{19.168}$$

Now we can determine matrices $\mathbf{H}_1 = \left[h_{i,j}^{(1)}\right]$, $\mathbf{H}_2 = \left[h_{i,j}^{(2)}\right]$, and $\mathbf{H}_3 = \left[h_{i,j}^{(3)}\right]$ as follows:

$$\begin{aligned} h_{1,1}^{(1)} &= -h_{4,1}^{(1)} = I_{1\zeta\zeta} \cos\alpha,\ h_{3,8}^{(1)} = -h_{8,3}^{(1)} = \tfrac{1}{2}\left(I_{2\eta\eta} + I_{2\zeta\zeta} - I_{2\xi\xi}\right), \\ h_{6,10}^{(1)} &= -h_{10,6}^{(1)} = \tfrac{1}{2}\left(I_{2\eta\eta} - I_{2\xi\xi} - I_{2\zeta\zeta}\right),\ \text{every other } h_{i,j}^{(1)} = 0\ . \end{aligned}$$

$$\begin{aligned} h_{1,1}^{(2)} &= -h_{4,1}^{(2)} = I_{1\zeta\zeta} \sin\alpha,\ h_{3,7}^{(2)} = -h_{7,3}^{(2)} = \tfrac{1}{2}\left(I_{2\xi\xi} - I_{2\eta\eta} - I_{2\zeta\zeta}\right), \\ h_{6,9}^{(2)} &= -h_{9,6}^{(2)} = \tfrac{1}{2}\left(I_{2\xi\xi} - I_{2\eta\eta} + I_{2\zeta\zeta}\right),\ \text{every other } h_{i,j}^{(2)} = 0 \end{aligned}$$

$$h_{2,5}^{(3)} = -h_{5,2}^{(3)} = I_{2\xi\xi},\ h_{7,8}^{(3)} = -h_{8,7}^{(2)} = I_{2\eta\eta} - I_{2\xi\xi},\ \text{every other } h_{i,j}^{(3)} = 0\ .$$

By partitioning of matrix $\mathbf{H}_j$ related to Eq. (19.147), we obtain sub-matrices $\mathbf{H}_{j1}$, $\mathbf{H}_{j2}$, $\mathbf{H}_{j3}$, $\mathbf{H}_{j4}$. Then, vectors $\mathbf{u}_j$ and $\mathbf{u}_j^*$, the matrices $\mathbf{S}_j$ and $\mathbf{S}_j^*$ can be formulated by using Eqs. (19.138), (19.139), (19.149), and (19.150), for example

$$u_2 = \left[-m_1 h \eta_{S1} \sin\alpha, 0, m_1 h \xi_{S1} \sin\alpha + m_2 h l_1 \left(1 - \frac{\xi_{S2}}{l_2}\right)\right.$$
$$\left. \sin\alpha, 0, 0,\ m_2 h \xi_2 \left(1 - \frac{\xi_{S2}}{l_2}\right),\ 0,\ 0,\ 0\right]^T$$

$$\mathbf{u}_2^* = \left[0,\ 0,\ 0,\ 0,\ 0,\ \frac{h}{2l_2}\left(I_{2\xi\xi} - I_{2\eta\eta} - I_{2\zeta\zeta}\right),\ 0,\ 0,\ 0\right]^T.$$

Finally, according to Eq. (19.152) we obtain the non-zero elements of vectors $\mathbf{u}_j + \mathbf{u}_j^*$ and matrices $\mathbf{S}_j + \mathbf{S}_j^*$ in the form

$$\begin{aligned}
k_1 &= m_1 h\eta_{S1}\cos\alpha - m_2 dl_1 \tfrac{\xi_{S2}}{l_2}\left(1 - \tfrac{\xi_{S2}}{l_2}\right) - \tfrac{dl_1}{2l_2^2}\left(I_{2\xi\xi} - I_{2\eta\eta} - I_{2\zeta\zeta}\right)\\
k_2 &= \left[m_1 h\xi_{S1} + m_2 hl_1\left(1 - \tfrac{\xi_{S2}}{l_2}\right)^2 - \tfrac{hl_1}{2l_2^2}\left(I_{2\xi\xi} - I_{2\eta\eta} - I_{2\zeta\zeta}\right)\right]\cos\alpha\\
k_3 &= \left[m_1 h\xi_{S1} + m_2 hl_1\left(1 - \tfrac{\xi_{S2}}{l_2}\right)\right]\sin\alpha\\
k_4 &= m_2 h\xi_{S2}\left(1 - \tfrac{\xi_{S2}}{l_2}\right) + \tfrac{h}{2l_2}\left(I_{2\xi\xi} - I_{2\eta\eta} - I_{2\varsigma\varsigma}\right)\\
k_5 &= m_2 dl_1 \tfrac{\xi_{S2}}{l_2}\sin\alpha + m_3 dl\sin\alpha,\ k_6 = m_2 d\tfrac{\xi_{S2}^2}{l_2} + m_3 dl_2 - \tfrac{d}{l_2}\left(I_{2\xi\xi} - I_{2\eta\eta}\right)\\
k_7 &= \left[m_1\left(\xi_{S1}^2 + \eta_{S1}^2\right) + m_2 l_1^2\left(1 - \tfrac{\xi_{S2}}{l_2}\right) - \tfrac{l_1^2}{2l_2^2}\left(I_{2\xi\xi} - I_{2\eta\eta} - I_{2\varsigma\varsigma}\right) + I_{1\varsigma\varsigma}\right]\cos\ \alpha\\
k_8 &= \left[m_1\left(\xi_{S1}^2 + \eta_{S1}^2\right) + m_2 l_1^2\left(1 - \tfrac{\xi_{S2}}{l_2}\right) + I_{1\varsigma\varsigma}\right]\sin\ \alpha\\
k_9 &= m_2 l_1\xi_{S2}\left(1 - \tfrac{\xi_{S2}}{l_2}\right) + \tfrac{l_1}{2l_2}\left(I_{2\xi\xi} - I_{2\eta\eta} - I_{2\varsigma\varsigma}\right)\\
k_{10} &= \left[m_2 l_1\xi_{S2}\left(1 - \tfrac{\xi_{S2}}{l_2}\right)\tfrac{l_1}{l_2}\left(I_{2\xi\xi} - I_{2\eta\eta}\right)\right]\cos\ \alpha\\
k_{11} &= I_{2\xi\xi} - I_{2\eta\eta} + I_{2\varsigma\varsigma},\ k_{12} = m_1 h\eta_{S1}\sin\ \alpha,\ k_{13} = I_{2\xi\xi}
\end{aligned} \tag{19.169}$$

Note that the above obtained expressions are original and can be further simplified. Now we choose $\alpha = \pi/2$, $\eta_{S1} = 0$ and let $I_{2\eta\eta} = I_{2\varsigma\varsigma}$, the expressions k_i in Eq. (19.169) are reduced as follows:

$$\begin{aligned}
f_4 &= m_2 l_2\xi_{S2} - m_2\xi_{S2}^2 - I_{2\varsigma\varsigma},\ f_5 = m_1\xi_{S1} + m_2 l_1\left(1 - \tfrac{\xi_{S2}}{l_2}\right)\\
f_6 &= \tfrac{m_2\xi_{S2}l_1}{l_2} + m_3 l_1,\ f_7 = m_1\xi_{S1}^2 + m_2 l_1^2\left(1 - \tfrac{\xi_{S2}}{l_2}\right) + I_{1\varsigma\varsigma},\ f_8 = I_{2\xi\xi}.
\end{aligned} \tag{19.170}$$

The shaking moment is completely balanced if the values of f_i $(i = 4, 5, \ldots, 8)$ in Eq. (19.170) vanish simultaneously. It is clearly shown that these conditions cannot be completely satisfied by adding counterweights, since the values of f_8 are not equal to zero in any case. These conditions are mainly of theoretical interest. However, Eq. (19.170) provide the necessary tool for the minimization of the shaking moment. Another way for solving the problem is the simultaneous minimization of the shaking force and shaking moment based on Eqs. (19.163) and (19.170). From the conditions

$$f_i \to \min\ \ (i = 1, 2, \ldots, 8)$$

one can choose a set of optimizing values for geometrical and inertia parameters of the links: $m_1, m_2, m_3, \xi_{S1}, \xi_{S2}, I_{1\varsigma\varsigma}, I_{2\xi\xi}, I_{2\eta\eta}, I_{2\varsigma\varsigma}$. This problem will be considered in the future investigation.

19.5 Conclusions

This chapter provided an approach to derive the dynamic balancing conditions of planar and spatial mechanisms. The following conclusions have been reached:

- Based on theory of multibody dynamics, the algebraic balancing conditions for the shaking force and shaking moment of planar and spatial mechanisms have been established.
- A specialized code has been developed on the MAPLE® environment for this study. It can be concluded that the proposed method is suitable for the application of the widely accessible computer algebra systems such as MAPLE®.
- The proposed method is illustrated for a planar 8R-eightbar mechanism having multi degrees-of-freedom and multi-links is an appropriate object to demonstrate the suggested procedure. Based on the obtained balancing conditions of the shaking force, a number of balancing schema with counterweights can be established by assigning the parameters of two arbitrary links and determining parameters of the other links.
- The proposed method is illustrated by using a spatial slider crank mechanism. In the application of balancing techniques using counterweights and supplementary links [35–37] for spatial mechanisms, the proposed method may provide a helpful tool to obtain exactly the balancing conditions and therefore we can get better balancing results. This will be the subject of future work.

Acknowledgment The work discussed in this chapter was completed with the financial support given by the National Foundation for Science and Technology Development of Vietnam.

References

1. Dresig, H, Vulfson, JI: Dynamik der Mechanismen. Deutscher Verlag der Wissenschaften, Berlin (1989)
2. Schiehlen, W, Eberhard, P: Technische Dynamik, 3rd edn. Vieweg + Teubner, Wiesbaden (2007)
3. Dresig, H, Holzweissig, F: Dynamics of Machinery. Springer, Berlin (2010)
4. Huston, RL: Multibody Dynamics. Butterworth-Heinemann, Boston (1990)
5. Haug, EJ: Computer Aided Kinematics and Dynamics of Mechanical Systems, vol. 1: Basic Methods. Allyn and Bacon, Boston, MA (1989)
6. Chaudhary, H, Saha, SK: Dynamics and Balancing of Multibody Systems. Springer, Berlin (2009)
7. Lowen, GG, Tepper, FR, Berkorf, RS: Balancing of linkages – an update. Mech Mach Theory **18**, 213–220 (1983)
8. Thümmel, T: Literaturbericht zum dynamischen Ausgleich schnellaufender Mechanismen. Wiss Schriftenreihe der TH Karl-Marx-Stadt. Mech Mater **7**, 57–92 (1983)
9. Arakelian, VH, Smith, MR: Shaking force and shaking moment balancing of mechanisms: a historical review with new examples. ASME J Mech Des **127**, 334–339 (2005)
10. Shchepetilnikov, VA: The determination of the mass centers of mechanisms in connection with the problem of mechanism balancing. J Mech **3**, 367–389 (1968)

11. Berkof, RS, Lowen, GG: A new method for completely force balancing simple linkage. Trans ASME J Eng Ind **91**(1), 21–26 (1969)
12. Kaufman, RE, Sandor, GN: Complete force balancing of spatial linkages. Trans ASME J Mech Des **93B**(2), 620–626 (1971)
13. Berkof, RS: Complete force and moment balancing of inline four-bar linkages. Mech Mach Theory **8**, 397–410 (1973)
14. Dresig, H, Rockhausen, L, Naake, S: Balancing conditions for planar mechanism. Flex Mech Dyn Anal ASME **47**, 67–73 (1992)
15. Dresig, H, Rockhausen, L, Naake, S: Vollständiger und harmonischer Ausgleich ebener Mechanismen, Fortschritt-Berichte VDI, Reihe 18, Nr. 155. VDI Verlag, Düsseldorf (1994)
16. Kochev, IS: General theory of complete shaking moment balancing of planar linkages: a critical review. Mech Mach Theory **35**, 1501–1514 (2000)
17. Kochev, IS: General method for active balancing of combined shaking moment and torque fluctuations in planar linkages. Mech Mach Theory **25**, 679–687 (1990)
18. Ye, Z, Smith, MR: Complete balancing of planar linkages by an equivalence method. Mech Mach Theory **29**(5), 701–712 (1994)
19. Esat, I, Bahai, H: A theory of complete force and moment balancing of planer linkage mechanisms. Mech Mach Theory **34**, 903–922 (1999)
20. Arakelian, VH, Smith, MR: Design of planar 3-DOF 3-RRR reactionless parallel manipulators. Mechatronics **18**(10), 601–606 (2008)
21. Wu, Y, Gosselin, CM: On the dynamic balancing of multi-DOF parallel mechanisms with multiple legs. ASME J Mech Des **129**(2), 234–238 (2007)
22. Nguyen, VK: Über den Massenausgleich in Mehrkörpersystemen. Tech Mech **14**(3–4), 231–238 (1994)
23. Nguyen, VK, Nguyen, PD, Pham, VS: Balancing conditions of planar mechanisms with multi-degree of freedom. Vietnam J Mech **27**, 204–212 (2005)
24. Nguyen, VK, Nguyen, PD: Balancing conditions for spatial mechanisms. Mech Mach Theory **42**, 1141–1152 (2007)
25. Nguyen, VK, Nguyen, PD: On the dynamic balancing conditions of planar multi-DOF parallel manipulators with revolute joints. In: Proceedings of the 1st IFToMM International Symposium on Robotics and Mechatronics, Hanoi, 21–23 Sept, 2009
26. Arakelian, VH, Smith, VH: Complete shaking force and shaking moment balancing of linkages. Mech Mach Theory **34**, 1141–1153 (1999)
27. Arakelian, V, Dahan, M: Partial shaking moment balancing of fully shaking force balanced linkages. Mech Mach Theory **36**, 1241–1252 (2001)
28. Arakelian, VH: Complete shaking force and shaking moment balancing of RSS'R spatial linkages. J Multibody Dyn **221**, 303–310 (2007)
29. Chaudhary, H, Saha, SK: Balancing of shaking forces and shaking moments for planar mechanisms using the equimomental systems. Mech Mach Theory **43**, 310–334 (2008)
30. Feng, G: Complete shaking force and shaking moment balancing of four types of six-bar linkages. Mech Mach Theory **24**(4), 275–287 (1989)
31. Feng, G: Complete shaking force and shaking moment balancing of 17 types of eight-bar linkages only with revolute pairs. Mech Mach Theory **26**, 197–206 (1991)
32. Moore, B, Schicho, J, Gosselin, CM: Determination of the complete set of shaking force and shaking moment balanced planar four-bar linkages. Mech Mach Theory **44**, 1338–1347 (2009)
33. Kaufman, RE, Sandor, GN: Complete force balancing of spatial linkages. Trans ASME J Eng Ind **93**, 620–626 (1971)
34. Bagci, C: Complete balancing of space mechanisms – shaking force balancing. J Mech Trans Automat Des **105**, 609–616 (1983)
35. Chen, N-X: The complete shaking force balancing of a spatial linkage. Mech Mach Theory **19**, 243–255 (1984)
36. Yue-Qing, Y: Complete shaking force and moment balancing of spatial irregular force transmission mechanisms using additional link. Mech Mach Theory **23**, 279–285 (1988)

37. Abdel-Rahman, TM, Elbestawi, MA: Synthesis and dynamics of statically balanced direct-drive manipulators with decoupled inertia tensors. Mech Mach Theory **26**, 389–402 (1991)
38. Wang, J, Gosselin, CM: Static balancing of spatial four-degree-of freedom parallel mechanisms. Mech Mach Theory **35**, 563–592 (2000)
39. Arakelian, V., Smith, M.R: Shaking moment minimization of fully force-balanced linkages. In: Proceedings of the 11th World Congress in Mechanism and Machine Science, Tianjin, China (2004)
40. Park, J: Principle of dynamical balance for multibody systems. Multibody Syst Dyn **14**, 269–299 (2005)
41. Russo, A, Sinatra, R, Xi, F: Static balancing of parallel robots. Mech Mach Theory **40**, 191–202 (2005)

Chapter 20
Static Balancing of Articulated Wheeled Vehicles by Parallelogram- and Spring-Based Compensation

Aliakbar Alamdari and Venkat Krovi

Abstract Articulated wheeled vehicles (AWVs) offer superior uneven terrain traversal capabilities by virtue of the superior reconfigurability within their articulated structure. However, this capability can be realized only at the price of increased actuation-based equilibration, oftentimes solely to support the gravitational loading. Hence, the simultaneous reduction of the overall actuation remains one of the critical challenges in such AWVs.

In this chapter, we address the static balancing of six degree-of-freedom AWVs with multiple leg-wheel subsystem. Static balancing is defined as a set of conditions on dimensional and inertial parameters of articulated vehicle components which ensure that the weight of the links and platform does not produce any torque/force at the actuators for any configuration of vehicle. In this study, elastic elements such as springs are employed in conjunction with parallelogram linkages to achieve the static balancing. The underlying principle is to realize an overall articulated system whose total potential energy including the elastic potential energy stored in springs and gravitational potential energy becomes constant.

Keywords Static balancing • Articulated wheeled vehicles • Parallelograms Reconfigurable vehicle

20.1 Introduction

Articulated wheeled vehicles (AWVs) are a class of wheeled vehicles where the terrain-contact wheels are attached to the chassis via an articulated-multibody chain. The resulting articulations allow the wheel-relocation with respect to chassis during locomotion, and provide the vehicle with significant reconfigurability and redun-

A. Alamdari (✉) • V. Krovi
The State University of New York at Buffalo, Buffalo, NY, USA
e-mail: aalamdar@buffalo.edu

D. Zhang, B. Wei (eds.), *Dynamic Balancing of Mechanisms and Synthesizing of Parallel Robots*, DOI 10.1007/978-3-319-17683-3_20

dancy [1]. Viewed from a traditional perspective, the intermediate-articulations are nothing more than the traditional suspension-mechanisms, commonly used in vehicle-designs. However, the use of alternate architectures (such as serial-, parallel-, or hybrid-blends) can unlock significant potential performance benefits. For instance, the articulated-suspensions (with adequate workspace) can permit the vehicle to change the location of center of mass by adjusting the linkages joints, so as to avoid the rollover when passing the uneven terrain [2].

Traditionally, the articulated linkages require the role of both the kinematic constraints and the support distribution of the load to the ground. Load support can be achieved by a combination of actuation equilibration and structural equilibration. Here, our focus is on a new class of the AWVs that allow a significant range of motion of the axle with respect to the chassis. In such cases, actuation-based equilibration becomes necessary but compensation for weight of the chassis and links requires very powerful actuators at the joints. However, a significant component of this load due to gravity is configuration dependent. We can choose semi-active actuation such as springs and dampers to modulate the actuation requirements. Maintaining the platform of articulated vehicle with multiple legs in static equilibrium requires considerable power because of their weights. Hence, reducing or eliminating these static forces needs careful consideration and can be achieved by kinetostatic design of parallel-articulated mechanism [3]. Undeniably, other than reducing the number of actuators which leads to reduction in calibration efforts and costs [4], using smaller size and less powerful actuators improves the efficiency. A balanced articulated mechanism is easier to move its platform and can be brought to static equilibrium in any of its configurations without exerting external forces.

To the knowledge of authors, static balancing of mobile articulated mechanism with variable ground-contact point has not been explicitly addressed in the literature. In the proposed mobile parallel structure with articulated legs, in addition to having high performance, torque minimization also is one of the biggest challenges. Therefore, the elimination of static torques due to gravity through static balancing reduces the torque requirements and provides much more efficient design (reduction of the actuator size).

Many gravity-compensated mechanisms have been designed using zero free-length springs [5, 6], counterweights [7], torsion springs [8], cam and pulleys [9] in literature. A general approach for gravity balancing of planar parallel mechanism using elastic elements such as springs is presented in [5]. Static balancing using counterweights [7] eliminates the static torques but may also degrade the dynamic properties due to an increase of the inertia. In [8], a new approach for static balancing optimization of a parallel medical robot is proposed to satisfy two objective functions, safety as well as static balancing. In this approach, torsion springs are added on the actuated and passive revolute joints to eliminate gravity effects.

In this study, the static balancing of six degree-of-freedom (6-DOF) spatial parallel mechanism with four articulated legs is addressed. First static balancing definition is defined, and then the formulation of balanced mechanism with elastic elements is presented. Next, the conditions for static balancing of three-dimensional articulated mechanism with springs and parallelograms are introduced.

Definition Parallel mechanisms are said to be statically balanced when the center of mass of the whole mechanism remains fixed for any arbitrary motion of the mechanism. In other words, in the gravity-compensated parallel mechanism, no torque will be produced by weight of the links at the actuators for any configuration of mechanism under static conditions. On the other hand, in mechanism with elastic elements (linear or torsion springs), static balancing can be defined as the set of conditions which the total potential energy of the mechanism including gravitational potential energy and potential energy stored in the springs constant for any configuration of mechanism. Without using springs, static balancing can be defined as the set of conditions, which the total gravitational potential energy of the mechanism constant for any arbitrary motion of the mechanism. It means that the center of mass of mechanism does not move in the direction of gravity vector. A popular approach to static balancing is the use of tension springs attached to each link to compensate the effect of gravity on any configuration within the workspace. Herder [10] proposed different spring configurations for the static balancing of the planar mechanism. Shin and Streit [11] provided a mathematical basis for comparing the complexity of equilibrators methodologies using spring. Fattah and Agrawal [12] proposed a passive gravity-balanced leg orthosis using pantograph to assist persons with hemiparesis to walk by eliminating the effects of gravity. A 6-DOF platform was statically balanced by Ebert-Uphoff et al. [13] using three parallelogram legs. Jean and Gosselin [14] placed a counterweight on each revolute joint to move the center of gravity on the rotation axis to statically balance the mechanisms. Simionescu and Ciupitu [15] used counterweights in combination with springs. Static balancing of a 6-DOF table was compared using springs and counterweights by Gosselin and Wang [7]. In another study, a hexapod was statically balanced with a counterweight attached to a pantograph by Russo et al. [16]. In these studies, the added counterweight (mass) or pantograph (links) increased the robot's complexity and manufacturing cost. The resulting balanced mechanism becomes too large, and thus less movable.

In a recent study [17], a design methodology from energy perspective for determining the spring configuration on statically balanced planar articulated manipulator is presented. Lin et al. [18] studied the theory of weight-balanced mechanism for the design of a class of spatial mobile arm support to facilitate the arm movement in space by the complete weight compensation of the upper limb at any possible posture.

As mentioned earlier, in this study, the static balancing of spatial 6-DOF parallel mechanism with articulated legs using springs and parallelograms (structural equilibration) is studied. The balanced mechanism has constant total potential energy (including elastic potential energy stored in springs as well as the potential energy due to gravitational potential energy). Therefore, in the gravity-eliminated mechanism the actuators do not have to support the weight of links and platform;

hence, they will support only the added payload. Consequently, in statically balanced mechanism, actuators will be reduced in size and energy consumption will be minimized.

20.2 Background

20.2.1 Static Balancing with Spring Assist

As mentioned earlier, the use of springs for static balancing is preferred as it adds very little mass and inertia to the system. The resulting mechanism is balanced for any orientation of its platform with respect to the polygon formed by wheel-ground contact points.

The position vector of the center of mass of the vehicle with respect to the fixed frame can be expressed as:

$$^{F}[\mathbf{r}_{CG}] = \frac{1}{M}\sum_{i=1}^{n_m} m_i \mathbf{r}_i \tag{20.1}$$

where m_i and $\boldsymbol{r}_i$ are defined as mass and position vector of ith moving part, respectively. n_m represents total number of moving parts and M is the total mass of moving parts, defined as:

$$M = \sum_{i=1}^{n_m} m_i \tag{20.2}$$

Generally, the position vector of the center of mass is configuration dependent. Therefore, it is presented as a function of configuration vector of the mechanism θ, as:

$$^{F}[\mathbf{r}_{CG}] = {}^{F}[\mathbf{r}_{CG}](\theta) \tag{20.3}$$

According to the condition for static balancing, the position vector of center of mass has to be independent from the configuration vector θ, the position vector $^{F}[\boldsymbol{r}_{CG}]$ can be written as

$$^{F}[\mathbf{r}_{CG}] = \text{constant} \tag{20.4}$$

In this study, the weight of the vehicle is one of the main concerns, for which the elastic elements (springs) are used instead of counterweights. The total potential energy of the vehicle is defined as the sum of the gravitational and elastic potential energy and can be written as:

$$V = M\mathbf{g}^{TF}[\mathbf{r}_{CG}] + \frac{1}{2}\sum_{j=1}^{n_s} K_j(S_j - S_j^0)^2 \tag{20.5}$$

where $\mathbf{g}$ is the gravitational acceleration vector and n_s is the number of springs in the vehicle, K_j is the stiffness of the jth elastic element, and S_j and S_j^0 are the length and free-length of jth elastic element, respectively.

The condition for static balancing when springs are used is that the total potential energy of the vehicle be constant for any configuration of the system which is written as:

$$V = \text{constant} \tag{20.6}$$

The gravity compensation by elastic elements usually requires the undeformed length of springs to be equal to zero. Zero free-length springs can physically satisfy this requirement.

20.3 6-DOF AWV

An AWV with 6-DOF platform is illustrated in Fig. 20.1. This vehicle consists of four identical legs connecting platform to wheels. Each of these articulated leg-wheels consists of 2-Link manipulator attached to the wheel-axle. The fixed coordinate frame {F}-$\{\boldsymbol{X}_F, \boldsymbol{Y}_F, \boldsymbol{Z}_F\}$ is attached to the ground with $\boldsymbol{Z}_F$ pointing vertically upward.

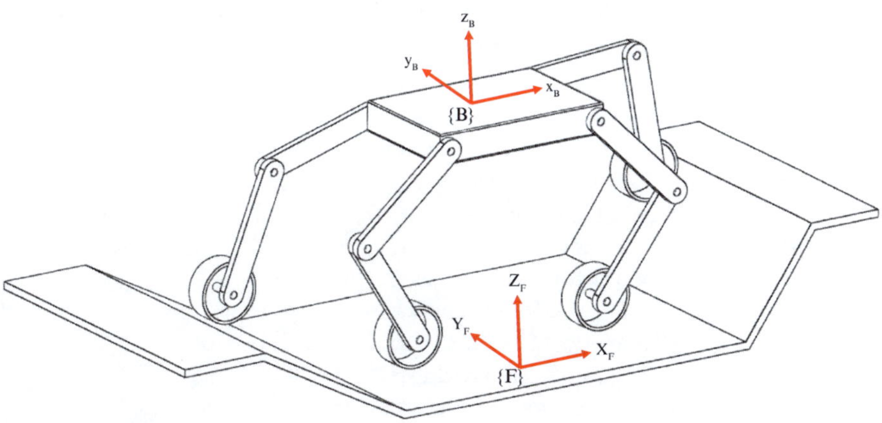

Fig. 20.1 Kinematic architecture of the wheeled vehicle with revolute actuators

Similarly, the body coordinate frame {B}-$\{\boldsymbol{x}_B, \boldsymbol{y}_B, \boldsymbol{z}_B\}$ is attached to the platform. The Cartesian coordinate of platform is given by position vector of frame {B} with respect to the fixed frame {F}, noted $^F[\boldsymbol{r}_B] = [\boldsymbol{X}_B\ \boldsymbol{Y}_B\ \boldsymbol{Z}_B]^T$ and the orientation of platform with respect to fixed frame can be expressed as function of Euler angles by comparing with conventional Yaw–Pitch–Roll (ZYX) representation $(\Phi_{BZ}, \Phi_{BY}, \Phi_{BX})$. The rotation matrix $^F R_B$ from inertial frame {F} to the body frame {B} can be written as:

$$
\begin{aligned}
{}^F R_B &= \mathrm{Rotz}(\Phi_{BZ})\mathrm{Roty}(\Phi_{BY})\mathrm{Rotx}(\Phi_{BX}) \\
&= \begin{bmatrix} c\Phi_{BZ}c\Phi_{BY} & r_{12} & r_{13} \\ s\Phi_{BZ}c\Phi_{BY} & r_{22} & r_{23} \\ -s\Phi_{BY} & c\Phi_{BY}s\Phi_{BX} & c\Phi_{BY}c\Phi_{BX} \end{bmatrix}
\end{aligned}
\tag{20.7}
$$

where

$$r_{12} = c\Phi_{BZ}s\Phi_{BY}s\Phi_{BX} - s\Phi_{BZ}c\Phi_{BX},\ r_{13} = c\Phi_{BZ}s\Phi_{BY}c\Phi_{BX} + s\Phi_{BZ}s\Phi_{BX}$$

$$r_{22} = s\Phi_{BZ}s\Phi_{BY}s\Phi_{BX} + c\Phi_{BZ}c\Phi_{BX},\ r_{23} = s\Phi_{BZ}s\Phi_{BY}c\Phi_{BX} - c\Phi_{BZ}s\Phi_{BX}$$

where 'c' and 's' stand for trigonometric functions 'cosine' and 'sine', respectively.

The Cartesian coordinate of first revolute joint, i.e. J_{1i}, in each leg-wheel subsystem relative to platform frame {B} is noted by $^B[\boldsymbol{r}_{J_{1i}}] = [a_i\ b_i\ c_i]^T$ with $i = \{1, 2, 3, 4\}$.

20.4 Static Balancing of 6-DOF AWV

20.4.1 Derivation of Center of Mass of the Wheeled Vehicle Without Springs

The notations and vectors for two-link of the ith leg-wheel subsystem vehicle are illustrated schematically in Fig. 20.2. A reference frame $\{W_i\}$-$\{\boldsymbol{x}_{W_i}, \boldsymbol{y}_{W_i}, \boldsymbol{z}_{W_i}\}$ is attached to the end of the second link at the wheel axle. The coordinates of origin $\{W_i\}$ expressed in fixed frame {F} are $[X_{W_i}\ Y_{W_i}\ Z_{W_i}]^T$ with $i = \{1, 2, 3, 4\}$. The unit vector $\mathbf{z}_{W_i}$ is defined along the second link, i.e., from the origin of $\{W_i\}$ toward joint J_{2i} and $\mathbf{y}_{Wi}$ is defined along the wheel axle. Also, points CG_{1i} and CG_{2i} denote the center of mass of first and second link, respectively.

Let γ_i be the angle of second link relative to the positive direction of the unit vector $\boldsymbol{X}_F$ and ζ_i be the angle between the positive direction of $\boldsymbol{x}_{Wi}$ of the wheel axle frame and the unit vector $\boldsymbol{X}_F$ in fixed frame from the top view. Therefore, the homogeneous transformation starting from the fixed frame to the wheel axle frame $\{W_i\}$ can be written as:

$$^F A_{W_i} = \mathrm{Trans}(X_{W_i}, Y_{W_i}, Z_{W_i})\mathrm{Rotz}(\zeta_i)\mathrm{Roty}\left(\frac{3\pi}{2} - \gamma_i\right)$$

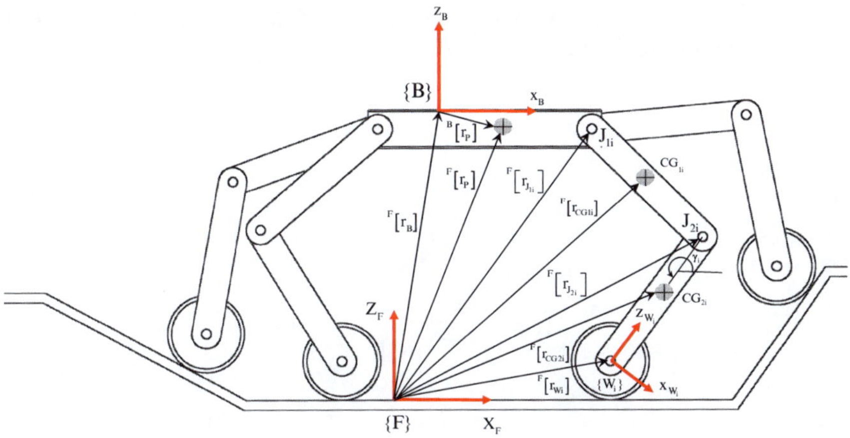

Fig. 20.2 Geometry of *i*th leg-wheel subsystem

$$= \begin{bmatrix} -\cos(\zeta_i)\sin(\gamma_i) & -\sin(\zeta_i) & -\cos(\zeta_i)\cos(\gamma_i) & X_{W,i} \\ -\sin(\zeta_i)\sin(\gamma_i) & \cos(\zeta_i) & -\sin(\zeta_i)\cos(\gamma_i) & Y_{W,i} \\ \cos(\gamma_i) & 0 & -\sin(\gamma_i) & Z_{W,i} \\ 0 & 0 & 0 & 1 \end{bmatrix} \tag{20.8}$$

Therefore, the center of mass of the second link of the *i*th leg-wheel subsystem can be expressed as

$$^{F}[\boldsymbol{r}_{CG_{2i}}] = {}^{F}A_{W_i}{}^{W_i}[\boldsymbol{r}_{CG_{2i}}] \tag{20.9}$$

where $^{F}[\boldsymbol{r}_{CG_{2i}}]$ and $^{W_i}[\boldsymbol{r}_{CG_{2i}}]$ are the position vectors of the center of mass of the second link with respect to the fixed {F} and wheel-axle frame$\{W_i\}$, respectively. As presented in Fig. 20.4 l_{CG2i} is the location of center of mass respect to the local coordinate frame $\{W_i\}$; thus

$$^{W_i}[\boldsymbol{r}_{CG_{2i}}] = \left[0\ 0\ l_{CG2i}\ 1\right]^T \tag{20.10}$$

Similarly, Eq. (20.9) can be written to find the position vector of joint J_{2i} with respect to fixed frame {F} as:

$$^{F}[\boldsymbol{r}_{J_{2i}}] = {}^{F}A_{W_i}\left[0\ 0\ l_{2i}\ 1\right]^T \tag{20.11}$$

where l_{2i} is the length of second link. Moreover, position vector $^{F}[\boldsymbol{r}_{J_{1i}}]$ can be expressed as a function of the position and orientation of platform, i.e.,

$$^{F}[\boldsymbol{r}_{J_{1i}}] = {}^{F}[\boldsymbol{r}_B] + {}^{F}R_B{}^{B}[\boldsymbol{r}_{J_{1i}}] \tag{20.12}$$

One can compute the position vector of center of mass of first link of ith leg-wheel subsystem from the position vectors ${}^{F}[\boldsymbol{r}_{J_{1i}}]$ and ${}^{F}[\boldsymbol{r}_{J_{2i}}]$ as:

$$ {}^{F}[\boldsymbol{r}_{CG_{1i}}] = {}^{F}[\boldsymbol{r}_{J_{2i}}] + \frac{l_{CG_{1i}}}{l_{1i}} \left({}^{F}[\boldsymbol{r}_{J_{1i}}] - {}^{F}[\boldsymbol{r}_{J_{2i}}] \right) \tag{20.13} $$

The position vector of center of mass of the platform in frame B is expressed as ${}^{B}[\boldsymbol{r}_P] = [x_P \; y_P \; z_P]^T$, then the position of mass center of the platform in fixed frame can be written as:

$$ \begin{bmatrix} {}^{F}[\boldsymbol{r}_P] \\ 1 \end{bmatrix} = {}^{F}A_B \begin{bmatrix} {}^{B}[\boldsymbol{r}_P] \\ 1 \end{bmatrix} \tag{20.14} $$

The global center of mass of the vehicle, noted ${}^{F}[\boldsymbol{r}_{CG}]$, can then be written as:

$$ {}^{F}[\boldsymbol{r}_{CG}] = \frac{1}{M} \left(m_p {}^{F}[\boldsymbol{r}_P] + \sum_{i=1}^{4} \left(m_{1i} {}^{F}[\boldsymbol{r}_{CG_{1i}}] + m_{2i} {}^{F}[\boldsymbol{r}_{CG_{2i}}] \right) \right) \tag{20.15} $$

Substituting Eqs. (20.9), (20.13), and (20.14) into Eq. (20.15), one then obtains

$$ {}^{F}[\boldsymbol{r}_{CG}] = \frac{1}{M} \begin{bmatrix} \boldsymbol{r}_{CGx} \\ r_{CGy} \\ r_{CGz} \end{bmatrix} \tag{20.16} $$

where r_{CGx}, r_{CGy}, and r_{CGz} are the x, y, and z components of center of mass of the vehicle. Here, the main attention is on the z component of ${}^{F}[\boldsymbol{r}_{CG}]$ which the gravitational potential energy is emerged from this component.

$$ r_{CGz} = m_p A + \sum_{i=1}^{4} \left(m_{1i} \left(\sin(\gamma_i)\, l_{2i} + Z_{Wi} + \frac{B}{l_{1i}} \right) + m_{2i} \left(\sin(\gamma_i)\, l_{CG2i} + Z_{Wi} \right) \right) $$

where

$$ A = \left(r_{31} x_p + r_{32} y_p + r_{33} z_p + r_{34} \right) $$
$$ B = l_{CG1i} \left(r_{31} a_i + r_{32} b_i + r_{33} c_i + r_{34} - \sin(\gamma_i)\, l_{2i} - Z_{Wi} \right) $$

where m_{1i} and m_{2i} are the first and second link's mass, respectively. Parameters r_{31}, r_{32}, and r_{33} are the components of third row of rotation matrix in Eq. (20.7) and r_{34} is equal to Z_B.

In order to show the variation of potential energy due to gravity, a dynamic simulation of the vehicle has been implemented in MapleSim software (see Fig. 20.3a). MapleSim is an advanced physical modeling and simulation tool that applies modern techniques to dramatically reduce model development time, provide

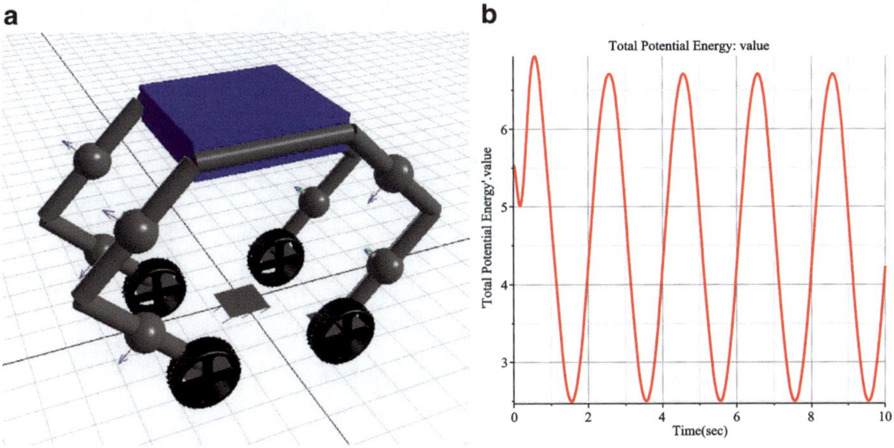

Fig. 20.3 (**a**) MapleSim model (**b**) total potential energy of the vehicle while the elevation of the platform is changing sinuously

greater insight into system behavior, and produce fast, high-fidelity simulations. Figure 20.3b shows the variation of potential energy while the platform elevation is changing sinuously.

20.4.2 Derivation of Static Balancing Conditions of Vehicle with Spring and Parallelogram Assist

For static balancing of AWV, a special architecture is proposed for the legs. A parallelogram linkage is employed instead of using single link with a revolute joint, and a spring is attached inside the parallelogram (see Fig. 20.4).

One can easily find the position vector of mass center of the added links in the parallelogram. The position vector of mass center of link 4 with length l_{4i}, and the position vector of joint 3 (J_{3i}) of the leg-wheel subsystem can be computed, respectively, as

$$^{F}[\boldsymbol{r}_{CG_{4i}}] = {}^{F}A_{W_i}\left[0\ 0\ l_{2i} + \tfrac{l_{4i}}{2}\ 1\right]^T \tag{20.17}$$

$$^{F}[\boldsymbol{r}_{J_{3i}}] = {}^{F}A_{W_i}\left[0\ 0\ l_{2i} + l_{4i}\ 1\right]^T \tag{20.18}$$

The center of mass of the link 3 with length l_{3i} can be computed from the position vector of J_{0i} and J_{3i}

$$^{F}[\boldsymbol{r}_{J_{0i}}] = {}^{F}A_{B}\left[x_{0i}\ y_{0i}\ z_{0i}\ 1\right]^T \tag{20.19}$$

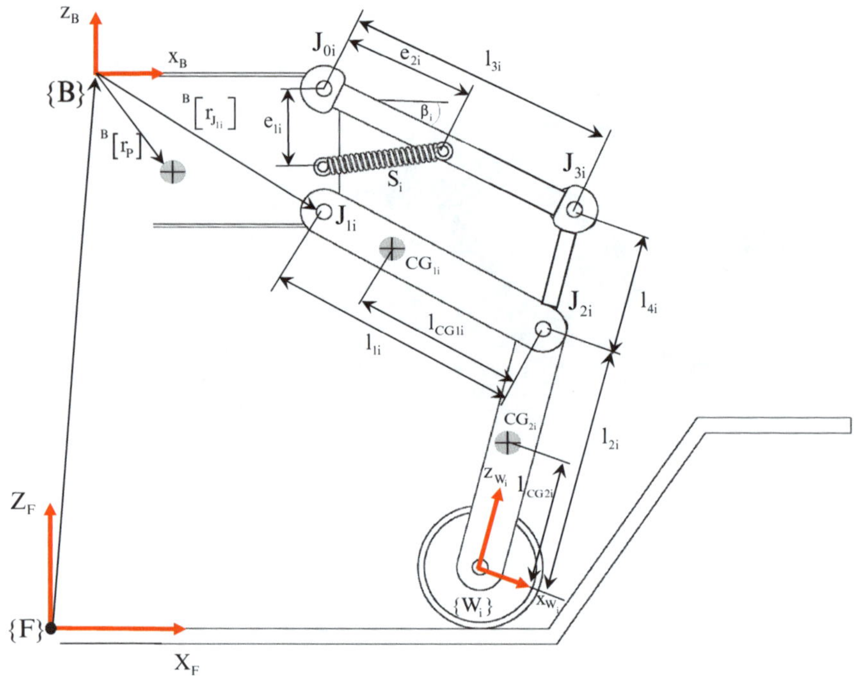

Fig. 20.4 Kinematic architecture of *i*th leg-wheel subsystem with springs

where $\left[x_{0i}\ y_{0i}\ z_{0i}\right]^T$ is the position vector of joint J_{0i} with respect to frame B; then

$$ {}^F[\boldsymbol{r}_{CG_{3i}}] = {}^F[\boldsymbol{r}_{J_{3i}}] + \frac{l_{CG3i}}{l_{3i}}\left({}^F[\boldsymbol{r}_{J_{0i}}] - {}^F[\boldsymbol{r}_{J_{3i}}]\right) \tag{20.20} $$

Therefore, the new global center of mass of vehicle, noted ${}^F[\boldsymbol{r}_{CG}]$, can then be written as:

$$ {}^F[\boldsymbol{r}_{CG}] = \frac{1}{M}\left(m_p{}^F[\boldsymbol{r}_P] + \sum_{i=1}^{4}\left(m_{1i}{}^F[\boldsymbol{r}_{CG_{1i}}] + m_{2i}{}^F[\boldsymbol{r}_{CG_{2i}}] + m_{3i}{}^F[\boldsymbol{r}_{CG_{3i}}] + m_{4i}{}^F[\boldsymbol{r}_{CG_{3i}}]\right)\right) \tag{20.21} $$

The total potential energy of the articulated vehicle including springs and links' weights can be written as

$$ V = V_w + V_s \tag{20.22} $$

where V_w and V_s are gravitational potential energy and elastic potential energy stored in the springs, respectively.

$$V_w = r_{CGz} g \tag{20.23}$$

$$V_s = \frac{1}{2} \sum_{i=1}^{4} K_i S_i^2 \tag{20.24}$$

where S_i is the length of springs, and it is assumed that springs are zero free-length to obtain complete balancing. One can then obtain the length of springs as:

$$S_i^2 = e_{1i}^2 + e_{2i}^2 - 2e_{1i}e_{2i}\cos\left(\frac{\pi}{2} - \beta_i\right) \tag{20.25}$$

where e_{1i} and e_{2i} are distances from joint J_{0i} to both ends of spring with length S_i, and β_i is the angle of link 3 with respect to horizon.

$$\sin(\beta_i) = \frac{z_{J_{0i}} - z_{J_{3i}}}{l_{3i}} = \frac{r_{31}x_{0i} + r_{32}y_{0i} + r_{33}z_{0i} + r_{34} + \sin(\gamma_i)(l_{2i} + l_{4i}) - Z_{Wi}}{l_{3i}} \tag{20.26}$$

where z_{J_0} and z_{J_3} are z component of joints J_0 and J_3 as shown in Fig. 20.4.

Substituting Eqs. (20.23), (20.24), and (20.25) into Eq. (20.22), one can obtain the total potential energy of the articulated vehicle.

$$\begin{aligned} V_T = {} & m_p g\left(r_{31}x_p + r_{32}y_p + r_{33}z_p + r_{34}\right) + \sum_{i=1}^{4} m_{2i} g\left(-\sin(\gamma_i)\, l_{CG2i} + Z_{Wi}\right) \\ & + \sum_{i=1}^{4} m_{1i} g\left(-\sin(\gamma_i)\, l_{2i} + Z_{Wi}\right) + \sum_{i=1}^{4} m_{3i} g\left(-\sin(\gamma_i)\, l_{5i} + Z_{Wi}\right) \\ & + \sum_{i=1}^{4} m_{4i} g\left(-\sin(\gamma_i)\left(l_{2i} + \frac{l_{4i}}{2}\right) + Z_{Wi}\right) + \sum_{i=1}^{4} \frac{K_i}{2}\left({e_{1i}}^2 + {e_{2i}}^2\right) \\ & + \sum_{i=1}^{4} m_{1i} g\left(\frac{l_{CG1i}\left(r_{31}a_i + r_{32}b_i + r_{33}c_i + r_{34} + \sin(\gamma_i)\, l_{2i} - Z_{Wi}\right)}{l_{1i}}\right) \\ & + \sum_{i=1}^{4} m_{3i} g\left(\frac{l_{CG3i}\left(r_{31}x_{0i} + r_{32}y_{0i} + r_{33}z_{0i} + r_{34} + \sin(\gamma_i)\, l_{5i} - Z_{Wi}\right)}{l_{3i}}\right) \\ & - \sum_{i=1}^{4} K_i\left(\frac{e_{1i}e_{2i}\left(r_{31}x_{0i} + r_{32}y_{0i} + r_{33}z_{0i} + r_{34} + \sin(\gamma_i)(l_{2i} + l_{4i}) - Z_{Wi}\right)}{l_{3i}}\right) \end{aligned} \tag{20.27}$$

where $l_{5i} = l_{2i} + l_{4i}$.

The total potential energy of the vehicle V_T remains constant when the coefficients of $r_{31}, r_{32}, r_{33}, r_{34}$, Z_{Wi}, and $sin(\gamma_i)$ be equal to zero. Thus, we rewrite Eq. (20.27) as follows:

$$V_T = H_1 r_{31} + H_2 r_{32} + H_3 r_{33} + H_4 r_{34} + H_{i+4} Z_{Wi} + H_{i+8} \sin(\gamma_i) + \sum_{i=1}^{4} \frac{K_i}{2}\left(e_{1i}^2 + e_{2i}^2\right) \tag{20.28}$$

where $i = \{1, 2, 3, 4\}$ and,

$$H_1 = m_p x_p g + \sum_{i=1}^{4}\left(\left(\frac{m_{1i} l_{CG1i} a_i}{l_{1i}} + \frac{m_{3i} l_{CG3i} x_{0i}}{l_{3i}}\right) g - \frac{K_i e_{1i} e_{2i} x_{0i}}{l_{3i}}\right)$$

$$H_2 = m_p y_p g + \sum_{i=1}^{4}\left(\left(\frac{m_{1i} l_{CG1i} b_i}{l_{1i}} + \frac{m_{3i} l_{CG3i} y_{0i}}{l_{3i}}\right) g - \frac{K_i e_{1i} e_{2i} y_{0i}}{l_{3i}}\right)$$

$$H_3 = m_p z_p g + \sum_{i=1}^{4}\left(\left(\frac{m_{1i} l_{CG1i} c_i}{l_{1i}} + \frac{m_{3i} l_{CG3i} z_{0i}}{l_{3,i}}\right) g - \frac{K_i e_{1i} e_{2i} z_{0i}}{l_{3i}}\right)$$

$$H_4 = m_p g + \sum_{i=1}^{4}\left(\left(\frac{m_{1i} l_{CG1i}}{l_{1i}} + \frac{m_{3i} l_{CG3i}}{l_{3i}}\right) g - \frac{K_i e_{1i} e_{2i}}{l_{3i}}\right)$$

$$H_{i+4} = \left(m_{1i}\left(1 - \frac{l_{CG1i}}{l_{1i}}\right) + m_{2i} + m_{3i}\left(1 - \frac{l_{CG3i}}{l_{3i}}\right) + m_{4i}\right) g + \frac{K_i e_{1i} e_{2i}}{l_{3i}}$$

$$H_{i+8} = m_{1i}\left(-l_{2i} + \frac{l_{CG1i} l_{2i}}{l_{1i}}\right) g - m_{2i}(l_{CG2i}) g + m_{4i}\left(-l_{2i} - \frac{l_{4i}}{2}\right) g$$
$$+ m_{3i}\left(-l_{2i} - l_{4i} + \frac{l_{CG3i} l_{5i}}{l_{3i}}\right) g - \frac{K_i e_{1i} e_{2i} l_{5i}}{l_{3i}}$$

From Eq. (20.28) one can obtain sufficient conditions for static balancing of the articulated vehicle to maintain total potential energy constant.

$$H_i = 0, \{i = 1, 2, \ldots, 12\} \tag{20.29}$$

The balanced mechanism is illustrated schematically in Fig. 20.5a with parameters given in Table 20.1. For verification of statically balanced vehicle a physical model of balanced system is implemented in MapleSim (see Fig. 20.5b) to show the roughly constant potential energy of the system while the platform elevation is changing sinuously. The result of simulation is presented in Fig. 20.6; the significant portion of the load is due to gravity which is configuration dependent. With proper selection of spring constants, spring attachment points, and mass center of the

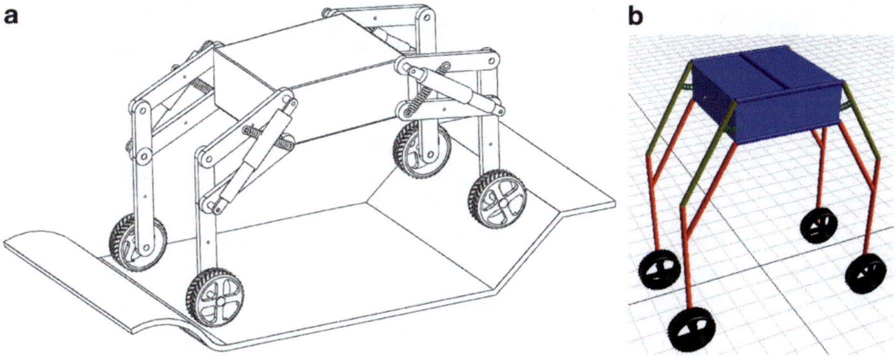

Fig. 20.5 (**a**) Balanced articulated vehicle with springs and parallelograms assist (**b**) MapleSim model with parallelogram and spring assist

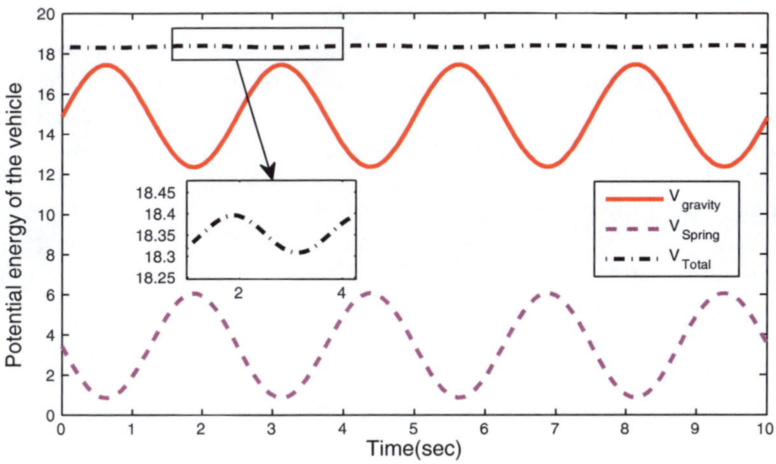

Fig. 20.6 Total potential energy remains fixed while the elevation of the platform is changing sinuously

Table 20.1 Statically balancing condition for six degrees of freedom vehicle

Spring constants (N/m)	Link parameters
$K_1 = 43.2$	$\boldsymbol{J}_{0i} = [\pm 0.2, \pm 0.2, 0]^T(m), \boldsymbol{J}_{1i} = [\pm 0.2, \pm 0.2, -0.15]^T$
$K_2 = 43.2$	$l_{1i} = 0.4, l_{2i} = 0.4, l_{3i} = 0.4, l_{4i} = 0.15(\text{m})$
$K_3 = 43.2$	$e_{1i} = 0.15, e_{2i} = 0.2(\text{m}), x_p = y_p = 0, z_p = -0.12(\text{m})$
$K_4 = 43.2$	$m_p = 0.5, m_{1i} = m_{2i} = m_{3i} = 0.2, m_{4i} = 0.1(\text{kg})$

platform, the total potential energy of the vehicle maintains constant even during vehicle elevation changes. In other words, in the gravity-compensated parallel mechanism, no torque will be produced by weight of the links at the actuators for any configuration of mechanism under static conditions. As we mentioned before, in parallel structures with articulated legs, in addition to having high performance, torque minimization also is one of the biggest challenges. Here, elimination of static torques due to gravity through static balancing reduces the torque requirements and provides much more efficient design (reduction of the actuator size). Now, a balanced articulated mechanism is easier to move its platform and can be brought to static equilibrium in any of its configurations without exerting external forces.

20.5 Conclusion

In AWVs, the terrain-contact wheels are attached to the chassis via an articulated-multibody chain. They offer superior uneven terrain traversal capabilities by virtue of the reconfigurability within their articulated structure. Changing the vehicle platform elevation with multiple legs requires considerable power because of their weights. Hence, reducing or eliminating these static forces needs careful consideration which is the main focus of this chapter. Undeniably, having balanced condition leads to smaller size and less powerful actuators and also it improves the efficiency. Finally, it is easier for vehicle to move its platform and can be brought to static equilibrium in any of its configurations without exerting external forces. In this chapter, the static balancing of articulated vehicles with leg-wheel subsystems has been addressed. Two methods of balancing, springs and parallelograms, are merged together to derive the total potential energy of the vehicle. The sets of conditions for static balancing of the vehicle have finally been obtained from these expressions.

References

1. Alamdari, A., Krovi, V.: Active reconfiguration for performance enhancement in articulated wheeled vehicles. In: ASME Dynamic Systems Controls Conference, San Antonio, 22–24 October 2014
2. Alamdari, A., Zhou, X., Krovi, V.: Kinematic modeling, analysis and control of highly reconfigurable articulated wheeled vehicles. In: Proceedings of the ASME 2013 International Design Engineering Technical Conferences and Computers in Engineering Conference, ASME IDETC/CIE 2013, Portland, 4–7 August 2013
3. Alamdari, A., Sovizi, J., Jun, S. K., Krovi, V.: Kinetostatic optimization for an adjustable four-bar based articulated leg-wheel subsystem. In: IEEE/RSJ International Conference on Intelligent Robots and Systems, Chicago, 14–18 September 2014
4. Bidarvatan, M., Shahbakhti, M.: Integrated HCCI engine control based on a performance index. In: ASME 2013 Internal Combustion Engine Division Fall Technical Conference, pp. V001T05A007–V001T05A007 (2013)

5. Laliberte, T., Gosselin, C., Jean, M.: Static balancing of 3-DOF planar parallel mechanisms. IEEE/ASME Trans. Mechatron. **4**(4), 363–377 (1999)
6. Alamdari, A., Krovi, V.: Parallel articulated-cable exercise robot (PACER): novel home-based cable-driven parallel platform robot for upper limb neuro-rehabilitation. In: Proceedings of the ASME 2015 International Design Engineering Technical Conferences and Computers in Engineering Conference, Boston, 2–5 August 2015
7. Gosselin, C., Wang, J.: Static balancing of spatial six-degree-of-freedom parallel mechanisms with revolute actuators. J. Rob. Syst. **17**(3), 159–170 (2000)
8. Lessard, S., Bigras, P., Bonev, I.: A new medical parallel robot and its static balancing optimization. Trans. ASME J. Med. Dev. **1**, 272–278 (2007)
9. Ulrich, N., Kumar, V.: Passive mechanical gravity compensation for robot manipulators. In: Proceeding of the 1991 IEEE International Conference on Robotics and Automation, pp. 1536–1541 (1991)
10. Herder, J.L.: Energy-free systems. Theory, conception and design of statically balanced mechanisms. Ph.D thesis, Delft University of Technology, Delft (2001)
11. Shin, E., Streit, D.A.: Spring equilibrator theory for static balancing of planar pantograph linkages. Mech. Mach. Theory **26**(7), 645–657 (1991)
12. Fattah, A., Agrawal, S.K.: On the design of a passive orthosis to gravity balance human legs. ASME J. Mech. Des. **127**(4), 802–808 (2005)
13. Ebert-Uphoff, I., Gosselin, C.M., Laliberte, T.: Static balancing of spatial parallel platform mechanisms-revisited. ASME J. Mech. Des. **122**(1), 43–51 (2000)
14. Jean, M., Gosselin, C.M.: Static balancing of planar parallel manipulators. In: IEEE International Conference on Robotics and Automation, Minneapolis, pp. 3732–3737 (1996)
15. Simionescu, I., Ciupitu, L.: Static balancing of the industrial robot arms. Part I: discrete balancing. Mech. Mach. Theory **35**(9), 1287–1298 (2002)
16. Russo, A., Sinatra, R., Xi, F.: Static balancing of parallel robots. Mech. Mach. Theory **40**(2), 191–202 (2005)
17. Lee, Y.Y., Chen, D.Z.: Determination of spring installation configuration on statically balanced planar articulated manipulators. Mech. Mach. Theory **74**, 319–336 ()2014
18. Lin, P., Shieh, W., Chen, D.: A theoretical study of weight-balanced mechanisms for design of spring assistive mobile arm support (MAS). Mech. Mach. Theory **61**, 156–167 (2013)